Sustainable Materials and Technology

Sustainable Materials and Technology (SMT) book series publishes research monographs (both edited and authored volumes) showcasing the latest developments in the field and comprehensively covering topics such as:

- Recycling of waste into useful material and their energy applications
- Catalytic action of Nano oxides for efficient carbon reforming process
- Sustainable technologies for plastic transformation
- Bifunctional nanoparticles for sustainable water splitting applications
- Sustainable dying and printing
- New materials from waste
- Sustainable Manure Management and Technology: Potentials, Uses and limitations
- Sustainable Mechanical Engineering Approach
- Sustainable biochemistry for the improvement of health
- Sustainable development of Mechanical recycling of automotive components
- Sustainable-waste recycling and conversion in useful materials for different applications
- Sustainable development of inexpensive Nano-photo catalysts
- Sustainable development of recycling of discarded lithium ion batteries
- Modern sustainable cement and concrete
- Sustainable adsorbent for hazardous removal
- Sustainable superior electromagnetic shielding materials
- Excellent sustainable nanostructured materials for energy storage device
- Sustainable development of heavy metal detoxification from water
- Carbon dioxide utilization for sustainable energy
- Sustainable development in green syntheses of materials
- Environment friendly and sustainable cloth for garments application
- Sustainable design and application of eco-materials
- Nanoparticles for sustainable environment applications
- Sustainable remediation of industrial contaminated water towards potential industrial applications
- Biomaterials for sustainable bioremediations

Rishabha Malviya · Sonali Sundram
Editors

Sustainable Cellulose Materials for Biomedical Applications

Springer

Editors
Rishabha Malviya
Department of Pharmacy
Galgotias University
Greater Noida, Uttar Pradesh, India

Sonali Sundram
Department of Pharmacy
Galgotias University
Greater Noida, Uttar Pradesh, India

ISSN 2731-0426 ISSN 2731-0434 (electronic)
Sustainable Materials and Technology
ISBN 978-981-96-9903-2 ISBN 978-981-96-9904-9 (eBook)
https://doi.org/10.1007/978-981-96-9904-9

This Springer imprint is published by the registered company Springer Nature Singapore Pte Ltd.
The registered company address is: 152 Beach Road, #21-01/04 Gateway East, Singapore 189721, Singapore

Foreword

The advancement of sustainable biomaterials has emerged as a transformative approach in modern biomedical science, addressing both ecological concerns and the need for safer, biocompatible materials in healthcare. Among the various biopolymers explored, cellulose, a naturally abundant, renewable, and biodegradable material, has shown exceptional potential in medical and pharmaceutical applications. Its unique physicochemical properties, including biocompatibility, non-toxicity, and mechanical strength, have paved the way for innovative developments in tissue engineering, wound healing, drug delivery, and medical device fabrication.

The book *Sustainable Cellulose Materials for Biomedical Applications*, edited by Dr. Rishabha Malviya, is a comprehensive and insightful compilation that brings together cutting-edge research on cellulose-based biomaterials and their applications in modern healthcare. It offers an interdisciplinary perspective, bridging the gap between material science, biomedical engineering, and clinical applications. The content presented in this book is timely and highly relevant, given the increasing global emphasis on sustainable healthcare solutions and the transition towards eco-friendly medical materials.

This book covers a broad spectrum of topics, from the chemical modification of cellulose and its safety to its functional applications in tissue engineering, pharmaceutical formulations, and medical devices. The chapters provide in-depth discussions on microbial cellulose synthesis, cellulose polyelectrolyte complexes, hybrid hydrogels, and nanofiber-based biomaterials, highlighting their significance in regenerative medicine and drug delivery systems. Furthermore, the book delves into advanced applications, including cellulose-derived materials for wound healing, hemostasis, antibacterial coatings, and ophthalmology, demonstrating the versatility of this biopolymer in medical innovation.

One of the standout features of this book is its focus on sustainability. With growing concerns about the environmental impact of synthetic biomaterials, this book provides a forward-thinking approach by emphasizing the role of cellulose as an eco-friendly and sustainable alternative. It not only explores scientific advancements but also aligns with the broader global vision of reducing biomedical waste and promoting green materials in healthcare.

The editors, Dr. Rishabha Malviya and Sonali Sundram, along with esteemed contributors from diverse scientific domains, have compiled a valuable resource that will benefit researchers, academicians, biomedical engineers, material scientists, pharmaceutical professionals, and clinicians. The multidisciplinary nature of this book ensures that it serves as a reference point for future innovations in cellulose-based biomedical applications.

As we move towards a more sustainable and responsible healthcare industry, this book plays a pivotal role in educating and inspiring the next generation of scientists and innovators. I am confident that Sustainable Cellulose Materials for Biomedical Applications will serve as a basis for further advancements in biomedical materials research and will contribute significantly to the future of eco-friendly medical solutions.

It is with great enthusiasm that I recommend this book to scientists, engineers, and healthcare professionals seeking to explore the potential of cellulose-based materials in biomedical applications.

Happy Reading and Best Wishes

Dr. Dhruv Galgotia
CEO
Galgotias University
Greater Noida, India

Preface

Advancements in biomedical science and material engineering have led to the development of sustainable biomaterials for healthcare applications. Cellulose, a renewable, biodegradable polymer, is gaining attention due to its physicochemical properties, biocompatibility, and versatility. This book, *Sustainable Cellulose Materials for Biomedical Applications*, provides a comprehensive exploration of chemical modifications of cellulose, structural and functional properties, and its integration into biomedical systems. Through the appropriate use of cellulose in the development of innovative medical solutions that are sustainable, the book seeks to fill the gap between material science and biomedical engineering.

The book is structured into twenty chapters, covering wide topics related to cellulose-based biomaterials, ranging from fundamental concepts to advanced applications. Chapters "Cellulose Derivatives: Chemical Modification and Biomedical Research Applications", "Safety and Biocompatibility of Cellulose-Based Materials", and "Advancements in Bacterial Cellulose Synthesis and Functional Applications" introduce the chemical modifications of cellulose, the safety and biocompatibility of cellulose-based materials, and microbial cellulose synthesis as novel approaches to obtaining high-purity cellulose for medical use. Chapters "Fundamentals and Applications of Cellulose Polyelectrolyte Complexes" and "Design, Characteristics, and Biomedical Applications of Cellulose-Based Hydrogel" explore cellulose polyelectrolyte complexes and the design and biomedical applications of cellulose-based hydrogels. Chapter "Utilizing Cellulose-Based Hybrid Smart Hydrogel for Tissue Engineering" explores hybrid smart hydrogels for tissue engineering and the utilization of sustainable sources such as grape stems for biomedical applications. Chapter "Modified Cellulose-Supported Catalysts and Reagents in Three-Component Reactions for Sustainable Medicinal Chemistry" highlights cellulose nanofibers derived from plant sources and their applications in medical science. Chapter "Cellulose-Derived Matrices for Enhanced Healing and Regeneration in Plastic Surgery" covers cellulose-derived matrices for enhanced healing and plastic surgery applications, cellulose-based pharmaceutical dosage forms, and its broader biomedical applications. Chapter "Development of Dosage Form Using Cellulose-Based Materials" discusses the cellulose-based dosage form for drug delivery applications while Chapter "Sustainable Cellulose Materials for Biomedical Application" explores the role of cellulose in biomedical applications. Chapter "Cellulose-Based Material as Antibacterial Agent" explores antibacterial cellulose-based materials and their role in preventing infections. The book presents recent advancements in bacterial nanocellulose-based biomaterial composites for tissue engineering (Chapter "Bacterial Nanocellulose-Based Biomaterial Composite for Tissue Engineering") and cellulose-based medical devices (Chapter "Cellulose-Based Medical Devices"), highlighting their role in medical diagnostics and therapeutic applications. Chapters also focus on cellulose-based haemostatic agents (Chapters "Cellulose-Based Hemostatic Agents: Innovations and Applications in Wound Healing" and "Wound Healing and Cellulose-Based Hemostatic Agent"), which are crucial for wound healing and emergency medicine. Chapter "Advances in Cellulose-Based Biomaterials for the Sustainable Drug Delivery Systems" discusses cellulose-based biomaterials in sustainable drug delivery, emphasizing their potential to revolutionize pharmaceutical formulations. The book concludes with an insightful discussion on cellulose-based materials in corneal regeneration (Chapter "Innovative Approaches to Corneal Regeneration: From Biomaterials to Clinical Application"), an emerging field that highlights the adaptability of cellulose in ophthalmology and vision restoration. Chapter "Cellulose in 3D Printing for the Development of Biomedical Devices" explores the applications of 3D-printed cellulose devices in biomedicine.

The book is not only a scientific compilation but also a visionary guide to the future of sustainable healthcare materials, encouraging innovation in eco-friendly and effective medical solutions. The integration of cellulose-based materials into modern medicine signifies a step towards a more sustainable and environmentally conscious future in healthcare.

Greater Noida, India

Rishabha Malviya
Sonali Sundram

About This Book

The book *Sustainable Cellulose Materials for Biomedical Applications* is an in-depth discussion of the most recent developments in cellulose-based biomaterials and their transformational role in healthcare. As the world gradually transitions to ecologically friendly and renewable biomaterials, cellulose has emerged as a promising choice for biomedical solutions. The book explores extensively cellulose-based materials, including their chemical modifications, structural characteristics, and several uses in drug delivery, tissue engineering, wound healing, and medical device development.

The book begins by discussing the chemical modifications of cellulose (Chapter "Cellulose Derivatives: Chemical Modification and Biomedical Research Applications") and its safety and biocompatibility (Chapter "Safety and Biocompatibility of Cellulose-Based Materials"), essential for its application in biomedical science. It further delves into microbial cellulose synthesis (Chapter "Advancements in Bacterial Cellulose Synthesis and Functional Applications") and polyelectrolyte complexes (Chapter "Fundamentals and Applications of Cellulose Polyelectrolyte Complexes"), highlighting the versatility of cellulose in biomedical engineering. A significant portion of the book is dedicated to hydrogels and hybrid materials, including their design, characteristics, and applications in tissue engineering and regenerative medicine (Chapters "Design, Characteristics, and Biomedical Applications of Cellulose-Based Hydrogel" and "Utilizing Cellulose-Based Hybrid Smart Hydrogel for Tissue Engineering"). The exploration of sustainable cellulose sources (Chapter "Grape Stems as a Sustainable Source of Cellulose for Skin Regeneration") and nanofibers derived from plants (Chapter "Cellulose Nanofibers from Plant Sources and Their Use in Biomedical Applications") reinforces the book's focus on renewable and green biomaterials.

The functional uses of cellulose in medicinal chemistry (Chapter "Modified Cellulose-Supported Catalysts and Reagents in Three-Component Reactions for Sustainable Medicinal Chemistry"), plastic surgery and tissue repair (Chapter "Cellulose-Derived Matrices for Enhanced Healing and Regeneration in Plastic Surgery"), and dosage form development (Chapter "Development of Dosage Form Using Cellulose-Based Materials") are covered in the book in addition to the fabrication of materials. The role of cellulose as an antibacterial agent (Chapter Cellulose-Based Material as Antibacterial Agent"), a biomaterial composite for tissue engineering (Chapter "Bacterial Nanocellulose-Based Biomaterial Composite for Tissue Engineering"), and its use in medical devices (Chapter "Cellulose-Based Medical Devices") are thoroughly investigated, demonstrating its diverse applications in biomedical sciences. The focus is on wound healing and haemostasis, with chapters on cellulose-based haemostatic agents (Chapters "Cellulose-Based Hemostatic Agents: Innovations and Applications in Wound Healing" and "Wound Healing and Cellulose-Based Hemostatic Agent") and advancements in sustainable drug delivery methods (Chapter "Advances in Cellulose-Based Biomaterials for the Sustainable Drug Delivery Systems"). The book also discusses ground-breaking ways of corneal regeneration (Chapter "Innovative Approaches to Corneal Regeneration: From Biomaterials to Clinical Application"), highlighting cellulose's potential in ophthalmology. Chapter "Cellulose in 3D Printing for the Development of Biomedical Devices" discusses the role of 3D-printed cellulose material in the development of biomedical devices.

The book *Sustainable Cellulose Materials for Biomedical Applications*, which includes contributions from leading experts in biomaterials, chemistry, and biomedical engineering, is an invaluable resource for researchers, scientists, healthcare professionals, and students in biomedical engineering, pharmaceutical sciences, regenerative medicine, and material science. It not only emphasizes the importance of cellulose as a sustainable biomaterial but also offers insight into its future potential in enhancing healthcare solutions while reducing environmental effects.

Contents

Cellulose Derivatives: Chemical Modification and Biomedical Research Applications

Murat Torun and Nursel Pekel Bayramgil

Abstract Cellulose is a natural polymer and is modified with chemical agents to obtain cellulose derivatives used in various industrial applications. These modifications make cellulose more functional by changing its physical and chemical properties. The main reactions used to derivatize cellulose are as follows: Acetylation, that is modification of cellulose with acetyl groups, produces cellulose acetate. This process increases the water solubility of cellulose and gives it biodegradable properties. Thus, cellulose acetate stands out as an environmentally friendly material. It is generally used in coating and film production. Carboxymethylation, which is the modification of cellulose with carboxymethyl groups, produces carboxymethyl cellulose (CMC), which is water-soluble and used as a stabilizer and thickener in the food, pharmaceutical and cosmetic industries. CMC prevents the formation of ice crystals in frozen products and increases the volume of baked goods. In the esterification reaction, cellulose is esterified with various acids, increasing the stress–strain properties and water resistance of cellulose. Cellulose derivatives obtained as a result of chemical modifications are applied in a wide range of pharmaceutical formulations, cosmetic products and biomedical applications. Chemical modification of cellulose derivatives increases the functionality of this natural polymer, allowing it an important role in modern industries with both environmentally friendly and functional properties. In addition to the derivatives mentioned here, other derivatives have also found a place in various applications. In this section, in addition to the chemical modifications applied to the products production of cellulose derivatives, biomedical applications in which these derivatives are used are also included.

Keywords Cellulose · Modification · Grafting · Etherification · TEMPO-oxidized · Methyl cellulose

M. Torun · N. P. Bayramgil (✉)
Faculty of Science, Department of Chemistry, Hacettepe University, Beytepe, Ankara, Turkey
e-mail: nursel@hacettepe.edu.tr

M. Torun
e-mail: mtorun@hacettepe.edu.tr

R. Malviya and S. Sundram (eds.), *Sustainable Cellulose Materials for Biomedical Applications*, Sustainable Materials and Technology,
https://doi.org/10.1007/978-981-96-9904-9_1

Abbreviations

3D	Three dimensional
CMC	Carboxymethyl cellulose
DAC	Dialdehyde cellulose
DS	Degree of substitution
NH_2-PEG-*b*-PLA	Poly (ethylene glycol-block-poly (*D,L*)-lactic acid)
PNIPAM	Poly (N-isopropyl-acrylamide
TEMPO	2, 2, 6, 6-Tetramethylpiperidine-1-oxyl

1 Introduction to Cellulose Derivatives

Cellulose derivatives are modified versions of cellulose via chemical reactions which is the greatest quantified natural polymer on earth. Cellulose is the main constituent of lignocelluloses in cell wall of plants [28]. It is one of the most consumed carbohydrate homopolymer repeating ß-anhydro-D- glucopyranose units with a formula of $(C_6H_{10}O_5)_n$ that consists of a linear chain linked via oxygen atom as shown in Fig. 1. Fahmy and Mobarak [29]. Cellulose is safe in use for human activities, as food or biomedical applications since its toxicity is very low and it is chemically stable [66].

Cellulose is one of the most modified biopolymers for bio-applications since its resistance on pH, time and temperature changes as well as it is stable in biological systems [63]. Bacterial cellulose is the one that is chemically synthesized, and its nanostructure with highest purity brings sometimes advantages depending upon the application that includes biomedicine [41]. Some bacteria, such as a vinegar bacterium called acetobacter, can convert small organic units to pure cellulose under specific conditions [107]. Bacterial cellulose, itself and its composites with other organic polymers or metals, are mainly used in wound dressing, tissue engineering, drug delivery and cancer treatment applications.

The natural abundance and renewable properties of cellulose make it a sought-after raw material for many industrial uses. The intrinsic properties of natural cellulose, including its insoluble behavior in water and most organic solvents, limit its direct applicability in many respects. In order to reduce these limitations and to reveal its

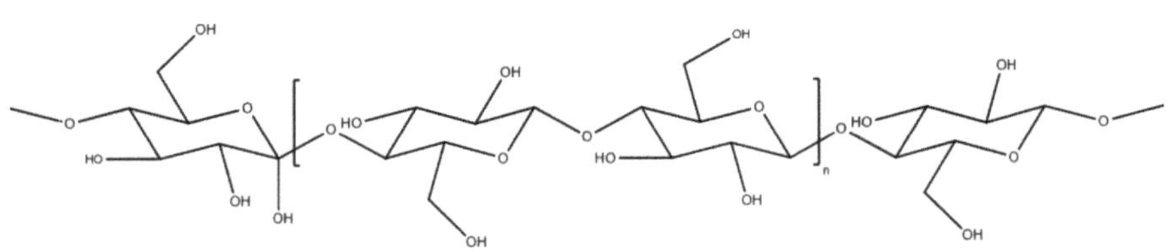

Fig. 1 The structure of the cellulose

potential in different areas of use, cellulose is obtained in the form of different derivatives by chemical modifications. Cellulose can be modified via hydroxyl groups by substitution or conversion of any other functional groups [103]. The steric hindrance and poor solubility in most solvents limit the regioselective modification of cellulose. Hydroxyl groups on cellulose are not active nucleophiles and hard reactions are necessary for modification. The hydroxyl groups substituted on different carbon atoms can bring reactivity difference as well, but in cellulose structure all hydroxyl groups attached on different carbon atoms have low reactivity, and, the modification of cellulose via hydroxyl groups is not selective [142]. The details of these modifications are given in the following section.

2 Chemical Modification of Cellulose or Cellulose Derivatives

It is possible to enrich the functionality and thus the areas of use by adding new functional groups or polymers to the cellulose structure through chemical modification reactions. These modifications can improve the solubility, thermal stability, mechanical strength and chemical reactivity of cellulose significantly. The most intensively studied chemical modification methods are esterification, etherification, oxidation, grafting, and cross-linking in the literature. Each of these methods gives cellulose unique properties, making it suitable for certain applications. In addition to these, bioactive species attached onto cellulose films enhances biocompatibility and functionality. There are also considerable studies in the literature on biomedical applications where biopolymers are used in coating materials alone or as blend [72, 90]. These biopolymers can be processed into cellulosic films with very different methods. Ultimately, non-toxic, biocompatible cellulose films were developed under mild processing conditions, and it was reported that these films reduced adverse immunogenicity and minimized side effects.

In the last few years, there have been intensive efforts to use natural cellulose fibers as reinforcements in polymer composites [48, 134]. The reason for this is that natural fibers have advantages over traditional reinforcements due to their economic costs, biodegradability, and high specific properties. However, the high -OH content of cellulose, and therefore its rigid structure due to its high H-bonding ability, leads to poor compatibility with the polymers that are used as matrices and to the obtainment of composites with poor mechanical properties. In order to improve the matrix-reinforcement interface properties, the matrix, the reinforcement, or both must be subjected to physical or chemical modifications [97, 139, 144].

2.1 Esterification

Esterification involves the reaction of the –OH groups of cellulosic materials with carboxylic acids, acid anhydrides, or acyl halide to form ester bonds [120, 130]. This reaction can be accomplished using a variety of reagents and conditions, depending on the expected properties of the cellulose ester. The carboxylic acid or acid anhydride is activated, usually using an acidic catalyst such as sulfuric acid or a base catalyst such as pyridine. The –OH group of cellulose acts as a nucleophile and attacks the activated carboxylic group to form an ester bond. A byproduct, usually water or acetic acid, is formed. This process improves the solubility, thermal stability, and mechanical properties of cellulose, making it advantageous for some applications such as textiles, photographic films, and biomedical devices. It also increases its compatibility with other materials, for use in composites, films, and coatings. (Examples of these reaction products include cellulose acetate and cellulose acetate butyrate.)

The high crystallinity resulting from the dense intramolecular and intermolecular hydrogen bonds present in the cellulose structure prevents cellulose from general type all chemical solvents, especially water. The absence of a solvent that perfectly dissolves cellulose has allowed esterification to be carried out only with heterogeneous methods. Studies have shown that the heterogeneous system is a disadvantage for the esterification of cellulose because it cannot control the reaction process well and that the degree of substitution is also low. However, the esterification of cellulose in a homogeneous system is a strategy that facilitates the entry of functional groups into the main chain and leads to different applications of cellulose esters. Increasing the esterification efficiency of cellulose and preventing breaks in the cellulose main chain can only be achieved with the existence of a homogeneous reaction system [117, 138]. The solution to this is possible by finding new solvent/solvent systems that will provide a homogeneous esterification environment. In the literature, ionic liquids as well as LiCl/DMAc solvent mixtures with good recycling and thermal stability have been used to provide a homogeneous reaction environment and their effectiveness in the reaction has been demonstrated.

The cellulose acetate is one of the most widely used ester form of the cellulose (see Fig. 2). The cellulose is immersed into the glacial acetic acid with stirring for half an hour. The sulfuric acid in glacial acetic acid is added on stirring the mixture in a manner and after that, let the mixture to stir for 15 min without heating. After addition of acetic anhydride on the final solution just stirring, allow stirring for half an hour. The solution is rested for one day and the distilled water is added on the solution carefully, then the stirring is started for one hour. The modified cellulose is washed with distilled water until neutralization is observed. The washed sample is kept in an oven at 50 °C to dry Homem and Amorim [39].

Cellulose acetate is an amorphous polymer that is transparent, colorless, shiny, fairly hard thermoplastic including good clarity but have moderate UV stability and chemical resistance. High moisture absorption of cellulose acetate can cause dimensional changes [135].

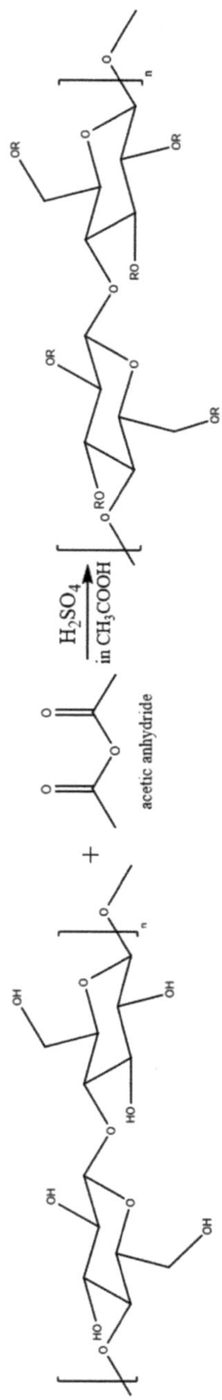

(a) R: H or CH_3CO

Fig. 2 General reaction mechanism for the synthesis of cellulose acetate

2.2 Etherification

Etherification is the reaction of cellulose with agents to form ether groups in the cellulose structure. Commonly used reagents are alkyl halides and epoxides. The alkylating agent is usually activated using a base such as sodium hydroxide. The –OH group of cellulose acts as a nucleophile and attacks the activated alkylating agent to form an ether bond. Halide ions or other small molecules are usually formed as byproducts. Reaction conditions can be tailored to control the substitution degree and the features of the final cellulose ether. Etherification increases the water solubility, chemical reactivity and thermal stability of cellulose. Cellulose ethers are widely used and create high added value due to their properties. They have critical applications in membrane science, and biomedicine [52, 125]. Cellulose ethers with excellent performance are produced by developing reaction media whose conditions can be adjusted as desired. Nagel et al. [75] synthesized methylcellulose reaction conditions that is homogeneous using LiOH/urea system as solvent mixture and dimethyl sulfate as etherifying agent. (Examples of this derivatization are methylcellulose, ethyl cellulose, hydroxyethyl cellulose, hydroxypropyl cellulose, carboxymethyl cellulose.)

The hydroxyl groups on cellulose can be converted to ether functional groups without any selection, partially or completely depending on the process. The common reagents for the etherization processes are epoxides, alpha halogenated carboxylic acids, and halogenoalkanes [49]. Ethyl and methyl celluloses are derived in a similar path (Fig. 3). The pure cellulose samples are kept in concentrated sodium hydroxide at room temperature. The alkali cellulose is separated from sodium hydroxide solution by filtration and poured into the solvent acetone. Dimethyl or diethyl sulfate is added into the mixture drop by drop at 50 °C to carry out the reaction. After filtration of cellulose, acetone and dimethyl or diethyl sulfate is again added on the filtrated cellulose. The same procedure is followed for several times, then cellulose is filtrated and washed in acetic acid to remove excess sodium hydroxide, at the end, to remove possible acetic acid on cellulose, the filtrated sample is washed with acetone for several times, and kept in oven for drying [122, 123]. The use of the alkyl halogens, such as methyl sulfate or methyl chloride to produce methyl cellulose and ethyl sulfate or ethyl chloride to produce ethyl cellulose is also on demand. In basic medium, the alkyl halides react with cellulose to form alkyl celluloses as well similar with the process in alkyl sulfate alkylation discussed above, but, differently, the elapsed time at about 6 h under reflux conditions in alkyl halide method brings hard reaction conditions for the reaction [93]. Ethyl methyl cellulose is also commercially possible, but the reaction conditions should be maintained very carefully since there are two etherization reagents which are methyl and ethyl groups. It is important to have the same stereochemistry, the attached methyl and ethyl distribution in eteric cellulose should be identical for all the alkylated cellulose samples. The change of the hydroxyl radicals with eteric functional groups reduce the solubility of the cellulose in water. Degree of substitution is the most important parameter for the solubility of alkyl celluloses in water and organic solvents.

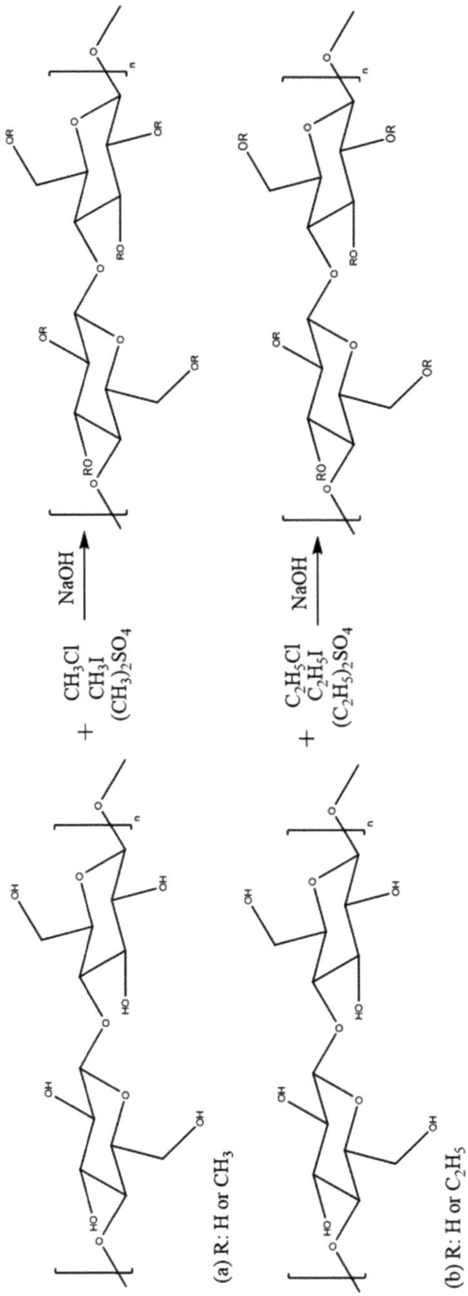

Fig. 3 General reaction mechanism for the synthesis of **a** methyl cellulose **b** ethyl cellulose

The methyl cellulose is odorless, it has no taste and appears as cotton. It is insoluble in most common organic solvents but swells in water. It is stable in acidic and basic pH values that is from 2 to 12. The viscosity of the methyl cellulose in water does not change with the change in pH [49]. Methyl cellulose is mainly used for drug delivery systems and in tissue engineering applications, as well as, it is in the list of Food and Drug Administration [77]. Methyl cellulose based systems are used in situ gel drug delivery systems because of its solubility and swelling properties that makes its release from the system easier [64]. The drug release from methyl cellulose is maintained successfully because gel network has optimum viscosity and physically stable as well, that resist on drug diffusion from the gel matrix [9, 10, 16].

Ethyl cellulose is an ether derivative of cellulose that can be derived from cellulose as well as can be prepared as a powder form from wood pulp or cotton with light color through light-tan depending on the origin, molecular weight, etc. It is insoluble in aqueous media at any pH but it swells in acidic media like gastric juice [59] as well as soluble in most organic solvents [82]. The swelling properties of ethyl cellulose and its non-toxic behavior makes it one of the good candidate to be applied at biomedical applications as well as ıt has improved mechanical properties [3]. The degree of etherification (substitution) determines the physical properties of ethyl cellulose, if degree of substitution is between 2.2 and 2.6, the solubility of the ethyl cellulose in organic solvents is expected [132].

The main hydroxyalkyl groups derived from cellulose are hydroxypropyl cellulose, hydroxyethyl cellulose and hydroxypropyl methyl cellulose. The conversion of cellulose to hydroxypropyl cellulose and hydroxyethyl cellulose are similar (Fig. 4). The synthesis of hydroxyethyl cellulose is carried out in heterogeneous media. Primarily dried cellulose is immersed in at about 20% (m/m) NaOH solution with stirring for 1 h without heating. The cellulose sample is swollen in the alkaline media and it is just alkali. The alkaline cellulose is treated with ethylene oxide under vacuum with a mechanical stirrer, the reaction should be carried out in an autoclave, at 40 °C at about 2 h, when the pressure dropped into the initial stage, and the reaction should be stopped. The modified cellulose is rinsed into acetic acid and water, respectively before drying [128]. The modification of hydroxypropyl cellulose follows the same procedure (Fig. 5), but, instead of ethylene oxide, propylene oxide is used. An addition of methyl chloride to the synthesis media, hydroxypropyl methyl cellulose is produced.

Hydroxypropyl methyl cellulose is an odorless and tasteless powder that can dissolve in water. The solubility of hydroxypropyl methyl cellulose does not change with pH as well as it is stable in acidic and basic solutions reported as pH from 1 to 10 [47, 69]. The neutrality for the permanent charges and non-ionic structure of the hydroxypropyl methyl cellulose makes this biopolymer important because the electrolytes, proteins and enzymes in the body fluid will not be affected in the presence of hydroxypropyl methyl cellulose [119]. Hydroxypropyl methyl cellulose is a promising cellulose derivative and a wide range of concentrations of the biopolymer can be prepared in water for the required application since its solubility in water [89].

The other etherification method of the cellulose is the carboxyalkylation (Fig. 6). The cellulose samples are immersed into the 30% (w/v) NaOH in ethyl alcohol

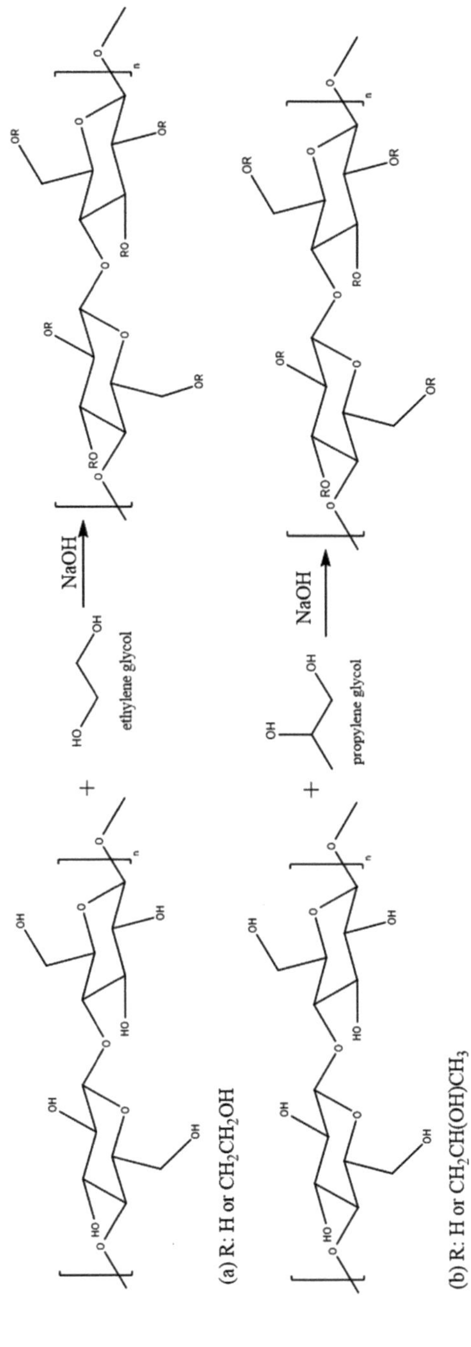

Fig. 4 General reaction mechanism for the synthesis of **a** hydroxyethyl cellulose **b** hydroxypropyl cellulose

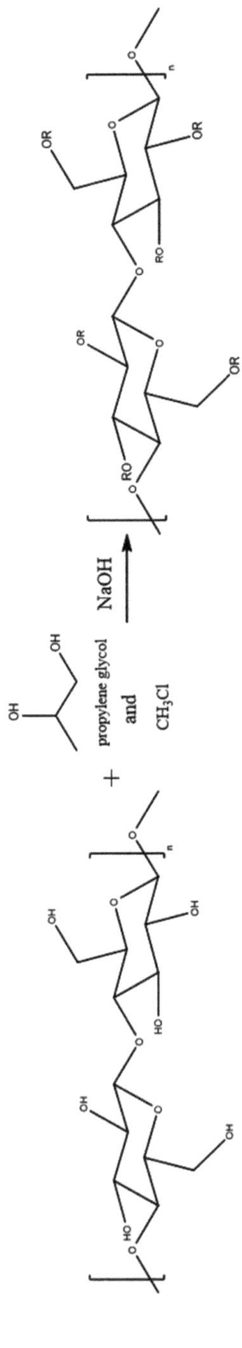

(b) R: H, CH_3 or $CH_2CH(OH)CH_3$

Fig. 5 General reaction mechanism for the synthesis of hydroxypropyl methyl cellulose

solution and stirred for 1 h without heating. After that, monochloroacetic acid is dropped into the solution while stirring for half an hour, then the mixture is heated up to 55 °C with stirring for 3.5 h. The modified cellulose is immersed into the methanol and then acetic acid for neutralization, respectively. The modified cellulose is washed with ethanol for several times then dried at 60 °C in an oven, respectively [5].

Carboxymethyl cellulose can dissolve in water via ionization to anions. It is light yellowish in color and odorless. It is a hexose containing the components carboxymethyl cellulose, acetic acid, sodium and mannose. It is a water-dispersible sodium salt of the carboxy-methyl ether of cellulose, forming a clear colloidal solution. It is a hygroscopic material with the ability to absorb more than 50% water in high humidity. It is one of the natural derivatives that can be used in the detergent, food and textile industries. Thanks to its thickener, binder, stabilizer, film forming and gelling functions, CMC is used in various sectors such as food, drilling, mining, construction, paper, detergent, paint, textile, pharmaceutical and personal care [21, 96].

2.3 Oxidation

Oxidation is a reaction used to add carboxyl groups to the cellulose structure. This reaction can be classified as selective and nonselective oxidation. In nonselective type, oxidants include H_2O_2, NaOCl, periodate, etc. Selective oxidants can oxidize the -OH groups of cellulose to obtain specific cellulose derivatives [14, 62]. The oxidizing agent is usually activated in an aqueous solution. This process can oxidize primary hydroxyl groups to carboxyl groups selectively. Depending on the oxidant used, byproducts such as water or small organic molecules are formed. Wen et al. [133] used a green oxidation process combining ultraviolet light, hydrogen peroxide, and ozone. In this study, nanocellulose fibers were prepared by homogenizing oxidized cellulose under high pressure. The average length–width values of nanocellulose fibers prepared with this green method were measured as 11 μm–22 nm, and these values showed that nanocellulose fibers are high-quality fibers and have great application potential. Oxidation increases the biodegradability and biocompatibility of cellulose, making it advantageous in biomedical applications such as wound dressings, drug delivery systems and tissue engineering scaffolds. It also increases the ability of cellulose to form hydrogels and interact with other biomolecules. Dialdehyde cellulose (DAC) is a cellulose derivative produced by periodate oxidation. DAC is used in biomedical applications due to its improved biodegradability and biocompatibility. (TEMPO-oxidized cellulose is cellulose derivatives obtained by dialdehyde cellulose oxidation.)

An example for this type modification is 2, 2, 6, 6-tetramethylpiperidine-1-oxyl (TEMPO)-oxidized cellulose (Fig. 7). The cellulose samples are immersed in aqueous solution of TEMPO and NaBr. The pH of the NaClO solution is adjusted to 10 with HCl. The NaClO solution that the pH adjusted to 10 is added on the solution carefully and stirred at ambient temperature. The pH of the solution is adjusted to

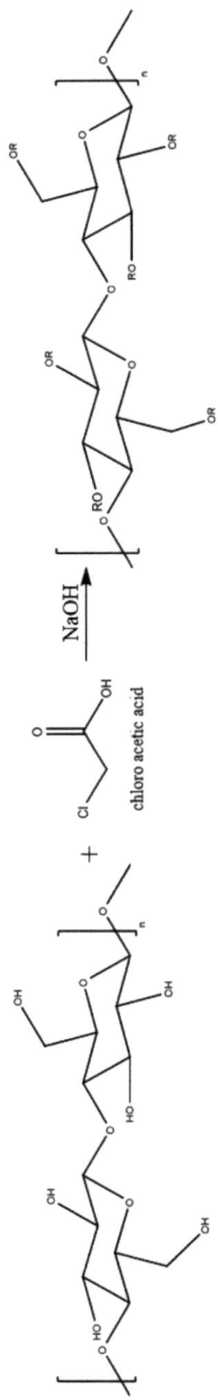

(a) R: H or CH_2CO_2H

Fig. 6 General reaction mechanism for the synthesis of carboxymethyl cellulose

10 with 0.5 M NaOH solution until all the NaOH is consumed. The oxidized sample is washed and stored in cold, i.e., in refrigerator at 4 °C [98].

2.4 *Grafting*

Grafting is a reaction that attaches polymer chains or functional groups to the cellulose backbone. Active –OH groups along the chain in the cellulose molecule are used as grafting points. Grafting can be achieved using various techniques, including primarily atom transfer radical polymerization, and click chemistry [18, 21]. Monomers or functional groups are added to the activated sites to form a covalent bond with cellulose. The reaction is terminated by deactivating the reactive sites. This reaction increases the mechanical strength, thermal stability, and chemical reactivity of cellulose. Grafting is a reaction that customizes cellulose for specific applications, particularly smart hydrogels [91], nanocomposites [57], and responsive coatings [50]. Grafted cellulose derivatives are also used in drug delivery and tissue engineering studies [37, 141]. (Poly(N-isopropylacrylamide) grafted cellulose, polyethylene glycol grafted cellulose are examples of this type of derivatization.)

2.5 *Crosslinking*

Crosslinking reaction is a reaction in which crosslinkers such as N, N'-methylene bisacrylamide, epichlorohydrin, glutaraldehyde or citric acid are used to form a 3-D network structure of cellulose or its derivatives with other polymeric species. Crosslinkers are usually activated by heat or a catalyst. The activated crosslinker reacts with –OH groups in different chains of cellulose to form covalent bonds. The final product has a three-dimensional network structure and has improved mechanical and thermal properties. Characterization of modified surfaces is evaluated by elemental analysis, contact angle measurements, SEM, ^{13}C-NMR Cross-Polarization/Magic-Angle Spinning, FTIR and XPS [15]. With this modification, cellulose, which has increased mechanical strength, thermal stability and water resistance, becomes suitable for use in superabsorbent materials and some biomedical applications including wound dressings, controlled release systems and tissue scaffolds [25, 118]. (Epichlorohydrin cross-linked cellulose, glutaraldehyde cross-linked cellulose can be given as examples of this type of derivatization.)

An example for this type of modification is crosslinking with glutaraldehyde (Fig. 8). There are three possible paths for the crosslinking of the cellulose dependent upon the cellulose functional groups. The soluble cellulose sample is first dissolved in aqueous solution containing glutaraldehyde that is stirred for 20 min, then the crosslinked cellulose samples were separated with centrifugation and washed with distilled water. The second path for the crosslinking of the cellulose samples is before gelation. The film form of the cellulose, obtained by centrifugation, is immersed in

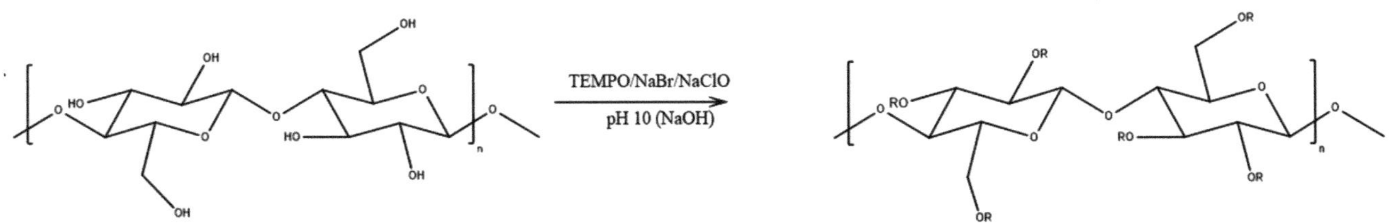

(a) R: H or COONa

Fig. 7 General reaction mechanism for the oxidation of cellulose with TEMPO

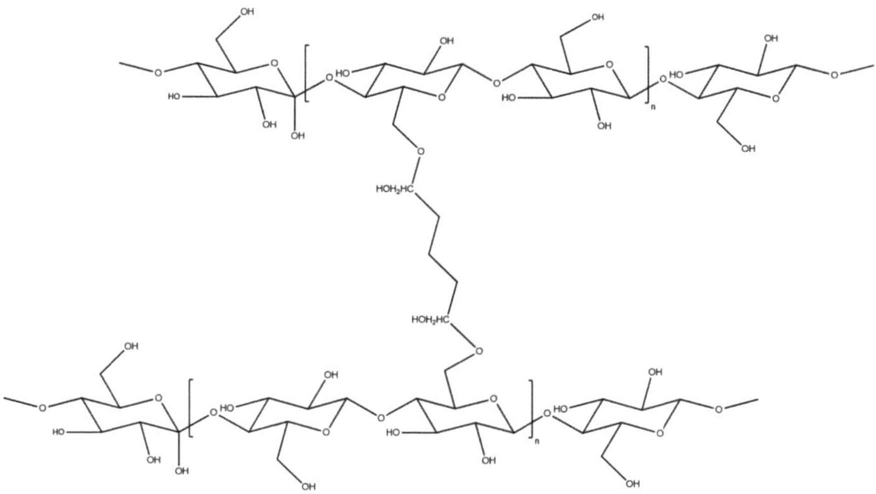

Fig. 8 Crosslinked cellulose with glutaraldehyde

glutaraldehyde solution that stands for 20 min at 60 °C, then the cellulose is immersed in H_2SO_4 solution, that the gels are formed after contact. Finally cellulose films are washed with distilled water. The last path for the crosslinking of cellulose is after gelation in H_2SO_4. The gelated cellulose is immersed in glutaraldehyde solution for 20 min at 60 °C, then crosslinked cellulose is washed and dried [33].

Cellulose graft derivatives, in which a polymer such as thermo-responsive poly(N-isopropyl-acrylamide) (PNIPAM) is grafted onto cellulose to create smart materials that respond to temperature changes, offer advantages in drug delivery and tissue engineering. Glutaraldehyde cross-linked cellulose is used in wound dressings and tissue scaffolds where increased stability and durability are required.

3 Biomedical Research Applications of Cellulose or Cellulose Derivatives

Cellulose derivatives obtained by chemical modification of cellulose and having improved properties are widely used in various fields. In tissue engineering, cellulose-based scaffolds provide a supportive matrix for cell growth and tissue regeneration. Their porous structures allow for the uptake of nutrients and the disposal of unused substances, which are very important for tissue development. Cellulose hydrogels and films are used especially in wound dressings because they create a moist environment, reducing the risk of infection in wounds and contributing to faster healing. Cellulose derivatives can be designed as controlled drug systems to control the release of drugs, increase therapeutic efficacy and minimize side effects. The ability to create a release profile specific to the drug used makes cellulose derivatives ideal for targeted drug release and sustained release formulations. Recent developments in cellulose research have focused on the development of nanocellulose, stimuli-responsive cellulose and green process techniques. Cellulose produced at nanoscale exhibits mechanical strength and large surface area, that brings advantageous for high-performance materials and biomedical devices. Cellulose derivatives respond to certain stimulants as pH, temperature and light, providing successful results in smart drug delivery systems and wound dressings. Green process techniques aim to reduce the environmental impact of harmful chemicals used in cellulose modification environments by using environmentally friendly methods and reagents.

Biomedical areas where cellulose derivatives are used are given in more detail under subheadings with examples from the literature.

3.1 Drug Delivery Systems

Cellulose is derivatized to form hydrogels and films that can release drugs in a controlled manner. These systems provide time-dependent slow release, increasing the efficacy of drugs and reducing their side effects [86, 87]. Studies have shown that a new modification method has been used to develop nanocrystalline cellulose with antibacterial properties. In this method, aldehyde groups are attached to nanocrystalline cellulose, followed by direct binding of antibacterial agents. The efficiency of this method is the ability of these agents to maintain their antibacterial activity for long periods of time, and this has been considered a promising approach for antibacterial applications. New techniques and chemicals used in cellulose derivatization for wound dressing applications have overcome the obstacles to obtain improved absorbent antibacterial and biofilm-disrupting hydrogels. Modified cellulose can be used as a drug carrier. The grafted polymer contains different functional groups and crosslinking agents in the cellulose graft copolymer matrix. Cellulose graft polymers have hydrophobicity which helps in drug interaction and contains different weaker forces like van der Waals forces, electrostatic force etc. Drug carriers with hydrophilic character show a faster rate of drug release compared to a hydrophobic carrier. Therefore, modified cellulose has great potential in future as a strong drug carrier [61].

3.1.1 Ophthalmic Drug Release

The cellulose derivatives can be used in ophthalmic drug release systems, since drug release is slower and more continuous in ophthalmic systems, the contact time of the active substance in cellulosic materials with the tissue in this drug form is longer and the total bioavailability is higher.

Taghe et al. [109] used nanofibers produced by electrospinning of polycaprolactone and cellulose acetate to use the drug azithromycin for the ocular treatment. The average diameters for the polycaprolactone and cellulose were 160 nm and 190 nm, respectively. The microbial behavior of the nanofibers was quantified with *Micrococcus luteus* microorganism as well as they showed antibacterial efficiency against *Escherichia coli* and *Staphylococcus aureus*. The polymer systems were identified as non-cytotoxic and not irritant. The polycaprolactone nanofibers released the model drug azithromycin with a high rate that follows Peppas kinetic model as well as cellulose acetate nanofibers released the model drug azithromycin with a controlled rate that follows first order model. In in-vitro applications, it was observed that all formulations can release effective concentrations of azithromycin about 6 days. Nilay et al. [79] examined on the ocular applications of chloramphenicol as a model drug. The cellulose acetate and cellulose acetate butyrate films were prepared in ocular film forms. The effect of the formulation on controlled drug release was investigated. The mechanical strength, thickness of the films and *in-vitro* studies were applied for different formulations. All different compositions enhanced the drug delivery by retaining about 12 h.

The patent for the methylcellulose-based solution for dry eye problems was obtained by Haddad and Loucas [36]. Methyl cellulose was used in the eye drops as viscosity modifier to increase the residence time of ofloxacin in cornea, but after the application, it is observed that, methyl cellulose precipitated [4]. The effect of composition and temperature of the ternary methyl cellulose-polyethylene glycol-citric acid ophthalmic drug delivery vehicle was determined to be useful for the application into the eyes and when compared with the commercial one, this vehicle system was determined to be more effective. Takeuchi et al. [110]. The drug delivery system based on methyl cellulose and carbapol was examined that resulted in a regular release of ocular drugs in nearly 12 h that is comparatively high [58]. In vitro and in vivo results of the timolol maleate formulation in 3% methyl cellulose improved ocular bioavailability as well as the temperature of the eye shifted the system into gel after instillation that maintains the optimum release capacity [27]. The effect of molecular weight of methylcellulose on gelation temperature of Poloxamer 407 was determined to examine the release of ketorolac tromethamine. The presence of –OH groups in methyl cellulose gums up the physical attractions between Poloxamer 407 and water. Researches showed that molecular weight of methyl cellulose retains the release of ketorolac tromethamine from the system. The drug release kinetics follow first order equation and controlled by the Fickian-diffusion that obeys Korsmeyer–Peppas equation [24]. The release of ketorolac tromethamine from 1% methyl cellulose complexes were also examined by Bain and co-researchers in several works [8]. They prepared a system from methyl cellulose, polyethylene glycol and sodium chloride. They examined the effect of molecular weight of polyethylene glycol on gelation of 1% methyl cellulose and observed that increase in molecular weight of polyethylene glycol decrease the gelation temperature as well as increase in the ketorolac-tromethamine release time. The other research examined for the release of ketorolac tromethamine by Bain et al. [7] is the effect of amount of citrate tribasic dihydrate on gelation temperature of methyl cellulose in the presence of fructose. They observed that an increase in amount of citrate tribasic dihydrate in constant 1% methyl cellulose and 10% fructose, a decrease in the sol–gel transition was observed through 1–5% citrate tribasic dihydrate composition that enhances the release of ketorolac tromethamine. The release of ketorolac tromethamine from 1% methyl cellulose, 10% polyethylene glycol, sodium tartarate and sodium citrate system was observed. Sodium citrate was determined to be more effective to decrease the gelation temperature of the methyl cellulose-polyethylene glycol system when compared with sodium tartarate. The release of ketorolac tromethamine duration increased more effectively in the presence of sodium citrate [11]. The effect of molecular weight of polyvinyl alcohol in 1% methyl cellulose and sodium chloride to the release of ketorolac tromethamine was examined. The increase in the molecular weight of the polyvinyl alcohol for hundred times decrease the gelation temperature as well as an increase in the ketorolac tromethamine release time, that was attributed to the higher viscosity of the system and effective hydrogen bonds with an increase of molecular weight of polyvinyl alcohol [9].

Obeidallah et al. [83] produced micro sponges from ethyl cellulose to formulate the antiglaucoma drug acetazolamide for ocular drug delivery in situ gel form with Pluronic F-127. The applications on rabbit's eye showed a good biocompatibility and the therapeutic effect of the drug was observed to be high when compared with free drug. Zhu et al. [145] designed a inner layer embedded contact lens and prepared a blend film of ethyl cellulose and Eudragit S100 as the inner layer and poly(2-hydroxyethyl methacrylate hydrogel as the outer layer by using diclofenac as a model drug. In vivo studies in rabbits showed an improved corneal residence time in rabbit eyes. Maulvi et al. [71] also developed a drug delivery system via contact lenses. Thymol maleate was loaded in an ethyl cellulose-acrylate based hydrogel contact lenses. The mean residence time was enhanced when compared with the drug alone. Nair et al. [76] examined the release of timolol maleate as well, through a corneal membrane by developing a gelation system from sodium alginate, ethyl cellulose and plasticizer polyethylene glycol. The results concluded that increase in the amount of the ethyl cellulose in the gelation system delayed the release of the drug from corneal membrane when compared with marketed thymol maleate eye drops. Balzus et al. [12] loaded dexamethasone on ethyl cellulose and Eudragit nanoparticles to understand the controlled release of the drug on the skin. The researchers examined on ex vivo drug release and observed that the penetration of the drug into the skin from the nanoparticles is slower when compared with commercial one, that improved system can be used for the delivery of corticosteroids.

Popa et al. [89] developed bupivacaine hydrochloride loaded chitosan and hydroxypropyl methyl cellulose in situ gelling systems for ocular applications. The mucoadhesive polymer chitosan is pH sensitive as well as other mucoadhesive polymer, hydroxypropyl methyl cellulose is thermosensitive and viscous, the researchers aimed to combine the distinct properties of two biopolymers. The optimum parameters of the combination of in situ gelling systems were determined to be based on response surface methodology, contact angle, viscosity and *in-vitro* release of bupivacaine hydrochloride. The gelling system containing 1% chitosan, hydroxypropyl methyl cellulose was the optimum formulated gelling system that was suitable for the range of optimal viscosity of the eye, the released bupivacaine hydrochloride after 6 h was 69.55%, which is the highest one. Liu et al. [65] examined on the formulations of nanomiscelles of poly (ethylene glycol-block-poly(*D,L*)-lactic acid) (NH_2-PEG-*b*-PLA) and hydroxypropyl methylcellulose to improve the penetration of hydrophobic drugs to the eye, such as tacrolimus. The nanomiscelles were investigated for stability, encapsulation efficiency, cytotoxicity, surface tension and in vitro release. The release of the bupivacaine hydrochloride was confirmed as higher concentration for a long time that shows the system is a promising bioavailable and efficient material. Cheng et al. [20] synthesized methacrylated hydroxypropyl methyl cellulose for the ophthalmic drug delivery of natural compound curcumin. The rheological analysis of the gelation system showed that methacrylation of hydroxypropyl methyl cellulose improved the bio adhesive forces, shear thinning processes and self-recovery behaviors. The *in-vivo* applications results concluded that the synthesized gelation system was biocompatible.

Mu et al. [73] examined on the collagen peptide modified carboxymethyl cellulose for a potential drug carrier. The protective effects of the modified carboxymethyl cellulose were investigated for an H_2O_2 induced injury on the retinal endothelial cells. The researchers concluded that there was a significant increase in the viability of the cells initially treated with H_2O_2 as well as the production of the reactive oxygen species and cell apoptosis were restricted in in vitro studies. After that, the modified carboxymethyl cellulose, alone and interleukin-10 loaded, was applied as drug carrier for the treatment of retinal ischemia/reperfusion in rats. The modification of the carboxymethyl cellulose with collagen peptide improved the treatment of retinal injury when compared with carboxymethyl cellulose alone. Yang et al. [136] developed a thermosensitive hydrogel based on oxidized carboxymethyl cellulose and aminated Poloxamer 407. The synthesized hydrogel system was determined to have high transparency, adjusted gelation temperature and time. The bone morphogenetic protein 4 was loaded on the hydrogel system for the treatment of corneal neovascularization. The duration of the drug in the target area is longer in the presence of hydrogel that improves the inhibition of corneal neovascularization. The animal experiments concluded the biocompatibility of hydrogel and its effective inhibition for the treatment of corneal neovascularization after ocular surface surgery. Zhang et al. [140] examined on the development of a novel formulation of the tacrolimus, a drug for the dry eyes. They first synthesized shapeless tacrolimus microcrystals and in the presence of poly (allylamine hydrochloride) and then carboxymethyl cellulose, the product was poly(allylamine hydrochloride)/carboxymethyl coated tacrolimus. The obtained drug matrix was in vivo applied to a mouse drug eye and the drug matrix improved the therapeutic performance when compared with the commercially available one.

3.1.2 Oral Drug Release

Zhou et al. [143] prepared carboxymethyl chitosan/cellulose acetate microspheres by emulsion solvent evaporation for drug delivery system. The prepared microspheres were spherical porous structure with small volumes. The prepared samples had an efficiency for the delayed release of acetaminophen as well as the optimum formulation contained 2.0% carboxymethyl chitosan and 1360 KD. The release of the drug was observed slower in acidic conditions when compared with relatively basic saline solution for 1 day, meaning that delivery of the drug is slower in simulated gastric fluid rather than simulated intestine environment. The kinetics of the drug release was found as first order meaning that this process strongly depends on drug amount in the chains of the polymers. Çetin et al. [23] developed a controlled release system of anti-inflammatory agents. They examined the efficiency of the cellulose acetate films for in vitro release of chlorhexidine gluconate, indomethacin, and meloxicam model drugs. The films were loaded with model drugs and the release of the active drug agents from the cellulose acetate films were investigated in 10 mL of simulated saliva at 37 °C. The indometacin release kinetics was higher when compared with other formulations.

Oral dosages of drugs generally based on the solid tablets, and sometimes to use these types of drugs, it is difficult for the patients that have a dysphagic problems. Viscous drug systems are easy to swallow when compared with solid ones [53, 54]. Methyl cellulose based liquid drugs have different properties to improve the release duration that is useful for the dysphagic patients. The thermally reversible methylcellulose solutions were prepared by Itoh et al. [42] in pectin for an ion-responsive gelation that forms in situ gelling agent for oral drug delivery. The gels produced by methyl cellulose (1–2) % and pectin (0.5–2.0) % were suitable for in vivo and in vitro paracetamol release with a suitable gel strength. When 20% d-sorbitol and calcium ions were complexed with 1.5% pectin and 2% cellulose, the complex gel became more stable with a viscosity that brings ease for swallowing of dysphagic problems. Itoh et al. [43] examined the effect of d-sorbitol on the gelation temperature and strength of methyl cellulose in liquid media. The dissolution of (25–30) % d-sorbitol in (1–2) % methyl cellulose decreased the gelation temperature from 53 °C to 37 °C, a body temperature, with an increase of the gel strength. The in vitro release of paracetamol from the above gel system was observed as diffusion control. Shimoyama et al. [105] formulated in situ gelling liquid composed of methyl cellulose and sodium alginate, that the gelation is thermally reversible as well as it is ion-responsive, suitable for dysphagic patients. They prepared solutions of sodium alginate (0.25–1.0) % in constant methyl cellulose (2.0) % to examine the rheological and gelation characteristics of the solutions for dysphagic patients. The paracetamol release was followed and compared for the different gelation systems, that is improved, in vitro release of paracetamol was diffusion controlled in 2.0% methyl cellulose and 0.5% alginate gels in acidic (1.2) and neutral (6.8) pH values. The release of paracetamol was also investigated by Sharma et al. [104] for methyl cellulose and alginate in situ gelling system. Aqueous sodium alginate (1.5% w/v) complexed with calcium ions to form a gel for the release of calcium ions in acidic media such as stomach liquid. The alginate–calcium system was improved by addition of methyl cellulose, sodium chloride and polyethylene glycol. The results of *in-vitro* studies concluded that the paracetamol release from the gel system was diffusion controlled. The developed *in-situ* gelation system containing of 1.5% sodium alginate, 1.5% methyl cellulose, 3% $CaCO_3$, 2% NaCl 0.05% polyethylene glycol was more stable and efficient in the stomach of rabbit when compared with the equal dose of commercially available suspension.

Wang et al. [126] examined on the controlled release of the model drugs zolmitriptan and etodolac from a film composed of chitosan, sodium alginate and ethyl cellulose for buccal mucosal applications. The chitosan/sodium alginate/methyl cellulose film was mechanically stable and showed optimum performance for the release of the model drugs when compared with chitosan/sodium alginate and ethyl cellulose/sodium alginate films. The permeability behavior of the discussed films was concluded in the TR146 cells model as well as in the rabbit buccal mucosa. The researchers observed an improvement of the drug release from the films when compared with the drug alone. Siew et al. [106] studied on film coatings from amylose and ethyl cellulose blends for colonic drug delivery, the effect of thickness of the film on the model drug 5-aminosalicylic acid pellets. The drug release was observed both in acidic gastric and basic small intestine conditions. The release of the model drug from the film was enhanced when the amylose content was relatively higher as well as the thicker films with low amylose content showed a poor release of the drug. The presence of the pepsin and pancreatin did not affect the drug release characteristics from the films. The researchers concluded that the film thickness and amylose content should be adjusted for a controlled delay of the drug release. El-Habashy et al. [26] studied on the ethyl cellulose-based nanoparticles that were loaded with piroxicam to control the release and to reduce the gastric irritation resulted from piroxicam. The nanoparticles stabilized with poloxamer brought well sized nanoparticles and enhanced the drug release as delaying in optimum duration. Sajeev et al. [100] examined on the controlled release of diclofenac sodium that is microencapsulated with ethyl cellulose in a form of compressed tablets. The in vitro release of drug from tablet form microcapsules in distilled water was determined to be zero order and a delay of release more than 24 h.

Bodini et al. [17] examined on the preparation of films that were formulated in different compositions of pregelatinized starch and hydroxypropyl methyl cellulose with improved mechanical and disintegrating properties. The amount of the sorbitol plasticizer is 20% of the total macromolecules. They observed that the increase in the amount of the hydroxypropyl methyl cellulose reduced the mechanical properties in units of tensile strength as well as the mucoadhesive behavior of the films decreased. The in vitro examinations on disintegration time was less than 30 s. The 80% starch and 20% hydroxypropyl methyl cellulose had the best taste in the mouth according to in vivo studies as well as this composition was found to be the best for the applications on oral release. Ravi et al. [96] investigated the effect of the amount of hydroxypropyl methyl cellulose with different viscosity grades on controlled *in-vitro* release of zidovudine from matrix tablets as well as the effect of compression force was also evaluated. They determined the kinetics of releasing by using different modeling equations. The delivery of the zidovudine was observed to be the least efficient in high hydroxypropyl methyl cellulose concentration and naturally with higher viscosity as well as the higher force for compression of tablets reduced the rate of delivery. Talukdar et al. [112] compared the drug release performance for xanthan gum and hydroxypropyl methyl cellulose as prepared in hydrophilic matrix-forming agents for the delivery of caffeine, indomethacin and sodium salt of indomethacin. The two gelling systems have similar characteristics when compared but the presence of salts in gastro-intestinal system, adversely brought disadvantage when compared with hydroxypropyl methyl cellulose, a polymer stable in ionic strength solutions as well as xanthan gum has better flowing abilities.

Barkhordari and Yadollahi [13] examined on the cephalexin as a model drug that was intercalated between pH sensitive carboxymethyl cellulose beads capsuled for layered double hydroxides via co-precipitation method, in nanohybrid structure and designed for gastrointestinal tract conditions. The produced nanohybrid structure was used to synthesize nanocomposite hydrogel beads with carboxymethyl cellulose. The gastrointestinal track in-vitro conditions were simulated to carry out the efficiency of the drug delivery system that showed a good protection against acidic stomach pH and improved release in gastrointestinal track conditions. Agarwal et al. [2] examined on the preparation of calcium alginate/carboxymethyl cellulose beads for oral drug delivery addressed to the colon. The drug was designed as pH sensitive, mucoadhesive and specific for being released in colons. The calcium alginate/carboxymethyl cellulose beads were synthesized by ionic gelation method. In the simulated colonic environment, the swelling and mucoadhesive properties of the polymeric system was determined as higher as well as the biodegradation of the composition is slower in the simulated environment when compared with colonic microflora. The anticancer drug, 5-fluorouracil was selected as a model drug for *in-vitro* release applications and, more than 90% of the drug was released from the polymeric system in colonic enzymes found environment, and the results concluded that carboxymethyl cellulose improves the duration of the controlled drug release. Javanbakht et al. [45] aimed to develop copper-based metal–organic nanocomposite for the encapsulated ibuprofen in pH sensitive carboxymethyl cellulose for oral delivery system. The *in-vitro* drug delivery tests were applied and it was determined that the encapsulation ibuprofen with carboxymethyl cellulose improved the controlled release of the gastrointestinal tract conditions as well as the nanocomposite showed lower toxicity to Caco-2 cells.

3.1.3 Injectable Gels

Some drugs are not useful orally for some patients via digestion and injection methods are improved for these types of patients. The thermosensitive gelation-based systems are developed in most studies. Khaing et al. [55] developed a cellulose acetate butyrate gelation system for the release of metronidazole and doxycycline hyclate. The authors observed the increase of mechanical strength of the gelation system as the amount of the polymer increased. The diffusion-controlled release of metronidazole and doxycycline hyclate was achieved for at about a week after in-situ gelling. Wang et al. [127] studied to develop a mixture of cellulose acetate and sodium alginate that contains laponite. The hydrogel system showed good resistance against *Staphylococcus aureus* and *Escherichia coli* bacteria as antimicrobial activity as well as it was biocompatible for the human that was concluded with osteosarcoma cell line

MG-63. *In-vivo* applications concluded that laponite improved the bone healings when compared with the free gels.

Kim et al. [56] aimed to produce low molecular weight methyl cellulose with Pluronic F127 gelation system to release docetaxel. The low molecular weight methyl cellulose with Pluronic F127 injectable thermoreversible gelation system was examined on its efficiency of subcutaneous tumor model. The gelation system combined from low molecular weight methyl cellulose, ammonium sulfate and Pluronic F127 system formed gels at (15–40) °C depending on ammonium sulfate concentration, an increase in salt concentration resulted in a gel formation temperature. The gelation system improved anti-cancer effects of docetaxel when compared with free docetaxel since the gelation system sustainably released docetaxel more than 30 days with the maintained structure of the docetaxel. The controlled release of Pluronic F127 from methyl cellulose-based gelation systems were also examined by Rangabhatla et al. [95]. The 4% methyl cellulose was prepared with different amounts Pluronic F127 to prepare injectable drug delivery system. The in vitro degradation test results confirmed that methyl cellulose improved the gel strength of Pluronic F127 and retained the degradation of the gelation that brought sustainable drug delivery for osteogenesis. Hyaluronan with methyl cellulose gelation system was developed by Gupta et al. [35] that the blend is ready for the gelation after just injection. The strength of the gel is maintained with temperature as well, in vivo studies in the rat concluded the biocompatibility of the gel system for 1 month. The results showed that hyaluronan with methyl cellulose gelation system is suitable for localized delivery of the therapeutic drug agents.

Adhikari et al. [1] mixed the ethyl cellulose with ethanol to obtain gel that is injected to destroy cancerogenic cells in which ethyl cellulose is used as a biopolymer that controls the release of ethyl alcohol. The tissue simulated phantoms in 1% agarose solution were applied during injection to follow the effect of injection parameters on the character of the resulting gel. The needle type and rate of the injection had minor effect on the distribution of the gel in the body as well as the depth of the injection greatly affected the distribution of the gel. The effect of the concentration of the ethyl cellulose became important when the injection pressure is (70–250) kPa. Mueller et al. [74] also examined the effect of injectable ethyl cellulose in ethanol on cervical dysplasia. The researchers investigated the effect of the relative amounts of the ethyl cellulose and ethanol with injection volume and parameters such as needle size, injection direction, rate, depth and infusion rate to the body. The researchers found the optimum efficient conditions via in vivo applications on large animals like swine. Thakur et al. [115] examined on the thermosensitive injectable gels in which carboplatin were loaded in ethyl cellulose nanoparticles. They prepared the system by using double emulsion solvent evaporation method and studied on the size, shape and efficiency of encapsulation. The release of the drug from nanoparticle system synthesized in optimum conditions at ambient temperature was for a week as well as the same is for 2 h from the pure drug and commercial one. The good anti-tumor activity of the nanoparticles in hydrogels system was reported in situ enrich ascitic carcinoma mice model as well as it was safe when compared with the commercial one.

Talasaz et al. [111] examined on the in situ gel combination consists of poloxamer 407 for the release of model drug vancomycin. They aimed to decrease the concentration of poloxamer by addition cellulose derivatives, such as hydroxypropyl methyl cellulose and hydroxypropyl cellulose to reduce the diverse effects of poloxamer. The cellulose additives decreased the poloxamer concentration from 18 to 10% that forms *in-situ* gelation. The addition hydroxypropyl methyl cellulose increased the gelation temperature more when compared with the hydroxypropyl cellulose. The higher viscosity above 27 °C concluded that all the compositions are possible candidates for the applications of drug delivery as an injectable local drug delivery. Okubo et al. [84] investigated a thermoresponsive injectable hydrogel from hydrophobically modified hydroxypropyl methyl cellulose and cyclodextrin that the release was completely depend on intermolecular interactions. The thermo-responsive injectable hydrogel consisting of hydrophobically modified hydroxypropyl methyl cellulose and cyclodextrin was in the gel form near body temperature. The addition of the cyclodextrin decreased the yield stress of the gelation system and it is necessary for the injectable form of the gelation of hydrophobically modified hydroxypropyl methyl cellulose. *In-vivo* applications of the indocyanine loaded gelation system on the mice concluded that the retained indocyanine was constant for 1 day that is longer when compared with indocyanine alone. Wang et al. [124] developed thermosensitive injectable material composed of hydroxypropyl methyl cellulose, polyvinyl alcohol and Fe_3O_4 to magnetic hyperthermia ablation of tumors. The *in-vitro* and *ex-vivo* experiment results of the injectable gelation system concluded that the temperature and necrotic volume of bovine liver were correlated with hydroxypropyl methyl cellulose/Fe_3O_4 weight, heating time and Fe_3O_4 content. 180 s heating duration for 0.06 mL hydroxymethyl cellulose and 60% Fe_3O_4 is enough for complete ablation of *in-vivo* tumors applications.

Varma et al. [121] developed an injectable modified carboxymethyl cellulose for soft tissue filler. The carboxymethyl cellulose was methacrylated and crosslinked with ammonium persulfate and ascorbic acid to form stable hydrogels for the treatment. The mechanical properties of the synthesized gel system were suitable when compared with native soft tissues and commercial surgical fillers currently used. The mechanical and swelling properties, mesh size were affected from polymer concentration and crosslinking density. Rheological analysis, hydrolytic degradation, cytocompatibility results concluded that crosslinked carboxymethyl cellulose hydrogels are suitable as a soft tissue filler in medical applications. Jiang et al. [46] reported a development of injectable hydrogels produced from modification of carboxymethyl cellulose with methacrylic anhydride. The methacrylic anhydride was grafted on carboxymethyl cellulose via UV radiation. The amount of grafted methacrylic anhydride on carboxymethyl cellulose improved the surface characteristics, swelling behaviors and mechanical properties. The gelation system was found to be *in-vitro* cytocompatibility. The researchers synthesized a light sensitive and injectable carboxymethyl cellulose derived hydrogel system for biomedical applications. Chen and Fan [19] synthesize thermosensitive chitosan/carboxymethyl cellulose injectable blends as a biomaterial for the culture of chondrocytes. The combinations have physiological pH and can be stored as liquid below room temperatures as well as after just *in-vivo* injection at the body temperature, these formulations turned out to be gel structures. The thermoset gelation system was approved as compatible with biological compounds.

3.2 Tissue Engineering and Wound Dressing

In tissue engineering, cellulose scaffolds provide a suitable support for cells to grow and form new tissues. These scaffolds are designed to mimic the extracellular matrix and promote cell attachment and proliferation [103].

Cellulose-based wound dressings are highly effective due to their ability to maintain a moist environment, promote healing, and prevent infections. Bacterial cellulose, in particular, is used due to its high purity and excellent water retention properties [68]. The inherent features of bacterial cellulose, which include high biocompatibility and biodegradability, enable it to be a potentially useful polymer for wound healing. In addition, bacteria naturally form hydrogels through biosynthesis, making cellulose an ideal material to facilitate the healing process in dry wounds where more moisture is required. However, since bacterial cellulose also lacks antibacterial properties, it is inevitable to modify it chemically. In one study, green chemistry processes were used to functionalize cellulose with antibacterial functional groups [85]. The material obtained by covalent derivatization of –OH groups of two different active ingredients through heterogeneous reaction in basic aqueous conditions was then subjected to biological evaluation using both bacterial cells and human keratinocytes. It was observed that the cellulose derivative, which was brought into contact with different bacterial species, caused a 50% decrease in the population within 24 h. Meanwhile, cellulose-based hybrid films and materials are attracting attention among recent developments, as cellulose offers new and effective antibacterial solutions with extraordinary versatility.

Nosar et al. [81] examined on the three-dimensional scaffolds based on cellulose acetate, produced via wet electrospinning technique. Different amounts of cellulose acetate were dissolved in acetone for fabrication of scaffolds, and, the highest porosity percentage was observed as 69.5%, that is relatively low for tissue engineering applications. The scaffolds were determined to be hydrophobic as well as the attachment of the cells was enhanced. Luo et al. [67] reported the synthesis of carbon nanotubes that contain nano fibrous scaffolds based on polysaccharides for tissue engineering applications. The cellulose acetate nano fibers were mixed with chitosan and multiwalled carbon nanotubes, or, sodium alginate. The presence of the multiwalled carbon nanotubes instead of sodium alginate in the scaffold system, enhanced the mechanical strength.

Methyl cellulose is one of the hydrogels extensively used for the tissue engineering [78]. Tate et al. [114] developed a methyl cellulose as a scaffold for the defects of the brain that is biocompatible and injectable. The blend of the methyl cellulose with polymethyl methacrylate was also used for tissue applications by Stalling et al. [34, 108] after some modifications. Stalling et al. modified methyl cellulose with methacrylate groups to increase the functionality by crosslinking with photolysis to produce hydrogels for the possible application in plastic and reconstructive surgery after determination of the optimum moduli of the gels. Gold et al. also modified methyl cellulose with methacrylate groups to increase the functionality by redox polymerization to obtain hydrogels. They measured the rheological data and moduli values of hydrogels that was determined as same with human adipose tissue.

Another research that includes organic polymer with methyl cellulose was studied by Sá-Lima et al. [101]. The researchers grafted poly(N-isopropylacrylamide) on methyl cellulose to prepare thermoreversible hydrogel that behaves as a three-dimensional support for cell encapsulation to regenerate articular cartilage. The grafted copolymer behaved to be a gel at about 32 °C or higher temperatures. The chondrogenic cells were encapsulated with the gelation system and it was observed that hydrogel scaffolds protected morphology with their round shapes as well as the biology of the chondrocytes worked well. The injectable chitosan-methyl cellulose blend hydrogels were examined as three-dimensional matrix for tissue repairing by Tang et al. [113]. The effect of the different salts on, NaCl, Na_3PO_4, $NaHCO_3$ and glycerophosphate, on gelation process was investigated. They concluded that NaCl included system had the highest gelation temperature as well as glycerophosphate included system had both highest gelation rate and gelation strength. The Na_3PO_4 included chitosan-methyl cellulose gelation systems behaved as most suitable for cell viability and proliferation when used as a scaffold for chondrocytes. An injectable hydrogel composed of hyaluronic acid and methyl cellulose were examined about the appropriate degradation rate. The gelation system was stabilized by polyethylene glycol as crosslinker, the rheological properties and stability of the gelation system increased the duration time of the hydrogel. The gelation system was observed to be fast-gelling, injectable and biocompatible as well as can have the shape of the injured tissue by *in-situ* forming [146].

Sajadi-Javan et al. [99] prepared a novel hydrogel blends of methyl cellulose and Persian gum, that the gelation is sensitive to temperature. The strength of gelation system was improved with taxifolin loaded halloysite nanotubes for the bone tissues. The mechanical and rheological tests were performed as well as injectability was investigated for different formulations. The optimum conditions for the gelation system was determined to be as 1% Persian gum and 3% halloysite nanotubes in methyl cellulose, that brings easier injection with rapid gelation at the ambient temperature with the highest mechanical strength. The thermal response of methyl cellulose for gelation was investigated as well for cell sheet engineering applications. The solution-gel analysis was performed via differential calorimetry and rheological tests to understand the effects of molecular weight and concentration of methyl cellulose as well as dissolved salts in the media. The hydrogel stability and water uptake capacity were investigated via swelling and degradation experiments of the hydrogel system in the cell culture solutions at the incubation temperature for several days. The optimum conditions were determined to be for (12–16) % methyl cellulose in water with a molecular weight of 15,000 at or above 32 °C. The gelation system was applied on the stromal/stem cells derived from human adipose tissue derived stem cells for a week to understand the efficiency. The addition of bovine collagen on the methyl cellulose hydrogel surface at 37 °C improved the adhesion and the cell growth on the hydrogel system [116].

Altomare et al. [6] also investigated the methyl cellulose based thermosensitive hydrogels for the preparation of cell sheets. The methyl cellulose-based hydrogel solutions were prepared by dispersion technique in the presence of different kinds of salts which are sodium sulfate, sodium phosphate, calcium chloride and phosphate buffers to understand the optimum conditions for gelation temperature and gel stability, which is determined for the hydrogel 8% (w/v) methyl cellulose in 0.05 M sodium sulfate at 37 °C thermoreversible temperature. At the optimum conditions, the cell sheet was generated effectively and separated from hydrogel surface successfully in the absence of enzymes. The methyl cellulose-based tissue engineering scaffolds can also be fabricated via 3D printing. The blends of methyl cellulose and alginates were studied in literature for 3D printed tissue engineering scaffolds. Schütz et al. [102] investigated a methyl cellulose and alginate-based blend material for production of 3D structures with defined geometry to be used for embedding cells. The 3% alginate and methyl cellulose blend solution was very viscous, that methyl cellulose enables printing and after plotting of scaffold, alginate was complexed with Ca^{2+} ions to separate it from noncomplexed methyl cellulose. The viability of embedded cells after 3 weeks cultivation show similar behaviors with the cells in pure unprinted alginate. Li et al. [60] also examined on the methyl cellulose and alginate blend 3D printing. The interfacial bonding between the printed layers were improved by trisodium citrate solution via breaking down of the crosslinks formed by calcium ions. The resulted system was concluded as an effective 3D printed biomaterial. Karavasili et al. [51] reported the physical and mechanical data of alginate-methyl cellulose hydrogels that can be 3D printed for wound dressing with some biological components which are Manuka honey, aloe vera gel, eucalyptus essential oil. The 3D printable gel inks had good swelling properties that can enhance the moisture on the injury and had antimicrobial actions.

Hokmabad et al. [38] studied on three dimensional scaffolds for bone damages treatment. They grafted ε-caprolactone on ethyl cellulose and mixed the grafted cellulose with alginate and different amounts of nano-hydroxyapatite via electrospinning and freeze-drying methods. The expected scaffolds after the combined synthesis methods were nanofibrous, microporous and mechanically stable scaffolds. The decrease in swelling properties and porosity of the structure was observed as well as the increase in *in-vitro* biomineralization and the mechanical strength was observed with the increase in amount of nano-hydroxyapatite amount in the system. Croitoru et al. [22] examined on the ethyl cellulose-based fibers for a potential wound dressing application. They loaded different concentrations of gallic acid on ethyl cellulose to synthesize fiber type polymeric materials. The optimum composition of the fibers was determined as 10% ethyl cellulose in 100 mg gallic acid solution after tensile strength, biocompatibility and antimicrobial tests. It was concluded that as the gallic acid amount increased, the cell viability increased. For 10% ethyl cellulose in 100 mg gallic acid solution, the microbial reduction capacity showed more than 90% against *Staphylococcus* (S.) *aureus*, *Escherichia* (E.) *coli*, *Pseudomonas* (Ps.), and *Candida* (C.) *albicans*.

In-vitro tests concluded that gallic acid was completely released from the system after 2880 min that demonstrated the possible application of the ethyl cellulose/gallic acid system for potential wound dressing applications. Yang et al. [131] *investigated the release of ibuprofen from 3D printed ethyl cellulose-based tablets. The release modifiers in this system as used as hydroxypropyl methyl cellulose, sodium alginate polyvinyl alcohol and xanthan gum. The methyl cellulose, model drug ibuprofen and release modifiers were mixed before hot melt extrusion in filaments. These filaments were printed as tablets* via *fused deposition model. The effect of the composition and printing parameters on the drug release were investigated.*

Fatimi et al. [30] examined the rheological properties of 3D synthetic matrix, that is silated hydroxypropyl methyl cellulose that was used as s tissue engineering. The silated hydroxypropyl methyl cellulose was grafted with 3-glycidoxypropyltrimethoxysilane. Silanization was observed to be no effect on the change of the rheological behaviors of concentrated solutions. The viscous silated hydroxypropyl methyl cellulose had a pH of 12.8, in an acid buffer, at acidic ranges, the gelation system was cross-linked that the solution is elastic. The cross-linking kinetics and elasticity was found to be depend on polymer concentration, pH and temperature. The results showed that the gelation system is suitable for tissue engineering. Hoo et al. [40] modified hydroxypropyl methyl cellulose with methacrylic anhydride that produced 3D hydrogel hydroxypropyl methyl cellulose methacrylate for adipose tissue applications. The three different concentrations of hydrogel scaffolds were prepared by using thermosensitive properties of hydroxypropyl methyl cellulose and temperature-initiated phase separation of 3D hydrogel hydroxypropyl methyl cellulose methacrylate. The pore sizes, porosity percentage, swelling ratio and storage modulus showed that scaffolds were biocompatible for adipose tissue regeneration as well as they were concluded as a good candidate for this application. Feroz and Dias [31] examined on the development of a hydroxypropyl methyl cellulose crosslinked keratin scaffold with an inorganic component hydroxyapatite for the use of regeneration of alveolar bone. The synthesized scaffold had a high pore sizes between (108.36–22.56) μm that the networks were interconnected. The maximum compressive strength was also recorded as similar with the human trabecular bone, as well as, in vitro biodegradation showed the stability with 26% weight loss in 4 weeks. The synthesized scaffold successfully mimics the trabecular part of alveolar bone, but, still there is a need for the improvement of the behavior of the scaffold.

Ninan et al. [80] synthesized a scaffold composed of pectin, carboxymethyl cellulose and micro fibrillated cellulose via lyophilization technique. The scaffold with an 0.1% micro fibrillated cellulose showed the highest mechanical stability and glass transition temperature as 103 °C. The pore size of the scaffold was reduced when the micro fibrillated was in the composition of the scaffold as well as it improved the thermal stability and degradation rate. The scaffold with 0.1% micro fibrillated cellulose showed the highest *in-vitro* cell viability toward NIH3T3 fibroblast cell line. The results confirmed the usability of the synthesized scaffolds in tissue engineering applications. Gaihre et al. [32] developed carboxymethyl cellulose microparticles in aqueous ionic complex of zirconium solutions via ionic crosslinking, then formation of complex with chitosan. The zirconium in the structure improved the roughness of the surfaces of the microparticles as well as it increased the stability of the microparticles in phosphate buffered saline solution. The microparticles showed no cytotoxic effects when the cells were attached as well as cells were interacted directly with the surface of microparticles. After 10 days culturing, it was observed that the surface of the microparticles were covered with the cells. The authors concluded that the synthesized microparticles were good candidates in bone tissue engineering. Pasqui et al. [88] introduced hydroxyapatite added carboxymethyl cellulose based hydrogel. The experiments confirmed that hydroxyapatite was not diffused into the water or saline solutions. The carboxymethyl cellulose and hydroxyapatite gel system was tested with osteoblast MG63 line which was compared with free hydrogels whether its biological efficiency is enough. The results confirmed that addition of hydroxyapatite to carboxymethyl cellulose improved the biological activity as well as an increase in the production rate of mineralized extracellular matrix. The developed biomaterial can be used safely in bone injury treatments.

3.3 Other Biomedical Applications

Surface modification is a critical approach to prepare blood-compatible biomedical devices that ideally do not require anticoagulants when in contact with blood or plasma [44]. Modifications such as grafting phospholipid polymers onto the cellulose membrane surface to enhance heparin immobilization have been noted as efforts to increase the blood compatibility of cellulose. Ishihara et al. studied the preparation and blood compatibility of cellulose membranes grafted with a phosphoryl choline derivative. Graft polymerization of zwitterion vinyl monomer onto cellulose significantly increased the blood compatibility of cellulose membranes. The –OH groups present on the membrane surface were modified by grafting with reactive polymers in heterogeneous media, particularly by covalent attachment of polar and/or ionic groups to the membrane surface. Membranes modified by incorporating sulfate, sulfonate, and carboxylic groups showed improved compatibility compared to pure cellulose [70].

With the advancement of electrospinning, fiber scaffolds at nm scale derived from cellulose have been used in transdermal systems, antibacterial drugs, wound dressings, cancer treatments, and carriers for growth factors [94].

Gold-doped bacterial cellulose nanocomposite has been used for enzyme immobilization and biosensor fabrication. Heme proteins in blood have been immobilized on the cellulose surface [129]. The sensitivity of the biosensor has increased with the increase of gold nanoparticles. You et al. [137] found that gold quartz crystal coating on cellulose facilitated high-resolution protein separation by electrophoresis.

4 Conclusion

Cellulose derivatives offer a versatile and sustainable solution for various biomedical applications, driven by ongoing research and technological advancements. Enzymatic and genetic engineering approaches represent cutting-edge techniques in cellulose modification, offering precise control over the structure and properties of cellulose derivatives. These methods are environmentally friendly, highly specific, and can be tailored to produce cellulose with enhanced functionalities for a wide range of applications. As research in these areas continues to advance, we can expect to see even more innovative and sustainable cellulose derivatives emerging in future.

Modifying cellulose and scaling up its production for industrial and biomedical applications present several technical challenges. These are heterogeneity of cellulose sources, solubility issues, control over degree of substitution (DS), environmental and safety concerns, reproducibility and scalability, cost of raw materials and reagents, environmental impact, etc. Addressing these technical challenges requires a multidisciplinary approach, combining advances in chemistry, engineering, and materials science. By overcoming these hurdles, it is possible to unlock the full potential of cellulose derivatives for a wide range of industrial and biomedical applications.

Finally, cellulose derivatives are chemically modified forms of cellulose that offer enhanced properties and versatility for a wide range of applications. The ongoing research and technological advancements in cellulose modification continue to expand the potential uses of cellulose derivatives, promising innovative solutions in the biomedical field and beyond.

References

1. Adhikari G, Sarojasamhita VP, Powell VR, Farooqui A, Budzinski M, Garvey DT, Yang J, Katz D, Crouch B, Ramanujam N, Mueller JL (2024) Impact of injection-based delivery parameters on local distribution volume of ethyl-cellulose ethanol gel in tissue and tissue mimicking phantoms. IEEE Trans Biomed Eng 71:1488–1498. https://doi.org/10.1109/TBME.2023.3340613
2. Agarwal T, Narayana GH, Pal K, Pramanik K, Giri S, Banerjee I (2015) Calcium alginate-carboxymethyl cellulose beads for colon-targeted drug delivery. Int J Biol Macromol 75:409–417. https://doi.org/10.1016/j.ijbiomac.2014.12.052
3. Ahmadi P, Jahanban-Esfahlan A, Ahmadi A, Tabibiazar M, Mohammadifar M (2020) Development of ethyl cellulose-based formulations: a perspective on the novel technical methods. Food Rev Int 38(4):685–732. https://doi.org/10.1080/87559129.2020.1741007
4. Ahuja M, Singh G, Majumdar DK (2008) Effect of formulation parameters on corneal permeability of ofloxacin. Sci Pharm 76:505–514. https://doi.org/10.3797/scipharm.0804-22505
5. Alabi FM, Lajide L, Ajayi OO, Adebayo AO, Emmanuel S, Fadeyi AE (2020) Synthesis and characterization of carboxymethyl cellulose from Musa paradisiaca and Tithonia diversifolia. Afr J Pure Appl Chem 14(1):9–23. https://doi.org/10.5897/AJPAC

6. Altomare L, Cochis A, Carletta A, Rimondini L, Fare S (2016) Thermo-responsive methylcellulose hydrogels as temporary substrate for cell sheet biofabrication. J Mater Sci Mater Med 27:95. https://doi.org/10.1007/s10856-016-5703-8
7. Bain MK, Bhowmik M, Ghosh SN, Chattopadhyay D (2009) In situ fast gelling formulation of methyl cellulose for in vitro ophthalmic controlled delivery of ketorolac tromethamine. J Appl Polym Sci 113(2):1241–1246. https://doi.org/10.1002/app.30040
8. Bain MK, Bhowmik M, Maity D, Bera NK, Ghosh S, Chattopadhyay D (2010) Control of thermo reversible gelation of methylcellulose using polyethylene glycol and sodium chloride for sustained delivery of ophthalmic drug. J Appl Polym Sci 118(2):631–637. https://doi.org/10.1002/app.32350
9. Bain MK, Bhowmick B, Maity D, Mondal D, Mollick MMR, Rana D (2012) Chattopadhyay, D, synergistic effect of salt mixture on the gelation temperature and morphology of methylcellulose hydrogel. Int J Biol Macromol 51(5):831–836. https://doi.org/10.1016/j.ijbiomac.2012.07.028
10. Bain MK, Bhowmick B, Maity D, Mondal D, Mollick MMR, Paul BK, Bhowmik M, Rana D, Chattopadhyay D (2012) Effect of PVA on the gel temperature of MC and release kinetics of KT from MC based ophthalmic formulations. Int J Biol Macromol 50(3):565–572. https://doi.org/10.1016/j.ijbiomac.2012.01.025
11. Bain MK, Maity D, Bhowmick B, Mondal D, Mollick MMR, Sarkar G, Bhowmik M, Rana D, Chattopadhyay D (2013) Effect of PEG–salt mixture on the gelation temperature and morphology of MC gel for sustained delivery of drug. Carbohydr Polym 91(2):529–536. https://doi.org/10.1016/j.carbpol.2012.08.040
12. Balzus B, Sahle FF, Hönzke S, Gerecke C, Schumacher F, Hedtrich S, Burkhard K, Bodmeier R (2017) Formulation and ex vivo evaluation of polymeric nanoparticles for controlled delivery of corticosteroids to the skin and the corneal epithelium. Eur J Pharm Biopharm 115:122–130. https://doi.org/10.1016/j.ejpb.2017.02.001
13. Barkhordari S, Yadollahi M (2016) Carboxymethyl cellulose capsulated layered double hydroxides/drug nanohybrids for cephalexin oral delivery. Appl Clay Sci 121–122:77–85. https://doi.org/10.1016/j.clay.2015.12.026
14. Baron RI, Biliuta G, Macsim AM, Dinu MV, Coseri S (2023) Chemistry of hydroxypropyl cellulose oxidized by two selective oxidants. Polymers 15(19):3930. https://doi.org/10.3390/polym15193930
15. Belgacem MN, Salon-Brochier MC, Krouit M, Bras J (2011) Recent advances in surface chemical modification of cellulose fibres. J Adhes Sci Technol 25(6–7):661–684. https://doi.org/10.1163/016942410X525867
16. Bhowmik M, Kumari P, Sarkar G, Bain MK, Bhowmick B, Mollick MM, Mondal D, Maity D, Rana D, Bhattacharjee D, Chattopadhyay D (2013) Effect of xanthan gum and guar gum on in situ gelling ophthalmic drug delivery system based on poloxamer-407. Int J Biol Macromol 62(1):117–123. https://doi.org/10.1016/j.ijbiomac.2013.08.024
17. Bodini RB, Guimarães JGL, Monaco-Lourenço CA, de Carvalho RA (2019) Effect of starch and hydroxypropyl methylcellulose polymers on the properties of orally disintegrating films. J Drug Deliv Sci Technol 51:403–410. https://doi.org/10.1016/j.jddst.2019.03.028
18. Bukharina D, Cauffiel K, Killingsworth LM, Brackenridge JA, Poliukhova V, Kim M, Tsukruk VV (2024) Click-chemistry-enabled functionalization of cellulose nanocrystals with single-stranded DNA for directed assembly. ACS Biomater Sci Eng 10(10):6155–6166. https://doi.org/10.1021/acsbiomaterials.4c01518
19. Chen H, Fan M (2008) Novel thermally sensitive ph-dependent chitosan/carboxymethyl cellulose hydrogels. J Bioact Compat Pol 23(1):38–48. https://doi.org/10.1177/0883911507085070
20. Cheng YH, Fung MP, Chen YQ, Chiu YC (2024) Development of mucoadhesive methacrylic anhydride-modified hydroxypropyl methylcellulose hydrogels for topical ocular drug delivery. J Drug Deliv Sci Technol 93:105450. https://doi.org/10.1016/j.jddst.2024.105450
21. Coskun M, Temuz MM (2005) Grafting studies onto cellulose by atom-transfer radical polymerization. Polym Int 54(2):342–347. https://doi.org/10.1002/pi.1684

22. Croitoru AM, Ayran M, Altan E, Karacelebi Y, Ulag S, Ali Sahin A, Guncu MM, Aksu B, Gunduz O, Tihăuan BM, Ficai D, Ficai A (2023) Development of gallic acid-loaded ethylcellulose fibers as a potential wound dressing material. Int J Biol Macromol 253(5):126996. https://doi.org/10.1016/j.ijbiomac.2023.126996
23. Çetin EÖ, Buduneli N, Atlıhan E, Kırılmaz L (2004) In vitro studies on controlled-release cellulose acetate films for local delivery of chlorhexidine, indomethacin, and meloxicam. J Clin Periodontol 31:1117–1121. https://doi.org/10.1111/j.1600-051X.2004.00620.x
24. Dewan M, Bhowmick B, Sarkar G, Rana D, Bain MK, Bhowmik M, Chattopadhyay D (2015) Effect of methyl cellulose on gelation behavior and drug release from poloxamer based ophthalmic formulations. Int J Biol Macromol 72(1):706–710. https://doi.org/10.1016/j.ijbiomac.2014.09.021
25. Elangwe CN, Morozkina SN, Olekhnovich RO, Krasichkov A, Polyakova VO, Uspenskaya MV (2022) A review on chitosan and cellulose hydrogels for wound dressings. Polymers 14(23):5163. https://doi.org/10.3390/polym14235163
26. El-Habashy SE, Allam AN, El-Kamel AH (2016) Ethyl cellulose nanoparticles as a platform to decrease ulcerogenic potential of piroxicam: formulation and in vitro/in vivo evaluation. Int J Nanomed 11:2369–2380. https://doi.org/10.2147/IJN.S93354
27. El-Kamel AH (2002) In vitro and in vivo evaluation of pluronic F127-based ocular delivery system for timolol maleate. Int J Pharm 241(1):47–55. https://doi.org/10.1016/S0378-5173(02)00234-X
28. El-Saied H, Basta AH, Hassanen ME, Korte H, Helal A (2012) Behaviour of rice-byproducts and optimizing the conditions for production of high performance natural fiber polymer composites. J Polym Environ 20:838–847. https://doi.org/10.1007/s10924-012-0439-0
29. Fahmy Y, Mobarak F (1971) Fine structure and reactivity of cellulose. J Polym Sci Part B Polym Lett 9(10):767–769. https://doi.org/10.1002/pol.1971.110091012
30. Fatimi A, Tassin JF, Quillard S, Axelos MAV, Weiss P (2008) The rheological properties of silated hydroxypropylmethylcellulose tissue engineering matrices. Biomater 29(5):533–543. https://doi.org/10.1016/j.biomaterials.2007.10.032
31. Feroz S, Dias G (2021) Hydroxypropylmethyl cellulose (HPMC) crosslinked keratin/hydroxyapatite (HA) scaffold fabrication, characterization and in vitro biocompatibility assessment as a bone graft for alveolar bone regeneration. Heliyon 7(11):e08294. https://doi.org/10.1016/j.heliyon.2021.e08294
32. Gaihre B, Jayasuriya AC (2016) Fabrication and characterization of carboxymethyl cellulose novel microparticles for bone tissue engineering. Mater Sci Eng C 69:733–743. https://doi.org/10.1016/j.msec.2016.07.060
33. Gao M, Xie X, Huang T, Zhang N, Wang Y (2022) Glutaraldehyde-assisted crosslinking in regenerated cellulose films toward high dielectric and mechanical properties. Cellulose 29:8177–8194. https://doi.org/10.1007/s10570-022-04785-2
34. Gold GT, Varma DM, Harbottle D, Gupta MS, Stalling SS, Taub PJ, Nicoll SB (2014) Injectable redox-polymerized methylcellulose hydrogels as potential soft tissue filler materials. J Biomed Mater Res 102(12):4536–4544. https://doi.org/10.1002/jbm.a.35132
35. Gupta D, Tator CH, Shoichet MS (2006) Fast-gelling injectable blend of hyaluronan and methylcellulose for intrathecal, localized delivery to the injured spinal cord. Biomater 27(11):2370–2379. https://doi.org/10.1016/j.biomaterials.2005.11.015
36. Haddad HM, Loucas SP (1996) US Patent 5,556,633 Accepted: 17 September 1996, barcode number: US005556633A, p 7
37. Hickey RJ, Pelling AE (2019) Cellulose biomaterials for tissue engineering. Front Bioeng Biotechnol 7:45. https://doi.org/10.3389/fbioe.2019.00045
38. Hokmabad VR, Davaran S, Aghazadeh M, Rahbarghazi R, Salehi R, Ramazani A (2019) Fabrication and characterization of novel ethyl cellulose-grafted-poly(ε-caprolactone)/alginate nanofibrous/macroporous scaffolds incorporated with nano-hydroxyapatite for bone tissue engineering. J Biomater Appl 33(8):1128–1144. https://doi.org/10.1177/0885328218822641
39. Homem NC, Amorim MTP (2020) Synthesis of cellulose acetate using as raw material textile wastes. Mater Today Proc 31(2):315–317. https://doi.org/10.1016/j.matpr.2020.01.494

40. Hoo SP, Loh QL, Yue Z, Fu J, Tan TTY, Choong C, Chan PPY (2013) Preparation of a soft and interconnected macroporous hydroxypropyl cellulose methacrylate scaffold for adipose tissue engineering. J Mater Chem B 1:3107–3117. https://doi.org/10.1039/C3TB00446E
41. Iqbal HMN, Kyazze G, Tron T, Keshavarz T (2014) Laccase-assisted grafting of poly(3-hydroxybutyrate) onto the bacterial cellulose as backbone polymer: development and characterisation. Carbohydr Polym 113:131–137. https://doi.org/10.1016/j.carbpol.2014.07.003
42. Itoh K, Hatakeyama T, Shimoyama T, Miyazaki S, D'Emanuele A, Attwood D (2011) In situ gelling formulation based on methylcellulose/pectin system for oral-sustained drug delivery to dysphagic patients. Drug Dev Ind Pharm 37(7):790–797. https://doi.org/10.3109/03639045.2010.541465
43. Itoh K, Hatakeyama T, Kimura T, Shimoyama T, Miyazaki S, D'Emanuele A, Attwood D (2010) Effect of D-sorbitol on the thermal gelation of methylcellulose formulations for drug delivery. Chem Pharm Bull 58(2):247–249. https://doi.org/10.1248/cpb.58.247
44. Ishihara K, Takayama R, Nakabayashi N, Fukumoto K, Aoki J (1992) Improvement of blood compatibility on cellulose dialysis membrane: 2. Blood compatibility of phospholipid polymer grafted cellulose membrane. Biomater 13(4):235–239. https://doi.org/10.1002/jbm.820290207
45. Javanbakht S, Pooresmaeil M, Hashemi H, Namazi H (2018) Carboxymethylcellulose capsulated Cu-based metal-organic framework-drug nanohybrid as a pH-sensitive nanocomposite for ibuprofen oral delivery. Int J Biol Macromol 119:588–596. https://doi.org/10.1016/j.ijbiomac.2018.07.181
46. Jiang X, Yang Z, Zhang J, Liang H, Wang H (2024) Lu J (2024) Preparation and characterization of photosensitive methacrylate-grafted sodium carboxymethyl cellulose as an injectable material to fabricate hydrogels for biomedical applications. Int J Biol Macromol 263:130190. https://doi.org/10.1016/j.ijbiomac.2024.130190
47. Joshi SC (2011) Sol-gel behavior of hydroxypropyl methylcellulose (HPMC) in ionic media including drug release. Mater Basel 4(10):1861–1905. https://doi.org/10.3390/ma4101861
48. Kamarudin SH, Mohd Basri MS, Rayung M, Abu F, Ahmad SB, Norizan MN, Abdullah LC (2022) A review on natural fiber reinforced polymer composites (NFRPC) for sustainable industrial applications. Polymers 14(17):3698. https://doi.org/10.3390/polym14173698
49. Kamel S, Ali N, Jahangir K, Shah S, El-Gendy A (2008) Pharmaceutical significance of cellulose: a review. Express Polym Lett 2:758–778. https://doi.org/10.3144/expresspolymlett.2008.90
50. Kan KH, Li J, Wijesekera K, Cranston ED (2013) Polymer-grafted cellulose nanocrystals as pH-responsive reversible flocculants. Biomacromol 14(9):3130–3139. https://doi.org/10.1021/bm400752k
51. Karavasili C, Tsongas K, Andreadis II, Andriotis EG, Papachristou ET, Papi RM, Tzetzis D, Fatouros DG (2020) Physico-mechanical and finite element analysis evaluation of 3D printable alginate-methylcellulose inks for wound healing applications. Carbohydr Polym 247:116666. https://doi.org/10.1016/j.carbpol.2020.116666
52. Keldibekova R, Suleimenova S, Nurgozhina G, Kopishev E (2023) Interpolymer complexes based on cellulose ethers: application. Polymers 15(15):3326. https://doi.org/10.3390/polym15153326
53. Kelly J, D'Cruz G, Wright D (2009) A qualitative study of the problems surrounding medicine administration to patients with dysphagia. Dysphagia 24:49–56. https://doi.org/10.1007/s00455-008-9170-3
54. Kelly J, D'Cruz G, Wright D (2010) Patients with dysphagia: experiences of taking medication. J Adv Nurs 66(1):82–91. https://doi.org/10.1111/j.1365-2648.2009.05145.x
55. Khaing EM, Lertsuphotvanit N, Thammasut W, Rojviriya C, Chansatidkosol S, Phattarateera S, Pichayakorn W, Phaechamud T (2024) Cellulose acetate butyrate-based in situ gel comprising doxycycline hyclate and metronidazole. Polym 16:3477. https://doi.org/10.3390/polym16243477
56. Kim JK, Won YW, Lim KS, Kim YH (2012) Low-molecular-weight methylcellulose-based thermo-reversible gel/pluronic micelle combination system for local and sustained docetaxel delivery. Pharm Res 29:525–534. https://doi.org/10.1007/s11095-011-0581-8

57. Kiriakou MV, Pakdel AS, Berry RM, Hoare T, Dubé MA, Cranston ED (2021) Incorporation of polymer-grafted cellulose nanocrystals into latex-based pressure-sensitive adhesives. ACS Mater Au 2(2):176–189. https://doi.org/10.1021/acsmaterialsau.1c00052
58. Kumar S, Haglund BO, Himmelstein KJ (1994) In-situ forming gels for ophthalmic drug delivery. J Ocular Pharmacol 10(1):47–56. https://doi.org/10.1089/jop.1994.10.47
59. Lane ME (2008) Modified-release drug delivery technology, Rathbone MJ, Hadgraft J, Roberts MS (eds), 2nd ed. Informa Healthcare: London, UK, vol 2
60. Li H, Tan YJ, Leong KF, Lin L (2017) 3D bioprinting of highly thixotropic alginate/methylcellulose hydrogel with strong interface bonding. ACS Appl Mater Interfaces 9(23):20086–20097. https://doi.org/10.1021/acsami.7b04216
61. Liang S (2023) Advances in drug delivery applications of modified bacterial cellulose-based materials. Front Bioeng Biotechnol 11:1252706. https://doi.org/10.3389/fbioe.2023.1252706
62. Lin F, You Y, Yang X, Jiang X, Lu Q, Wang T, Lu B (2017) Microwave-assisted facile synthesis of TEMPO-oxidized cellulose beads with high adsorption capacity for organic dyes. Cellulose 24:5025–5040. https://doi.org/10.1007/s10570-017-1473-9
63. Lin N, Dufresne A (2014) Nanocellulose in biomedicine: current status and future prospect. Eur Polym J 59:302–325. https://doi.org/10.1016/j.eurpolymj.2014.07.025
64. Lin Y, Sun J, Jiang G, Zan J, Ding F (2007) In vitro evaluation of lysozyme-loaded microspheres in thermosensitive methylcellulose-based hydrogel. Chin J Chem Eng 15:566–572. https://doi.org/10.1016/s1004-9541(07)60125-6
65. Liu D, Wu Q, Chen W, Lin H, Zhu Y, Liu Y, Liang H, FangMing Zhu F (2019) A novel FK506 loaded nanomicelles consisting of amino-terminated poly(ethylene glycol)-block-poly(D, L)-lactic acid and hydroxypropyl methylcellulose for ocular drug delivery. Int J Pharm 562:1–10. https://doi.org/10.1016/j.ijpharm.2019.03.022
66. Liu W, Du H, Zhang M, Liu K, Liu H, Xie H, Zhang X, Si C (2020) Bacterial cellulose-based composite scaffolds for biomedical applications: a review. ACS Sustain Chem Eng 8(20):7536–7562. https://doi.org/10.1021/acssuschemeng.0c00125
67. Luo Y, Wang S, Shen M, Qi R, Fang Y, Guo R, Cai H, Cao X, Tomás H, Zhu M, Xiangyang Shi X (2013) Carbon nanotube-incorporated multilayered cellulose acetate nanofibers for tissue engineering applications. Carbohydr Polym 91(1):419–427. https://doi.org/10.1016/j.carbpol.2012.08.069
68. Madhushree M, Vairavel P, Mahesha GT, Bhat KS (2024) A Comprehensive review of cellulose and cellulose-based materials: extraction, modification, and sustainable applications. J Nat Fibers 21(1):2418357. https://doi.org/10.1080/15440478.2024.241835
69. Maderuelo C, Zarzuelo A, Lanao JM (2011) Critical factors in the release of drugs from sustained release hydrophilic matrices. J Control Release 154(1):2–19. https://doi.org/10.1016/j.jconrel.2011.04.002
70. Malsch G, Paul D, Krämer S, Wolf H, Thomaneck U, Brown GS, Falkenhagen D (1991) Chemical modification of cellulose membranes with improved blood compatibility. Clin Mater 8(1–2):65–69. https://doi.org/10.1016/0267-6605(91)90011-4
71. Maulvi FA, Lakdawala DH, Shaikh AA, Desai AR, Choksi HH, Vaidya RJ, Ranch KM, Koli AR, Vyas BA, Shah DO (2016) In vitro and in vivo evaluation of novel implantation technology in hydrogel contact lenses for controlled drug delivery. J Controlled Release 226:47–56. https://doi.org/10.1016/j.jconrel.2016.02.012
72. Moohan J, Stewart SA, Espinosa E, Rosal A, Rodríguez A, Larrañeta E, Donnelly RF, Domínguez-Robles J (2020) Cellulose nanofibers and other biopolymers for biomedical applications. A Review. Appl Sci 10(1):65. https://doi.org/10.3390/app10010065
73. Mu H, Wang Y, Wei H, Lu H, Feng Z, Yu H, Xing Y, Wang H (2018) Collagen peptide modified carboxymethyl cellulose as both antioxidant drug and carrier for drug delivery against retinal ischaemia/reperfusion injury. J Cell Mol Med 22:5008–5019. https://doi.org/10.1111/jcmm.13768
74. Mueller JL, Morhard R, DeSoto M, Chelales E, Yang J, Nief C, Crouch B, Everitt J, Previs R, Katz D, Ramanujam N (2021) Optimizing ethyl cellulose-ethanol delivery towards enabling ablation of cervical dysplasia. Sci Rep 11:16869. https://doi.org/10.1038/s41598-021-96223-9

75. Nagel MC, Koschella A, Voiges K, Mischnick P, Heinze T (2010) Homogeneous methylation of wood pulp cellulose dissolved in LiOH/urea/H_2O. Eur Polym J 46(8):1726–1735. https://doi.org/10.1016/j.eurpolymj.2010.05.009
76. Nair RV, Shefrin S, Suresh A, Anoop KR, Nair SC (2018) Sustained release timolol maleate loaded ocusert based on biopolymer composite. Int J Biol Macromol 110:308–317. https://doi.org/10.1016/j.ijbiomac.2018.01.029
77. Nasatto PL, Pignon F, Silveira JL, Duarte ME, Noseda MD, Rinaudo M (2015) Methylcellulose, a cellulose derivative with original physical properties and extended applications. Polymers 7:777–803. https://doi.org/10.3390/polym7050777
78. Niemczyk-Soczynska B, Gradys A, Kolbuk D, Krzton-Maziopa A, Sajkiewicz P (2019) Crosslinking kinetics of methylcellulose aqueous solution and its potential as a scaffold for tissue engineering. Polymers 11(11):1772. https://doi.org/10.3390/polym11111772
79. Nilay S, Patıl UA, Dinesh BM, Desai BG (2008) Drug-loaded cellulose acetate and cellulose acetate butyrate films as ocular inserts. East Central Afr J Pharm Sci 11:9–15. https://doi.org/10.4314/ecajps.v11i1.44761
80. Ninan N, Muthiah M, Park I, Elain A, Thomas S, Grohens Y (2013) Pectin/carboxymethyl cellulose/microfibrillated cellulose composite scaffolds for tissue engineering. Carbohydr Polym 98(1):877–885. https://doi.org/10.1016/j.carbpol.2013.06.067
81. Nosar MN, Salehi M, Ghorbani S, Beiranvand SP, Goodarzi A, Azami M (2016) Characterization of wet-electrospun cellulose acetate based 3-dimensional scaffolds for skin tissue engineering applications: influence of cellulose acetate concentration. Cellulose 23:3239–3248. https://doi.org/10.1007/s10570-016-1026-7
82. Notario-Pérez F, Cazorla-Luna R, Martín-Illana AJ, Galante J, Ruiz-Caro R, das Neves J, Veiga MD (2020) Design, fabrication and characterisation of drug-loaded vaginal films: State-of-the-art. J Controlled Release 327: 477-479. https://doi.org/10.1016/j.jconrel.2020.08.032
83. Obeidallah MM, Abdel-Mageed AM, Elfaham TH (2018) Ocular administration of acetazolamide microsponges in situ gel formulations. Saudi Pharm J 26(7):909–920. https://doi.org/10.1016/j.jsps.2018.01.005
84. Okubo M, Iohara D, Anraku M, Higashi T, Uekama K, Hirayama FA (2020) Thermoresponsive hydrophobically modified hydroxypropylmethylcellulose/cyclodextrin injectable hydrogel for the sustained release of drugs. Int J Pharm 575:118845. https://doi.org/10.1016/j.ijpharm.2019.118845
85. Orlando I, Basnett P, Nigmatullin R, Wang W, Knowles JC, Roy I (2020) Chemical modification of bacterial cellulose for the development of an antibacterial wound dressing. Front Bioeng Biotechnol 8:557885. https://doi.org/10.3389/fbioe.2020.557885
86. Pandey A, Singh MK, Singh A (2024) Bacterial cellulose: a smart biomaterial for biomedical applications. J Mater Res 39(1):2–18. https://doi.org/10.1557/s43578-023-01116-4
87. Pandey A, Singh MK, Singh A (2025) Low-cost bacterial cellulose production from agricultural waste for antibacterial applications. J Bioact Compat Polym 40(1):94–116. https://doi.org/10.1177/08839115241293464
88. Pasqui D, Torricelli P, De Cagna M, Fini M, Barbucci R (2014) Carboxymethyl cellulose—hydroxyapatite hybrid hydrogel as a composite material for bone tissue engineering applications. J Biomed Mater Res Part A 102A:1568–1579. https://doi.org/10.1002/jbm.a.34810
89. Popa L, Ghica MV, Popescu R, Irimia T, Dinu-Pîrvu CE (2021) Development and optimization of chitosan-hydroxypropyl methylcellulose in situ gelling systems for ophthalmic delivery of bupivacaine hydrochloride. Processes 9(10):1694. https://doi.org/10.3390/pr9101694
90. Qi S, Kiratzis I, Adoni P, Tuekprakhon A, Hill HJ, Stamataki Z, Zhang ZJ (2023) Porous cellulose thin films as sustainable and effective antimicrobial surface coatings. ACS Appl Mater Interfaces 15(17):20638–20648. https://doi.org/10.1021/acsami.2c23251
91. Qiu X, Hu S (2013) “Smart” materials based on cellulose: a review of the preparations, properties, and applications. Materials 6(3):738–781. https://doi.org/10.3390/ma6030738
92. Rahman MS, Hasan MS, Nitai AS, Nam S, Karmakar AK, Ahsan MS, Shiddiky MJA, Ahmed MB (2021) Recent developments of carboxymethyl cellulose. Polym 13:1345. https://doi.org/10.3390/polym13081345

93. Rahminar L, Irawan E, Athifah S, Sudiarti T (2020) Synthesis and characterization of methylated cellulose as biocomposite hydrogel. Int J Sci Technol Res 9(1):3993–3996
94. Ramezani G, Stiharu I, van de Ven TG, Nerguizian V (2024) Advancements in hybrid cellulose-based films: Innovations and applications in 2D nano-delivery systems. J Funct Biomater 15(4):93. https://doi.org/10.3390/jfb15040093
95. Rangabhatla ASL, Tantishaiyakul V, Oungbho K, Boonrat O (2016) Fabrication of pluronic and methylcellulose for etidronate delivery and their application for osteogenesis. Int J Pharm 499(1–2):110–118. https://doi.org/10.1016/j.ijpharm.2015.12.070
96. Ravi PR, Ganga S, Saha RN (2008) Design and in vitro evaluation of zidovudine oral controlled release tablets prepared using hydroxypropyl methylcellulose. Chem Pharm Bull 56(4):518–524. https://doi.org/10.1248/cpb.56.518
97. Saedi S, Garcia CV, Kim JT, Shin GH (2021) Physical and chemical modifications of cellulose fibers for food packaging applications. Cellulose 28(14):8877–8897. https://doi.org/10.1007/s10570-021-04086-0
98. Saito T, Kimura S, Nishiyama Y, Isogai A (2007) Cellulose nanofibers prepared by TEMPO-mediated oxidation of native cellulose. Biomacromol 8(8):2485–2491. https://doi.org/10.1021/bm0703970
99. Sajadi-Javan ZS, Varshosaz J, Mirian M, Manshaei M, Aminzadeh A (2022) Thermo-responsive hydrogels based on methylcellulose/persian gum loaded with taxifolin enhance bone regeneration: an in vitro/in vivo study. Cellulose 29:2413–2433. https://doi.org/10.1007/s10570-021-04383-8
100. Sajeev C, Vinay G, Archna R, Saha RN (2002) Oral controlled release formulation of diclofenac sodium by microencapsulation with ethyl cellulose. J Microencapsul 19(6):753–760. https://doi.org/10.1080/02652040210000022734
101. Sá-Lima H, Tuzlakoglu K, Mano JF, Reis RL (2011) Thermoresponsive poly(N-isopropylacrylamide)-gmethylcellulose hydrogel as a three-dimensional extracellular matrix for cartilage-engineered applications. J Biomed Mater Res Part A 98:596–603. https://doi.org/10.1002/jbm.a.33140
102. Schütz K, Placht AM, Paul B, Brüggemeier S, Gelinsky M, Lode A (2017) Three-dimensional plotting of a cell-laden alginate/methylcellulose blend: towards biofabrication of tissue engineering constructs with clinically relevant dimensions. J Tissue Eng Regen Med 11:1574–1587. https://doi.org/10.1002/term.2058
103. Seddiqi H, Oliaei E, Honarkar H, Jin J, Geonzon LC, Bacabac RG, Klein-Nulend J (2021) Cellulose and its derivatives: towards biomedical applications. Cellulose 28(4):1893–1931. https://doi.org/10.1007/s10570-020
104. Sharma S, Gunjan Sarkar G, Srestha B, Chattopadhyay D, Bhowmik M (2019) In-situ fast gelling formulation for oral sustained drug delivery of paracetamol to dysphagic patients. Int J Biol Macromol 134:864–868. https://doi.org/10.1016/j.ijbiomac.2019.05.092
105. Shimoyama T, Itoh K, Kobayashi M, Miyazaki S, D'Emanuele A, Attwood D (2012) Oral liquid in situ gelling methylcellulose/alginate formulations for sustained drug delivery to dysphagic patients. Drug Dev Ind Pharm 38(8):952–960. https://doi.org/10.3109/03639045.2011.634809
106. Siew LF, Basit AW, Newton JM (2000) The potential of organic-based amylose-ethylcellulose film coatings as oral colon-specific drug delivery systems. AAPS Pharm Sci Tech 1(3):22. https://doi.org/10.1208/pt010322
107. Son HJ, Heo MS, Kim YG, Lee SJ (2001) Optimization of fermentation conditions for the production of bacterial cellulose by a newly isolated Acetobacter sp. A9 in shaking cultures. Biotechnol Appl Biochem 33(1):1–5. https://doi.org/10.1042/BA20000065
108. Stalling SS, Akintoye SO, Nicoll SB (2009) Development of photocrosslinked methylcellulose hydrogels for soft tissue reconstruction. Acta Biomater 5(6):1911–1918. https://doi.org/10.1016/j.actbio.2009.02.020
109. Taghe S, Mirzaeei S, Pakdaman N, Kazemi A, Nokhodchi A (2024) Macrolide-loaded nanofibrous inserts with polycaprolactone and cellulose acetate base for sustained ocular delivery: pharmacokinetic study in Rabbit's eye. Int J Pharm 665:124699. https://doi.org/10.1016/j.ijpharm.2024.124699

110. Takeuchi M, Kageyama S, Suzuki H, Wada T, Notsu Y, Ishii F (2003) Rheological properties of reversible thermo-setting in situ gelling solutions with the methylcellulose–polyethylene glycol–citric acid ternary system. Colloid Polym Sci 281:1178–1183. https://doi.org/10.1007/s00396-003-0893-2
111. Talasaz AHH, Ghahremankhani AA, Moghadam SH, Malekshahi MR, Atyabi F, Dinarvand R (2008) In situ gel forming systems of poloxamer 407 and hydroxypropyl cellulose or hydroxypropyl methyl cellulose mixtures for controlled delivery of vancomycin. J Appl Polym Sci 109:2369–2374. https://doi.org/10.1002/app.28163
112. Talukdar MM, Michoel A, Rombaut P, Kinget PR (1996) Comparative study on xanthan gum and hydroxypropylmethyl cellulose as matrices for controlled-release drug delivery I. Compaction and in vitro drug release behaviour. Int J Pharm 129(1–2):233–241. https://doi.org/10.1016/0378-5173(95)04355-1
113. Tang Y, Wang X, Li Y, Lei M, Du Y, Kennedy JF, Knill CJ (2010) Production and characterisation of novel injectable chitosan/methylcellulose/salt blend hydrogels with potential application as tissue engineering scaffolds. Carbohydr Polym 82(3):833–841. https://doi.org/10.1016/j.carbpol.2010.06.003
114. Tate MC, Shear DA, Hoffman SW, Stein DG, LaPlaca MC (2001) Biocompatibility of methylcellulose-based constructs designed for intracerebral gelation following experimental traumatic brain injury. Biomater 22(10):1113–1123. https://doi.org/10.1016/S0142-9612(00)00348-3
115. Thakur S, Singh H, Singh A, Kaur S, Sharma A, Singh SK, Kaur S, Kaur G, Jain SK (2020) Thermosensitive injectable hydrogel containing carboplatin loaded nanoparticles: a dual approach for sustained and localized delivery with improved safety and therapeutic efficacy. J Drug Deliv Sci Technol 58:101817. https://doi.org/10.1016/j.jddst.2020.101817
116. Thirumala S, Gimble JM, Devireddy RV (2013) Methylcellulose based thermally reversible hydrogel system for tissue engineering applications. Cells 2(3):460–475. https://doi.org/10.3390/cells2030460
117. Tosh B, Saikia CN, Dass NN (2000) Homogeneous esterification of cellulose in the lithium chloride–N, N-dimethylacetamide solvent system: effect of temperature and catalyst. Carbohydr Res 327(3):345–352. https://doi.org/10.1016/S0008-6215(00)00033-1
118. Tudoroiu EE, Dinu-Pîrvu CE, Albu Kaya MG, Popa L, Anuța V, Prisada RM, Ghica MV (2021) An overview of cellulose derivatives-based dressings for wound-healing management. Pharm 14(12):1215. https://doi.org/10.3390/ph14121215
119. Tundisi LL, Mostaço GB, Carricondo PC, Petri DFS (2021) Hydroxypropyl methylcellulose: physicochemical properties and ocular drug delivery formulations. Eur J Pharm Biopharm 159:105736. https://doi.org/10.1016/j.ejps.2021.105736
120. Vaca-Garcia C, Thiebaud S, Borredon ME, Gozzelino G (1998) Cellulose esterification with fatty acids and acetic anhydride in lithium chloride/N, N-dimethylacetamide medium. J Am Oil Chem Soc 75:315–319. https://doi.org/10.1007/s11746-998-0047-2
121. Varma DM, Gold GT, Taub PJ, Nicoll SB (2014) Injectable carboxymethylcellulose hydrogels for soft tissue filler applications. Acta Biomater 10(12):4996–5004. https://doi.org/10.1016/j.actbio.2014.08.013
122. Vieira JG, Oliveria GC, Rodrigues G, de Assuncao RMN, Meireles CD, Cerqueira DA, Silva WG, Motta LAD (2009) Production, characterization, and evaluation of methylcellulose from sugarcane bagasse for applications as viscosity enhancing admixture for cement based material. Carbohydr Polym 78:779–783. https://doi.org/10.1016/j.carbpol.2009.06.016
123. Vieira RGP, Rodrigues FG, Assuncao RMN, Meireles CD, Vieira JG, Oliveria GS (2007) Synthesis and characterization of methylcellulose from sugar cane bagasse cellulose. Carbohydr Polym 67:182–189. https://doi.org/10.1016/j.carbpol.2006.05.007
124. Wang F, Yang Y, Ling Y, Liu J, Cai X, Zhou X, Tang X, Liang B, Chen Y, Chen H, Chen D, Li C, Wang Z, Hu B, Zheng Y (2017) Injectable and thermally contractible hydroxypropyl methyl cellulose/Fe3O4 for magnetic hyperthermia ablation of tumors. Biomater 128:84–93. https://doi.org/10.1016/j.biomaterials.2017.03.004

125. Wang J, Abbas SC, Li L, Walker CC, Ni Y, Cai Z (2024) Cellulose membranes: synthesis and applications for water and gas separation and purification. Membr 14(7):148
126. Wang L, Yang T, Zhao G (2022) An injectable cellulose acetate/sodium alginate hydrogels-loaded laponite microsphere as a potential wound healing in nursing care in perioperative period. Mater Res Express 9(3):035402. https://doi.org/10.1088/2053-1591/ac565f
127. Wang S, Gao Z, Liu L, Li M, Zuo A, Guo J (2022) Preparation, in vitro and in vivo evaluation of chitosan-sodium alginate-ethyl cellulose polyelectrolyte film as a novel buccal mucosal delivery vehicle. Eur J Pharm Biopharm 168:106085. https://doi.org/10.1016/j.ejps.2021.106085
128. Wang W, Li F, Yu J, Navard P, Budtova T (2015) Influence of substitution on the rheological properties and gelation of hydroxyethyl cellulose solution in NaOH–water solvent. Carbohydr Polym 124:85–89. https://doi.org/10.1016/j.carbpol.2015.01.065
129. Wang W, Zhang TJ, Zhang DW, Li HY, Ma YR, Qi LM, Zhang XX (2011) Amperometric hydrogen peroxide biosensor based on the immobilization of heme proteins on gold nanoparticles–bacteria cellulose nanofibers nanocomposite. Talanta 84(1):71–77. https://doi.org/10.1016/j.talanta.2010.12.015
130. Wang Y, Wang X, Xie Y, Zhang K (2018) Functional nanomaterials through esterification of cellulose: a review of chemistry and application. Cellulose 25:3703–3731. https://doi.org/10.1007/s10570-018-1830-3
131. Yang Y, Wang H, Li H, Ou Z, Yang G (2018) 3D printed tablets with internal scaffold structure using ethyl cellulose to achieve sustained ibuprofen release. Eur J Pharm Sci 115:11–18. https://doi.org/10.1016/j.ejps.2018.01.005
132. Wasilewska K, Winnicka K (2019) Ethylcellulose–a pharmaceutical excipient with multidirectional application in drug dosage forms development. Mater 12:3386. https://doi.org/10.3390/ma12203386
133. Wen Y, Yuan Z, Qu J, Wang C, Wang A (2020) Evaluation of ultraviolet light and hydrogen peroxide enhanced ozone oxidation treatment for the production of cellulose nanofibrils. ACS Sustain Chem Eng 8(7):2688–2697. https://doi.org/10.1021/acssuschemeng.9b06028
134. Westman MP, Fifield LS, Simmons KL, Laddha S, Kafentzis TA (2010) Natural fiber composites: a review. U.S. Deparment of Energy. https://doi.org/10.2172/989448
135. Wu S, Qin X, Li M (2014) The structure and properties of cellulose acetate materials: a comparative study on electrospun membranes and casted films. J Ind Text 44(1):85–98. https://doi.org/10.1177/1528083713477443
136. Yang Y, Nan W, Zhang R, Shen S, Wu M, Zhong S, Zhang Y, Cui X (2024) Fabrication of carboxymethyl cellulose-based thermo-sensitive hydrogels and inhibition of corneal neovascularization. Int J Biol Macromol 261(2):129933. https://doi.org/10.1016/j.ijbiomac.2024.129933
137. You J, Zhao L, Wang G, Zhou H, Zhou J, Zhang L (2014) Quaternized cellulose-supported gold nanoparticles as capillary coatings to enhance protein separation by capillary electrophoresis. J Chromatogr A 1343:160–166. https://doi.org/10.1016/j.chroma.2014.03.079
138. Zhang J, Chen W, Feng Y, Wu J, Yu J, He J, Zhang J (2015) Homogeneous esterification of cellulose in room temperature ionic liquids. Polym Int 64(8):963–970. https://doi.org/10.1002/pi.4883
139. Zhang J, Yingping QI, Yongfeng SHEN, Hua LI (2022) Research progress on chemical modification and application of cellulose: a review. Mater Sci 28(1):60–67. https://doi.org/10.5755/j02.ms.25485
140. Zhang C, Zheng Y, Li M, Zhang Z, Chang L, Ai M, Wang J, Zhao S, Li C, Zhou Z (2020) Carboxymethyl cellulose-coated tacrolimus nonspherical microcrystals for improved therapeutic efficacy of dry eye. Macromol Biosci 20:2000079. https://doi.org/10.1002/mabi.202000079
141. Zhao J, Lu C, He X, Zhang X, Zhang W, Zhang X (2015) Polyethylenimine-grafted cellulose nanofibril aerogels as versatile vehicles for drug delivery. ACS Appl Mater Interfaces 7(4):2607–2615. https://doi.org/10.1021/am507601m

142. Zheng X, Gandour RD, Edgar KJ (2014) Remarkably regios—elective deacylation of cellulose esters using tetraalkylammonium salts of the strongly basic hydroxide ion. Carbohydr Polym 111:25–32. https://doi.org/10.1016/j.carbpol.2014.04.014
143. Zhou HY, Cao PP, Li JB, Zhang FL, Ding PP (2015) Preparation and release kinetics of carboxymethyl chitosan/cellulose acetate microspheres as drug delivery system. J Appl Polym Sci 132:42152. https://doi.org/10.1002/app.42152
144. Zhou L, Ke K, Yang MB, Yang W (2021) Recent progress on chemical modification of cellulose for high mechanical-performance poly (lactic acid)/Cellulose composite: a review. Compos Commun 23:100548. https://doi.org/10.1016/j.coco.2020.100548
145. Zhu Q, Liu C, Sun Z, Zhang X, Liang N, Mao S (2018) Inner layer-embedded contact lenses for pH-triggered controlled ocular drug delivery. Eur J Pharm Biopharm 128:220–229. https://doi.org/10.1016/j.ejpb.2018.04.017
146. Zhuo F, Liu X, Gao Q, Wang Y, Hu K, Cai Q (2017) Injectable hyaluronan-methylcellulose composite hydrogel crosslinked by polyethylene glycol for central nervous system tissue engineering. Mater Sci Eng C 81:1–7. https://doi.org/10.1016/j.msec.2017.07.029

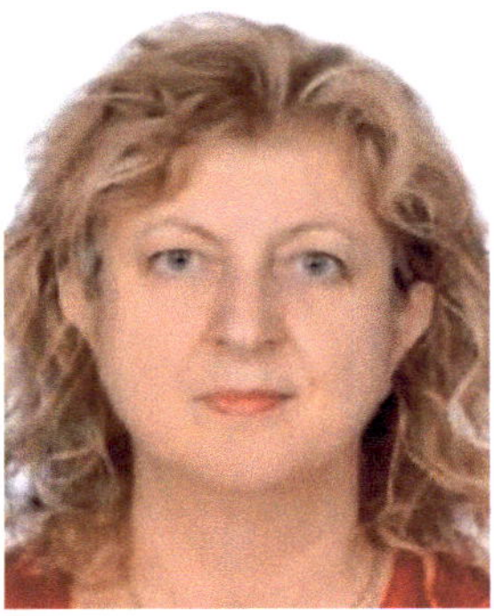

Murat Torun is a lecturer and researcher in the Department of Chemistry, Faculty of Science at Hacettepe University, Ankara, Türkiye. He received MSc and PhD degrees in Chemistry of Physical Chemistry Section from Hacettepe University in 2004 and 2008, respectively. He visited Institute of Nuclear Chemistry and Technology as a researcher (Warsaw, Poland) in 2009 for four months. Dr. Torun has experiences on natural polymers, chemical toxicology, advanced oxidation processes for organic contaminants in water, nanoparticles, polymer synthesis and characterization, development of drug delivery systems that all are carried out in his own research laboratory. He has some interdisciplinary projects that are partnered with pharmacy, biology and nuclear energy departments. He has projects, articles, book chapters and some presentations about his research area.

Nursel Pekel Bayramgil is currently working in the Department of Chemistry, Faculty of Science at Hacettepe University, Ankara, Türkiye. She received her MSc, and Ph.D. degrees in Physical Chemistry from Hacettepe University in 1994 and 1998, respectively. She worked as a visiting researcher at the Takasaki-Japan Atomic Energy Research Institute in 2002–2003. Dr. Pekel Bayramgil has research experience in cellulose and its derivatives, nanoparticles, nanocomposites, nanofibers, polymer synthesis and characterization, hydrogels, composite films, graphene and its derivatives, drug delivery systems, wound healing systems, personal protective equipment production, and in addition to her own department, she continues her interdisciplinary studies with departments such as medicine, pharmacy, dentistry, and biology. She has numerous projects and scientific articles in her field.

Safety and Biocompatibility of Cellulose-Based Materials

Lailla Daianna Soltau Missio Pinheiro, Gabriela Geraldo Sangoi, Alencar Kolinski Machado, and William Leonardo da Silva

Abstract The safety and biocompatibility of a material is determined by its ability to interact with the biological system without causing adverse reactions. Regarding safety, microbiological safety stands out, when applied to cells, chemical modification, when its derivatives are used with some modification in the cellular chain. When referring to biocompatibility, some care such as degradability, when modified for this purpose, with in vivo and in vitro studies are necessary to evaluate the real biocompatibility of the materials. Cellulose, in general, generates biocompatible and safe materials for use in the medical field. This chapter will discuss points relating to cellulose, briefly discussing how it is obtained and the challenges it faces in terms of possible contamination, its toxicity and biocompatibility, and how to proceed with assessing its safety in in vitro and in vivo models, as well as environmental toxicity analyses and regulatory guidelines on its production and use. Briefly, its application, especially in the area of biomedical engineering, has been gaining prominence in product development and its use is increasingly promising, as most studies indicate that the use of cellulose is very promising due to its high biocompatibility with organisms and its good biodegradability in the environment, including in tests with aquatic animals. However, although it is normally considered safe, whenever a new product is produced with cellulose and its derivatives, safety assessments must be carried out, as the origin of the cellulose and the method of obtaining it can influence the toxicity of the material.

Keywords Cellulose biocompatibility · Microbiological safety · Medical biomaterials · Chemical modification · Health device quality control

L. D. S. M. Pinheiro · G. G. Sangoi · A. K. Machado · W. L. da Silva (✉)
Graduate Program in Nanoscience, Franciscan University (UFN), Santa Maria, RS, Brazil
e-mail: williamleonardo_silva@hotmail.com

L. D. S. M. Pinheiro · W. L. da Silva
Applied Nanomaterials Research Group (GPNAp), Franciscan University (UFN), Santa Maria, RS, Brazil

G. G. Sangoi · A. K. Machado
Laboratory of Cell Culture and Bioactive Effects, Franciscan University (UFN), Santa Maria, RS, Brazil

R. Malviya and S. Sundram (eds.), *Sustainable Cellulose Materials for Biomedical Applications*, Sustainable Materials and Technology,
https://doi.org/10.1007/978-981-96-9904-9_2

1 Introduction

In recent years, there has been a growing interest in sustainable and biodegradable materials for various applications, such as domestic use, food, pharmaceuticals, cosmetics, and medical fields [59, 69]. Among these materials, cellulose and its derivatives have gained prominence because they are natural polysaccharides composed of long chains of glucose molecules. They possess unique properties, such as biodegradability, biocompatibility, and versatility [135]. Cellulose is mainly found in plants, wood, and other natural sources, where it plays a fundamental role in plant structure and rigidity. It is abundantly available and of great industrial importance, used in products ranging from paper to biomedical fabrics and bioenergy (Fig. 1) [36, 48, 124]. Cellulose and its derivatives have a fibrous and durable structure, making them ideal for applications that require strength and durability.

Cellulose-based materials for medical applications have been extensively studied due to their biodegradability, biocompatibility, and highly organized structure. Their applications span regenerative medicine, tissue engineering, medical devices, and drug delivery systems [1, 20, 114, 115]. In tissue engineering and regeneration, they are used as scaffolds to support cell growth and tissue regeneration, providing a favorable environment for cell adhesion and proliferation due to their porous structure and the possibility of chemical modifications [90, 103, 118, 141]. In medical devices, cellulose and cellulose-based materials are used in dressings, adhesives, and films for controlled drug release, as they provide an ideal moist environment for healing and preventing infections [35, 97]. In drug delivery, they are used due to their fibrous structure and high surface area, as chemical modifications of cellulose allow for

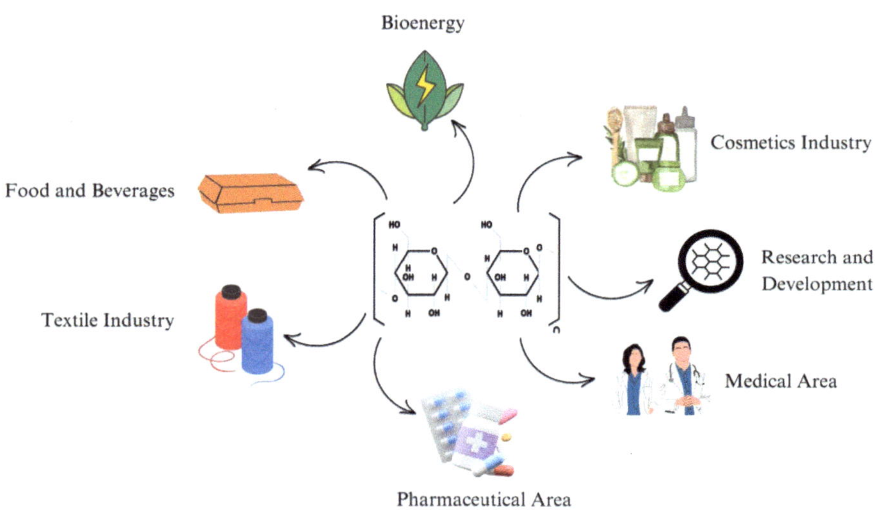

Fig. 1 Areas of application of cellulose-based materials

controlling the drug release rate, ensuring more effective therapies with fewer side effects [61, 64, 71].

However, for use in biomaterials, it is important to highlight the safety and biocompatibility of these materials. Regarding safety, it is crucial to emphasize the absence of toxic or harmful effects during the material's use, requiring a rigorous evaluation of these properties, which is essential to ensure the efficacy and safety of these materials in clinical applications [41, 156]. For biocompatibility, it is important to mention the material's ability to interact with biological tissues without causing adverse responses, such as inflammation activation or rejection [129, 175].

Aiming at the best applicability of cellulose-based materials, this chapter aims to explore the safety and biocompatibility of cellulose-based materials, discussing the most recent advances, the challenges faced, and the evaluation methods used to ensure that these materials can be widely applied in innovative solutions for medicine and other areas of biotechnology.

2 Overview of Security Aspects

Broadly speaking, the safety of cellulose-based materials is a fundamental aspect, especially when related to medical and/or pharmacological devices. Being a natural material, cellulose is widely recognized for its low toxicity potential and excellent biodegradability. However, to be considered safe, a series of precautions must be taken to evaluate the material and avoid any health risks [42, 158]. Some of these precautions include: (i) origin, (ii) impurities and contaminations during production; (iii) reactivity and potential toxic effects; (iv) biodegradability and degradation of the products; (v) allergic or immunological reactions; (vi) safety testing and toxicological evaluation; and (vii) accelerating clinical acceptance.

2.1 Origin, Extraction, and Contamination

Cellulose is a natural biopolymer abundantly found in nature, consisting of long chains of glucose (D-glucose) molecules that form rigid fibers and structures resistant to degradation by organisms [43]. However, its origin and extraction process directly influence the resulting product, thereby affecting its quality [11]. Some of its primary sources are plant-based, such as: trees, the most common sources for cellulose production; cotton, notable for its high purity, making it suitable for textiles, pharmaceuticals, and tissue engineering; bamboo, which offers good strength; and agricultural residues, a sustainable and low-cost alternative [38, 155]. In addition to plants, cellulose can be extracted from algae, characterized by their crystalline structure and potential biomedical applications [135, 147]. Cellulose can also be derived from microbial sources, which exhibit excellent biocompatibility, high crystallinity, and mechanical strength, and are free from lignin and hemicellulose [30, 88]. Figure 2 shows the primary sources of cellulose and their respective benefits.

The cellulose extraction process depends on factors such as the origin, purity level, and final application of raw material. It involves separating cellulose from components such as lignin, hemicellulose, proteins, and minerals [130]. The extraction of plant-based cellulose can be carried out using three main methods: chemical, mechanical, and biological, depending on the process and source [50, 89]. Table 1 summarizes these three main methods, along with their characteristics and steps.

For agricultural residues, the process differs as it requires a pre-treatment step to remove impurities, followed by chemical processing and subsequent purification [128]. On the other hand, bacterial cellulose extraction involves bacterial fermentation, the collection of cellulose films, and then purification [28].

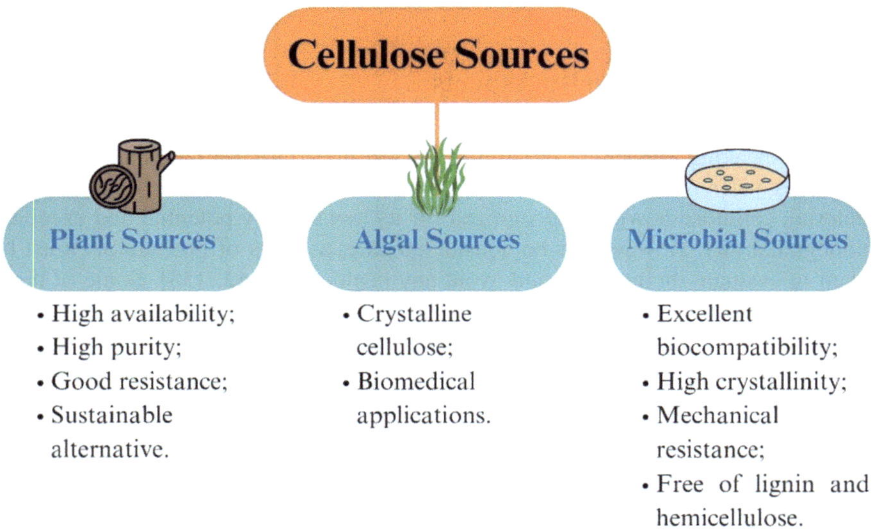

Fig. 2 Main sources and their respective benefits

Table 1 Overview of the main extraction methods and their characteristics

Method	Steps	Characteristic
Chemical	Pre-treatment. Chemical cooking (Kraft or sulfite process); bleaching; purification	High yield and purity cellulose; need for effluent treatment as chemical compounds is released
Mechanical	Grinding under high pressure; mechanically released fibers	Fast and economical; lower purity
Biological	Exposed to fungi or enzymes; degradation of non-cellulosic components	Sustainable, without intensive use of chemicals; time-consuming and expensive on a large scale

However, the production of cellulose-based materials faces certain challenges, such as impurities and contamination, since many of their applications require materials with high purity, especially for food packaging, medical, and pharmaceutical products [133, 139, 169]. Undesirable substances can significantly alter the functionality and safety of these materials. Impurities may have various origins, including: (i) chemical impurities: lignin, which affects the material's biocompatibility; (ii) hemicellulose, which impacts the thermal and chemical stability of the material; (iii) chemical residues, leading to material contamination and heavy metals, known for their toxicity at high concentrations [107, 170]; (iv) microbial contamination: from bacteria and fungi, posing infection risks in biomedical applications; and v physical contamination: involving foreign particles such as dust and silica fibers [106, 113, 157, 172]. Thus, controlling these contaminants requires good manufacturing practices, proper storage, and strict inspection of materials.

3 Toxicity and Biocompatibility

Toxicity and biocompatibility are critical factors in the evaluation of materials intended for medical and biological applications. Toxicity refers to the potential for a material to cause adverse effects on biological systems, such as cellular damage, inflammation, or systemic toxicity [145]. These effects can arise from the material itself, residual chemicals from its synthesis, or degradation by product during its use in biological environments [173]. To ensure safety, extensive in vitro and in vivo studies are conducted to evaluate the cytotoxicity, genotoxicity, and immune responses of materials [73]. Regulatory standards require that medical devices and biomaterials meet stringent toxicity thresholds before they can be approved for clinical use [55].

Biocompatibility, on the other hand, focuses on the ability of a material to interact with living tissues without causing harmful reactions. A biocompatible material should integrate seamlessly into biological systems, supporting processes such as cell adhesion, proliferation, and tissue regeneration while avoiding inflammatory or immune responses [77]. For example, materials used in implants, drug delivery systems, or wound dressings must demonstrate high biocompatibility to ensure their efficacy and safety [47, 166]. Factors such as surface properties, degradation rate, and chemical composition play important roles in determining biocompatibility [56, 137]. Moreover, advances in materials engineering, such as surface functionalization and nanotechnology, have enabled researchers to increase biocompatibility while minimizing toxicity, paving the way for innovative applications in regenerative medicine and biomedical devices [19, 65].

3.1 *Toxicity*

Toxicity assesses the potential of materials to cause harm to live organisms, making it a critical factor in biomedical applications. While cellulose is a naturally non-toxic biopolymer, chemical residues from the synthesis process, additives, natural impurities, and/or contaminants can impart toxic characteristics to its derivatives [154].

Toxicity with chemical residues occurs in the extraction process, where many chemicals are used; for example, sodium hydroxide (NaOH) and sodium sulfide (Na_2S) are utilized in the Kraft process, and in the bleaching process chlorine dioxide (ClO_2), hypochlorite (ClO^-) and peroxides (H_2O_2) are applied. These components release toxic byproducts, such as dioxins, which are harmful to the environment and human health, resulting in respiratory or systemic toxicity and skin irritation. In addition, by receiving some active ingredients to improve mechanical properties and resistance, these cellulose-based materials end up receiving plasticizers, stabilizers, or retardants, and also fluorochemicals or silicones, which causes them to act as endocrine disruptors, which means that they are mechanisms that replace or block hormones and their natural actions, as well as presenting bioaccumulate and toxic action [15, 138].

Heavy materials such as lead, cadmium, mercury, and/or arsenic may appear due to contamination of the raw material (soil) or induced by catalysts used in some of the production stages, causing neurotoxicity, kidney, and liver damage, in addition to being carcinogenic [98]. Another important point that leads to toxicity is the degradation byproducts, which occur when cellulose is degraded, releasing chemical compounds, causing irritation, allergic reactions, and systemic toxic effects [29]. Figure 3 shows the main contamination methods mentioned, chemical residues, additives, heavy metals, degradation byproducts, and biological contamination.

3.2 *Biocompatibility*

Biocompatibility is a critical factor when cellulose and its derivatives are used in medical devices, biomedical applications, and/or pharmaceuticals. Pure cellulose is naturally biocompatible; however, residues and chemical modifications can alter this property [29]. Therefore, rigorous testing is essential to ensure quality, safety, and efficacy characteristics.

Pure cellulose exhibits inherent biocompatibility and is hypoallergenic and biodegradable. When used in healthcare, its degradation within the body does not produce toxic byproducts, and it is absorbed without causing allergic reactions [114]. However, modifications to enhance certain properties can sometimes reduce its biocompatibility. For example:

- *Regenerated Cellulose (RC)*: This involves dissolving cellulose in solvents and then regenerating it, which can introduce solvents or chemical residues into the

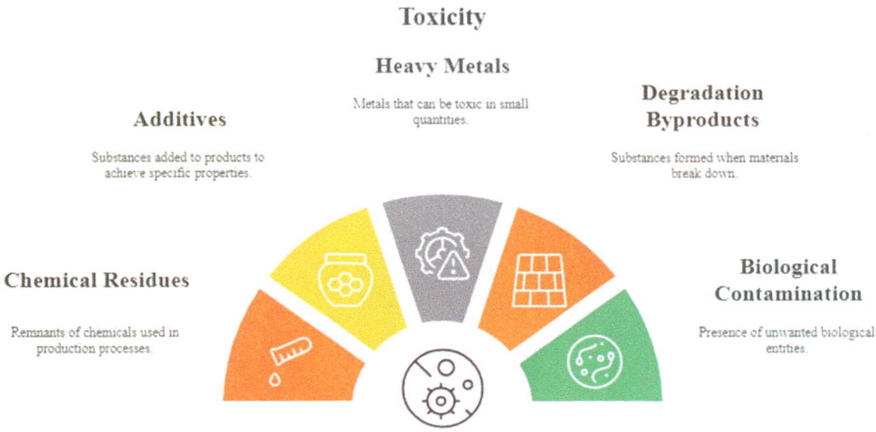

Fig. 3 Main methods of contamination during the process of extracting cellulose and its derivatives: chemical residues, additives, heavy metals, degradation byproducts, and biological contamination

material. Its biocompatibility depends on the purity of the process and the residual chemicals. For instance, processes using N-methylmorpholine-N-oxide (NMMO) are more biocompatible than the viscose process, which uses carbon disulfide (CS_2) [7].

- *Modified Cellulose (MC):* This involves chemically altering cellulose to achieve specific properties, such as increased hydrophobicity or flexibility, as seen in cellulose acetate, carboxymethylcellulose, or cellulose esters. However, chemical modifications can introduce compounds that lead to cytotoxicity and/or allergic reactions. Thus, selecting appropriate chemicals for cellulose modification is essential [108].

Bacterial cellulose exhibits excellent biocompatibility, making it highly interested in medical applications. This is primarily due to its high purity, derived from its bacterial origin rather than plant sources. Additionally, it offers a high absorption capacity and the ability to retain moisture, which is advantageous for medical dressings. Its porous structure also facilitates oxygen exchange, promoting wound healing [60, 164].

Because of these properties, cellulose has been widely researched for applications in tissue engineering, wound dressings, implants, and contact lenses [20, 21, 81, 148]. Several factors influence the biocompatibility of cellulose and its derivatives, such as: (i) material purity: impurities and chemical residues can impact biocompatibility; (ii) particle size and form: for instance, nanocellulose enhances integration with biological tissues due to its high surface area and nanoscale dimensions; and (iii) chemical modifications: adding functional groups or biomolecules can improve biocompatibility for biomedical devices [45, 100, 149]. Figure 4 shows the summary of the main factors influencing and enhancing biocompatibility.

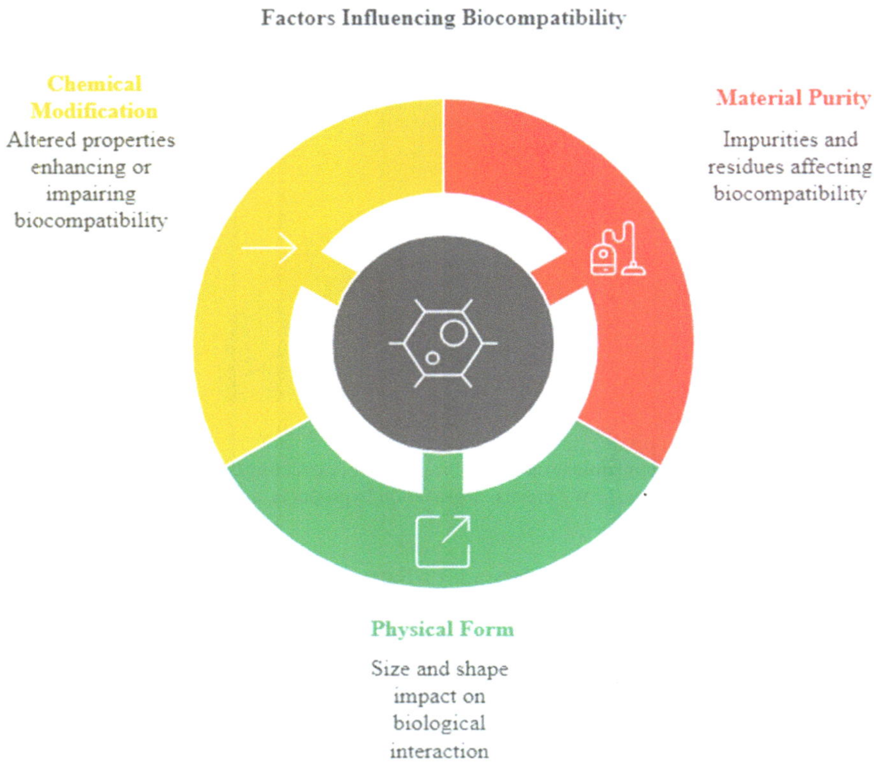

Fig. 4 Main factors influencing and enhancing biocompatibility

4 Allergic and Immunological Reactions

An important aspect to be considered in the use of cellulose and its derivatives is the potential of causing allergic and immunological reactions, as these materials should not cause adverse effects in the body. While cellulose has excellent biodegradability, the factors contributing to its toxicity can also lead to infections and allergic responses.

Allergic reactions are immune system responses triggered when a substance is identified as harmful to the body. Some allergic sources related to cellulose and its derivatives can stem from natural origins or clinical manifestations [45]. Table 2 outlines these two aspects and their primary reactions. The importance of testing the allergenic potential of biomaterials cannot be overstated. Tests such as patch tests and, particularly, vitro tests are critical. Typical symptoms include swelling, pain, or redness at the material's contact site.

Immunological reactions occur when the body identifies a "foreign body," thereby activating the immune system. This can result on: local inflammatory responses: inflammation occurs at the site where the cellulose-based biomaterial has contacted

Table 2 Summary of natural and clinical allergy sources and associated reactions

Sources of reactions	Characteristics
Allergic potential	Impurities natural; chemical waste; additives
Clinical manifestation	Contact dermatitis; respiratory reactions; eye irritation

the body; and foreign body reactions: more severe responses, such as rejection of cellulose-based implants, can happen due to toxic incompatibility or improper sterilization [102, 122, 160]. In cases of rejection, the immune system may be activated at a systemic level, leading to broader complications. Figure 5 shows some immunological reactions that can occur from contact with cellulose and its derivatives.

Safety tests are also extremely necessary to assess the biocompatibility of cellulose. The main tests include in vitro tests, conducted on cells to assess the material's cytotoxicity; in vivo tests, performed to evaluate the interaction of the material with living organisms in complex environments; sensitivity and irritation tests, designed to determine potential irritations on the skin or mucosa; and biodegradability tests, used to assess whether the material degrades in a controlled manner, either within the body or in the environment.

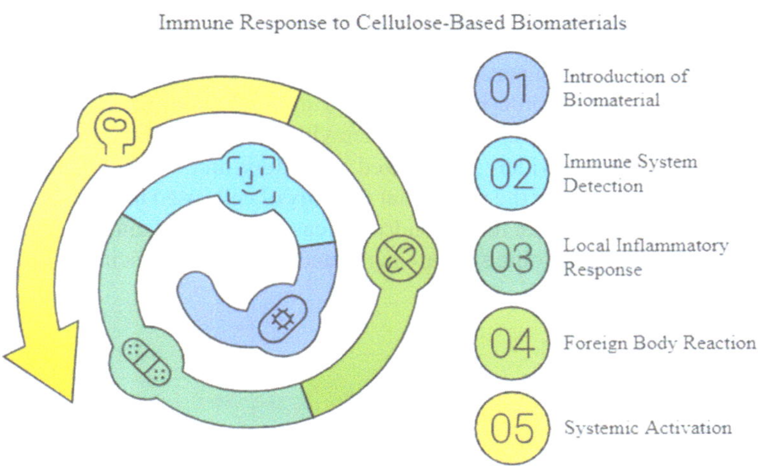

Fig. 5 Illustration of the primary immunological reactions

5 Safety Tests

Considering toxicity, biocompatibility and adverse reactions, safety tests are essential for the use of cellulose-based materials, as they ensure their safety during use. These tests aim to evaluate the chemical, physical, biological, and mechanical properties, thus identifying possible risks. Initially, the main tests to be performed are biological safety tests, which include cytotoxicity, dermal sensitivity, irritation, acute systemic toxicity, and genotoxicity. The cytotoxicity test evaluates the toxicity of the cellulose-based material to human cells and can be performed directly on specific cell lines for each application, while genotoxicity tests evaluate the damage caused to DNA by the biomaterials used [108]. Dermal sensitivity tests will evaluate, both in vivo and in vitro, the immune response after contact with the biomaterial [122]. Irritation tests evaluate the potential irritability of the biomaterial, both to the skin and the eyes or mucous membrane [40], while the acute systemic toxicity test will determine whether toxic reactions may occur in internal organs caused by the release of certain compounds by the cellulose-based material [63]. Figure 6 demonstrates possible safety evaluations of cellulose.

5.1 Detection of Cellular Viability

Prior to testing cellulose agents in animal experimental models, it is necessary to investigate their safety profile in cells. The primordial assay for cellular viability analysis is the MTT (tetrazoline 3-(4, 5-dimethylthiazol-2yl)-2, 5-diphenyl bromide) assay. This is an assay widely recognized for safety and security. This assay is responsible for determining the percentual viability of cells used in the exposure. The reagent interacts with cellular mitochondria, releasing the formazan crystal produced by viable cells [87]. Tetrazolite salts are generated through the succinate-tetrazolium

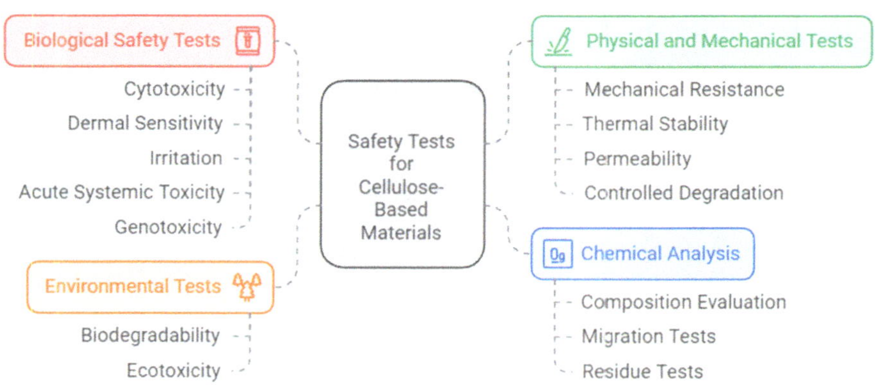

Fig. 6 Possible safety evaluation of the cellulose

reductase system of the electron transport chain and are only activated in viable mitochondria. Despite MTT, other tetrazolite salt reagents can be used to determine cell viability, such as WST-1 and XTT. These reagents also form formazan crystals, the former being related to the glycotylic production of NAD(P)H while the XTT assay consists of a reaction that generates electron coupling, a reaction that is only possible in viable cells due to the need for mitochondrial metabolism [74, 134].

There are various tests for determining cell viability, as this analysis is crucial for determining the safety of compounds and materials [58]. Another test that can be applied to measure cellular viability is the neutral red assay. Neutral red is a reagent that is metabolized by lysosomes of alive cells.

In addition to metabolism by organelles, other tests can be used to determine cell viability, such as measuring the production of molecules, such as measuring the ATP produced by cells, since only viable cells can produce ATP—its production is directly proportional to the number of alive cells. In these assays, the principle of the technique consists of the conversion of luciferin into oxyluciferin, by the action of the luciferase enzyme and cellular ATP, generating the luminescence that is detected in the test [116].

Techniques using fluorescent reagents, which by impregnating antibodies differentiate viable cells from dead cells, can also be used to analyze cytotoxicity. Fluorescent markers are attached to antibodies that end up binding to antigens and/or cell receptors expressed by the cells. The differentiation between viable and non-viable cells is due to the difference in the expression of the markers. Assays such as these can generate images captured by fluorescence microscopy. Examples include assays using the dyes calcein and propidium iodide [99, 167]. In addition, another technique that can be used with fluorescent reagents is the CFSE assay, a kit that marks cells by mitotic generations, so that exposure to the material can be analyzed for more days, determining how much of the cells have proliferated exposed to the material in test. Flow cytometry is required to carry out this test, so it ends up costing more financially. However, it is excellent for differentiating viability in lymphocytes, for example, and is useful in immunological tests [151].

Still in relation to the use of antibodies to evaluate cellular viability, it can be mentioned the BrdU assay (bromodeoxyuridine reagent). This is an assay in which the reagent used binds genomic DNA recently synthesized in the S phase of the cell cycle, with the joint binding of an antibody to this complex. This analysis can be carried out both in cell culture and in vivo models, making it easier for researchers to quantify cells in analyses of hematopoietic stem cells, for example [4, 171]. Very similar to this assay, there is the EdU technique, where the reagent also binds to synthesized DNA; however, this one uses a click method, i.e., click chemistry, i.e., with the formation of new molecules, without the addition of antibodies.

Protein expression analyses can also be used to determine cell viability. A good example is the measurement of KI67 protein. This immunohistochemical assay leads to a direct quantification of cell proliferation, as this protein is expressed in all the active phases of the cell cycle, except for phase zero (GO), so by quantifying its levels, the rate of cell proliferation [9] can be obtained. This test is commonly used for cancer diagnosis or as a cancer biomarker, however, due to its importance in the cell proliferative phase, the KI67 protein can be used as a cell proliferation marker, including, for example, analyses that use cellulose in healing investigations.

Although the assessment of cell viability is paramount, other markers can and should also be analyzed to determine the biocompatibility and safety of cellulose, such as genotoxicity analyses and determination of biomarkers related to the oxidative metabolism.

5.2 Genotoxicity Assessment

Genotoxicity analyses assess possible damages to genomic DNA, which is an important factor to analyze. To perform this type of analysis, the comet assay can be a great alternative. The comet assay is a test that detects damage to the genetic material according to the "comet tail" formed after applying an electrophoresis condition. Through the electrophoresis, damaged material ends up being dragged along, forming the "comet tail" that can be microscopically analyzed [152]. The comet assay is one of the most common options to perform genotoxicity assessments [22]. It has been worldwide used since it is a low-cost method that provides consistent results quickly, making it more popular than other genotoxic tests such as micronucleus, for example [72]. The analysis identifies different types of DNA damage, such as breaks in single and double stranded genomic DNA, and the potential for damage to the material depends on a classification of the damage analyzed via optical microscopy. One important limitation of this technique is the possibility of different classification between professionals [31], and for this reason, there are already methods that can facilitate the interpretative homogeneity of the analysis, such as the development of quantitative analysis imaging techniques, which aim to reduce interference between comet assay analyzers [16].

5.3 Determination of Oxidative Stress Levels

Cellular assays that report on cellular oxidative metabolism are also widely used to determine the safety profile of different materials. Determining the levels of reactive oxygen species (ROS) is important because the exacerbated production of these molecules can lead to oxidative stress, which is extremely toxic to the cells. Oxidative stress can be related to pro-inflammatory activation and cell death, since the imbalance between levels of ROS and enzymatic antioxidants can result in a cellular lack of homeostasis [49]. Reactive nitrogen species (RNS) are also a crucial factor in establishing cellular oxidative stress [125] and end up being generated after metabolic processes such as oxidative phosphorylation, cell differentiation, and generation of transcriptional factors and apoptosis. On the other hand, the presence of ROS and RNS in the environment also causes these factors to be manufactured by the organism, thus creating a loop [125].

However, the direct determination of ROS is extremely difficult, due to the instability of the molecules and their short half-life [84]. For this reason, it is necessary to carry out tests that, using reagents, can proportionally measure the quantification of these molecules, as is the case of the DCFH-DA (2, 7'-dichlorofluorescein diacetate) test. DCFH-DA is an assay that estimates the levels of ROS produced by the cells. DCFH-DA reagent, in the presence of ROS, ends up being deacetylated into another non-fluorescent compound (2' 7'-dichlorofluorescin), which remains inside the cells and, in contact with hydrogen peroxide, ends up generating 2', 7'-dichlorofluorescein, a fluorescent compound. Thus, the higher the fluorescence obtained, the higher the levels of ROS in the environment [44, 110]. However, the probe used in the DCFH-DA assay is not specific for determining peroxide radicals, and studies show that it can also be sensitive to pH and oxygen in the environment, so the fluorescence emitted may not be specifically proportional to the increase in ROS levels only [105].

However, the half-life of hydrogen peroxide and peroxyl radicals, when compared to other ROS, are relatively stable, with a half-life of around seconds to minutes, whereas other reactive molecules have a half-life of only a few nanoseconds, making them extremely difficult to measure. Alkoxyl, oxygen, and hydroxyl anions are examples in this case [76, 131].

Oxygen can be measured using a fluorescent probe called dihydroethidium (DHE), which inside the cell is oxidized to 2-hydroxyethidium ($2-OH-E+$) by the presence of oxygen; this final compound is then fluorescent, and its measurement is proportional to intracellular oxygen levels. However, DHE is not only sensitive to oxygen, as it can be oxidized by radicals such as hydroxyl and hydrogen peroxide [76, 131].

Techniques capable of directly measuring ROS can be electron spin resonance, spin capture or pulse radiolysis. Nevertheless, these tests are more expensive and require more elaborate equipment's, as well as being extremely labor-intensive, which is the reason why they are not widely used by researchers [131].

The difficulty of directly determining the total levels of ROS has led researchers to look for techniques that indirectly measure these markers. We can measure oxidative stress through tests that identify molecules originating from the reaction between oxidizing species and cellular components, such as lipid peroxidation products, protein carbonyls, as well as markers of protein and even nuclear damage [79].

Due to the susceptibility of lipids to structural damage from oxidation, the end of the lipid peroxidation cascade generates markers that are used to indicate an environment with exacerbated oxidative stress. The process of lipid peroxidation is subdivided into three stages: initiation, propagation, and termination. Firstly, it is very likely that the hydroxyl radical initiates the process of the lipid peroxidation cascade, as it is the most active radical. These radical attacks the methylene hydrogen molecule of a fatty acid, resulting in a carbon-centered lipid radical, which in turn ends up reacting with an oxygen molecule to form a lipid peroxyl radical where it forms endoperoxides through cyclization [101].

At the end of the sequence of the lipid peroxidation cascade, toxic molecules are generated that are related to DNA and protein damage, malondialdehyde and

4-hydroxy-2-nonenal, considered biomarkers of lipid peroxidation, along with other molecules such as 4-hydroxynonenalmercapturic acid and 4-oxo-2-nonenal [101].

The measurement techniques are mentioned below, in vivo tests (5.5) as they are more specific to this type of analysis, since the samples used for indirect measurement of these biomarkers come from animal metabolism and excretion.

5.4 Irritability Tests

In addition to cell tests, other in vitro evaluations can also be carried out to determine the safety of cellulose, such as irritability tests such as the HAT-CAM (Chorioallantoic Membrane) test. This test assesses the potential vasodilation of the materials, i.e., how much they will irritate the membrane. The technique can be carried out on bovine eyes or chicken eggs, for example, as the principle of the technique requires the membranes to be exposed to the material. In this regard, macroscopic changes in the membrane are assessed, such as hyperemia, coagulation, and hemorrhage [52]. The use of this test can then determine cellulose's ability to promote skin or membrane irritation, for example, when cellulose is used in dermal dressings.

5.5 In Vivo Assay

Once safety has been established with cellular and other in vitro tests, the investigations can be moved to in vivo analysis, as this will allow the reduction of the concentrations tested and consequently the "n" number of animals used in the experiments.

In vivo tests allow us to evaluate the systemic action of cellulose, as well as dermal sensitivity. Assessing acute systemic toxicity allows the detection of health risks of acute exposure to the material, in this case cellulose, about its route of action [136]. In addition, analyses of the animal's weight loss or gain, the appearance of anomalies and other morphological or clinical alterations are also evaluated during the experiments [136].

However, attention must also be paid to the analysis of systemic oxidative stress, since once established it can lead to inflammatory activation and even cell death in some cases, since an environment with high oxidative stress ends up becoming a toxic environment [37, 109, 110, 112].

5.5.1 Levels of Reactive Oxygen Species: Lipidic Peroxidation

As a systemic evaluation of the use of cellulose, indirect biological markers of oxidative stress can be assessed, as mentioned above, by targeting the products of metabolism, depending on the route of action and/or contact of the cellulose in the animal's organism.

Thus, urine, plasma or serum, cerebrospinal fluid (CSF), saliva, and organ tissue can be evaluated and the techniques for determining lipid peroxidation biomarkers include detection by high performance liquid chromatography (HPLC) chemiluminescence, iodometric assay, ferrous oxidation of xylenol, thiobarbituric acid reactive substances (TBARS) assay, gas chromatography coupled to mass spectrometry (GC–MS assay), enzyme-linked immunosorbent assay (ELISA), high performance liquid chromatography coupled to sequential mass spectrometry (HPLC–MS), isotope dilution mass spectrometry, immunohistochemistry (IHC) [70, 86, 96, 142, 143]. While the biomarkers range from:

- Lipid hydroperoxide which can be detected by HPLC, chemiluminescence, iodometric assay and xylenol ferrous oxidation, and can be measured from samples such as plasma, supernatant, tissue, and serum lipoproteins [142],
- Malondialdehyde measured by TBARS assay, HPLC, and GC–MS, which in addition to the samples mentioned above can also be detected in erythrocytes, saliva, and CSF [86],
- 4-hydroxy-2-nonenal measured by ELISA, HPLC–MS, GC–MS, and IHC detected by serum, plasma, urine, tissue, and CSF samples [80, 86, 96],
- 4-hydroxynonenalmercapturic acid measured by HPLC–MS from urine 9 [96];
- 4-oxo-2-nonenal detected on urine by isotope dilution mass spectrometry [86] [114];
- 8-isoprostaglandin detected on urine by GC–MS, ELISA, and HPLC–MS [86, 96],
- 11-β-prostaglandin F2α (11-PGF2α), 15-prostaglandin F2α (15-PGF2α), and 8-Iso-15(R)-prostaglandin F2α both detected in urine by HPLC–MS [96].

5.5.2 Levels of Reactive Oxygen Species: Damage to Nucleic Acids

Another biomarker of oxidative stress that can be analyzed from the systemic analysis of materials is the percentage of damage to nucleic acids, because ROS, especially hydroxyl groups, act directly on genetic material to stabilize it [49]. At the end of the cascade of chemical reactions of free radical attack on DNA, 8-hydroxy-2'-deoxyguanosine (8-OHdG) is generated, an important biomarker for determining levels of oxidative stress since its formation is predominant at levels of damage to genetic material [121, 159], and is even mutagenic[14], since it leads to irreversible damage to DNA, being able to change the order of base pairs from Guanine-Cytosine (GC) to Thymine-Adenine (TA) [144].

Furthermore, the isolation of mitochondrial DNA for this type of analysis must be carried out primarily because mitochondrial DNA is more susceptible to damage by this molecule in question. As such, 8-OHdG levels are higher in mitochondrial DNA than in genomic DNA [163].

Additionally, another biomarker related to damage to genetic material is 8-hydroxyguanine (8-OHG), which results from the oxidation of the Guanine base by ROS. This marker is a lesion found in equal proportions in both genomic and mitochondrial genetic material [144]. The Guanine base can also be oxidized by RNS, especially peroxynitrite, resulting in 8-nitroguanine (8-NO2-G). The presence of this molecule acts on DNA synthesis, promoting negative events in synthesis because 8-NO2-G can bind to Adenosine resulting in GT transversions. For this reason, 8-NO2-G is known to be mutagenic [121]. Thus, 8-NO_2-G is considered a marker of nitrative damage induced by RNS [96].

Despite the DNA damage, ROS can cause damage to RNA, the latter being even more susceptible to damage than DNA, due to its single-stranded structural form and the absence of a repair mechanism and its proximity to mitochondria [163].

The oxidation of RNA by ROS leads to altered protein synthesis, and the protein formed is then functional or has a different property to what it should have and can even lead to cell death or degradation [92]. There are two biomarkers of RNA damage: 8-hydroxyguanosine or 7, 8-dihydro-8-oxo-guanosine (8-oxoG), the main product of RNA damage generated by the attack of RNS [75], and 8-nitroguanosine (8-NdG, produced by nitrogen dioxide (NO_2) [168].

To measure the markers, urine, saliva, plasma, tissue, serum, and peripheral lymphocytes can be used as samples and the techniques consist of: HPLC, ELISA, HPLC with electrochemical detector (ECD), HPLC coupled with mass spectrometry (LC–MS) [26, 78, 84, 111, 112, 140, 159, 165].

Biomarkers can be measured as follows:

- 8-hydroxy-2'-deoxyguanosine through saliva, tissue, plasma or serum, and urine using LC–MS, ELISA, and HPLC techniques [26, 165],
- 8-hydroxyguanine through saliva, urine, and serum using HPLC and HPLC with ECD techniques [78, 96, 140],
- 8-nitroguanine through urine samples or isolation of lymphocytes from peripheral blood, the techniques to be used being HPLC with ECD and HPLC–MS [14, 131],
- 8-hydroxyguanosine and 8-nitroguanosine both through urine using the HPLC–MS technique [96].

5.5.3 Levels of Reactive Oxygen Species: Protein Damage

So far, we have seen the damage caused by RNS at the cellular level, but their action can also cause damage to the protein itself. There are several processes that proteins can undergo as a result of oxidative stress, such as carbonylation, oxidation of sulfur-containing amino acids, and oxidation of aromatic fractions and glycolization. The free radicals that can cause protein damage are the superoxide anion, hydroxyl group, peroxyl radical, alkoxyl, and hydroperoxyl [84, 121].

Briefly, the most widely used marker of protein damage is carbonyl, a product of oxidative damage to amino acids such as lysine, proline, threonine, and arginine [79]. Carbonyls circulate in the body for longer than other markers, such as lipid peroxidation products, which take around minutes to degrade, while carbonyl takes up to days to be metabolized and excreted [34].

In addition to carbonyl, other markers are associated with protein damage, such as cystine, which is the oxidized form of cysteine, which is not used as a marker of oxidative stress due to its high reactivity [80, 119]. Cysteine is susceptible to oxidation by reactive aminothiol residues which will interact with ROS or RNS to form cysteine [80].

Methionine is also an amino acid that is oxidized by ROS or RNS, producing methionine sulfoxide or methionine sulfone. However, the former is considered a marker of oxidative stress due to protein damage, as it is produced first and on a larger scale than the latter [54]. In addition, methionine sulfoxide, among most thiol-oxidized products, is considered the most stable [84].

Amino acids that have aromatic rings in their composition also undergo oxidation and the products of these oxidations are markers of oxidative or nitrative stress [39]. Called advanced oxidation protein products, these products contain dityrosine, a cross-reaction of two amino acid tyrosines, which is accelerated in the presence of the hydroxyl radical or nitrative species of peroxynitrite and nitrosoperoxycarbonate [39]. Thus, the levels of advanced oxidation protein products are biomarkers of oxidative and nitrative stress [162], including 3-nitrotyrosine, produced by the reaction between tyrosine and nitrogen dioxide [3].

Levels of carboxymethyl lysine are markers of oxidative stress from the glycoxidation pathway [56, 79]. Carboxymethyl lysine is an advanced degradation product normally formed in the body by the oxidative process; however, its high production is an indicator of an environment with established oxidative stress [80].

Allantoin is a non-enzymatic oxidation product of uric acid that is a biomarker of systemic oxidative stress [39].

The samples that can be used for the analysis of protein damage markers and their respective techniques can be:

- Protein carbonylation content can be detected in plasma or serum, tissue, aqueous humor, and saliva. The techniques to be used can be spectrophotometry, 2, 4-dinitrophenylhydrazine assay (DNPH), IHC, HPLC–MS, ELISA or immunoblot [34, 79],
- Cystine detected in serum or plasma samples by HPLC [31, 80];
- Methionine sulfoxide detected in serum or plasma by western blotting or liquid chromatography coupled to mass spectrometry (LC–MS) [84],
- Dityrosine found in urine, serum or plasma, which can be detected by LC–MS, spectrophotometry, spectrofluorimetry or HPLC–MS [80, 96]
- Advanced oxidation protein products detected in saliva, serum or plasma by spectrophotometry [80, 150],
- Nitrotyrosine found in urine, serum, or plasma by mass spectrometry, IHC, ELISA, HPLC or LC–MS [80, 96]
- Carboxymethyl lysine through tissue, serum, or plasma by ELISA, IHC, HPLC–MS, spectrophotometry or immunoblot techniques [80, 96],
- Pentosidine through tissue, serum, or plasma using the same techniques as above [56, 80]

- 3-bromotyrosine detected in urine using LC–MS or HPLC–MS techniques [113, 140];
- 3-chlorotyrosine found in serum, plasma or whole blood and urine using HPLC–MS [27, 33],
- Acrolein detected in urine or tissue using ELISA or IHC [146],
- Allantoin present in urine by the specific Rimini-Schryver reaction method, by HPLC–MS or LC–MS [66, 96].

5.5.4 Histological Analysis

The histological analysis of animal tissues after exposure to cellulose is of great importance, as this determines the local reaction generated by the product.

Acute inflammation after contact with cellulose can be tested. Inflammatory activation can be confirmed when there is a large population of neutrophils and macrophages at the site near the graft [57]. Calcification can also be found [127]. However, this inflammatory reaction depends on the production of the cellulose and, in the case of bacterial cellulose, the species of bacteria from which it was extracted. For this reason, it is more common to find in vivo studies that indicate good biocompatibility of cellulose in animal tissue, without the presence of inflammation [5, 57, 123, 127, 153].

In addition to in vitro and in vivo tests, it is recommended to carry out environmental tests on cellulose-based biomaterials. Biodegradability tests assess the degradation of the biomaterial by microorganisms, and also ecotoxicity, which assesses whether the biomaterial ends up affecting living organisms in the environment [2, 117, 132, 161].

5.6 Environmental Tests

The biodegradability of cellulose is an extremely important parameter in environmental assessments, as its accumulation in the environment can be harmful, and it is also dependent on factors such as microbial activity, environmental conditions, and chemical modifications in the environment in which it will be degraded [17, 18, 83, 126].

The coexistence of various microbial colonies (fungi and bacteria) in the soil facilitates the degradation of cellulose [17, 18, 24]. In addition, heat treatment methods with alkaline pH can also increase the biodegradability of cellulose and its derivatives by reducing ester groups and consequently increasing microbial accessibility to cellulose [126]. Factors such as the amount of lignin and pectin (structural components of cellulose) in cellulose fibers can also alter their biodegradability [17].

The biodegradability of cellulose-based pharmaceutical products can be considered limited and for this reason there is a concern about their use and accumulation, especially in aquatic environments [13, 154]. Its transformation products, including hydroxypropylmethylcellulose, hydroxypropylcellulose, methylcellulose, ethylcellulose and hydroxyethylcellulose, can be toxic, which is the reason why it is recommended that, from the outset of production, it is known which products will be generated after the degradation of cellulose and whether or not they will be toxic to the environment and animals [154].

Regarding environmental toxicity, environmental tests on aquatic animals have shown that cellulose, especially nanocellulose, is biocompatible and is considered to be of minimal environmental risk [12]. Among the analyses that can be carried out with aquatic animals, we can mention the use of zebra fish, as this animal has around 70% genetic compatibility with the human species [94], in acute or chronic toxicity tests due to its rapid larval development [91].

Despite the use of zebra fish for environmental assessment, the reproduction test with Ceriodaphnia, a small crustacean commonly found in freshwater. This crustacean has a rapid reproductive cycle and is highly sensitive to contaminants, so its use to determine environmental toxicity has been increasingly discussed [93, 109].

Multitrophic microassays are tests that can evaluate various trophic levels in the environment and can mimic the food chain and ecosystem interactions, especially in aquatic environments, including various species such as Colpidium, Daphnia, and Hyalella, providing a greater understanding of their response to environmental stressors [104].

5.7 Chemical and Physical Analysis

When chemically analyzing cellulose compounds, the composition of the biomaterials must first be evaluated, as this will identify the substances present, which may indicate possible contaminants and additives [46]. In addition, tests to evaluate migration are important, in which the release of chemical substances by the biomaterial will be tested, and residue tests, in which the presence of residual solvents or catalysts in the biomaterials will be verified [32].

Chemical analysis consists of microscopy techniques, Fourier Transform Infrared Spectroscopy (FTIR), Nuclear Magnetic Resonance (NMR), RAMAN spectroscopy and quantitative analysis of cellulose components [25, 62, 95].

Atomic force microscopy is one of the techniques used to characterize cellulose. It gives a detailed image of the cellulose, providing topographical data of the sample, allowing us to understand the surface of the cellulose and its derivatives [25]. The FTIR method identifies the functional groups of the molecules present in cellulose, a crucial factor in understanding the reactivity of the material [95] and RAMAN spectroscopy complements this test, as it helps to determine the molecular bonds present in the sample [25] and in addition to these, NMR evaluates the chemical environment of cellulose, also helping, like FTIR, in the chemical determination of the material [25].

For quantitative analysis of chemical compounds, standard methods such as GB/T5889-865 are used to evaluate the quantity of chemical compounds through a calculation relating to the weights of pectin, hemicellulose, and lignin, both compounds of cellulose, to the percentage of 100% cellulose [62].

The physical and mechanical tests of cellulose-based biomaterials consist of mechanical resistance tests, in which the durability and elasticity of the material are evaluated, in addition to thermal stability, in which the behavior of the biomaterials at high temperatures will be evaluated [10, 23] as well as permeability and controlled degradation, which will assess the permeability of moisture, gases, and/or liquids for biomaterials in certain uses, and the degradation rate in different environments, respectively [51].

Cellulose has significant potential as a reinforcing material in various applications, given its tensile and flexural strength, especially cellulose from sisal, which has shown an increase in strength of around 300% compared to fibers not made from or treated with sisal [174].

Cellulose nanocrystals are effective in increasing the elasticity and deformation mechanisms of the material, a characteristic of nanotechnology [53]. The difference in tensile strength is also related to the difference in thickness and weight of the cellulose and its original matrix [120]. In addition, changes in the physical state of cellulose can also alter its mechanical performance; studies show that gelled cellulose, for example, has greater resistance than solid cellulose [82].

6 Standards and Regulations

To ensure that cellulose and its derivatives are safe for medical and environmental applications, it is necessary to follow regulations that establish their biosafety and biocompatibility. These standards cover everything from biological and chemical tests for use in medical devices to environmental guidelines that ensure these materials do not cause negative impacts on the environment after disposal.

The biological evaluation of medical devices is carried out by ISO 1099 [67]. It is the most relevant for the biological safety assessment of materials that present contact to the human body, such as medical devices. It covers a series of tests that consider the interaction between biological systems (skin, blood, tissue, and organs). It is crucial to ensure that cellulose-based biomaterials used in implants or medical devices are safe for human use, without causing adverse reactions such as irritation, allergies, or toxicity.

The definition of biodegradability of plastic materials is defined by ASTM D6400 [8], which defines the requirements for the composting of plastic materials and products in contact with food and the environment ranging from initial contact to possible contact with microorganisms after disposal. This is important when considering the use of biodegradable materials, including certain cellulose derivatives, to ensure that they do not cause adverse environmental impacts. In addition to ASTM D6400 standards, other standards for environmental impacts and biodegradability must be observed, such as ISO 14855 [68]. This standard regulates tests for the aerobic biodegradability of plastics and composting materials, including cellulose derivatives. It also defines how the material should decompose under controlled conditions and how byproducts should be analyzed to assess their environmental impact.

As for products that come into contact with food, in this case cellulose, there are regulations that guide production and tests to be carried out before sale to the public. Materials, including cellulose and its derivatives, are regulated regarding the presence of toxic substances, as well as the use of additives in their production, such as chlorination. The production process must then be monitored to ensure that there is no contamination of the food products that will come into contact with the cellulose-based product [25]. Jurisdictions in Australia-New Zealand, Canada, China, the EU, Japan, Korea, MERCOSUR, Switzerland, Turkey, the UK, and the USA have regulations on regenerated cellulose for food contact, which aim to guarantee safety and prevent the migration of cellulose components into food [6].

About sustainability-related regulations, the EU's Single-Use Plastics Directive regulates and prioritizes the biodegradability and compostability of cellulose, its derivatives and all plastic material, in order to reduce waste [85].

Another regulatory aspect that needs to be addressed is operational regulations, as pulp production involves chemical and physical risks [6] and on a larger scale, in Brazil, for example, mills must comply with occupational safety legislation, following regulations such as Regulatory Standard for Unhealthy Activities (NR-15). In other countries, such as the USA and countries in the European Union, chemical safety and biodegradability protocols are required for products entering the market, to guarantee both the safety of consumers and the environment in general [6].

7 Challenges and Futures Perspectives

The challenges and future perspectives of cellulose and its derivatives are directly linked to the demand for sustainable raw materials and technological advancements in various applications. When related to toxicity and biocompatibility, these perspectives and challenges become even more critical, as they involve health-related applications such as biomedical uses, pharmaceuticals, cosmetics, and environmental concerns. The challenges related to toxicity and biocompatibility include the purity and contaminants of cellulose matrices, biological viability (in vitro and in vivo), and degradability. Nevertheless, these factors can still generate harmful residues that impair cellular function, leading to undesired inflammatory responses.

Future perspectives involve: Safe and efficient functionalization to reduce the material's toxicity and make processes more sustainable; advanced biocompatibility Assessment, with new methods to evaluate the biocompatibility and toxicity of cellulose derivatives in a more precise and representative manner in cellular environments; nanocellulose and toxicity, aiming for a better understanding of the interaction mechanisms between nanocellulose and biological systems to minimize associated risks; and certifications and regulations, where specific international regulatory standards for cellulose derivatives could accelerate their acceptance in health-related applications, including guidelines for toxicity and biocompatibility in medical and pharmaceutical uses. This approach ensures the functionality, biocompatibility, and low toxicity of cellulose-based materials, thereby expanding their applications in the healthcare field.

8 Final Remarks

Cellulose and its derivatives are widely known as excellent biodegradable materials with great biocompatibility. These characteristics stem from their natural origin, as they can come from plants and bacteria. They are also resistant and durable due to their fibrous structure.

The applications of cellulose range from paper production to regenerative engineering in the production of dressings and drug delivery systems, for example, in other words, its applicability is great and for this reason it is one of the reasons why it is in high demand in the scientific field.

However, as with any material, even if it is natural, its toxicity and safety must be assessed. Thinking of a biomedical application for tissue regeneration, for example, cellulose will be a "foreign body," so due to the body's defense pathways, it is possible that there will be a reaction against cellulose. Similarly, its use, like that of any other material, can promote localized or even systemic oxidative stress, given that the blood support from the area in contact with the cellulose runs throughout the body.

Therefore, in vitro tests evaluating cytotoxicity, production of oxidative stress, irritation potential and genotoxic profile should be carried out prior to in vivo analysis. In vivo tests, in turn, make it possible to detect morphological and/or histological changes at the site of contact with cellulose, in addition to the systemic toxicity profile.

In addition to the biological tests, chemical and physical analyses of the material should be carried out to confirm its structure and purity, as well as assessing its elasticity, permeability, degradability, and thermal stability. These parameters should be analyzed with a view to the material's applicability and subsequent tests carried out to determine its safety.

Another aspect to be analyzed is its environmental toxicity, because although its applicability may be to humans, for example, it is very important not to forget that after use, the cellulose will come into contact with the environment and consequently with aquatic animals. Therefore, analyses using marine animals or fresh water are widely used to determine environmental toxicity.

The toxicity of cellulose is generally minimal; however, there are studies that indicate that some concentrations in animals can cause acute inflammation and this heterogeneity of results may be related to the method of cellulose extraction and its production. Similarly, and perhaps for the same reason, its biodegradability, which is considered by most studies to be good, is not reported by some.

In short, although most studies point to cellulose as an excellent alternative for various areas, it is necessary to analyze its safety in vitro, in vivo and environmentally, as these parameters are required, as in the case of environmental tests by ISO 14855, which regulates the biodegradability of plastic materials and compost products, including cellulose derivatives.

Acknowledgements We would like to thank the Franciscan University (UFN, Brazil) for the support and assistance to carry out the present work. This study was financed in part by the Coordenação de Aperfeiçoamento de Pessoal de Nível Superior-Brazil (CAPES)-Finance Code 001.

References

1. Abdelhamid H, Mathew A (2022) Cellulose-based nanomaterials advance biomedicine: a review. Int J Mol Sci 23:5405–5440
2. Abe M, Branciforti M, Montagnolli R et al (2021) Production and assessment of the biodegradation and ecotoxicity of xylan—and starch-based bioplastics. Chemosphere 287(3):132290–132305
3. Ahsan H (2013) 3-Nitrotyrosine: a biomarker of nitrogen free radical species modified proteins in systemic autoimmunogenic conditions. Hum Immunol 74(10):1392–1399
4. An N, Kang Y (2013) In vivo BrdU incorporation assay for murine hematopoietic stem cells. Bio Protocol 3(21):e960–e966
5. Andrade FK, Alexandre N, Amorin I et al (2012) Studies on the biocompatibility of bacterial cellulose. J Bioact Compat Polym 28(1):97–112
6. Ariosti A (2022) Global legislation for regenerated cellulose materials in contact with food. In: Global legislation for food contact materials. Woodhead Publishing, pp 51–98
7. Armir N, Zulkifli A, Gunaseelan S et al (2021) Regenerated cellulose products for agricultural and their potential: a review. Polymers 13(20):3586–3629
8. ASTM International (2021) ASTM D6400—Standard specification for labeling of plastics designed to be aerobically composted in municipal or industrial facilities. West Conshohocken, PA: ASTM International
9. Aung TN, Acs B, Warrell J et al (2021) A new tool for technical standardization of the Ki67 immunohistochemical assay. Mod Pathol 34:1261–1270
10. Avérous L, Boquillon N (2004) Biocomposites based on plasticized starch: thermal and mechanical behaviours. Carbohydr Polym 56:111–122
11. Aziz T, Farid A, Haq F et al (2022) A review on the modification of cellulose and its applications. Polymers 14(15):3206–3239
12. Bading M, Olsson O, Kümmerer K (2024) Analysis of environmental biodegradability of cellulose-based pharmaceutical excipients in aqueous media. Chemosphere 352:141298–141305
13. Bartesaghi S, Radi R (2018) Fundamentals on the biochemistry of peroxynitrite and protein tyrosine nitration. Redox Biol 14:618–625
14. Birben E, Sahiner UM, Sackesen C et al (2012) Oxidative stress and antioxidant defense. World Allergy Organ J 5(1):9–19
15. Bou-Maroun E, Dahbi L, Dujourdy L et al (2023) Migration studies and endocrine disrupting activities: chemical safety of cosmetic plastic packaging. Polymers 15:4009–4028
16. Brianezi G, De Camargo JLV, Miot HÁ (2009) Development and validation of a quantitative image analysis method to evaluate comet assay (silver staining). J Bras Patol Med Lab. 45(4):213–217

17. Brunšek R, Kopitar D, Schwarz I et al (2023) Biodegradation properties of cellulose fibers and PLA biopolymer. Polymers 15(17):3532–3551
18. Bueno F, Fultz L, Husseneder C et al (2023) Biodegradability of bacterial cellulose polymer below the soil and its effects on soil bacteria diversity. Polymer Degrad Stab 217:110535–110543
19. Bullock C, Busy C (2019) Biocompatibility considerations in the design of graphene biomedical materials. Adv Mater Interfaces 6(11):1900229–1900315
20. Carvalho T, Guedes G, Sousa F et al (2019) Latest advances on bacterial cellulose-based materials for wound healing, delivery systems, and tissue engineering. Biotechnol J 14(12):e1900059–e1900068
21. Carvalho T, Guedes G, Sousa F et al (2019) Latest advances on bacterial cellulose-based materials for wound healing, delivery systems, and tissue engineering. Biotechnol J 14:1900059–1900069
22. Cavallo D, Ursini CL, Rondinone B et al (2009) Evaluation of a suitable DNA damage biomarker for human biomonitoring of exposed workers. Environ Mol Mutagen 50:781–789
23. Cazón P, Velázquez G, Vázquez M (2019) Characterization of bacterial cellulose films combined with chitosan and polyvinyl alcohol: evaluation of mechanical and barrier properties. Carbohydr Polym 216:72–85
24. Cazón P, Vázquez M (2021) Bacterial cellulose as a biodegradable food packaging material: a review. Food Hydrocolloids 113:106530–106541
25. Chakraborty I, Rongpipi S, Govindaraju I et al (2022) An insight into microscopy and analytical techniques for morphological, structural, chemical, and thermal characterization of cellulose. Microsc Res Tech 85(5):1990–2015
26. Chao MR, Evans MD, Hu CW et al (2021) Biomarkers of nucleic acid oxidation: a summary state-of-the-art. Redox Biol 42:101872–101893
27. Chen HJ, Chiu WL (2008) Simultaneous detection and quantification of 3-nitrotyrosine and 3-bromotyrosine in human urine by stable isotope dilution liquid chromatography tandem mass spectrometry. Toxicol Lett 181(1):31–39
28. Cheng W, Zhu Y, Jiang G et al (2023) Sustainable cellulose and its derivatives for promising biomedical applications. Prog Mater Sci 138:101152–101162
29. Chimpibul W, Nakaji-Hirabayashi T, Yuan X (2020) Controlling the degradation of cellulose scaffolds with Malaprade oxidation for tissue engineering. J Mater Chem B 8:7904–7913
30. Choi SM, Rao KM, Zo SM et al (2022) Bacterial cellulose and its applications. Polymers 14(6):1080–1123
31. Cortés-Gutiérrez EI, Dávila-Rodríguez MI (2018) López-Fernández C (2018) The Comet Assay. In: Zini A, Agarwal A (eds) A clinician's guide to sperm DNA and chromatin damage. Springer, pp 147–213
32. Costa A, Almeida F, Vinhas G et al (2017) Production of bacterial cellulose by Gluconacetobacter hansenii using corn steep liquor as nutrient sources. Front Microbiol 8:1–12
33. Crow BS, Quiñones-González J, Pantazides BG et al (2016) Simultaneous measurement of 3-chlorotyrosine and 3,5-dichlorotyrosine in whole blood, serum and plasma by isotope dilution HPLC-MS-MS. J Anal Toxicol 40(4):264–271
34. Dalle-Donne I, Rossi R, Giustarini D et al (2003) Protein carbonyl groups as biomarkers of oxidative stress. Clin Chim Acta 329(1–2):23–38
35. Das M, Zandraa O, Mudenur C et al (2022) Composite scaffolds based on bacterial cellulose for wound dressing application. ACS Appl Bio Mater 5:3722–3733
36. De Amorim J, De Souza K, Duarte C et al (2020) Plant and bacterial nanocellulose: production, properties and applications in medicine, food, cosmetics, electronics and engineering a review. Environ Chem Lett 18:851–869
37. De Muro P, Zhang S, Li S et al (2024) The emerging role of oxidative stress in inflammatory bowel disease. Front Endocrinol 15:1390351–1390373
38. Debnath B, Haldar D, Purkait M (2021) A critical review on the techniques used for the synthesis and applications of crystalline cellulose derived from agricultural wastes and forest residues. Carbohydr Polym 273:118537–118542

39. Di Marco T, Giulivi C (2007) Current analytical methods for the detection of dityrosine, a biomarker of oxidative stress, in biological samples. Mass Spectrom Rev 26(1):108–120
40. Dourado F, Gama M, Rodrigues A (2017) A review on the toxicology and dietetic role of bacterial cellulose. Toxicol Rep 4:543–553
41. Durnev A (2023) Pharmaceutical toxicology is the most important component of preclinical studies. Bull Sci Cent Expert Eval Med Prod. 13(1):8–13
42. Ede J, Ong K, Goergen M et al (2019) Analysis of cellulose nanomaterials by inhalation: current state of science. Nanomaterials 9:337–354
43. Emenike EC, Iwuozor KO, Saliu OD et al (2023) Advances in the extraction, classification, modification, emerging and advanced applications of crystalline cellulose: a review. Carbohydr Polym Technol Appl. 6:100337–100357
44. Engelbrecht I, Horn S, Giesy JP et al (2024) A method to determine reactive oxygen species production in intestinal and liver cell cultures using the 2'-7'dichlorodihydrofluorescein diacetate assay. MethodsX. 15:102615–102620
45. Erdal N, Hakkarainen M (2022) Degradation of cellulose derivatives in laboratory, man-made, and natural environments. Biomacromol 23:2713–2729
46. Fatema N, Ceballos R, Fan C (2022) Modifications of cellulose-based biomaterials for biomedical applications. Front Bioeng Biotechnol 10:1–8
47. Feron K, Lim R, Sherwood C et al (2018) Organic bioelectronics: materials and biocompatibility. Int J Mol Sci 19(8):2382–2421
48. Figueiredo J, Ismael M, Anjo C et al (2010) Cellulose and derivatives from wood and fibers as renewable sources of raw-materials. Top Curr Chem 294:117–128
49. Forman HJ, Zhang H (2021) Targeting oxidative stress in disease: promise and limitations of antioxidant therapy. Nat Rev Drug Discov 20(9):689–709
50. Gabriel T, Belete A, Syrowatka F et al (2020) Extraction and characterization of celluloses from various plant byproducts. Int J Biol Macromol 164:3071–3077
51. Gaidukovs S, Platnieks O, Gaidukova G et al (2021) Understanding the impact of microcrystalline cellulose modification on durability and biodegradation of highly loaded biocomposites for woody-like materials applications. J Polym Environ 30:1435–1450
52. Garcia P, Wang Y, Viallet J et al (2021) The chicken embryo model: a novel and relevant model for immune-based studies. Front Immunol 12:1–12
53. Geng L, Shi Y, Fang W et al (2023) Analysis of transverse elastic modulus and compressive deformation mechanism of cellulose crystals based on molecular dynamics. Mater Today Commun 35:106414–106426
54. Ghesquière B, Jonckheere V, Colaert N et al (2011) Redox proteomics of protein-bound methionine oxidation. Mol Cell Proteomics 10(5):M110.006866–M110.006877
55. Ghosh M, Hartmann H, Jakobi M et al (2020) The impact of biomaterial cell contact on the immunopeptidome. Front Bioeng Biotechnol 8:571294–571313
56. Gillery P (2006) Oxidative stress and protein glycation in diabetes mellitus. Ann Biol Clin 64(4):309–314
57. Girard VD, Chaussé J, Borduas M et al (2024) In vitro and in vivo biocompatibility of bacterial cellulose. J Biomed Mater Res 112(10):1–17
58. Gomez-Gutierrez G, Bhutiani N, McNally M et al (2020) The neutral red assay can be used to evaluate cell viability during autophagy or in an acidic microenvironment in vitro. Biotech Histochem 96:302–308
59. González-Arancibia F, Mamani M, Valdés C et al (2024) Biopolymers as sustainable and active packaging materials: fundamentals and mechanisms of antifungal activities. Biomolecules 14:1224–1247
60. Gorgieva S, Trček J (2019) Bacterial cellulose: production, modification and perspectives in biomedical applications. Nanomaterials 9(10):1352–1371
61. Gupta R, Raghav N (2019) Nano-crystalline cellulose: preparation, modification and usage as sustained release drug delivery excipient for some non-steroidal anti-inflammatory drugs. Int J Biol Macromol 147:921–930

62. Song H, Wu H, Li X (2018) Quantitative analysis of chemical components of plant fiber. Appl Chem Eng 1(1):35–46
63. Hadrup N, Knudsen KB, Berthing T et al (2019) Pulmonary effects of nanofibrillated celluloses in mice suggest that carboxylation lowers the inflammatory and acute phase responses. Environ Toxicol Pharmacol 66:116–125
64. Hujaya S, Lorite G, Vainio S et al (2018) Polyion complex hydrogels from chemically modified cellulose nanofibrils: structure-function relationship and potential for controlled and pH-responsive release of doxorubicin. Acta Biomater 75:346–357
65. Iaquinta M, Mazzoni E, Manfrini M et al (2019) Innovative biomaterials for bone regrowth. Int J Mol Sci 20(3):618–634
66. Il'yasova D, Scarbrough P, Spasojevic I (2012) Urinary biomarkers of oxidative status. Clin Chim Acta 413(19–20):1446–1453
67. International Organization for Standardization (2009) ISO 10993–5: Biological evaluation of medical devices—Part 5: Tests for in vitro cytotoxicity. ISO, Geneva
68. International Organization for Standardization (2012) ISO 14855–1: Determination of the ultimate aerobic biodegradability of plastic materials under controlled composting conditions—Method by analysis of evolved carbon dioxide—Part 1: General method. ISO, Geneva
69. Irastorza A, Zarandona I, Andonegi M et al (2021) The versatility of collagen and chitosan: from food to biomedical applications. Food Hydrocoll 116:106633–106643
70. Ito F, Sono Y, Ito T (2019) Measurement and clinical significance of lipid peroxidation as a biomarker of oxidative stress: oxidative stress in diabetes, atherosclerosis, and chronic inflammation. Antioxidants 8(3):72–99
71. Javanbakht S, Shaabani A (2019) Carboxymethyl cellulose-based oral delivery systems. Int J Biol Macromol 133:21–29
72. Jiang N, Naz S, Ma Y et al (2023) An overview of comet assay application for detecting DNA damage in aquatic animals. Agriculture 13(3):623–642
73. Jong W, Carraway J, Geertsma R (2020) In vivo and in vitro testing for the biological safety evaluation of biomaterials and medical devices. In: Biocompatibility and performance of medical devices, pp 123–144
74. Jost L, Kirkwood J, Whiteside T (1992) Improved short—and long-term XTT-based colorimetric cellular cytotoxicity assay for melanoma and other tumor cells. J Immunol Methods 147:153–165
75. Kaneko K, Akuta T, Sawa T et al (2008) Mutagenicity of 8-nitroguanosine, a product of nitrative nucleoside modification by reactive nitrogen oxides, in mammalian cells. Cancer Lett 262(2):239–247
76. Katerji M, Filippova M, Duerksen-Hughes P (2019) Approaches and methods to measure oxidative stress in clinical samples: research applications in the cancer field. Oxid Med Cell Longev 2019:1279250–1279278
77. Kaushik S, Thungon P, Goswami P et al (2020) An emerging biocompatible material for application of enzymes and whole cells in bioelectronics and bioanalytical sciences. ACS Biomater Sci Eng 6(8):4337–4418
78. Kawai K, Kasai H, Li YS et al (2018) Measurement of 8-hydroxyguanine as an oxidative stress biomarker in saliva by HPLC-ECD. Genes Environ 40(5):1–4
79. Kehm R, Baldensperger T, Raupbach J et al (2021) Protein oxidation—formation mechanisms, detection and relevance as biomarkers in human diseases. Redox Biol 42:101901–101923
80. Kehm R, Baldensperger T, Raupbach J et al (2021) Protein oxidation—formation mechanisms, detection, and relevance as biomarkers in human diseases. Redox Biol 42:101901–101923
81. Khan R, Khan M, Stojanović G et al (2024) Fabrication of bilayer nanofibrous-hydrogel scaffold from bacterial cellulose, PVA, and gelatin as advanced dressing for wound healing and soft tissue engineering. ACS Omega 9:6527–6536
82. Kim T, Song Y, Ahn J et al (2021) Intermediate physical state of gel and LC in cellulose solution and its synergetic effects. Res Square 2021:1–17

83. Kovacs T, Naish V, O'Connor B et al (2010) An ecotoxicological characterization of nanocrystalline cellulose (NCC). Nanotoxicology 4(3):255–270
84. Krishnamurthy HK, Pereira M, Rajavelu I et al (2024) Oxidative stress: fundamentals and advances in quantification techniques. Front Chem 12:1–12
85. Kryvets OV, Krechyk AS, Savchuk OO (2024) Legal regulation of restriction of use of disposable plastic products in the European Union. Uzhhorod Natl Univ Herald Law Ser 2(85):23–28
86. Kuiper HC, Langsdorf BL, Miranda CL et al (2010) Quantitation of mercapturic acid conjugates of 4-hydroxy-2-nonenal and 4-oxo-2-nonenal metabolites in a smoking cessation study. Free Radic Biol Med 48(1):65–72
87. Kumar P, Nagarajan A, Uchil PD (2018) Analysis of cell viability by the MTT assay. Cold Spring Harb Protoc 6:1–4
88. Lahiri D, Nag M, Dutta B et al (2021) Bacterial cellulose: production, characterization, and application as antimicrobial agent. Int J Mol Sci 22(23):12984–13001
89. Li H, Shi H, He Y et al (2020) Preparation and characterization of carboxymethyl cellulose-based composite films reinforced by cellulose nanocrystals derived from pea hull waste for food packaging applications. Int J Biol Macromol 164:4104–4112
90. Liang X, Duan P, Gao J (2018) Bilayered PLGA/PLGA-HAp composite scaffold for osteochondral tissue engineering and tissue regeneration. ACS Biomater Sci Eng 4(10):3506–3521
91. Lima VA, Dos Santos ABN, De Morais ESG et al (2024) Zebrafish: a versatile model organism for toxicological research and applications. Adv Res 25(5):387–396
92. Liu Z, Chen X, Li Z et al (2020) Role of RNA oxidation in neurodegenerative diseases. Int J Mol Sci 21(14):5022–5035
93. Lydy VR, Regn OZ, Bouldin JL (2023) Toxicant responses and culturing characteristics of long-term laboratory-reared and field populations of Ceriodaphnia dubia. Environ Toxicol Chem 43(1):159–169
94. Madivalar V, Shalavdi M, Teli S et al (2024) Zebrafish: a versatile animal model in biomedical research. Int J Multidiscip Res 6(5):1–22
95. Mahmoud AA, Osman O, El-hotaby W et al (2014) Modeling and molecular spectroscopic analyses of cellulose. J Appl Sol Chem Model 3(3):159–163
96. Martínez-Moral MP, Kannan K (2022) Analysis of 19 urinary biomarkers of oxidative stress, nitrative stress, metabolic disorders, and inflammation using liquid chromatography-tandem mass spectrometry. Anal Bioanal Chem 414(6):2103–2116
97. Maver T, Maver U, Mostegel F et al (2015) Cellulose based thin films as a platform for drug release studies to mimic wound dressing materials. Cellulose 22:749–761
98. Medda N, Patra R, Ghosh T et al (2020) Neurotoxic mechanism of arsenic: synergistic effect of mitochondrial instability, oxidative stress, and hormonal-neurotransmitter impairment. Biol Trace Elem Res 198:8–15
99. Mehnath S, Karthikeyan K, Jeyaraj M (2024) Mechanical force on hydrogel implication on enhanced drug release, antibacterial, and M2 macrophage polarization: new insights alleviate diabetic wound healing. ACS Appl Mater Interfaces 16:55166–55180
100. Meng X, Ragauskas A (2014) Recent advances in understanding the role of cellulose accessibility in enzymatic hydrolysis of lignocellulosic substrates. Curr Opin Biotechnol 27:150–158
101. Milne GL, Yin H, Hardy KD et al (2011) Isoprostane generation and function. Chem Rev 111(10):5973–5996
102. Modulevsky D, Cuerrier C, Pelling A (2016) Biocompatibility of subcutaneously implanted plant-derived cellulose biomaterials. PLoS ONE 11(6):1–19
103. Molina M, Malollari K, Komvopoulos K (2021) Design challenges in polymeric scaffolds for tissue engineering. Front Bioeng Biotechnol 9:1–29
104. Munawar M, Munawar IF, Mayfield CI et al (1989) Probing ecosystem health: a multi-disciplinary and multi-trophic assay strategy. In: Munawar M, Dixon G, Mayfield CI, Reynoldson T, Sadar MH (eds). Environmental bioassay techniques and their application. Developments in Hydrobiology, p 119–34

105. Murphy MP, Bayir H, Belousov V et al (2022) Guidelines for measuring reactive oxygen species and oxidative damage in cells and in vivo. Nat Metab 4(6):651–662
106. Naser H, Czinkota I, Dorkota A (2023) A review of the advancements of potentially toxic element adsorption by various cellulose-based materials and the used adsorbents' fate. Prog Agric Eng Sci 19:1–13
107. Nasution H, Yahya EB, Khalil HPSA et al (2022) Extraction and isolation of cellulose nanofibers from carpet wastes using supercritical carbon dioxide approach. Polymers 14:326–339
108. Nazir F, Iqbal M (2020) Synthesis, characterization and cytotoxicity studies of aminated microcrystalline cellulose derivatives against melanoma and breast cancer cell lines. Polymers 12(11):2634–2653
109. Nogueira LR, Lira VS, Carvalho MM et al (2024) Ecotoxicity of the chlorantraniliprole insecticide on Ceriodaphnia dubia. Stud Environ Anim Sci. 5(2):1–15
110. Nova Z, Skovierova H, Strnadel J et al (2020) Short-term versus long-term culture of A549 cells for evaluating the effects of lipopolysaccharide on oxidative stress, surfactant proteins, and cathelicidin LL-37. Int J Mol Sci 21(3):1148–11164
111. Ohshima H, Sawa T, Akaike T (2006) 8-nitroguanine, a product of nitrative DNA damage caused by reactive nitrogen species: formation, occurrence, and implications in inflammation and carcinogenesis. Antioxid Redox Signal 8(5–6):1033–1045
112. Olaniyan MF, Muhibi MA, Olaniyan TB (2023) Oxidative stress and inflammatory response interplay. J Prev Diagn Treat Strat Med 2(2):94–100
113. Olmos L, Delgado P, Barbosa W (2023) Chemical and microbiological risks identified in operating rooms of Santander health institutions and their prevention strategies. Int Vis Cult Rev 14:1–7
114. Oprea M, Voicu Ș (2020) Recent advances in composites based on cellulose derivatives for biomedical applications. Carbohydr Polym 247:116683–116692
115. Pandey A (2021) Pharmaceutical and biomedical applications of cellulose nanofibers: a review. Environ Chem Lett 19:2043–2055
116. Park JS, Kim WO, Kim H et al (2023) Development of a novel ATP bioluminescence assay based on engineered probiotic Saccharomyces boulardii expressing firefly luciferase. J Microbiol Biotechnol 33(11):1506–1512
117. Patel RS, Ghasemzadeh N, Eapen DJ et al (2016) Novel biomarker of oxidative stress is associated with risk of death in patients with coronary artery disease. Circulation 133(4):361–369
118. Patil S, Dhyani V, Kaur T et al (2020) Spatiotemporal control over cell proliferation and differentiation for tissue engineering and regenerative medicine applications using silk fibroin scaffolds. ACS Appl Bio Mater 3(6):3476–3493
119. Paulsen CE, Carroll KS (2013) Cysteine-mediated redox signaling chemistry, biology, and tools for discovery. Chem Rev 113(7):4633–4679
120. Pego MF, Bianchi ML, Yasumura PK (2022) Physical and mechanical properties of papers produced from different cellulosic pulps blending. Res Square 2022:1–25
121. Phaniendra A, Jestadi DB, Periyasamy L (2015) Free radicals: properties, sources, targets, and their implication in various diseases. Indian J Clin Biochem 30(1):11–26
122. Piasecka-Zelga J, Zelga P, Szulc J et al (2018) An in vivo biocompatibility study of surgical meshes made from bacterial cellulose modified with chitosan. Int J Biol Macromol 116:1119–1127
123. Piasecka-Zelga J, Zelga P, Szulc J et al (2018) An in vivo biocompatibility study of surgical meshes made from bacterial cellulose modified with chitosan. Int J Biol Macromol 116:119–127
124. Picot-Allain M, Emmambux M (2021) Isolation, characterization, and application of nanocellulose from agro-industrial by-products: a review. Food Rev Int 39:941–969
125. Pizzino G, Irrera N, Cucinotta M et al (2017) Oxidative stress: harms and benefits for human health. Oxid Med Cell Longev 1:1–13

126. Prabhaharan D, Park H, Choi O et al (2024) Enhancing cellulose acetate biodegradability in cigarette filters: an in-depth analysis of thermal alkaline pretreatment, microbial dynamics, and breakdown pathway prediction. Microb Cell Fact 23:199–210
127. Pértile RA, Moreira S, Gil da Costa RM et al (2012) Bacterial cellulose: long-term biocompatibility studies. J Biomater Sci Polym Ed 23(10):1339–1354
128. Ramakrishnan R, Kim JT, Roy S et al (2024) Recent advances in carboxymethyl cellulose-based active and intelligent packaging materials: a comprehensive review. Int J Biol Macromol 259:129194–129205
129. Raut H, Das R, Ziqian L, Xiaoling L et al (2020) Biocompatibility of biomaterials for tissue regeneration or replacement. Biotechnol J 15:2000160–2000169
130. Rovera C, Carullo D, Bellesia T et al (2023) Extraction of high-quality grade cellulose and cellulose nanocrystals from different lignocellulosic agri-food wastes. Front Sustain Food Syst 6:1–16
131. Rubio C, Cerón J (2021) Spectrophotometric assays for evaluation of reactive oxygen species (ROS) in serum: general concepts and applications in dogs and humans. BMC Vet Res 17:226–238
132. Salehpour S, Jonoobi M, Ahmadzadeh M et al (2018) Biodegradation and ecotoxicological impact of cellulose nanocomposites in municipal solid waste composting. Int J Biol Macromol 111:264–270
133. Santos M, Durval I, De Medeiros A et al (2024) Biotechnology in food packaging using bacterial cellulose. Foods 13:3327–3352
134. Sari C, Kolayli S, Eyüpoğlu FC (2021) A comparative study of MTT and WST-1 assays in cytotoxicity analysis. Haydarpasa Numune Train Res Hosp Med J 61:281–288
135. Seddiqi H, Oliaei E, Honarkar H et al (2021) Cellulose and its derivatives: towards biomedical applications. Cellulose 28:1893–1931
136. Shanmugam PST, Sampath T, Jagadeeswaran I (2023) Acute systemic toxicity. In: Srinivasan P, Shanmugam T, Sampath T, Jagadeeswaran I (eds) Biocompatibility protocols for medical devices and materials. Academic Press, pp 67–76
137. Sharma K, Mujawar M, Kaushik A (2019) State-of-art functional biomaterials for tissue engineering. Front Mater 6:1–10
138. Sheikh I, Beg M (2021) Structural binding perspectives of common plasticizers and a flame retardant, BDE-153, against thyroxine-binding globulin: potential for endocrine disruption. J Appl Toxicol 42:841–851
139. Shen Y, Seidi F, Ahmad M et al (2023) Recent advances in functional cellulose-based films with antimicrobial and antioxidant properties for food packaging. J Agric Food Chem 71:16469–16487
140. Shin CS, Moon BS, Park KS et al (2021) Serum 8-hydroxy-guanine levels are increased in diabetic patients. Diabetes Care 24(4):733–737
141. Sivashankari P, Prabaharan M (2016) Prospects of chitosan-based scaffolds for growth factor release in tissue engineering. Int J Biol Macromol 93:1382–1389
142. Sochor J, Ruttkay-Nedecky B, Babula P et al (2012) Automation of methods for determination of lipid peroxidation. In: Catala A (ed). Lipid Peroxidation
143. Stopforth A, Burger BV, Crouch AM et al (2006) Urinalysis of 4-hydroxynonenal, a marker of oxidative stress, using stir bar sorptive extraction-thermal desorption-gas chromatography/mass spectrometry. J Chromatogr B Anal Technol Biomed Life Sci 834(1–2):134–140
144. Suzuki T, Kamiya H (2016) Mutations induced by 8-hydroxyguanine (8-oxo-7, 8-dihydroguanine), a representative oxidized base, in mammalian cells. Genes Environ 39(2):2–7
145. Taboureau O, M'Selmi W, Audouze K (2020) Integrative systems toxicology to predict human biological systems affected by exposure to environmental chemicals. Toxicol Appl Pharmacol 405:115210–115218
146. Tamura S, Tsukahara H, Ueno M et al (2006) Evaluation of a urinary multi-parameter biomarker set for oxidative stress in children, adolescents and young adults. Free Radic Res 40(11):1198–1205

147. Tarchoun A, Trache D, Klapötke T (2019) Microcrystalline cellulose from posidonia oceanica brown algae: extraction and characterization. Int J Biol Macromol 138:837–845
148. Tarrah R, Khataee A, Karimi A et al (2021) The latest achievements in plant cellulose-based biomaterials for tissue engineering focusing on skin repair. Chemosphere 288:132529–132536
149. Tavakolian M, Jafari S, Van De Ven T (2020) A review on surface-functionalized cellulosic nanostructures as biocompatible antibacterial materials. Nano-Micro Lett. 12:1–23
150. Taylor EL, Armstrong KR, Perrett D et al (2015) Optimization of an advanced oxidation protein products assay: its application to studies of oxidative stress in diabetes mellitus. Oxid Med Cell Longev 2015:496271–496280
151. Terrén I, Orrantia A, Vitallé J et al (2020) CFSE dilution to study human T and NK cell proliferation in vitro. Methods Enzymol 631:239–255
152. Tice RR, Agurell E, Anderson D et al (2000) Single cell gel/comet assay: guidelines for in vitro and in vivo genetic toxicology testing. Environ Mol Mutagen 35:206–21510
153. Torres FG, Commeaux S, Troncoso OP (2024) Biocompatibility of bacterial cellulose based biomaterials. J Funct Biomater 3(4):864–878
154. Tudoroiu E, Dinu-Pîrvu C, Kaya M et al (2021) An overview of cellulose derivatives-based dressings for wound-healing management. Pharmaceutics 14(12):1215–1244
155. Vallejo M, Cordeiro R, Dias P et al (2021) Recovery and evaluation of cellulose from agroindustrial residues of corn, grape, pomegranate, strawberry-tree fruit and fava. Bioresour Bioprocess 8:25–36
156. Van Norman G (2019) Limitations of animal studies for predicting toxicity in clinical trials. JACC Basic Transl Sci 4:845–854
157. Ventura C, Pinto F, Lourenço A et al (2020) On the toxicity of cellulose nanocrystals and nanofibrils in animal and cellular models. Cellulose 27:5509–5544
158. Ventura C, Pinto F, Lourenço A et al (2023) Assessing the genotoxicity of cellulose nanomaterials in a co-culture of human lung epithelial cells and monocyte-derived macrophages. Bioengineering 10:986–997
159. Verigos KE, Sagredou S, Orfanakos K et al (2020) 8-Hydroxy-2'-deoxyguanosine and 8-nitroguanine production and detection in blood serum of breast cancer patients in response to postoperative complementary external ionizing irradiation of normal tissues. Dose-Response 8(4):155932582098217–155932582098226
160. Vieyra H, Juárez E, López U et al (2018) Cytotoxicity and biocompatibility of biomaterials based in polyhydroxybutyrate reinforced with cellulose nanowhiskers determined in human peripheral leukocytes. Biomed Mater 13(4):45011–45013
161. Vikman M, Vartiainen J, Tsitko I et al (2015) Biodegradability and compostability of nanofibrillar cellulose-based products. J Polym Environ 23:206–215
162. Villaverde AISB, Netherton J, Baker MA (2019) From past to present: the link between reactive oxygen species in sperm and male infertility. Antioxidants 8(12):616–634
163. Vona R, Pallotta L, Cappelletti M et al (2021) The impact of oxidative stress in human pathology: focus on gastrointestinal disorders. Antioxidants 10(2):201–226
164. Wahid F, Huang L, Zhao X et al (2021) Bacterial cellulose and its potential for biomedical applications. Biotechnol Adv 53:107856–107931
165. Watters JL, Satia JA, da Costa KA et al (2009) Comparison of three oxidative stress biomarkers in a sample of healthy adults. Biomark Biochem Indic Expo Response Susceptibility Chem 14(8):587–595
166. Williams D (2014) There is no such thing as a biocompatible material. Biomaterials 35(38):10009–10014
167. Wischmann B, Toft MM, Malten M et al (2012) Biomass conversion determined via fluorescent cellulose decay assay. Methods Enzymol 510:19–36
168. Wu J, Li Z (2008) Human polynucleotide phosphorylase reduces oxidative RNA damage and protects HeLa cells against oxidative stress. Biochem Biophys Res Commun 372(2):288–292
169. Yadav M, Kaushik B, Rao GK et al (2023) Solubility and stability of chitosan derivatives in aqueous systems: a review. Carbohydr Polym Technol Appl 5:100323–100344

170. Yang J, Li J (2018) Self-assembled cellulose materials for biomedicine: a review. Carbohydr Polym 181:264–274
171. Yu J, Wang Z, Wang Y (2022) BrdU incorporation assay to analyze the entry into S phase. Methods Mol Bio 2022:209–226
172. Zaam N, Benssabri S, Bajjou T (2021) Evaluation of infectious risk among biomedical agents in Morocco. Int J Commun Med Publ Health 8:550–557
173. Zhang X, Xia P, Wang P (2018) Advances in ecotoxicology. Environ Sci Technol 52(7):3842–3851
174. Zidi S, Miraoui I (2024) Mechanical and morphological analysis of cellulose extracted from sisal fibers and their effect on bio-based composites mechanical properties. Funct Compos Struct. 6(1):015008–015015
175. Zor F, Selek F, Orlando G et al (2019) Biocompatibility in regenerative nanomedicine. Nanomedicine 14:2763–2775

Lailla Daianna Soltau Missio Pinheiro holds a degree in Biomedical Engineering from the Universidade Franciscana (UFN, 2017), as well as a master's and a PhD. in Nanosciences from the same institution (2019 and 2024, respectively). She is currently a member of the Applied Nanomaterials Research Group (GPNAp). She has experience in the green synthesis of nanomaterials, with an emphasis on nanotechnology applied to health, seeking innovative and sustainable solutions for biomedical applications. Her practical and theoretical expertise in green nanotechnology, focused on the synthesis of metallic nanoparticles, aims to understand their biological effects and explore their potential uses in areas such as biomedical diagnosis and therapy, always with a focus on innovation and sustainability.

Gabriela Geraldo Sangoi holds a bachelor's degree in Biomedicine from the Franciscan University (UFN, 2021), where she also obtained her master's degree in Nanosciences (2023). She is currently a PhD student in Graduate Program Nanoscience at UFN, in addition to carrying out her PhD exchange at the Institute of Plant Biochemistry (IPB) in Germany. As a member of the Cell Culture and Bioactive Effects Laboratory (LabCultBio), her area of expertise involves nanotechnology combined with natural products and cell culture, with applications in various areas of healthcare. At present, she is researching natural extracts with a focus on neurotherapy, exploring the correlation between nanoparticles, oxidative stress, and mitochondrial dysfunction.

Alencar Kolinski Machado obtained his bachelor degree in Biomedicine from the Franciscan University (Brazil), and Master and PhD in Pharmacology from the Federal University of Santa Maria (Brazil). He completed a postdoctoral fellow period at the Department of Pharmacology and Toxicology from the University of Toronto (Canada). Dr. Machado is currently an adjunct professor (level II) at the Franciscan University, acting as the coordinator of the Graduate Program in Nanosciences and developing research with natural health products, bioactive molecules, and nanotechnology. His research group works mainly with cell culture methods to investigate new potential therapeutic agents against inflammation, oxidative metabolism impairments, and mitochondrial dysfunction.

William Leonardo da Silva holds a degree in Chemical Engineering from the Federal University of Santa Maria (UFSM, 2010), a master's degree (2012) and a PhD (2016) in Chemical Engineering from the Federal University of Rio Grande do Sul (UFRGS). He developed research as a sandwich PhD student abroad (PDSE / CAPES-2015) at Ulster University at Nanotechnology and Integrated Bioengineering Centre (NIBEC). Currently, he is an adjunct professor at the Franciscan University (UFN) and a permanent professor of the Graduate Program in Nanosciences, coordinating the Applied Nanomaterials Research Group (GPNAp). He has experience in green nanotechnology, with emphasis on green synthesis of metallic nanoparticles, chemical kinetics and catalysis, working mainly on the following topics: removal of organic pollutants in wastewater by advanced oxidative processes (heterogeneous photocatalysis), adsorption process, biomaterials, machine learning, and synthesis and characterization of nanocatalysts.

Advancements in Bacterial Cellulose Synthesis and Functional Applications

Deepak Kumar, Rishabha Malviya, Sonali Sundram, and Sathvik Belagodu Sridhar

Abstract Microbial cellulose, a biopolymer synthesized by bacteria, is gaining significant attention for its superior features related to plant-derived cellulose. Its higher purity, crystallinity, tensile strength, and tunable structure enable it highly useful for biomedical, food, and environmental applications. This study explores the production, properties, along with wide-ranging utilization of microbial cellulose, emphasizing its potential for industrial and healthcare innovations. Microbial cellulose exhibits remarkable physical and chemical characteristics, including excellent mechanical strength, WHC, and biocompatibility. Unlike plant cellulose, bacterial cellulose lacks contaminants like lignin and hemicellulose, offering enhanced purity. Its porous structure enables its use in wound dressings, where it promotes healing and prevents infections. In tissue engineering, bacterial cellulose mimics the extracellular matrix, supporting cell proliferation and regeneration. The food industry benefits from microbial cellulose as a stabilizer and fat replacer, while its biodegradable nature makes it ideal for sustainable packaging solutions. Microbial cellulose is also effective in environmental remediation, such as removing dyes, heavy metals, and oils from wastewater. Advances in fermentation techniques, including static, shaking, and agitated cultures, have improved its production efficiency. However, challenges like high costs and limited scalability necessitate further research, including genetic engineering and the use of inexpensive substrates. Microbial cellulose holds immense potential across diverse sectors, providing sustainable solutions for healthcare, food, and environmental challenges. Continued advancements in production and application development will drive its broader industrial adoption.

Keywords Bacterial cellulose · Biopolymer · Tissue engineering · Wound dressing · Food industry · Nanofibers · Biomedical applications

D. Kumar · R. Malviya (✉) · S. Sundram
Department of Pharmacy, School of Medical and Allied Sciences, Galgotias University, Greater Noida, Uttar Pradesh, India
e-mail: rishabhamalviya19@gmail.com

S. B. Sridhar
RAK College of Pharmacy, RAK Medical and Health Sciences University, Ras Al Khaimah, UAE

R. Malviya and S. Sundram (eds.), *Sustainable Cellulose Materials for Biomedical Applications*, Sustainable Materials and Technology,
https://doi.org/10.1007/978-981-96-9904-9_3

1 Introduction

Bio-based materials are gaining popularity for their direct uses in biological sectors such as the healing of wounds, tissue engineering, and delivery of drugs. It is made up of a linear chain of β-1,4-linked units of D-glucose called polysaccharides [1]. It is mostly composed of units of monosaccharide, which are bio-based substances with greater possibility for use in medical uses that combine the unique benefits of polymers made from nature over synthetic ones [2]. Cellulose and chitin are the major biopolymers from nature among a variety of polysaccharide types because of their widespread occurrence in nature [3, 4]. The most prevalent natural polymer in the biosphere is cellulose. It is generated at an average rate of 1.5×10^{12} tons per annually worldwide which is equivalent to the primary planetary reserves fossil along with mineral sources [5]. Cellulose primarily comes from plants, animals, fungi, and algae, but in 1986, A. J. Brown released the first document regarding the recognition of cellulose synthesized by the bacterium *Acetobacter xylinum* [6].

Bacterial cellulose $(C_6H_{10}O_5)_n$ has a similar molecular formula as cellulose from plants but shows distinct physical as well as chemical characteristics [7]. Bacterial cellulose (BC) is chosen above plant cellulose because of its high purity, crystallinity index, and polymerization. It has a stronger tensile strength along with water-holding volume in comparison to plant cellulose, enabling it an ideal raw substance for creating paper, speakers, and desserts [8]. BC fibers are 100 times thinner as related to cellulose from plants, making them very porous. This permits antibiotics and additional treatments to enter the wound though acting as a physical barrier versus exterior infection. It is often utilized for healing of wound [9]. BC offers advantages due to its purity. It lacks wax, lignin, pectin, as well as hemicelluloses, unlike plant-derived cellulose. BC can be effectively regulated for repeating units and molecular weight through fermentation [10].

2 Structure and Features of BC

Bacterial cellulose possesses a chemical composition comparable to cellulose that is derived from plant sources but lacks hemicellulose, pectin, and lignin [11]. BC fibers vary from plant-derived cellulose due to their mechanical characteristics, lack of contaminating polymers, and increased crystallinity [12].

Acetobacter xylinum generates cellulose I, which is a ribbon-like polymer as well as cellulose II, which is a thermodynamically stable polymer, as seen in Fig. 1. Protofibrils of glucose chains produced via bacterium cell walls combine to generate nanofibrils and cellulose ribbons, resulting in a web-shaped network structure. The plenty of hydroxyl groups on the layer of cellulose contributes to its hydrophilic nature, biodegradability as well as chemical alteration capability [13].

The accessible surface area along with the pore volume of the BC determines its water-holding capacity (WHC) along with the water release rate (WRR). WHC and

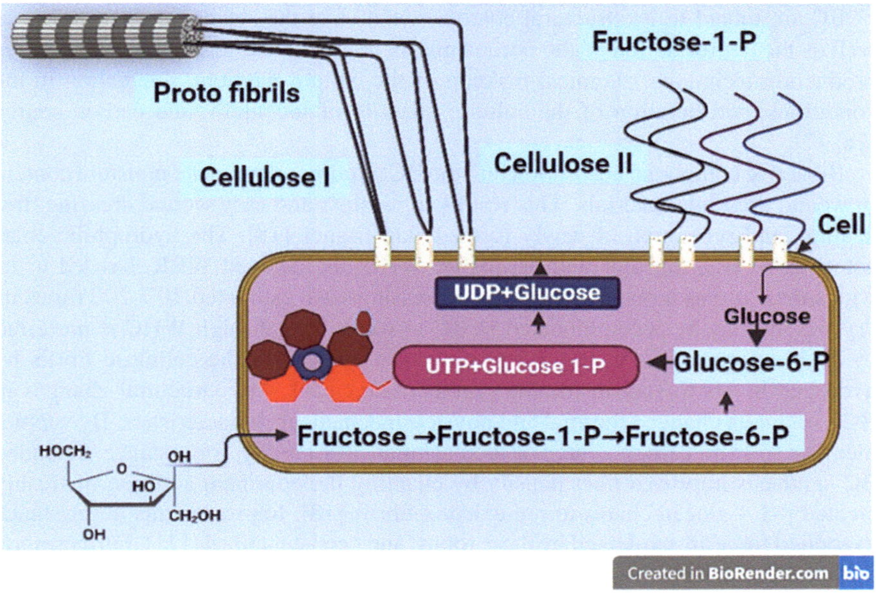

Fig. 1 Schematic depicting the synthesis of BC type I as well as II

WRR values decrease when the pore, as well as the surface area of the BC structure, decreases. Water evaporation is prevented by the denser microfibrils because they create hydrogen bonds that keep more water in the system and less free bulk water [14].

Numerous investigations into the relationship between the structure and WHC of BC have been conducted in recent years. These findings demonstrate that WHC falls and WRR rises as the size of the pores along with the surface area of BC are reduced [15]. Single sugar-linked glucuronic acid-based oligosaccharides can be supplied to the culture media to decrease the size of the pores as well as the surface. Furthermore, hydroxypropyl methylcellulose can decrease BC porosity, which lowers the WHC value. When carboxymethylcellulose is employed in the culture media, it adheres to the BC surface and forms a fibrillary network with bigger ribbons, which raises the WHC. BC is a capable material for wound dressings because of its inherent characteristics and web-shaped network framework, which enable the regulated release of chemicals [12].

2.1 Physiological Properties

The physiological characteristics of BC, including WHC, WRR, mechanical features, and thermal properties, are crucial to its wide range of uses. These characteristics of BC are linked to its structural characteristics, like the arrangement of fibrils as well as the conformation of the porous matrix that is linked to the microbial strain, production technique, chemical makeup of the culture medium, fluctuation in the conditions, and duration of the culture, quantity of inoculum, and carbon source [16, 17].

Bioactive compound adsorption into the BC requires appropriate moisture content in wound dressing materials. This results in painless and easy wound dressing, fast healing, and protection of newly formed skin tissues [18]. The hydrophilic characteristic of BC, together with its higher WHC and delayed WRR, has led to its widespread use as a material for wound dressings. An estimated 100–200 times its dry weight may be accommodated in BC in water [19]. A high WHC is preferred by highly porous BC, where water molecules connect to the cellulose fibrils by hydrogen bonds to stay inside the porous matrix [20]. Any structural changes in BC explain its changed thermal and physico-mechanical characteristics. By supplementing an extra carbon source or increasing its availability, for instance, extended BC synthesis improves fiber density by enabling the continual addition of freshly created β-1,4-glucan chains to pre-existing fibrils [18]. Improvements in resistance to applied force are explained by these robust and persistent fibrils [21]. Furthermore, a compact structure gives BC thermal resilience, which is a crucial characteristic of its industrial uses.

2.2 *Porous Structure and Absorptive Capacity of BC in Composite Synthesis*

The synthesis technique and pre- and post-synthesis treatment procedures greatly influence the porosity of BC, which is porous in nature. Its porous structure offers the perfect conditions for the adsorption of a variety of substances, including liquids and solid particles, such as wound exudates, nanoparticles, polymer solutions, water, cells, solutions, and medium components. From a medical standpoint, BC's modest natural porosity is a drawback because it hinders the entry of bigger particles and mammalian cells deeply into its matrix, but it is also beneficial since it stops microbial cell invasion [22].

Presently, research is being done to develop artificially porous BC using a variety of substances known as "porogens," which include gelatin, ice crystals, salts, paraffin particles, and other carbohydrates. A porogen is any substance that is applied as a space holder at first and then carefully removed to avoid harm or interfering with the geometry of the pores designed [23, 24]. Depending on the goals of the investigation, porogens are adsorbed into the BC matrix physically, not chemically, during varying periods. The needs of the cultured cells or additional substance to be added to the BC matrix also influence the sizes and forms of the porogens. Bäckdahl et al. found that potato starch along with paraffin wax particles were used as porogens to create BC tubes with linked pores of different sizes. Additionally, they successfully managed the pore size and produced partial particle fusion by heating the original paraffin wax particles to certain temperatures. The smooth muscle cell development inside the pores was efficiently supported by the structures that developed [23].

2.3 Crystallization and Self-Assembly

BC has distinct structural along with physico-mechanical properties due to the extrusion as well as self-assembly of β-1,4-glucan chains outside the cell. Once the produced fibrils are extruded via terminal complex (TCs), they eventually form bundles, 80 × 4 nm ribbon-shaped microfibrils, and 2–4 nm protofibrils [25].

Cell membrane electron micrographs showed that there were between 50 and 80 pore-like TCs along the axis of a microbial cell and in connection with the extracellular ribbon [26]. The β-1,4-glucan chains that form the first cellulose framework in the form of aggregates are thought to be extruded from these TCs. Such creation of aggregates indicates that a common characteristic of cellulose microfibril assembly in both higher and lower species is the simultaneous production of multiple β-1,4-glucan chains [27]. The initial arrangement of TCs controls the mutual orientation and connection of β-1,4-glucan chains, as well as their aggregation and the creation of bundles, microfibrils, and ribbons, whereas cellulose self-assembly along with crystallization are cell-directed processes [28]. *G. xylinum* specifically produces two different physical forms of cellulose: cellulose I as well as cellulose II [29]. In contrast to cellulose II, which comprises β-1,4-glucan chains arranged randomly, cellulose I has parallel, uniaxially aligned chains. Higher tensile strength, high WHC, higher crystallinity, slow WRR rate, and enhanced thermal stability are intriguing and distinctive qualities of BC, which are mostly caused by the microfibrillar fiber configurations and their extraordinary purity as related to cellulose from plants.

2.4 Regulation and Alignment of Macrostructure

The structural characteristics of BC are intriguing in both wet and dry forms, and their uses differ appropriately [30]. The physical structure of BC is sufficiently adaptable to be managed at the macro, micro, and nanoscales. Additionally, BC may be made in a variety of forms by altering the microbial strains and growth conditions (static, shaking, or agitation) [31]. Furthermore, at the micro, macro, and nanoscales, a number of its distinguishing characteristics, including porosity, surface chemistry, and fiber arrangement as well as orientation, may be adjusted. There are benefits to this microstructure control for its wide range of applications [9, 23].

Anisotropic patterns are seen in the culturing plane of BC in nature. BC fibers are random, which can be clarified by the mobility of microbial strains in the culture media. Controlling the motion of microbial cells allows for the manipulation of fiber organization. For instance, Uraki et al. used microgrooves with a honeycomb pattern in an agarose film to create a honeycomb pattern of BC [32]. Wang et al. used microfluidic channels to regulate microbial cell migration in a different investigation, and they were successful in obtaining aligned BC fibers [33]. To produce patterned BC fibers, Shi et al. employed a magnetic field to regulate the movement of *G. xylinum* [34]. Although BC fibers naturally exhibit random activity, their overall structure is robust and compact. BC may be applied to a diversity of fields and obtain equivalent isotropic alignment.

2.5 Biocompatibility

The term "biocompatibility" describes BC's capacity to stay in touch with live tissue without producing any adverse allergic or toxic reactions [35, 36]. High WHC and various other liquid-absorbing capacities of BC are directly related to its biocompatibility, which supports the long-term development and proliferation of live cells along with tissues.

According to reports thus far, BC exhibits moderate to high degrees of biocompatible characteristics with many mammalian cell and tissue types [37–39]. For instance, a subcutaneous BC implant in vivo trial has demonstrated biocompatibility for up to 12 weeks. There were no large cells or fibrotic capsules visible under a microscope, proving that there had been no foreign body response. Additionally, redness, edema, and exudate production or buildup surrounding the BC implant were prevented by BC implantation [40]. In a study, Khan et al. established the development of a microporous BC scaffold that is highly biocompatible for the application of skin tissue regeneration. The scaffold showed that human keratinocytes (HaCaT) adhered and proliferated in vitro and that experimental skin tissues of mice completely recovered in two weeks [41]. However, in some situations, BC's biocompatibility falls short of expectations and has to be improved. Combining it with other biocompatible materials to create composites is one way to increase its biocompatibility. The manufacturing of composites using different materials, for instance, has been proven to improve cell adhesion, biocompatibility, and proliferation for the application of dressing of wounds and the development of scaffolds in tissue engineering [42, 43].

2.6 Biodegradability

It has long been known that biodegradable polymers offer promise [44]. A biomaterial needs to be able to elicit the proper host response in a particular application to be considered biodegradable. This response to the implant form and structure may be explained in terms of the physical, chemical, and biological characteristics of the materials. The most intriguing and challenging aspect of BC is its ability to suit the demands of different biological applications through biodegradation. BC is difficult to biodegrade because of its compact structure and high degree of crystallinity. Alternative methods for breaking down BC implants must be developed because mammalian cells or tissues lack the cellulase enzyme. To do this, Li et al. described improved periodate oxidation-based BC degradation in vivo, which demonstrated accelerated degradation in water, and phosphate-buffered saline, along with simulated body fluid without affecting the initial structure of the fibrous network [45].

3 Biosynthesis of Bacterial Cellulose

Numerous specialized enzymes and regulatory proteins control the intricate process of BC biosynthesis. It is an aerobic activity that is closely related to the catabolism of the cells; nonetheless, it does not obstruct additional anabolic processes, such as the production of proteins in bacterial cells. BC is synthesized anaerobically by a cell-free system employing cellulose-producing enzymes as well as cofactors in addition to aerobically by microbial cells [46]. Although the mechanism of uridine-di-phosphoglucose synthesis is well established, the glucose polymerization step of BC production has not yet been well investigated.

3.1 Biochemistry of BC Biosynthesis

Sutherland characterized the synthesis of BC by microbial cells as a four-step procedure that includes (i) the formation of glucose nucleotides, which activates monosaccharides; (ii) the sequential accumulation of glucose repeating units, which polymerizes glucose; (iii) the simultaneous addition of acyl groups, if present, to specific glucose units; and (iv) the excretion of cellulose fibers into the extracellular environment through the wall/membrane complex. UDP-glucose, a precursor to BC synthesis, is produced throughout the BC synthesis process. Following this procedure, specific glucose units are polymerized to generate β-1 → 4-glucan chains. Hundreds of thousands of separate chains combine to create ribbon-like structures that are expelled through TCs. In turn, fibrils are formed by these ribbon-like structures [47]. The accessibility of certain substrates, enzymes, and a cofactor controls each biochemical step in the BC synthesis route, if they are not available, the metabolic pathway is eliminated.

Sugar nucleotides produce monosaccharides in the first stage of BC synthesis by interconversion via the catalyzed processes of epimerization, dehydrogenation, and decarboxylation. The second stage involves the sequential addition of UDP-glucose and d-glucose-1-phosphate, which are done by cellulose synthase and UDP-glucose pyrophosphorylase, respectively. This generates the cellulose backbone. It is then followed by the acylation of cellulose, which is mediated by certain sugar transferases such as 1-acyl-sn-glycerol-3-phosphate acyltransferase and involves the movement of the acyl group to create acetyl CoA [48]. This process produces acetyl CoA, a precursor to the tricarboxylic acid cycle. It is important to note that very specialized sugar transferases control the sequential movement of acyl groups and glucose monomers to the appropriate donors, which determines the structure of BC. Finally, in a process driven by cellular energy and regulated by adenosine triphosphate (ATP), β-1 → 4-glucan chains synthesized within the cytoplasm are organized into crystalline forms near or within the outer membrane and transported to the extracellular environment through systematically arranged terminal complexes (TCs) [49]. This indicates the need for a well-structured mechanism on the microbial cell surface to guarantee effective cellulose fibril excretion. A cell-free system does not have a cell wall or membrane to act as a barrier, therefore it avoids the complex process of β-1,4-glucan chain extrusion to the extracellular environment that microbial cell systems do. Moreover, it circumvents the periplasm's cellulose internalization and energy consumption, preventing any fatal consequences [50].

3.2 Biochemical Pathway for the Production of BC

LC–MS/MS LT Orbitrap investigation of the crude cell-free lysate of *G. hansenii* PJK demonstrated the occurrence of many major enzymes participating in the metabolism of cellulose [51, 52]. The glucokinase, phosphoglucomutase, cellulose synthase, and UDP-glucose pyrophosphorylase all directly participate in the synthesis of BC. Similarly, glucose-6-phosphate is a common intermediary for the TCA cycle, pentose phosphate pathway, and primary cellulose synthesis. The amount of phosphorylated glucose that passes via the two potential pathways dictates the quantity of BC production. Cellulose synthase and UDP-glucose dehydrogenase both use UDP-glucose as a mutual substrate. Once more, in both microbial as well as cell-free systems, the amount of synthesized BC is determined by the flow of UDP-glucose via the two potential pathways. A lesser quantity of nicotinamide adenine dinucleotide in the cell-free lysate may promote a higher degree of cellulose synthesis by partially inhibiting the action of UDP-glucose dehydrogenase. Crucially, 2 phosphodiesterases and diguanylic cyclase work together to control cellulose synthase activity via a particular activator called bi-(3′ → 5′)-cyclic diguanylic acid. Notably, the activator's production and degradation are mediated by PDE-A and PDE-B, respectively [53]. It could help speed up UDP-glucose polymerization into cellulose. In a similar vein, glucose can potentially enter microbial metabolism through additional mechanisms [54]. Pyruvate can be formed from a small amount of triose sugar generated by the Entner Doudorouff and PP pathways. Pyruvate dehydrogenase can then transform this pyruvate into acetyl-CoA. To increase the effectiveness of BC synthesis, acetyl-CoA is used to generate bulk ATPs [55, 56]. High ATP concentrations generally inhibit G6P dehydrogenase, which lowers its activity by a factor of many. Therefore, it may boost BC synthesis via the main route and slow down the PP pathway in microbial or cell-free environments.

3.3 Regulatory Mechanisms of BC Biosynthesis

The intricate process of producing BC and extruding it into the extracellular environment can be easily impacted through multiple mutations in the genes controlling the synthesis-related enzymes. Any changes to important genes cause cellulose to build up in the periplasm, which will inevitably have fatal consequences [57].

A minimum of four distinct genes control the generation of BC by microbial cells that is axcess A, axcess B, axcess C, and axcess D [58, 59]. These genes are organized in the form of an operator and encode specific proteins, AxCESB, AxCESA, and AxCESD, respectively, which conduct specific functions [26]. The polymerization of individual β-1,4-glucan chains is catalyzed as well as regulated by AxCESA along with AxCESB. In particular, UDP-glucose is bound by AxCESA, which also has amino acid patterns indicative of processive β-glycosyltransferase [60]. Likewise, the AxCES subunit binds cyclic di-guanosine monophosphate in high amounts reversibly, activating AxCESB. The salt bridge, which ties a conserved gating loop that regulates access to substrate coordination at the active site, is broken by the released c-di-GMP, which auto-inhibits the state of the enzyme [61]. Mutagenesis produces a constitutively active cellulose synthase by disrupting the salt bridge.

Furthermore, the c-di-GMP-activated BcsA-B complex has been identified to include a growing cellulose polymer, with its terminal glucose unit positioned near a novel site above the active region of BcsA, facilitating catalysis. During cellulose synthesis, AxCESC and AxCESD are considered potential rate-limiting components, as they are believed to regulate the extrusion of β-1,4-glucan chains and the crystallization of cellulose sub-elementary fragments, respectively [62, 63]. According to reports, the octamer that forms the functional unit of AxCESD is a cylinder-shaped protein. All four β-1,4-glucan chains are extruded separately along the dimer interface in a twisted fashion to the extracellular environment from the four passageways created by the octamer's N-termini, which are positioned inside the AxCeSD cylinder. This implies the presence of the AxCESD protein at the TC's extracellular side. These four subunits can be seen, along with perhaps additional elements from the complexes that produce cellulose [64].

4 Bacterial Cellulose Production: Methods of Synthesis

BC synthesis requires a chemically specified medium that includes glucose, yeast extract, peptone, disodium phosphate, as well as citric acid. BC synthesis occurs under acidic circumstances (pH 5–6). Variations in other growing parameters, like temperature, pH, and carbon source concentration, can impact quality along with quantity. Different growing methods produce BC with varying structures and characteristics [65]. Attempts have been made to establish a cost-effective and effective process for producing BC with higher yield and enhanced characteristics. The sections below provide a summary of popular fermentation tactics for producing

BC, such as static, shaking, and agitation, each with unique structural as well as physiological features and yields [66].

4.1 Static Fermentation for Bacterial Cellulose Membrane, Film, or Sheet Production

Static fermentation is the utmost prevalent technique for producing BC. Static cultivation involves inoculating cellulose-synthesizing microbial cells into the culture medium at optimal pH and incubating them at a suitable temperature. After 5–10 days of undisturbed incubation, microbial cells multiply and proliferate, using accessible carbon sources and medium components to make BC. During static culture, BC develops as reticulated crystalline ribbons at the air-medium interface, eventually forming a gel, membrane, film, or sheet as incubation duration increases [67, 68]. The BC pellicle eventually entraps all microbial cells, causing them to become dormant or die owing to nutrition or oxygen deficiency [31].

The static culture of BC produces a sheet with a 3D-linked reticulate structure, making it ideal for building scaffolds in biological applications [69]. Static cultivation's poor yield and extended incubation time make it unsuitable for large-scale production and commercial uses. Efforts have been undertaken to address constraints in static fermentative culture. Jung et al. developed a new approach for static culture that used intermittent feeding to produce BC. After 7–10 days of static cultivation, the BC pellicles detach from the medium because of nutrient depletion and restricted oxygen supply, resulting in a thickness of around 2 mm. Intermittent feeding resulted in a considerable increase in BC thickness after 30 days of cultivation to 30 mm under the same circumstances [70].

4.2 Shaking Fermentation for Bacterial Cellulose Pellet Production

It is easy to distinguish shaking fermentative cultivation from agitation that is the synthesis of BC in a reactor with a regulated speed, by referring to the production of BC in an incubator with a rotor or shaker [67]. Shaking cultivation involves inoculating the cellulose-producing microbial cells to the culture medium at the ideal pH and then shaking the mixture while incubating at the proper temperature. The standard unit of measurement for shaking is revolutions per minute (rpm). Microbial cells develop and multiply during this sort of culture, which typically lasts 24–36 h, producing BC by partially consuming the accessible carbon source and additional medium components. Small pellets of BC are created in the culture media during shaking cultivation. Depending on the microbial strain type, incubation time, and shaking speed, the pellet's form changes. According to research by Hu and Catchmark, when the shaking speed was maintained at 125 rpm, the BC pellet took on a sphere-like form. When the shaking speed was raised further, the pellets showed a tail-like structure that seemed to get bigger as the particle size shrank or the shaking speed rose. It was discovered that the tail-like BC pellets were joined to one another at even higher rotating speeds [71].

4.3 *Agitated Fermentation for Bacterial Cellulose Granule Production*

Granules of BC are produced during agitation cultivation. Several parameters, including bioreaction, movement of oxygen along with carbon supplies, and elimination of cellulose fibrils, have been optimized to provide a complete model for BC production by agitation culture [72]. Compared to static as well as shaking fermentative cultivation, agitation cultivation produces BC at a substantially higher pace. In agitated cultures, the increased cell density and improved interaction with accessible oxygen lead to extremely high volumetric production. Furthermore, the high density of microbial cells is also linked to increased production. The fermenters provide regulated pH, temperature, nutrition delivery, and oxygen availability, and inhibit foam formation [73].

Even though agitation cultivation produces a lot of BC, there are several drawbacks to this type of farming. For instance, a high-power supply is needed, which typically results in cellulose pellets rather than pellicles during aerated and submerged culture, limiting its range of applications [74]. Additionally, BC produced by agitated culture has a lesser degree of polymerization, and overall crystallinity than BC produced in static or shaking cultivations. Occasionally, agitation or stirring can also result in the production of irregular or sphere-like cellulose particles (SCPs), which are less crystalline, have low mechanical strength, and exhibit a low degree of polymerization [75]. Furthermore, the BC granules are not suitable for direct utilization in biomedicine or several other applications. Persistent agitation also changes the cellulose-producing microbial cultures in more enriched cellulose-negative (Cel−) mutants, that cultivate quicker than the wild type and generate less BC [76].

Several modified reactors have been created to improve the synthesis of microbial cellulose, comprising a direct oxygen and glucose feeding reactor, a cylindrical silicone membrane vessel, and a spinning disk bioreactor. In the sections that follow, a few of these are briefly explained.

4.3.1 Rotating Disk Reactor

This bioreactor is meant to increase the production of BC by keeping half of its disk at the air surface and the other half inside the growth media. BC synthesis requires direct interaction with the medium and oxygen in the environment for cells on the disk's surface to get nutrients [77]. The use of horizontal fermenters with spinning disks has been shown in prior studies to enhance BC output and culture conditions significantly. *G. xylinum* ATCC 700178 retained its key structural characteristics over a 5-day culture period and synthesis BC production (0.24 g/l/day) by adhering to the plastic composite surface (PCS) [78, 79].

4.3.2 Trickling Bed Reactor

This type of reactor, which has inlets to aid in air circulation, is typically used to produce vinegar. In this type of reactor, a pump moves the fermenting liquid from the collecting reservoir to the top of the tank till the intended product is produced. BC manufacturing has also made good use of such a reactor. This is accomplished by adsorbing cellulose-producing microbial cells, such as *G. xylinum*, onto packing materials, like rice or corncobs, and exposing them to the fermentation medium and air space. The capacity of the system is further increased by the oxygen delivery to the fermenter. By increasing oxygen availability, this culture technique lowers sheer force, which raises BC production [80].

In these reactors, BC films either adhere to the corncobs' surface or grow between husk gaps. Bacteria may readily adhere and develop in the semisolid as well as liquid–solid microenvironments because the reactor's many packings increase the surface-to-volume ratio and improve the oxygen amount to a sufficient level. The BC generated by trickling bed reactors seemed to be uneven sheet films that varied in thickness from 1 to 5 mm. In comparison to that produced in static and trembling cultivations, BC generated in such reactors has superior levels of purity, porosity, polymerization, and thermal stability. Therefore, BC generated by trickling reactors opens the door to several industrial uses [81]. Several kinds of bacterial cellulose production patents are mentioned in Table 1.

5 Application

5.1 Food Processing Sector

BC membrane is a solid contender for several food industry applications owing to its smooth texture and gelatinous consistency. It can develop edible when treated with substances like sugar or calcium chloride, giving it the consistency of fruits or molluscs [88]. BC is currently recognized as a food component or food additive after the US FDA classified it as a safe kind of dietary fiber. The food sector has shown promise in using BC as a thickening, gelling, and water-binding agent, which enhances the rheological properties of BC-stabilized emulsions [89, 90]. From a dietary perspective, BC can reduce cholesterol and replace fats [91]. It can also be

Table 1 Numerous patents associated with the development of the synthesis of bacterial cellulose

Patent number	Inventor	Publication date	Title	Note	References
CA1327536C	BYROM, DAVID	08.03.1994	Process for the production of microbial cellulose	The process involves cultivating a novel *Acetobacter* strain in an aqueous medium to produce extracellular microbial cellulose, which is claimed to be capable of producing cellulose	[82]
WO2019109133A1	Jinzarli, Madian Mohamad Cass, Gary Andrew Moursounidis, John Best, Wayne Morris	13.06.2019	Methods for producing a viscose dope from microbial cellulose	The method involves exposing microbial cellulose to water to create microbial cellulose pulp for viscose dope production, having a cellulose concentration below 0.040 g per mL	[83]
EP0186495B1	Roberts, Eric M. Hardison, Linda K. Brown, R. Malcolm	02.07.1986	Production of microbial cellulose	The invention outlines a technique for producing microbial cellulose using a liquid culture and a structure permeable to gases, allowing for continuous cellulose production and modification of cellulose products	[84]
US20140080184A1	Rajendra Kumar Saxena, Firdaus Jahan	20.03.2014	Isolated bacterial strain of gluconacetobacter oboediens and an optimized economic process for microbial cellulose production therefrom	The development presents a new bacterial species, *Gluconacetobacter oboediens*, for producing microbial cellulose. It optimizes production, develops an efficient drying method, and offers various shapes and sizes for burn and injured patients	[85]

(continued)

Table 1 (continued)

Patent number	Inventor	Publication date	Title	Note	References
CA1289494C	Hardison, Linda K., Brown, R. Malcolm	24.09.1991	Production of microbial cellulose	The development outlines a procedure for microbial cellulose production, involving a liquid culture and a structure permeable to gases, allowing for continuous cellulose production and altering configuration	[86]
US5955326A	Bungay Henry R., Serafica Gonzalo C.	21.09.1999	Production of microbial cellulose using a rotating disk film bioreactor	The process of creating new microbial cellulose with high water content involves submerging disks in a biological medium, forming cellulose on the surface, and harvesting it	[87]

used to decrease the fat content of ice creams and exhibit noticeable enhancements in stability as well as rheological characteristics, including greater resistance to melting [92]. Finally, it can be utilized as a fat substitute in mayonnaise with satisfactory sensory qualities and additional physical characteristics [93].

BC is appropriate for use in food packaging because of its robust network and barrier qualities. BC is used to produce cellulose nanofiber, which provides nanopaper its great strength, optical clarity, and thermal stability. BC is an excellent substitute for conventional micro-sized pulp sheets, which are currently used for bags and other packaging because it too has high oxygen barrier qualities that successfully reduce the entry of oxygen [94].

BC has recently been used in active packing, where the packaging elements can include ingredients that improve the shelf-life environments by entrapping or releasing certain compounds into the environment or into the food that is packaged [95]. To prevent the development and activity of *Listeria monocytogenes*, BC, impregnated with lyophilized *Lactobacillus plantarum* bacteria are employed as an antimicrobial nanopaper wrapping for minced beef [96]. Another study found that food packaging materials made of BC along with potato peel films treated with the phenolic component curcumin improved tensile strength and significantly decreased oxygen and water vapor permeability [97]. Various applications of BC are mentioned in Fig. 2.

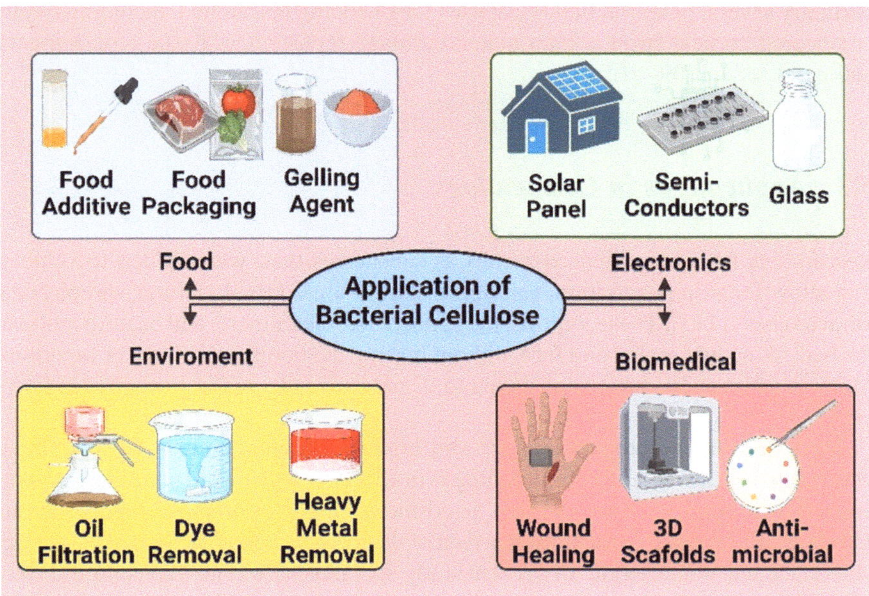

Fig. 2 Diagram illustrating the use of BC in the food, electronics, environment, and biomedical sectors

5.2 *Biomedical Applications*

The distinct physico-mechanical characteristics of BC, including its high tensile strength, crystalline structure, excellent permeability, HWA capacity, and biocompatibility, have made it valuable in a diversity of biomedical uses, including tissue engineering as well as wound dressing. BC is a perfect material for dressing wounds since it may hasten the healing process, stop microbial infections, and help the skin regain its structure and function [98]. One advantage of BC fibers is their capability to mirror the structure and shape of the natural extracellular matrix, which promotes the production of epithelial cells and aids in the development of new tissues. This is due to their random alignment with a nanometer-size dispersion [99]. Additionally, BC is a great membrane for wound healing because it can upsurge the sorption of liquids from the wound, which wipes the exudates from the wound, lets cells breathe properly, and makes changing dressings easier and less painful while maintaining the integrity of the skin's newly formed epithelial lining [100–102].

BC was examined for biocompatibility in rats having dural defects after 120 days. Results showed mechanical stability, no inflammation, and comparable characteristics to local tissues [103]. The BC has been successfully shaped into nanofibers that closely resemble the body's collagen fibers, which makes it appealing for applications in tissue engineering [104]. Additionally, BC has been utilized in dental implants, artificial skin, artificial blood vessels, and scaffold materials [105–108]. There are currently several potential biological uses for BC being researched, including tissue-engineered corneal stroma, dura mater, cartilage replacement tissue replacements, and nasal septa [109–112].

5.3 *Applications in Cosmetology*

Researchers have identified cosmetics as substances that, when added to vehicles that allow for skin penetration, can improve a person's visual features and physical attractiveness [113]. These improve the appearance, cleanliness, and beautification of the hair, skin, eyes, nails, and face without hurting the bodily structures or functions. Many benefits of BC have contributed to its extensive use in the cosmetics industry [114].

Modern cosmetic preparations are shifting away from chemicals like parabens, which can have adverse effects including skin allergies, and toward natural treatments derived from botanical sources [115]. In cosmetics, BC is a common biopolymer that stabilizes oil-in-water emulsions deprived of the need for extra skin-irritating surfactants since it is non-allergic. In a recent study, BC and carboxymethyl cellulose were effectively employed to replace two commonly used chemical surfactants in oil-in-water emulsions to create cosmetic creams [116]. Additionally, BC has been utilized in contact lenses, washing solutions, and face masks and cleansers [117]. Currently, BC is favored over botanical cellulose due to its higher crystalline structure, advanced tensile strength, improved WHC, and chemical purity, as it lacks hemicellulose or lignin [77, 118, 119].

The addition of more natural elements has enhanced BC's overall qualities in the cosmetics industry. For example, propolis concentrate has been added to BC films to improve their anti-inflammatory and moisturizing qualities, particularly for sheet masks used to treat and cure skin prone to inflammation and acne [113]. Through the transfer of 1,3-dihydroxy-2-propanone in the corneum extract, BC has also been utilized to treat skin pigmentation, changing the color of the skin. According to Stasiak-Różańska et al., this treatment produced a real skin tone that was thought to be the closest to the effect of a natural tan and may likely be used as a substitute for patients with vitiligo [120]. In a further study, both in situ and *ex-situ* techniques were employed to successfully functionalize BC with silk sericin and hyaluronic acid for usage in the cosmetic industry to improve skin texture and preserve moisture [121].

BC offers several benefits that strongly encourage its involvement in the cosmetics industry. It can hold liquids about ten times longer than nonwoven masks and 100 times longer than its dry weight because of its strong absorption capacity [122]. It can reach all of the facial contours, fine lines, wrinkles, and other areas that typical masks would normally miss because of its thin thickness, which adds flexibility for decent adherence to uneven surfaces of the treated skin [123]. Additionally, because of its hydrating properties, its gentle touch supports the skin [124]. More research is required to add active ingredients to BC in order to improve its anti-aging, cleaning, or whitening qualities, hence expanding its use in the cosmetics industry [119].

5.4 Applications in Environmental Science

BC is a sustainable and biodegradable green bio-based substance that meets modern environmental laws. It is therefore seen as a viable substitute for the wastewater treatment methods in use today. According to recent research conducted in our research facility, BC loaded by graphite or charcoal complexes has successfully eliminated cationic colors like methylene blue from tainted water. BC membrane was only able to remove 53% of the methylene blue when tested alone; however, when graphite and charcoal were added, this proportion increased to 98.7% and 100%, respectively [125]. Additionally, polydopamine and TiO_2 nanoparticles have been added to BC membranes to increase their surface area and active sites, which will boost dye photocatalytic degradation [126]. According to studies, BC can act as a structural support for the Cu and Ni nanoparticles that are utilized to reduce 4-nitophenol, a dangerous chemical found in wastewater from the paper, dye, and pharmaceutical sectors [127].

BC membranes are used in conjunction with absorbent materials like chitosan, graphene oxide, magnetic nanoparticles, and EDTA to remove harmful heavy metals that have been free into the environment as a result of industrial processes (such as factories that make batteries, fertilizers, or tanneries) and that are harmful to living things, including people. Zhu et al. found that combining spherical BC with Fe_3O_4 effectively removed heavy metal ions as Cr^{3+}, Mn^{2+}, and Pb^{2+} [128]. According to previous research, BC has also demonstrated a strong efficiency template for the removal of additional heavy metals from polluted environments, including Sr, Sb, Fe, Pb, Cr, and As [129–133].

Industrial procedures that produce oil-contaminated wastewater local wildlife, endanger marine life, and people. A sphere-like BC/graphene composite with a honeycomb-like surface and a three-dimensional interconnected porous structure thoroughly absorbed oils and organic solvents from contaminated water [72]. With an 88% oil recovery rate in mixed water along with oil emulsions, SiO_2-functionalized BC membranes have demonstrated excellent separation efficiency [134, 135]. Additionally, Galdino et al. established the viability of using BC membrane as a filter for oil elimination [136].

5.5 Electronic Applications

Researchers have been looking for transparent, flexible, and strong glass substitute materials for electrical devices [113]. Cellulose nanofibers that are optically transparent and appropriate for utilization in electronic devices like solar cells, antennas, organic light-emitting diodes, along transistors, with flexible displays can be developed, according to a new study [137]. Utilizing nanocellulose, the triboelectric nanogenerator famous for its capacity to change mechanical energy to electrical energy was also developed. As electronic technology advances, there is a growing request for devices that can store energy deprived of compromising their functionality. Biodegradable polymers are one such technology that can replace non-renewable materials that are currently in use. A BC gel electrolyte with 1-ethyl-3-methylimidazolium tetrafuoroborate and chitosan and alginate covering it worked best for use in electric capacitors [138]. Polyaniline was applied to the surface of BC/graphene nanosheets to create an electrically conductive nanocomposite that may be used effectively in electromagnetic shielding and flexible electrode materials [139]. Although it costs more to produce, BC can also be utilized as loudspeaker membranes due to it can withstand sound speed and frequency and respond to sound power efficiently. Various application of BC in the healthcare sector is mentioned in Table 2.

6 Current Problems and Future Prospects of BC Research

The production of BC has significantly improved in recent years due to the integration of several fermentation techniques and contemporary molecular biology methods, while certain areas are still undiscovered and hence require further research. To meet the needs of large-scale commercial production, these include: (1) using new and inexpensive wastes as substitute carbon and nitrogen substrates to reduce the time and price of fermentation and increase the BC yield; and (2) examining the synthesis

Table 2 Applications of BC in various healthcare industries

Field	Application	Highlight	References
Dermal applications	Wound dressing for burns and cuts	Maintains moisture, reduces pain, and accelerates healing compared to conventional dressings	[140]
Ophthalmology	Corneal regeneration	Flexibility, high hydrophilicity, and mechanical stability make BC suitable for use in corneal replacements	[141]
Tissue engineering	Artificial cartilage and orthopedic scaffolds	Elasticity close to natural cartilage; used in intervertebral disk replacements	[142]
Blood vessels	Artificial vascular grafts	High breaking pressure (800 mmHg), suitable for small and large vascular grafts; biocompatible when combined with graphene oxide	[143]
Drug delivery	Transdermal systems and controlled drug release	Used as carriers for drugs like lidocaine; systems enable controlled API release	[122, 144]
Biomedical implants	Scaffolds for cell growth in organ reconstruction	Allows proliferation of muscle and endothelial cells for organ tissue engineering	[145]
Burn treatments	Dressing systems for burn wounds	BC/ZnO nanocomposites promote wound healing and antimicrobial activity	[146]
Cosmetic industry	Hydrating and anti-aging masks	BC films retain high water content, delivering sustained hydration and enhancing skin elasticity	[147]
Skin substitutes	Temporary skin replacements	Biofill® BC dressings provide pain relief, tight adhesion, and fast healing; low elasticity limits certain uses	[148]
Periodontal medicine	Guided tissue regeneration (GTR) for dental applications	BC scaffolds aid in tissue recovery and serve as durable, biocompatible membranes	[149]
Environmental uses	Water filtration membranes	BC-based membranes filter impurities while maintaining high mechanical and chemical stability	[150, 151]

of BC from commonly utilized Gram-negative strains and the usage of Gram-positive bacteria and extremophiles. In this regard, the number of BC producers may rise with the use of modern gene-editing technologies like CRISPR as well as metagenomics investigation of numerous strains; (3) extensive research should be done to optimize the cultivation situations for BC generation, including the impact of oxygen tensions, particularly under hyperxic and hypoxic circumstances. Similarly, co-culturing is a new strategy for increased yield that needs more research; and (4) broadening the use of BC as a sustainable and green biopolymer to address urgent ecological issues.

7 Conclusion

Microbial cellulose, synthesized primarily by *Acetobacter xylinum*, stands out due to its unparalleled purity, high crystallinity, and mechanical strength, distinguishing it from plant-derived cellulose. Its biocompatibility and structural versatility enable diverse applications, particularly in healthcare, where it excels as a scaffold for tissue engineering applications and as an advanced wound dressing material. The porous structure as well as the high WHC of microbial cellulose facilitate controlled drug release, promote proliferation of cells, and accelerate healing of wound.

In the food sector, microbial cellulose serves as a stabilizer, fat replacer, and component of active packaging, enhancing food quality and sustainability. Its role in environmental remediation, including efficient removal of dyes and heavy metals, highlights its potential as a green material. Moreover, its integration into cosmetics and electronics underscores its adaptability across industries. While advancements in fermentation techniques have improved production efficiency, challenges such as high costs and limited scalability remain. Future efforts must focus on optimizing culture conditions, employing cost-effective substrates, and leveraging genetic modifications to enhance yield. Research into composite materials and functionalization will expand its utility in high-demand sectors like drug delivery and regenerative medicine.

Microbial cellulose embodies a sustainable and versatile biopolymer with transformative potential. Continued innovation and strategic research will unlock its full capacity, paving the way for widespread adoption in healthcare, food, environmental science, and beyond.

References

1. Di X, Liang X, Shen C et al (2022) Carbohydrates used in polymeric systems for drug delivery: from structures to applications. Pharmaceutics 14
2. Moon RJ, Martini A, Nairn J et al (2011) Cellulose nanomaterials review: structure, properties and nanocomposites. Chem Soc Rev 40. https://doi.org/10.1039/c0cs00108b
3. Barikani M, Oliaei E, Seddiqi H et al (2014) Preparation and application of chitin and its derivatives: a review. Iran Polym J (English Edition) 23

4. Seddiqi H, Oliaei E, Honarkar H et al (2021) Cellulose and its derivatives: towards biomedical applications. Cellulose 28
5. Heinze T (2015) Cellulose: structure and properties. Adv Polym Sci 271. https://doi.org/10.1007/12_2015_319
6. Esa F, Tasirin SM, Rahman NA (2014) Overview of bacterial cellulose production and application. Agric Agric Sci Procedia 2. https://doi.org/10.1016/j.aaspro.2014.11.017
7. Yoshinaga F, Tonouchi N, Watanabe K (1997) Research progress in production of bacterial cellulose by aeration and agitation culture and its application as a new industrial material. Biosci Biotechnol Biochem 61. https://doi.org/10.1271/bbb.61.219
8. Shoda M, Sugano Y (2005) Recent advances in bacterial cellulose production. Biotechnol Bioprocess Eng 10
9. Czaja W, Krystynowicz A, Bielecki S et al (2006) Microbial cellulose—the natural power to heal wounds. Biomaterials 27. https://doi.org/10.1016/j.biomaterials.2005.07.035
10. Ummartyotin S, Manuspiya H (2015) A critical review on cellulose: from fundamental to an approach on sensor technology. Renew Sustain Energy Rev 41
11. Gallegos AMA, Carrera SH, Parra R et al (2016) Bacterial cellulose: a sustainable source to develop value-added products—a review. Bioresources 11. https://doi.org/10.15376/biores.11.2.Gallegos
12. Portela R, Leal CR, Almeida PL et al (2019) Bacterial cellulose: a versatile biopolymer for wound dressing applications. Microb Biotechnol 12
13. Vandamme EJ, De Baets S, Vanbaelen A et al (1998) Improved production of bacterial cellulose and its application potential. Polym Degrad Stab 59. https://doi.org/10.1016/s0141-3910(97)00185-7
14. Lupaşcu RE, Ghica MV, Dinu-Pîrvu CE et al (2022) An overview regarding microbial aspects of production and applications of bacterial cellulose. Materials 15
15. Lin SP, Loira Calvar I, Catchmark JM et al (2013) Biosynthesis, production and applications of bacterial cellulose. Cellulose 20
16. Kaewnopparat S, Sansernluk K, Faroongsarng D (2008) Behavior of freezable bound water in the bacterial cellulose produced by Acetobacter xylinum: an approach using thermoporosimetry. AAPS PharmSciTech 9. https://doi.org/10.1208/s12249-008-9104-2
17. Tang W, Jia S, Jia Y et al (2010) The influence of fermentation conditions and post-treatment methods on porosity of bacterial cellulose membrane. World J Microbiol Biotechnol 26. https://doi.org/10.1007/s11274-009-0151-y
18. Ullah MW, Khattak WA, Ul-Islam M et al (2016) Metabolic engineering of synthetic cell-free systems: strategies and applications. Biochem Eng J 105
19. Schrecker ST, Gostomski PA (2005) Determining the water holding capacity of microbial cellulose. Biotechnol Lett 27. https://doi.org/10.1007/s10529-005-1465-y
20. Gelin K, Bodin A, Gatenholm P et al (2007) Characterization of water in bacterial cellulose using dielectric spectroscopy and electron microscopy. Polymer (Guildf) 48. https://doi.org/10.1016/j.polymer.2007.10.039
21. Ul-Islam M, Ha JH, Khan T et al (2013) Effects of glucuronic acid oligomers on the production, structure and properties of bacterial cellulose. Carbohydr Polym 92. https://doi.org/10.1016/j.carbpol.2012.09.060
22. Hutchens SA, Benson RS, Evans BR et al (2006) Biomimetic synthesis of calcium-deficient hydroxyapatite in a natural hydrogel. Biomaterials 27. https://doi.org/10.1016/j.biomaterials.2006.04.032
23. Bäckdahl H, Esguerra M, Delbro D et al (2008) Engineering microporosity in bacterial cellulose scaffolds. J Tissue Eng Regen Med 2. https://doi.org/10.1002/term.97
24. Khan S, Ul-Islam M, Ikram M et al (2016) Three-dimensionally microporous and highly biocompatible bacterial cellulose-gelatin composite scaffolds for tissue engineering applications. RSC Adv 6. https://doi.org/10.1039/C6RA18847H
25. Iguchi M, Yamanaka S, Budhiono A (2000) Bacterial cellulose—a masterpiece of nature's arts. J Mater Sci 35. https://doi.org/10.1023/A:1004775229149

26. Endler A, Sánchez-Rodríguez C, Persson S (2010) Cellulose squeezes through. Nat Chem Biol 6(12):883–884. https://doi.org/10.1038/nchembio.480
27. Manan S, Ullah MW, Ul-Islam M et al (2022) Bacterial cellulose: molecular regulation of biosynthesis, supramolecular assembly, and tailored structural and functional properties. Prog Mater Sci 129
28. Ross P, Mayer R, Benziman M (1991) Cellulose biosynthesis and function in bacteria. Microbiol Rev 55. https://doi.org/10.1128/mr.55.1.35-58.1991
29. Yu X, Atalla RH (1996) Production of cellulose II by Acetobacter xylinum in the presence of 2,6-dichlorobenzonitrile. Int J Biol Macromol 19. https://doi.org/10.1016/0141-8130(96)01109-9
30. Shi Z, Gao X, Ullah MW et al (2016) Electroconductive natural polymer-based hydrogels. Biomaterials 111
31. Shah N, Ul-Islam M, Khattak WA et al (2013) Overview of bacterial cellulose composites: a multipurpose advanced material. Carbohydr Polym 98
32. Uraki Y, Nemoto J, Otsuka H et al (2007) Honeycomb-like architecture produced by living bacteria, Gluconacetobacter xylinus. Carbohydr Polym 69. https://doi.org/10.1016/j.carbpol.2006.08.021
33. Wang G, Chen X, Shi X et al (2008) Bio-fabrication of patterned cellulose nano-fibers. Adv Mater Res 47–50:1359–1362
34. Shi X, Shi Z, Wang D (2016) Microbial cells with a Fe_3O_4 doped hydrogel extracellular matrix: manipulation of living cells by magnetic stimulus. Macromol Biosci. https://doi.org/10.1002/mabi.201600143
35. Ul-Islam M, Khan S, Ullah MW et al (2015) Bacterial cellulose composites: synthetic strategies and multiple applications in bio-medical and electro-conductive fields. Biotechnol J 10. https://doi.org/10.1002/biot.201500106
36. Ul-Islam M, Shehzad A, Khan S et al (2014) Antimicrobial and biocompatible properties of nanomaterials. J Nanosci Nanotechnol 14
37. Khan S, Ul-Islam M, Ullah MW et al (2015) Engineered regenerated bacterial cellulose scaffolds for application in in vitro tissue regeneration. RSC Adv 5. https://doi.org/10.1039/c5ra16985b
38. Klemm D, Schumann D, Udhardt U et al (2001) Bacterial synthesized cellulose - Artificial blood vessels for microsurgery. Prog Polym Sci (Oxford) 26
39. Schumann DA, Wippermann J, Klemm DO et al (2009) Artificial vascular implants from bacterial cellulose: preliminary results of small arterial substitutes. Cellulose 16. https://doi.org/10.1007/s10570-008-9264-y
40. Helenius G, Bäckdahl H, Bodin A et al (2006) In vivo biocompatibility of bacterial cellulose. J Biomed Mater Res A 76. https://doi.org/10.1002/jbm.a.30570
41. Khan S, Ul-Islam M, Ikram M et al (2018) Preparation and structural characterization of surface modified microporous bacterial cellulose scaffolds: a potential material for skin regeneration applications in vitro and in vivo. Int J Biol Macromol 117. https://doi.org/10.1016/j.ijbiomac.2018.06.044
42. Ciechańska D (2004) Multifunctional bacterial cellulose/chitosan composite materials for medical applications. Fibres Textiles Eastern Eur 12
43. Kim J, Cai Z, Lee HS et al (2011) Preparation and characterization of a Bacterial cellulose/Chitosan composite for potential biomedical application. J Polym Res 18. https://doi.org/10.1007/s10965-010-9470-9
44. Ul-Islam M, Khan S, Ullah MW et al (2015) Structure, chemistry and pharmaceutical applications of biodegradable polymers. In: Handbook of polymers for pharmaceutical technologies
45. Li J, Wan Y, Li L et al (2009) Preparation and characterization of 2,3-dialdehyde bacterial cellulose for potential biodegradable tissue engineering scaffolds. Mater Sci Eng C 29. https://doi.org/10.1016/j.msec.2009.01.006
46. Ullah MW, Ul-Islam M, Khan S et al (2015) Innovative production of bio-cellulose using a cell-free system derived from a single cell line. Carbohydr Polym 132. https://doi.org/10.1016/j.carbpol.2015.06.037

47. Sutherland IW (2001) Microbial polysaccharides from gram-negative bacteria. Int Dairy J 11(9):663–674
48. Ullah MW, Manan S, Kiprono SJ et al (2019) Synthesis, structure, and properties of bacterial cellulose. In: Nanocellulose: from fundamentals to advanced materials, p 81–113
49. Shi Z, Zhang Y, Phillips GO et al (2014) Utilization of bacterial cellulose in food. Food Hydrocoll 35
50. Ullah MW, Ul Islam M, Khan S et al (2017) Recent advancements in bioreactions of cellular and cell-free systems: a study of bacterial cellulose as a model. Korean J Chem Eng 34
51. Li G, Wang L, Deng Y et al (2022) Research progress of the biosynthetic strains and pathways of bacterial cellulose. J Ind Microbiol Biotechnol 49
52. Singhania RR, Patel AK, Tsai ML et al (2021) Genetic modification for enhancing bacterial cellulose production and its applications. Bioengineered 12
53. Ross P, Weinhouse H, Aloni Y et al (1987) Regulation of cellulose synthesis in Acetobacter xylinum by cyclic diguanylic acid. Nature 325. https://doi.org/10.1038/325279a0
54. Ha JH, Shah N, Ul-Islam M et al (2011) Bacterial cellulose production from a single sugar α-linked glucuronic acid-based oligosaccharide. Process Biochem 46. https://doi.org/10.1016/j.procbio.2011.05.024
55. Naritomi T, Kouda T, Yano H et al (1998) Effect of ethanol on bacterial cellulose production from fructose in continuous culture. J Ferment Bioeng 85. https://doi.org/10.1016/S0922-338X(98)80012-3
56. Wang HM, Loganathan D, Linhardt RJ (1991) Determination of the pK(a) of glucuronic acid and the carboxy groups of heparin by 13C-nuclear-magnetic-resonance spectroscopy. Biochem J 278. https://doi.org/10.1042/bj2780689
57. Buldum G, Mantalaris A (2021) Systematic understanding of recent developments in bacterial cellulose biosynthesis at genetic, bioprocess and product levels. Int J Mol Sci 22
58. Kim Y, Ullah MW, Ul-Islam M et al (2019) Self-assembly of bio-cellulose nanofibrils through intermediate phase in a cell-free enzyme system. Biochem Eng J 142. https://doi.org/10.1016/j.bej.2018.11.017
59. Ullah MW, Ul-Islam M, Shahzad A et al (2021) Cell-free nanocellulose synthesis: mechanism, characterization, and applications. In: Bacterial cellulose: synthesis, production, and applications
60. Lin FC, Brown RM, Drake RR et al. (1990) Identification of the uridine 5′-diphosphoglucose (UDP-Glc) binding subunit of cellulose synthase in Acetobacter xylinum using the photoaffinity probe 5-Azido-UDP-Glc. J Biol Chem 265. https://doi.org/10.1016/s0021-9258(19)34039-6
61. Morgan JLW, McNamara JT, Zimmer J (2014) Mechanism of activation of bacterial cellulose synthase by cyclic di-GMP. Nat Struct Mol Biol 21. https://doi.org/10.1038/nsmb.2803
62. Benziman M, Haigler CH, Brown RM (1980) Cellulose biogenesis: Polymerization and crystallization are coupled processes in Acetobacter xylinum. Proc Natl Acad Sci USA 77. https://doi.org/10.1073/pnas.77.11.6678
63. Saxena IM, Kudlicka K, Okuda K et al (1994) Characterization of genes in the cellulose-synthesizing operon (acs operon) of Acetobacter xylinum: implications for cellulose crystallization. J Bacteriol 176. https://doi.org/10.1128/jb.176.18.5735-5752.1994
64. Hu SQ, Gao YG, Tajima K *et al.* (2010) Structure of bacterial cellulose synthase subunit D octamer with four inner passageways. Proc Natl Acad Sci USA 107. https://doi.org/10.1073/pnas.1000601107
65. Castro C, Cordeiro N, Faria M et al (2015) In-situ glyoxalization during biosynthesis of bacterial cellulose. Carbohydr Polym 126. https://doi.org/10.1016/j.carbpol.2015.03.014
66. Tanskul S, Amornthatree K, Jaturonlak N (2013) A new cellulose-producing bacterium, Rhodococcus sp. MI 2: screening and optimization of culture conditions. Carbohydr Polym 92. https://doi.org/10.1016/j.carbpol.2012.09.017
67. Ruka DR, Simon GP, Dean KM (2014) Bacterial cellulose and its use in renewable composites. Wiley Blackwell 6

68. Ullah MW, Ul-Islam M, Khan S et al (2016) Structural and physico-mechanical characterization of bio-cellulose produced by a cell-free system. Carbohydr Polym 136. https://doi.org/10.1016/j.carbpol.2015.10.010
69. Shezad O, Khan S, Khan T et al (2010) Physicochemical and mechanical characterization of bacterial cellulose produced with an excellent productivity in static conditions using a simple fed-batch cultivation strategy. Carbohydr Polym 82. https://doi.org/10.1016/j.carbpol.2010.04.052
70. Jung JY, Khan T, Park JK et al (2007) Production of bacterial cellulose by Gluconacetobacter hansenii using a novel bioreactor equipped with a spin filter. Korean J Chem Eng 24. https://doi.org/10.1007/s11814-007-5058-4
71. Hu Y, Catchmark JM (2010) Formation and characterization of spherelike bacterial cellulose particles produced by acetobacter xylinum JCM 9730 strain. Biomacromolecules 11. https://doi.org/10.1021/bm100060v
72. Wang J, Tavakoli J, Tang Y (2019) Bacterial cellulose production, properties and applications with different culture methods—a review. Carbohydr Polym 219
73. Lee KY, Buldum G, Mantalaris A et al (2014) More than meets the eye in bacterial cellulose: biosynthesis, bioprocessing, and applications in advanced fiber composites. Macromol Biosci 14. https://doi.org/10.1002/mabi.201300298
74. Kralisch D, Hessler N, Klemm D et al (2010) White biotechnology for cellulose manufacturing—the HoLiR concept. Biotechnol Bioeng 105. https://doi.org/10.1002/bit.22579
75. Shi QS, Feng J, Li WRU et al (2013) Effect of different conditions on the average degree of polymerization of bacterial cellulose produced by gluconacetobacter intermedius BC-41. Cellul Chem Technol 47
76. Kim YJ, Kim JN, Wee YJ et al (2007) Bacterial cellulose production by Gluconacetobacter sp. PKY5 in a rotary biofilm contactor. In: Applied biochemistry and biotechnology, pp 529–537
77. Chawla PR, Bajaj IB, Survase SA et al (2009) Microbial cellulose: fermentative production and applications. Food Technol Biotechnol 47
78. Cheng KC, Catchmark JM, Demirci A (2011) Effects of CMC addition on bacterial cellulose production in a biofilm reactor and its paper sheets analysis. Biomacromolecules 12. https://doi.org/10.1021/bm101363t
79. Lin SP, Hsieh SC, Chen KI et al (2014) Semi-continuous bacterial cellulose production in a rotating disk bioreactor and its materials properties analysis. Cellulose 21. https://doi.org/10.1007/s10570-013-0136-8
80. Lu H, Jiang X (2014) Structure and properties of bacterial cellulose produced using a trickling bed reactor. Appl Biochem Biotechnol 172. https://doi.org/10.1007/s12010-014-0795-4
81. Blanco Parte FG, Santoso SP, Chou CC et al. (2020) Current progress on the production, modification, and applications of bacterial cellulose. Crit Rev Biotechnol 40
82. CA1327536C (2025) Process for the production of microbial cellulose—Google Patents. https://patents.google.com/patent/CA1327536C/en?oq=CA1327536. Accessed 23 Jan 2025
83. WO2019109133A1 (2025) Methods for producing a viscose dope from microbial cellulose—Google Patents. https://patents.google.com/patent/WO2019109133A1/en?oq=WO2019109133. Accessed 23 Jan 2025
84. EP0186495B1 (2025) Production of microbial cellulose—Google Patents. https://patents.google.com/patent/EP0186495B1/en?oq=EP0186495. Accessed 23 Jan 2025
85. US20140080184A1 (2025) Novel isolated bacterial strain of gluconacetobacter oboediens and an optimized economic process for microbial cellulose production therefrom—Google Patents. https://patents.google.com/patent/US20140080184A1/en?oq=US20140080184. Accessed 24 Jan 2025
86. CA1289494C (2025) Production of microbial cellulose—Google Patents. https://patents.google.com/patent/CA1289494C/en?oq=CA1289494. Accessed 23 Jan 2025
87. US5955326A (2025) Production of microbial cellulose using a rotating disk film bioreactor—Google Patents. https://patents.google.com/patent/US5955326A/en?oq=US5955326. Accessed 24 Jan 2025

88. Okiyama A, Motoki M, Yamanaka S (1992) Bacterial cellulose II. Processing of the gelatinous cellulose for food materials. Top Catal 6. https://doi.org/10.1016/S0268-005X(09)80033-7
89. Paximada P, Koutinas AA, Scholten E et al (2016) Effect of bacterial cellulose addition on physical properties of WPI emulsions. Comparison with common thickeners. Food Hydrocoll 54. https://doi.org/10.1016/j.foodhyd.2015.10.014
90. Paximada P, Tsouko E, Kopsahelis N et al (2016) Bacterial cellulose as stabilizer of o/w emulsions. Food Hydrocoll 53. https://doi.org/10.1016/j.foodhyd.2014.12.003
91. Dourado F, Gama M, Rodrigues AC (2017) A review on the toxicology and dietetic role of bacterial cellulose. Toxicol Rep 4
92. Guo Y, Zhang X, Hao W et al (2018) Nano-bacterial cellulose/soy protein isolate complex gel as fat substitutes in ice cream model. Carbohydr Polym 198. https://doi.org/10.1016/j.carbpol.2018.06.078
93. Akoğlu A, Cakir I, Karahan AG et al (2017) Effects of bacterial cellulose as a fat replacer on some properties of fat-reduced mayonnaise. Rom Biotechnol Lett 23
94. Samyn P, Barhoum A, Öhlund T et al (2018) Review: nanoparticles and nanostructured materials in papermaking. J Mater Sci 53. https://doi.org/10.1007/s10853-017-1525-4
95. Urbina L, Corcuera MÁ, Gabilondo N et al (2021) A review of bacterial cellulose: sustainable production from agricultural waste and applications in various fields. Cellulose 28
96. Shafipour Yordshahi A, Moradi M, Tajik H et al (2020) Design and preparation of antimicrobial meat wrapping nanopaper with bacterial cellulose and postbiotics of lactic acid bacteria. Int J Food Microbiol 321. https://doi.org/10.1016/j.ijfoodmicro.2020.108561
97. Xie Y, Niu X, Yang J (2020) Active biodegradable films based on the whole potato peel incorporated with bacterial cellulose and curcumin. Int J Biol Macromol 150:480–491. https://doi.org/10.1016/J.IJBIOMAC.2020.01.291
98. Abrigo M, McArthur SL, Kingshott P (2014) Electrospun nanofibers as dressings for chronic wound care: advances, challenges, and future prospects. Macromol Biosci 14
99. Frone AN, Panaitescu DM, Nicolae CA et al (2020) Bacterial cellulose sponges obtained with green cross-linkers for tissue engineering. Mater Sci Eng C 110. https://doi.org/10.1016/j.msec.2020.110740
100. Shalumon KT, Anulekha KH, Nair SV (2011) Sodium alginate/poly(vinyl alcohol)/nano ZnO composite nanofibers for antibacterial wound dressings. Int J Biol Macromol 49:247–254 https://doi.org/10.1016/J.IJBIOMAC.2011.04.005
101. Ul-Islam M, Khan T, Khattak WA et al (2013) Bacterial cellulose-MMTs nanoreinforced composite films: novel wound dressing material with antibacterial properties. Cellulose 20. https://doi.org/10.1007/s10570-012-9849-3
102. Zou B, Liu Y, Luo X (2012) Electrospun fibrous scaffolds with continuous gradations in mineral contents and biological cues for manipulating cellular behaviors. Acta Biomater 8:1576–1585. https://doi.org/10.1016/J.ACTBIO.2012.01.003
103. de Lima FDMT, Pinto FCM, Andrade-da-Costa BLDS, da Silva JGM, Campos Júnior O, de Aguiar JLA (2017) Biocompatible bacterial cellulose membrane in dural defect repair of rat. J Mater Sci Mater Med 28. https://doi.org/10.1007/s10856-016-5828-9
104. Torgbo S, Sukyai P (2018) Bacterial cellulose-based scaffold materials for bone tissue engineering. Appl Mater Today 11
105. Keskin Z, Sendemir Urkmez A, Hames EE (2017) Novel keratin modified bacterial cellulose nanocomposite production and characterization for skin tissue engineering. Mater Sci Eng C 75:1144–1153. https://doi.org/10.1016/J.MSEC.2017.03.035
106. Kumbhar JV, Jadhav SH, Bodas DS et al (2017) In vitro and in vivo studies of a novel bacterial cellulose-based acellular bilayer nanocomposite scaffold for the repair of osteochondral defects. Int J Nanomed 12. https://doi.org/10.2147/IJN.S137361
107. Lee SE, Park YS (2017) The role of bacterial cellulose in artificial blood vessels. Mol Cell Toxicol 13
108. Voicu G, Jinga SI, Drosu BG (2017) Improvement of silicate cement properties with bacterial cellulose powder addition for applications in dentistry. Carbohydr Polym 174:160–170 https://doi.org/10.1016/J.CARBPOL.2017.06.062

109. Binnetoglu A, Demir B, Akakin D et al (2020) Bacterial cellulose tubes as a nerve conduit for repairing complete facial nerve transection in a rat model. Eur Arch Oto-Rhino-Laryngol 277. https://doi.org/10.1007/s00405-019-05637-9
110. Mandour YMH, Fahmy Shendy M, Badie S et al (2019) Use of bacterial cellulose in closure of nasal septal perforation. Int J Otorhinolaryngol Head Neck Surg 5. https://doi.org/10.18203/issn.2454-5929.ijohns20192701
111. Pang M, Huang Y, Meng F et al (2020) Application of bacterial cellulose in skin and bone tissue engineering. Eur Polym J 122
112. Zhang C, Cao J, Zhao S et al (2020) Biocompatibility evaluation of bacterial cellulose as a scaffold material for tissue-engineered corneal stroma. Cellulose 27. https://doi.org/10.1007/s10570-020-02979-0
113. de Amorim JDP, de Souza KC, Duarte CR et al (2020) Plant and bacterial nanocellulose: production, properties and applications in medicine, food, cosmetics, electronics and engineering: a review. Environ Chem Lett 18
114. Hasan N, Biak DRA, Kamarudin S (2012) Application of bacterial cellulose (BC) in natural facial scrub. Int J Adv Sci Eng Inf Technol 2. https://doi.org/10.18517/ijaseit.2.4.201
115. Darbre PD, Harvey PW (2008) Paraben esters: review of recent studies of endocrine toxicity, absorption, esterase and human exposure, and discussion of potential human health risks. J Appl Toxicol 28
116. Martins D, Rocha C, Dourado F (2021) Bacterial cellulose-carboxymethyl cellulose (BC:CMC) dry formulation as stabilizer and texturizing agent for surfactant-free cosmetic formulations. Colloids Surf A Physicochem Eng Asp 617:126380https://doi.org/10.1016/J.COLSURFA.2021.126380
117. Ullah H, Santos HA, Khan T (2016) Applications of bacterial cellulose in food, cosmetics and drug delivery. Cellulose 23
118. Jonas R, Farah LF (1998) Production and application of microbial cellulose. Polym Degrad Stab 59. https://doi.org/10.1016/S0141-3910(97)00197-3
119. Mbituyimana B, Liu L, Ye W et al (2021) Bacterial cellulose-based composites for biomedical and cosmetic applications: Research progress and existing products. Carbohydr Polym 273
120. Stasiak-Różańska L, Płoska J (2018) Study on the use of microbial cellulose as a biocarrier for 1,3-dihydroxy-2-propanone and its potential application in industry. Polymers (Basel) 10. https://doi.org/10.3390/polym10040438
121. Wang X, Tang J, Huang J et al (2020) Production and characterization of bacterial cellulose membranes with hyaluronic acid and silk sericin. Colloids Surf B Biointerfaces 195. https://doi.org/10.1016/j.colsurfb.2020.111273
122. Trovatti E, Freire CSR, Pinto PC et al (2012) Bacterial cellulose membranes applied in topical and transdermal delivery of lidocaine hydrochloride and ibuprofen: in vitro diffusion studies. Int J Pharm 435. https://doi.org/10.1016/j.ijpharm.2012.01.002
123. Wei B, Yang G, Hong F (2011) Preparation and evaluation of a kind of bacterial cellulose dry films with antibacterial properties. Carbohydr Polym 84. https://doi.org/10.1016/j.carbpol.2010.12.017
124. Bianchet RT, Vieira Cubas AL, Machado MM et al (2020) Applicability of bacterial cellulose in cosmetics—bibliometric review. Biotechnol Rep 27
125. AK Saleh H El-Gendi JB Ray 2023 A low-cost effective media from starch kitchen waste for bacterial cellulose production and its application as simultaneous absorbance for methylene blue dye removal Biomass Convers Biorefin 13 12437 12449 https://doi.org/10.1007/s13399-021-01973-1
126. Yang L, Chen C, Hu Y et al (2020) Three-dimensional bacterial cellulose/polydopamine/TiO_2 nanocomposite membrane with enhanced adsorption and photocatalytic degradation for dyes under ultraviolet-visible irradiation. J Colloid Interface Sci 562. https://doi.org/10.1016/j.jcis.2019.12.013
127. Song L, Shu L, Wang Y et al (2020) Metal nanoparticle-embedded bacterial cellulose aerogels via swelling-induced adsorption for nitrophenol reduction. Int J Biol Macromol 143. https://doi.org/10.1016/j.ijbiomac.2019.09.152

128. Zhu H, Jia S, Wan T et al (2011) Biosynthesis of spherical Fe3O4/bacterial cellulose nanocomposites as adsorbents for heavy metal ions. Carbohydr Polym 86. https://doi.org/10.1016/j.carbpol.2011.06.061
129. Cheng R, Kang M, Zhuang S et al (2019) Adsorption of Sr(II) from water by mercerized bacterial cellulose membrane modified with EDTA. J Hazard Mater 364. https://doi.org/10.1016/j.jhazmat.2018.10.083
130. Hassan A, Sorour NM, El-Baz A et al (2019) Simple synthesis of bacterial cellulose/magnetite nanoparticles composite for the removal of antimony from aqueous solution. Int J Environ Sci Technol 16. https://doi.org/10.1007/s13762-018-1737-4
131. Meng C, Zhong N, Hu J et al (2019) The effects of metal elements on ramie fiber oxidation degumming and the potential of using spherical bacterial cellulose for metal removal. J Clean Prod 206. https://doi.org/10.1016/j.jclepro.2018.09.072
132. Mensah A, Lv P, Narh C et al (2019) Sequestration of Pb(II) ions from aqueous systems with novel green bacterial cellulose graphene oxide composite. Materials 12. https://doi.org/10.3390/ma12020218
133. Stoica-Guzun A, Stroescu M, Jinga SI et al (2016) Box-Behnken experimental design for chromium(VI) ions removal by bacterial cellulose-magnetite composites. Int J Biol Macromol 91. https://doi.org/10.1016/j.ijbiomac.2016.06.070
134. He J, Zhao H, Li X et al (2018) Superelastic and superhydrophobic bacterial cellulose/silica aerogels with hierarchical cellular structure for oil absorption and recovery. J Hazard Mater 346. https://doi.org/10.1016/j.jhazmat.2017.12.045
135. Hou Y, Duan C, Zhu G et al (2019) Functional bacterial cellulose membranes with 3D porous architectures: conventional drying, tunable wettability and water/oil separation. J Memb Sci 591. https://doi.org/10.1016/j.memsci.2019.117312
136. Galdino CJS, Maia AD, Meira HM et al (2020) Use of a bacterial cellulose filter for the removal of oil from wastewater. Process Biochem 91. https://doi.org/10.1016/j.procbio.2019.12.020
137. Nogi M, Iwamoto S, Nakagaito AN et al (2009) Optically transparent nanofiber paper. Adv Mater 21. https://doi.org/10.1002/adma.200803174
138. Kotatha D, Morishima K, Uchida S et al (2018) Preparation and characterization of gel electrolyte with bacterial cellulose coated with alternating layers of chitosan and alginate for electric double-layer capacitors. Res Chem Intermed 44. https://doi.org/10.1007/s11164-018-3348-6
139. Wan Y, Li J, Yang Z et al (2018) Simultaneously depositing polyaniline onto bacterial cellulose nanofibers and graphene nanosheets toward electrically conductive nanocomposites. Curr Appl Phys 18. https://doi.org/10.1016/j.cap.2018.05.008
140. Swingler S, Gupta A, Gibson H et al (2021) Recent advances and applications of bacterial cellulose in biomedicine. Polymers (Basel) 13
141. Anton-Sales I, D'Antin JC, Fernández-Engroba J et al. (2020) Bacterial nanocellulose as a corneal bandage material: A comparison with amniotic membrane. Biomater Sci 8. https://doi.org/10.1039/d0bm00083c
142. Akaraonye E, Filip J, Safarikova M et al (2016) Composite scaffolds for cartilage tissue engineering based on natural polymers of bacterial origin, thermoplastic poly(3-hydroxybutyrate) and micro-fibrillated bacterial cellulose. Polym Int 65. https://doi.org/10.1002/pi.5103
143. Scherner M, Reutter S, Klemm D et al (2014) In vivo application of tissue-engineered blood vessels of bacterial cellulose as small arterial substitutes: proof of concept? J Surg Res 189. https://doi.org/10.1016/j.jss.2014.02.011
144. Amorim J, Liao K, Mandal A (2024) Impact of carbon source on bacterial cellulose network architecture and prolonged lidocaine release. Polymers (Basel) 16:3021 https://doi.org/10.3390/POLYM16213021/S1
145. Lv X, Yang J, Feng C et al (2016) Bacterial cellulose-based biomimetic nanofibrous scaffold with muscle cells for hollow organ tissue engineering. ACS Biomater Sci Eng 2. https://doi.org/10.1021/acsbiomaterials.5b00259

146. Khalid A, Khan R, Ul-Islam M et al (2017) Bacterial cellulose-zinc oxide nanocomposites as a novel dressing system for burn wounds. Carbohydr Polym 164. https://doi.org/10.1016/j.carbpol.2017.01.061
147. Oliveira TJ, Segato TCM, Machado GP et al (2022) Evolution of bacterial cellulose in cosmetic applications: an updated systematic review. Molecules 27
148. Czaja WK, Young DJ, Kawecki M et al (2007) The future prospects of microbial cellulose in biomedical applications. Biomacromolecules 8
149. Kraisuriyawong P, Kornsuthisopon C, Pavasant P et al (2024) Functionalizable bacterial cellulose composite membrane for guided tissue regeneration. Int J Biol Macromol 268:131655. https://doi.org/10.1016/J.IJBIOMAC.2024.131655
150. AA Alves WE Silva MF Belian 2020 Bacterial cellulose membranes for environmental water remediation and industrial wastewater treatment Int J Environ Sci Technol 17 3997 4008 https://doi.org/10.1007/S13762-020-02746-5/METRICS
151. Boonyaporn W, Phisalaphong M, Painmanakul P (2024) Nanoporous membranes from bacterial cellulose derivatives for water treatment. ACS Appl Nano Mater 7:24508–24521 https://doi.org/10.1021/acsanm.4c04152

Deepak Kumar has completed B. Pharmacy from Galgotias University, Greater Noida. Currently he is pursuing his M. Pharmacy from Galgotias University, Greater Noida. His area of interest includes nano formulations, formulation optimization, cancer nanomedicine, targeted drug delivery system, pharmaceuticals, and green nanotechnology. He has authored 3 book chapters. He has published many good Scopus index papers with reputed publishers like Elsevier and Bentham Science. He has attended more than 10 international conferences.

Rishabha Malviya completed B. Pharmacy from Uttar Pradesh Technical University and M. Pharmacy (Pharmaceutics) from Gautam Buddha Technical University, Lucknow, Uttar Pradesh. His PhD (Pharmacy) work was in the area of Novel formulation development techniques. He has 15 years of research experience and is presently working as Professor in the Department of Pharmacy, School of Medical and Allied Sciences, Galgotias University for past 12 years. He has edited/editing 110 books with Springer Nature, CRC Press, Taylor and Francis, Apple Academic Press, Wiley, etc. His area of interest includes formulation optimization, nanoformulation, targeted drug delivery, artificial intelligence in healthcare, localized drug delivery and characterization of natural polymers as pharmaceutical excipients. He has authored more than 150 research/review papers for national/international journals of repute. He is an editorial board member/reviewer of more than 40 journals. More than 25 patent grants from different countries are in his credit, while 40 patents are published/under evaluation.

Sonali Sundram completed B. Pharm & M. Pharm (pharmacology) from AKTU, Lucknow. She has worked as a research scientist on the project of ICMR at King George's Medical University, Lucknow, after that she joined BBDNIIT, and currently, she is working at Galgotias University, Greater Noida. Her PhD (Pharmacy) work was in the area of Neurodegeneration and Nano formulation. Her area of interest is Neurodegeneration, clinical research, and artificial Intelligence. She has edited 28 books (Wiley, Springer Nature, CRC Press/Taylor and Francis, River Publisher, and IOP Publishing). She has attended as well, and organized more than 15 national and international seminars/conferences/workshops. She has more than 8 patents, national and international, in her credit.

Fundamentals and Applications of Cellulose Polyelectrolyte Complexes

Bruno Solis-Cruz, Raquel López-Arellano, Jorge L. Mejía-Méndez, Abraham Méndez-Albores, and Daniel Hernandez-Patlan

Abstract Cellulose, one of the most bountiful biopolymers worldwide since it is found in plant cell walls, is considered an environmentally friendly and low-cost material compared to other natural, semi-synthetic, or synthetic biopolymers. This biopolymer has been extensively studied in biomedical applications such as tissue engineering (3D printing) and wound dressing manufacturing, as well as in the pharmaceutical industry for drug delivery due to its excellent physicochemical properties, as well as its safety, biocompatibility, biodegradability, biomimetic, and low toxicity. However, polyelectrolyte complexes, which are macromolecular structures obtained from the joining of two or more positively or negatively charged polyelectrolytes, have emerged as a strategy to solve the limitations presented by cellulosic materials and their derivatives by themselves. In this sense, one of the main advantages of using cellulose polyelectrolyte complexes lies in the fact that the possibility of toxicity is

B. Solis-Cruz · R. López-Arellano · D. Hernandez-Patlan (✉)
Laboratory 5: LEDEFAR, Multidisciplinary Research Unit, National Autonomous University of Mexico-Superior Studies Faculty at Cuautitlan (UNAM-FESC), Cuautitlan Izcalli, Mexico State, Mexico
e-mail: danielpatlan@comunidad.unam.mx; daniel.hernandez@upvm.edu.mx

B. Solis-Cruz
e-mail: brunosc@comunidad.unam.mx; bruno.solis@upvm.edu.mx

R. López-Arellano
e-mail: lopezar@unam.mx

B. Solis-Cruz · D. Hernandez-Patlan
Nanotechnology Engineering Division, Polytechnic University of the Valley of Mexico, Tultitlan, Mexico State, Mexico

J. L. Mejía-Méndez
Colegio de Postgraduados, Campus Montecillo, Programa de Edafología, Montecillo, México
e-mail: mejia.jorge@colpos.mx

A. Méndez-Albores
Unidad de Investigación Multidisciplinaria L14-A1 (Ciencia y Tecnología de Materiales), Facultad de Estudios Superiores Cuautitlán (FESC), Universidad Nacional Autónoma de México (UNAM), Cuautitlán Izcalli, Mexico
e-mail: albores@unam.mx

R. Malviya and S. Sundram (eds.), *Sustainable Cellulose Materials for Biomedical Applications*, Sustainable Materials and Technology,
https://doi.org/10.1007/978-981-96-9904-9_4

reduced by using chemical cross-linking agents, as well as improvements in their physicochemical properties. Therefore, this book chapter describes the fundamentals and applications of different cellulose polyelectrolyte complexes in the biomedical and pharmaceutical fields, mainly.

Keywords Cellulose · Polyelectrolytes · Polyelectrolyte complexes · Drug delivery systems · Tissue engineering

Abbreviations

CAc	Cellulose acetate
CAcB	Cellulose acetate butyrate
CD	Charge density
CMC	Carboxymethyl cellulose
CN	Cellulose nitrate
CP	Cellulose propionate
CPC	Cellulose polyelectrolyte complexes
CS	Cellulose sulfate
EC	Ethyl cellulose
HEC	Hydroxyethyl cellulose
HEC-E	Hydroxyethyl cellulose ethoxylate
HPC	Hydroxypropyl cellulose
HPMC-AcS	Hydroxypropyl methylcellulose acetate succinate
HPMC-P	Hydroxypropyl methylcellulose phthalate
IS	Ionic strength
MC	Methyl cellulose
MTT	Thiazolyl blue tetrazolium bromide
MW	Molecular weight
NaCMC	Sodium carboxymethylcellulose
SC	Sodium cellulose
SCS	Sodium cellulose sulfate

1 Introduction

Cellulose is a biopolymer that is structurally formed by repeated glucose residues linked by β-1,4-glycosidic bonds that is characterized by excellent biocompatibility, biodegradation properties, and reduced or zero toxicity [12]. This biopolymer is one of the most bountiful in nature since it is the structural polysaccharide of the cell wall of plants [114]. However, cellulose biosynthesis is not limited to plants (Fig. 1), but it is also known that its production is carried out by a wide variety of bacteria,

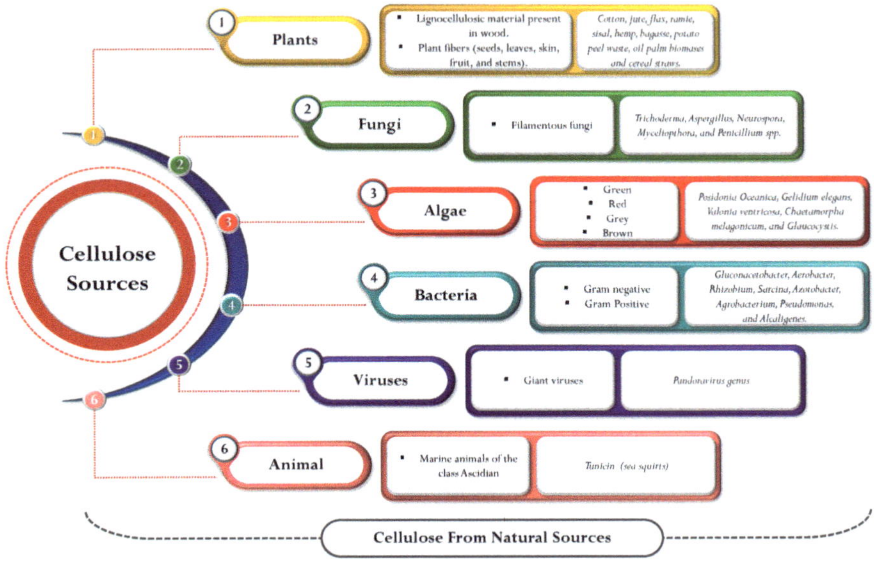

Fig. 1 Classification of the different sources of cellulose biosynthesis

fungi, algae, giant viruses, and even animals [99]. Cellulose can also be obtained synthetically from the polymerization of the 6-di-O-benzyl-α-d-glucopyranose 1,2,4-orthopivalate ring or through chemical reactions with 1-allyl-2,6-di-O-acetyl-3-benzyl-4-O-(p-methoxybenzyl)-β-d-glucopyranoside. Furthermore, a large number of cellulosic derivatives (semi-synthetic cellulose) can be synthesized from chemical modifications of natural cellulose [50].

Despite plants/agro-industrial waste are the main sources of cellulose production due to advantages such as profitability derived from lower production costs, biocompatibility, biomimicry, and lower immunogenicity and toxicity compared to synthetic cellulosic materials [7], it is a fact that ecological production of cellulose, considering global trends, has promoted the employ of green methods that contemplate the use of environmentally friendly solvents [N-methylmorpholine-N-oxide, dimethylacetamide, ionic liquids (IL), deep eutectic solvents, and thiourea], with the objective of reducing not only the environmental impact but also because higher yields have been reported, as well as better biodegradability, stability, and easy recycling characteristics [1, 41, 97].

In addition to the use of green methods for obtaining cellulose, interest in cellulose biosynthesis from biotechnological processes using bacteria has recently increased due to high yields, and its importance is such that a growth of 38% was estimated during 2019–2027, equivalent to USD 777 million [99]. Among the advantages of cellulose biosynthesized by bacteria are higher purity (> 90%), crystallinity (80–90%), high mechanical strength, water retention, chemical stability and degree of polymerization (7000–16,000 glucose residues), biodegradability, biocompatibility, non-cytotoxic, and non-genotoxic compared to cellulose of plant origin, allowing a broader range of industrial applications [2, 111]. The isolation and purification processes are also considered simple and green as they do not require chemical treatments [142].

Even though cellulose obtained from biotechnological processes has better properties compared to that obtained from plants, cellulose is characterized by being non-water soluble and in a large amount of organic solvents, which delimits its employment in the biomedical and pharmaceutical fields [142, 143]. However, chemical modifications in the structure of cellulose, specifically in its hydroxyl groups (C2, C3, and C6 positions) through interaction with some reagents that introduce functional groups have allowed obtaining cellulose ethers [methyl cellulose (MC), carboxymethyl cellulose (CMC), ethyl cellulose (EC), hydroxyethyl cellulose (HEC), and hydroxypropyl cellulose (HPC)] and cellulose esters [cellulose acetate (CAc), cellulose acetate butyrate (CAcB), cellulose nitrate (CN), cellulose propionate (CP), and cellulose sulfate (CS)], but many of these latter derivatives are characterized by being insoluble in water [1].

Therefore, another strategy that has been used to solve the limitations of cellulose solubility is also the use of ionic liquids, which must have low toxicity, good thermal stability and, reduced volatility [90, 142]. Furthermore, in the last two decades, much attention has been paid to supramolecular structures constituted by the interaction of two or more polymeric/biopolymeric compounds (polyelectrolyte complexes or interpolyelectrolyte complexes), especially due to their unique physicochemical characteristics and technofunctionalities compared to those they present by themselves [161]. These polyelectrolyte compounds are formed by the association of polyelectrolytes with opposite charges, that is, must have a positive charge and the other a negative charge (self-assembly), as well as external conditions such as pH, ionic strength (IS), and temperature [64, 83, 102]. Some examples of anionic polyelectrolytes that are characterized by presenting carboxylic, sulfate, or phosphate functional groups are acrylic acid derivatives (carbomer, carbopol), alginic acid, sodium carboxymethyl cellulose (NaCMC), and hyaluronic acid. Polyelectrolytes that can be positively charged include chitosan and synthetic acrylic polymers (Eudragits). These are characterized by having protonated nitrogen atoms in their structures [150]. However, among the most technologically relevant polymers/biopolymers with the potential to form polyelectrolyte complexes are cellulose derivatives [46]. In this sense, this book chapter aims to show the fundamental principles of cellulose polyelectrolyte complexes (CPC) and their main applications in the pharmaceutical and biomedical fields.

2 Cellulose Polyelectrolyte Complexes in the Pharmaceutical Field

Cellulose and its spinoffs are highly used in pharmaceutical products due to their low or non-toxic properties, biodegradability, and biocompatibility. These biopolymers are used as excipients since they allow to improve the flow and compression properties of powders. Likewise, they are also used as super disintegrants and viscous/suspending agents, as well as to control drug release, improve drug solubility by forming solid dispersions, film coatings, and as the main component of capsules (Table 1). Like cellulose derivatives, CPC, which have gained significant ground in recent years in the pharmaceutical field, have been used for film coating, development of micro/nanoparticulate systems (micro/nanospheres and micro/nanocapsules) for the release and protection (increased stability) of drugs, proteins and nucleic acids, and the formation of hydrogels, mainly [31]. Furthermore, some of the advantages of CPC include better controlled release of drugs, greater encapsulation properties which improve therapeutic efficiency and, by presenting excellent mechanical and water resistance properties, they form long-lasting films with a high protective capacity [77].

Table 1 Examples of cellulosic materials applied in the pharmaceutical field

Cellulose derivatives	Pharmaceutical application	References
MCC and cellulose ethers	Improvement in fluidity and compressibility of powders	[135]
Croscarmellose sodium	Super disintegrants	[59]
Cellulose ethers: MC, NaCMC, HEC, and HPMC	Viscous/suspending agents	[74]
Cellulose ethers: EC, HPMC, and HPC	Controlled drug release	[29]
Cellulose esters: CAc, CN, CAcP, HPMC-P, and HPMC-AcS	Film coating	[29]
CAcP and HPMC-P	Development of amorphous solid dispersions (improve drug solubility)	[89]
HPMC, MC CMC, and HEC-E	Main component of the capsules	[42]

MCC microcrystalline cellulose, *MC* methyl cellulose, *NaCMC* carboxymethyl cellulose, *HEC* hydroxyethyl cellulose, *HPMC* hydroxypropyl methylcellulose, *EC* ethyl cellulose, *HPC* hydroxypropyl cellulose, *CAc* cellulose acetate, *CN* cellulose nitrate, *CAcP* cellulose acetate phthalate, *HPMC-P* hydroxypropyl methylcellulose phthalate, *HPMC-AcS* hydroxypropyl methylcellulose acetate succinate, *CMC* carboxymethyl cellulose, *HEC-E* hydroxyethyl cellulose ethoxylate

3 Fundamentals of Cellulose Polyelectrolyte Complexes

3.1 Polyelectrolyte Complex Formation Mechanism

As mentioned previously, polyelectrolyte complexes are higher structures formed by the interaction between polymers/biopolymers with opposite charges, which facilitates their self-assembly given the strong, but reversible, electrostatic interactions [39]. In the specific case of cellulose polyelectrolyte complexes, one of the biopolymers that form the complex is a cellulose derivative. These types of macromolecular structures are obtained spontaneously in aqueous media without requiring the addition of chemical reagents or cross-linking agents [156]. In addition to electrostatic interactions, the formation of polyelectrolyte complexes involves interactions between polymers such as dipolar forces, Van der Waals forces, hydrogen bonds, and hydrophobic interactions. However, factors such as charge density (CD), molecular weight (MW), amount/concentration of polymers (polyelectrolytes), pH, IS, salt concentration (SC), and mode of addition/mixing can affect the stability of these complexes [83].

As shown in Fig. 2, the mechanism by which cellulose polyelectrolyte complexes are formed consists of three main stages: (1) interaction between the polyelectrolyte polymers/biopolymers to give rise to the first primary complex, (2) formation of a diffuse cloud of the polyelectrolytes, which adopt a combination between a disordered structure similar to a scrambled egg and a highly ordered ladder-like structure, and (3) coacervation/agglomeration of the polyelectrolyte chains to give rise to the secondary complex. In this last stage, hydrophobic interactions allow the formation of large fibrils, entanglement, and networks [124, 126]. Furthermore, in the late-stage interactions of polyelectrolyte complex formation, the formation of nanoparticle systems with a core–shell structure can occur., that is, one can serve as a core former and another as a polymer shell former [91].

The method of forming CPC from the mixture of complementary with another polymer/biopolymer in aqueous medium is one of the most frequently used due to its feasibility and reliability, but other methods such as matrix polymerization, interactions at the boundary of liquid compounds and application on solid and flexible surfaces have also been developed [77]. In the case of matrix polymerization, this method considers the conventional synthesis of one of the polymers and offers a viable alternative when other methods are difficult or unattainable. The complexes obtained present excellent stability, and only the polymers can be separated if the conditions favor the modification of the intermolecular bonds [16, 17]. Meanwhile, the method of interactions at the boundary of liquid compounds considers reactions between two immiscible liquids, which contain the complementary polymers separately, and once associated form a thin film [9]. Unlike the previous methods, the solid surface application method involves the deposition of a polymer/biopolymer that is in aqueous solution on a solid surface that may be flexible. The complexes formed in this last method can be soluble or insoluble in water, allowing the thickness of coatings to be controlled to a greater extent [32].

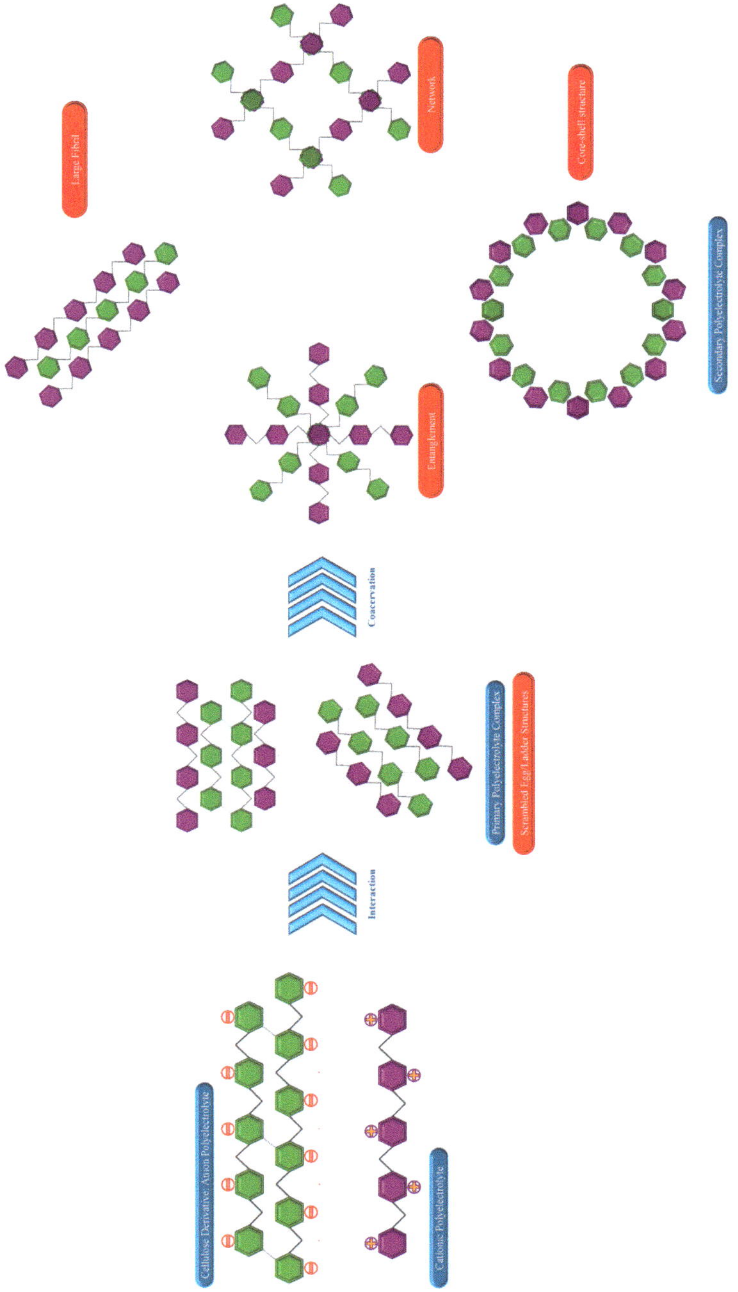

Fig. 2 Stages involved in the formation of cellulose polyelectrolyte complexes (CPC)

Like CPC, other intermacromolecular complexes can be categorized, but these mainly involve mechanisms such as the formation of hydrogen bonds (Complexes of hydrogen bonds), stereoisomerism (complementation of stereoregular polymers), and the electronic transition to an excited electronic state by the polymers (charge transfer complex), in which one acts as an electron donor (Lewis base) and the other as an electron acceptor (Lewis acid), that is, electrostatic type interactions are not involved [19, 77, 134].

3.2 *Factors Affecting the Formation of Polyelectrolyte Complexes*

It has already been described in this chapter that CPC have completely different properties than the polyelectrolytes alone and, in addition, some of their properties can be enhanced (synergistic effect) [125]. However, it has already been described that the formation of polyelectrolyte complexes is affected by aspects such as charge-to-charge stoichiometry, CD, MW, polyelectrolyte amount/concentration, IS, mixing mode/stirring speed, pH, addition order, mixing ratio and salt concentration [73, 83]. Figure 3 shows the structural, preparation, and reaction medium parameters that have a significant influence on the formation of polyelectrolyte complexes.

3.2.1 Charge-to-Charge Stoichiometry

Charge-to-charge stoichiometry can be defined as the total proportion of positive to negative charges of the electrolytes involved [121]. Thus, when the charge of

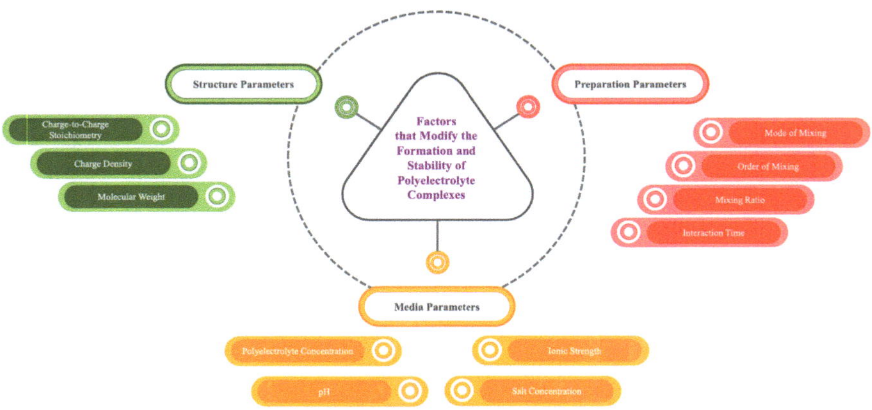

Fig. 3 Parameters at the level of structure, preparation conditions, and reaction medium that influence the formation and stability of polyelectrolyte complexes

both polyelectrolytes is high, the strong interactions lead to a phase transition from liquid to solid, that is, aggregates are formed spontaneously and provide a viable alternative to selectively encapsulate not only small molecules but also biomolecules [88]. Otherwise, when the interactions are very weak, derived from weak charges, polyelectrolyte complexes are formed that can remain in solution [118]. However, it has been described that some weakly charged polyelectrolytes can influence the dissociation behavior of others [3]. If stoichiometry is not considered, polyelectrolyte complexes are produced in the form of colloids with solvation, charge, and hydrophobicity that vary depending on the surface charge [88], but soluble non-stoichiometric complexes can be formed when: (1) the charge on one of the polyelectrolytes is weak, (2) the MW are significantly different, (3) the higher MW component is in excess, (4) the presence of salts, and (5) a narrow range of mix proportions [144].

3.2.2 Charge Density (CD)

Charge density is one of the parameters that has recently generated interest because it can significantly affect the phase behavior and properties of polyelectrolyte complexes [62]. In this sense, it has been reported that when the CD of polyelectrolytes is reduced, the resistance of the complexes to salt is reduced and their relaxation dynamics is accelerated [61]. Furthermore, structural, compositional, and stoichiometry changes of the polyelectrolyte complexes occur [61], but when the charge density is doubled, a liquid-to-solid phase transition can be promoted [146]. However, such behavior is frequently observed in polyelectrolytes with hydrophilic side chains, since for polymers with hydrophobic side chains there is a critical CD that is dominated by hydrophobic interactions rather than CD [61]. In addition to salts, pH plays an important role in the interaction between polyelectrolytes due to the effect it has on charge density, and in fact, it has been reported that the interaction of polyelectrolyte complexes with biomolecules such as proteins to achieve higher encapsulation efficiency depends largely on the surface charge distribution [118].

3.2.3 Molecular Weight (MW)

Like the type of polyelectrolyte of choice for the development of polyelectrolyte complexes, the MW of the polyelectrolytes has a great influence on the formation of the complexes and their surface properties [149]. In this sense, the MW can control whether the formation of the polyelectrolyte complex is exponential or linear, so, while high MWs favor a linear growth of the polyelectrolyte complexes, low MWs do so exponentially because their diffusion coefficients are higher, which allows them to interpenetrate at a higher speed and even the formation of surface complexes [129, 149]. Furthermore, it is known that low MW polyelectrolytes favor the formation of more stable polyelectrolyte complexes with smaller particle size compared to high MW polyelectrolytes, but it has been reported that low MW polyelectrolytes tend to present lower adsorption levels [125], obviously with their exceptions since the effect of MW depends on the IS [137].

Although the effects of polyelectrolyte molecular weight on the properties of the formed polyelectrolyte complexes have been described previously, factors such as MW and CD directly affect the stability of polyelectrolyte complex flocs since the interaction of low MW polyelectrolytes with high CD promotes the formation of fragile flocs due to shearing and are sensitive to the polyelectrolyte concentration. Otherwise, low CDs and high MW of the polyelectrolytes lead to flocs that are more resistant to shear force because they can bind more particles by forming bridges [136].

3.2.4 Polyelectrolyte Concentration

Polyelectrolyte concentration can promote phase changes, as well as affect the formation of polyelectrolyte complexes by reducing the dissociation capacity of polyelectrolyte functional groups, which results in an incomplete formation of complexes or aggregation phenomena is minimized when the polyelectrolyte concentration is relatively low [102]. High polyelectrolyte concentration can lead to high electrostatic repulsion between molecules, thus the homogeneity/uniformity of polyelectrolyte complexes can be affected [81], but the adsorption of polyelectrolytes in relatively short times is favored [118].

3.2.5 Ionic Strength (IS)

It has been reported that IS promotes the formation of secondary aggregates and the higher the IS, the size of the aggregates increases significantly, as well as multilayer thickness and roughness of the polyelectrolyte complex given the strong interdigitation of the coiled chains [38, 80]. However, with this parameter, the thickness of the layers of the polyelectrolyte complex and its absorption capacity can be controlled [63]. Furthermore, the increase in IS promotes the precipitation of soluble polyelectrolyte complexes due to salting-out effect, so turbidity increases and viscosity decreases. On the contrary, when the ionic strength decreases, so do the electrostatic interactions and the formation of complexes is affected, as well as the turbidity and viscosity, which decrease and increase, respectively [34].

3.2.6 Mode of Mixing

Mixing cationic and anionic polyelectrolyte solutions to form polyelectrolyte complexes may seem straightforward, but it is necessary to consider the type and protocol of mixing, and the devices to achieve its association. However, it has been described that the molecular weight of the polyelectrolytes plays an important role; therefore, if the molecular weight is kept constant and only the association/stirring speed varies, smaller polyelectrolyte complexes are obtained when the stirring speed is high and when the association/stirring speed is low, two stages are carried out: (1) the particle size is reduced to a minimum (precomplex formation), and (2) equilibrium is achieved and the size increases, being able to favor aggregation in some cases [83, 84].

3.2.7 pH

According to their ionization capacity, polyelectrolytes can be classified as weak or strong, that is, strong polyelectrolytes ionize easily and the pH is not so critical, but the ionization of weak polyelectrolytes depends largely on the pKa [103]. Therefore, modifying the pH of the medium produces changes in the charge, solubility, conformation of the molecules, and the intermolecular interactions of the polyelectrolytes, which can affect or favor the formation of polyelectrolyte complexes [109, 118].

3.2.8 Order of Addition

One of the most sensitive and critical preparation parameters of polyelectrolyte complexes is the order of addition of the polyelectrolytes since it affects the particle size and stability [91]. In theory, the smallest particle size is achieved by adding the minor component to the major component. Furthermore, the fast addition rate also favors the formation of smaller and more stable polyelectrolyte complexes compared to slow addition rates (drop by drop), so it could be said that the particle formation process is kinetically controlled [83].

3.2.9 Mixing Ratio

The mixing ratio of polyelectrolytes is another parameter of preparation of polyelectrolyte complexes that is used to qualitatively evaluate their particle size considering the turbidity of the medium [73].

3.2.10 Salt Concentration

The salt concentration significantly influences the formation of polyelectrolyte complexes, so their particle size is affected. However, particle size can be modified according to the type of polyelectrolyte complex formed, since the increase in salt concentration can favor the aggregation of primary complexes, giving rise to larger secondary complexes, or it can also favor smaller particle sizes [83, 118].

4 Preparation Methods for Polyelectrolyte Complexes

Some factors that have a significant influence on the preparation and even stability of polyelectrolyte complexes were described above, but it is a fact that the selection of the preparation method is also essential to obtain polyelectrolyte complexes with particular characteristics. Polyelectrolyte complexes are generally formed at room temperature starting mainly from independent solutions of the polyelectrolytes (cationic and anionic), followed by the addition of the anionic polyelectrolyte solution in a constant and slow manner (drop or drop) to the solution containing the cationic polyelectrolyte or in a rapid manner, that is, in a single stage under constant stirring, but both polyelectrolyte solutions must have the same ionic strength [83]. However, the above procedure, which is part of the bottom-up methodologies (emulsions and emulsion polymerizations, amphiphiles self-assembly and polymerization induced self-assembly) [30], is difficult to standardize and scale up to an industrial level, so top-down methods, such as ultrasound, high-pressure homogenization, and ultra-high pressure homogenization, which lead to the reduction of particle size through the application of high energy, are most frequently used because the particle size and its homogeneity (polydispersity index) are consistent; in addition, they are scalable at an industrial level and technology transfer is not complicated [106]. Although top-down methodologies have advantages compared to bottom-up ones, the excess energy applied in top-down methodologies has the disadvantage of inducing the chemical degradation of polyelectrolyte complexes [106].

Then, the preparation methods of polyelectrolyte complexes consist of: (1) bottom-up methods as self-assembly, dip coating, spray coating, spin coating, and electrodeposition, which are considered as deposition methods; (2) top-down methods as jet mixing, high-pressure homogenization, ultra-high pressure homogenization, ultrasound; and (3) hot-melt extrusion [5, 129].

4.1 Bottom-Up Methods

4.1.1 Self-assembly

This method is mainly used to form multilayer polyelectrolyte complexes such as membranes, microcapsules, and nanoparticles through electrostatic interactions, hydrogen bonding, hydrophobic interactions, and Van der Waals forces [157]. Extremely stable structures with regular geometry due mainly to non-covalent interactions, but this method is sensitive to parameters of the reaction medium such as polyelectrolyte concentration, pH, and IS, as well as temperature [60]. In this method, it is necessary to prepare the polyelectrolyte solutions separately and then associate them through mixing processes [92]. Some of the advantages of this technique are that it allows the thickness of the layers to be controlled, and it is low cost and can be scaled industrially [11].

4.1.2 Dip Coating

The dip coating method is one of the first-choice methods for obtaining multilayer polyelectrolyte complexes and involves immersion of the substrate in a polyelectrolyte solution, followed by several rinses and then immersion in a polyelectrolyte solution with the opposite charge to the first solution [160]. The deposition of polyelectrolytes is produced through electrostatic interactions and entropic driving forces that allow the formation of thin layers of the polyelectrolyte (0.5–5 nm) on layers with opposite charges [130]. Furthermore, dip coating is a scalable process that allows thin films to be formed on a wide variety of substrates and, although large volumes of polyelectrolyte solutions are required, its performance is high and does not require specialized equipment [148, 160].

4.1.3 Spray Coating

The spray coating method is another polyelectrolyte deposition method that can be easily scaled to industrial level for coating application since it is not limited to the substrate size, its manufacturing is low cost and requires low amounts of polyelectrolyte [6, 51]. This method consists of spraying the polyelectrolyte on a substrate, followed by a period of rest, washing and rest again, and then repeating the same operation with the oppositely charged polyelectrolyte [160]. The formation of polyelectrolyte complex is rapid due to the strong interaction of the charges involved in each layer and does not necessarily depend on diffusion processes [57]. Spray coating is also faster and more efficient than immersion coating because it covers a larger surface area. However, there is a risk of forming non-uniform coatings with a lower yield, which can even be 1% [160].

4.1.4 Spin Coating

Polyelectrolyte complexes obtained by precipitation or layer-by-layer obtained in aqueous media have driven technological advances in the medical/pharmaceutical field. As briefly described, multilayer polyelectrolyte complexes are characterized by forming thin and continuous coatings. However, the speed of formation of the layers often depends on the combination of polyelectrolytes, and can be exponential and achieve multilayer polyelectrolyte compounds in the order of micrometers or linear when they are formed below the glass transition temperature of the complex, so their sizes are smaller [78]. In this sense, the deposition rate can be significantly improved with other methods such as spin coating, where the transport of the polyelectrolyte to the surface is carried out by spinning the substrate upside down in the polyelectrolyte solution and then spinning the substrate upside down while the solution is dropped or sprayed on it. This type of spin coating is characterized by rapidly producing narrow, uniform, and highly elastic quality multilayers, but is limited to small substrates with flat surfaces and produces more waste than dip coating [69, 78].

4.1.5 Electrodeposition

Currently, special emphasis has been placed on materials that can modify their properties in response to stimuli such as changes in pH, IS, temperature, application of an electric field or ultraviolet light, as well as electrochemical stimuli, and thus have important fields of application in materials science, biomedicine/pharmaceuticals, sensing, and electronics [47]. The electrodeposition method is based on the application of an electrochemical potential that modifies the interfacial environment of the polyelectrolyte in solution derived from oxidation–reduction reactions that cause its precipitation, followed by its electrophoretic deposition on the working electrode [47, 107]. This electrochemical method is attractive because it is easy to achieve an adequate deposition rate and control the thickness of the films, but the high potentials applied to the electrode can lead to electrolysis of water, which can affect the deposition [133]. Furthermore, the deposition method is applied when the substrate surface is non-planar or porous, while on planar surfaces, spin coatings of molecular precursors and their subsequent thermal or photonic cross-linking are preferred [153].

4.2 Top-Down Methods

4.2.1 Jet Mixing

Mixing is one of the most important operations when process kinetics are faster than mixing time because the mixing/stirring rate directly impacts particle size and homogeneity as well as the formation of intermediates. Jet mixing is a method for the obtaining of polyelectrolyte complexes that is based on the collision of two collinear jets at high velocities within a small volume [35]. This method has superior advantages to conventional methods where stirring tanks, static mixers, or microfluidic mixers are used, since its performance is superior and it is also highly scalable at an industrial level, making it one of the most viable options for obtaining polyelectrolyte complexes [35, 141]. In addition, in this method, the most important parameters to control are: mixing time, jet velocity, jet diameter, jet location, jet angle, tank, and liquid characteristics [154]. Larger particle sizes of polyelectrolyte complexes are obtained when the mixing time is too short, so the mixing time is critical to achieve smaller particle sizes [5].

4.2.2 High-Pressure/Ultra-High Pressure Homogenization

This methodology was based on high-energy dispersion processes that lead to the disaggregation of coarse systems as they pass through a micrometric diameter nozzle, causing high turbulence, shear, and cavitation [139]. The advantages of this methodology include its easy reproducibility and scalability at an industrial level, its relatively low price, and eco-friendly, as well as the obtaining of more homogeneous micrometric and even nanometric systems [138]. However, one of its major disadvantages is the chemical degradation of the polyelectrolytes/components with biological activity derived from the excess energy applied to the system [54, 106].

4.2.3 Ultrasound

Another methodology used to increase the stability of polyelectrolyte complexes is the use of ultrasound (non-thermal physical processing), which is based on the mechanical effects of sound waves and bubbles in liquids to cause sonochemical reactions (acoustic cavitation) that result in a reduction in particle size [26]. Although the cavitation performance of ultrasound at a single frequency allows obtaining reproducible and homogeneous polyelectrolyte complexes in terms of size, it has been described that the use of ultrasound at multiple frequencies has a synergistic effect in reducing particle size [26]. However, it is important to consider the fields of application of polyelectrolyte complexes since the frequency of the ultrasonic waves and the power have a direct impact on the propagation medium [105]. Furthermore, it is known that ultrasound can cause structural and chemical changes on polyelectrolyte complexes derived from the degradation of polyelectrolytes or the breaking of polymeric chains, the latter resulting in the formation of macroradicals, which can be useful to promote the coupling of polymers from radicals [10]. The advantages of this method include its high efficiency, energy savings, and eco-friendly [26].

4.3 Hot-Melt Extrusion

Hot-melt extrusion is a pharmaceutical process that can be used in continuous mode, which is carried out in solid phase at temperatures higher than the glass transition temperatures of polyelectrolyte materials and consists of a rotating screw. In some cases, melting temperatures can even be used to achieve mixtures not only of polyelectrolytes but also of other excipients such as binders and active components, at a molecular level [119]. The technology used for hot-melt extrusion is considered robust, effective and environmentally friendly since it does not require solvents, making it one of the first options at the industrial level for the development of conventional (granules, tablets, capsules, films) and innovative (inhalable powders, nanoextrudates, complexation, and amorphous solid dispersions) pharmaceutical forms containing low-solubility active components [120]. However, the polymers that can be pharmaceutically used in hot-melt extrusion are restricted to Carbopol, NaCMC, HPMC-AcS, some Eudragits, or their combinations, derived from their excellent thermoplastic and physicochemical properties (melt viscosity, glass transition temperature, thermostability), so they can improve the properties of polymeric matrices given the new interactions that can be taken with other excipients or active components [150]. Although it is a fairly popular method in industrial terms, one of its disadvantages is that the polyelectrolytes must have high thermal stability (thermostable) and good physical and chemical stability during processing and prolonged storage [119].

5 Methods for Characterization of Polyelectrolyte Complexes

The characterization of polyelectrolyte complexes must always be comprehensive, that is, it must be supported by techniques that allow establishing whether their formation was satisfactory and whether they present improved physicochemical and mechanical properties compared to the polyelectrolytes separately. Table 2 shows some of the most frequently used techniques for the characterization of polyelectrolyte complexes [5, 36, 77, 83, 102, 116].

6 Applications of Cellulose Polyelectrolyte Complexes

One of the main applications of cellulose polyelectrolyte complexes in the pharmaceutical field is in drug delivery since they can be used to develop systems that allow the association/encapsulation of small molecules with therapeutic activity, as well as therapeutic biomolecules of low stability such as genes, peptides, proteins and antigens [65, 128]. Likewise, this type of complexes can facilitate the transport of molecules with therapeutic activity and control the release rate, increase their lifetime, and even direct/make specific a pharmacological treatment [39]. Therefore, through CPC, different types of conventional and non-conventional delivery systems as microparticles, nanoparticles, hydrogels, nanogels, multilayer films, capsules, and tablets can be developed [83]. Below are summarized some cellulose polyelectrolyte complexes considering their method of production and application in the pharmaceutical field (Table 3).

7 Cellulose-Based Polyelectrolyte Complex for Biomedical Applications

Cellulosic polymers are a very encouraging group of biomaterials in the biomedical area mainly because of their diversity and number of applications. Cellulose is a biopolymer whose properties such as biocompatibility, biodegradability, water holding capacity, renewability, abundance, renewable, zero toxicity, excellent

Table 2 Techniques for characterization of polyelectrolyte complexes and applications

Characterization technique	Importance/application
Viscometry	Indicative of the formation of the polyelectrolyte complex considering the pH
Potentiometric methods	These methods estimate ionic activity by using electrodes; then pH, pKa, and degree of ionization can be evaluated
Ultracentrifugation	Useful method for estimating the molecular weight of polyelectrolyte complexes in a range of 10^2–10^8 g/mol, as well as the sedimentation rate
Turbidity measurement	Simple evaluation that allows to assess the dispersion of polyelectrolyte complexes, as well as their stability
Fluorometric methods	Monitoring polyelectrolyte complex formation
Dynamic light scattering	Determination of the hydrodynamic size of the polyelectrolyte complex and its homogeneity through the polydispersity index
Electrophoretic light scattering	Estimation of the surface charge of the polyelectrolyte complex
Microscopic techniques: Scanning electron microscope (SEM) Transmission electron microscopy (TEM) Scanning force microscopy (SFM)	Evaluation of the morphology of polyelectrolyte complexes
X-ray diffraction	They indicate the degree of crystallinity of the polyelectrolyte complex and even the homogeneity of the polyelectrolytes
Fourier transform infrared spectroscopy	From the molecular vibrations of the characteristic functional groups of each polyelectrolyte, it can be established whether the polyelectrolyte complex was formed as a result of their interaction, which is why changes in the intensity or pronounced broadening of the corresponding bands are observed
Nuclear magnetic resonance (NMR) spectroscopy	Evaluates the chemical composition of the polyelectrolyte complex and provides information on its molecular dynamics
Thermal methods: Differential scanning calorimetry (DSC) Thermogravimetric analysis	Useful for evaluating the thermal stability of polyelectrolyte complexes. Furthermore, information such as glass transition temperature, crystallization temperature, melting temperature, and degradation or decomposition temperature can be obtained to establish the type of application

mechanical properties, economical production (compared to synthetic biopolymers), and tunability which have made it an ideal biomaterial for biomedical applications [12, 142]. Therefore, cellulosic biomaterials and their spinoffs have been employed in wound dressings, artificial blood vessels, blood purification, tissue engineering, 3D bioprinting, and biosensors [8, 27, 49, 58].

However, for a polymeric material to be used efficiently in the biomedical field, certain characteristics become indispensable, such as the biocompatibility of these biomaterials, and the effect produced when they come into contact with different cell

Table 3 Pharmaceutical applications of different cellulose polyelectrolyte complexes

Polyelectrolyte 1	Polyelectrolyte 2	Active compound	Method of obtaining	Application
NaCMC	Chitosan	–	Magnetic stirring/slow addition rates (drop by drop)	Antimicrobial agent (gram-positive and negative) [159]
CMC	Partially reacetylated chitosan	–	Titration	Biocompatible and biodegradable drug delivery systems [15]
NaCMC	Gelatin	Isoniazid	–	Controlled-release microparticles [73]
CS	Poly(L-lysine)	Rifampicin and bisphosphonate risedronate	–	Drug delivery system with controlled loading and enhanced release performance for bone therapies [73]
CMC	Quaternized cellulose	Plasmid DNA	Magnetic stirring	Vaccine delivery [145]
CMC	Chitosan and alginate	Doxorubicin	Self-assembly	Drug delivery vehicle based on multilayer microcapsules [164]
NaCMC	Chitosan	–	Self-assembly	Controlled-release of drugs from microcapsules [132]
HPMC	Eudragit L100/ Eudragit S100	Griseofulvin	Co-spray-drying	Solid dispersion for improved dissolution profile [115]
HPMC-AcS	Soluplus®	Carbamazepine	Hot-melt extrusion	Improves solubility and physical stability [4]
HPMC-AcS	Eudragit® EPO	Acetaminophen	Physical mixing in organic medium	Novel controlled release tablets [70]
HPMC-P	Eudragit® EPO	Acetaminophen	Physical mixing in organic medium	Novel controlled release tablets [70]

(continued)

Table 3 (continued)

Polyelectrolyte 1	Polyelectrolyte 2	Active compound	Method of obtaining	Application
NaCMC	Lysine/ Meglumine	Fenofibrate	Solvent evaporation—hot-melt extrusion	Increased solubility of poorly water-soluble drugs [37]
EC	Hydroxypropyl methylcellulose/ Gelucire® 44/14	–	Hot-melt extrusion	Matrix system for sustained release of zero-order drugs [101]
HPC	Hydroxypropyl methyl cellulose	Lidocaine	Hot-melt extrusion	Controlled release of drugs from films [131]
CMC	Chitosan	Vancomycin	Spray-drying	Microparticles for colon drug delivery [22]
SCS	Calcium alginate/ chitosan hydrochloride	–	Self-assembly	Vehicles for sustained drug release [157]
CMC	Magnetic chitosan	Penicillins, macrolides, aminoglycosides, rifampicins	Precipitation	Antibiotic delivery vehicles [52]
CMC/cellulose nanocrystals	Hydroxypropyl β cyclodextrin	Neohesperidin/ copper (II)	Magnetic stirring	Drug delivery from hydrogels [158]
SC	Chitosan	–	–	Films as drug carriers for oral controlled release [102]
CMC	Chitosan	Clarithromycin	–	Gastric-specific delivery [102]
CMC	Chitosan	Chlorhexidine digluconate	Magnetic stirring/spin coating	Vaginal delivery system with antimicrobial activity [14]
CMC	Quaternized hydroxyethyl cellulose ethoxylate	–	Magnetic stirring (layer-by-layer assembly)	Controlled drug release systems [42]

(continued)

Table 3 (continued)

Polyelectrolyte 1	Polyelectrolyte 2	Active compound	Method of obtaining	Application
CMC	Chitosan	Dexamethasone sodium phosphate	Magnetic stirring	Nanoparticulate ocular delivery system for sustained drug release and increased bioavailability [112]
CMC	Chitosan	–	Magnetic stirring (layer-by-layer assembly)	Sustained-release drug delivery microspheres to reduce fluctuations in blood drug levels [48]
CMC	Chitosan	Curcumin	Magnetic stirring	Cryogels for sustained delivery of anti-inflammatory drugs [40]
CMC	Chitosan	Simvastatin	Magnetic stirring	Increased dissolution rate due to improved wettability [96]

NaCMC sodium carboxymethyl cellulose, *CMC* carboxymethyl cellulose, *CS* cellulose sulfate, *HPMC* hydroxypropyl methylcellulose, *HPMC-AcS* hydroxypropyl methylcellulose acetate succinate, *HPMC-P* hydroxypropyl methylcellulose phthalate, *EC* ethyl cellulose, *HPC* hydroxypropyl cellulose, *SCS* sodium cellulose sulfate, *SC* sodium cellulose

lines, since for many biomedical applications, the biomaterials remain in contact for a long time with different biological tissues and should not cause cytotoxic or other adverse effects.

In this regard, cellulose, as well as most biopolymers of biological origin, has demonstrated unique characteristics concerning biological environments [56]. These characteristics including low cytotoxicity, biodegradability, absorbability, non-accumulative and negligible immunogenicity (which results in non-allergic and non-rejectable properties), in addition to its lack of reactivity with biological fluids, have allowed cellulosic polymers and their spinoffs to be one of the main choices of biomaterials for biomedical applications [110, 117]. All these characteristics depend largely on the physicochemical properties of the cellulosic biomaterials, such as pH, temperature, mechanical properties, hydrophobicity, and hydrophilicity, but also on the characteristics of the biological environment with which they interact [33, 93]. their physical properties, surface chemistry, and biological properties can be improved to develop cellulose-based biomaterials with potential biomedical applications. [45, 147]. Additionally, cellulose-based biomaterials can be found in multiple structural types, from universal macroscopic forms such as membranes, hydrogels, scaffolds, liquids, fibers, and powders, to nanoscale structures such as nanofibers, nanocrystals, and bacterial cellulose [27, 45, 110, 159].

There are various methods for the synthesis of cellulose-based biomaterials for biomedical applications, however, it is the polyelectrolyte complexes formation by chemical cross-linking one of the most prevalent methods [65, 102, 110, 159]. CPC, like many other polyelectrolytes, have been used for the development of polymeric biomaterials with promising applications in the biomedical area. In recent years, CPC have been one of the most studied biomaterials because they structurally contain reactive and ionizable groups, which favors their interaction with biological systems such as cells, proteins, and biological fluids [69, 86]. In addition to the functional groups of the polyelectrolytes, there are more properties related to their chemical structure, such as MW and stereochemical fitting, and other conditions that also influence their behavior, such as concentration before mixing, ratio of mixing, and mixing order. Furthermore, to form a polyelectrolyte complex it is important to know the degree of polymerization (important for compatibility between the associated polyelectrolytes), the IS of the medium, the reaction temperature, and the pH of the medium, since the reaction is influenced by all these conditions [18, 127, 140]. With all these conditions under control, the employment of different chemical and synthetic cross-linking agents is circumvented, which reduces the possible toxicity effect and other harmful effects that these cross-linking agents can cause, and it becomes a key advantage of polyelectrolyte complexes in the biomedical field since they hold superior biocompatibility concerning other biomaterials. As mentioned above, many cellulosic biomaterials have been used for biomedical purposes, and cellulose polyelectrolyte complexes also have been widely used in this field since they have multiple applications according to their ionic interactions, from tissue engineering to gene therapy [71, 123, 145], nevertheless, the applications of polyelectrolyte complexes differ dramatically depending on the properties and morphologies of associated polyelectrolytes, the balance of water, type of polymer, salt ions within the complex, among others.

It is well known that cellulose is a natural polysaccharide insoluble in water, so it is necessary to cross-link it to form a hydrogel network [28, 79]. Polyelectrolyte complexes can be classified into different categories according to the properties of the constituent polyelectrolytes; hence, cellulose polyelectrolyte complexes can be considered natural and amphoteric polyelectrolytes that can be used as hydrogels, films, hydrocolloids, microparticles and nanoparticles for biomedical applications, given the biocompatibility and biological activities of cellulose, as well as the ease of responding to external stimuli that can be chemical or physical such as pH, IS, solvent polarity, and external electric field [18, 65]. Furthermore, CPC have been employed in controlled drug release systems, cell delivery carriers, wound dressings, tissue adhesives, and tissue engineering scaffolds [71, 100, 113, 123, 145].

Undoubtedly, the most common form of CPC for biomedical applications is as hydrogel, since these are biocompatible with living cells and growth factors, as well as viable to prepare under mild conditions, non-toxic to cells and tissues, biodegradable, hydrophilic, and thermally and chemically stable [76, 151]. Hydrogels are characterized by absorbing large amounts of water or biological fluids, providing a microenvironment with biomimetic properties that serve as scaffolds and support the recovery of damaged tissue [20, 23].

The use of CPC has evolved the biomedical field of functional hydrogel systems. In this regard, several scientific research have investigated the use of CPC for tissue engineering. CPC are insoluble in water, have a remarkable tissue biocompatibility, and can be obtained by the association of cellulose derivatives with other oppositely charged polyelectrolytes in aqueous solution. Nevertheless, it has been shown that the water content of CPC is not so high as compared to those of polyelectrolyte complexes formed between cellulose derivatives and synthetic polymers [104].

One of the first uses of CPC for biomedical purposes was CPC formed from two types of quaternary ammonium cellulose spinoffs (cellulose and HEC) as polycation components, while CMC and CS were employed as anionic derivatives [66, 67]. In this research, the authors examined the blood compatibility of these CPC and the effect of their excess charge on the blood compatibility both in vivo and in vitro blood tests. The results showed excellent blood compatibility for these polyelectrolyte complexes, however, excess charge had an effect independent of the molar ratio of the polycation to polyanion component on the relative blood clotting time, but platelet adhesion increased with increasing molar ratio of polycation/polyanion in the polyelectrolyte complexes, and activation of the intrinsic coagulation system increased with decreasing molar ratio.

As mentioned above, it is difficult to achieve the best performance of the polyelectrolyte complexes for biomedical purposes only using cellulose derivatives. So, many of the biomedical applications for cellulose polyelectrolyte complexes have been developed by the combination of cellulose polyelectrolytes and other types of polyelectrolytes of different natures. Such is the case of the development of a compatible mixture of polyelectrolyte complexes based on CMC and chitosan, to achieve a biomaterial that meets various requirements to be used in cartilage tissue engineering. In this study, ultrasound was applied as a method to prepare polyelectrolyte complex mixtures to improve the stability and mechanical properties through the coupling of interpolymer macroradicals produced by sonochemical reactions [10]. The results of this research demonstrated that the scaffolds obtained by applying this methodology turned out to have improved properties compared to the simple mixture of the two polysaccharides, that is, applying ultrasound during the polyelectrolyte complexes preparation resulted in a more homogenous surface and pore size distribution. Moreover, no evidence of cytotoxicity was determined by macrophage cells, which suggests that this biomaterial could be useful in the regeneration of the connective tissue that forms cartilage.

On the other hand, CPC also have been investigated as biomaterials for wound dressing and skin repair. An example of this is a study carried out using a green aqueous process to develop hydrogel membranes composed of a carboxymethyl-functionalized cellulose derivative, chemically cross-linked with citric acid and modified with polyethylene glycol to fine-tune its properties [21]. The results of the study demonstrated that superabsorbent hydrogels were produced with distinct morphological features and different degrees of swelling, ranging from 100 to 5000%, depending on the degree of CMC substitution, the extent of citric acid cross-linking and the addition of polyethylene glycol. Furthermore, the nanomechanical properties of the surface were drastically affected due to the formation of carboxymethyl-polyethylene glycol hybrid nanostructures. These cellulose-based hydrogels did not exhibit cytotoxicity considering the in vitro cell viability responses of more than 95% toward human embryonic kidney cells using the in vitro MTT (Thiazolyl blue tetrazolium bromide) assay. Therefore, they are considered high-efficiency adsorbent hydrogels with key properties to help skin wound healing and regeneration. Cross-linked polyelectrolyte complex membranes based on gelatin and CMC have been also developed. For this, different conditions were evaluated to determine the proper mix ratio of these polymers, the concentration of the cross-linking agent, and film manufacturing conditions, while membrane properties and permeability of solutes were assessed. The cross-linked polyelectrolyte complex membranes showed suitable mechanical strength, high solute permeability, hydrophilicity, biocompatibility, non-toxicity, and non-pyrogen for dialysis applications [85].

Bacterial cellulose is an organic compound produced by certain types of bacteria, that has shown different morphology, structure, properties, and applications [152]. For biomedical purposes, this cellulosic polymer has also been used as a cross-linked polyelectrolyte complex for wound dressing applications due to its excellent water retention capacity, great conformability, good porosity, high crystallinity, and high mechanical strength. In recent research, bacterial cellulose hydrogel dressings were developed in combination with silver nanoparticles with curcumin for wound-management applications [53]. Results of this study demonstrate that the bacterial cellulose hydrogel, loaded with silver nanoparticles and curcumin, turned out to have a broad-spectrum antimicrobial activity along with antioxidant properties, and cytocompatibility with several cell lines.

Although bacterial cellulose has unique physical and mechanical properties, it lacks natural antibacterial properties; thus, to widen its biomedical applications, it is almost always used in combination with other biomaterials and bioactive molecules. Silver sulfadiazine, copper (II) sulfate, zinc oxide, collagen, and curcumin also have been incorporated into bacterial cellulose hydrogels, generally in combination with other polymers to enhance their performance and functionality [43, 75, 108, 155, 165].

Cellulose polyelectrolyte hydrogels are considered as up-and-coming candidates for the development of wound dressings due to their ability to maintain a moist microclimate at the wound site, which has proven effective as a conductive cell and facilitator of the healing process. In another investigation, a bacterial cellulose hydrogel enriched with thymol was developed to address the complications in third-degree burns [72]. The burn wound healing efficiency of the cellulose hydrogel was evaluated in a rodent model, observing faster wound closure, and suggesting that this hydrogel can be used as a resourceful and natural burn wound dressing material. Cellulose polyelectrolyte hydrogels have demonstrated high biocompatibility and low cytotoxicity, as well as bio-integration, stability, and resistance to degradation and elimination routes in different tissues [33, 82, 122].

Hydrogels based on cellulose also have been used as cell culture scaffolds for different purposes. Although cellulose is usually combined with proteins, other polysaccharides, or both in combination [55, 95, 163], it. Has been also used in combination with inorganic compounds such as hydroxyapatite, to be applied for bone and cartilage tissue engineering [44, 94, 113]. The structural properties of these scaffolds, which have a similar structure to natural extracellular matrices, have shown appropriate rheological properties, cellular biocompatibility, cellular polarization, and differentiation to several cell lines [24, 56, 57, 68]. Recently, a study to demonstrate the feasibility of plant-derived native nanofibrillar cellulose hydrogels as 3D-cell culture scaffolds was carried out [13]. In this study, the synthesized hydrogels had the desired functionalities, such as rheological properties that allow the formation of a 3D-scaffold in situ after facile injection, cellular biocompatibility without added growth factors, cellular polarization, and differentiation of human hepatic cell lines.

In summary, several CPC have been developed and evaluated for a wide variety of biomedical purposes. All these investigations have shown that the biocompatibility, the simple preparation, the mechanical properties, and the unique characteristics of these biomaterials, so that they become one of the main options for biomedical applications, receiving increasing attention in recent years. Below are summarized some cellulose polyelectrolyte complexes considering their method of production and application in the biomedical field (Table 4).

8 Conclusion

Cellulose and its derivatives are extensively used in the biomedical/pharmaceutical field due to their properties of zero toxicity, biodegradability, and biocompatibility. However, the use of CPC emerged as a strategy to improve the physicochemical and mechanical properties of these materials, thus allowing to extend the applications of cellulosic base materials in: (1) tissue engineering; (2) cell regeneration; (3) development of implants and scaffolds; (4) modified drug release systems; (5) stabilization of dispersed systems (emulsions and suspensions) and hydrogels; (6) improvement in the stability of biomolecules; and (7) enzymatic immobilization.

Although the number of applications of cellulose polyelectrolyte complexes has expanded, there is still an important field to be explored, starting with microbial methods for obtaining cellulose, since it has been described that this type of cellulose has very different properties to those obtained from sources such as plants or synthetically. Therefore, the fields of application of cellulose polyelectrolyte complexes can be extended. Finally, it is important to consider that nanotechnology also offers new applications for cellulose polyelectrolyte complexes since their physicochemical and mechanical properties are modified mainly due to their low particle size.

Table 4 Biomedical applications of different cellulose polyelectrolyte complexes (CPC)

Polyelectrolyte 1	Polyelectrolyte 2	Active compound	Method of obtaining	Application
Quaternary ammonium derivatives of HEC and cellulose	Aminoethyl cellulose Diethylaminoethyl cellulose Carboxymethyl cellulose Cellulose sulfate	–	By mixing oppositely charged cellulose derivatives in formic acid solutions at pH 7.4	Blood applications [66]
CMC	Chitosan	–	Sonochemically produced interpolymeric macroradical coupling	Cartilage tissue engineering [10]
CMC	Polyethylene glycol	No active compound	Green aqueous process using citric acid as cross-linking agent	Skin wound healing and regeneration [21]
DCMC	Gelatin–polyethylene glycol composite	No active compound	Gel-spinning process	Wound dressings [87]
CMC	Gelatin	No active compound	Mixture of CMC and gelatin, followed by cross-linking with glutaraldehyde	Dialysis applications [85]
Bacterial cellulose	Chitosan	Zinc oxide	Immersing the oxidized bacterial cellulose hydrogels in chitosan solution and then combining zinc oxide in alkaline environment	Antibacterial dressing [75]
Bacterial cellulose	Alginate chitosan	Copper (II) sulfate	Bacterial cellulose was immersed in the mixtures of alginate and chitosan. Then they were freeze-drying and complexing using ion adsorption	Antibacterial wound dressing [155]
NaCMC	Sodium alginate	No active compound	Physical cross-linking	Artificial skin [162]
Regenerated cellulose nanofiber	Chitosan	No active compound	Chemical cross-linking	Bone tissue engineering [98]
CMC	Chitosan	No active compound	Freeze-dry process	Scaffold for pulp tissue engineering [25]

References

1. Abolore RS, Jaiswal S, Jaiswal AK (2024) Green and sustainable pretreatment methods for cellulose extraction from lignocellulosic biomass and its applications: a review. Carbohydr Polym Technol Appl 7:100396. https://doi.org/10.1016/j.carpta.2023.100396
2. Akhtar S, Mandal SK, Singh L et al (2024) Cellulose for biomedical and pharmaceutical applications. Biopolym Pharm Food Appl 2:759–792
3. Allegri G, Huskens J, Martinho RP, Lindhoud S (2024) Distribution of polyelectrolytes and counterions upon polyelectrolyte complexation. J Colloid Interface Sci 672:654–663. https://doi.org/10.1016/j.jcis.2024.06.062
4. Alshahrani SM, Lu W, Park J-B et al (2015) Stability-enhanced hot-melt extruded amorphous solid dispersions via combinations of Soluplus® and HPMCAS-HF. AAPS PharmSciTech 16:824–834. https://doi.org/10.1208/s12249-014-0269-6
5. Arya R, Singh V, Juyal D, Rawat G (2017) A brief review of polyelectrolyte complex: an update. INDO Am J Pharm Sci 4:1026–1033
6. Aziz F, Ismail AF (2015) Spray coating methods for polymer solar cells fabrication: a review. Mater Sci Semicond Process 39:416–425
7. Aziz T, Farid A, Haq F et al (2022) A review on the modification of cellulose and its applications. Polymers (Basel) 14:3206
8. Bacakova L, Pajorova J, Bacakova M et al (2019) Versatile application of nanocellulose: from industry to skin tissue engineering and wound healing. Nanomaterials 9:164
9. Bekturov EA, Bimendina LA (1997) Complexes of water-soluble polymers. J Macromol Sci Part C Polym Rev 37:501–518
10. Belluzo MS, Medina LF, Cortizo AM, Cortizo MS (2016) Ultrasonic compatibilization of polyelectrolyte complex based on polysaccharides for biomedical applications. Ultrason Sonochem 30:1–8. https://doi.org/10.1016/j.ultsonch.2015.11.022
11. Berbezier I, De Crescenzi M (2015) Self-assembly of nanostructures and nanomaterials. Beilstein J Nanotechnol 6:1397–1398. https://doi.org/10.3762/bjnano.6.144
12. Bhaladhare S, Das D (2022) Cellulose: a fascinating biopolymer for hydrogel synthesis. J Mater Chem B 10:1923–1945
13. Bhattacharya M, Malinen MM, Lauren P et al (2012) Nanofibrillar cellulose hydrogel promotes three-dimensional liver cell culture. J Control Release 164:291–298. https://doi.org/10.1016/j.jconrel.2012.06.039
14. Bigucci F, Abruzzo A, Vitali B et al (2015) Vaginal inserts based on chitosan and carboxymethylcellulose complexes for local delivery of chlorhexidine: preparation, characterization and antimicrobial activity. Int J Pharm 478:456–463. https://doi.org/10.1016/j.ijpharm.2014.12.008
15. Blagodatskikh IV, Vyshivannaya OV, Tishchenko NA et al (2023) Complexation between chitosan and carboxymethyl cellulose in weakly acidic, neutral, and weakly alcaline media. Int J Biol Macromol 243:125277
16. Blumstein A, Kakivaya SR, Salamone JC (1974) Matrix polymerization on an ionene template. The formation of an ordered polyelectrolyte complex. J Polym Sci Polym Lett Ed 12:651–658
17. Boeva Z, Pyshkina O, Sergeyev V (2010) Matrix polymerization of aniline in the presence of polysulfonic acids. J Riga Tech Univ 21:104–108
18. Buriuli M, Verma D (2017) Polyelectrolyte complexes (PECs) for biomedical applications. Adv Biomater Biomed Appl 45–93
19. Calatayud JM, Zamora LL (2013) Spectrophotometry | pharmaceutical applications. In: Worsfold P, Poole C, Townshend A, Miró M (eds) Encyclopedia of analytical science. Academic Press, Oxford, pp 249–262
20. Cao H, Duan L, Zhang Y et al (2021) Current hydrogel advances in physicochemical and biological response-driven biomedical application diversity. Signal Transduct Target Ther 6:426

21. Capanema NSV, Mansur AAP, de Jesus AC et al (2018) Superabsorbent crosslinked carboxymethyl cellulose-PEG hydrogels for potential wound dressing applications. Int J Biol Macromol 106:1218–1234
22. Cerchiara T, Abruzzo A, Parolin C et al (2016) Microparticles based on chitosan/carboxymethylcellulose polyelectrolyte complexes for colon delivery of vancomycin. Carbohydr Polym 143:124–130. https://doi.org/10.1016/j.carbpol.2016.02.020
23. Chai Q, Jiao Y, Yu X (2017) Hydrogels for biomedical applications: their characteristics and the mechanisms behind them. Gels 3:6
24. Chen C, Xi Y, Weng Y (2022) Recent advances in cellulose-based hydrogels for tissue engineering applications. Polymers (Basel) 14
25. Chen H, Fan M (2007) Chitosan/carboxymethyl cellulose polyelectrolyte complex scaffolds for pulp cells regeneration. J Bioact Compat Polym 22:475–491. https://doi.org/10.1177/0883911507081329
26. Chen X, Wang Z, Kan J (2021) Polysaccharides from ginger stems and leaves: effects of dual and triple frequency ultrasound assisted extraction on structural characteristics and biological activities. Food Biosci 42:101166. https://doi.org/10.1016/j.fbio.2021.101166
27. Cheng W, Zhu Y, Jiang G et al (2023) Sustainable cellulose and its derivatives for promising biomedical applications. Prog Mater Sci 138:101152
28. Cheng Y, Luo X, Payne GF, Rubloff GW (2012) Biofabrication: programmable assembly of polysaccharide hydrogels in microfluidics as biocompatible scaffolds. J Mater Chem 22:7659–7666
29. Ciolacu DE, Nicu R, Ciolacu F (2020) Cellulose-based hydrogels as sustained drug-delivery systems. Materials (Basel, Switzerland) 13. https://doi.org/10.3390/ma13225270
30. Cook AB, Clemons TD (2022) Bottom-up versus top-down strategies for morphology control in polymer-based biomedical materials. Adv NanoBiomed Res 2:2100087. https://doi.org/10.1002/anbr.202100087
31. Dakhara SL, Anajwala CC (2010) Polyelectrolyte complex: a pharmaceutical review. Syst Rev Pharm 1
32. De Geest BG, De Koker S, Sukhorukov GB et al (2009) Polyelectrolyte microcapsules for biomedical applications. Soft Matter 5:282–291
33. de Oliveira Barud HG, Da Silva RR, da Silva BH et al (2016) A multipurpose natural and renewable polymer in medical applications: bacterial cellulose. Carbohydr Polym 153:406–420
34. De Vasconcelos CL, Bezerril PM, Dos Santos DES et al (2006) Effect of molecular weight and ionic strength on the formation of polyelectrolyte complexes based on poly(methacrylic acid) and chitosan. Biomacromolecules 7:1245–1252
35. Devos C, Mukherjee S, Inguva P et al (2024) Impinging jet mixers: a review of their mixing characteristics, performance considerations, and applications. AIChE J e18595
36. Ditzinger F, Dejoie C, Sisak Jung D, Kuentz M (2019) Polyelectrolytes in hot melt extrusion: a combined solvent-based and interacting additive technique for solid dispersions. Pharmaceutics 11:174
37. Ditzinger F, Wieland R, Statelova M et al (2020) In vivo performance of innovative polyelectrolyte matrices for hot melt extrusion of amorphous drug systems. Mol Pharm 17:3053–3061
38. Dodoo S, Steitz R, Laschewsky A, von Klitzing R (2011) Effect of ionic strength and type of ions on the structure of water swollen polyelectrolyte multilayers. Phys Chem Chem Phys 13:10318–10325
39. Dogaris I, Pylypchuk I, Henriksson G, Abbadessa A (2024) Polyelectrolyte complexes based on a novel and sustainable hemicellulose-rich lignosulphonate for drug delivery applications. Drug Deliv Transl Res 1–15
40. Dragan ES, Dinu MV, Ghiorghita CA (2022) Chitosan-based polyelectrolyte complex cryogels with elasticity, toughness and delivery of curcumin engineered by polyions pair and cryostructuration steps. Gels 8:240

41. Dutta SD, Patel DK, Lim K-T (2019) Functional cellulose-based hydrogels as extracellular matrices for tissue engineering. J Biol Eng 13:55
42. Eivazi A, Medronho B, Lindman B, Norgren M (2021) On the development of all-cellulose capsules by vesicle-templated layer-by-layer assembly. Polymers (Basel) 13:589
43. Faisul Aris FA, Mohd Fauzi FNA, Tong WY, Syed Abdullah SS (2019) Interaction of silver sulfadiazine with bacterial cellulose via ex-situ modification method as an alternative diabetic wound healing. Biocatal Agric Biotechnol 21:101332. https://doi.org/10.1016/j.bcab.2019.101332
44. Fang B, Wan Y-Z, Tang T-T et al (2009) Proliferation and osteoblastic differentiation of human bone marrow stromal cells on hydroxyapatite/bacterial cellulose nanocomposite scaffolds. Tissue Eng Part A 15:1091–1098. https://doi.org/10.1089/ten.tea.2008.0110
45. Fatema N, Ceballos RM, Fan C (2022) Modifications of cellulose-based biomaterials for biomedical applications. Front Bioeng Biotechnol 10:993711
46. Ferreira DCM, Ferreira SO, de Alvarenga ES et al (2022) Polyelectrolyte complexes (PECs) obtained from chitosan and carboxymethylcellulose: a physicochemical and microstructural study. Carbohydr Polym Technol Appl 3:100197. https://doi.org/10.1016/j.carpta.2022.100197
47. Gersdorf S, Schildknecht V, Schumann E et al (2023) Aqueous polyelectrolyte electrodeposition: the effects of alkyl substitution and varying supporting electrolyte concentrations on the deposition efficiency. ChemElectroChem 10:e202300217. https://doi.org/10.1002/celc.202300217
48. Ghiorghita CA, Dinu MV, Dragan ES (2022) Burst-free and sustained release of diclofenac sodium from mesoporous silica/PEI microspheres coated with carboxymethyl cellulose/chitosan layer-by-layer films. Cellulose 29:395–412. https://doi.org/10.1007/s10570-021-04282-y
49. Golmohammadi H, Morales-Narváez E, Naghdi T, Merkoçi A (2017) Nanocellulose in sensing and biosensing. Chem Mater 29:5426–5446
50. Gopi S, Balakrishnan P, Chandradhara D et al (2019) General scenarios of cellulose and its use in the biomedical field. Mater Today Chem 13:59–78. https://doi.org/10.1016/j.mtchem.2019.04.012
51. Gribkova OL, Melnikova EI, Teplonogova MA et al (2024) Conductive polyaniline-based composition for spray coating. Colloids Surf A Physicochem Eng Asp 692:133903
52. Grumezescu AM, Andronescu E, Ficai A et al (2012) Synthesis, characterization and in vitro assessment of the magnetic chitosan–carboxymethylcellulose biocomposite interactions with the prokaryotic and eukaryotic cells. Int J Pharm 436:771–777. https://doi.org/10.1016/j.ijpharm.2012.07.063
53. Gupta A, Briffa SM, Swingler S et al (2020) Synthesis of silver nanoparticles using curcumin-cyclodextrins loaded into bacterial cellulose-based hydrogels for wound dressing applications. Biomacromolecules 21:1802–1811. https://doi.org/10.1021/acs.biomac.9b01724
54. Han K, Wu Y, Peng X et al (2024) Effect of high-pressure homogenization on maize starch-stearic acid and maize starch-stearic acid-whey protein complexes. Front Sustain Food Syst 8:1324217
55. Haroun AA, Gamal-Eldeen A, Harding DRK (2009) Preparation, characterization and in vitro biological study of biomimetic three-dimensional gelatin–montmorillonite/cellulose scaffold for tissue engineering. J Mater Sci Mater Med 20:2527–2540. https://doi.org/10.1007/s10856-009-3818-x
56. Hasanin MS (2022) Cellulose-based biomaterials: chemistry and biomedical applications. Starch-Stärke 74:2200060
57. Hautmann A, Hedtke T, Sislema-Muñoz S et al (2024) Design of a composite wound dressing: combining an electrospun fleece with a free-standing multilayer film. Next Mater 2:100060
58. Hickey RJ, Pelling AE (2019) Cellulose biomaterials for tissue engineering. Front Bioeng Biotechnol 7:45
59. Hiremath P, Nuguru K, Agrahari V (2019) Material attributes and their impact on wet granulation process performance. In: Handbook of pharmaceutical wet granulation. Elsevier, pp 263–315

60. Hu L, Zhou S, Zhang X et al (2024) Self-assembly of polymers and their applications in the fields of biomedicine and materials. Polymers (Basel) 16
61. Huang J, Laaser JE (2021) Charge density and hydrophobicity-dominated regimes in the phase behavior of complex coacervates. ACS Macro Lett 10:1029–1034
62. Huang J, Morin FJ, Laaser JE (2019) Charge-density-dominated phase behavior and viscoelasticity of polyelectrolyte complex coacervates. Macromolecules 52:4957–4967. https://doi.org/10.1021/acs.macromol.9b00036
63. Huang W-P, Hu J-Q, Qian H-L et al (2022) Dynamic structural controlment for the functionalization of polyelectrolyte multilayer films. Supramol Mater 1:100016. https://doi.org/10.1016/j.supmat.2022.100016
64. Hubbe MA (2021) Contributions of polyelectrolyte complexes and ionic bonding to performance of barrier films for packaging: a review. BioResources 16
65. Ishihara M, Kishimoto S, Nakamura S et al (2019) Polyelectrolyte complexes of natural polymers and their biomedical applications. Polymers (Basel) 11:672
66. Ito H, Miyamoto T, Inagaki H et al (1986) In vivo and in vitro blood compatibility of polyelectrolyte complexes formed between cellulose derivatives. J Appl Polym Sci 32:3413–3421
67. Ito H, Shibata T, Miyamoto T et al (1986) Formation of polyelectrolyte complexes between cellulose derivatives and their blood compatibility. J Appl Polym Sci 31:2491–2500
68. Jacob S, Nair AB, Shah J et al (2021) Emerging role of hydrogels in drug delivery systems, tissue engineering and wound management. Pharmaceutics 13
69. Jagtap P, Patil K, Dhatrak P (2021) Polyelectrolyte complex for drug delivery in biomedical applications: a review. IOP Conf Ser Mater Sci Eng 12007
70. Jeganathan B, Prakya V (2015) Interpolyelectrolyte complexes of Eudragit® EPO with hypromellose acetate succinate and Eudragit® EPO with hypromellose phthalate as potential carriers for oral controlled drug delivery. AAPS PharmSciTech 16:878–888. https://doi.org/10.1208/s12249-014-0252-2
71. Jiang L, Li Y, Zhang L, Wang X (2009) Preparation and characterization of a novel composite containing carboxymethyl cellulose used for bone repair. Mater Sci Eng C 29:193–198
72. Jiji S, Udhayakumar S, Rose C et al (2019) Thymol enriched bacterial cellulose hydrogel as effective material for third degree burn wound repair. Int J Biol Macromol 122:452–460. https://doi.org/10.1016/j.ijbiomac.2018.10.192
73. Joshi KS, Sonawane RO, Ige PP, Bhavsar SK (2019) Polyelectrolyte complex for pharmaceutical aid. Int J All Res Writings 1:69–99
74. Jujare VK, Dantuluri A, Liu Y, Durig T (2023) Hydroxyethylcellulose as a versatile viscosity modifier in the development of sugar-free, elegant oral liquid formulations. Int J Curr Res Chem Pharm Sci 10:1–23. https://doi.org/10.22192/ijcrcps.2023.10.04.001
75. Kai J, Xuesong Z (2020) Preparation, characterization, and cytotoxicity evaluation of zinc oxide–bacterial cellulose–chitosan hydrogels for antibacterial dressing. Macromol Chem Phys 221:2000257. https://doi.org/10.1002/macp.202000257
76. Karoyo AH, Wilson LD (2021) A review on the design and hydration properties of natural polymer-based hydrogels. Materials (Basel) 14(5):1095
77. Keldibekova R, Suleimenova S, Nurgozhina G, Kopishev E (2023) Interpolymer complexes based on cellulose ethers: application. Polymers (Basel) 15
78. Kelly KD, Schlenoff JB (2015) Spin-coated polyelectrolyte coacervate films. ACS Appl Mater Interfaces 7:13980–13986. https://doi.org/10.1021/acsami.5b02988
79. Klemm D, Kramer F, Moritz S et al (2011) Nanocelluloses: a new family of nature-based materials. Angew Chem Int Ed 50:5438–5466
80. Kovačević D, Borković S, Požar J (2007) The influence of ionic strength, electrolyte type and preparation procedure on formation of weak polyelectrolyte complexes. Colloids Surf A Physicochem Eng Asp 302:107–112
81. Krishna BA, Willott JD, Lindhoud S, de Vos WM (2022) Hot-pressing polyelectrolyte complexes into tunable dense saloplastics. Polymer (Guildf) 242:124583. https://doi.org/10.1016/j.polymer.2022.124583

82. Kucińska-Lipka J, Gubanska I, Janik H (2015) Bacterial cellulose in the field of wound healing and regenerative medicine of skin: recent trends and future prospectives. Polym Bull 72:2399–2419. https://doi.org/10.1007/s00289-015-1407-3
83. Kulkarni AD, Vanjari YH, Sancheti KH et al (2016) Polyelectrolyte complexes: mechanisms, critical experimental aspects, and applications. Artif Cells Nanomed Biotechnol 44:1615–1625
84. Lappan U, Naas C, Scheler U (2021) Influence of the mixing ratio on the dynamics of polymer segments in polyelectrolyte complexes. Macromol Chem Phys 222:2000445. https://doi.org/10.1002/macp.202000445
85. Lee SH (2000) Preparation and evaluation of crosslinked polyelectrolyte complex membranes. Polym J 32:716–721
86. León-Campos MI, Claudio-Rizo JA, Caldera-Villalobos M et al (2023) Polyelectrolyte-based hydrogels for biomedicine. Mediterr J Basic Appl Sci 7:58–74
87. Li D, Ye Y, Li D et al (2016a) Biological properties of dialdehyde carboxymethyl cellulose crosslinked gelatin–PEG composite hydrogel fibers for wound dressings. Carbohydr Polym 137:508–514. https://doi.org/10.1016/j.carbpol.2015.11.024
88. Li H, Fauquignon M, Haddou M et al (2021) Interfacial behavior of solid- and liquid-like polyelectrolyte complexes as a function of charge stoichiometry. Polymers (Basel) 13
89. Li J, Yu D, Zeng C et al (2025) Hydrolysis of cellulose acetate phthalate and hydroxypropyl methylcellulose phthalate in amorphous solid dispersions. J Pharm Sci 114:89–97. https://doi.org/10.1016/j.xphs.2024.04.006
90. Li X, Zhang Y, Tang J et al (2016b) Efficient preparation of high concentration cellulose solution with complex DMSO/ILs solvent. J Polym Res 23:1–8
91. Lim C, Shin Y, Oh KT (2024) Polyelectrolyte complex for drug delivery. Drug Targets Ther 3:62–73
92. Lin J, Jiao G, Scott AJ et al (2025) Green synthesis of self-assembly, self-healing, and injectable polyelectrolyte complex hydrogels using chitosan, sulphated polysaccharides, hydrolyzed collagen and nanocellulose. Int J Biol Macromol 288:138566. https://doi.org/10.1016/j.ijbiomac.2024.138566
93. Lin N, Dufresne A (2014) Nanocellulose in biomedicine: current status and future prospect. Eur Polym J 59:302–325
94. Liu Y, Wang G, Cai Y et al (2009) In vitro effects of nanophase hydroxyapatite particles on proliferation and osteogenic differentiation of bone marrow-derived mesenchymal stem cells. J Biomed Mater Res Part A 90A:1083–1091. https://doi.org/10.1002/jbm.a.32192
95. Liuyun J, Yubao L, Chengdong X (2009) Preparation and biological properties of a novel composite scaffold of nano-hydroxyapatite/chitosan/carboxymethyl cellulose for bone tissue engineering. J Biomed Sci 16:65. https://doi.org/10.1186/1423-0127-16-65
96. López-Manzanara Pérez C, Torres-Pabón NS, Laguna A et al (2023) Development of chitosan/sodium carboxymethylcellulose complexes to improve the simvastatin release rate: polymer/polymer and drug/polymer interactions' effects on kinetic models. Polymers (Basel) 15:4184
97. Magalhães S, Fernandes C, Pedrosa JFS et al (2023) Eco-friendly methods for extraction and modification of cellulose: an overview. Polymers (Basel) 15
98. Maharjan B, Park J, Kaliannagounder VK et al (2021) Regenerated cellulose nanofiber reinforced chitosan hydrogel scaffolds for bone tissue engineering. Carbohydr Polym 251:117023. https://doi.org/10.1016/j.carbpol.2020.117023
99. Martirani-VonAbercron S-M, Pacheco-Sánchez D (2023) Bacterial cellulose: a highly versatile nanomaterial. Microb Biotechnol 16:1174–1178. https://doi.org/10.1111/1751-7915.14243
100. Märtson M, Viljanto J, Hurme T, Saukko P (1998) Biocompatibility of cellulose sponge with bone. Eur Surg Res 30:426–432
101. Mehuys E, Vervaet C, Remon JP (2004) Hot-melt extruded ethylcellulose cylinders containing a HPMC–Gelucire® core for sustained drug delivery. J Control Release 94:273–280
102. Meka VS, Sing MKG, Pichika MR et al (2017) A comprehensive review on polyelectrolyte complexes. Drug Discov Today 22:1697–1706

103. Meneses-Juárez E, Márquez-Beltrán C, González-Melchor M (2019) Influence of pH on the formation of a polyelectrolyte complex by dissipative particle dynamics simulation: from an extended to a compact shape. Phys Rev E 100:12505
104. Miyamoto T, Takahashi S, Ito H et al (1989) Tissue biocompatibility of cellulose and its derivatives. J Biomed Mater Res 23:125–133
105. Mokhova E, Gordienko M, Menshutina N et al (2023) Influence of ultrasound on the properties of polysaccharide complexes and materials based on them. Polysaccharides 4:189–207
106. Montero N, Alhajj MJ, Sierra M et al (2020) Development of polyelectrolyte complex nanoparticles-PECNs loaded with ampicillin by means of polyelectrolyte complexation and ultra-high pressure homogenization (UHPH). Polymers (Basel) 12
107. Moore D, Arcila JA, Saraf RF (2020) Electrochemical deposition of polyelectrolytes is maximum at the potential of zero charge. Langmuir 36:1864–1870
108. Moraes PRFS, Saska S, Barud H et al (2016) Bacterial cellulose/collagen hydrogel for wound healing. Mater Res 19:106–116. https://doi.org/10.1590/1980-5373-MR-2015-0249
109. Morariu S, Avadanei M, Nita LE (2024) Effect of pH on the poly(acrylic acid)/poly(vinyl alcohol)/lysozyme complexes formation. Molecules 29
110. More RV, Antanitta SV, Khonde R, Kandasubramanian B (2024) Cellulose and derivatives serving as natural, versatile and biocompatible polymers in biomedical applications. Int J Polym Mater Polym Biomater 1–15
111. Moreno-Díaz C, González-Arranz S, Martínez-Cerezo C (2024) Bacterial cellulose production within a circular economy framework: utilizing organic waste. Polymers (Basel) 16
112. Mujtaba MA, Desai H, Ambekar A et al (2024) Development of chitosan/sodium carboxymethyl cellulose-based polyelectrolyte complex of dexamethasone for treatment of anterior uveitis. Biomed Mater 19:65016
113. Müller FA, Müller L, Hofmann I et al (2006) Cellulose-based scaffold materials for cartilage tissue engineering. Biomaterials 27:3955–3963
114. Navya PV, Gayathri V, Samanta D, Sampath S (2022) Bacterial cellulose: a promising biopolymer with interesting properties and applications. Int J Biol Macromol 220:435–461. https://doi.org/10.1016/j.ijbiomac.2022.08.056
115. Ohyagi N, Ueda K, Higashi K et al (2017) Synergetic role of hypromellose and methacrylic acid copolymer in the dissolution improvement of amorphous solid dispersions. J Pharm Sci 106:1042–1050. https://doi.org/10.1016/j.xphs.2016.12.005
116. Olea AF, Carrasco H, Santana F et al (2024) Formation and characterization of chitosan-based polyelectrolyte complex containing antifungal phenylpropanoids. Polymers (Basel) 16
117. Oprea M, Voicu SI (2020) Recent advances in composites based on cellulose derivatives for biomedical applications. Carbohydr Polym 247:116683
118. Otoni CG, Queiros MVA, Sabadini JB et al (2020) Charge matters: electrostatic complexation as a green approach to assemble advanced functional materials. ACS Omega 5:1296–1304
119. Patil H, Tiwari RV, Repka MA (2016) Hot-melt extrusion: from theory to application in pharmaceutical formulation. AAPS PharmSciTech 17:20–42. https://doi.org/10.1208/s12249-015-0360-7
120. Patil H, Vemula SK, Narala S et al (2024) Hot-melt extrusion: from theory to application in pharmaceutical formulation—where are we now? AAPS PharmSciTech 25:37. https://doi.org/10.1208/s12249-024-02749-2
121. Pergushov DV, Müller AHE, Schacher FH (2012) Micellar interpolyelectrolyte complexes. Chem Soc Rev 41:6888–6901
122. Pita PCC, Pinto FCM, Lira MMM et al (2015) Biocompatibility of the bacterial cellulose hydrogel in subcutaneous tissue of rabbits. Acta Cir Bras 30:296–300. https://doi.org/10.1590/S0102-865020150040000009
123. Pötzinger Y, Rabel M, Ahrem H et al (2018) Polyelectrolyte layer assembly of bacterial nanocellulose whiskers with plasmid DNA as biocompatible non-viral gene delivery system. Cellulose 25:1939–1960

124. Putro JN, Lunardi VB, Soetaredjo FE et al (2021) A review of gum hydrocolloid polyelectrolyte complexes (PEC) for biomedical applications: their properties and drug delivery studies. Processes 9:1796
125. Qian L, Dong C, Liang X et al (2014) Polyelectrolyte complex containing antimicrobial guanidine-based polymer and its adsorption on cellulose fibers. Holzforschung 68:103–111
126. Quiñones JP, Peniche H, Peniche C (2018) Chitosan based self-assembled nanoparticles in drug delivery. Polymers (Basel) 10:235
127. Raj S, Kumar Sharma P, Malviya R (2018) Pharmaceutical and tissue engineering applications of polyelectrolyte complexes. Curr Smart Mater 3:21–31
128. Rao SS, Venkatesan J, Yuvarajan S, Rekha P-D (2022) Self-assembled polyelectrolyte complexes of chitosan and fucoidan for sustained growth factor release from PRP enhance proliferation and collagen deposition in diabetic mice. Drug Deliv Transl Res 12:2838–2855
129. Regenspurg JA, Martins Costa AF, Achterhuis I, de Vos WM (2022) Influence of molecular weight on the performance of polyelectrolyte multilayer nanofiltration membranes. ACS Appl Polym Mater 4:2962–2971. https://doi.org/10.1021/acsapm.1c00826
130. Regenspurg JA, Martins Costa AF, Achterhuis I, de Vos WM (2021) Influence of molecular weight on the performance of polyelectrolyte multilayer nanofiltration membranes. ACS Appl Polym Mater 4:2962–2971
131. Repka MA, Gutta K, Prodduturi S et al (2005) Characterization of cellulosic hot-melt extruded films containing lidocaine. Eur J Pharm Biopharm Off J Arbeitsgemeinschaft Pharm Verfahrenstech eV 59:189–196. https://doi.org/10.1016/j.ejpb.2004.06.008
132. Roy JC, Ferri A, Giraud S et al (2018) Chitosan–carboxymethylcellulose-based polyelectrolyte complexation and microcapsule shell formulation. Int J Mol Sci 19
133. Sadman K, Wang Q, Chen SH et al (2017) pH-controlled electrochemical deposition of polyelectrolyte complex films. Langmuir 33:1834–1844
134. Saklani R, Domb AJ (2024) Peptide and protein stereocomplexes. ACS Omega 9:17726–17740. https://doi.org/10.1021/acsomega.4c00178
135. Salatin S, Yari Khosroushahi A (2017) Overviews on the cellular uptake mechanism of polysaccharide colloidal nanoparticles. J Cell Mol Med 21:1668–1686. https://doi.org/10.1111/jcmm.13110
136. Salmi J, Österberg M, Stenius P, Laine J (2007) Surface forces between cellulose surfaces in cationic polyelectrolyte solutions: the effect of polymer molecular weight and charge density. Nord Pulp Pap Res J 22:249–257
137. Sampl C, Schaubeder J, Hirn U, Spirk S (2023) Interplay of electrolyte concentration and molecular weight of polyDADMAC on cellulose surface adsorption. Int J Biol Macromol 239:124286
138. Sanchez-Ballester NM, Sciortino F, Mir SH, Rydzek G (2022) Weak polyelectrolytes as nanoarchitectonic design tools for functional materials: a review of recent achievements. Molecules 27
139. Satam CC, Meredith JC (2021) Increasing efficiency of the homogenization process for production of chitin nanofibers for barrier film applications. Carbohydr Polym 274:118658. https://doi.org/10.1016/j.carbpol.2021.118658
140. Sato H, Nakajima A (1975) Formation of a polyelectrolyte complex from carboxymethyl cellulose and poly(ethylenimine). Polym J 7:241–247
141. Scott D, Prud'Homme R, Priestley R (2019) Confined impingement jet mixing of charged polymers for functional structured colloids and encapsulation. In: APS march meeting abstracts, pp B52–012
142. Seddiqi H, Oliaei E, Honarkar H et al (2021) Cellulose and its derivatives: towards biomedical applications. Cellulose 28:1893–1931
143. Seiler ERD, Takeoka Y, Rikukawa M, Yoshizawa-Fujita M (2020) Development of a novel cellulose solvent based on pyrrolidinium hydroxide and reliable solubility analysis. RSC Adv 10:11475–11480. https://doi.org/10.1039/d0ra01486a
144. Shovsky AV, Varga I, Makuška R, Claesson PM (2009) Formation and stability of soluble stochiometric polyelectrolyte complexes: effects of charge density and polyelectrolyte concentration. J Dispers Sci Technol 30:980–988

145. Song Y, Zhou Y, Chen L (2014) Cellulose-based polyelectrolyte complex nanoparticles for DNA vaccine delivery. Biomater Sci 2:1440–1449
146. Stevens KC, Tirrell MV (2024) Impact of a lightly branched star polyelectrolyte architecture on polyelectrolyte complexes. ACS Macro Lett 13:688–694
147. Tarrahi R, Khataee A, Karimi A, Yoon Y (2022) The latest achievements in plant cellulose-based biomaterials for tissue engineering focusing on skin repair. Chemosphere 288:132529
148. Thukral TS, Rabbi KF, Khodakarami S et al (2024) Scalable and durable polydimethylsiloxane coating for anti-corrosion, anti-fouling, and condensation applications. Cell Rep Phys Sci 5
149. Towle EG, Ding I, Peterson AM (2020) Impact of molecular weight on polyelectrolyte multilayer assembly and surface properties. J Colloid Interface Sci 570:135–142. https://doi.org/10.1016/j.jcis.2020.02.114
150. Tsiaxerli A, Karagianni A, Ouranidis A, Kachrimanis K (2021) Polyelectrolyte matrices in the modulation of intermolecular electrostatic interactions for amorphous solid dispersions: a comprehensive review. Pharmaceutics 13. https://doi.org/10.3390/pharmaceutics13091467
151. Wang C, Surat'man NEB, Chang JJ et al (2022) Polyelectrolyte hydrogels for tissue engineering and regenerative medicine. Chem Asian J 17:e202200604
152. Wang J, Tavakoli J, Tang Y (2019) Bacterial cellulose production, properties and applications with different culture methods—a review. Carbohydr Polym 219:63–76. https://doi.org/10.1016/j.carbpol.2019.05.008
153. Wang W, Resing AB, Brown KA, Werner JG (2024) Electrodeposition of polymer networks as conformal and uniform ultrathin coatings. Adv Mater 36:2409826
154. Wasewar K (2006) A design of jet mixed tank. Chem Biochem Eng Q 20(1):20
155. Wichai S, Chuysinuan P, Chaiarwut S et al (2019) Development of bacterial cellulose/alginate/chitosan composites incorporating copper (II) sulfate as an antibacterial wound dressing. J Drug Deliv Sci Technol 51:662–671. https://doi.org/10.1016/j.jddst.2019.03.043
156. Wu D, Zhu L, Li Y et al (2020) Chitosan-based colloidal polyelectrolyte complexes for drug delivery: a review. Carbohydr Polym 238:116126
157. Wu Q-X, Wang D-D, Su T et al (2017) Self-assembly of polyelectrolyte complexes microcapsules with natural polysaccharides for sustained drug release. Cellulose 24:4949–4962
158. Xia N, Wan W, Zhu S, Liu Q (2020) Preparation of crystalline nanocellulose/hydroxypropyl β cyclodextrin/carboxymethyl cellulose polyelectrolyte complexes and their controlled release of neohesperidin-copper (II) in vitro. Int J Biol Macromol 163:1518–1528. https://doi.org/10.1016/j.ijbiomac.2020.07.272
159. Yang D, Gong L, Li Q et al (2023) Preparation of a biobased polyelectrolyte complex from chitosan and sodium carboxymethyl cellulose and its antibacterial characteristics. Int J Biol Macromol 227:524–534
160. Yang H, Peterson AM (2017) Inkjet printed drug-releasing polyelectrolyte multilayers for wound dressings. Growth Factors 32:36
161. Zhang H, Wang C, Zhu G, Zacharia NS (2016) Self-healing of bulk polyelectrolyte complex material as a function of pH and salt. ACS Appl Mater Interfaces 8:26258–26265
162. Zhang K, Wang Y, Wei Q et al (2021) Design and fabrication of sodium alginate/carboxymethyl cellulose sodium blend hydrogel for artificial skin. Gels 7
163. Zhao L, Mitomo H, Yosh F (2008) Synthesis of pH-sensitive and biodegradable CM-cellulose/chitosan polyampholytic hydrogels with electron beam irradiation. J Bioact Compat Polym 23:319–333. https://doi.org/10.1177/0883911508092302
164. Zhao Q, Han B, Wang Z et al (2007) Hollow chitosan-alginate multilayer microcapsules as drug delivery vehicle: doxorubicin loading and in vitro and in vivo studies. Nanomedicine 3:63–74. https://doi.org/10.1016/j.nano.2006.11.007
165. Zikmundova M, Vereshaka M, Kolarova K et al (2020) Effects of bacterial nanocellulose loaded with curcumin and its degradation products on human dermal fibroblasts. Materials (Basel) 13

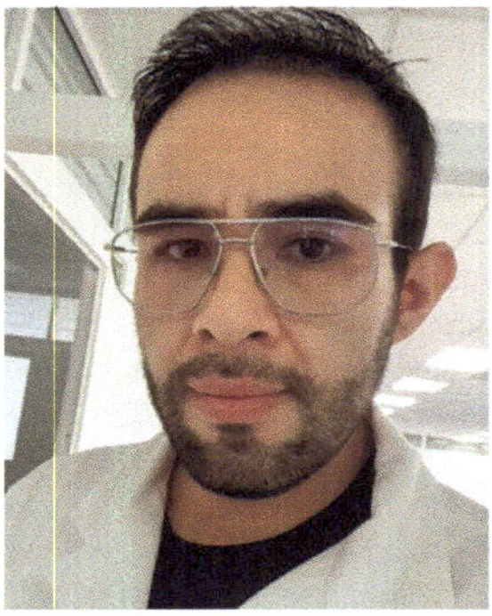

Bruno Solis-Cruz in 2011 obtained the title of Biological Pharmaceutical Chemist from the National Autonomous University of Mexico, in 2014 he obtained the degree of Master's in Animal Production and Health Sciences, and in 2020 he graduated as a Doctor of Sciences from the same institution, where he was recognized with three university merit medals. He has conducted and collaborated on research at the Department of Poultry Sciences at the University of Arkansas in the USA, as well as at the Faculty of Pharmacy at the University Clermont-Auvergne in France. He has published research articles in international peer-reviewed journals and international book chapters; and he has presented papers at national and international conferences.

Since 2010 he has been part of the working group of the Pharmaceutical Development Testing Laboratory (LEDEFAR) at the FESC, where he has carried out and collaborated in projects related to the development and validation of analytical methods, development of dissolution tests for the evaluation and characterization of drugs and medicines, permeability and toxicity tests in cell cultures, as well as in the search for new administration and release systems that allow improving the biopharmaceutical properties of active ingredients with low solubility, both in the human and veterinary areas. He has also participated in various projects for SAGARPA (now SADER)/SENASICA, regarding the regulation of chemical-pharmaceutical and biotechnological products for veterinary use. He is currently a subject professor at the Universidad Politécnica del Valle de México and an Analytical Development Chemist in the Research and Development laboratory of Bausch Health Companies.

Raquel López-Arellano is Full-Time Professor "C" at the Faculty for Higher Studies (Facultad de Estudios Superiores, FES) Cuautitlán of the National Autonomous University of Mexico (UNAM). She is Specialist in industrial pharmacy and Doctor in quality systems and optimization of pharmaceutical processes from the Claude Bernard University Lyon 1, France, and chemical pharmaceutical biology graduated from the Faculty of Chemical Sciences of the Benemérita Universidad Autónoma de Puebla, Mexico. Dr. López Arellano is Member of the National System of Researchers (SNII), level 2, and is Level "C" in the Full-Time Academic Staff Performance Bonus Program (PRIDE). Furthermore, she participates in the committee of experts in general methods of analysis of the Pharmacopoeia of the United Mexican States and is part of the National College of Pharmaceutical Biological Chemists of Mexico A.C. (CNQFBM A.C.). She has 7 years of experience in the pharmaceutical industry, has collaborated in 17 joint projects with the pharmaceutical industry, and has directed 7 doctoral theses, 25 master's theses, 90 bachelor's theses in the pharmaceutical area. Likewise, she is Tutor for the postgraduate program in Animal Health and Production Science and Chemical Sciences at UNAM, as well as Accredited Evaluator for

CONAHCYT and in the Program to Support Research Projects and Technological Innovation (PAPIIT) at UNAM. Finally, Dr. López Arellano is responsible for the Laboratory of Tests of Pharmaceutical Development (LEDEFAR) at FES Cuautitlán, and her research lines focus on the development of human and veterinary drugs, as well as the development and validation of analytical/bioanalytical methods.

Dr. Jorge L. Mejía-Méndez graduated with honors from the Bachelor of Pharmaceutical Sciences Program of the Universidad de las Americas Puebla (UDLAP), Mexico. Recently, he received the degree of Doctor in Molecular Biomedicine from the same institution together with the Magna Cum Laude distinction. Over the last few years, the research interests of Dr. Mejía-Méndez have focused on evaluating the biological activities of extracts from medicinal plants and the synthesis and characterization of polymeric and metallic nanomaterials. The results retrieved from his research projects have been published in international journals such as biomaterials science, antioxidants, and the Journal of Nanobiotechnology. Finally, Dr. Mejía-Méndez has also participated in international and national microbiological, pharmacological, phytochemical, and nanotechnological research conferences.

Abraham Méndez-Albores is Chemical Engineer from the Benemerita Autonomous University of Puebla (BUAP), Master's in Food Science and Technology from the Autonomous University of Querétaro (UAQ), and Doctor of Science with a specialty in materials from the Center for Research and Advanced Studies of the National Polytechnic Institute (CINVESTAV-IPN). He has completed a two-year postdoctoral stay at UNAM in the area of mycotoxins and mycotoxicosis. He is Permanent Full-Time Professor B and Head of the Food, Mycotoxins, and Mycotoxicosis Laboratory at UNAM Cuautitlan. He has published 81 articles in national and international peer-reviewed journals, one book, and three book chapters (index $h = 24$). He is Member of the National System of Researchers of Mexico (Level 3).

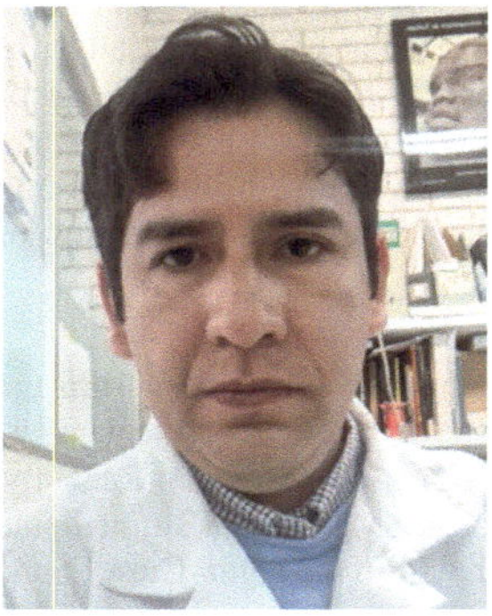

Dr. Daniel Hernandez-Patlan obtained the degree of biological pharmaceutical chemist he Faculty for Higher Studies (Facultad de Estudios Superiores, FES) Cuautitlán of the National Autonomous University of Mexico (UNAM) in 2011. Subsequently, in 2014 he obtained the degree of Master of Science, and in 2019 he graduated as Doctor of Science, in the Animal Production and Health program, UNAM. As part of his academic/professional training, he carried out research stays at the Department of Poultry Sciences at the University of Arkansas, United States, and at the Faculty of Pharmacy at the University of Clermont-Auvergne, France. Since 2010, Dr. Hernandez-Patlan has collaborated with the working group of the Laboratory of Tests of Pharmaceutical Development (LEDEFAR) at FES Cuautitlán, developing projects related to the development and validation of analytical methods, development of dissolution tests and dissolution studies, and in strategies for increasing the solubility and permeability of drugs through solid dispersions and micro/nanoparticulate systems. Furthermore, he has participated in various projects for the Secretary of Agriculture and Rural Development (SADER)/the National Service for Agrifood Health, Safety and Quality (SENASICA), regarding the regulation of chemical-pharmaceutical and biotechnological products for veterinary use.

He is currently a full-time professor "A" at the Polytechnic University of the Valley of Mexico, belongs to the National System of Researchers (SNII) Level 1 of CONAHCYT, and has 4 years of experience in the pharmaceutical industry. Finally, his line of research includes the development of solid dispersions and micro/nanoparticle systems to encapsulate/associate phytobiotics, probiotics, and enzymes as an alternative to the use of antibiotics to reduce bacterial resistance problems.

Design, Characteristics, and Biomedical Applications of Cellulose-Based Hydrogel

Md. Aliahsan Bappy, Md. Nizam Uddin, Md. Moniruzzaman, Nelson Irizarry, and Md. Mahbubur Rahman

Abstract Cellulose-based hydrogels have emerged as versatile materials with significant potential in biomedical applications due to their biocompatibility, biodegradability, and tunable physical and chemical properties. Derived from one of the most abundant natural polymers, cellulose, these hydrogels offer a sustainable and cost-effective alternative to synthetic materials. Cellulose hydrogels, synthesized by forming a three-dimensional network through either physical or chemical crosslinking of cellulose or its derivatives, are well-regarded for their high-water absorption capacity and excellent biocompatibility. The design of cellulose-based hydrogels involves modifications at the molecular level, allowing control over parameters like porosity, water retention, mechanical strength, and responsiveness to environmental stimuli. Their unique characteristics include high water content, mechanical resilience, and the capacity to incorporate bioactive molecules, making them ideal for diverse biomedical uses. Applications span wound healing and drug delivery systems to tissue engineering scaffolds and biosensors. In wound healing, cellulose hydrogels provide a moist environment that facilitates tissue repair, while in drug delivery, they allow for controlled release of therapeutic agents. In tissue engineering, they offer scaffolding structures that mimic the extracellular matrix (ECM), supporting cell adhesion and proliferation. Advances in cellulose functionalization have further expanded their applicability, enhancing cell interaction and material stability. Furthermore, recent advancements in nanocellulose-based hydrogels for tissue engineering have been extensively evaluated and documented. This

Md. Aliahsan Bappy
Department of Mechanical Engineering, Lamar University, Beaumont, TX, USA

Md. Nizam Uddin (✉) · N. Irizarry
James C. Morriss Division of Engineering, Texas A&M University-Texarkana, Texarkana, TX, USA
e-mail: muddin@tamut.edu

Md. Moniruzzaman
Department of Industrial and Systems Engineering, Lamar University, Beaumont, TX, USA

Md. Mahbubur Rahman
Department of Mechanical Engineering, Khulna University of Engineering and Technology, Khulna, Bangladesh

R. Malviya and S. Sundram (eds.), *Sustainable Cellulose Materials for Biomedical Applications*, Sustainable Materials and Technology,
https://doi.org/10.1007/978-981-96-9904-9_5

chapter explores the design strategies, intrinsic properties, and wide-ranging biomedical applications of cellulose-based hydrogels, highlighting recent developments and future directions.

Keywords Cellulose-based hydrogels · Biocompatibility · Scaffold · Wound healing · Tissue engineering

Abbreviations

BCNCs	Bacterial cellulose nanocrystals
BCNFs	Bacterial cellulose nanofibrils
CGLH	Cellulose/γ-PGA hydrogel
CMC	Carboxymethyl cellulose
CNCs	Cellulose nanocrystals
CNFs	Cellulose nanofibrils
ECM	Extracellular matrix
HEC	Hydroxyethyl cellulose
HPC	Hydroxypropyl cellulose
HPMC	Hydroxypropyl methylcellulose
MC	Methylcellulose
PEG	Polyethylene glycol
PLA	Polylactic acid
PVA	Polyvinyl alcohol
SAPs	Superabsorbent polymers
TiO_2	Titanium dioxide

1 Introduction

Biopolymers, or natural polymers, are derived from living things and are typically biodegradable. Among these, cellulose, a polymer generated by some microbes and present in plant cell walls, is notable for its mechanical strength, biocompatibility, and abundance. Bacterial cellulose is released by bacteria such as *Acetobacter xylinum* as a very pure nanofibrous network, whereas plant cellulose usually forms microfibrils with crystalline and amorphous areas. Both bacterial cellulose and plant-derived nanocellulose function as reinforcing fillers in composites, enhancing their qualities for biomedical applications like bone tissue engineering and wound healing. Pure cellulose presents processing challenges due to its strong hydrogen bonding and consequent insolubility in most organic solvents; however, modified forms, such as cellulose acetate and carboxymethyl cellulose, get around this restriction and find use in everything from bioadhesives to drug delivery systems. Cellulose is a strong contender for "green" polymer composites in the medical industry due to its adaptable structure and biodegradability [25]. In Fig. 1, we can see that the first step in creating bacterial cellulose is the formation of protofibrils from glucose chains. The cell wall's synthetic sites are subsequently used to secrete these protofibrils. The protofibrils combine to form microfibrils, which are then arranged into bundles after being secreted. These bundles ultimately form the bacterial cellulose pellicle, a highly organized and robust substance bacterial cellulose's special qualities, including its high purity, mechanical strength, and biocompatibility, are made possible by this complex procedure, which makes it useful for a range of applications.

The most prevalent bio-inspired polymer on earth is cellulose, mostly derived from plants but made by bacteria and sea invertebrates called tunicates. Despite having the

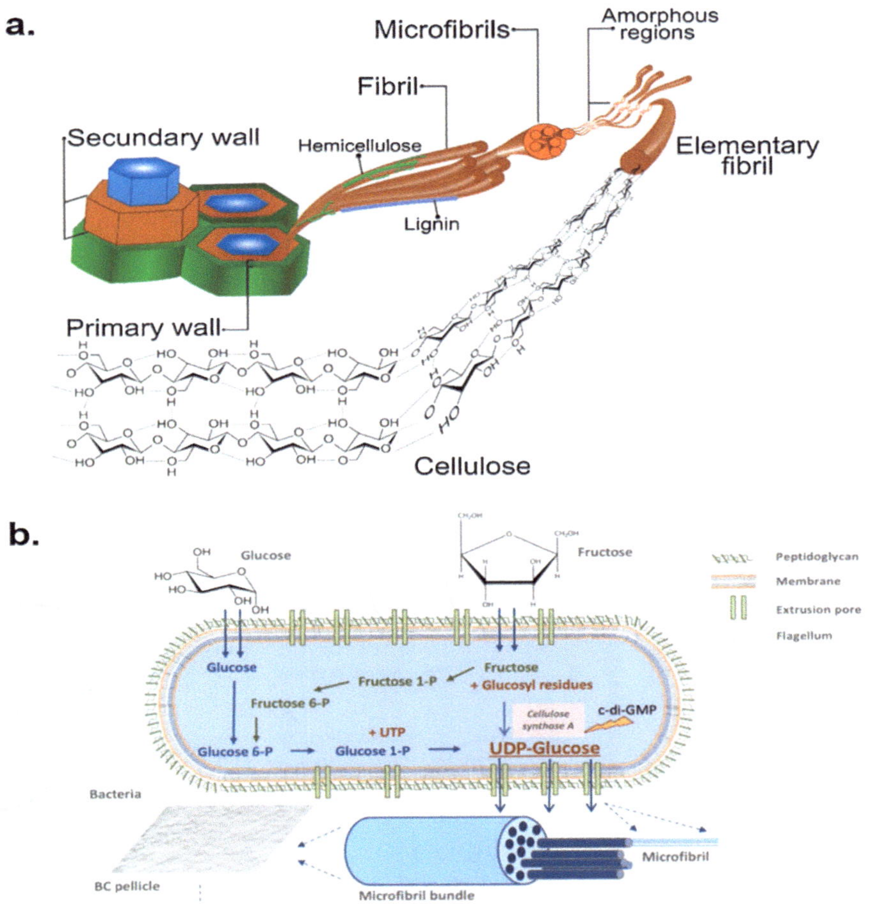

Fig. 1 **a** Hierarchical structure of cellulose extracted from plants. **b** Schematic representation of the steps involved in bacterial cellulose production by *Acetobacter xylinum* [25]

same core β-1,4-D-glucopyranose structure, celluloses differ in how their fibrils are organized, which results in different characteristics and uses. Cellulose I, the natural form, can be irrevocably changed into cellulose II, which has increased porosity and is easier to digest. Because of its crystalline stability, cellulose is mostly insoluble in moderate environments, but it can undergo chemical modification to create a variety of "isomorphs," each with a unique chain configuration. Because of their adaptable surface, low toxicity, and strong mechanical qualities, cellulose nanofibrils frequently separated by acid hydrolysis are highly valued in biological applications. Cellulose-based hydrogels have excellent biocompatibility, structural stability, and swelling capacity which is explained in Fig. 2 [21].

Biomedical materials have been utilized for illness prevention and injury repair since the dawn of human civilization. Bio-renewable materials including gelatin, silk fibroin, chitosan, and cellulose have drawn a lot of interest in the biomedical area as a means of enhancing illness diagnosis and therapy due to the quick advancement of science and technology. However, due to difficulties with material selection, molecular design, and structural regulation, biomaterials' functional design and medicinal applications are still in their infancy. The most prevalent natural biomaterial is cellulose, which is derived from bacteria, algae, and plants. It is perfect for usage in a variety of biomedical applications due to its abundance and biocompatibility. Furthermore, cellulose is easily converted into derivatives that are both soluble and processable, resulting in a variety of useful materials such hydrogels, carbon dots, scaffolds, and membranes. In terms of ionic conductivity, mechanical strength, and thermal stability, these cellulosic materials outperform synthetic polymers. Cellulose and its derivatives are therefore commonly utilized in tissue

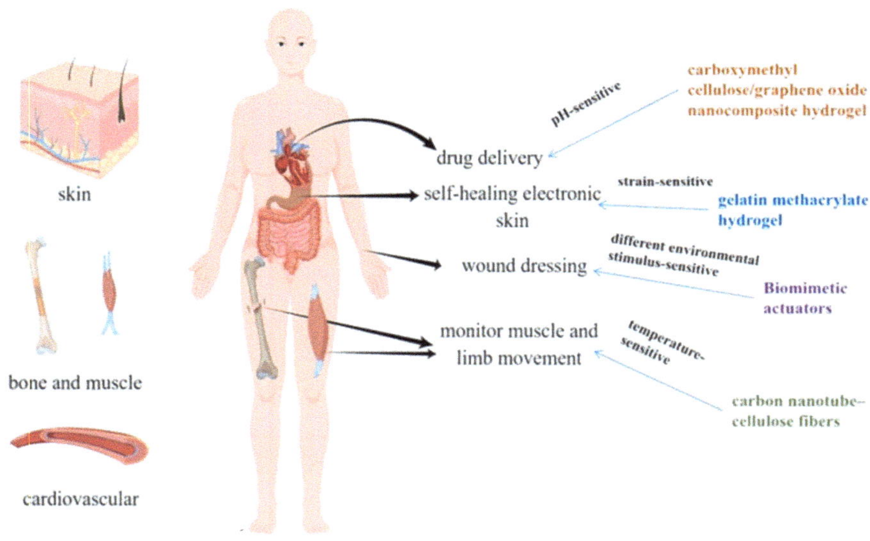

Fig. 2 Illustration of potential applications for various smart cellulose-based hydrogels [21]

engineering, wound dressings, artificial blood vessels, and biosensors, among other applications. As evidenced by the rising number of published papers, research on the biological use of cellulose has been expanding continuously. Materials such as bacterial cellulose, cellulose nanocrystals, and cellulose derivatives like cellulose acetate are frequently the subject of studies. Cellulose is a desirable material for a variety of biomedical applications due to its important characteristics, which include mechanical strength, rheological behavior, antibacterial qualities, and barrier performance. Notwithstanding the advancements, more research is required to fully comprehend the connection between material performance and molecular structure. The present status of cellulose-based materials in the biomedical industry is summed up in this study, with particular attention paid to their structural traits, techniques of modification, and their uses in tissue engineering, medication delivery, wound healing, health products, and diagnostic technologies [9].

Cellulose is extensively utilized in sectors including pulp production, textiles, and paper. Apart from plant sources, some marine species, including tunicates, also biosynthesis cellulose, which is known as tunicin in these animals. In vitro enzymatic synthesis and chemosynthesis from glucose derivatives are two more important cellulose manufacturing techniques. Microbial production is one of the most significant processes for cellulose synthesis. Other microbes that can create cellulose include fungi, algae, and non-pathogenic bacteria from genera such as Agrobacterium, Sarcina, Rhizobium, and Gluc acetobacter [40]. One of the main structural polysaccharides, cellulose, is unique in that its reactive functional groups allow for a variety of alterations, and its derivatives, particularly carboxymethyl cellulose (CMC), are being thoroughly researched for potential applications in biomedicine. Despite having a great swelling capacity due to its superabsorbent nature, CMC's strong hydrophilicity may make it less stable in watery settings. Aldehydes, carboxylic acids, urea derivatives, and epichlorohydrin are examples of crosslinkers that can improve water stability and mechanical strength. The goal of the current study is to create a crosslinked CMC hydrogel with enhanced stability under physiological settings, building on earlier research that used CMC membranes for bone regeneration and wound healing. The study aims to maximize crosslinking and expand the potential of cellulose-based hydrogels in regenerative medicine and other biomedical domains by employing a crosslinker immersion approach followed by drying—and investigating the effects of equilibration in distilled water [8]. Growing environmental concerns and the need for eco-friendly materials have led to the study of cellulose, a renewable, biocompatible, non-toxic, and biodegradable biopolymer found in plant cell walls. The high absorption efficiency and stimuli-responsiveness of cellulose-based hydrogels, which are crosslinked macromolecular networks that can absorb and release huge volumes of water and other fluids, have drawn attention. These hydrogels are very useful for flexible electronics, biomedicine, and agriculture. For researchers examining the potential of cellulose-based hydrogels as biomaterials, this study offers a comprehensive overview of the synthesis, design, architectures, physicochemical characteristics, and various uses of these materials [36]. It is anticipated that cellulose, a naturally occurring substance that is plentiful, adaptable, and sustainable, would replace non-renewable polymers. Natural cellulose's physicochemical properties are associated with the isolation process that gives it a specific chain length, crystallinity, and quantity of intramolecular and intermolecular hydrogen bonds. For specific cellulosic materials, these properties may be different from others, giving cellulose special qualities [14].

Strong β 1 → 4-glycosidic bonds bind repeated glucose units to form cellulose, an unbranched natural polymer that is one of the most prevalent organic compounds on Earth. Cellulose, which is mostly found in the cell walls of plants and wood, as well as in the tissues of algae and bacterial nanofiber networks, exhibits a hierarchical structure that extends from macro- to nanoscale fibrils with both crystalline and disordered domains. The polymer's intrinsic biodegradability, biocompatibility, flexibility, and cost-effectiveness, along with its highly ordered structure, make it appealing for use in biomedical applications. Cellulose can be chemically altered at its hydroxyl groups to customize characteristics like porosity, moisture sensitivity, and mechanical strength, despite its low water solubility and susceptibility to microbial attacks. Since their tunable porosity and sturdy physical characteristics allow for the creation of materials that promote cell growth, offer a protective interface for wound healing, and permit controlled therapeutic release, cellulose, and its derivatives have thus demonstrated promise in tissue engineering, wound dressing, and drug delivery. The properties, manufacturing processes, and biomedical uses of cellulose-based materials are covered in this review, which highlights their increasing significance in cutting-edge healthcare solutions [30].

Cellulose, a biopolymeric carbohydrate found in large quantities in bacteria and plants, has great potential for use in biomedical applications because of its remarkable mechanical and physical properties, biocompatibility, biodegradability, and natural abundance. However, because of its structure's strong intramolecular and intermolecular hydrogen bonding, it has limitations in biomedicine, including low microbiological resistance, moisture sensitivity, and water insolubility. Cellulose can be chemically altered to overcome these restrictions by substituting different functional groups for its hydroxyl groups, enhancing its qualities for use in biomedicine. To increase its use in medical disciplines, a variety of cellulose derivatives have been produced, such as cellulose ethers like methylcellulose (MC), carboxymethyl cellulose, and hydroxyethyl cellulose (HEC), as well as cellulose esters like cellulose acetate [27]. Green chemistry and sustainability have been more prominent in materials science in recent decades, producing novel nanomaterials with a wide range of applications. Because of its renewability, biocompatibility, and biodegradability, cellulose-one of the most prevalent natural polymers on earth—plays a crucial part in these developments. Its active hydroxyl groups enable a wide range of chemical interactions and changes, making it easier to create cellulose hydrogels using chemical, physical, or radiation crosslinking techniques. These hydrogels are strengthened by nanocellulosic materials like cellulose nanofibrils (CNFs) and cellulose nanocrystals (CNCs), derived from cellulose's molecular structure and later technological alterations. An overview of hydrogels highly porous, three-dimensional networks of polymers and their development from their initial manufacturing in 1960 is given in this text. It discusses how they are divided into two categories: physical and chemical crosslinking, traditional and functional/smart response, and their uses in biomedical sectors (e.g., drug delivery, wound dressings) and larger industries (e.g., sensors, batteries). The discussion then turns to cellulose-based hydrogels, highlighting their benefits (such as affordability, biocompatibility, and renewability) but also pointing out that their low mechanical strength may constrain their application. Developing synthetic or hybrid hydrogels with improved toughness, self-healing, or other capabilities is one recent tactic to enhance these qualities. The study summarizes the most recent developments in cellulose hydrogel research emphasizing new uses, difficulties, and potential paths forward [45].

Hydrogels can absorb and retain large volumes of water, which are three-dimensional networks of hydrophilic polymers. Using a variety of crosslinking processes, including ionic, chemical, physical, and hydrogen bonding, they can be created from synthetic and natural polymers. They are perfect for use in fields including drug delivery, tissue engineering, and agriculture because of their high-water content, biocompatibility, and adjustable qualities. Hydrogels based on polysaccharides (such as cellulose derivatives) have drawn interest due to their biocompatibility, sustainability, and responsiveness to environmental stimuli. These "green" solutions can optimize water resource management and replace synthetic materials [26]. Three-dimensional networks called hydrogels are made of hydrophilic polymers that can hold a lot of water while keeping their shape. Because of their inherent biodegradability, biocompatibility, and environmental friendliness, natural polymers like cellulose, chitosan, collagen, alginate, hyaluronic acid, starch, guar gum, agarose, and dextran are particularly appealing for hydrogel synthesis. Nevertheless, a lot of these hydrogels made from natural polymers have poor mechanical strength and toughness, which leads to the use of techniques like physical or chemical crosslinking and the addition of artificial polymers or inorganic particles to increase their resilience. These materials have a wide range of biological uses, such as tissue engineering, wound healing, and drug delivery, because of their bio-friendly qualities. The structures, extraction processes, processing methods, current advancements, and prospects of regularly used natural polymers are all covered in their hydrogel-based applications [44].

Hydrogels present interesting options for biomedical applications, especially tissue engineering, because of their high-water-retention capacity, biocompatibility, and biodegradability. Both manufactured (such as polyacrylic acid and polyvinyl alcohol) and naturally occurring (like collagen, chitosan, and cellulose derivatives) polymers can be used to make them; however, organically derived polymers are typically chosen because of their abundance, renewability, and biocompatible breakdown products. Despite these benefits, naturally occurring hydrogels frequently have low mechanical strength and stability, which can be improved by crosslinking by chemical (covalent) or physical (hydrogen bonding or hydrophobic interactions). Cellulose hydrogels and their hybrid composites have been used in various applications, from energy harvesting and environmental protection to biomedicine and personal care, because of their exceptional mechanical strength, biodegradability, and adjustable properties. These hydrogels' structural and functional qualities have been greatly enhanced by advanced preparation methods such as ultrasonic assistance, acid-treated mesh processes, in situ polymerization, Janus hybrid design, and

3D printing. As a result, their applications in drug delivery, tissue engineering, and environmental remediation have increased. This review highlights how crucial it is to summarize fundamental ideas, cutting-edge preparation techniques, and optimization strategies as cellulose hydrogel research advances. It also examines existing uses and projects future possibilities in this rapidly developing field [32].

Hydrogels are used in various applications and are distinguished by their great capacity to absorb water (e.g., contact lenses, wound dressings, and medication capsules). Unlike many synthetic hydrogels, natural hydrogels, or hydrogels based on biopolymers, have attracted attention due to their biocompatibility, biodegradability, and low toxicity. Because cellulose is abundant, renewable, and can undergo a variety of chemical changes to produce ethers (like carboxymethyl cellulose) or esters (like cellulose acetate), cellulose-based hydrogels stand out among them. Methods of preparation include creating composites and hybrid hydrogels or dissolving pure cellulose in certain solvents (such as ionic liquids). Cellulose-based hydrogels have been investigated for a variety of biomedical applications, including tissue engineering, wound dressings, and drug delivery, as well as non-biomedical uses, such as sensors, agriculture (e.g., water retention, pesticide carriers), and hygiene products, due to their desirable properties, including porosity, tunable mechanical strength via crosslinking, and intrinsic biocompatibility. Cellulose-based hydrogels' growing potential is highlighted by ongoing research and patent activity; however, there are still few thorough studies that cover the whole range of these applications. To demonstrate how customized functionalization and crosslinking techniques can maximize performance for end uses, the current work focuses on the structure, physicochemical characteristics, and various application domains of cellulose-based hydrogels [15].

The two most common techniques for creating cellulose-based hydrogels are chemical and physical crosslinking. By using hydrogen bonds between cellulose molecules, physical crosslinking produces a stable three-dimensional structure without the use of extra chemicals, guaranteeing high biocompatibility for use in biomedical applications. The use of these hydrogels is advantageous in settings where toxicity is an issue. In contrast to physically crosslinked hydrogels, chemical crosslinking produces hydrogels with greater mechanical strength, elasticity, wear resistance, and extended service life through a chemical interaction between cellulose and other substances. Depending on the intended use, each variety has unique benefits [42].

2 Design and Characteristics

Several techniques are used in the manufacture of cellulose-based gels to alter cellulose to produce the required gel-like structure; the method selected will affect the final characteristics and uses. These techniques fall into three general categories: physical, chemical, and hybrid. Physical crosslinking is biocompatible and environmentally benign because it uses non-covalent interactions, such as hydrogen bonds and van der Waals forces, to create a gel network without the use of chemical chemicals. Conversely, chemical crosslinking strengthens the gel's structural integrity and mechanical resilience by covalently bonding cellulose molecules with chemical agents. By varying the crosslinking agent's concentration, temperature, and reaction time, this technique enables exact control over gel strength and durability. By adding additional polymers or nanoparticles, hybrid approaches combine physical and chemical processes, frequently improving mechanical qualities and providing new capabilities, such as antibacterial properties. Furthermore, cellulose's inherent structure may be maintained during gel formation by solvent exchange procedures, which makes it perfect for use in biomedical applications such as drug delivery systems and wound dressings. These adaptable synthesis techniques enable cellulose-based gels to be customized to satisfy functional needs in a range of applications [35]. Bacterial cellulose scaffolds are interesting for biomedical applications because of their 3D network structure, which offers benefits such as high purity, high porosity, large specific surface area, non-toxicity, and good biocompatibility. However, they have some drawbacks, such as modest mechanical strength that limits their usage in bone tissue engineering applications and a lack of antimicrobial qualities that are essential for wound-dressing applications. To address these shortcomings and improve the mechanical and biological qualities of bacterial cellulose scaffolds, scientists have added certain polymers or nanoparticles to the scaffolds, resulting in composite materials with better qualities. Furthermore, bacterial cellulose nanofibrils (BCNFs) and bacterial cellulose nanocrystals (BCNCs), which are produced from bacterial cellulose by mechanical homogenization and acid hydrolysis, respectively, can be disseminated in aqueous solutions and employed as reinforcing agents [22].

Bacterial cellulose has a better purity, crystallinity, and water-holding capacity than its plant-derived counterpart. Both bacteria and plants naturally contain cellulose. Physical and chemical changes can be made to produce cellulose derivatives (such as cellulose ethers and esters) to overcome the difficulties that both forms encounter, such as poor solubility and challenging processing. By altering structure without changing chemical composition, physical methods such as mechanical grinding, swelling, and ion exchange produce nanocellulose with improved mechanical and antibacterial qualities as well as adjustable optical and thermal characteristics. Chemical changes such as graft copolymerization, etherification, and esterification can add new functional groups for enhanced mechanical strength, solubility, and biocompatibility, opening a range of uses from drug administration and biomedicine to sensors and energy storage showed in Fig. 3. Notably, cellulose-based hydrogels have exceptional conductivity, mechanical performance, thermal stability, and anisotropic optical qualities, which make them appropriate for intelligent, responsive devices like flexible electronics and force sensors [45].

Mostly found in plants (where it is entangled with lignin and hemicellulose), cellulose is the most prevalent biodegradable polymer on Earth. It is also found in some bacteria, algae, fungi, and marine creatures. Its fundamental repeating unit is D-glucopyranose, which is connected by β-1,4-glycosidic linkages. Each glucose residue has three hydroxyl groups, which promote strong hydrogen interactions

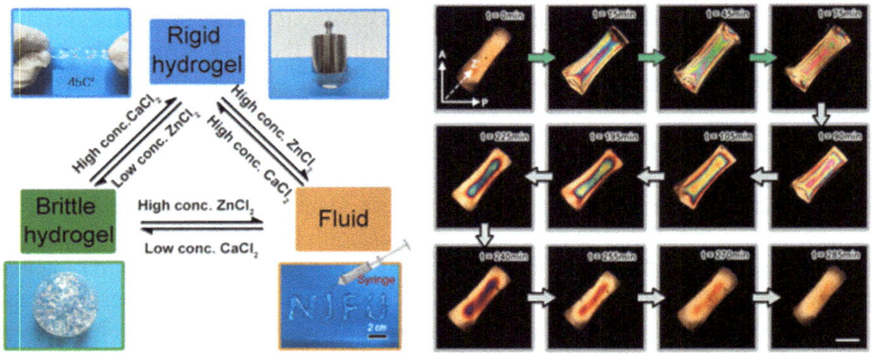

(a)

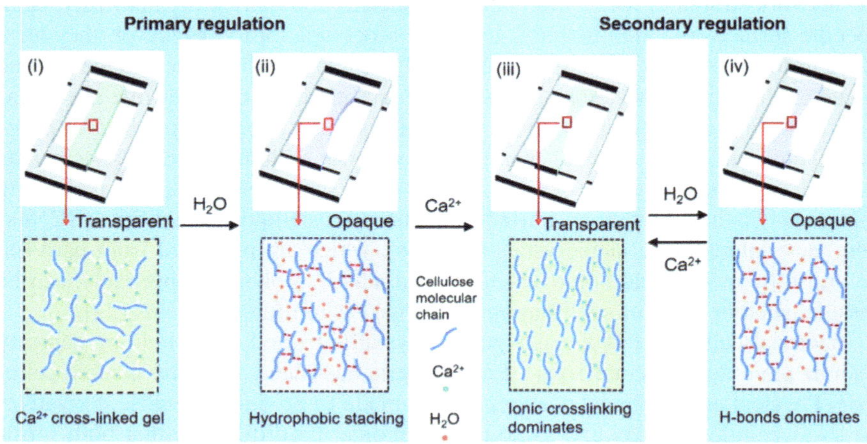

(b)

Fig. 3 New method for preparing anisotropic cellulose hydrogels induced by calcium ions. **a** Reversible multiphase transformation of cellulose hydrogels based on Ca^{2+}/Zn^{2+} exchange and color change of the gel interference under orthogonally polarized light during ion exchange. **b** The schematic diagram of the preparation of a highly oriented cellulose hydrogel by H_2O/Ca^{2+} exchange shows that the flexible switch between ion coordination/hydrogen bond dominance is achieved, thus achieving continuous regulation of a high-oriented structure. **c** Mechanism diagram and microstrain test diagram of gradient anisotropic hydrogel prepared by directional diffusion of cellulose solution in $CaCl_2$ solution [45]

both inside and between molecules. In addition to giving cellulose its high mechanical strength, crystallinity, and fibrous, porous structure, this strong hydrogen-bond network also renders it almost insoluble in common solvents. The ultra-high purity (free of lignin and hemicellulose), remarkable water-retention capacity (up to 98%), and highly crystalline nanofiber network of bacterial cellulose, which is produced by a variety of Gram-positive and Gram-negative bacteria, make it particularly appealing for biomedical applications like hydrogels. Cellulose is frequently chemically altered

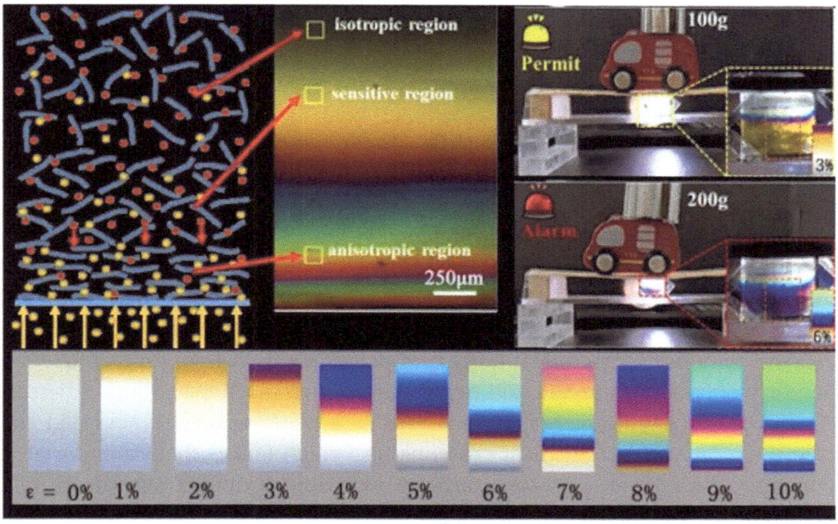

(c)

Fig. 3 (continued)

to increase its solubility and customize its functionality. This results in two primary groups of derivatives: water-soluble cellulose ethers (such as carboxymethyl cellulose and methyl cellulose) and mostly water-insoluble cellulose esters (such as cellulose acetate) [8].

Composed of D-glucose units joined by β (1 → 4) glycosidic linkages, cellulose is an unbranched polymer with a high degree of polymerization and substantial intra- and intermolecular hydrogen bonding. Its main sources are plants, wood, bacteria, algae, and tunicates. Depending on where it comes from and how it is processed, it can have different levels of mechanical performance and crystallinity. Although cellulose's robust hydrogen-bond network makes it insoluble in most solvents, chemical changes, especially to the hydroxyl groups, produce soluble or processable derivatives (such as MC, CMC, and cellulose acetate), each with unique qualities that allow for a wide range of uses. High strength, a huge surface area, and adjustable functionality are further benefits of nanocellulose forms (such as cellulose nanocrystals and nanofibrils). Because of its chemical diversity, mechanical strength, biocompatibility, and biodegradability, cellulose and its derivatives collectively serve a wide range of biomedical applications, from tissue engineering scaffolds to drug delivery and wound healing [30].

Composed of β-D-glucose units connected by β-(1 → 4)-glycosidic bonds, cellulose is a widely distributed organic polymer with noteworthy properties like non-toxicity, biodegradability, and renewability. Through hydrogen bonds and van der Waals interactions, its linear polymer chains create both crystalline and amorphous areas, providing chemical stability and mechanical strength. Natural wood cellulose's inherent porosity can cause brittleness even as its water insolubility supports structural stability. Researchers can improve cellulose's solubility, hydrophilicity, and toughness by chemically altering it, especially in its hydroxyl groups. One excellent example is "elastic memory wood," which increases flexibility by making use of the reversible flow of water through cell walls and lumens. These specific alterations highlight cellulose's versatility and increase its potential uses in hydrogel engineering and biomaterials [32].

When exposed to different environmental stimuli, such as heat, light, humidity, pH, or electrical fields, cellulose-based hydrogels can display customized color, form, or dimensional alterations. These "smart" hydrogels are typically made using three primary techniques: electrospinning, chemical crosslinking, and grafting. Grafting preserves cellulose's biocompatibility and degradability while combining it with functional polymers to create additional capabilities (such as pollution detection and oil–water separation). Stable networks with water-responsiveness and shape memory are created by chemical crosslinking, which makes them appropriate for use in biomedical devices, sensors, and actuators. Some synthetic substitutes, on the other hand, degrade more quickly and produce more hazardous byproducts. Lastly, the ability to precisely adjust fiber orientation by electrospinning improves mechanical qualities and makes pH-responsive membranes suitable for filtering and separation applications. Combined, these techniques broaden the possibilities for creating cellulose-based hydrogels with adjustable properties and various uses, ranging from sophisticated medical treatment to environmental cleanup.

The structure, functionality, and applicability of cellulose hydrogels for a variety of applications are greatly influenced by environmental and procedural factors, including solvent type, temperature, pH, and crosslinking technique. While temperature and pH regulate swelling, mechanical strength, and compatibility, particularly in biomedical contexts, solvent selection is crucial for dissolving cellulose and determining hydrogel qualities. While physical and radiation crosslinking prioritizes reversibility, biocompatibility, and safety, chemical crosslinking creates robust, durable networks but necessitates a careful selection of low toxicity crosslinkers. Assessing the mechanical characteristics, swelling behavior, thermal stability, and biocompatibility of the resultant hydrogels is essential for various applications, from industrial processes and environmentally responsive materials to tissue scaffolds and wound dressings. Researchers may create cellulose hydrogels with optimal strength, flexibility, and degradability by adjusting these properties, increasing their applicability in materials science, biomedicine, and environmental sciences, and the preparation and performance are shown in Fig. 4 [32].

Although collagen, a crucial protein in the extracellular matrix (ECM), is essential for tissue regeneration, its poor mechanical qualities and rapid rate of degradation limit its use in biomedical applications. Collagen is the source of gelatin, which is biocompatible but needs more crosslinking to increase its mechanical strength. To create hydrogels that are appropriate for cell transplantation, non-toxic enzymes such as tyrosinase and transglutaminase are frequently employed as crosslinking agents. It has been demonstrated that combining collagen with other polymers, such as chitosan, can improve its mechanical qualities and improve cell encapsulation,

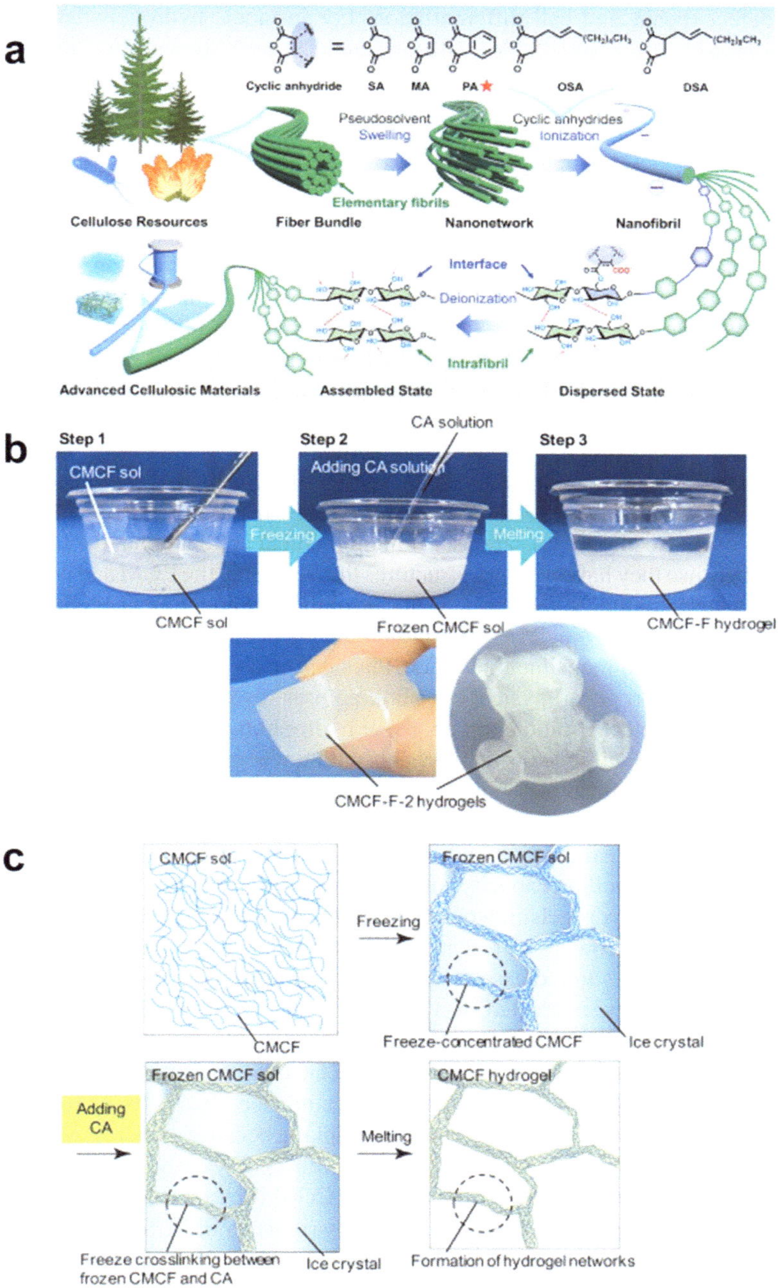

Fig. 4 Preparation techniques for cellulose hydrogels: **a** solvent systems and cellulose dissolution, chemical crosslinking techniques, exfoliation of cellulose elementary fibrils through a dual-step method, and sophisticated cellulose material assembly through surface deionization [32]

bone defect repair, and cell differentiation. For instance, hydrogels derived from glycerophosphates, acid-soluble collagen, and CS have shown good biocompatibility and promote cell viability [41]. Hydrogels based on natural polymers, which come from sources like proteins (like collagen and gelatin) and polysaccharides (like cellulose, chitosan, dextran, alginate, hyaluronic acid, starch, guar gum, and agarose), have high biocompatibility and biodegradability, which makes them great options for pharmaceutical and biomedical applications. Particularly noteworthy is cellulose, the most prevalent natural polymer in the world. Its structure, which is made up of glucose units connected by β-(1 → 4)-glycosidic bonds, can be altered chemically or using solvents to improve its low solubility. Further improving processability and utility, cellulose derivatives (such as MC and CMC) produce hydrogels with desired qualities such as mechanical robustness, flexibility, and reactivity to environmental stimuli. Physical or chemical crosslinking helps other polymers (such as chitosan's antibacterial properties, alginate's chelating ability, and hyaluronic acid's ECM-mimicking properties) create stable networks. When combined, these hydrogels based on natural polymers allow for a variety of uses, such as scaffolds for tissue engineering, drug delivery systems, and wound dressings [44].

Cellulose-based hydrogels are useful in biomedical applications like tissue engineering because they have network structures that resemble the ECM and retain much water. However, cellulose's intrinsic water insolubility frequently necessitates either chemical modification (such as adding functional groups) or nanocellulose fibers to stabilize the three-dimensional network. Physical or chemical crosslinking can be used to create hydrogels. Physical crosslinking uses reversible interactions, such as hydrogen bonds, electrostatic interactions, or hydrophobic interactions, and is easier to process, but it usually results in hydrogels with lower mechanical strength. Chemical crosslinking, on the other hand, uses covalent bonds to create more stable, robust hydrogels. The kind of cellulose derivative utilized, the degree of crosslinking, and the makeup of the crosslinking agents all affect the characteristics, including mechanical performance, swelling capacity, and stimulus responsiveness. The potential uses of cellulose-based hydrogels in wound dressings, medication delivery, and regenerative medicine are growing as researchers look for more environmentally friendly crosslinkers and ways to increase mechanical toughness and biocompatibility [8].

CMC hydrogels were created by vacuum-drying 7% (w/v) CMC at 45 °C after it had been submerged in 1 M citric acid for 1–24 h. While CG02 samples were equilibrated in water for 24 h prior to drying, CG01 samples were dried right away after immersion. Successful esterification was confirmed by FTIR-ATR (peak at ~ 1718 cm^{-1}), and CG01 displayed greater crosslinking because of increased citric acid retention. Greater esterification in CG01 was confirmed by carboxyl group titration, which showed a higher total carboxyl content in CG01 (up to ~ 1350 mEq/100 g) than in CG02 (~ 190 mEq/100 g). Studies on swelling (in water or PBS at 37 °C) revealed that CG02 swelled more than CG01, which was more crosslinked, probably because CG02 had fewer residual crosslinking agents. Even after 48 h in PBS, SEM images showed solid structures and rough surfaces. Due to the abundance of crosslinked citric acid, CG01 started to break down at slightly lower temperatures, according to thermogravimetric analysis, which revealed two main degradation stages. Although both hydrogels remained mechanically stable and swellable for possible biomedical or other applications, compression tests showed that CG01 had a higher compressive strength (up to 3.55 MPa dry, ~ 1.3 MPa in PBS) than CG02, demonstrating its denser crosslinked network [3].

Because of their biocompatibility, biodegradability, and adjustable swelling/ mechanical properties, cellulose-based hydrogels—made from native cellulose or its derivatives (such as MC, CMC, and cellulose acetate) and frequently combined with other natural or synthetic polymers—are appealing for a variety of applications. While cellulose derivatives add ether or ester groups to increase solubility, tunability, and crosslinking potential, native cellulose hydrogels rely on hydrogen bonds between cellulose strands. To create new structural designs and functional performances, composite hydrogels combine extra polymers (such as chitosan, starch, and polyvinyl alcohol (PVA)) with inorganic reinforcements (like silica) that can improve mechanical or bioactive qualities. The type of hydrophilic groups and the crosslinking density of these hydrogels determine their swelling capacity, whereas modified rubber elasticity theories can be used to describe their mechanical strength. This means that it is cellulose based. Consequently, hydrogels based on cellulose can be engineered to satisfy the needs of applications that range from scaffolds for tissue engineering to controlled drug delivery and wound dressings [17].

Composed of β-(1 → 4)-linked D-glucopyranose units, cellulose is a carbohydrate homopolymer that has both crystalline (densely packed, low reactivity) and amorphous (readily reactive) portions. Its numerous hydroxyl groups provide a great deal of hydrogen bonding between and among molecules, which affects accessibility, crystallinity, and the ease of chemical modification. Common cellulose derivatives include ether derivatives and ester derivatives, each of which has unique solubility and functional characteristics based on the type and degree of substitution. Because cellulose and its derivatives are typically hydrophilic, non-toxic, biocompatible, and possibly easily biodegradable, they serve as biomaterials in biomedical settings. These characteristics enable the use of cellulose-based materials in tissue engineering, drug delivery, wound dressing, and medical devices, regardless of whether they are membranes, hydrogels, scaffolds, fibers, or powders. Their application in individualized and cutting-edge healthcare solutions is increased by chemical modification, which further customizes their mechanical strength, swelling behavior, and reactivity to pH and temperature (e.g., adding functional groups or mixing with other polymers or inorganic components) [16, 46]. The use of biomass resources, which provide biodegradability, biocompatibility, and advantageous biological features, has been emphasized by recent developments in hydrogel production. For example, chitosan, which is derived from chitin, has been used to generate chitosan-based hydrogels modified with dextran by grafting procedures, while cotton linter pulp has been used to create cellulose-based hydrogels using a single crosslinking agent. These hydrogels have a porous structure, good swelling capacity, and encourage cell proliferation, all of which help with wound healing, especially when used for drug delivery applications. Additionally, CMC-based hydrogels have been created using a two-step molding and acidification method, which makes them perfect for self-healing applications and appropriate for usage in sealants and stomach mucosa regeneration. These examples highlight the expanding range of biomedical applications for hydrogels made from biomass [46].

Because of their capacity to replicate the ECM and promote cell connections, hydrogels—especially those made from natural biopolymers like cellulose—have drawn a lot of interest in biomedical applications. The lack of antibacterial qualities and moderate mechanical strength are two drawbacks of bacterial cellulose scaffolds, despite their excellent purity, porosity, and biocompatibility. To address these issues, scientists have improved the mechanical and biological characteristics of bacterial cellulose scaffolds by adding nanoparticles and certain polymers. Furthermore, composite scaffolds employ BCNCs and BCNFs, which are dispersible in aqueous solutions, as reinforcing agents.

Controlled drug release and stem cell applications are made possible by the ability to design these hydrogels to react to environmental stimuli including pH, temperature, and electrical or magnetic fields. Though natural hydrogels like collagen, fibrin, and hyaluronic acid are frequently employed, synthetic materials like polyethylene glycol (PEG) hydrogels have adjustable qualities that can be adjusted for increased cell adherence and biodegradability. Biocompatibility is still a critical problem. To hold out hope for improving tissue engineering and scientific research [20]. The three main categories of cellulose—regenerated cellulose, nanocellulose, and cellulose derivatives—each have unique characteristics and uses. While nanocellulose, such as bacterial cellulose, CNCs, and CNFs, has a high specific surface area, exceptional mechanical strength, and distinct surface chemistry that makes it appropriate for use in nanotechnology, materials science, and biomedicine, regenerated cellulose preserves the advantageous properties of natural cellulose. Cellulose derivatives include cellulose ethers, esters, and nitrates, which are created by esterification or etherification reactions. Both chemical and physical crosslinking can be used to make cellulose-based hydrogels; chemical crosslinking creates stable covalent bonds for stronger, quicker-curing materials, while physical crosslinking uses molecular interactions like hydrogen bonds. These hydrogels are being utilized more and more in batteries, supercapacitors, and flexible sensors [47]. Bamboo has drawn a lot of interest as a raw material for cellulose-based hydrogels because of its high cellulose content and its lightweight, biodegradable, and biocompatible qualities. By using physical, chemical, or hybrid crosslinking techniques, these hydrogels are produced, giving rise to a three-dimensional network structure with adjustable chemical and physical characteristics. Bamboo can be treated with bleach and alkali to extract cellulose, or it can be treated with green solvents such as ionic liquids to create cellulose and nanocellulose. These substances can then be combined to create hydrogels, which have a lot of potential for use in fields like environmental remediation, biomedicine, agriculture, and food packaging. With the help of nanoparticles like titanium dioxide (TiO_2), cellulose-based hydrogels have been developed with improved mechanical strength, biodegradability, and antibacterial qualities, which make them appropriate for application in tissue engineering, drug delivery, and wound healing. This adaptable substance is still being investigated for its potential in a range of environmental and biomedical applications, making use of its sustainable qualities and inherent advantages [5].

3 Applications

Because of its biocompatibility, film-forming ability, water absorption qualities, and adjustable solubility, cellulose and its derivatives—in particular, cellulose esters (such as cellulose acetate, cellulose acetate phthalate) and cellulose ethers (such as MC, HPC)—are widely used. Cellulose ethers, which are frequently soluble and chemically inert, can expand in water and create hydrogels appropriate for matrix tablets and other pharmaceutical uses. Typically, cellulose esters form water-insoluble coatings used in membranes, filters, and controlled-release systems. To achieve certain performance features, such as superabsorbent qualities for hygiene products and agriculture or advanced biomedical uses in wound dressings, drug delivery, and tissue engineering, cellulose-based hydrogels are created by chemical or physical crosslinking. Occasionally, these hydrogels are further enhanced by adding synthetic or natural polymers. Cellulose hydrogels and their derivatives adaptability enable a broad range of uses, from complex excipients and scaffolds in regenerative medicine to moisture-control materials in arid soils [26]. Advanced, multipurpose hydrogels and aerogels are increasingly made from cellulose, a plentiful and mechanically strong natural polymer. Because of their distinct physicochemical characteristics, nanocellulose can be mixed with other materials to create hybrid, "smart" systems that are suited for a variety of biological applications. These hybrids made of cellulose offer improved mechanical strength, multisession capability, and reactivity to different stimulus. Recent research has taken advantage of cellulose's natural biocompatibility, renewability, and affordability to create materials with better performance in a variety of applications, from medication administration and wearable medical devices to tissue engineering and wound care.

3.1 Superabsorbent for Personal Hygiene Products

Crosslinked hydrophilic compounds called superabsorbent polymers (SAPs) are used extensively in medical supplies, feminine hygiene products, and disposable nappies because of their capacity to absorb enormous amounts of fluids, such as blood or pee. Many SAPs have historically been synthetic (polyacrylic acid, for example), which presents disposal and environmental issues. Since cellulose fluff pulp is already utilized in commercial absorbent products and new methods, like micro fibrillated cellulose crosslinked with agents like glutaraldehyde, demonstrates superior absorbency and retention capacities, cellulose-based hydrogels present a promising biodegradable substitute. While additives (such as enzymes) might stop infections and odors, the addition of microporous structures can improve swelling rates and water absorption even more. The market is still dominated by synthetic SAPs, although cellulose-based superabsorbents are becoming more popular because of their eco-friendliness, biodegradability potential, and use in fields other than personal hygiene (such as waste management and agriculture).

In personal care and hygiene goods like nappies and feminine napkins, cellulose-based hydrogels show promise as SAPs. Cellulose hydrogels have advantages over conventional synthetic SAPs, which are frequently based on acrylic or acrylamide. These advantages include less toxicity, biocompatibility, and biodegradability. They can be chemically altered (e.g., crosslinked carboxymethyl cellulose or HEC) to improve swelling capacity and mechanical strength. They also have exceptional liquid absorption and retention, which is crucial for controlling bodily fluids without leaking or irritating the skin. To create SAPs with enhanced absorbency, better environmental profiles, and possible antibacterial qualities, current research also investigates more environmentally friendly and sustainable methods, such as gamma-irradiation crosslinking and employing cellulose generated from waste. These developments in cellulose-based SAPs offer practical, environmentally responsible substitutes for synthetic polymers in the personal care sector [17].

The ability of cellulose-based superabsorbent hydrogels to absorb and retain moisture makes them essential in personal care products, especially nappies. They provide comfort and lessen nappy rash or skin irritation by keeping the skin dry and reducing dampness. Additionally, by preventing leaks, their use in nappies reduces the danger of sickness transmission and fecal contamination in childcare facilities. The growing demand for training pants and adult incontinence products highlights the adaptability and health advantages of these cellulose-based hydrogels [15].

3.2 Skin Applications

Tissue engineering uses natural or synthetic materials that promote cell adhesion, development, and differentiation to replace or repair damaged tissues. Scaffolds composed of synthetic polymers (like polylactic acid (PLA) and PEG) or natural biopolymers (like cellulose, collagen, and chitosan) offer a suitable 3D platform that replicates the natural ECM in skin tissue engineering. While synthetic polymers can provide controlled degradation and customizable mechanical properties, natural polymers frequently encourage cell adhesion, proliferation, and biocompatibility. Particularly noteworthy are materials based on cellulose because of their exceptional cytocompatibility, mechanical durability, and ability to create advantageous hydrogen bonds. By increasing mechanical strength, surface area, and cell–scaffold interactions, nanocellulose significantly improves scaffold performance. Notwithstanding certain drawbacks, such as the poor or extremely slow biodegradability of cellulose and the possible brittleness of some synthetic polymers, hybrid techniques that combine natural and synthetic materials can provide scaffolds with the best qualities for skin and other tissue regeneration applications. In the end, these biomaterials' ECM-like environment promotes stronger tissue remodeling, improved integration, and quicker wound healing [33].

Biomaterials based on cellulose, especially nanocellulose or composite hydrogels, offer a flexible foundation for wound healing and skin tissue engineering. While absorbing wound exudate, they can resemble the ECM, encourage cell adhesion and proliferation, keep the environment wet, and permit gas exchange. Improved mechanical integrity, controlled drug release, and antibacterial qualities are demonstrated by cellulose hydrogels that include ions or bioactive compounds (such as growth factors or antibiotics). Despite its low in vivo biodegradability, cellulose's strong mechanical durability and hydrophilic surface promote advantageous cell responses. To improve wound-healing results like quicker re-epithelialization, lower risk of infection, and better tissue regeneration, researchers have mixed cellulose with different polymers (such as chitosan, polyvinylpyrrolidone, and PLA), or inorganic elements (like calcium ions, and silver). With characteristics like adjustable porosity, robust mechanical qualities, and cytocompatibility that promote the formation of new, healthy skin, scaffolds and dressings based on nanocellulose generally show promise for treating serious wounds and burns. A crucial barrier of defense, the skin can sustain superficial (epidermal) or deep (dermoepidermal) lesions from physical stress and burns. Regenerative tissue engineering is being used more and more in modern wound care, using biomaterials, especially hydrogels based on polysaccharides like cellulose, to speed up healing. These hydrogels can absorb exudate, maintain moisture, allow gas exchange, stabilize temperature, and provide infection protection. Nanocellulose is notable for its exceptional mechanical strength, biocompatibility, and ECM-mimicking capabilities. By encouraging vascularization, regulating bacterial growth, and boosting cell adhesion, adding ions (such as strontium or zinc), antibiotics, or substances like silver to cellulose-based dressings improve wound healing even more. According to studies, these composites can increase the rate of wound closure, lower infection rates, and even remove dressings painlessly, which makes cellulose-based scaffolds a desirable choice for managing challenging wounds, including diabetic ulcers and extensive burns [33].

Hydrogels based on cellulose are becoming more and more adaptable and "smart" dressings that react to pH, temperature, and other changes in the wound microenvironment to speed up wound healing. Cellulose-based dressings have antibacterial efficacy against both Gram-positive and Gram-negative bacteria by adding antimicrobial agents such as copper coatings or silver nanoparticles. Furthermore, on-demand drug release and real-time monitoring of wound parameters (such as pH and uric acid levels) are made possible by functional alterations, such as the addition of conductive materials, stimuli-responsive polymers, or biosensors. These intelligent hydrogel systems are therefore intriguing instruments for enhanced wound care and skin regeneration since they provide an ideal environment for cell development and tissue regeneration, guard against infection, and offer useful diagnostic input. Hydrogels based on cellulose are especially well-suited for wound dressings because they offer a moist environment that promotes quick healing, guards against infection, and frequently gives patients cooling comfort. To treat various wound types, such as burns and chronic ulcers, they can be made into sheets, gels, or gauze-like dressings. Cellulose or its derivatives (like CMC) are used in a lot of commercial goods (including Woundtab®, Granugel®, Aquacel® Ag, and XCell®), and they are occasionally mixed with antimicrobials like silver. Because of its capacity to control wound exudates, preserve the ideal moisture content, and promote tissue regeneration, bacterial cellulose is also becoming more and more well-known. According to studies, cellulose-based hydrogels can help second-degree burns and other difficult-to-heal wounds heal more quickly, with less discomfort, and with better results. Research is also being done to fulfill clinical needs for individualized wound care plans, incorporate active chemicals, and modify the swelling and antibacterial qualities of polymeric hydrogels.

Recent studies show that by preserving a moist environment, absorbing wound fluids, and promoting nutrient exchange, hydrogels based on natural polymers can successfully enhance wound healing. Numerous sophisticated hydrogel formulations, which frequently contain materials like zinc, calcium, MXene, or bioactive compounds, demonstrate extra features including improved mechanical strength, antibacterial qualities, and stimuli-responsiveness (e.g., photo/magnetic). These features facilitate enhanced vascularization, less inflammation, quicker tissue repair, and precise medication or factor release. Notably, chemicals (such as deferoxamine, exosomes, or growth factors) that target the difficulties of chronic wounds by controlling oxidative stress, fostering angiogenesis, and encouraging M2 macrophage polarization are included in specialized hydrogels for diabetic wounds. All things considered, these adaptable hydrogel technologies present encouraging approaches to more effective, focused, and biocompatible wound care [44].

Because they offer a moist, biocompatible, and structurally stable three-dimensional environment for cell growth, cellulose-based hydrogels have great potential to accelerate and promote wound healing. Various formulations with improved mechanical qualities, electrical conductivity, antibacterial/antioxidant activities, and controlled drug release are frequently made with CNFs, CNCs, or composite additives like MXene. These hydrogels can enhance cell adhesion and proliferation, promote angiogenesis and epithelialization, and lessen excessive wound contraction, which minimizes scarring. Their adaptable architectures also allow for painless removal or on-demand breakdown when necessary, underscoring their value as cutting-edge scaffolding materials for wound care and skin tissue engineering applications and the details are shown in Fig. 5 [8].

For wound-dressing applications, cellulose-based hydrogels—and chitosan in particular—have attracted a lot of interest because of their chemically easy modification, biocompatibility, and porous three-dimensional structure. While pure cellulose usually necessitates the addition of antibacterial agents (such as ZnO or Ag nanoparticles), chitosan is notable for its inherent hemostatic and antimicrobial qualities. In addition to acting as transporters for medicinal substances, these hydrogels can be shaped into a variety of wound-dressing forms, including fibers, membranes, and sponges. Notably, the creation of dressings that react to temperature and stimuli (such as polypropylene nonwoven textiles that have been grafted with acrylic acid and NIPAAm and then loaded with chitosan) is an example of the continuous advancements in this area. These dressings are potential options for safe, multipurpose wound care materials because they can control permeability and water vapor transport rates while inhibiting bacterial development [17]. For both sheet and

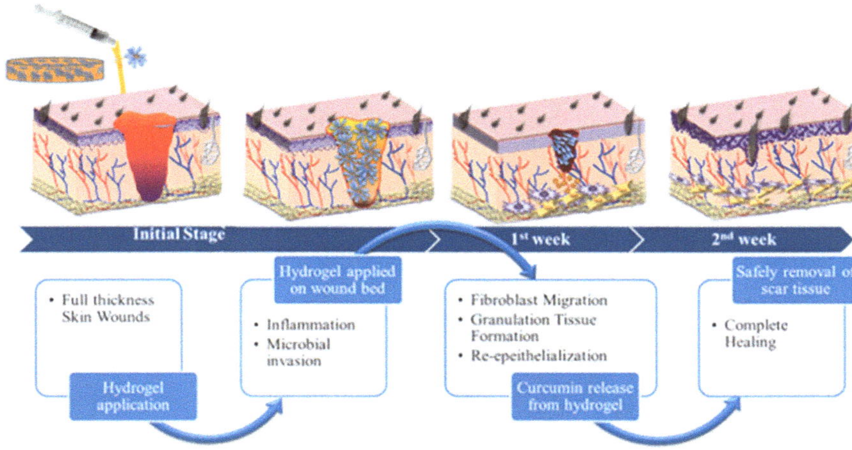

Fig. 5 Wound-healing mechanism at the same stage after hydrogel application [8]

amorphous wound dressings—the latter having non-adherent qualities that facilitate removal without damaging the wound hydrogels are perfect because they create a moist environment that speeds up healing. Many studies have created cellulose-based hydrogels for wound care such as HPC hydrogels to maintain optimal moisture with low adhesiveness. As shown in Fig. 6, combining HPMC and HPC with Gemifloxacin-loaded sodium alginate membranes could sustain drug release and combined HPMC, ofloxacin, and biodegradable microfibers to increase collagen deposition. While antimicrobial modifications, such as those made with ZnO nanoparticles or gentamicin-loaded films, improve the hydrogel's ability to fight infections and improve patient outcomes, formulations containing CMC, either by itself or in combination with additives like zinc oxide, mesoporous silica, or chestnut honey, have demonstrated promise in promoting healing in a variety of wound types, including diabetic ulcers and mucosal injuries [15].

The tunable physical, chemical, mechanical, and biological properties of cellulose-based biomaterials, along with their accessibility, affordability, and biocompatibility, make them extremely attractive choices for wound-healing and dressing applications. These substances perform vital tasks including preventing infections, retaining hydration, absorbing exudate, reducing inflammation, and accelerating the healing process. Active wound dressings derived from cellulose, as opposed to passive biomaterials, can provide advantageous biological activities, such as proliferative, antioxidant, and antibacterial effects. The outcomes of in vitro and in vivo investigations, bioactive compounds added to cellulose-based wound dressings, and different manufacturing techniques are all carefully examined in this review. With the ability to administer bioactive chemicals and improve wound-healing processes, the results demonstrate the efficacy of cellulose-based biomaterials as advanced, adaptable wound dressings or grafts [1].

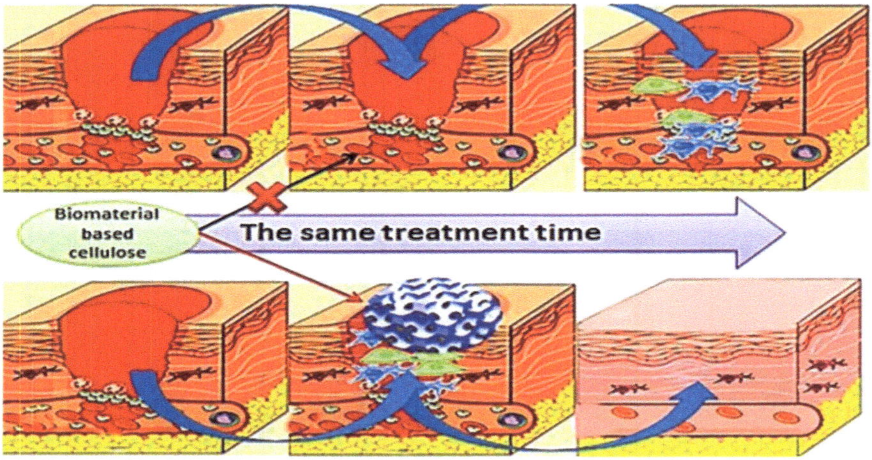

Fig. 6 Different stages of the wound-healing process with and without cellulose-based biomaterial [16]

Cellulose-based materials can mimic skin, encourage rapid cell regeneration, and reduce scarring; they are being used more and more in wound-healing applications. Nanocellulose is a perfect bio-ink component for bioprinting, which produces some of the most cutting-edge materials for wound dressings. The production of porous nanostructures that can stabilize and release functionalized molecules is made possible by the reduction of viscosity and improvement of rheological characteristics of bio-inks by nanocellulose fibrils, particularly those obtained from TEMPO-mediated oxidation. For instance, 3D porous structures that can distribute microbicides have been created using carboxymethylated periodate nanocellulose [13].

3.3 Personal Care Products

Having superior water absorption, non-irritating qualities, and a wide range of biocompatibility, cellulose hydrogels have become more and more popular in a variety of applications. Innovations in personal care such as CNC and crosslinked sodium alginate offer superior absorption for adult incontinence products [1], while environmentally friendly NaCMC/starch membranes take the role of non-biodegradable materials in hygiene products. PVA and TEMPO-oxidized CNFs combine to create mechanically strong, self-healing, antibacterial hydrogels that stick to human skin well and don't cause discomfort in wound-dressing applications shown in Fig. 7. Additional functionalization with bioactive substances, such as resveratrol, bestows antibacterial and antioxidant qualities. These developments highlight the adaptability of cellulose hydrogels in the food, personal care, environmental engineering, and healthcare sectors, demonstrating their great promise for secure, effective, and long-lasting solutions [32].

Nowadays, acrylate-based superabsorbent hydrogels are widely employed to absorb fluids in personal care products. These hydrogels serve to promote skin health and customer comfort by keeping moisture and wetness away from the skin. Disposable nappies with a SAP are thought to be used by thousands of hospitals and nursery facilities worldwide, as well as by most parents in developed nations. Additionally, this material is being used more and more in adult incontinence goods and training pants globally [12].

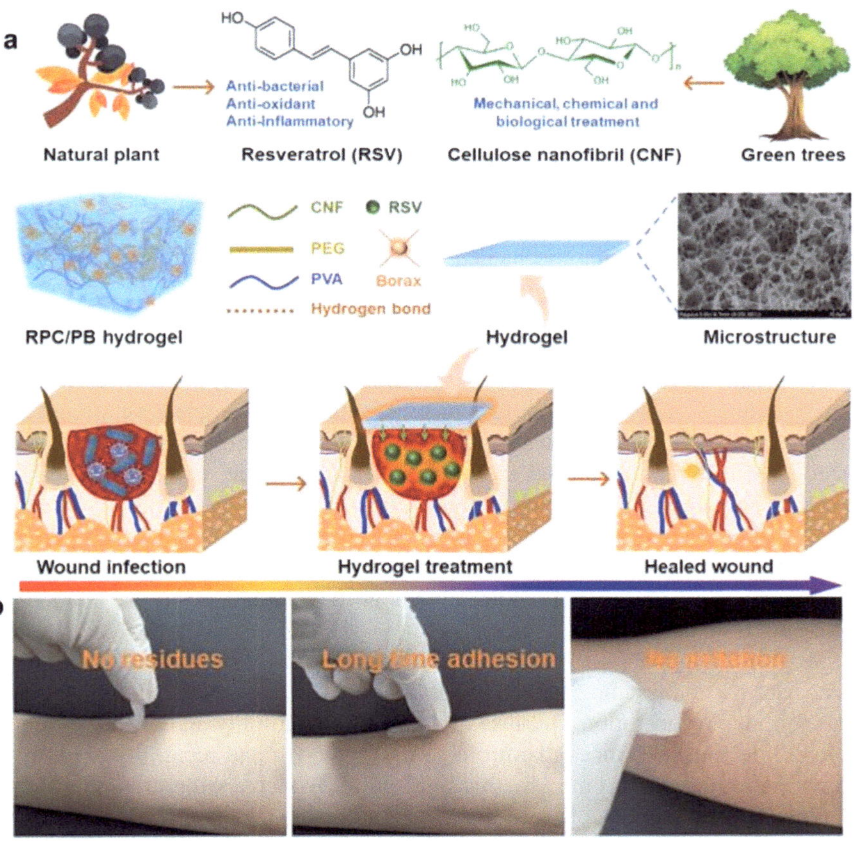

Fig. 7 **a** Cellulose nanofibril-reinforced hydrogel, and the application of the hydrogel in wound healing. **b** RPC/PB-0.5: the hydrogel is removed from human skin without residue [32]

3.4 Bone Applications

Smart hydrogels made of cellulose can be used as sophisticated drug delivery systems and greatly promote bone repair. They lessen the chance of rejection by using the body's own stem cells to promote the creation of specific tissue. Cell migration, proliferation, and vascularization are supported by hydrogel scaffolds that closely resemble natural bone or cartilage environments when temperature-responsive or shape-memory polymers are included with cellulose. Furthermore, scaffolds made of cellulose–nanocomposite materials (such as those containing carbon nanotubes or cellulose nanofibers) enable enhanced mechanical performance and the possibility of patient monitoring in the field of rehabilitation. Beyond bone mending, cellulose hydrogels are useful drug carriers for a range of medications (such as anti-inflammatory, antibacterial, or anticancer treatments) because of their swelling capacity and reactivity to internal or external stimuli. This improves controlled-release precision. These materials have limits in their capacity to completely match the mechanical strength of natural bone and in minimizing oxidative stress during cartilage and bone healing, despite their great promise in tissue regeneration and drug delivery. However, continued advancements in smart materials based on cellulose are probably going to hasten the development of regenerative medicine and treatments. Because they replicate the mechanical characteristics and hierarchical structure of real bone, cellulose-based hydrogels exhibit great promise for bone tissue synthesis. Because cellulose fibers can be aligned, chemically altered, and/or mineralized to increase mechanical strength, bioactivity, and cell compatibility, they offer improved osteoconduction, osteoinduction, and osteointegration. The use of regenerated cellulose nanofibers in chitosan matrices to greatly increase compressive strength, the incorporation of nano-hydroxyapatite to modify gelation temperatures and promote osteogenic differentiation, and the creation of anisotropic, mineralized scaffolds (such as sodium alginate in wood frameworks) that closely mimic the structural complexity of native bone are some strategies. In addition to encouraging cell adhesion, proliferation, and osteogenic differentiation, these cellulose-based scaffolds show promise as substitutes for conventional grafting techniques in the treatment of extensive bone lesions [8].

Millions of individuals worldwide suffer from bone illnesses like osteoporosis, osteoarthritis, and fractures, which are driving up demand for cutting-edge biomaterials for bone regeneration. Because of their biocompatibility and capacity to incorporate bioactive substances that promote tissue growth and repair, hydrogel-based composites hold great promise for this application. To increase the strength, stiffness, and mechanical qualities of these hydrogels, inorganic nanoparticles such as silica, gold, or TiO_2 are frequently added. Studies have demonstrated that the combination of synthetic polymers like PVA and natural polysaccharides like cellulose and chitosan produces a hydrogel that is both flexible and durable. TiO_2 nanoparticles are added to improve mechanical stability, encourage cell division, and have the potential to serve as a drug delivery vehicle. With potential advantages like antibacterial qualities and prolonged drug release that can hasten bone regeneration, this multi-component hydrogel technique offers a novel approach to bone tissue creation. A unique method that addresses past difficulties in hydrogel manufacturing is the incorporation of chitosan, cellulose, and PVA with TiO_2 in a single hydrogel matrix. This approach offers mechanical and biological enhancements for bone regeneration applications [19].

3.5 Skeletal Muscle Regeneration

To encourage healing and muscle growth, self-healing hydrogels are being utilized more and more in skeletal muscle regeneration. They frequently contain skeletal muscle cells like C2C12. For instance, a hydrogel composed of N-carboxyethyl chitosan and dextran-graft-aniline tetramer showed efficient release of C2C12 cells, facilitating muscle regeneration. Likewise, in mice, a hyaluronic acid-chitosan hydrogel containing C2C12 cells promoted blood vessel and neuron development and myofiber regeneration. Other hydrogels that promote muscle fiber growth include GelMA containing poly(3,4-ethylenedioxythiophene) nanoparticles and Gelatin Methacryloyl (GelMA). Because of their proven biocompatibility and capacity to promote the growth and regeneration of skeletal muscle tissues, these hydrogels offer a viable treatment option for volumetric muscle loss [23].

3.6 Oral Drug Delivery

Smart cellulose-based hydrogels offer versatile drug delivery options through various routes oral, nasal, ocular, and beyond by leveraging their responsiveness to environmental stimuli such as pH, temperature, or redox conditions. In oral delivery, cellulose-based materials can undergo coil-to-globule transitions and modulate drug release in response to gastrointestinal pH changes. For nasal administration, thermosensitive cellulose derivatives can improve drug retention and permeation across nasal mucosa, while reducing enzymatic degradation. Similarly, in ocular drug delivery, gentle cellulose gels prolong drug residence time and help overcome tear-induced losses. Moreover, dual-responsive systems (e.g., combining pH and redox sensitivity) further refine the control over release profiles. Collectively, these intelligent platforms enable precise, site-specific drug delivery with minimized side effects and enhanced therapeutic efficacy. Because of their exceptional biocompatibility, which reduces adverse effects and supports vital physiological processes, cellulose hydrogels have become extremely promising materials for drug delivery and wound-healing applications. The creation of a -PL-modified cellulose/γ-PGA hydrogel (CGLH) using a double-network approach is one recent development. It has potent antibacterial qualities and can hasten wound healing by promoting collagen deposition, vascularization, and cell proliferation. A pH-responsive, injectable, self-healing hydrogel made of carboxymethyl chitosan and oxidized hydroxypropyl cellulose is another breakthrough. It uses reversible imine linkages to improve drug release in mildly acidic environments, such as those prevalent in tumor microenvironments. These results demonstrate the potential of hydrogels based on cellulose for clinical application in targeted therapies and regenerative medicine [32].

Hydrogels have been investigated by researchers as possible drug delivery systems to enhance the dose–response relationship, biodistribution, and bioavailability of drugs shown in Fig. 8. Because of their three-dimensional structure and hydrophilic chemicals, hydrogels are perfect for drug administration since they can swell and absorb a lot of water. Non-toxicity, the capacity to include poorly soluble medicines, good bioadhesive characteristics, and control, and delayed drug release are some of the benefits of these systems, particularly those made from natural polymers. Their high porosity and swelling behavior enhance pharmacokinetics by facilitating effective drug loading and prolonged release. Hydrogels can be administered by a variety of methods since they are biocompatible, flexible, and biodegradable. MC, CMC, HEC, and HPMC are cellulose derivatives that have proven especially useful in the production of hydrogels for regulated drug release [28].

3.7 Controlled Drug Delivery

Because of its biocompatibility, ability to swell, and resilience to acidic conditions, cellulose and its derivatives—especially cellulose ethers—are frequently utilized as excipients in medicinal formulations. When water is present, cellulose-based polymers (such as HPMC, CMC, and ethyl cellulose) in controlled-release tablets form a hydrogel layer that serves as a diffusion barrier and maintains drug release. They are useful for oral, nasal, and even injectable routes due to their mucoadhesive qualities, which prolong residence duration and improve the bioavailability of a variety of medications, including those that break down in the gastrointestinal tract. The antiviral and enteric coating uses of cellulose acetate phthalate make it unique. Injectable cellulose hydrogels can transport proteins and cells in tissue engineering settings in addition to oral delivery. Even though cellulose has been used for more complex delivery methods (such as targeting, zero-order release, and bioactive encapsulation), there are still obstacles to overcome to increase their clinical adaptability, especially in terms of achieving release profiles, improving biodegradability, and resolving some drug stability issues. Hydrogels made of natural polymers, occasionally mixed with synthetic or inorganic materials, have exceptional drug delivery capabilities by enclosing and shielding both big and small drug molecules while permitting regulated and stimulus-responsive release. Recent developments include polyelectrolyte complex hydrogels to lessen the gastrointestinal side effects of medications like piroxicam, multidomain peptide hydrogels for sustained cyclic dinucleotide release, and micellar hydrogels based on chitosan that can deliver hydrophilic and hydrophobic drugs at varying rates for stroke therapy. Targeted drug carriers with enhanced responsiveness, extended-release durations, and decreased toxicity are produced by using carbon dots or magnetic nanoparticles. These hydrogels show

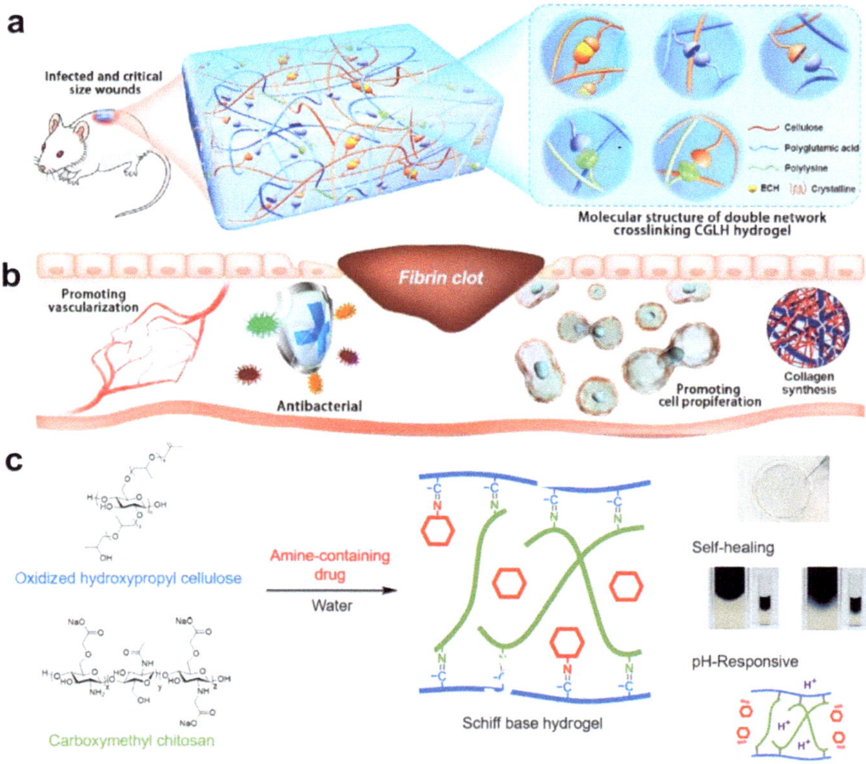

Fig. 8 Cellulose hydrogels in medical and drug delivery fields: **a** schematic illustration depicting the method of preparation and biomedical utilization of double-network CGLH. **b** CGLH exhibits potential in accelerating wound healing by eliminating bacteria and promoting vascularization, cell proliferation, and collagen deposition. **c** Cellulose-based hydrogels are pH responsive to targeted drug delivery [32]

great promise for a range of therapeutic uses, such as targeted intravascular delivery, anticancer therapies, and multipurpose systems that can release drugs and images at the same time.

In pharmaceutical applications, cellulose ethers are essential, especially when used as excipients in medication compositions. These materials are commonly employed in pharmacokinetics and nano pharmacology research due to their biocompatibility and flexibility. The creation of "smart" drug carriers using cellulose-based hydrogels that can react to many stimuli, including pH, temperature, and magnetic fields, and produce regulated drug release is a major area of interest [15].

Since vaginal and rectal drug delivery routes avoid first-pass metabolism and enable direct drug absorption, they provide substantial advantages for both local and systemic treatments. But there are issues like pain, leakage, and decreased bioavailability, especially for medications that are subject to widespread first-pass elimination. The use of thermo-reversible mucoadhesive in situ gels, which lengthen residence time, lessen leakage, and enhance drug efficacy, can help with these problems. Systemic toxicity and low anti-tumor medication absorption sometimes limit the use of chemotherapy in cancer treatment, requiring frequent and high dosages. By enhancing drug delivery to tumor locations, minimizing side effects, and regulating localized drug release, stimuli-sensitive hydrogels—as thermosensitive hydrogels—offer a possible remedy [10].

3.8 *Bulking Agents for Stomach*

One in three adults and almost one in ten people worldwide suffer from obesity, which contributes to several chronic illnesses including type 2 diabetes, heart disease, and some types of cancer. Obesity has negative effects on psychological health in addition to physical health. Exercise, supervised diets, and, in extreme situations, surgery or medication are usually part of the treatment. Dietary supplements, such as bulking agents that decrease hunger by causing stomach bloating, have recently been promoted as weight-loss tools. Though several bulking agents exhibit promise, there is uncertainty regarding their efficacy, and side effects such as gastrointestinal problems have been reported. The potential of new superabsorbent hydrogels, especially those based on cellulose, to expand in the stomach and create a feeling of fullness without the need for invasive surgery is being investigated. Although more research is required to demonstrate their long-term efficacy and safety, these hydrogels are biocompatible and can be customized for used in weight reduction therapy [12].

3.9 *Regenerative Medicines*

Cellulose-based hydrogels exhibit great promise as scaffolds in tissue engineering since they are frequently created using biocompatible crosslinkers (like carbodiimides) and mixed with other natural polymers (like hyaluronic acid). They provide a bioactive milieu that can promote cell adhesion, proliferation, and differentiation, as well as high-water content and mechanical qualities akin to rubber. These materials' surface chemistry can easily be altered using biomolecules to direct cell functions, and they can be designed to break down gradually to match the rate at which new tissue forms. Cellulose scaffolds have been investigated in practice for the regeneration of bone, cartilage, neurons, arteries, and hearts often with negligible foreign-body responses. Researchers can further customize mass movement, cell penetration, and ultimately scaffold resorption by adjusting porosity and adding degradable segments. To better replicate natural extracellular matrices, enable regulated drug or enzyme delivery, and provide carefully tailored three-dimensional architectures for cutting-edge regenerative medicine applications, research is still being done on cellulose-based hydrogels. Because natural polymer-based hydrogels closely mimic the ECM, they are extremely adaptable scaffolds for regenerative medicine. Their high-water content and three-dimensional hydrophilic networks facilitate the effective interchange of metabolites, nutrients, and oxygen, supporting tissue development and cellular activity. To improve osteogenesis, angiogenesis, and inflammation control, recent advancements have concentrated on adjusting mechanical qualities and adding bioactive elements, such as growth factors, magnetically sensitive nanoparticles, and bioactive ceramics. The alignment requirements of tissues like muscle and cartilage are met by hydrogels with anisotropic cell configurations (such as GelMA-based scaffolds). Cells and other components can be precisely positioned using methods like magneto-patterning and vertical 3D cryo-bioprinting to create intricate, gradient-structured tissues. These developments highlight how hydrogels can offer strong, flexible, and biomimetic substrates for tissue repair and the engineering of functional, heterogeneous tissues [44].

As hydrogel supports and guide cells throughout the synthesis of new tissue, smart cellulose-based superabsorbent hydrogels—especially those based on MC—are perfect for regenerative medicine. Because of their hydrophilic (–OH) and hydrophobic (–OCH_3) groups, MC hydrogels are thermosensitive and show reversible changes in water affinity, making them appropriate as injectable, biodegradable scaffolds for tissue engineering and cell treatments. In cell sheet engineering, studies have shown its usefulness in improving cell adhesion and proliferation, creating spatial gradients for endothelial differentiation, and enabling the enzyme-free detachment of constantly growing cell sheets. Furthermore, composite hydrogels that combine MC with other substances, such as collagen for mesenchymal stem cell encapsulation and hyaluronan for tissue engineering of the central nervous system, have demonstrated encouraging outcomes in preserving cell viability and encouraging proliferation [15]. Because of their high-water content and rubbery mechanical characteristics, which mimic soft tissues, hydrogels are extremely biocompatible. Additionally, they permit the integration of bioactive chemicals and cells during their development. The chemistry of hydrogels can be altered using ECM domains to enhance cell adhesion and function, even if cells do not readily adhere to extremely hydrophilic surfaces. Hydrogels are perfect for making biomimetic scaffolds for tissue regeneration because of their characteristics [12].

3.10 Other Biomedical Applications

Recent developments emphasize the value of 3D hydrogel systems in stem cell engineering, cancer research, and drug discovery. Surface plasmon resonance imaging combined with a dextran hydrogel chip allows for consistent, label-free, and high-throughput small-molecule drug identification. In the meantime, 3D alginate–collagen hydrogel beads replicate tumor cell–ECM interactions and enable homogeneous spheroid development, offering more biologically realistic tumor spheroid models than conventional 2D cultures. By triggering mechanosensitive signaling pathways, magneto-responsive GelMA-based hydrogels in regenerative medicine use realigned magnetic nanorods to dynamically modify stiffness and affect stem cell development. Furthermore, spatial control of stem cell proliferation and differentiation is made possible by dual-crosslinked alginate hydrogels with micropatterned growth factor immobilization, highlighting the potential of hydrogel-based platforms to closely resemble physiological environments and direct tissue engineering procedures.

Because of their biocompatibility, adjustable swelling behavior, redox-, pH-, and temperature responsiveness, cellulose-based hydrogels provide a flexible substrate for targeted and regulated drug administration. Researchers can improve pore structure and stimulus sensitivity by altering cellulose at different response levels (such as carboxymethylation or the addition of nanoparticles), which allows for precise control over drug loading and release. The kinetics of drug release can be controlled by these hydrogels' ability to swell or contract in response to environmental stimuli like chemical agents, temperature changes, or pH shifts. To deliver a variety of therapies (such as insulin, proteins, and small-molecule medications) to target regions like the stomach, intestines, or wound areas, different formulations, such as semi-interpenetrating and interpenetrating polymer networks, have been investigated [17].

3.11 Cellulose-Based Hydrogels as Biosensor

Additionally, cellulose-based sensors, such as those incorporating glucose oxidase or urease enzymes, have been developed for monitoring glucose, urea, and other metabolites in biological samples. These cellulose-based platforms offer high sensitivity, flexibility, reusability, and low cost, making them ideal for point-of-care diagnostics and other rapid testing applications. To monitor glucose, urea, and other metabolites in biological samples, cellulose-based sensors have also been developed. Examples of these sensors include those that incorporate urease or glucose oxidase enzymes. These cellulose-based platforms are perfect for point-of-care diagnostics and other quick testing applications because of their high sensitivity, versatility, reusability, and affordability [18].

3.12 Other Tissue Applications

In addition to skin and bone, smart cellulose-based hydrogels have the potential to regenerate other tissues. By supplying the electrical cues necessary for neuron growth and function, electrically conductive or piezoelectric cellulose scaffolds can aid in directing cell migration and fostering neural tissue repair in nerve regeneration. Similarly, cellulose hydrogels that replicate the lubricating and mechanical characteristics of the ECM can help cartilage tissue by promoting chondrocyte differentiation and proliferation. Additionally, through improved cellular support and nutrient exchange, cellulose-based hydrogels with customized responsiveness—such as dual pH/ionic sensitivity—assist cartilage healing. These hydrogels can improve angiogenesis and heart function after myocardial infarction by delivering growth factors (such as bFGF) in a way that responds to reactive oxygen species in cardiac tissue regeneration. Therefore, developments in stimulus-responsive cellulose hydrogels present a wide range of opportunities. In tissue engineering, cellulose-based hydrogels are useful scaffolds because they produce three-dimensional, extremely porous matrices that resemble the ECM. To regenerate tissues including cartilage, bone, muscle, and skin, these scaffolds encourage cell adhesion, proliferation, and differentiation. Furthermore, CB hydrogels can be functionalized to either promote or prevent cell adherence, which makes them appropriate for a variety of biomedical supports and implants. When the temperature changes, thermos-responsive systems (such as chitosan-grafted PNIPAAm) transform from an adherent to a non-adherent state, allowing for gentle cell harvesting. Because it prevents damage from forceful detachment techniques, this property is very beneficial for cultivating and transporting stem cells. Additionally, injectable CB hydrogel systems have demonstrated potential in vivo applications when paired with living cells and bioactive compounds, including cartilage repair, offering a minimally invasive approach to tissue regeneration [17].

Because hydrogels may replicate the ECM, they create the perfect milieu for cellular adhesion, development, and migration, making them indispensable in tissue engineering. Hydrogels made of cellulose, which comes from natural sources like microorganisms or plants, are very useful due to their mechanical stability, biodegradability, and biocompatibility. These hydrogels can be customized for a variety of uses, such as the regeneration of skin, cartilage, bone, and nerve tissue. Their high porosity promotes tissue growth by facilitating the dispersion of bioactive molecules and cell distribution. Hydrogels with adjustable qualities are made from cellulose derivatives such MC, HEC, and HPMC. The functioning of these scaffolds is further enhanced by the introduction of bacterial cellulose, which has a higher tensile strength and is more hydrophilic [2].

Effective treatment methods are necessary to encourage tissue regeneration since diseases, trauma, and traumas frequently result in tissue and organ failure. Traditional techniques like autografts and allografts are constrained by problems including donor shortages, high infection risk, and several procedures. Tissue engineering with biomimetic 3D scaffolds seeded with cells has therefore become a viable substitute. Because of their structural resemblance to the ECM, hydrogels are especially appealing for tissue engineering because they offer appropriate three-dimensional microenvironments for cell survival, proliferation, and function. Scaffolds must encourage cell recruitment and regulate the rate at which bioactive molecules are released to maximize tissue regeneration. This will help maintain local concentration and preserve the bioactivity of the molecules. Polymeric scaffolds of various topologies have been created using a variety of methods, including solvent casting, electrospinning, and freeze-drying. Both natural and synthetic polymers are frequently utilized in the construction of scaffolds; natural polymers, such as plant-based biopolymers (cellulose, hemicellulose, and starch), provide biocompatibility and resemblance to the components of mammalian ECM. Along with bioactive substances from plants like Stevia and Rubia Philippines that have antibacterial and antioxidant qualities, plant protein-based hydrogels like those made from soy and zein also exhibit potential. The potential of plant-based hydrogels in tissue engineering and drug delivery is highlighted in this review, which focuses on current developments in their biomedical applications [24].

3.13 Cartilage

Cellulose-based hydrogels show tremendous potential for the production of cartilage tissue because they replicate the key characteristics of real cartilage, including a highly hydrated extracellular matrix, significant mechanical strength, and biocompatibility. Scaffolds that promote cell adhesion, proliferation, and differentiation are essential since cartilage has restricted circulatory networks and hence weak self-repair capacities. Several techniques include the development of multi-zonal or layered hydrogels that imitate the superficial, intermediate, and deep sections of cartilage, as well as the reinforcement of hydrogels with substances such as hydroxyapatite, silk fibroin, PVA, and bacterial cellulose. These combinations are ideal for load-bearing cartilage repair because they produce better compressive strength, elasticity, low friction coefficients, and strong fatigue resistance. Overall, the use of cellulose in hydrogel formulations highlights the potential of the material by creating a more cartilage-like milieu that can successfully encourage tissue regeneration [8].

3.14 Water Reservoirs in Agriculture

Superabsorbent hydrogels based on cellulose are becoming more popular in agriculture as biodegradable and sustainable substitutes for traditional acrylate-based solutions. They can greatly swell in the soil, which improves water retention, lowers the frequency of irrigation, and increases soil porosity and aeration (which promotes root growth). These hydrogels also serve as controlled-release nutrient platforms by controlling water release, increasing fertilizer effectiveness, and lessening environmental impact. The efficacy of cellulose-based superabsorbent hydrogels is further improved by crosslinking with biodegradable agents or adding additional renewable polymers, such as starch. In the end, cellulose-derived superabsorbent hydrogels can be a useful tactic for maximizing irrigation and soil health in water-scarce areas, providing a sustainable substitute for petroleum-based, non-biodegradable superabsorbent [26].

It is commonly recognized that a hydrogel undergoes a transformation from a glassy to a rubber-like state during the swelling process, which allows the material to retain water even when severely compressed. However, if there is a difference in humidity between the material's inside and exterior, the inflated hydrogel can gradually release water by a diffusion-driven mechanism [12]. Since the 1960s, hydrogel networks of hydrophilic polymers that can absorb huge amounts of water—have been employed in biological applications. They are being marketed for a variety of uses. The hydrogel-based polymers utilized in tissue engineering, medication delivery, wound dressing, contact lens manufacturing, and hygiene products are thoroughly examined in this review. It categorizes the materials according to their chemical makeup and lists the essential specifications for every application. The commercial name, manufacturer, chemical makeup, and salient features of each product are listed in detail, providing a thorough rundown of hydrogels that have achieved commercial maturity in the biomedical industry [6].

3.15 Applications for Cellulose Hydrogels in the Food Industry

As stabilizers, thickeners, and packaging materials that improve food quality and shelf life while lowering environmental pollution, cellulose hydrogels are important to the food industry. Recent studies have demonstrated their wide range of applications, which include strong antifungal and mild antibacterial capabilities against common infections. These are made possible by designed hydrogel films that contain cationic polymers and TEMPO-oxidized cellulose nanofibers. Furthermore, cellulose hydrogels exhibit two stimuli-responsive behaviors: temperature and pH. This allows for real-time colorimetric freshness indicators and regulated preservative delivery in packaging which is shown in Fig. 9. The potential of cellulose-based hydrogels as environmentally acceptable substitutes for conventional plastic packaging is further highlighted by sustainable production techniques, such as ionically crosslinked hydrogels made from wheat straw or cotton linter pulp [32].

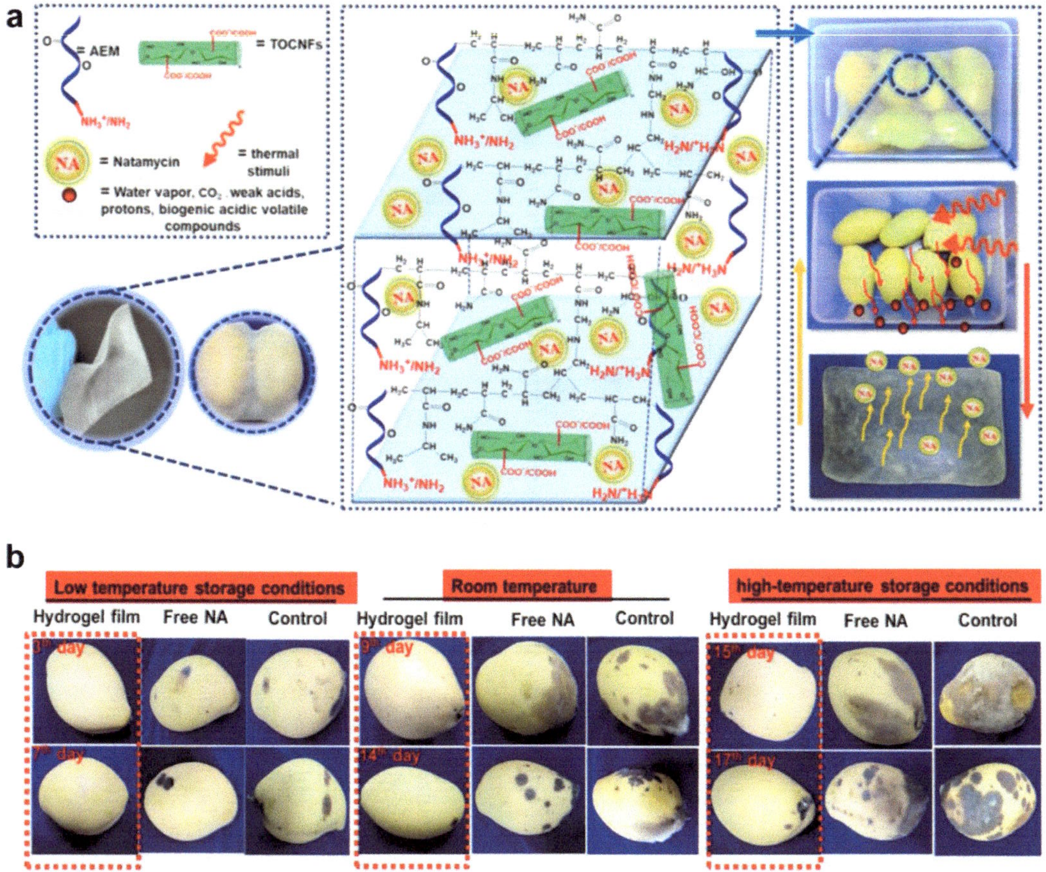

Fig. 9 Cellulose hydrogels in the environmental food industry. **a** Thermo-/pH-responsive preservative delivery based on TEMPO cellulose nanofiber/cationic copolymer hydrogel film in fruit packaging, and **b** the appearance development of mangos from two cultivars, subjected to different in vivo treatments and stored under the conditions [32]

3.16 Applications for Cellulose Hydrogels in Environmental Engineering

In environmental engineering, cellulose hydrogels show exceptional adaptability and sustainability as efficient adsorbents for organic contaminants, heavy metals, and colors in wastewater. While novel nanocomposite designs, including incorporating tungsten oxide nanodots allow solar-driven water evaporation for seawater desalination, their adsorption mechanisms include hydrophobic, ion, and electrostatic interactions. In addition to purifying water, cellulose hydrogels improve soil quality by retaining moisture and regulating fertilizer release. They are also used to purify the air by capturing harmful gases and particle matter. Together, these diverse uses highlight the potential of cellulose hydrogels as environmentally benign, multipurpose solutions for resource management and environmental rehabilitation which is shown in Fig. 10 [32].

3.17 Agricultural Use

SAP hydrogels based on cellulose have shown in Fig. 11 great promise in agriculture as water reservoirs and as vehicles for agrochemicals such as insecticides, fertilizers, and herbicides. They are particularly helpful in dry or drought-prone areas because they absorb and hold onto huge amounts of water, reducing the need for frequent watering and preventing water loss through evaporation. Additionally, their ability to swell enhances soil aeration and structure, which promotes the growth of plant roots. Additionally, hydrogels based on cellulose are renewable and biodegradable, making them more environmentally benign than the widely used SAPs based on synthetic acrylate, which can build up and contaminate the soil. To provide plants with a consistent and focused supply of nutrients or manage the release of herbicides more sustainably, several formulations are utilized, frequently incorporating CMC, chitosan, starch, and other polysaccharides [17].

Hydrogels based on cellulose are attracting a lot of interest in the field of agriculture. This is mostly because horticulture and agriculture need to optimize water resources and minimize water usage. Moreover, these resources are the main tactic to develop the proper way to use water to save it for later [15].

3.18 Food Industry

Their adaptability, biodegradability, and potential to improve food quality and safety, cellulose-based hydrogels are receiving a lot of attention in a variety of food-related applications. Cellulose-based hydrogels are employed in biosensors in the food sector to identify analytes like acids, alcohol, and carbohydrates in fermented foods to check

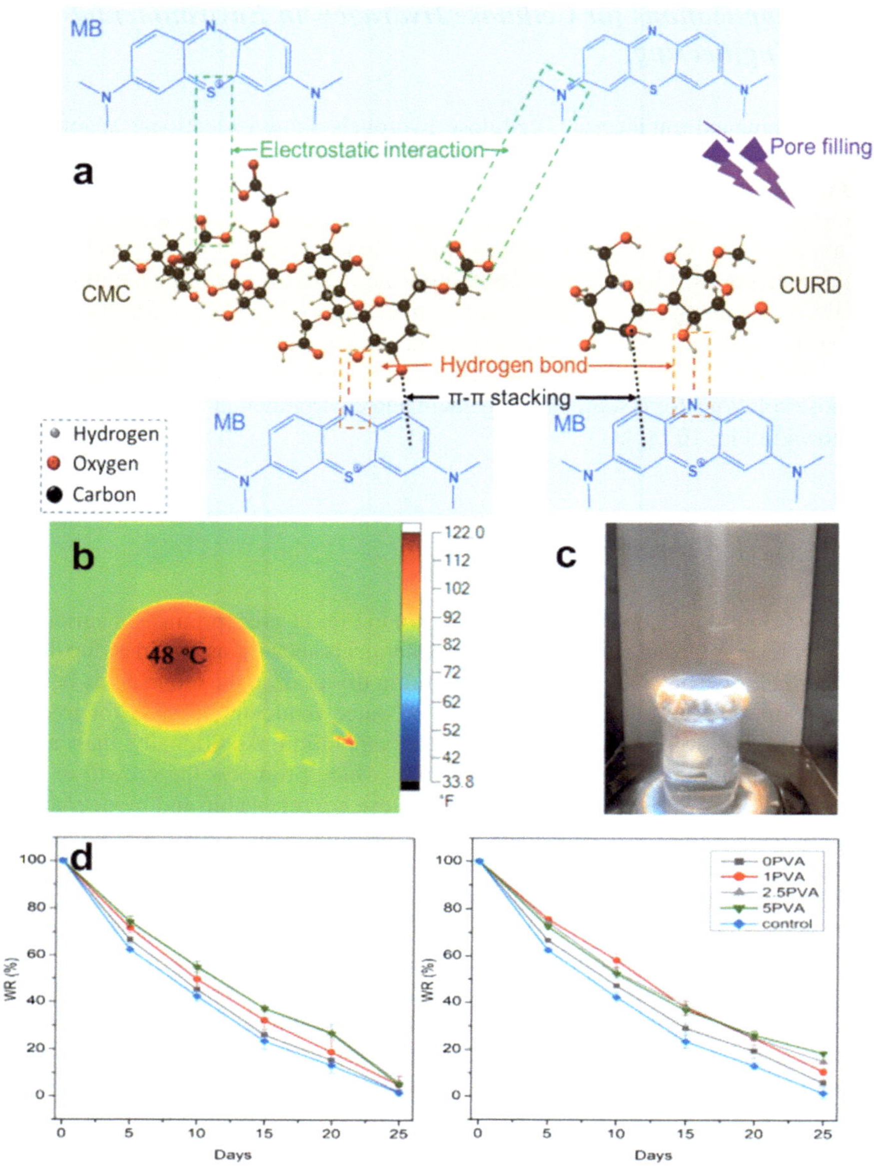

Fig. 10 Cellulose hydrogels in environmental engineering fields. **a** A biodegradable solution to remove organic dyes from water via the mechanism of adsorption of MB onto CURD/CMC adsorbent. **b**, **c** Picture of practical application of water vapor evaporation of composite materials under sunlight irradiation. **d** Soil water-retention experiment: comparison of swelling properties of hydrogel particle size ≥ 2 mm and hydrogel particle size ≤ 2 mm in soil [32]

Fig. 11 Change of soil porosity with the swelling of hydrogels [17]

food quality. By eliminating pollutants like dyes and heavy metals from wastewater, they also significantly contribute to environmental conservation. Cellulose-based hydrogels are used to produce biodegradable polymers with enhanced barrier qualities, antibacterial capabilities, and shelf-life extension in food packaging. Cellulose-based hydrogels are also utilized in the creation of functional foods, which offer regulated release of bioactive substances for positive health effects. They are a vital component of packaging, functional food delivery systems, and food quality control because of their versatility, adaptability, and environmental friendliness [34, 37, 38].

3.19 Cellulose Hydrogels for Enzyme Sensing

Using luminous nano cellulosic hydrogels containing sulfur and nitrogen-doped graphene quantum dots, Palomera et al. created a novel, inexpensive fluorescence-based platform for enzyme sensing that targets laccase. With graphene quantum dots acting as the luminophore, this method makes use of the hydrogels' fluorescence response rather than conventional techniques that depend on catalytic activities influenced by environmental variables. Increased fluorescence in the hydrogels prevented self-quenching and increased the sensitivity of laccase detection. Through non-covalent interactions between the sensor and the analyte, this straightforward and affordable technique enables stable detection and possible enzyme recycling [4].

3.20 Applications for Textile Processing

Because of their high absorbency, porous structure, and abundance of reactive functional groups, cellulose-based hydrogels are being used more and more in the textile industry. They can be used in heat and moisture management textiles, textile wet processing (dying and printing), and self-fitting disposable personal care items. Furthermore, by using interactions like complexation, ion exchange, and electrostatic attraction, these hydrogels—especially in hybrid forms—are very successful at eliminating different contaminants (such as metal ions and colors) from textile wastewater. As a result, cellulose-based hydrogels exhibit great potential as environmentally friendly adsorbents. At the right pH, they can bind and extract harmful substances from wastewater using their reactive amine or hydroxyl groups, assisting in the resolution of important environmental issues facing the textile sector [17]. To stop odors and microbes from growing, antimicrobial treatment is becoming a common finish for textiles, especially in hygienic, medical, and institutional settings. It is also used in women's clothes, sportswear, and fashionable apparel. Because of their hydrophilic and porous nature, which facilitates the movement of moisture, natural fibers like cellulose and textiles made of proteins are more vulnerable to microbial attack. To stop bacterial growth and enhance the fabric's durability and hygienic qualities, antibacterial chemicals are now frequently used in textile manufacture [29].

3.21 Ionic Conduction/Battery

Because of their exceptional elasticity, transparency, mechanical properties that can be adjusted, and stable ionic conductivity, cellulose-based conductive hydrogels hold great promise for a variety of applications, from flexible electronics and sustainable energy devices to wearable sensors and tissue engineering. Recent developments include the creation of nanocellulose-graphene composites for a variety of sensing platforms, antifreeze conductive hydrogels that continue to function in subzero temperatures and the addition of cellulose to alkaline-resistant double-network hydrogels that allow for ultra-stretchable zinc-air batteries. Additionally, the possibility for biocompatible, self-powered implantable devices is highlighted by transitory zinc-ion batteries with totally biodegradable cellulose aerogel-gelatin electrolytes. When combined, these cellulose-derived materials show the synergistic advantages of enhanced ionic conductivity, toughness, and mechanical strength, paving the way for energy storage systems, transient electronic technologies, and next-generation sensors [45].

3.22 *Thermal Insulation*

Because of their superior mechanical performance, low heat conductivity, and lightweight nature, cellulosic polyphonous materials have a lot of promise to replace traditional thermal insulation materials. One prominent example is a mixed aerogel made of BC and methyltrimethoxysilane, which has a lightweight, layered porous structure that offers great thermal insulation across a wide temperature range (− 20 to 150 °C), high compression resilience (99%), and super hydrophobicity (≈ 168°). Even at high humidity (30–90%), its aerogel's moisture-proof qualities guarantee constant heat conductivity. Additionally, it may be scaled up and freeze-formed into a variety of shapes, indicating bright futures for industrial uses shown in Fig. 12 [45].

3.23 *Separation of Water and Oil*

Ecosystems and human health are seriously threatened by the discharge of oily wastewater, which membrane separation technology is thought to be an efficient way to address. For oil–water separation, cellulose-based hydrogels work incredibly well, especially those having super hydrophilic and underwater super oleophobic qualities. Membranes with antimicrobial, self-healing, and self-cleaning qualities are being developed to enhance their performance. Over time, these membranes' ability to recover from mechanical injury allows them to continue to separate materials effectively. For instance, an Ag/AgCl nanoparticle-enhanced cellulose-based hydrogel was developed, and even after 50 cycles, it showed outstanding oil–water separation with 99.5% efficiency. Ag/AgCl nanoparticles are added to the hydrogel to improve its mechanical characteristics and photocatalytic self-cleaning capabilities. They also have antibacterial activity, which increases the membrane's sustainability and durability [43].

3.24 *Diagnostics*

In contemporary medicine, diagnostic tools are essential for measuring vital signs, identifying microbial or viral agents, and evaluating physiological characteristics. Particularly in isolated or underdeveloped regions, these devices need to be accurate, reliable, affordable, and easy to use. The price and efficacy of paper-based diagnostics in tests including blood type, HIV, dengue, and cancer biomolecule detection have drawn attention. Although less research has been done on hydrogel-based diagnostics, they provide benefits such as improved biomolecule interactions and optimized enzyme function. For example, research exhibiting enzyme immobilization for color change detection and fluorescence emission via nanoparticle integration has

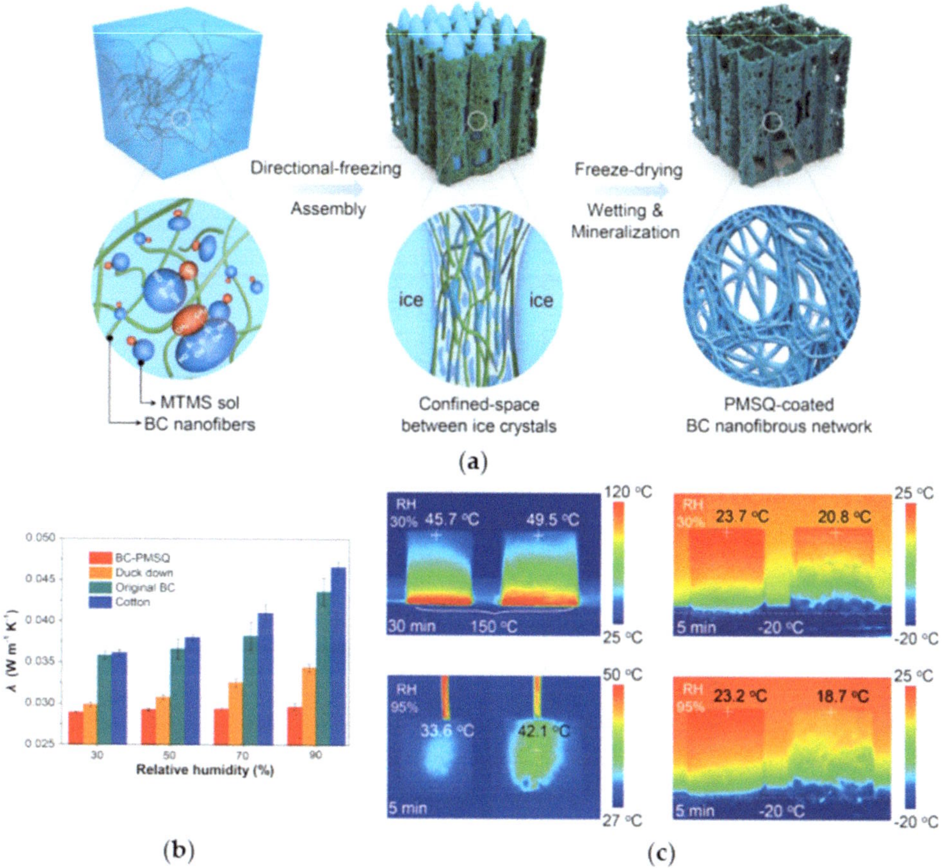

Fig. 12 Nanocellulose hybrid aerogels for thermal insulation. **a** Processing principles and synthesis of the BC–PMSQ hybrid aerogels. **b** Thermal conductivities λ of BC–PMSQ, down feathers, and pure BC membranes. **c** Durable thermal insulation performance was evaluated by optical and infrared images [45]

demonstrated the potential of cellulose-based hydrogels in biosensing applications, underscoring their potential in the creation of novel diagnostic technologies [11].

3.25 Cellulose-Based Hydrogels for Magnetic Hyperthermia Applications

The adaptability of materials based on HPMC in cutting-edge biomedical applications has been highlighted by recent investigations. For instance, in vitro and ex vivo tests have indicated that a composite of HPMC, PVA, and Fe_3O_4 that is thermally contractible, injectable, and biodegradable has potential for magnetic hyperthermia. Another invention employs HPMC in freeze-dried vaginal inserts, which offer appropriate mucoadhesive Ness and regulated probiotic release, potentially helping to prevent and treat vaginal infections [15].

3.26 Cellulose-Based Hydrogels in Filler Technology

Formulations comprising polymethylmethacrylate, dextran, and HPMC have been demonstrated to promote safe and effective facial contouring in both animal models and clinical applications. Additionally, MC, an FDA-approved cellulose derivative, has become a promising soft tissue filler because it effectively supports fibroblast adhesion and proliferation. In addition, cellulose ethers are widely used in pharmacokinetics and nano pharmacology research and are important excipients in medicinal formulations. By creating "smart" hydrogels that react to different stimuli, including pH, temperature, or magnetic fields, these materials can be used to enable regulated drug release and open new possibilities for contemporary drug delivery systems [15].

3.27 Applications of Hydroxypropyl Methylcellulose (HPMC), Hydroxypropyl Cellulose (HPC), and Carboxymethyl Cellulose (CMC)

Because of their exceptional film-forming or superabsorbent qualities, biocompatibility, and environmental responsiveness, HPMC, HPC, and CMC are versatile cellulose derivatives that are widely used in tissue engineering and drug delivery. HPMC has been used in several formulations, including films for better oral disintegration and bio adhesion, pH-sensitive hydrogels for regulated oral medication release, self-setting injectable hydrogels for cartilage tissue engineering, and highly porous scaffolds for bone healing. Because of its temperature responsiveness, HPC has been used to produce sophisticated drug delivery systems like nanocarriers and floating tablets, biodegradable 3D hydrogels that assist stem cell development, and photo-patternable scaffolds that promote cell proliferation. Likewise, CMC's chelating and deforming properties have been used to create hybrid hydrogels, microparticles, and composite scaffolds for targeted cancer treatments, neuroma treatment, and bone and dental tissue creation. All this research shows how cellulose-based biomaterials have a great deal of promise for creating intelligent, multipurpose systems for regenerative medicine and regulated drug delivery [15].

4 Conclusions and Future Perspectives

Because of its hydrophobic nature in water, poor wettability, sensitivity to condensation, and vulnerability to bacterial contamination, cellulose has not found extensive use in pharmaceutical goods. Strong intermolecular hydrogen bonds and electrostatic interactions inside its fibril network are the main causes of its insolubility in water and common solvents, which is a considerable obstacle. Because of these interactions, stiff bundles with uniformly spaced hydroxyl groups are formed, giving cellulose its great stiffness and crystallinity. Despite its potential, this limits its adaptability in medical applications by making it challenging to change its intrinsic structure and straight backbone [31]. Collagen, chitosan, and particularly cellulose are examples of natural polymers that are used in tissue engineering to make scaffolds that mimic the ECM and promote skin regeneration. Cellulose is very useful for creating hydrogels or composite scaffolds that hold moisture, permit gas exchange, and promote cell growth because of its sustainability, mechanical strength, and simplicity of functionalization. There are still many obstacles to overcome, though, such as the requirement for "green" synthesis techniques that steer clear of hazardous solvents, practical ways to lower production costs, and improved storage procedures to preserve scaffold performance. The wider use of stem cells and growth factors is also limited by ethical and technical challenges, and problems like immunological rejection, infection or cancer risk, and the intricate criteria for completely regenerating skin appendages still exist. Though promising, innovations such as 3D bioprinting and the use of nanoscale additives must get over these obstacles before perfect, completely functional tissue-engineered solutions may be made possible.

Since their inception in the early 2000s, smart cellulose-based hydrogels have significantly increased the use of cellulose in biomedical applications by including dynamic, stimuli-responsive capabilities (such as redox, pH, temperature, and magnetic sensitivity). Recent developments involve shaping these materials into intricate designs using methods like 3D printing and integrating graphene or carbon nanotubes to impart electronic properties that are helpful in wearable sensors and other bionic applications. Although cellulose is biocompatible, there are still issues with in-body deterioration because the human body does not have the cellulases necessary for effective breakdown. Research on the incorporation of cellulase into cellulose hydrogels suggests that this problem may be solved. Future directions include developing multifunctional platforms that combine medicinal advantages, intelligent reaction, and mechanical support, as well as improving production techniques using cell-based or additive manufacturing techniques. All things considered, the adaptability of smart cellulose-based hydrogels is expected to spur additional development in drug delivery, tissue engineering, and other cutting-edge biomedical applications.

With their many renewable resources, eco-friendliness, high biocompatibility, and multifunctional tunability, cellulose-based hydrogels have great potential for use in biomedicine and other cutting-edge fields such as flexible wearable sensors, drug administration, and wound dressings. Self-healing, high strength, adhesion, conductivity, and stimulus–response are some of their intriguing qualities, but there are still issues with them, such as poor antifreeze performance, signal loss, restricted chemical compatibility, foreign-body responses, and limitations in large-scale manufacture. Cellulose-based hydrogels are set to advance from laboratory experiments to industrial manufacturing and daily uses as research into these challenges continues. This will allow them to fully utilize their special qualities. Even though cellulose-based hydrogels are known for being environmentally benign, biocompatible, and adaptable, they nonetheless confront several difficulties. Their dependence on hazardous solvents, their low mechanical strength and stability in physiological settings, and the challenge of accurately regulating biodegradation profiles are the main ones. Large-scale production is also hampered by costly and complicated modification procedures, and market and regulatory acceptability necessitate thorough testing and cost-effectiveness. The development of green solvents, sophisticated crosslinking techniques to improve mechanical performance, and innovative approaches to designing responsive, "smart" hydrogels will be the main areas of future study. Additionally, efforts will be made to explore multifunctional applications (such as drug delivery, wound healing, and biosensing), ensure environmentally sustainable production and disposal practices to support wider commercial and biomedical adoption and customize in vivo biocompatibility and biodegradation rates to meet specific application needs.

Because of their structural resemblance to the ECM, biocompatibility, and biodegradability, hydrogels based on natural polymers hold great promise for use in biomedical applications. To increase their mechanical strength, researchers have investigated several techniques, including the development of nanocomposite, double-network, and dual-crosslinking hydrogels. In the meantime, novel alterations-through enzymatic or biological processes—improve the structural characteristics of natural polymers, producing materials with improved performance, injectability, and responsiveness for uses such as tissue regeneration, medication administration, and wound healing. By providing exact control over their structure and activity, advanced fabrication techniques like 3D and 4D printing further increase the potential of hydrogels and open the door for next-generation solutions in drug development, regenerative medicine, and other healthcare domains. Along with the discovery of new cellulose solvents such as Lithium chloride/N, N-dimethylacetamide (LiCl/DMAc), N-methylmorpholine oxide (NMMO), ionic liquids, and NaOH/urea (or thiourea) aqueous solution, cellulose hydrogels have also been documented. Certain stimuli-responsive hydrogels have been created from MC, HPC, HPMC, and CMC using both chemical and physical processes because of the synthesis of cellulose derivatives, primarily cellulose ether. This has resulted in hydrogels that are sensitive to pH and temperature. Cellulose and other polymers have been combined to create composite hydrogels [7, 38, 39].

References

1. Abazari MF et al (2021) Recent advances in cellulose-based structures as the wound-healing biomaterials: a clinically oriented review. Appl Sci 11(17):7769
2. Asmatulu R, Veisi Z, Uddin MN, Mahapatro A (2019) Highly sensitive and reliable electrospun polyaniline nanofiber based biosensor as a robust platform for COX-2 enzyme detections. Fibers Polym 20(5):966–974
3. Aswathy SH, NarendraKumar U, Manjubala I (2022) Physicochemical properties of cellulose-based hydrogel for biomedical applications. Polymers 14(21):4669
4. Billah SR, Mondal MIH, Somoal SH, Pervez MN (2019) Cellulose-based hydrogel for industrial applications. In: Cellulose-based superabsorbent hydrogels, pp 1–41
5. Cao X, Li F, Zheng T, Li G, Wang W, Chen S, Li Y, Lu Y (2022) Cellulose-based functional hydrogels derived from bamboo for product design. Front Plant Sci 13:958066
6. Cascone S, Lamberti G (2020) Hydrogel-based commercial products for biomedical applications: a review. Int J Pharm 573:118803
7. Chang C, Zhang L (2011) Cellulose-based hydrogels: present status and application prospects. Carbohydr Polym 84(1):40–53
8. Chen C, Xi Y, Weng Y (2022) Recent advances in cellulose-based hydrogels for tissue engineering applications. Polymers 14(16):3335
9. Cheng W, Zhu Y, Jiang G, Cao K, Zeng S, Chen W, Zhao D, Yu H (2023) Sustainable cellulose and its derivatives for promising biomedical applications. Prog Mater Sci 138:101152
10. Ciolacu DE, Nicu R, Ciolacu F (2020) Cellulose-based hydrogels as sustained drug-delivery systems. Materials 13(22):5270
11. Curvello R, Raghuwanshi VS, Garnier G (2019) Engineering nanocellulose hydrogels for biomedical applications. Adv Colloid Interface Sci 267:47–61
12. Dhanasekaran PS, Uddin MN, Wooley P, Asmatulu R (2020) Fabrication and biological characterization of highly porous PEEK bio-nanocomposites incorporated with carbon and hydroxyapatite nanoparticles for scaffold applications. Molecules 25(16):3572
13. Fatema N, Ceballos RM, Fan C (2022) Modifications of cellulose-based biomaterials for biomedical applications. Front Bioeng Biotechnol 10:993711
14. Firmanda A, Syamsu K, Sari YW, Cabral J, Pletzer D, Mahadik B, John F, Farah F (2022) 3D printed cellulose-based product applications. Mater Chem Front 6(3):254–279
15. Ghorbani S et al (2018) Hydrogels based on cellulose and its derivatives: applications, synthesis, and characteristics. Polym Sci Ser A 60:707–722
16. Hasanin MS (2022) Cellulose-based biomaterials: chemistry and biomedical applications. Starch-Stärke 74(7–8):2200060
17. Kabir SF, Sikdar PP, Haque B, Bhuiyan MR, Ali A, Islam M (2018) Cellulose-based hydrogel materials: chemistry, properties and their prospective applications. Prog Biomater 7:153–174
18. Kamel S, Khattab TA (2020) Recent advances in cellulose-based biosensors for medical diagnosis. Biosensors 10(6):67
19. Kasi PB, Azar MG, Dodda JM, Bělský P, Kovářík T, Šlouf M, Dobrá JK, Babuška V (2023) Chitosan and cellulose-based composite hydrogels with embedded titanium dioxide nanoparticles as candidates for biomedical applications. Int J Biol Macromol 243:125334
20. Laurén P (2018) Biomedical applications of nanofibrillar cellulose. Helsinki University
21. Liu H, Hu Y, Liu Y, Hu R, Wu X, Li B (2023) A review of recent advances in biomedical applications of smart cellulose-based hydrogels. Int J Biol Macromol 253:127149

22. Liu W, Du H, Zhang M, Liu K, Liu H, Xie H, Hongxiang Z, Xinyu S, Si C (2020) Bacterial cellulose-based composite scaffolds for biomedical applications: a review. ACS Sustain Chem Eng 8(20):7536–7562
23. Mahmood A, Patel D, Hickson B, DesRochers J, Hu X (2022) Recent progress in biopolymer-based hydrogel materials for biomedical applications. Int J Mol Sci 23(3):1415
24. Mohammadinejad R et al (2019) Status and future scope of plant-based green hydrogels in biomedical engineering. Appl Mater Today 16:213–246
25. Oprea M, Voicu SI (2020) Recent advances in composites based on cellulose derivatives for biomedical applications. Carbohydr Polym 247:116683
26. Onofrei M, Filimon A (2016) Cellulose-based hydrogels: designing concepts, properties, and perspectives for biomedical and environmental applications. In: Mendez-Vilas A, Solano A (eds) Polymer science: research advances, practical applications and educational aspects, pp 108–120
27. Pourmadadi M, Rahmani E, Shamsabadipour A, Samadi J, Esmaeili R, Arshad A, Rahdar F, Pandey TS (2023) Novel carboxymethyl cellulose-based nanocomposite: a promising biomaterial for biomedical applications. Process Biochem 130:211–226
28. Ribeiro AM, Magalhães M, Veiga F, Figueiras A (2018) Cellulose-based hydrogels in topical drug delivery: a challenge in medical devices. In: Cellulose-based superabsorbent hydrogels, pp 1–29
29. Sabbagh F, Muhamad II, Pa'e N, Hashim Z (2019) Strategies in improving properties of cellulose-based hydrogels for smart applications. In: Cellulose-based superabsorbent hydrogels, pp 887–908
30. Seddiqi H, Oliaei E, Honarkar H, Jin J, Geonzon LC, Bacabac RG, Klein-Nulend J (2021) Cellulose and its derivatives: towards biomedical applications. Cellulose 28(4):1893–1931
31. Selvaraj S, Chauhan A, Dutta V, Verma R, Rao SK, Radhakrishnan A, Ghotekar S (2024) A state-of-the-art review on plant-derived cellulose-based green hydrogels and their multifunctional role in advanced biomedical applications. Int J Biol Macromol 265:13099
32. Tang Y, Fang Z, Lee HJ (2024) Exploring applications and preparation techniques for cellulose hydrogels: a comprehensive review. Gels 10(6):365
33. Tarrahi R, Khataee A, Karimi A, Yoon Y (2022) The latest achievements in plant cellulose-based biomaterials for tissue engineering focusing on skin repair. Chemosphere 288:132529
34. Thivya P, Akalya S, Sinija V (2020) A comprehensive review on cellulose-based hydrogel and its potential application in the food industry. Appl Food Res 2(2):100161
35. Tofanica BM, Mikhailidi A, Samuil C, Ungureanu OC, Fortună ME, Ungureanu E (2024) Advances in cellulose-based hydrogels: current trends and challenges. Gels 10(12):842
36. Uddin MN, Dhanasekaran PS, Asmatulu R (2019) Synthesis, characterization, and applications of polymer-based biomaterials. In: Bartul Z, Trenor J (eds) Advances in nanotechnology, vol 22. Nova Science Publishers Inc., New York
37. Uddin MN, Desai F, Asmatulu E (2020) Engineered nanomaterials in the environment: bioaccumulation, biomagnification, and biotransformation. Environ Chem Lett 18:1073–1083
38. Uddin MN, Desai F, Asmatulu E (2021) Review of bioaccumulation, biomagnification, and biotransformation of engineered nanomaterials. In: Kumar V, Guleria P, Ranjan S, Dasgupta N, Lichtfouse E (eds) Nanotoxicology and nanoecotoxicology, vol 2. Environmental chemistry for a sustainable world, vol 67. Springer, Cham
39. Uddin MN, Rab MF, Islam AKMN, Asmatulu E, Rahman MM, Asmatulu R (2022) Nanostructured hybrid hydrogels for solar-driven clean water harvesting from the atmosphere. Materials 15:7538
40. Ullah H, Wahid F, Santos HA, Khan T (2016) Advances in biomedical and pharmaceutical applications of functional bacterial cellulose-based nanocomposites. Carbohydr Polym 150:330–352
41. Valle LJD, Díaz A, Puiggalí J (2017) Hydrogels for biomedical applications: cellulose, chitosan, and protein/peptide derivatives. Gels 3(3):27
42. Wang C, Bai J, Tian P, Xie R, Duan Z, Lv Q, Tao Y (2021) The application status of nanoscale cellulose-based hydrogels in tissue engineering and regenerative biomedicine. Front Bioeng Biotechnol 9:732513

43. Yang L, Wang H, Yang Y, Li Y (2014) Self-healing cellulose-based hydrogels: From molecular design to multifarious applications. Carbohydr Polym 347:122738
44. Zhao L, Zhou Y, Zhang J, Liang H, Chen X, Tan H (2023) Natural polymer-based hydrogels: from polymer to biomedical applications. Pharmaceutics 15(10):2514
45. Zou P, Yao J, Cui YN, Zhao T, Che J, Yang M, Li Z, Gao C (2022) Advances in cellulose-based hydrogels for biomedical engineering: a review summary. Gels 8(6):364
46. Zainal SH, Mohd NH, Suhaili N, Anuar FH, Lazim AM, Othaman R (2021) Preparation of cellulose-based hydrogel: a review. J Mater Res Technol 10:935–952
47. Zhang M, Xu T, Liu K, Zhu L, Miao C, Chen T, Gao M, Wang J, Si C (2024) Modulation and mechanisms of cellulose-based hydrogels for flexible sensors. SusMat e255

Md. Aliahsan Bappy currently pursuing an M.Eng. in Mechanical Engineering at Lamar University, Texas, USA, is focused on applying machine learning and nanoelectronics processes to optimize renewable energy technologies. With manufacturing and quality management expertise, he aims to advance solar cell efficiency by developing novel materials and enhancing layer interfaces to reduce energy loss. His research interests are machine learning, product design, development and modification, nanoparticles, energy-efficient engines, solar energy, etc.

Dr. Md. Nizam Uddin is Assistant Professor of Mechanical Engineering, at Texas A&M University-Texarkana, TX, USA. His research interests are design and advanced manufacturing, nanomaterials, polymer-based nanocomposites, and renewable energy and environmental systems. Throughout his studies, he has published more that 80 peer reviewed articles and conference proceedings received 20 honors/awards, and reviewed several manuscripts in international journals and conference proceedings. He has been serving Editorial Board Member and Guest Editor of several journals.

Md. Moniruzzaman is currently pursuing a Master of Engineering in Industrial Engineering at Lamar University, TX USA. His expertise includes mechanical product development, thermal systems, and manufacturing process optimization, with hands-on experience in CAD modeling. His contributions also include the development of advanced thermal imaging systems, robotics integration in manufacturing, and Lean Six Sigma-driven process improvements.

Dr. Nelson Irizarry has been Instructor at Texas A&M University Texarkana since 2020. He holds a Bachelor of Science in Civil Engineering from the University of Puerto Rico and Master of Science and Ph.D. from Texas A&M University. He has vast experience teaching and consulting in the private and public sectors.

Dr. Md. Mahbubur Rahman is currently Associate Professor in the Department of Mechanical Engineering at Khulna University of Engineering and Technology (KUET), Bangladesh. He earned his Ph.D. from the Department of Science and Advanced Technology of Saga University, Japan. He has published numerous articles in internationally recognized journals, contributing to the academia community. His research focuses include, but are not limited to, thermal engineering, energy science, materials science, and manufacturing processes.

Utilizing Cellulose-Based Hybrid Smart Hydrogel for Tissue Engineering

Pankay Sharma and Vinay Jain

Abstract The novel cellulose-based hybrid smart hydrogels have been identified as dynamic in the field of tissue engineering because of their impressive biocompatibility, robust mechanical properties, and capacity to adapt. Hydrogels designed for these tissue architectures are composed of nanoscale components and responsive polymers to achieve structural and biological properties that are relevant to the complexity of the tissue microenvironment. In this chapter, the current state of smart hydrogels based on cellulose hybrids covering the design of such hydrogels and functionalization steps in connection with the application of resultant hydrogels on various types of tissues would be discussed. This chapter describes how alteration of biomaterial properties, multi-stimuli-responsive hydrogels for tissue engineering, and cellular response to the engineered matrix are explored to understand how these synthetic materials can be designed in a manner that resembles the natural tissue milieu and support regenerative medicine application. A focus has been made on composites of hydrogel/nanoparticles, stimuli responsiveness, and bioactivity for enhanced cell growth and differentiation to enhance clinical outcomes involving tissue repair/regeneration applications.

Keywords Cellulose · Hydrogel · Smart material · Hybrid hydrogel · Biocompatibility · Regenerative medicine

P. Sharma (✉)
Department of Pharmaceutics, ShriRam College of Pharmacy, Banmore, Morena, Madhya Pradesh, India
e-mail: pankajsharma223@gmail.com

V. Jain
Department of Pharmacognosy, ShriRam College of Pharmacy, Banmore, Morena, Madhya Pradesh, India

R. Malviya and S. Sundram (eds.), *Sustainable Cellulose Materials for Biomedical Applications*, Sustainable Materials and Technology,
https://doi.org/10.1007/978-981-96-9904-9_6

List of Abbreviations

ECM	Extracellular Matrix
BMP	Bone Morphogenetic Protein
VEGF	Vascular Endothelial Growth Factor
PNIPAM	Poly(N-isopropylacrylamide)
LCST	Lower Critical Solution Temperature
CMC	Carboxymethyl Cellulose
NIR	Near-Infrared
MMPs	Matrix Metalloproteinases
PEG	Polyethylene Glycol
GIT	Gastrointestinal Tract
4D	Four-Dimensional
MRI	Magnetic Resonance Imaging
CT	Computed Tomography
PVA	Polyvinyl Alcohol
LOC	Lab-on-a-Chip
NPs	Nanoparticles
PCL	Polycaprolactone
RGD	Arginine-Glycine-Aspartic Acid (Cell Adhesion Peptide)
PLA	Polylactic Acid
HA	Hyaluronic Acid

1 Introduction

Tissue engineering is an interdisciplinary field that combines biology, engineering, and material sciences to respond to the need for organ and tissue repair around the world. These constraints cannot be easily overcome by traditional treatments that include transplants and prosthetic implants which are scarce, have immune rejection factors, and are limited in their functional capacity. Tissue engineering represents a dynamic approach by providing the use of cells, scaffolds, and signaling molecules to reconstruct replace or repair damaged tissue or organs [1].

This field has recently improved due to changes that have come in biomaterials used, cell biology, and ways of fabricating tissues. For example, the transplantation of skin substitutes for burn patients and the cartilage regeneration in osteoarthritis sufferers gives a hint at what tissue engineering can do. Additionally, combining patient-derived stem cells with bioengineered scaffolds avoids the problem of immune rejection and opens the way to personalized regenerative medicine [2].

In light of the growing de generative diseases, aging population, and rising instances of trauma across the global populace, there is a demand for tissue-engineered products. These approaches can be rightly said to harbor the potential of transforming health care and enhancing the quality of life and reducing the costs of health care. However, there are limitations such as obtaining adequate blood supply for the engineered tissue, biocompatibility and going up a level of tissue engineering manufacturing to meet the existing need of the clinical world [3]. Smart hydrogels, which is an example of an emerging material, is quickly proving to be useful in meeting these challenges.

Superabsorbent and biocompatible hydrogels resembling the structure and properties of natural ECMs are suitable for application in tissue engineering. These properties include: Mechanical property that can easily be tuned, biocompatibility, and ability to respond to stimuli from the environment making them suitable for tissue engineering [4].

For instance, hydrogels has been used as cartilage substitute, thanks to their stimuli-responsive properties that would allow for stimulation of chondrocyte spreading and matrix accumulation. Likewise, in wound healing, we have hydrogels not just keeping the wound moist but can also be incorporated with the growth factors or even antibiotics to hasten tissue healing [5].

In addition, hydrogels are also used as a carrier of drug delivery systems and even as cellular therapies. Sponsored hydrogels which can gel in situ are more preferred in cardiac and neural tissue engineering. These systems permit targeted delivery of therapeutic agents or cells to the defect site thus avoiding extensive surgery interventions [6].

Hydrogels also have one of the major strengths of flexibility. The property of hydrogels can be altered through changing the type of polymer used in its synthesis, the number of links that are formed between the polymer chains or the addition of certain functional groups. For instance, polyethylene glycol (PEG)-based hydrogels are preferred as drug delivery systems because of the hydrophilic character and chemical inertness of PEG while alginate hydrogels are favored in cartilage engineering because of their bioactivity [7]. However, general hydrogels have exhibited some drawbacks, including low mechanical strength and no reversible properties, and as a result, the hybrid smart hydrogels have been widely studied.

Smart hydrogels are a novel class of hydrogels that are half natural and half synthetic—they improve on both while bringing in a new level of responsiveness to chemical stimulus. They are made to be sensitive to the environment, temperatures, pH, light, or electric fields and can change with the condition of a body [8].

It is possible to mention the following considerable progress in the field: Practical applications of nanomaterials in hydrogels' usage are used. For instance, the use of graphene oxide-reinforced hydrogels improve their mechanical properties and electrical conductivity to support neural tissue engineering [9, 10]. In the same regard, antimicrobial activity is achieved through incorporating silver nanoparticles into hydrogels based on cellulose for use in wound healing since the area is prone to get infected [10].

The synthesis of hybrid hydrogels also includes molecular level modification aiming at targeted biomedical application. For example, hydrogels based on cellulose and containing peptides or growth factors will introduce stimuli specific to cell adhesion and proliferation, enhancing tissue repair [11]. In cancer therapy, pH-sensitive hydrogels enable targeted drug delivery at the tumor site and hence, minimal off-target toxicity [12].

Another milestone compounding the discoveries is application of 3D bioprinting where hybrid hydrogels are employed as bio-inks. These hydro gels enable realizing three-dimensional precise cell and biomolecules deposition in highly structurally complex system as tissue. For instance, 3D printed vascularized tissue construct using cellulose-alginate hydrogels had been effective in a preclinical experiment [13].

However, there are the following unsolved issues that need to be solved in order to improve the properties of hybrid hydrogels. The problems that arise include long-term biocompatibility, degradation control, and the possibility to produce the material at the industrial level. However, nanotechnology, synthetic biology, and newer fabrication techniques can be expected to address these barriers, making hybrid smart hydrogels as fundamental components in the future tissue engineering field [11].

2 Basic Concept of Cellulose Incorporated Hybrid Hydrogels

2.1 Structure and Properties of Cellulose

Cellulose, a natural biopolymer obtained from plant cell walls, is the most abundant and adaptable of all natural polymers. It is formed from β-D-glucose linked by β-1,4-glycosidic linkages that gives linear chains that arrange themselves by non-covalent interactions like hydrogen bonding, van der Waals interactions giving a region of crystalline, and other regions which are amorphous. This special anisotropic arrangement endows cellulose with affordable mechanical strength, high thermal stability, and great compatibility with living tissue, thus enabling its biomedical uses [14].

The repetitive structures of cellulose on the nanoscopic level, and micro-fibers or micro-scale level add to this versatility. At the nanoscale, cellulose nanocrystals (CNC) and cellulose nanofibrils (CNF) have received interest for their superior characteristics, including mechanical reinforcement capability, high surface zone, and surface chemistry versatility. Such nanoscale features make cellulose an ideal reinforcement material in the hybrid hydrogel systems [15].

Further, cellulose can be chemically activated and contain other functional groups as for example carboxyl, hydroxyl, and aldehyde groups in order to crosslink or to attach bioactive molecules. For instance, the oxidized cellulose was employed to form thermo and Ph-sensitive hydrogels for use in drug delivery systems [11]. In addition, cellulose is hydrophilic and biodegradable, which are favorable characteristics for the use of a biomaterial in tissue engineering and presented in Table 1.

Table 1 Properties of cellulose-based hybrid hydrogels to be used in tissue engineering

Property	Cellulose	Hybrid hydrogel (with chitosan, PEG, etc.)	Improvement for tissue engineering	References
Mechanical strength	Low	Moderate to high	Improved by crosslinking to increase the mechanical strength of the scaffold	[16]
Biodegradability	Biodegradable	Tunable	Controllable disintegration rate depending on the tissue	[17]
Hydrophilicity	High	High	Used in the maintenance of the cell integrity for improved cell adhesion	[18]
Biocompatibility	Excellent	Excellent	It is biocompatible for in vivo implantation with low immunogenicity	[19]
Porosity	Low to moderate	High	Enhanced nutrient and waste transfer characteristics for cell development	[20]
Electrical conductivity	Low	Enhanced with nanoparticles (but still moderate)	Most important in nerve and muscle tissue regeneration	[21]

2.2 Hybrid Hydrogel Design Overview

The multi-component hydrogels can be referred as advanced materials encompassing both the natural as well as synthetic polymers, which have enhanced biocompatibility, mechanical properties and functionability. In cellulose-based hybrid hydrogels, cellulose is the natural cellulose base polymer, while the second component is provided by synthetic polymers or nanomaterials with structural reinforcement functions and stimuli-responsive properties [22].

Hybrid hydrogels are built by select incorporation of some of the components which are desirable in achieving specific properties. For instance, introduction of cellulose into the PEG matrix improves elasticity to compensate brittleness needed for soft tissues hydrogel. Likewise, the additions of nanoparticles for example graphene oxide or silica nanoparticles into cellulose hydrogels enhance considerably the mechanical strength as well as conductivity [20].

Hybrid hydrogel structure is usually controlled by crosslinking density which determines the stability, porosity and responsivity of the hydrogel. Covalent crosslinking by agents including glutaraldehyde or Carbodiimides, form structures that are rigid while physical crosslinking through ionic gelation or hydrogen bonding techniques presents a reversible and stimuli-responsive nature. For instance, Rubber-like calcium-ion-crosslinked cellulose-alginate hydrogels for cartilage tissue engineering are has been used due to the biomechanical characteristics and cell bio potency [23].

The other factor that should be given a considerable amount of thought is the architecture of the hydrogel. Hybrid hydrogels may therefore be created as cellular supportive scaffolds that have interconnected porous networks, injectable gels, or laminated structures. For example, cellulose-based hydrogels generated by 3DP can permit the cell direction at the sub-micron level where scaffold architecture can be controlled to mimic that of natural tissue in applications such as in bone and vascular tissue regeneration [24].

2.3 Smart Hydrogel Mechanisms: Sensitivity to Environmental Cues

Smart hydrogels can be described as a relatively new generation of materials that are capable of significant changes in response to the effects of stimuli toward their chemical and/or physical surroundings, including pH, temperature, light, magnetic fields, and biological signals. These adaptive responses systems are most beneficial in biomedical aides where timed drug delivery, tunable stiffness, and changeable cell matrix are desirable [25].

pH-Responsive Mechanisms

Cellulose-based hydrogels might be prepared in such a way that they possess sites which are sensitive to the pH of the surrounding. For instance, carboxymethyl cellulose hydrogels swell/ shrink depending on the pH of the environment with the property that makes them suitable to be used for controlled drug delivery through the gastrointestinal system. These hydrogels do not degrade in the stomach owing to acidic pH but disintegrate in the more alkaline environment of the intestines [26].

Temperature-Responsive Mechanisms

The thermoresponsive cellulose hybrid hydrogels are developed using polymers such as poly(N-isopropylacrylamide) (PNIPAM) where the hydrogels undergo phase change at a particular temperature. These hydrogels are soluble at the room temperature and become gel at body temperature it can be used to deliver some cells or drugs through a minimally invasive method. This property has been utilized in cartilage regeneration whereby the gelation process favors cell confinement and deposition of matrix at the tender location [27].

Light-Responsive Mechanisms

The use of photo sensitive molecules for instance azobenzene or spiropyran into photoresponsive hydrogels occur in such a way that the molecules changes conformation when exposed to light. Hydrogels based on cellulose and these molecules have been utilized for drug delivery to control the release or for the reversible change of mechanical characteristics to suit tissue engineering with remarkable space and time-responsive features [13].

Magnetic and Electric Field Responsiveness

The process of loading magnetic particles including iron oxide to cellulose hydrogels allows for the application of external magnetic fields to control the material. Many of these materials are useful in directing the delivery of drugs to particular locations or in the alignment of cells in tissues. Also, electrically conductive cellulose hydrogels, therefore having frameworks such as graphene or carbon nanotubes, enhance neural tissue engineering by providing better encouragement for nerve cell development and direction of signals [28].

Biological Signal Responsiveness

Biomolecule-sensitive hydrogels can be synthesized for enzymes or other biomolecules which are present at the target tissue site. For example, cellulose hydrogels covalently crosslinked with peptide bonds degrade when exposed to matrix metalloproteinases, which can effectively deliver encapsulated therapeutic agents only in cancerous or inflamed tissues [13].

There are certain features where hydrogel mechanisms show great potential and that are smart hydrogels across the various area of practice such as wound healing, tissue regeneration, and drug delivery. A new development in the hybrid hydrogels will shift toward the ability of the hydrogels to incorporate numerous features of stimuli sensitivity in order to develop multipurpose hydrogel systems to tackle multifaceted biomedical issues [29].

3 Preparation and Modification of Cellulose-Based Composite Hydrogels

The development of hybrid hydrogels by incorporating cellulose with other materials and functionalizing cellulose-based hydrogels are significant in improving properties and performance of hydrogels used for tissue engineering applications and its summarized display illustrated in Fig. 1. These processes allow some adjustments to the mechanical, chemical, and biological characteristics of the hydrogels for better use in the treatment. The synthesis method addresses the right structure and function of the cellulose-based hydrogels while the functionalization strategies are tailored to afford specific biological activities like cell adhesion and differentiation, drug delivery, etc., and presented in Table 2 [30].

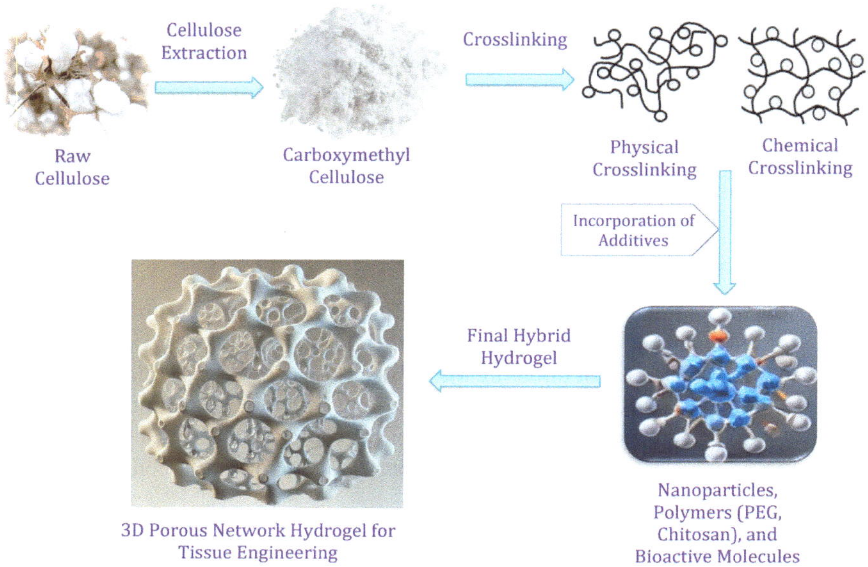

Fig. 1 Synthesis of cellulose-based hybrid hydrogels

3.1 Chemical Crosslinking and Physical Crosslinking Methods

Crosslinking strategies therefore still remain critical toward the preparation of cellulose-based hybrid hydrogels to form desired three-dimensional crosslinked polymer structures with appropriate mechanical and functional requirements. These methods can be further subdivided according to chemical and physical crosslinking mechanisms which both have merits [35].

Chemical Crosslinking

Chemical cross link is formed by covalent bonds that relate cellulose to other polymers or functional groups. This approach can produce well-characterized hydrogels with mechanical and degradation characteristics that are reproducible and predictable. Some of the usually used chemical crosslinking agents are glutaraldehyde, epichlorohydrin, and carbodiimides. For instance, crosslinked cellulose-PVA hydrogels with glutaraldehyde possess sufficient mechanical strength to prevent the deformation of the hydrogel structure under load and thus can be used for load-bearing applications such as cartilage tissue regeneration. Some of the enhanced chemical crosslinking techniques are based on click chemistry, which mainly provide low reaction interference. For example, cellulose ester derivatives containing azide groups can react with alkyne polymers via a conventional copper-catalyzed alkyne–azide cycloaddition (CuAAC) reaction to form highly stimuli-responsive hydrogels [36].

Table 2 Methods for synthesis and functionalization of cellulose-based hybrid hydrogels

Method	Description	Hydrogel composition	Advantages	References
Physical crosslinking	Hydrogen bonding, ionic and Van Der Waal's forces	Cellulose, chitosan, PVA	It is very easy, non-toxic crosslinker is not required	[31]
Chemical crosslinking	Crosslinking by covalent bonding, for example by means of glutaraldehyde and genipin	Cellulose, PEG, gelatin	Improved mechanical physics and stability of the polymer	[32]
Nanoparticle incorporation	Nanoparticles introduction to improve characteristics	Cellulose, silver nanoparticles, hydroxyapatite	Enhances the mechanical property, antibacterial property, and biological property	[33]
Biofunctionalization	Hence, use of peptides, growth factors, or ECM proteins	Laminates, RGD peptides, collagen	Promotes cell adhesion, cell proliferation and differentiation	[34]
Electrospinning	Manufacture of nanofiber scaffolds by employing electrostatic force	Cellulose acetate, PCL, Collagen	Copies the native structure of ECM and promotes cell movement	[13]

Physical Crosslinking

Physically crosslinked hydrogels involve crosslink density since they undergo physical crosslinking through methods such as hydrogen bonding, ionic interactions, or hydrophobicity. This also reduces the use of chemical crosslinker which may be toxic, thereby improving the accessibility of the cell matrix. For instance, usingención of friendly polyelectrolyte complexation of carboxymethyl cellulose (CMC) with calcium ions results in physically entangled hydrogels with high biodegradability. Another case in point is the freeze–thaw cycles in systems based on cellulose and polyvinyl alcohol (PVA). This procedure triggers formation of crystalline structures and hydrogen bonds that lead to improved elasticity of hydrogels and their water absorption capacity. One advantage of physical crosslinking is for reversible or stimuli-responsive hydrogels; this framework of the network can change with ease depending on stimuli [37].

3.2 Nanocomposites, Polymer, and Bioactive Elements

The introduction of nanoparticles, synthetic polymers, and bioactive agents into cellulose-based hydrogels increases the mechanical functional and biological characteristics.

Nanoparticles

Nanoparticles are used to enhance the mechanical properties, conductivity or biologic activity of hydrogels. For instance, when incorporating silver nanoparticles on cellulose hydrogels to produce a new product, the dressing acquires antimicrobial characteristics [38]. Likewise, in hydrogels, graphene oxide improves the electricity and mechanical strength necessary in neural tissue engineering. Other supplemental materials include, magnetic nanoparticles for example, iron oxide. These particles allow the manipulation of the hydrogel mechanical properties or delivery of the drug via magnetic field from outside. Such systems are especially applicable to targeted cancer therapy [37].

Synthetic Polymers

Cellulose combined with synthetic polymers formations yield hybrid gels with improved elasticity and mechanical stability. For instance, cellulose-PNIPAM hydrogel is thermosensitive, switching between gel state and sol state at physiological temperature and thus ideal for localized drug delivery treatments. Polyethylene glycol (PEG) is widely employed here to enhance the hydrophilicity and to minimize immunogenicity of cellulose-based hydrogels. Self-assembled PEG-cellulose hydrogels have found use in cartilage and bone tissue engineering because of the ability to tailor mechanical characteristics [39].

Bioactive Agents

The combination of growth factors or peptides as well as other bioactive molecules into the hydrogels increases their biological activity. For instance, the cellulose hydrogels containing VEGF can foster angiogenesis in the treatment of wound healing. In the same way, hydrogels containing osteogenic peptides have been employed for bone tissue engineering in orthopedic science [40].

3.3 Methods of Functionalizing the Surface to Improve Biocompatibility

It is, therefore, necessary to attach cell-compatible functional groups on the cellulose hybrid hydrogels to effectively support cell adhesion, proliferation, and tissue incorporation. To realize these enhancements, several techniques are used as earlier defined.

Surface Modification

Also known as surface functionalization, surface modification involves the attachment of other bioactive molecules or functional groups on cellulose chains. For example, in the case of RGD (Arg-Gly-Asp) peptide which imitates the natural cell adhesion motifs, it is fairly easy to graft these motifs onto cellulose hydrogels so as to enhance cell adhesion and proliferation. This approach has been invaluable in the design of hydrogels for vascular and neural tissue repair [41].

Chemical Functionalization

Chemical modification of cellulose entails the addition of functional groups at the cellulose chain to improve either its affinity to water or reactivity. Two of the most typical modification processes involving carboxylation and sulfonation in cellulose hydrogels are used to enhance their water adsorption and ion exchange properties, which is useful for medical purposes such as drug delivery and wound healing [42].

Incorporation of Bioinspired Molecules

Introducing biomolecules such as hyaluronic acid or collagen into cellulose hydrogels replicates the ECM and encompasses a suitable context for cell cultivation. For instance, cellulose collagen hydrogels have been employed in application for skin regeneration, and they allow for aggregation of cells and improved wound healing [43].

Stimuli-Responsive Functionalization

Cellulose hydrogels can be designed by incorporating stimuli-sensitive groups, so that they may reversibly respond to changes in the external conditions. For example, the pH-sensitive groups of cellulose hydrogels enhance the drug release in the acidic tumor environment and thus have better therapeutic outcomes [44].

4 Functional Improvements and Evolution for Versatile Applications

4.1 Antibacterial Sensitivity and Radical Scavenging Activity

Functional cellulose-based hybrid hydrogels are developed more often for the biomedical applications such as infection, oxidative stress, and inflammation which require antimicrobial as well as antioxidant properties.

Antimicrobial Properties

The integration of antimicrobial agents into cellulose hydrogels is a crucial approach to control infections in the case of wounds and implantable medical devices. For example, silver nanoparticles (AgNPs) are incorporated into cellulose hydrogels because they possess, among others, wide-spectrum antimicrobial properties. Cellulose hydrogels incorporating AgNP have displayed high efficiency in wound healing applications to decrease the microbial adhesion and enhance tissue repair. Others include chitosan and zinc oxide nanoparticles which are incorporated into cellulose matrices. Based from the overwhelming literature of both chitosan and zinc oxide nanoparticles, the practical approach and action of chitosan in destabilizing bacterial cell membrane and the sustained antimicrobial property of ZnO in releasing reactive oxygen species enhances the antibacterial activity toward the pathogens. They have been applied successfully in treating burns to preventing surgical site infection [45].

Antioxidant Properties

Reactive oxygen species (ROS) induced injury can hampering the healing process and promote chronic inflammation. In response, cellulose hydrogels are chemically modified to contain antioxidants like vitamin E, curcumin, or vegetables and fruits-derived polyphenols. For instance, the hydrogels prepared from cellulose with curcumin derived from turmeric exhibit both antioxidant and anti-inflammatory properties to promote wound healing in vitro and in vivo [46]. Also, enzymatic antioxidants, which are superoxide dismutase or catalase, are encapsulated or incorporated in hydrogels for a sustained delivery. These systems proactively capture ROS in affected tissues and thus minimize the degree of tissue injury while assisting cellular rehabilitation [47].

4.2 Drug Delivery Systems: A Synergistic Approach

Novel cellulose-based hybrid hydrogels can be further applied in the drug delivery system as they provide controlled release, targeted delivery and little impact on the overall side effects.

Controlled Release Mechanisms

These hydrogels are stable structures due to the presence of the hydrophilic cellulose and are especially suited for encapsulating and releasing drugs over time. New functional crosslinking densities or other stimuli-sensitive moieties can readily adjust the swelling concentrations of the gel to deliver the drug into the specified location at the desired rate. For example, the characteristic of cellulose-PVA hydrogels with encapsulated anticancer drugs shows controlled and sustained drug release enhancing the drugs' therapeutic efficiency and reducing their side effects [48].

Stimuli-Responsive Drug Delivery

Smart hydrogels provide the controlled release of medications by means of response to stimuli, such as pH, temperature, or enzymes. For instance, the cellulose hydrogels crosslinking with pH-sensitive polymers to deliver the drug selectively at the acidic tumor site for improving the cancer therapy outcome. Likewise, cellulose-PNIPAM hydrogels respond to physiological temperature and deliver drugs for chronic diseases without a frequent dosage application [49].

Nanoparticle-Enhanced Systems

Incorporating drug-loaded nanoparticles into cellulose hydrogels allows for dual functionality: Biodegradable hydrogel can serve as the carrier, and nanoparticles as drug delivery and targeting system. For example, cellulose hydrogels with loaded doxorubicin-containing liposomes have been used to enhance the localized cancer therapy by enhancing drug stand time at the tumor site [50].

Combination Therapies

As cellulose hybrid hydrogels can encapsulate multiple drugs, therapies can be provided synergistically at once. For instance, antibiotic and growth factor-containing hydrogels contribute to wound healing through the efficiency of antibacterial action and stimulation acceleration. The same two-in-one strategy has been used in osteoarthritis therapy where hydrogels release anti-inflammatory agents and also cartilage-building peptides [51].

4.3 Cell Adhesion and Proliferative Promoting

Therefore, enhancing cell adhesion and proliferation are important aspects for tissue engineering applications. CM-mimicking and bioactive cellulose-based hybrid hydrogels are designed to incorporate the cellular cues that promote cellular growth.

Bioactive Peptide Functionalization

The covalent immobilizations of cell attachment peptides like RGD (Arg-Gly-Asp) onto cellulose hydrogels improve cell adhesion because it imitates the natural integrin-binding domain of the peptides. For instance, the surfaces modified with RGD-functionalized cellulose hydrogels were revealed to enhance the adhesion and proliferation of osteoblasts in bone tissue engineering [52].

Collagen and Gelatin Incorporation

Natural ECM components such as collagen and gelatin are usually mixed with cellulose in the development of scaffolds to intensify its bioactivity. These materials present interactive sites for integrins that let cells adhere and migrate. The cellulose collagen hydrogels have exhibited excellent potential for skin regeneration including the promotion of keratinocytes' proliferation and wound healing [53].

Topographical and Mechanical Cues

It demonstrates that the micro- and nanoscale architecture of cellulose hydrogels can be designed to display suitable physical signs that can dictate cell actions. For instance, the hydrogels are displaying nanofibrils that help in directing the orientations of cells like neurons or muscle fibers in a better way that offers well- organized tissues. As well, stiffness-tuned cellulose hydrogels guide stem cell differentiation, with softer matrices promoting neural differentiation and stiffer matrices promoting osteogenesis.

Growth Factor Delivery

Cellulose hydrogels may be incorporated with factors, including vascular endothelial growth factor (VEGF) or bone morphogenetic proteins (BMPs) to promote cell division and differentiation. Their progressive release from hydrogels also provides a beneficial topographical climate for endogenous tissue healing. For example, VEGF incorporated in cellulose hydrogels has been applied in vascular tissue engineering for enhancement of angiogenesis and increase of endothelial cell number [54].

5 Multi-stimuli-Responsive Cellulose Incorporated Hydrogels

Multi-responsive cellulose-based hybrid hydrogels represent a leap forward in the production of biomedical polymers. Hydrogels described herein use at least two responsiveness elements to enable well-defined tuning of all properties and functions in response to external stimuli. It can be seen that this flexibility is especially beneficial for drug delivery, tissue engineering, and biosensing.

5.1 Temperature, pH, and Responsive Mechanisms of Enzyme

Temperature-Responsive Mechanisms

Thermo-responsive hydrogels are designed to change phase at certain temperatures. Cellulose-based hybrid hydrogels containing temperature-sensitive materials, including PNIPAM, have a LCST of approximately 32 °C, allowing potential applications in drug delivery and tissue engineering. For instance, cellulose-PNIPAM hydrogels are in liquid form at room temperature facilitating their delivery but transforms to gel form at body tem at and results in localized drug release. Also, dual-temperature-sensitive systems in which cellulose is mixed with polymers having different LCSTs have been synthesized for applications that may include systems with temporally separate release of drugs or biofactors [55].

pH-Responsive Mechanisms

Carboxyl, amino, and phosphate functionalized cellulose hydrogels show pH dependent swelling or deswelling properties. These hydrogels are useful for controlled drug delivery in conditions including that of different pH surroundings such as in the G.I.T or the tumor micro environment. For instance, CMC-hydrogels containing anticancer agents discharge their cargoes mainly in acidic tumor milieu and thus reduce systemic toxicity. pH and temperature responsive hydrogels, which combine both properties, have also been developed, to aid in altering the drug release rate based on the conditions within the body [38].

Enzyme-Responsive Mechanisms

Self-assembly responsive hydrogels rely on natural enzymatic activities to induce changes in their structure or disintegration of the hydrogel network. For instance, cellulosic hydrogels based on crosslinking that involves peptide linkers sensitive to matrix metalloproteinases (MMPs) release drugs in areas of escalated MMP activity such as cancer zones. As result, integration of cellulase-sensitive bonds enables the capability for selective cleavage of hydrogels in microbial systems and of added value in ant/ microbial therapies [56].

5.2 *Photoresponsive and Magnetic Field-Sensitive Hydrogels*

5.2.1 Light-Responsive Hydrogels

Both light-responsive hydrogels involve the incorporation of photoactive molecules like azobenzene, spiropyran, or graphene oxide within cellulose matrix. When subjected to stimuli of certain wavelengths of light, these photoresponsive molecules cause changes in the conformation of the hydrogel and the related changes in its swelling and/or mechanical properties. For example, chemically responsive hydrogels composed of cellulose material that incorporates azobenzene derivatives are promising for drug release on-demand in the treatment of wounds [57]. Near infrared (NIR) light-responsive hydrogels containing NIR absorbent materials including gold nanorods or carbon dots can be used for in-depth use due to high penetration depth of NIR light. These hydrogels have been used in therapy of cancers for targeted therapy and drug delivery and for photothermal treatment of tumor tissues [58].

Magnetic Field-Responsive Hydrogels

Magnetic hydrogels are obtained by dispersing cellulose matrices with magnetic nanoparticles like iron oxide. These nanoparticles allow for fine-tuning of hydrogel characteristics from an external magnetic field. Some uses are drug delivery where the hydrogel can be directed toward particular areas and dynamic tissue engineering where magnetic field causes mechanical movement for cell growth. For instance, cellulose-magnetic nanoparticle hydrogels can be used as carriers for chemotherapeutic agents so that by applying a magnetic field, it helps focus the hydrogel at the tumor site thereby enhancing drug concentration at the specific site while at the same time minimizing toxicity effects associated with chemotherapy [28].

Combined Light and Magnetic Responsiveness

Light and magnetic responsiveness at the same place in a single hydrogel system are unbeaten by any other hydrogel system. For instance, cellulose derived hydrogels containing magnetic nanoparticles as well as photoresponsive dyes allow precise spatiotemporal regulation of hydrogels that can be used in settings such as multimodal cancer treatment and sophisticated biosensing [59].

5.3 Uses in Targeted and Stimuli-Responsive

Drug delivery system recent stepped cellulose-based dual-active hydrogels for multiple-responsive drug delivery systems are more accurate and effective in therapeutic use.

Targeted Drug Delivery

Tissue or disease site targeted multi-responsive hydrogels are highly applicable particularly when the desired release is supposed to occur at the specific tissue site. For example, pH- and enzyme-sensitive cellulose hydrogels deliver drugs selectively at the desired site since they are activated only under the conditions characteristic of a tumor: low pH and high protease activity. Magnetic responsiveness can be added to this to thus enhance the targeting even further as the hydrogel can be remotely controlled through an applied magnetic field. A silver example is the synthesis of thermo- and magnetically sensitive cellulose hydrogels with potential for application in delivering anti-inflammatory drugs to the affected joint in osteoarthritis treatment. Ibuprofen of hydrogels prevents the inflammation locally in the joint without quick systemic effect besides extend the lasting effectiveness of the drug [60].

Controlled Drug Release

Multi-stimuli sensitivity allows for hydrogels having the ability to give precise control of the rate of drug delivery. For instance, a hydrogel made from cellulose with incorporated pH-sensitive polymers and NIR-absorbing nanoparticles, can deliver an antibiotic at a steady rate in the case of an acute infection and increase release upon irradiation with near-infrared light. Some of the other uses include Sequential drug delivery. Some cellulose hydrogels are pH and temperature sensitive so that they can release one drug in an acidic environment and another at a high or low temperature, respectively. This approach is used in cancer treatment where a first drug prepares the tumor microenvironment then a second drug exponentially increases its toxicity [61].

Theranostic Applications

This paper shows that multi-responsive cellulose hydrogels enable theranostic functionalities by incorporating diagnostic and therapeutic capabilities. For instance, magnetic and light-responsive cellulose hydrogels containing imaging agents and therapeutic agents make it possible for tracking the position of the hydrogels apart from drug release. Such systems have been proved to be useful in targeted cancer treatment that can also be used for imaging [62].

6 Mechanics and Biomechanics of Hybrid Hydrogels

The physical and biological characteristics of the cellulose-based hybrid hydrogels are important in tissue engineering and regeneration. These properties affect the ability of cells to perform their functions, how scaffolds function, and the results for tissue regeneration. In this chapter, we discuss mechanical property profiles, tissue-specific modifications, and characterization methods of hydrogel responses ex vivo and in vivo.

6.1 Mechanical Properties for Tissue Engineering Applications

Stiffness, elasticity and tensile strength are critical requirements presented by hybrid hydrogels in order to replicate the ECM structure and offer a mechanical background for cultured cells.

Stiffness and Strength

Hybrid hydrogels made from cellulose need to have adequate rigidity to offer mechanical array but also need to be somewhat elastic in order to allow for tissue motion. Hydrogels that are to be employed to repair cartilage need to exhibit a high compressive modulus to support mechanical loads, while those intended for neural tissue applications need to be very soft to prevent damaging neural cells. These reinforcing agents including graphene oxide or carbon nanotubes appreciably improve the strength of cellulose hydrogels. These nanomaterials have good compatibility with cellulose chains, enhancing tensile strength and elasticity though the strength of the composite is slightly reduced [63].

Resilience and Fatigue Resistance

Reusability is a crucial property in hydrogels for those applications where the hydrogels incorporate repetitive stresses, including cardiovascular uses. The copolymers of cellulose with other materials, such as elastomers, including polycaprolactone (PCL), show improved rebound elasticity and fatigue strength if used in such active contexts as a blood vessel [64].

Viscoelastic Behavior

Hydrogels need to possess viscoelastic properties since various tissues including muscles and tendons have this kind of property. Hydrogels derived from cellulose and capable of forming reversible covalent cross-linkage bonds includable hydrogen or ionic bonds have controllable viscoelasticity characteristics for load-bearing tissue applications [65].

6.2 Tunable Stiffness and Compliance for Different Tissues

The porosity and swelling behavior of cellulose-based hybrid hydrogels must be controlled in order to meet tissue mechanical demands.

Soft Tissues

Such tissues include the brain, skin, and liver, where therefore, hydrogels with low stiffness (elastic modulus of between 0.1–10 kPa) should be used. The addition of gelatin to cellulose hydrogels significantly decreases the gel stiffness but increases the cell compatibility when hyaluronic acid is added to the mix. These soft hydrogels have enhanced the cellular growth and differentiation without the generation of mechanical force [66].

Hard Tissues

Soft tissues like muscles, blood, and nerve require hydrogels that possess greater elasticity (elastic modulus > 1 MPa) while hard tissues like bone and cartilage need hydrogels of greater stiffness. Cellulose-based hydrogels with HA or bioactive glass create the stiffness needed for the hydrogels, while at the same time promoting osteogenic differentiation of stem cells. Specifically, cellulose-HA hybrid hydrogels exhibited promising applications in the field of bone tissue engineering where synergistic effects such as hydrogel mineralization and mechanical reinforcement have been recorded [67].

Dynamic Modulus Tuning

While there are situations including wound healing where; hydrogels with dynamic modularity are required. Photocrosslinkable or pH-sensitive cellulose-based hydrogels have the ability to change their mechanical properties on command. For instance, hydrogels containing light-responsive polymers such as spiropyran can transition between being compliant to support cell motion and hard to provide sufficient rigidity and stability needed during repair of a wound [68].

Gradient and Layered Designs

Structurally, several tissues and cells in the body interface each other like cartilage-bone interface, and this necessitates the need for hydrogels with mechanically tuned gradient. Indeed, the hydrogels containing both layers of cellulose with graded mechanical properties allow creating constructive tissue-like transitions which improve the integration and performance. Most significantly these designs have been useful especially in the repair of osteochondral defects [69].

6.3 *In Vitro and In Vivo Characterization of Mechanical Behavior*

It is therefore important to evaluate the mechanical properties of cellulose-based hybrid hydrogels in order to predict their behavior in biological systems.

In Vitro Characterization Techniques: Rheology

By rheological tests, the viscoelastic properties of hydrogels such as storage modulus and loss modulus can be determined. Rheometry has maximum application in examining the changes in hydrogel mechanics that occur due to temperature or pH.

Compression and Tensile Testing: These tests quantify Bulk modulus of elasticity, Tensile strength, and compressive strength of Hydrogels. For instance, mechanical assessments of the cellulose-PEG hydrogels have suggested their possible use in applications such as tendon repair [70].

Atomic Force Microscopy (AFM): AFM can perform assessment of hydrogel stiffness and surface roughness at nanoscale level which are important in the cells behavior [71].

Swelling and Degradation Analysis: The swelling properties are linked to the hydrogel mechanical strength when exposed to aqueous conditions and degradation investigations examine the compromised structural characteristics over a given period [72].

In Vivo Characterization: Implantation Studies

In vivo experiments determine the degree of compatibility of hydrogels with native tissue through mechanical testing. For example, the cellulose-chitosan hydrogels, when used to fill the cartilage defect, possess good load-carrying capacity and tissue compatibility [73].

Biomechanical Testing Post-Implantation: Following implantation, mechanical tests involve determining the values of the restored tissue such as the stiffness, elasticity, and resilience. These tests affirm the ability of a hydrogel to work in dynamic physiological environments [73].

Imaging Techniques: Ultrasound imaging elastography and MRI based mechanical resonance imaging offer a method of testing hydrogel mechanics in a live organism, allowing real-time evaluation of the hydrogel performance [74].

Correlating Mechanics with Cellular Responses: The degree of crosslinking in cellulose-based hybrid hydrogels determines the mechanical properties that affect cell attachment, division, and formative operations. For instance, soft hydrogel substrate favors the neural commitment of stem cells, on the contrary, rigid hydrogel matrix boost the osteogenic differentiation. Some of the features are now measurable in terms of cell mechanics, utilizing advanced characterization approaches like traction force microscopy to analyze cells' attachment to hydrogel matrices [75].

7 Bio-medical Applications in Tissue Engineering

Cellulose derived hybrid hydrogels have gained interest in Tissue engineering applications because of their biocompatibility, tailorable mechanical characteristics and multiple functionalities as illustrated in Fig. 2. Table 3 attempts to discuss how stem cells work in regenerating/repairing tissues such as bones, cartilage, soft tissues, cardiovascular tissues, neural tissues, and skins.

7.1 Bone Regeneration and Repair

Tissue engineering scaffolds for bone tissue must be characterized by high mechanical properties and biologic activity. These requirements are achieved by cellulose-based hybrid hydrogels through their biomimetic properties and osteogenic capacity.

Hydroxyapatite-Integrated Cellulose Hydrogels

The combined hydrogels containing hydroxyapatite (HA) improves osteoconductivity and mechanical properties. These hydrogels offer an appropriate platform for both stem cell adhesion and differentiation with respect to bone tissue engineering. For instance, cellulose-HA hydrogels have been used to treat critical-sized bone defects in rats [81].

Bioactive Glass Additives

Bioactive glass nanoparticles when incorporated into cellulose hydrogels enhance the mineralization to support bone regeneration. These hybrid hydrogels are then ideal where there is a higher rate of bone formation is desirable such as spinal fusion or dental implantation [82].

Growth Factor Delivery

Self-assembled cellulose hydrogels can support the controlled release of growth factors such as bone morphogenetic protein-2 (BMP-2) and improve the differentiation of osteoblasts and formation of new bone tissue. pH and enzyme-sensitive cellulose hydrogels help deliver these factors selectively in the bone microenvironment [83].

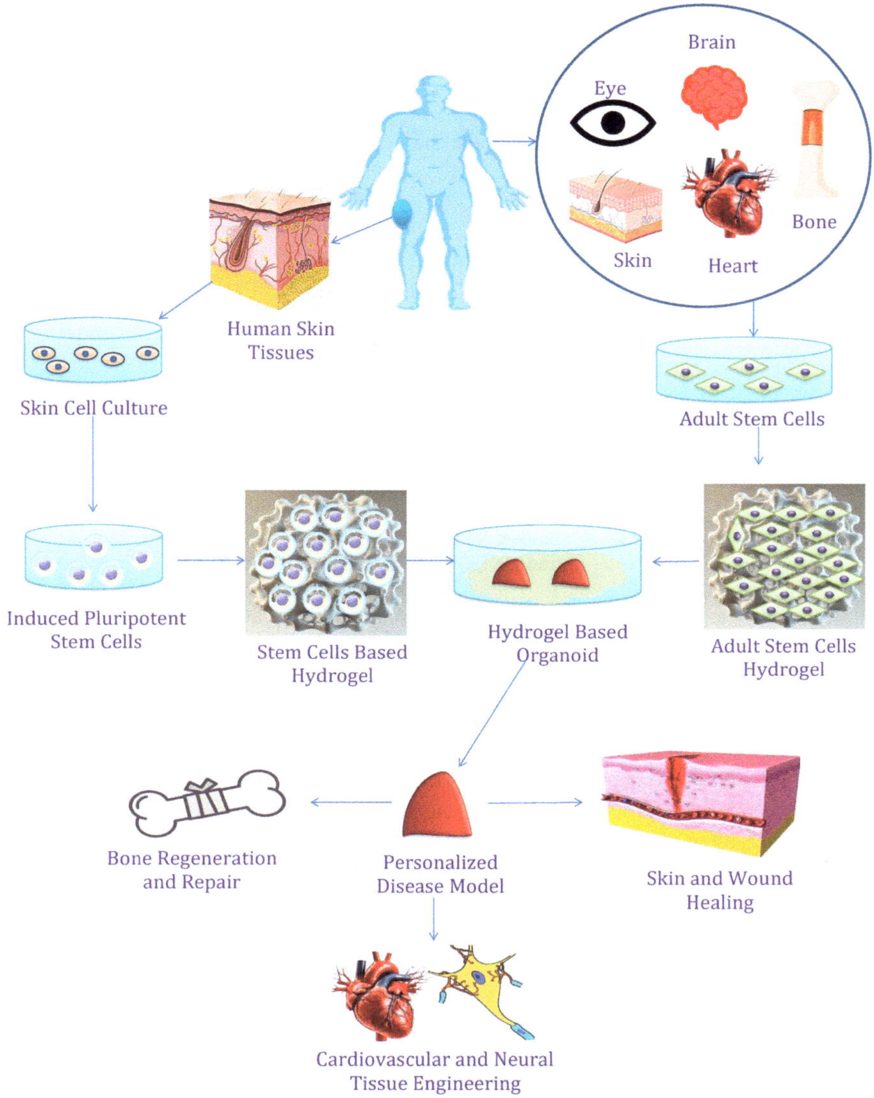

Fig. 2 Applications of cellulose-based hybrid hydrogels in tissue engineering

7.2 Cartilage and Soft Tissue Engineering

The restoration of cartilage is further complicated by the tissue's biomechanical characteristics, such as its comparatively high compressive modulus and extensibility. Cellulose-based hybrid hydrogels appear to provide such a solution.

Table 3 Uses of cellulose-based composite hydrogels in tissue regeneration

Tissue type	Application	Hydrogel composition	Key features	References
Bone	Bone regeneration and repair	Cellulose, hydroxyapatite and collagen	Stimulates osteoblast differentiation and promotes osteoid mineralization	[76]
Cartilage	Cartilage tissue engineering	Cellulose, chondroitine, hyaluronic acid	Stimulates chondrocyte differentiation with corresponding ECM production	[77]
Skin	Skin repair and wound healing	Cellulose, alginate, gelatin	Stimulates cell migration, Proliferation and promotes rapid healing of the wound	[78]
Neural tissue	The neural tissue repair and regeneration	The constituents of the material are cellulose, PEG, and graphene oxide	Enhances cell attachment, differentiation, and electrical conductivity to primary neural cells	[79]
Vascular	All forms of new blood vessel formation (angiogenesis)	Cellulose, VEGF, fibrinogen	Promotes proliferation of both endothelial cell lines as well as the formation of new blood vessels	[80]

Cartilage Repair

It has been postulated that, cellulose hydrogels and chitosan or polyvinyl alcohol (PVA) show enhanced performance in cartilage tissue engineering. These hydrogels possess the necessary mechanical characteristics and support favorable conditions for chondrocyte proliferation and the deposition of the extracellular matrix [83].

Soft Tissue Engineering

In ligament or tendon repair, new cellulose hybrid hydrogels containing silk fibroin or collagen fibers have been designed. Most of these hydrogels replicate the architectural and biomechanical characteristics of soft tissues, together with supporting cell attachment and alignment [84].

Gradient Hydrogels for Osteochondral Interfaces

Hydrogel engineering with gradient characteristics is essential in the reparative process of osteochondral junctions, where cartilage gradually merges with bone. There is evidence that cellulose-based hydrogels can form hydrogels with structures as a double-layer with promising mobility in rebuilding complicated tissue interface [85].

7.3 Cardiovascular and Neural Tissue Engineering

Cardiovascular and neural tissues are vulnerable and continuously changing; therefore, hydrogels require particular mechanical, electrical, and biological characteristics.

Cardiovascular Applications

Myocardial Regeneration: Conductive cellulose hybrid hydrogels containing elements such as polypyrrole or carbon nanotubes provide support to the regeneration of myocardial tissues through enhanced conductivity and synchronized function of cardiac cells. Some of these hydrogels have shown increased performance in preclinical applications for myocardial infarction therapy [86].

Vascular Tissue Engineering: Cellulose hydrogels can be crosslinked with elastomers such as polyurethane to have the flexibility and mechanical strength needed in vascular application. This is due to their characteristics of mimicking native blood vessels compliance to enable efficient integration and functionality [87].

Neural Tissue Engineering

Neurogenesis and Nerve Repair: Cultivated soft cellulose hydrogels have been shown to enhance the regeneration of peripheral and central nervous tissues when they are impregnated with nerve growth factor. These hydrogels mimic the nature ECM and thus supports differentiation of the neural stem cells into neurons [88].

Electrical Stimulation: Conductive polymers in cellulose hydrogels stimulate the neural signals required for nerve conduits in spinal cord injuries. Conductive cellulose-based scaffolds have also demonstrated potential for nerve regeneration in injury of sciatic nerve, in rat model [89].

7.4 Skin and Wound Healing Applications

Cellulose-based hybrid hydrogels biocompatibility and hydrophilicity contribute to their fitness for use in skin regeneration and wound healing applications.

Wound Dressings

It may also effectively deliver water, nutrients, and other required essential to the wound site, and at the same time, it acts as a barrier to microorganisms. Composite hydrogels with silver nanoparticle show very high antimicrobial activity thus minimizing infections on the chronic wound site [90].

Burn Treatment

Cellulose-based hydrogels of controlled swelling behavior thus embedded with growth factors like epidermal growth factor has a potential to reactivate burn injuries because of their inherent ability to enhance angiogenesis and promote the migration of keratinocytes. Some of those hydrogels that are used also minimize scar formation and enhance skin quality [91].

Skin Grafts

Cellulose hydrogels have a role of providing structural support to a skin construct with potential to support skin cell growth such as the keratinocytes and fibroblasts. Layered structure hydrogels mimic the dermis and epidermis, make grafts better integrated in host tissues [68].

Chronic Wound Management

In diabetic ulcers and other chronic wounds recurring cellulose hydrogels with curcumin or honey improve inflammation modulation and collagen synthesis. The dual-responsive hydrogels are also useful for the controlled release of therapeutics to match the treatment to wound conditions [92].

8 Cell-Signaling and Biocompatibility

The specific and presumably close relation between cells and the biomaterials is the most critical factor of the success of the tissue engineering applications. Cellulose-based hybrid hydrogels are advantageous for supporting cell attachment, spreading, proliferation, differentiation and subsequent re-establishment with native tissues on account of their structural flexibility and functional modularity. This chapter focuses on the migration, proliferation and differentiation of cells within hydrogels and reviews the immune reactions and biodegradation issues of the hydrogels as well as the approach to improve the cell–matrix interactions [93].

8.1 Cell Attachment, Growth, and Differentiation in Hydrogels

Cell Adhesion

Cell adhesion is a general requirement for essential cell activities such as proliferation and cell differentiation. Native cellulose does not bear cell-adhesive sites; nevertheless, when cellulose is combined with bioactive molecules like RGD peptides, CB-ECM exhibit improved cell adhesion. These modifications resemble the natural ECM and support integrin-mediated cell adhesion [94]. For example, if hydrogels from cellulose are used and each one is functionalized with collagen or fibronectin, the mesenchymal stem cells (MSCs) can adhere to it and when used on bone as well as cartilage tissue engineering, the hydrogels promote osteoblast and chondrocyte differentiation [95].

Cell Proliferation

We have engineered hybrid hydrogels which then establish a three-dimensional architecture for cell attachment leading to increased cell viability due to similarity with the native ECM. The above growth factors include vascular endothelial growth factor (VEGF) or transforming growth factor-beta (TGF-β) which in addition to the above methods also increases proliferation. Some research has demonstrated that the endothelial cell proliferation for the vascular tissue regeneration by the cellulose-chitosan hydrogels loaded with VEGF [96].

Cell Differentiation

Stem cell specialization marks is indispensable in the development of various tissues for repair. Bioactive ions (such as calcium or magnesium), or nanoparticles (such as hydroxyapatite) embedded within the hydrogel matrix present chemical signals to guide MSCs into osteogenic or chondrogenic differentiation. In a like manner, cellulose hydrogels that display neural growth factors have been effective in the process of converting neural progenitor cells into functional neurons useful in nerve repair [97].

8.2 Immunity and Biodegradation

Immune Compatibility

This is because cellulose-based hydrogels having to be biocompatible and are not allowed to elicit any contrary reactions to the immune system. The native cellulose is immunogenic, but the application of chemical changes or setting cross-linkage may elicit inflammation. To overcome this problem, crosslinking agents like genipin, enzymes are used instead of glutaraldehyde. For instance, cellulose-PEG hybrid hydrogels of the wound healing applications do not stimulate macrophage activation and cytokine generation [98].

Biodegradability

That is why biodegradability of hydrogels has to be controlled to ensure that fabricated hydrogel structures can be easily incorporated into the host tissue environment. Cellulose hydrogels also break down by enzyme (such as cellulase), and the products are non-hazardous. Nevertheless, it is possible to control the degradation rate by changing the crosslink density or by using biodegradable polymers as a blend with PLA or gelatin. A faster rate of degradation is useful where the scaffold is needed for only a few days or weeks such as in the case of wound healing but a slower rate of degradation is highly desirable where the scaffold is required for months or even years, for the purpose of bone or cartilage repair. Community reports show that hydrogels of both cellulose-SF and cellulose-SF-TP mixtures display tunable degradation kinetics compatible with tissue engineering applications [17].

8.3 Biological Techniques to Increase Cell–Matrix Attachment

For enhanced conformance to ideal cellular function, various approaches are used to enhance the cell-hydrogel comunicativity.

Incorporation of Bioactive Molecules

Presently some of the cell adhesion peptides, growth factors and other ECM components such as laminin or hyaluronic acid have been incorporated into the cellulose hydrogels. For example, RGD conjugated cellulose hydrogels improve adhesion and spreading of endothelial cells that are essential for blood vessel formation during tissue remodeling [99].

Nanomaterials for Cellular Cues

By immobilizing biochemical and mechanical stimuli on the cellulose hydrogels in the form of nanomaterials like graphene oxide, carbon nanotubes, or gold nanoparticles, cells can be controlled. These hybrid materials have been identified as helpful for neural and cardiac tissue engineering as they advanced electrical conductivity and signal transduction [97].

Dynamic and Stimuli-Responsive Properties

Reflecting the condition and functions of natural tissues, intelligent hydrogels that can react to the stimulus such as temperature, pH, or light provide dynamic conditions. For instance, cellulose-based hydrogels used in combination with light sensitive polymers have allowed for the localized control of cell behavior such as cell division and differentiation allowing for the formation of intricate tissue constructs [100].

Micro- and Nano-patterning

Other methods, for example, lithography or electrospinning, make a specific surface relief on a hydrogel, inducing cell orientation and movement. Aligned cellulose nanofiber hydrogels have shown superiormyocyte alignment and contraction in cardiac tissue engineering [101].

3D Bioprinting

Polymer-cellulose hybrid hydrogels are used in 3D bioprinting as a scaffold with appropriate designs of the architecture. Manufactured and natural bio-inks have featured cellulose derivatives and alginate for the formation of vascularized tissue constructs; it has also promoted the cell viability and integration [102].

9 Future Prospects and Problems

Since tissue engineering demands hydrogels to simulate a multifaceted and dynamic natural tissue micro environment; thus, the modification of cellulose-based hybrid hydrogels is relevant for the enhancement of their performance or versatility. Advanced properties in the material domain, stimuli sensitivity, and functions will expand the applications of hydrogels in regenerative medicine.

9.1 Clinical Translation: The Challenges and Regulation

Recent years witnessed much interest in the development of cellulose-based hybrid hydrogels as drug carriers in preclinical models, but several hurdles lie ahead of their clinical application. Such challenges are; manufacturing challenges, using regulatory bodies and safety and efficacy for the long-term concerns [103].

Manufacturing and Scalability

The poor comparability between laboratories and particularly the broad variations of parameters were a major concern when it comes to the large scaled up application of cellulose-based hydrogels in clinical settings. Successful hydrogels formation is a critical step that requires high accuracy, uniformity, and batch-to-batch reproducibility characteristic of clinical applications. However, the manufacture of hydrogels containing bioactive elements or nanoparticles at an industrial scale may be challenging and necessitate the use of techniques such as controlled polymerization or even microfluidics [104].

Regulatory Hurdles

The U.S Food and Drug Administration (FDA) and the European Medicines Agency (EMA) have set high standards for biomaterials used in any clinical practice. Hydrogels come under either the medical device category or the drug delivery system category, hence they demand biocompatibility, sterility and safety measures. When it comes to hydrogels, the absence of a clear understanding of the regulatory approval procedure is a major problem, e.g., hydrogels with nanomaterials or biosactive agents. As of now, there are no standard tests for cytotoxicity, immunogenicity, and degradation and these aspects need to be standardized for an engineered scaffold to be approved clinically [104].

Safety and Long-Term Efficacy

A second issue in translating cellulose-based hydrogels into clinical practice is the stability and biocompatibility of the long-term use of these hydrogels. Short-term in vitro and in vivo investigations result in highly favorable findings but little is known about the hydrogels' performance in chronic and implantation designs. There is a risk that the immune system might mount a response to the implanted scaffold that is either inflammatory or fibrotic encapsulation that needs to be assessed in chronic animal models and clinical trials. In addition, the degradation products of cellulose-based hydrogels must be non-toxic and biologically compatible [105].

9.2 Emerging Trends

4D Bioprinting and Biofabrication of Hybrid Hydrogels As the field of tissue engineering advances, new enabling technologies appear and 4D bioprinting and biofabrication of cellulose-based hybrid hydrogels present possible solutions for the construction of complex tissue structures. These innovations envisage the potential of bringing about radical changes in the process of constructing functional tissues or organs.

4D Bioprinting

4D bioprinting is a combination of 3D bioprinting technology in which time-dependent changes in the created structures can be induced. This kind of property enables the design of hydrogels that can be fabricated to alter their shape, rigidity, or utility based on an outside stimulus. These properties as well as their biocompatibility make cellulose-based hydrogels good candidates for use in 4D bioprinting owing to the fact that they can be functionalized to incorporate switchable elements. It, therefore, becomes possible to control the stimuli-responsive components to respond to some of the dynamic behaviors of native tissues, e.g., the contraction of muscles in the heart or the dilation of blood vessels; this is done through 4D bioprinting of hydrogels that can either expand or contract depending on the change in temperature [106].

Biofabrication and Organs-on-a-Chip

Tissue engineering and three-dimensional bioprinting methods—biofabrication methods—continue to create functional tissues and organ-like structures. For instance, the organs-on-a-chip, which replicate the human physiological conditions of organs, employ the hybrid hydrogels as an ECM milieu for culture of cells. The advantages of using cellulose-based hybrid hydrogels as a scaffold for constructing these models derive from the fact that the hydrogels can be loaded with bioactive molecules. They have the potential to provide a clearer explanation of cell behavior while interacting in an artificial environment typical of in vitro conditions before their application to treat patients [107].

Personalized Medicine and Custom Scaffold Design

There is an increased demand for tissue scaffolds for personalized medicine since the fields of tissue engineering have advanced to a level that calls for engineered tissues specific to patients. 3D bioprintable hybrid hydrogels enable generation of individualized scaffolds from data obtained through CT or MRI scans of patients' tissues. This trend means, the design of scaffolds that identical to the geometry and mechanical characters of the targeted tissue hence promoting tissue regeneration and integration [108].

The potential uses of cellulose-based hybrid hydrogels in tissue engineering appear to hold great prospects for the future, but they are still riddled with considerable challenges. Future developments in hydrogel technology are expected from the unlimited possibility of tweaking hydrogel properties through new formulations, clinics approval, and the newly available fabrication techniques including 4D bioprinting. Alleviating these challenges, these advanced materials could transform the field of regenerative medicine and provide answers for demanding tissue reconstruction and organ scaffolding [109].

10 Conclusion

Cellulose-based hybrid hydrogels are a subset of smart biomaterials well suited for, tissue engineering, because of their light weight, biocompatibility, and adjustable mechanical and chemical properties. As will be shown below, their usage in engineering materials covers applications in areas such as bone and cartilage, skin, and neural tissues. Of all the materials used to construct scaffolds, the hydrogels are most preferred because they can easily be designed to mimic the natural ECM hence promoting appropriate cell response. Advanced synthesis of hybrid hydrogels has advanced, with improving of responsiveness of the hydrogels to stimuli, addition of bioactive species in the hydrogel network, and introduction of nanoparticles in the hydrogel matrix for the improvement of mechanical characteristics, electrical conductivity and biological performance. Functions such as stimuli sensitivity in terms of pH, temperature, light, and mechanical stress suggest that there is great potential for its use in the course of controlled release, tissue generation, and phosphorylated therapy.

This notwithstanding, there remain some challenges that need to be addressed to make cellulose-based hybrid hydrogels useful in clinical practice. These are factors ranging from technological challenges in the manufacture of hydrogels, its ability to be reproduced in large scale and meeting regulators' specifications for release to the market as well as the requirement on safety and efficacy tests. While they also showed potential application of these hydrogels, long-term degradation and immune assessment of these hydrogels should have been conducted in vivo studies. Also, the complexity of the human tissues and organs provide major challenges to simulate the architecture and mechanics accurately.

The current developments of advanced manufacturing methods including 4D bioprinting and biofabrication will determine the future prospects of cellulose-based hybrid hydrogels in tissue engineering applications. In view of this assertion, these technologies will promote development of more luxurious and functional tissue construct, which signifies native tissue construct nature and dynamic performance. Besides, the use of these hydrogels for integration with personalized medicine will offer the patient tailor made solutions for tissue remodeling and healing.

Acknowledgements The authors acknowledge in particular ShriRam College of Pharmacy, Banmore, Morena, Madhya Pradesh, India, for encouraging us and providing necessary facilities.

Funding There was no funding for this project from any source.

References

1. Han F, Wang J, Ding L, Hu Y, Li W, Yuan Z, Guo Q, Zhu C, Yu L, Wang H, Zhao Z (2020) Tissue engineering and regenerative medicine: achievements, future, and sustainability in Asia. Front Bioeng Biotechnol 8:83
2. Sharma P, Rathore B, Jain S, Sharma M, Tiwari S, Kaushal R, Akram W, Jain V (2025) Principle characteristics and specification of pharmaceutical polymers. In: Polymers for oral drug delivery technologies. Elsevier Science Ltd, pp 197–231
3. De Chiara F, Ferret-Miñana A, Fernández-Costa JM, Ramón-Azcón J (2024) The tissue engineering revolution: from bench research to clinical reality. Biomedicines 12(2):453
4. Zhao L, Zhou Y, Zhang J, Liang H, Chen X, Tan H (2023) Natural polymer-based hydrogels: from polymer to biomedical applications. Pharmaceutics 15(10):2514
5. Ribeiro M, Simões M, Vitorino C, Mascarenhas-Melo F (2024) Hydrogels in cutaneous wound healing: insights into characterization, properties, formulation, and therapeutic potential. Gels 10(3):188
6. Jacob S, Nair AB, Shah J, Sreeharsha N, Gupta S, Shinu P (2021) Emerging role of hydrogels in drug delivery systems, tissue engineering and wound management. Pharmaceutics 13(3):357
7. Raina N, Pahwa R, Bhattacharya J, Paul AK, Nissapatorn V, de Lourdes PM, Oliveira SM, Dolma KG, Rahmatullah M, Wilairatana P, Gupta M (2022) Drug delivery strategies and biomedical significance of hydrogels: translational considerations. Pharmaceutics 14(3):574
8. Protsak IS, Morozov YM (2025) Fundamentals and advances in stimuli-responsive hydrogels and their applications: a review. Gels 11(1):30
9. Guo J, Cao G, Wei S, Han Y, Xu P (2023) Progress in the application of graphene and its derivatives to osteogenesis. Heliyon
10. Hussain S, Maktedar SS (2023) Structural, functional and mechanical performance of advanced graphene-based composite hydrogels. Results Chem 101029
11. Rana AK, Gupta VK, Hart P, Thakur VK (2024) Cellulose-alginate hydrogels and their nanocomposites for water remediation and biomedical applications. Environ Res 243:117889
12. Solanki R, Bhatia D (2024) Stimulus-responsive hydrogels for targeted cancer therapy. Gels 10(7):440
13. Sharma P, Jain V, Tailang M (2023) Advancement of nanocarrier-based engineering for specific drug delivery for cancer therapy. In: Targeted cancer therapy in biomedical engineering. Springer Nature Singapore, pp 465–486
14. Jabeen N, Atif M (2024) Polysaccharides based biopolymers for biomedical applications: a review. Polym Adv Technol 35(1):e6203
15. Jin SA, Spontak RJ (2023) Fundamentals of and advances in nanocellulose and nanochitin systems. Adv Ind Eng Polym Res 6(4):356–381
16. Ojha C, Sharma P, Jain V (2024) Design, optimization, and evaluation of topical gel of Cardiospermum halicacabum and *Ricinus communis L.* leaves extract for the treatment of rheumatoid arthritis. J Biomater Sci Polym Ed 1–22
17. Garcia-Garcia A, Muñana-González S, Lanceros-Mendez S, Ruiz-Rubio L, Alvarez LP, Vilas-Vilela JL (2024) Biodegradable natural hydrogels for tissue engineering, controlled release, and soil remediation. Polymers 16(18):2599
18. Xuanshu Z, Zongjian L, Xue G, Lin Y, Zengguo F, Jianing X (2022) Regulating cell adhesion by material surface properties. Prog Chem 34(5):1153

19. Hernandez JL, Woodrow KA (2022) Medical applications of porous biomaterials: features of porosity and tissue-specific implications for biocompatibility. Adv Healthc Mater 11(9):2102087
20. Sharma P, Akram W, Jain V, Tailang M (eds) (2024) Smart nanocarrier for effective drug delivery. CRC Press
21. Tamo AK (2024) Nanocellulose-based hydrogels as versatile materials with interesting functional properties for tissue engineering applications. J Mater Chem B
22. Kawadkar M, Mishra P, Mishra A, Sharma P (2024) Protein-based nanocarrier for targeted delivery of drug for cancer therapy. In: Sharma P, Akram W, Jain V, Tailang M (eds) Smart nanocarrier for effective drug delivery. CRC Press, pp 36–60
23. Reddy MS, Ponnamma D, Choudhary R, Sadasivuni KK (2021) A comparative review of natural and synthetic biopolymer composite scaffolds. Polymers 13(7):1105
24. Tevlek A, Çetin EA (2024) Smart hydrogels in lab-on-a-chip (LOC) applications. React Funct Polym 106023
25. Kasiński A, Zielińska-Pisklak M, Oledzka E, Sobczak M (2020) Smart hydrogels–synthetic stimuli-responsive antitumor drug release systems. Int J Nanomedicine 15:4541–4572
26. Batool N, Sarfraz RM, Mahmood A, Rehman U, Zaman M, Akbar S, Almasri DM, Gad HA (2023) Development and evaluation of cellulose derivative and pectin based swellable pH responsive hydrogel network for controlled delivery of cytarabine. Gels 9(1):60
27. Motloung B, Pfukwa R, Klumperman B (2024) Ion-mediated gelation of thermo-responsive cellulose nanofibril/poly (N-isopropylacrylamide) hybrid hydrogels with tunable de-swelling kinetics. Macromol Mater Eng 2300457
28. Londhe PV, Londhe MV, Salunkhe AB, Laha SS, Mefford OT, Thorat ND, Khot VM (2025) Magnetic hydrogel (MagGel): an evolutionary pedestal for anticancer therapy. Coord Chem Rev 522:216228
29. Choi H, Choi WS, Jeong JO (2024) A review of advanced hydrogel applications for tissue engineering and drug delivery systems as biomaterials. Gels 10(11):693
30. Tofanica BM, Mikhailidi A, Samuil C, Ungureanu OC, Fortună ME, Ungureanu E (2024) Advances in cellulose-based hydrogels: current trends and challenges. Gels 10(12):842
31. Hong F, Qiu P, Wang Y, Ren P, Liu J, Zhao J, Gou D (2024) Chitosan-based hydrogels: from preparation to applications, a review. Food Chem X 21:101095
32. Yao CH, Liu BS, Chang CJ, Hsu SH, Chen YS (2004) Preparation of networks of gelatin and genipin as degradable biomaterials. Mater Chem Phys 83(2–3):204–208
33. Xu L, Wang YY, Huang J, Chen CY, Wang ZX, Xie H (2020) Silver nanoparticles: synthesis, medical applications and biosafety. Theranostics 10(20):8996
34. Klimek K, Ginalska G (2020) Proteins and peptides as important modifiers of the polymer scaffolds for tissue engineering applications—a review. Polymers 12(4):844
35. Palantöken S, Bethke K, Zivanovic V, Kalinka G, Kneipp J, Rademann K (2020) Cellulose hydrogels physically crosslinked by glycine: synthesis, characterization, thermal and mechanical properties. J Appl Polym Sci 137(7):48380
36. Sannino A, Pappadà S, Madaghiele M, Maffezzoli A, Ambrosio L, Nicolais L (2005) Crosslinking of cellulose derivatives and hyaluronic acid with water-soluble carbodiimide. Polymer 46(25):11206–11212
37. Sharma P, Jain V, Tailang M (2022) Selection and role of polymers for designing of a drug carrier. In: Drug carriers. IntechOpen
38. Ahmad Z, Salman S, Khan SA, Amin A, Rahman ZU, Al-Ghamdi YO, Akhtar K, Bakhsh EM, Khan SB (2022) Versatility of hydrogels: from synthetic strategies, classification, and properties to biomedical applications. Gels 8(3):167
39. Ren Y, Wang Q, Xu W, Yang M, Guo W, He S, Liu W (2024) Alginate-based hydrogels mediated biomedical applications: a review. Int J Biol Macromol 279:135019
40. Fang W, Yang M, Liu M, Jin Y, Wang Y, Yang R, Wang Y, Zhang K, Fu Q (2023) Review on additives in hydrogels for 3D bioprinting of regenerative medicine: from mechanism to methodology. Pharmaceutics 15(6):1700

41. Kumar VB, Tiwari OS, Finkelstein-Zuta G, Rencus-Lazar S, Gazit E (2023) Design of functional RGD peptide-based biomaterials for tissue engineering. Pharmaceutics 15(2):345
42. Aziz T, Farid A, Haq F, Kiran M, Ullah A, Zhang K, Li C, Ghazanfar S, Sun H, Ullah R, Ali A (2022) A review on the modification of cellulose and its applications. Polymers 14(15):3206
43. Öztürk AB, Alarcin E, Yaşayan G, Avci-Adali M, Khosravi A, Zarepour A, Iravani S, Zarrabi A (2024) Innovative approaches in skin therapy: bionanocomposites for skin tissue repair and regeneration. Mater Adv
44. Miao L, Zhang M, Tu Y, Lin S, Hu J (2019) Stimuli-responsive cellulose-based hydrogels. Polymers and polymeric composites: a reference series, pp 269–308
45. Pangli H, Vatanpour S, Hortamani S, Jalili R, Ghahary A (2021) Incorporation of silver nanoparticles in hydrogel matrices for controlling wound infection. J Burn Care Res 42(4):785–793
46. Maeso L, Antezana PE, Hvozda Arana AG, Evelson PA, Orive G, Desimone MF (2024) Progress in the use of hydrogels for antioxidant delivery in skin wounds. Pharmaceutics 16(4):524
47. Dong Y, Wang Z (2023) ROS-scavenging materials for skin wound healing: advancements and applications. Front Bioeng Biotechnol 11:1304835
48. Thivya P, Akalya S, Sinija VR (2022) A comprehensive review on cellulose-based hydrogel and its potential application in the food industry. Appl Food Res 2(2):100161
49. Li Z, Zhou Y, Li T, Zhang J, Tian H (2022) Stimuli-responsive hydrogels: fabrication and biomedical applications. View 3(2):20200112
50. Babaei-Ghazvini A, Patel R, Vafakish B, Yazdi AF, Acharya B (2024) Nanocellulose in targeted drug delivery: a review of modifications and synergistic applications. Int J Biol Macromol 135200
51. Sun Y, Wang J, Li D, Cheng F (2024) The recent progress of the cellulose-based antibacterial hydrogel. Gels 10(2):109
52. Saraf S, Jain A, Tiwari A, Verma A, Panda PK, Jain SK (2020) Advances in liposomal drug delivery to cancer: an overview. J Drug Deliv Sci Technol 56:101549
53. Malandain N, Sanz-Fraile H, Farré R, Otero J, Roig A, Laromaine A (2023) Cell-laden 3D hydrogels of type I collagen incorporating bacterial nanocellulose fibers. ACS Appl Bio Mater 6(9):3638–3647
54. Cao H, Duan L, Zhang Y, Cao J, Zhang K (2021) Current hydrogel advances in physicochemical and biological response-driven biomedical application diversity. Signal Transduct Target Ther 6(1):426
55. Schönmeyr BH, Soares M, Avraham T, Clavin NW, Gewalli F, Mehrara BJ (2010) Vascular endothelial growth factor inhibits bone morphogenetic protein 2 expression in rat mesenchymal stem cells. Tissue Eng Part A 16(2):653–662
56. Pankaj S, Mukul T (2019) PH-dependent release potential of natural polymers in sustained release of ornidazole from colon targeted delivery system. Magnesium 10(10):10
57. Jiang J, Xu S, Ma H, Li C, Huang Z (2023) Photoresponsive hydrogel-based soft robot: a review. Mater Today Bio 20:100657
58. Zhan X, Yan J, Xiang D, Tang H, Cao L, Zheng Y, Lin H, Xia D (2024) Near-infrared light responsive gold nanoparticles coating endows polyetheretherketone with enhanced osseointegration and antibacterial properties. Mater Today Bio 25:100982
59. Gao W, Zhang Y, Zhang Q, Zhang L (2016) Nanoparticle-hydrogel: a hybrid biomaterial system for localized drug delivery. Ann Biomed Eng 44:2049–2061
60. Hawthorne D, Pannala A, Sandeman S, Lloyd A (2022) Sustained and targeted delivery of hydrophilic drug compounds: a review of existing and novel technologies from bench to bedside. J Drug Deliv Sci Technol 78:103936
61. Yu H, Gao R, Liu Y, Fu L, Zhou J, Li L (2024) Stimulus-responsive hydrogels as drug delivery systems for inflammation-targeted therapy. Adv Sci 11(1):2306152
62. Askari E, Seyfoori A, Amereh M, Gharaie SS, Ghazali HS, Ghazali ZS, Khunjush B, Akbari M (2020) Stimuli-responsive hydrogels for local post-surgical drug delivery. Gels 6(2):14

63. Karvinen J, Kellomäki M (2024) 3D-bioprinting of self-healing hydrogels. Eur Polym J 112864
64. Ranjha NM, Mudassir J, Majeed S (2011) Synthesis and characterization of polycaprolactone/acrylic acid (PCL/AA) hydrogel for controlled drug delivery. Bull Mater Sci 34:1537–1547
65. Devi VKA, Shyam R, Palaniappan A, Jaiswal AK, Oh TH, Nathanael AJ (2021) Self-healing hydrogels: preparation, mechanism and advancement in biomedical applications. Polymers 13(21):3782
66. Carvalho T, Bártolo R, Pedro SN, Valente BF, Pinto RJ, Vilela C, Shahbazi MA, Santos HA, Freire CS (2023) Injectable nanocomposite hydrogels of gelatin-hyaluronic acid reinforced with hybrid lysozyme nanofibrils-gold nanoparticles for the regeneration of damaged myocardium. ACS Appl Mater Interfaces 15(21):25860–25872
67. Carton F, Rizzi M, Canciani E, Sieve G, Di Francesco D, Casarella S, Di Nunno L, Boccafoschi F (2024) Use of hydrogels in regenerative medicine: focus on mechanical properties. Int J Mol Sci 25(21):11426
68. Olteanu G, Neacşu SM, Joiţa FA, Musuc AM, Lupu EC, Ioniţă-Mîndrican CB, Lupuliasa D, Mititelu M (2024) Advancements in regenerative hydrogels in skin wound treatment: a comprehensive review. Int J Mol Sci 25(7):3849
69. Gao Y, Zhang X, Zhou H (2023) Biomimetic hydrogel applications and challenges in bone, cartilage, and nerve repair. Pharmaceutics 15(10):2405
70. Alonso JM, Andrade del Olmo J, Perez Gonzalez R, Saez-Martinez V (2021) Injectable hydrogels: from laboratory to industrialization. Polymers 13(4):650
71. Alsteens D, Gaub HE, Newton R, Pfreundschuh M, Gerber C, Müller DJ (2017) Atomic force microscopy-based characterization and design of biointerfaces. Nat Rev Mater 2(5):1–6
72. Wang R, Cheng C, Wang H, Wang D (2024) Swollen hydrogel nanotechnology: advanced applications of the rudimentary swelling properties of hydrogels. ChemPhysMater
73. Xiao Y, Friis EA, Gehrke SH, Detamore MS (2013) Mechanical testing of hydrogels in cartilage tissue engineering: beyond the compressive modulus. Tissue Eng Part B Rev 19(5):403–412
74. Li Q, Feng Z, Song H, Zhang J, Dong A, Kong D, Wang W, Huang P (2020) 19F magnetic resonance imaging enabled real-time, non-invasive and precise localization and quantification of the degradation rate of hydrogel scaffolds in vivo. Biomater Sci 8(12):3301–3309
75. Sharma P, Dixit A, Rathore D, Rajpoot Y, Joshi R, Singh K (2015) To formulate and evaluate orodispersible tablets of primaquine. Am J Pharm Res 5(5)
76. Ielo I, Calabrese G, De Luca G, Conoci S (2022) Recent advances in hydroxyapatite-based biocomposites for bone tissue regeneration in orthopedics. Int J Mol Sci 23(17):9721
77. Chen M, Jiang Z, Zou X, You X, Cai Z, Huang J (2024) Advancements in tissue engineering for articular cartilage regeneration. Heliyon
78. Hsieh CF, Chen CH, Kao HH, Govindaraju DT, Dash BS, Chen JP (2022) PLGA/gelatin/hyaluronic acid fibrous membrane scaffold for therapeutic delivery of adipose-derived stem cells to promote wound healing. Biomedicines 10(11):2902
79. Aydin T, Gurcan C, Taheri H, Yilmazer A (2018) Graphene-based materials in neural tissue regeneration. In: Cell biology and translational medicine, vol 3: stem cells, biomaterials, and tissue engineering. Springer, pp 129–142
80. Everts PA, Lana JF, Onishi K, Buford D, Peng J, Mahmood A, Fonseca LF, van Zundert A, Podesta L (2023) Angiogenesis and tissue repair depend on platelet dosing and bioformulation strategies following orthobiological platelet-rich plasma procedures: a narrative review. Biomedicines 11(7):1922
81. Liu J, Yang L, Liu K, Gao F (2023) Hydrogel scaffolds in bone regeneration: their promising roles in angiogenesis. Front Pharmacol 14:1050954
82. Sousa AC, Alvites R, Lopes B, Sousa P, Moreira A, Coelho A, Rêma A, Biscaia S, Cordeiro R, Faria F, da Silva GF (2025) Hybrid scaffolds for bone tissue engineering: integration of composites and bioactive hydrogels loaded with hDPSCs. Biomater Adv 166:214042
83. Li H, Wu C, Yu X, Zhang W (2023) Recent advances of PVA-based hydrogels in cartilage repair application. J Mater Res Technol 24:2279–2298

84. Ruiz-Alonso S, Lafuente-Merchan M, Ciriza J, Saenz-del-Burgo L, Pedraz JL (2021) Tendon tissue engineering: cells, growth factors, scaffolds and production techniques. J Control Release 333:448–486
85. Karami P, Laurent A, Philippe V, Applegate LA, Pioletti DP, Martin R (2024) Cartilage repair: promise of adhesive orthopedic hydrogels. Int J Mol Sci 25(18):9984
86. Ghovvati M, Kharaziha M, Ardehali R, Annabi N (2022) Recent advances in designing electroconductive biomaterials for cardiac tissue engineering. Adv Healthc Mater 11(13):2200055
87. Chen J, Zhang D, Wu LP, Zhao M (2023) Current strategies for engineered vascular grafts and vascularized tissue engineering. Polymers 15(9):2015
88. Omidian H, Chowdhury SD, Cubeddu LX (2024) Hydrogels for neural regeneration: exploring new horizons. Materials 17(14):3472
89. Qin C, Qi Z, Pan S, Xia P, Kong W, Sun B, Du H, Zhang R, Zhu L, Zhou D, Yang X (2023) Advances in conductive hydrogel for spinal cord injury repair and regeneration. Int J Nanomedicine 7305–7333
90. Nandhini J, Karthikeyan E, Rani EE, Karthikha VS, Sanjana DS, Jeevitha H, Rajeshkumar S, Venugopal V, Priyadharshan A (2024) Advancing engineered approaches for sustainable wound regeneration and repair: harnessing the potential of green synthesized silver nanoparticles. Eng Regen 5(3):306–325
91. Kaur H, Gogoi B, Sharma I, Das DK, Azad MA, Pramanik DD, Pramanik A (2024) Hydrogels as a potential biomaterial for multimodal therapeutic applications. Mol Pharm 21(10):4827–4848
92. Firlar I, Altunbek M, McCarthy C, Ramalingam M, Camci-Unal G (2022) Functional hydrogels for treatment of chronic wounds. Gels 8(2):127
93. Kasoju N, Sunilkumar A (2024) Convergence of tissue engineering and sustainable development goals. Biotechnol Sustain Mater 1(1):20
94. Zhao X, Hu DA, Wu D, He F, Wang H, Huang L, Shi D, Liu Q, Ni N, Pakvasa M, Zhang Y (2021) Applications of biocompatible scaffold materials in stem cell-based cartilage tissue engineering. Front Bioeng Biotechnol 9:603444
95. Kang Y, Guan Y, Li S (2024) Innovative hydrogel solutions for articular cartilage regeneration: a comprehensive review. Int J Surg 110(12):7984–8001
96. Norahan MH, Pedroza-González SC, Sánchez-Salazar MG, Álvarez MM, de Santiago GT (2023) Structural and biological engineering of 3D hydrogels for wound healing. Bioact Mater 24:197–235
97. Zare I, Mirshafiei M, Kheilnezhad B, Far BF, Hassanpour M, Pishbin E, Vaghefi SS, Yazdian F, Rashedi H, Hasan A, Wang X (2024) Hydrogel-integrated graphene superstructures for tissue engineering: from periodontal to neural regeneration. Carbon 118970
98. Lu P, Ruan D, Huang M, Tian M, Zhu K, Gan Z, Xiao Z (2024) Harnessing the potential of hydrogels for advanced therapeutic applications: current achievements and future directions. Signal Transduct Target Ther 9(1):166
99. Isik M, Okesola BO, Eylem CC, Kocak E, Nemutlu E, Emregul E, D'Este M, Derkus B (2022) Tuning the cell-adhesive properties of two-component hybrid hydrogels to modulate cancer cell behavior, metastasis, and death pathways. Biomacromol 23(10):4254–4267
100. Chakrapani G, Zare M, Ramakrishna S (2022) Intelligent hydrogels and their biomedical applications. Mater Adv 3(21):7757–7772
101. Martin A, Nyman JN, Reinholdt R, Cai J, Schaedel AL, van der Plas MJ, Malmsten M, Rades T, Heinz A (2022) In situ transformation of electrospun nanofibers into nanofiber-reinforced hydrogels. Nanomaterials 12(14):2437
102. Yoon J, Han H, Jang J (2023) Nanomaterials-incorporated hydrogels for 3D bioprinting technology. Nano Converg 10(1):52
103. Ramezani G, Stiharu I, van de Ven TG, Nerguizian V (2024) Advancements in hybrid cellulose-based films: innovations and applications in 2D nano-delivery systems. J Funct Biomater 15(4):93
104. Karchoubi F, Ghotli RA, Pahlevani H, Salehi MB (2024) New insights into nanocomposite hydrogels; a review on recent advances in characteristics and applications. Adv Ind Eng Polym Res 7(1):54–78

105. Zöller K, To D, Bernkop-Schnürch A (2024) Biomedical applications of functional hydrogels: innovative developments, relevant clinical trials and advanced products. Biomaterials 122718
106. Lai J, Liu Y, Lu G, Yung P, Wang X, Tuan RS, Li ZA (2024) 4D bioprinting of programmed dynamic tissues. Bioact Mater 37:348–377
107. Mierke CT (2024) Bioprinting of cells, organoids and organs-on-a-chip together with hydrogels improves structural and mechanical cues. Cells 13(19):1638
108. Mirshafiei M, Rashedi H, Yazdian F, Rahdar A, Baino F (2024) Advancements in tissue and organ 3D bioprinting: current techniques, applications, and future perspectives. Mater Des 112853
109. Chen C, Xi Y, Weng Y (2022) Recent advances in cellulose-based hydrogels for tissue engineering applications. Polymers 14(16):3335

Pankaj Sharma Professor and Head of Pharmaceutics, ShriRam College of Pharmacy, Morena, M.P., India, with over 13 years of teaching and research experience. Dr. Sharma specializes in drug delivery systems, particularly nanotechnology and transdermal therapeutics. He has authored 10 books, 20 book chapters, and 48 research and review papers, with 34 indexed in Scopus/SCI journals. Awarded the "Best Doctorate Award" by IPES and holds a UK design patent. He is an M.Pharm. guide, having supervised several postgraduate projects. Dr. Sharma is dedicated to advancing pharmaceutical education and research through innovative methodologies and collaborative initiatives.

Vinay Jain is a highly accomplished academic and administrative professional, currently serving as the Principal of ShriRam College of Pharmacy. He holds an M.Pharm. and Ph.D. in Pharmacognosy, and has amassed 20 years of experience in teaching and administration. He has made significant contributions to the field through his research and publications, including peer-reviewed articles, edited and indexed books, and guidance for over 32 research scholars. Dr. Jain completed his M.Pharm. in Pharmacognosy from Barkatullah University in 2007 and began his professional career as Assistant Professor at ShriRam College of Pharmacy, Banmore. In 2017, he successfully defended his Ph.D. thesis entitled "Phytochemical and Pharmacological investigation of ***Arisaema leschenaultii*** from Berhampur University, Odisha, India. In addition to his academic pursuits, Dr. Jain has been an active member of the pharmacy community, serving as Co-convener of the Indo-Caribbean conference and Organizing Secretary of the national conference on "The Future Prospect in Pharma-Marketing." He has also filed 2 patents and published 50 + papers in international and national journals. In recognition of his excellence, Dr. Jain has received various awards and honors, including the **Dhanvantari Awarded** during SRIJAN 2021, and has been granted an allowance of INR 2.5 lac for conducting an international conference by AICTE and a grant of INR 8.5

lac by AICTE under MODROB Scheme for laboratory upgradation. He has also been invited as a guest of honor at several conferences and workshops held in MP. His recent achievements include **Best Principal award in 2022** by INSC publishers and **Pharma Rising Star Award in 2022**, which he received at New Delhi.

Grape Stems as a Sustainable Source of Cellulose for Skin Regeneration

Raquel Fernandes, Cláudia Botelho, and Ana Novo Barros

Abstract Skin, the largest organ of mammals, provides critical functions including internal organ protection, temperature control, and electrolyte/fluid balance maintenance. Loss of skin integrity impacts overall health, requiring its timely restoration. Skin regeneration, or wound healing, is a physiological process comprising hemostasis, inflammation, proliferation, and remodeling. Dysregulation of these phases, often exacerbated by bacterial infections that prolong inflammation, can result in chronic wounds. The medical community and researchers are gathering efforts to develop highly efficient and novel treatment strategies for managing chronic wounds. Researchers cannot disregard an important factor sustainability and reduced environmental impact of the new therapeutical approaches. So, natural materials have been investigators main choice to advance the creation of innovative medical devices designed for wound recovery, being an example cellulose. Cellulose, a naturally abundant polymer, has emerged as a versatile biomaterial with remarkable potential in the management of wounds, owing its biodegradable, biocompatible, and mechanical properties. This chapter will explore the physiology and dysregulation of wound healing, highlighting the impact of a microbial infection. It will delve into the characterization of cellulose, its derivatives, and cellulose-based biomaterials with improved antibacterial and skin regeneration properties. Special emphasis will be placed on natural-based cellulose sources, particularly grape stems as a sustainable and underutilized resource. This comprehensive analysis underscores the role of cellulose and its derivatives as innovative solutions for wound care, particularly addressing its microbial resistance and reduce carbon print.

R. Fernandes · A. N. Barros (✉)
Centre for the Research and Technology of Agroenvironmental and Biological Sciences, CITAB, Inov4Agro, Universidade de Trás-Os-Montes e Alto Douro, UTAD, Vila Real, Portugal
e-mail: abarros@utad.pt

R. Fernandes
e-mail: rgfernandes@utad.pt

C. Botelho
Centre of Biological Engineering (CEB), University of Minho, Braga, Portugal

LABBELS—Associate Laboratory, Braga, Guimarães, Portugal

R. Malviya and S. Sundram (eds.), *Sustainable Cellulose Materials for Biomedical Applications*, Sustainable Materials and Technology,
https://doi.org/10.1007/978-981-96-9904-9_7

Keywords Cellulose derivatives · Biomaterials · Wound healing · Antibacterial · Sustainable cellulose

1 Introduction

The skin, being the body's largest organ, serves as the primary defense, protecting internal structures from outside insults. Extensive research has shown skin's involvement in several other processes, namely the synthesis, processing, and metabolism comprising signaling molecules, proteins, glycans, and lipids [16, 31, 49, 69, 74, 119, 126]. Furthermore, skin has gained recognition as a crucial constituent of different systems including the nervous and endocrine [120].

The composition of human skin includes several layers, all of them presenting in their composition distinct cell types, conferring each layer specific functions. As expected, the specificities of each layer will contribute to distinct forms for the healing process. When skin integrity is breached, the dynamic process of tissue restoration begins [111].

The healing process of wounds is essential for reconstructing skin integrity after injury. It involves a well-orchestrated sequence of events, beginning with hemostasis and inflammation, followed by proliferation and tissue remodeling [37]. However, factors underlying health conditions, environmental stressors, and infections can disrupt this process, leading to delayed healing and formation of chronic wounds. Bacterial infections represent a significant impairment to wound closure, promoting inflammation and tissue damage while fostering antibiotic resistance [112].

In parallel with the growing demand for advanced wound care solutions, the field of biomaterials has considerable expanded. As the development of new treatment approaches must take in consideration the environmental impact and sustainability, the particular interest in natural materials has grown. Among them, cellulose, a natural polymer, which can be extracted from plants or synthetized by bacteria, has shown high potential to promote tissue repair, stemming from its distinctive functional and structural design [18]. Derivatives of cellulose, including carboxymethyl, methyl, ethyl, hydroxyethyl, and hydroxypropyl cellulose, cellulose acetate/sulfate exhibit enhanced mechanical strength, moisture retention, and antimicrobial properties, positioning them highly suitable for wound care products, such as dressings [107].

The continuous focus on sustainability and circular economy has driven the field of biomaterial to uncover newer sources of materials for skin regeneration. Agricultural residues, such as grape stems, are a sustainable and affordable resource for extraction of several bioactive components including cellulose. The ability to use by-products from the winery industry has several advantages. They are a natural source of highly bioactive products, such as phenolic compounds and cellulose. Moreover, the usage of these sources will give value to otherwise waste material, contributing for environmentally sustainability and bioeconomy [114].

This chapter seeks to offer an in-depth analysis of cellulose's and its derivative's role in wound healing, from their fundamental properties to their application in advanced biomaterials. Additionally, the exploration of grape stems as a sustainable cellulose source highlights the intersection of biomedical applications, healthcare, and environmental sustainability.

2 Skin Structure

The skin consists of the epidermis and dermis layers, along with the hypodermis, each characterized by its unique cell composition, anatomical features, roles, and thickness (Fig. 1).

Epidermis is mainly characterized by keratin-producing cells, the keratinocytes, pigment cells mainly melanocytes, immune cells (Langerhans), and Merkel cells. Keratinocytes are mature cells crucial for immune modulation and communication within the skin. These cells interact with stem and skin cells, as well as with immune cells of the epidermis, promoting epithelial cell and tissue stability [76]. Melanocytes, through toll like receptors (TLRs), stimulate the transport of melanin and the production of pro-inflammatory cytokines including interleukins and transforming growth factor (TGF)-β1 [53]. Langerhans cells play a key role in immune regulation and contribute to immune homeostasis and stimulation of T-cell population [108]. Resting epidermal Langerhans cells selectively induce the activation and proliferation of skin

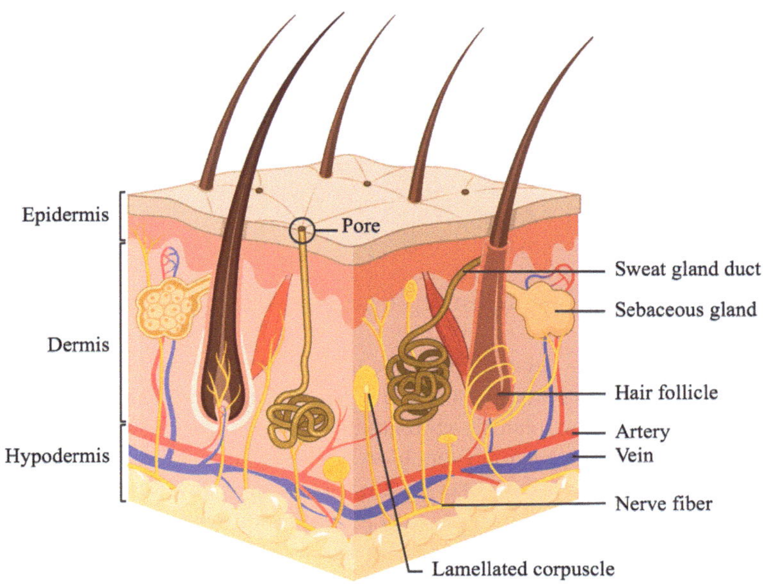

Fig. 1 Skin structure

resident regulatory T cells. However, when encountering pathogens, they activate effector memory T cells, while limiting the stimulation of Treg cells [108]. Within the epidermis, Merkel cells have endocrine and nervous functions. These mechanosensory cells interact with derma neurons, forming Merkel/neurite/Aβ nerve complexes [44]. Their proximity to nerve fibers suggests that Merkel corpuscles can release endocrine molecules through autonomic neural regulation [2]. Beyond that, they are important players in mechanotransduction, converting mechanical energy into electrical signals [36].

The dermal layer is mainly composed by fibroblasts and mast cells. Fibroblasts secrete extracellular matrix (ECM) and basement membrane proteins, important for skin health, contributing to homeostasis and the morpho functional organization of the skin [97]. Mast cells are known to secrete several pro-inflammatory mediators located at secretory granules and be involved in host defense and immune-mediated diseases [54, 127]. As a secretory cell, mast cells release several inflammatory and growth mediators including IL-4 [127] and fibroblast growth factors [54] that allow the communication with different cells, particularly fibroblasts. Endothelial cells are essential components of blood vessels [86]. They establish a barrier separating the bloodstream from the surrounding tissues and regulate vascular equilibrium [58]. Moreover, they serve as pivotal contributors to angiogenesis, essential to the formation of new microvasculature, thereby fostering the development of a well-coordinated network.

Adipocytes, located in the hypodermis, are responsible for storing fat as an energy reserve, providing insulation against heat loss, and acting as a protective cushion for underlying structures against trauma. Moreover, recent findings suggest that these cells are important skin immune mediators through the secretion of cytokines, adipokines, as well as antimicrobial peptides (AMPs) [41].

3 Wound Healing

3.1 Physiology

Skin repair occurs across four distinct stages, hemostasis, inflammation, proliferation, and maturation.

Hemostasis typically occurs within the first day's post-injury, typically ranging from 1 to 3 days and includes vasoconstriction, primary hemostasis and secondary hemostasis. At this stage, platelet aggregation and fibrinogen factor 1, produced by hepatocytes, are key players to prevent blood loss and microbial contamination [37]. At primary hemostasis, occurs platelet aggregation and development of a platelet plug. Secondary hemostasis initiates the clotting event, wherein soluble fibrinogen transforms into insoluble fibers forming the fibrin network. These structures collaborate to create the clot, stopping hemorrhaging. The release of immune and growth factors, leads to the formation of a temporary framework for infiltrating cells critical for tissue healing [72].

Inflammation, which generally spans days 3–20, often overlaps with the hemostasis phase. Immune cells including lymphocytes, neutrophils, and monocytes, are drawn to the damaged area. Moreover, signal molecules released at this stage lead to the activation and maturation of monocytes into macrophages. During the late phase of inflammation, macrophages are activated into a M2-like phenotype, promoting efferocytosis to clear expended neutrophils and contributing to new vessel formation [42].

Proliferation and cellular migration occur between day 7–40, and they are characterized by increased cell activity. M2-like macrophages induce the activation of fibroblasts and their differentiation into myofibroblasts. These cells increase the deposition of fibronectin, collagen, and other ECM components essential for re-epithelization. Concurrently, local microvascular endothelial cells are activated, and neovascularization ensures the supply of oxygen and nutrients, critical for effective healing [42].

Remodeling, the final stage of tissue repair, can span from 40 days to up to 2 years [37]. This phase includes unipotent epidermal stem cell proliferation, reversion of fully differentiated epidermal cells to a less differentiated state, and downregulation of the previously activated repair mechanisms. Neovasculature regresses, the production of collagen III decreases as collagen I increases, and myofibroblasts increase the production of matrix metalloproteases to target ECM for degradation (Fig. 2) [96, 112].

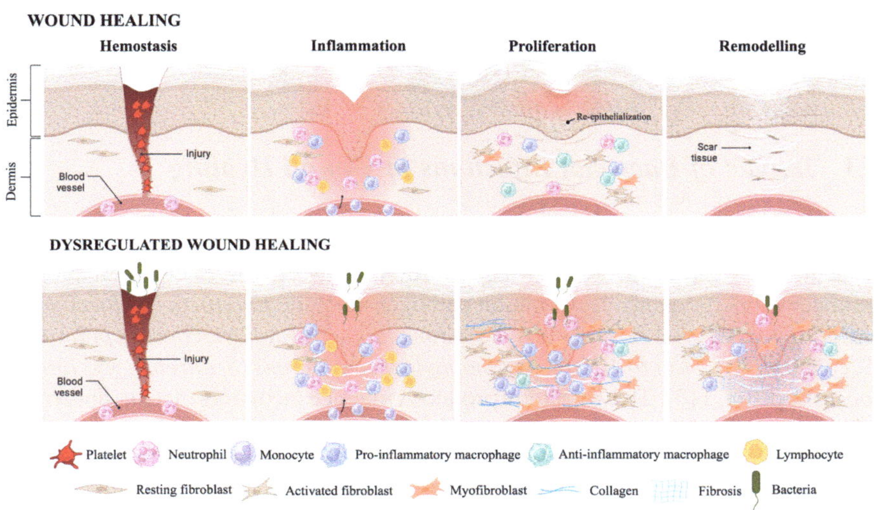

Fig. 2 Phases of wound healing and their corresponding time frame in homeostasis conditions or after bacterial infection

3.2 Wound Healing Dysregulation

Impaired neutrophil clearance or dysregulated production of pro-inflammatory macrophages at the inflammatory phase, as well excessive deposition of ECM, can delay the onset of the proliferative phase and establish an aberrant inflammatory response, leading to chronic wounds. Moreover, increased survival of myofibroblasts and increased TGF-β levels can induce a feedback loop of persistent ECM accumulation and fibrotic activity [103]. These abnormal wound repair with excessive fibrosis and scarring, does not follow the typical healing process (Fig. 2).

Chronic wounds, referred to as wounds that not heal within a month with excessive fibrosis and scarring, affects approximately 10.5 million United State (U.S.) Medicare beneficiaries and 6.5 million individuals in the U.S. annually. In Europe, the prevalence of wounds is estimated at 3–4 cases *per* 1000 people, translating to approximately 1.5–2.0 million individuals out of a population of 491 million, with a yearly incidence of 4 million newly diagnosed cases. Surgical wound infections are estimated to affect 30–40 patients per 1000 surgical procedures, posing potentially life-threatening risks, especially for older individuals. As a result, patients with wounds occupy 25–50% of acute hospital beds, significantly contributing to the overall economic burden on healthcare systems ([CSL STYLE ERROR: reference with no printed form.]). These wounds are linked to healthcare costs exceeding $50 billion each year, along with significant morbidity, functional impairment, and reduced quality of life, often culminating in amputations [96]. By 2050–2060, the aging populations of Europe and the U.S. are projected to exceed 129.8 million and 94.7 million, respectively, indicating a likely rise in the prevalence of chronic wounds ([CSL STYLE ERROR: reference with no printed form.], [CSL STYLE ERROR: reference with no printed form.]).

3.3 Impact of Bacterial Infections in Wound Healing

Infections by bacteria present a major challenge to the healing process of wounds, particularly in chronic wounds, which are highly susceptible to contamination by pathogenic microorganisms. Common pathogens include *Enterococcus faecium* (*E. faecium*), *Escherichia coli* (*E. coli*), *Pseudomonas* spp., *Staphylococcus epidermidis* (*S. epidermidis*), *Streptococcus* spp., *Staphylococcus aureus* (*S. aureus*), and *Enterococcus* spp. [98, 113]. These bacteria produce virulence mediators that disrupt critical cellular and physiological processes, including cell adhesion, nutrient acquisition, leukocyte function, and cytokine production. These disruptions exacerbate inflammation, impair granulation tissue formation, and significantly delay the healing process [112].

There are several antibiotics available in the market, although a significant low percentage of them are used in the clinic due to their toxicity, low specificity, and low efficiency due to the growing antibiotic resistance [115]. Antibiotic resistance not only impairs the treatment but also leads to increased morbidity and mortality rates associated with wound infections. These concerns underscore the urgent need for alternative solutions for the treatment of infect wounds, which are effective and sustainable [81]. In this context, bio-based materials derived from natural matrices are gaining increasing attention for medical uses, particularly in tissue regeneration. These materials, including cellulose and cellulose derivatives, offer unique benefits like compatibility with biological systems, degradability, and the capacity to integrate antimicrobial agents [18, 25]. Their use in treating infected wounds and prevention of further bacterial contamination represents a hopeful approach for reduce the effect of infections in wound care.

By leveraging the characteristics of bio-based compounds, it is possible to address the limitations of conventional antibiotics while promoting efficient wound healing reducing the risk of infection-related complications, meeting the required goals of sustainability and low carbon imprint.

4 Characterization of Cellulose

Cellulose is a straight-chain macromolecule $(C_6H_{10}O_5)_n$ connected by β-(1 − 4)-glycosidic linkages, where n can vary between 1000 and 30,000 glucose units depending on their source. It is the main constituent of the cell walls in marine invertebrates, algae, plants, and bacteria, contributing to their rigidity and strength [40, 107]. Hydrogen bounds within cellulose chain, the presence of oxygens in the rings, and the hydrogen bound between C_5 hydroxyl hydrogen and $C_{2'}$ hydroxyl oxygen confer high stability and crystallinity to the molecule, resulting in additional mechanical properties (Fig. 3) [40, 116].

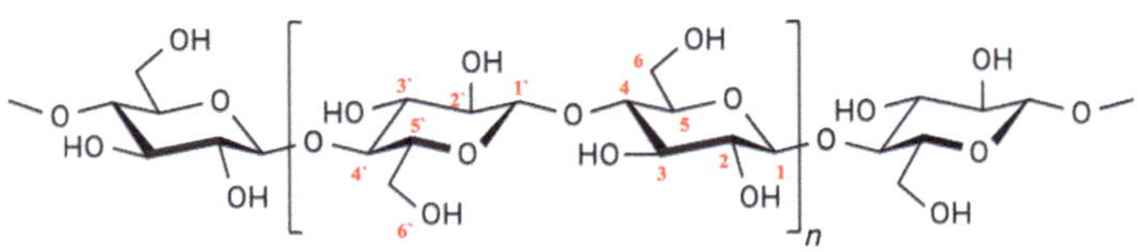

Fig. 3 Cellulose molecular composition Due to cellulose low toxicity, low density, high adaptability, and biodegradability, its structure can undergo chemically and physically modifications. These characteristics demonstrate its high potential in the field of biomaterials and biomedicine. Moreover, the possibility to obtain cellulose from distinct natural sources further increases its attractiveness as a biomaterial [18]

5 Cellulose Derivate Biomaterials

The use of chemical treatments and functionalization of cellulose expands its value and versability. The chemical hydroxyl derivatization of cellulose generally includes etherification generating carboxymethyl cellulose (CMC), methyl cellulose (MC), ethyl cellulose (EC), hydroxyethyl cellulose (HEC), hydroxypropyl cellulose (HPC), or esterification including cellulose acetate (CA) and cellulose sulfate (CS) (Fig. 4).

These cellulose derivatives can then be functionalized to increase their solubility, reactivity, and compatibility (Table 1). Their hydrophilic property supports moisture retention, an essential factor for wound dressings, while their mechanical strength provides support and stability [121]. The versatility of cellulose derivatives in producing different biomaterials, including films, fibers, hydrogels, scaffolds, and

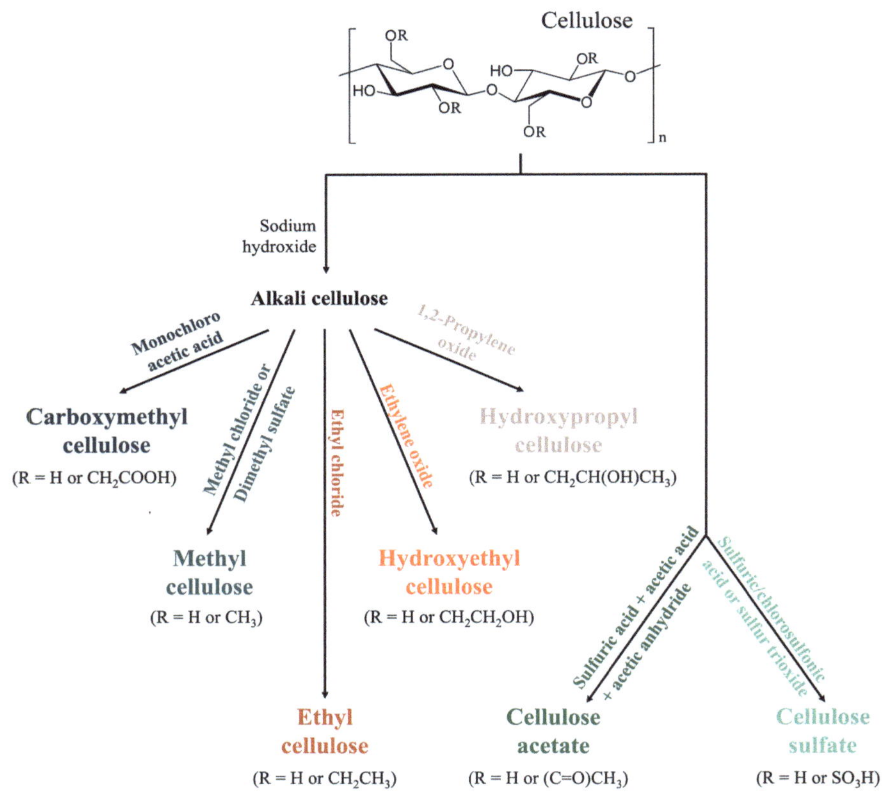

Fig. 4 Chemical reactions of cellulose involved in the formation of CMC, MC, EC, HEC, HPC by etherification as well as CA and CS by esterification. CA: cellulose acetate, CMC: carboxymethyl cellulose, CS: cellulose sulfate, EC: Ethyl cellulose, HEC: hydroxyethyl cellulose, HPC: hydroxypropyl cellulose, ME: methyl cellulose

nanostructures, underscores their potential to be used in biomedical applications (Table 1).

Carboxymethyl cellulose (CMC) is an anionic, water-dissolvable polymer synthetized as a result of the reaction between cellulose and chloroacetate in alkaline medium. In this reaction occurs carboxymethyl substitutions in C_2, C_3, or C_6 positions of glucose [55]. It is recognized by its solubility and stability in water, lack of toxicity, compatibility with biological systems, hydrophilicity, and degradability [99]. Studies demonstrated the potential of this cellulose derivate to form hydrogels and biofilms for wound treatment, pharmaceutical release, and regenerative medicine. Studies demonstrated the synthesis of non-toxic stable hydrogels of crosslinked CMC with citric acid and polyethylene glycol (PEG) [17], CMC films with nitrate ester [33], CMC gels with sodium hydroxide (NaOH) and monochloroacetic acid in ethanol with a high stability and purity (99.99%), [77] and a cellular non-toxic CMC biomaterial with incorporation of tyramine [85].

In methyl cellulose (MC), the hydroxyl groups are partially replaced by methoxyl with a common degree of substitution between 1.2 and 2.0. This results from alkali cellulose reacting with methyl chloride or dimethyl sulfate. MC thermosensitive gelation capacity, rheological features, biocompatibility and structural simplicity increase the commercial interest in this derivate. Studies synthetized MC films prepared with *Murta* fruit extracts, glutaraldehyde, and PEG [65]. Non-toxic 3D scaffold of alginate promoted no alterations on differentiation of human mesenchymal stem cells (hMSC) [106].

Ethyl cellulose (EC) derivate exhibits water insolubility and can be altered in C_2 and C_6 positions. It has residual free hydroxyl groups on C_2 and C_3 conferring it a low reactivity. EC is produced by reacting alkali cellulose with ethyl chloride and has a degree of substitution between 2.0 and 2.6. Due to its mechanical properties, capacity to prevent breakdown in alkali solutions and salt, thermo-plasticity, and stability, it is commonly incorporated in bio formulations [133]. Wang et al. [129] generated biocompatible triboelectric nanogenerators with a high output using EC films.

Hydroxyethyl cellulose (HEC) is obtained by reacting ethylene oxide with alkali cellulose. Its structure is easily altered owing to the reactivity of the hydroxyl groups and water and organic solvent solubility. HEC has compression features, is non-toxic and compatible with cellulose being interested to cosmetic applications [125]. A study demonstrated the potential of HEC incorporated with collagen and gelatin to form stable gel composites for dermatological and cosmetic applications [56].

Hydroxypropyl cellulose (HPC) results from the partial or full substitution of free hydroxyl groups with hydroxypropyl groups. It results from the reaction between alkaline cellulose and 1,2-propylene oxide forming secondary hydroxyl groups able to react. This characteristic attribute to HPC a high interest to be explored in cellular repair, wound healing, and selective delivery [135]. Studies described the importance of HPC to produce biosensors with an effective humidity-sensitive [63] and as innovative methods for immune organ engineering and immunodeficiency diseases [135].

Table 1 Properties of major cellulose derivatives, including their functional groups, primary synthesis methods, and key results achieved in the development of biomaterials utilizing these polymers

Cellulose derivate	Functional group	Type of bio material	Method used	Main results	References
Carboxy methyl cellulose (CMC)	H or CH_2COOH	Hydrogels and biofilms	CMC conjugated with polyethylene glycol (PEG) and citric acid	Stable hydrogels with no toxicity in HEK297T cell line	[17]
			CMC with different degrees of etherification (DE) functionalized with nitrate ester	Proper formation of CMC films with 0.38 DE	[33]
			CMC generated by reacting sodium hydroxide and monochloroacetic acid in an ethanol solution	CMC gels with 99.99% purity, and 2.4 (g/g) yield	[77]
			CMC synthetized by covalently incorporation of tyramine using aqueous-phase carbodiimide activation	80% viability of mammalian cells after 24 h with incubation of 1.5% (w/v) conjugated solution	[85]
Methyl cellulose (MC)	H or CH_3	Hydrogels and biofilms	MC films prepared with *Murta* fruit extracts, glutaraldehyde and PEG	Films with lower glutaraldehyde percentages more efficient for Gram-positive *Listeria innocua*	[65]
		Scaffolds	30 mg/mL alginate/MC (1:3 ratio)	3D scaffold of alginate and MC improved printability of a hydrogen membrane; non-toxic for cells; maintained differentiation of hMSCs	[106]
Ethyl cellulose (EC)	H or CH_2CH_3	Electric nano-generators	Triboelectric charging and electrostatic induction between the 317L stainless steel plate and the EC film	Increased power output of biocompatible triboelectric nanogenerators	[129]

(continued)

Table 1 (continued)

Cellulose derivate	Functional group	Type of bio material	Method used	Main results	References
Hydroxy ethyl cellulose (HEC)	H or CH_2CH_2OH	Gel composites	0.15 g of collagen/gelatin microspheres + 15 mL of 0.5% (w/v) collagen and gelatin solutions and 0.15 g HEC	Collagen/gelatin/HEC composites containing microspheres improved release performance of extracts; promising for biomedical/dermatological/cosmetic applications	[56]
Hydroxy propyl cellulose (HPC)	H or $CH_2CH(OH)–CH_3$	Biosensors	Proposed sensor composed of a section of 100 μM hollow-core fiber (HCF) connected with single-mode fiber and a layer of 10 μM HPC film covered on the endface of HCF	The sensor developed is an effective humidity-sensitive material with a binomial humidity response, hydrophilicity and biocompatibility	[64]
		Biofilms	hADSCs with 13% (w/v) HPC injected into BALB/c *nude* mice	hADSCs differentiated into lymphocyte-like cells forming lymphoid nodes; HPC as an alternative approach for engineering immune organs and immunodeficiency diseases	[135]
Cellulose acetate (CA)	H or $(C{=}O)CH_3$	Biofilms	CA/triethyl citrate nano-biocomposite films prepared with silver nanoparticles with incorporation of *Curcuma longa*/gelatin-modified montmorillonite nanofiller and thymol	Films with thermal stability, UV radiation protection and antimicrobial activity (*Salmonella* enterica, *E. coli*, *S. aureus* and *P. aeruginosa*	[24]
		Nanotubes	Carbon nanotubes functionalized with CA, ammonia and graphene oxide (CA/CNT/GO)	Stable nanomaterial against hydrodynamic forces, repellent interactions with BSA and Hb, high cytocompatibility with 3T3E1 cell line	[46]

(continued)

Table 1 (continued)

Cellulose derivate	Functional group	Type of bio material	Method used	Main results	References
		Nano-capsules	Nanosystems combining CA and peppermint, cinnamon and lemongrass essential oils	Optimized nanocapsules with CA prevented *S. aureus*, *P. aeruginosa* and *E. coli* growth, microbial colonization and biofilm development	[61]
Cellulose sulfate (CS)	H or SO_3H	Nanofiber scaffolds	Bacterial cellulose functionalized with sulfate groups (BCS) through acetosulfation	BCS/polyvinyl alcohol nanofiber scaffolds non-toxic for cells, with antioxidant and hemocompatibility properties; promoted cell adhesion and proliferation in L929 fibroblast cells	[87]

BSA—bovine serum albumin, *BSC*—bacterial cellulose functionalized with sulfate groups, *CA*—cellulose acetate, *CMC*—carboxymethyl cellulose, *CNT*—carbon nanotubes, *CS*—cellulose sulfate, *DE*—degrees of etherification, *EC*—ethyl cellulose, *hADS*—human adipose-derived stromal cells, *Hb*—hemoglobin, *HCF*—hollow-core fiber, *HEC*—hydroxyethyl cellulose, *hMSCs*—human mesenchymal stem cells, *HPC*—hydroxypropyl cellulose, *MC*—methyl cellulose, *PEG*—polyethylene glycol

The biodegradable cellulose acetate (CA) is synthesized through the acetylation of cellulose's hydroxyl groups with acetic anhydride, sulfuric acid, and acetic acid. Its synthesis involves ring-opening esterification and transesterification under homogeneous or heterogeneous conditions [51]. CA can be easily functionalized with functional groups to targeted approaches including oxidation, etherification, hydrolysis, and copolymerization. Moreover, as it has high stability, water affinity, availability, and film-forming features is a polymer with a widely properties to diverse biomedical applications including wound repair. However, its use has limited application due to its high cost [123]. Dairi et al. [24] optimized the synthesis of CA/triethyl citrate nano-biocomposite films prepared using silver nanoparticles with extract incorporation and demonstrated their stability, UV protection, and antibacterial activity. Ioniţă et al. [46] developed carbon nanotubes functionalized with CA with a cytocompatibility in 3T3E1 cell lines. Moreover, a study demonstrated that the combination of CA with essential oils can form nanosystems that prevent growth and colonization of *S. aureus*, *P. aeruginosa*, and *E. coli* [61].

Reaction of cellulose with sulfuric acid, chlorosulfonic acid or sulfur trioxide produces cellulose sulfate (CS). This derivate can be produced in large-scale, it has high biocompatibility, and film-forming capacity for potential use in tissue engineering, drug delivery, and cell encapsulation. Moreover, due to the presence of sulfate groups and the abroad degree of substitution, it has antibacterial and anticoagulant properties [87, 117]. Bacterial CS and polyvinyl alcohol nanofiber scaffolds developed demonstrated high antioxidant, hemocompatibility, and promoted proliferation in in vitro studies [87, 117].

6 Cellulose Derivatives with Antibacterial and Wound Healing Properties

As described, proper wound healing requires not only tissue regeneration but also moisture retention and protection against microbial contamination, which is a critical challenge while developing a wound dressing. Cellulose derivatives with antimicrobial properties provide a promising solution to address this issue. Considering this, researchers have been actively developing cellulose-based biomaterials and dressings with abilities to promote wound healing, support tissue regeneration, as well as with antibacterial activity (Table 2).

Studies using functionalized micro and nano crystalline cellulose with catechol/silver [66], 3-aminopropyltrimethoyl-silane/ethylenediamine/triethylenetetramine [30] or combined with bioactive plant extracts [90] presented antimicrobial activity in bacteria including *S. aureus*, *P. aeruginosa*, *E. coli*, *E. faecalis*, *S. aureus*, *S. epidermidis*, and *B. subtilis*. Moreover, the use of cellulose nanocrystals-based hydroxyapatite nanoparticles enhanced 50% healing in mice [118].

The use of micro/nanofibrils of cellulose also demonstrated potential in tissue regeneration, wound healing, and antibacterial activity. Nanofibril cellulose

Table 2 Wound healing, tissue regeneration, and antibacterial activity properties of cellulose and cellulose derivatives

Cellulose/ cellulose derivate	Biomedical application	Main results	References
Micro/nano crystalline cellulose	Antibacterial	Cellulose nanocrystal/catechol/silver (Ag) hydrogels promoted neovascularization and tissue repair owing to their antibacterial and anti-inflammatory capacity	[66]
		Amino-functionalized microcrystalline celluloses with 3-aminopropyltrimethoyl-silane, ethylenediamine, and triethylenetetramine inhibited *S. aureus*, *P. aeruginosa*, and *E. coli* growth	[30]
		Loading bioactive plant extracts within cellulose microfiber had antibacterial potential for *Bacillus subtilis*, *E. faecalis*, *S. epidermidis*, and *S. aureus*	[90]
	Wound healing	A solution mixture of chitosan and gelatin containing cellulose nanocrystals and hydroxyapatite nanoparticles (NPs) accelerated wound healing in mice by 50%	[118]
Cellulose micro/ nanofibrils	Tissue and skin regeneration	Calcium crosslinked nanofibrillated cellulose (NFC) hydrogels promoted fibroblast and keratinocyte proliferation in vitro and keratinocyte differentiation stimulating healing in vivo	[13]
		Combination of cellulose nanofibers and chitin promoted cell attachment, proliferation, and fibroblast adhesion and migration enhancing skin repair	[83]
		Faster epithelialization and healing of wounds using NFC-based dressings comparing to other dressing of the market	[43]
	Wound healing and antibacterial	Mupirocin-loaded α-cellulose nanopaper accelerated wound healing while limiting growth of *P. aeruginosa*, *E. coli*, and *S. aureus*	[91]
	Antibacterial	Cellulose reaction by eliminating the C_{12} to react with acrylamide induced antibacterial and antifungal activity against *E. coli*, *S. aureus*, and *Candida albicans*	[45]
Bacterial cellulose (BC)	Wound healing	Cellulose-vaccarin membranes promoted epithelialization and regeneration of wounds induced in rat skin	[94]

(continued)

Table 2 (continued)

Cellulose/ cellulose derivate	Biomedical application	Main results	References
		Cellulose/polyacrylamide hydrogel microparticles accelerated epithelial and fibroblast growth promoting tissue recovery	[89]
	Antibacterial	Cellulose-based polycaprolactone membranes polycaprolactone exhibited activity against *E. coli* and *S. aureus* pathogens	[25]
		BC/graphene oxide (GO)-Ag NPs inhibited *P. aeruginosa*, *S. aureus*, and *E. coli*	[68]
		BC/sodium alginate (SA)–Ag sulfadiazine composites exhibited antibacterial properties against *E. coli* and *S. aureus*	[110]
		BC/coniferyl alcohol composite hydrogel inhibited *L. monocytogenes*, *P. aeruginosa*, and *S. aureus*	[137]
		BC/bioactive glass nanocomposites impacted *E. coli, P. aeruginosa, S. typhimurium, Klebsiella pneumonia*; and *Proteus vulgaris, B. subtilis*, and *S. aureus* growth	[1]
		BC/methylglyoxal composite showed zone inhibition diameter from 12.3 to 17.1 mm against *Micrococcus luteus*, *P. aeruginosa*, S. *aureus*, and *E. coli*	[132]
		23% poly-lactic acid composites with cellulose nanocrystals and reduced GO had antibacterial effect against *E.coli* and *S. aureus*	[88]
	Tissue Regeneration and antibacterial	A composite of electrically modified BC with allantoin had antibacterial properties against *S. aureus*, supporting wound repair by promoting reepithelization and collagen I deposition	[3]
		Cellulose-chitosan membranes epithelialized and regenerated faster skin wounds *in* vivo inhibiting *E.coli* and *S. aureus*	[64]
Carboxymethyl cellulose (CMC)	Wound healing and skin regeneration	CMC/polycaprolactone hydrogel absorbed wound exudate, improved skin cell proliferation, cell migration and production of collagen I increasing anti-inflammatory cytokines	[75]

(continued)

Table 2 (continued)

Cellulose/ cellulose derivate	Biomedical application	Main results	References
		CMC/gelatin/cooper (Cu)-doped zinc oxide (ZnO) dressing increased adhesion and of fibroblasts	[80]
		Films of CMC and polyethylene glycol (PEG) promoted faster wound healing an collagen formation in healthy and diabetic mice	[14]
	Wound healing and antibacterial	CMC films controlled systemic inflammation, wound healing, and growth of *P. aeruginosa* and *S. aureus*	[130]
		Turmeric extract-loaded BC-CMC presents high swelling degrees and antimicrobial activity against *E. coli*, *S. aureus*, and *C. albicans*	[48]
	Antibacterial	Functionalization of CMC with tetrabutylammonium moieties increased antibacterial activity against *S. aureus* and *E. coli*	[134]
	Wound repair	Chitosan-CMC-g-PF127 injectable hydrogels enhanced fibroblast, keratinocyte and collagen deposition	[109]
Methyl cellulose (MC)	Antibacterial	MC cross-linked foams with Manuka honey inhibited *E.coli* and *S. aureus*	[104]
	Antibacterial and wound healing	Alginate/MC hydrogel inks with bioactive compounds stimulated human dermal fibroblast growth and antibacterial activity	[50]
		MC hydrogel containing Ag oxide NPs decreased growth of *S. aureus*, *K. pneumoniae*, and *E. coli* and promoted wound healing	[52]
	Tissue regeneration	A thermoresponsive hydrogel with keratin, MC, and sodium sulfate increased fibroblast migration and blood vessel branching	[29]
	Wound healing	Electrospun nanofibers with MC and polyvinyl alcohol accelerated wound closure by promoting cell migration and epithelialization	[131]
		Poly(ε-caprolactone)-MC biopolymer supported wound closure in human keratinocyte cells	[105]

(continued)

Table 2 (continued)

Cellulose/ cellulose derivate	Biomedical application	Main results	References
Ethyl cellulose (EC)	Wound healing and antibacterial	Hydrophobic EC nanofibers displayed improved fibroblast growth and proliferation and inhibited growth of *E.coli* and *S. aureus*	[62]
		Poly3-hydroxybutyrate grafted bio-composites improved infection and wound and exerted antibacterial activity in *E.coli* and *B. subtilis*	[47]
		EC/poly lactic acid/collagen embedded with Ag sulfadiazine inhibited *Bacillus* and *E.coli* bacteria and stimulated NIH 3T3 fibroblasts	[5]
		A pH-sensitive bilayer electrospun nanofibrous formed by EC on top and EC/ Eudragit S-100 on bottom with antibiotics decreased microbial burden while promoted a faster skin regeneration	[7]
		Curcumin-loaded EC-based carriers decreased *S. aureus* colony and fibroblast attachment after 168 h	[57]
	Wound healing	Ag NPs with EC cream enhanced healing properties improving wound granulation and epithelial cell proliferation	([CSL STYLE ERROR: reference with no printed form.])
		Fibers with EC and poly(vinyl pyrrolidone) loaded with naproxen attenuated wound pain and inflammation	[38]
		EC loaded with gallic acid using electrospinning showed a > 90% microbial reduction of *S. aureus*, *E. coli*, *P. aeruginosa* and *C. albicans*	[22]
Hydroxyethyl cellulose (HEC)	Wound healing and antibacterial	Polymeric scaffolds of HEC/hyaluronic acid/gelatin/tetracalcium phosphate increased wound closure and healing and reduced bacterial activity	[26]
		Hyaluronic acid/oxidized HEC hydrogel possessed a high absorbent capacity, creating a moist wound healing environment	[67]

(continued)

Table 2 (continued)

Cellulose/ cellulose derivate	Biomedical application	Main results	References
		HEC/mesocellular silica foam hydrogel exhibited hemorrhage control and antibacterial activities facilitating tissue repair in vivo	[128]
		HEC-based hydrogel with ferulic acid-grafted chitosan incorporating selenium NPs demonstrated antioxidant and antimicrobial activities promoting complete *in* vivo diabetic wound healing in 14 days	[93]
Hydroxypropyl cellulose (HPC)	Wound healing	Biodegradable ofloxacin-loaded HPC hydrogels promoted 95% wound size reduction and high collagen deposition	[4]
		Sodium HPC hydrogen films with high antibacterial after 3 days	[28]
		Coumestrol and hydroxypropyl-β-cyclodextrin incorporated within HPC hydrogel induced *in* vitro and *in* vivo cell proliferation and migration	[15]
Cellulose acetate (CA)	Wound healing	CA/gelatin/nanohydroxyapatite nanocomposites increased collagen synthesis, re-epithelialization, neovasculature and accelerated wound healing	[100]
	Antibacterial	CA/zinc sulfide/Ag oxide/GO had antibacterial activity against *E.coli*	[6]
		E. coli and *S. aureus* growth completely inhibited with AlOOH/Cu modified CA fibers for 1 h	[12]
		Cu NPs impregnation on CA nanofibers with significant inhibitory effect on *E. coli*	[102]
Cellulose sulfate (CS)	Wound healing	CS-coating quaternary ammonium salt (QAS) quickly exposed QAS promoting higher wound sealing and homeostasis	[136]
		CS/polyvinyl alcohol nanofiber scaffolds promoted fibroblast cell adhesion and proliferation	[87]

Ag—silver, *AlOOH*—aluminum hydroxide oxide, *BC*—bacterial cellulose, *CA*—cellulose acetate, *CMC*—carboxymethyl cellulose, *CS*—cellulose sulfate, *Cu*—cooper, *EC*—ethyl cellulose, *HEC*—hydroxyethyl cellulose, *HPC*—hydroxypropyl cellulose, *MC*—methyl cellulose, *MIC*—minimum inhibitory concentration, *NFC*—nanofibrillated cellulose, *NP*—nanoparticles, *PEG*—polyethylene glycol, *QAS*—quaternary ammonium salt, *SA*—sodium alginate, *ZnO*—zinc oxide crosslinked with calcium and chitin promoted fibroblast and keratinocyte proliferation and migration [13, 83]. Interestingly, nanofibrillated cellulose-based dressings stimulated faster epithelialization compared with other commercial dressings [43]. Pant et al. [91] and Ibrahim et al. [45] showed that mupirocin-doped α-cellulose nanopaper and cellulose reaction with acrylamide inhibited *E. coli*, *S. aureus*, *P. aeruginosa*, and *C. albicans* growth.

Moreover, functionalization of bacterial cellulose also demonstrated positive impact in fostering wound and tissue repair while attenuated bacterial burden. Cellulose-vaccarin membranes [94] and cellulose/polyacrylamide hydrogel microparticles [89] exhibited enhanced epithelialization and regeneration. Cellulose-based polycaprolactone membranes [25], bacterial cellulose/graphene oxide-silver nanoparticles [68], bacterial cellulose/sodium alginate-silver sulfadiazine composites [110], bacterial cellulose/coniferyl alcohol hydrogel [137], bacterial cellulose/bioactive glass nanocomposites [1], bacterial cellulose/methylglyoxal composite [132], and 23% poly-lactic acid/cellulose nanocrystals attenuated or inhibited *E. coli*, *S. aureus*, *P. aeruginosa*, *L. monocytogenes*, *S. typhimurium*, *K. pneumonia*, *P. vulgaris*, *B. subtilis*, and *M. luteus* colonial formation.

Complexation of CMC with polycaprolactone [75], gelatin/cooper-doped zinc oxide [80], and PEG [14], natural extracts [48], tetrabutylammonium moieties [134], and chitosan-g-PF127 [109], also fostered skin cell proliferation, collagen I production, fibroblast and keratinocyte deposition, and decreased *E. coli*, *S.* aureus, *P. aeruginosa*, and *C. albicans* growth in both wildtype rats and diabetic Wistar albino rats. Wong et al. [130] demonstrated that CMC films can effectively control systemic inflammation, a critical process for proper wound healing.

Foams of MC crosslinked with honey [104], incorporation of bioactive compounds in alginate/MC hydrogel [50], and MC hydrogels with silver oxide nanoparticles [52] decreased bacteria growth, while promoted activation of human dermal fibroblasts and wound healing. Electrospun nanofibers with MC [131] and polyvinyl alcohol and poly(ε-caprolactone)-MC biopolymer [105] supported cell epithelialization and wound closure in keratinocytes. Remarkably, a thermoresponsive hydrogel with keratin, MC, and sodium sulfate was shown to enhance blood vessel branching, which is crucial for tissue regeneration and accurate wound repair [29].

Biomaterials impregnated with EC also demonstrated proper antibacterial and wound healing properties. Hydrophobic EC nanofibers [62], poly3-hydroxybutyrate bio-composites with EC [47], EC/poly lactic acid/collagen incorporated with silver sulfadiazine [5], and EC loaded with gallic acid using electrospinning [22], inhibited *E. coli*, *S.* aureus, *P.* aeruginosa, *B. subtilis*, and *C. albicans*. Amani et al. [7] developed a pH-sensitive bilayer electrospun nanofibrous formed by EC on top and EC/Eudragit S-100 on bottom with antibiotics with ability to decrease microbial burden, while promoted a rapid skin regeneration. Additionally, curcumin-loaded EC-based carriers attenuated *S. aureus* proliferation and stimulated fibroblast attachment in 168 h [57]. Silver nanoparticles-EC creams improved wound granulation ([CSL STYLE ERROR: reference with no printed form.] and fibers incorporating EC and poly(vinyl pyrrolidone containing naproxen mitigated wound pain and inflammation [38].

Polymeric scaffolds of HEC/hyaluronic acid/gelatin/tetra calcium phosphate demonstrated benefits in reducing bacterial activity increasing wound closure [26]. Hydrogel containing HEC/mesocellular silica exhibited hemorrhage control and antibacterial activity in an in vivo full-thickness skin defect mouse model [128]. Ferulic acid-HEC grafted chitosan hydrogel loaded with selenium nanoparticles revealed antioxidant and antimicrobial activities promoting complete wound healing after 14 days in an in vivo diabetic model [93]. Also, hyaluronic acid/oxidized HEC hydrogel showed a high absorbent capacity, creating a moist wound healing environment [67].

Dharmalingam et al. [28] presented that sodium HPC hydrogen films have inhibited bacteria growth after three days after release. Biodegradable ofloxacin-loaded HPC hydrogels stimulated an accumulation of collagen leading to a 95% wound size reduction [4]. Likewise, coumestrol and hydroxypropyl-β-cyclodextrin incorporated within HPC hydrogel induced cell proliferation and migration in vitro and increased 50% wound repair in vivo [15].

Cellulose esterification producing CA also demonstrated antibacterial activities and wound healing. CA/gelatin/nanohydroxyapatite nanocomposites accelerated wound repair by stimulating collagen synthesis, re-epithelialization, and neovasculature [100]. Modified CA with aluminum hydroxide oxide and cupper for 1 h [12] and cupper nanoparticle impregnation on CA nanofibers [102] demonstrated an antibacterial activity against *E. coli* and *S. aureus*. A polymer of CA/zinc sulfide/silver oxide/graphene oxide also inhibited *E.coli* growth with a zone inhibition of 17.7 ± 0.5 mm [6].

Regarding CS, fewer studies are available as it is not commonly used for wound healing. However, one study showed that CS-coating quaternary ammonium salt improved wound sealing and homeostasis [136]. Additionally, Palaninathan et al. [87] produced CS/polyvinyl alcohol nanofiber scaffolds with a strong capacity to stimulate adhesion and proliferation of fibroblasts.

7 Natural-Based Sources of Cellulose

Based on information from the World Health Organization, antimicrobial resistance is expected to emerge as the primary cause of death globally by 2050. To address these challenges, the One Health concept aligns with the need to substitute synthetic compounds by natural alternatives [124]. Alongside, Directive (EU) 2018/851 of the European Parliament and the Council of Ministers (May 30, 2018), alongside the Sustainable Development Goals of the 2030 Agenda, supports Europe's transition to a circular economy. These initiatives promote sustainable production and consumption, emphasizing the recovery and regeneration of resources, including agrifood waste.

Plants serve as the main origin of cellulose, which is found in all plant species cell wall. In the context of a circular economy, plants offer significant advantages as a cellulose source due to their renewability, biodegradability, and capacity for integration into sustainable production systems. Moreover, plant and agriculture residues are rich in cellulose and often remain underused [101]. Several studies have been demonstrated that these residues, such as crop stalks, wood pulp, rice husk, corn waste, wheat waste, coffee waste, sisal fibers, sugarcane bagasse can be valorized to produce natural cellulose-based materials [101]. The major steps to extract cellulose from plants involve a pre-hydrolysis, pulping and bleaching. Pre-hydrolysis treatment typically uses mineral acids to break down and open the lignocellulosic matrix, facilitating subsequent processing steps.

Pulping involves incubation of fibers with alkaline agents such as NaOH to further separate the cellulose from lignin and hemicellulose. Bleaching is employed to obtain a purified and bleached form of cellulose. This can be achieved using agents such as hydrogen peroxide (H_2O_2), sodium chlorite, or ozone. Sulfuric acid (H_2SO_4) hydrolysis is used to increase the purity of the extracted cellulose under reflux conditions [19]. Commonly, hexane/ethanol is used to dewax dried rice husk, followed by delignification with NaOH at 121 °C. Cellulose is then separated from lignin and hemicellulose by ultrasound. Bleaching takes place using H_2O_2 and concentrated nitric acid [21]. Obtention of cellulose from wood pulp conventional includes semi-chemical degradation of hemicellulose and lignin from the matrix using NaOH and Na_2S at high pressure and temperature [23]. In the extraction of cellulose from coffee, NaOH is used as an alkali treatment and the residual solid is bleached with sodium chlorite, acetate buffer and water [20]. Sisal fibers are processed for 6 h with in toluene/ethanol (2:1 v/v) and pre-treatment process involves the use of 0.1 M NaOH in 50% ethanol for 3 h at 45 °C, followed by H_2O_2 treatment [92]. Delignification of sugarcane bagasse is carried out using sodium chlorite in acidic conditions for 2 h and then filtered, obtaining a residue rich in cellulose [34].

Interestingly, due to their high cellulose content and significant environmental impact, studies have been exploring methods to extract cellulose from winery by-products, particularly grape stems [123, 124]. Compared to bacterial cellulose, which is time-consuming, resource-intensive, and limited in scalability [95], grape stems present a cost-effective, abundant, and sustainable alternative for cellulose extraction. Additionally, grape stem-derived cellulose offers high biocompatibility with human tissues, positioning it as a favorable option for biomedical uses [8, 32]. Oxygen plays a crucial role in wound healing, and a lack of oxygen in tissues can trigger angiogenesis. The pores within nanocellulose from plant-derived cellulose serve as crucial pathways for oxygen molecules to permeate, enabling its functionality in wound healing applications [78]. Moreover, a study revealed that the subcutaneous implantation of bacterial cellulose could trigger a mild inflammatory response and fibroblast infiltration, potentially contributing to chronic wound formation through excessive ECM production and fibrosis [79]. These findings highlight the need to develop innovative cellulose-based biomaterials from plants to address challenges in chronic wound management.

7.1 Grape Stems for Sustainable Extraction of Cellulose

Vitis Vinifera (grapes) production stands as one of the most significant economic drivers in the global agrifood sector, with an annual output exceeding 74 million tones ([CSL STYLE ERROR: reference with no printed form.]). Wine production, a key economic and traditional industry particularly in Europe, generates significant amount of solid residues, accounting for approximately 30% (w/w) of the total processed material, amounting to over 20 million tons annually [35]. Currently, these by-products are either deposited in landfills or used in low-value applications, such as fertilizers and animal feed. In the past, wineries paid fees of around €3000 for waste disposal, however, this cost has recently risen to €30,000–€40,000. Large wineries face annual waste management costs exceeding $50,000, with potential increases up to $150,000 due to stricter environmental regulations [27, 82]. These financial pressures, coupled with environmental challenges, threaten the economic sustainability and competitiveness of the winery industry.

Among winery by-products, grape stems constitute around 25% (w/w) of the organic winery by-products and are mainly composed of cellulose (19.6–37.9%), lignin (12.8–47.3%), hemicellulose (9.8–35.3%), proteins (4.9–11.2%) [84], and bioactive compounds including phenolics (1.0–3.0%) [39] (Fig. 5).

Using grape stems for cellulose extraction aligns with efforts to comply with the EU Landfill Directive 1999/31/EC, which imposes strict limits on the organic carbon content of waste sent to landfills (< 3% or 30,000 mg/kg) to reduce greenhouse gas emissions and environmental pollution. By repurposing grape stems for cellulose extraction, waste is diverted from landfills, creating a valuable resource and promoting circular economy principles. This approach provides dual benefits including the mitigation of the environmental footprint of winery waste while providing a sustainable source of cellulose for various applications.

Additionally, grape stems contain high levels of bioactive compounds including polyphenols [39]. These secondary metabolites, predominantly produced by plants, exhibit strong antioxidant, anti-inflammatory, as well as antimicrobial activity [39, 59, 60]. In contrast, native bacterial cellulose lacks inherent antimicrobial properties and relies on the incorporation of antibacterial compounds or nanoparticles for use in wound healing. The bioactive properties of grape stem-derived cellulose make it particularly appropriate for wound repair by controlling bacteria growth and proliferation [39, 59, 60]. This enhances its advantages over bacterial cellulose, supporting its use as a sustainable alternative in biomedical applications.

Several studies have evaluated the efficacy of methods to extract cellulose from grape stems. Araújo et al. [9] first hydrolyzed the by-product with 5% H_2SO_4 (1:20 m/V) at 121 °C for 45 min, followed by washing until the liquid ran clear and drying at 50 °C for 24 h. Then, samples were treated with 4 mol/L NaOH (1:20 m/V) at room temperature with agitation for 24 h, followed by 5% H_2O_2 (pH 11.5) at 45 °C for 8 h. The samples were then cooled for 24 h, washed, and dried. Spigno et al. [114] tested the impact of acid–alkaline/oxidation using 2% (w/v) H_2SO_4 solution at 90 °C and alkaline-acidic hydrolysis using 2.9 M NH_4OH for 2 h, 7 h, 15 h, and 24 h at 26 °C to

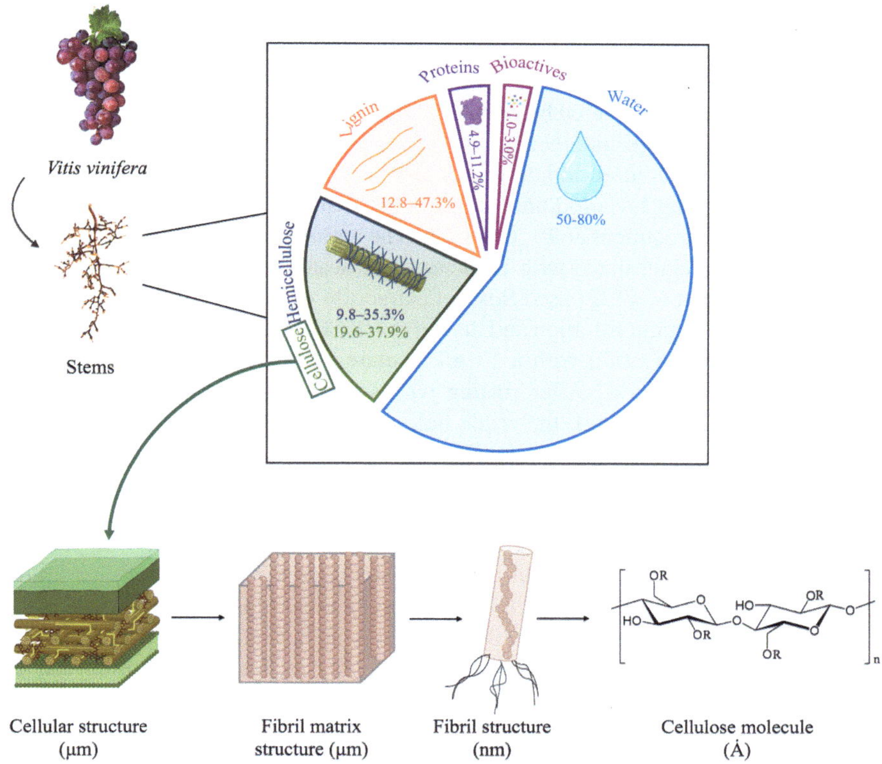

Fig. 5 Morphology and composition of grape stems: a sustainable source of cellulose [84]

extract lignocellulosic material. Through analyses of fiber content and composition using the Goering-Van Soest procedure, the modified Kurschner-Hanak method, and structural carbohydrate and lignin quantification techniques, they observed that direct hydrolysis under high temperatures and low acid conditions was more effective for hemicellulose extraction from grape stems [114]. In another study, grape stem raw material was dried at 105 °C for 24 h, ground, and subjected to subcritical water under a solid-to-water mass ratio of 1:10 for 30 min at 180 °C. Cellulose fibers from the residue were bleached using hydrogen peroxide solutions, pH 12, in four consecutive bleaching cycles (1 h each) at 40 °C. The fibers were then filtered, washed, and air-dried at room temperature for three days [71]. Atatoprak et al. [11] pioneered the extraction of cellulose nanocrystals from grape stems. In this study, grape stem extracts insoluble in water or ethanol were pretreated with a 1% NaOH solution for 15 min to 60 min at 100 °C. The delignified material, rich in cellulose and hemicellulose, was dried at 100 °C for 24 h, and the delignification percentage was calculated. Enzymatic hydrolysis was performed at 50 °C for 24 h, followed by a second hydrolysis for 48 h to increase sugar yield. The pellet was treated with 64% (w/w) H_2SO_4 at 50 °C for 30 min with constant stirring. Bleaching was carried out using an acetate buffer and 1.7% (m/v) sodium chlorite solution [11]. In another study, 72% H_2SO_4 and 0.35 g of the extracted sample were incubated at 30 °C for 1 h and then diluted to 3% sulfuric acid. Hydrolysis was performed at 120 °C for 1 h, and the solid residue was recovered by filtration [70]. Another study dried grape stems at 25 ± 1 °C for 7 days, ground them, and performed Soxhlet extraction with toluene/ethanol (2:1) for 24 h to eliminate waxes and pigments. Lignin and hemicellulose were eliminated using 1.4% sodium chlorite at 70 °C and 5% potassium hydroxide for 24 h, followed by treatment at 90 °C for 2 h. The cellulose was concentrated trough vacuum filtration, neutralized with acetic acid and water, and dried at 25 ± 1 °C for 72 h [73]. Vallejo et al. [122] used Soxhlet extraction with a 2:1 toluene-ethanol for 6 h, followed by vacuum filtration and drying at 105 °C. The dewaxed materials were bleached for delignification with a 17 g/L acetate buffer/sodium chlorite aqueous solution or 4 h at 100 °C. After rinsing with sodium hydrogen sulfate (20 g/L), deionized water, and ethanol, the crude holocellulose was dried at 105 °C. It was then treated with 175 g/L sodium hydroxide for 45 min at room temperature with agitation and subsequently rinsed with 10% acetic acid and deionized water. A final acidic treatment removed impurities using a 10:1 acetic acid:nitric acid at 120 °C for 15 min [122]. Lastly, Araújo et al. [10] extracted cellulose from vine pruning residues by immersing samples in ethanol (1:10 w/v) for 16 h to remove lipids, followed by drying at 50 °C for 24 h. Lignin and hemicellulose were removed using a 4% NaOH solution (1:20 w/v) at 80 °C for 4 h in an orbital shaker. After filtration and washing, the material underwent four bleaching cycles using sodium acetate buffer and a sodium chlorite aqueous solution at 60 °C for 4 h [10].

8 Conclusion

The incorporation of cellulose and cellulose derivatives in wound care for chronic wounds represents a significant advancement for the development of effective and sustainable biomedical solutions for this pathology. By addressing key challenges such as bacterial infections and chronic wound management, cellulose-based biomaterials offer innovative approaches to enhance healing outcomes. Furthermore, the valorization of agricultural by-products like grape stems underscores the potential of sustainable resource utilization in biomaterial development. This review highlights the dual benefits of cellulose in wound healing including its intrinsic properties that support tissue repair and its role for the delivery of antimicrobial agents. The increasing interest in biodegradable and renewable materials led to exploitation of cellulose-derived biomaterials, combining efficacy with sustainability. Future research should focus on optimizing grape stem cellulose extraction methods, expanding functionalization techniques, and conducting clinical studies to translate these materials into scalable and effective wound care products. By harnessing the synergy between natural resources and biomedical innovation, grape stem cellulose-based technologies can significantly contribute to improved patient care and environmental protection.

Acknowledgements The authors acknowledge CITAB—UTAD for providing the necessary conditions for the development of this chapter.

Funding This work was supported by National Funds by FCT—Portuguese Foundation for Science and Technology, under the projects UIDB/04033/2020 (https://doi.org/10.54499/UIDB/04033/2020) and LA/P/0126/2020 (https://doi.org/10.54499/LA/P/0126/2020), as well as the strategic funding of the UIDB/04469/2020 unit. Additionally, support was provided by LABBELS—Associate Laboratory in Biotechnology, Bioengineering, and Microelectromechanical Systems, LA/P/0029/2020. This research was also supported by the European Regional Development Fund (ERDF) under the Competitiveness Factors Operational program—Norte 2020, COMPETE.

Conflict of Interest The authors declare that they have no conflicts of interest.

References

1. Abdelraof M, Hasanin MS, Farag MM, Ahmed HY (2019) Green synthesis of bacterial cellulose/bioactive glass nanocomposites: effect of glass nanoparticles on cellulose yield, biocompatibility and antimicrobial activity. Int J Biol Macromol 138:975–985. https://doi.org/10.1016/j.ijbiomac.2019.07.144
2. Abraham J, Mathew S (2019) Merkel cells: a collective review of current concepts. Int J Appl Basic Med Res 9:9. https://doi.org/10.4103/ijabmr.IJABMR_34_18
3. Adhikari M, Bakadia BM, Wang L et al (2025) Electricallymodified bacterial cellulose tailored with plant based green materials for infected wound healing applications. Biomater Adv 167:214087. https://doi.org/10.1016/j.bioadv.2024.214087
4. Agubata CO, Okereke C, Nzekwe IT et al (2016) Development and evaluation of wound healing hydrogels based on a quinolone, hydroxypropyl methylcellulose and biodegradable microfibres. Eur J Pharm Sci 89:1–10. https://doi.org/10.1016/j.ejps.2016.04.017
5. Ahmadian S, Ghorbani M, Mahmoodzadeh F (2020) Silver sulfadiazine-loaded electrospun ethyl cellulose/polylactic acid/collagen nanofibrous mats with antibacterial properties for wound healing. Int J Biol Macromol 162:1555–1565. https://doi.org/10.1016/j.ijbiomac.2020.08.059
6. Alharthi AF, Gouda M, Khalaf MM et al (2023) Cellulose-acetate-based films modified with Ag_2O and ZnS as nanocomposites for highly controlling biological behavior for wound healing applications. Materials 16:777. https://doi.org/10.3390/ma16020777
7. Amani M, Rakhshani A, Maghsoudian S et al (2023) PH-sensitive bilayer electrospun nanofibers based on ethyl cellulose and Eudragit S-100 as a dual delivery system for treatment of the burn wounds; preparation, characterizations, and in-vitro/in-vivo assessment. Int J Biol Macromol 249:126705. https://doi.org/10.1016/j.ijbiomac.2023.126705
8. Apostolou A, Stagos D, Galitsiou E et al (2013) Assessment of polyphenolic content, antioxidant activity, protection against ROS-induced DNA damage and anticancer activity of Vitis vinifera stem extracts. Food Chem Toxicol 61:60–68. https://doi.org/10.1016/j.fct.2013.01.029
9. Araújo L, Machado AR, Pintado M et al (2023a) Toward a circular bioeconomy: extracting cellulose from grape stalks. In: ECP 2023. MDPI, Basel Switzerland, p 86
10. Araújo L, Machado AR, Sousa S et al (2023) Implementation of a circular bioeconomy: obtaining cellulose fibers derived from Portuguese vine pruning residues for heritage conservation, oxidized with TEMPO and ultrasonic treatment. Agriculture 13:1905. https://doi.org/10.3390/agriculture13101905

11. Atatoprak T, Amorim MM, Ribeiro T et al (2022) Grape stalk valorization for fermentation purposes. Food Chem Mol Sci 4:100067. https://doi.org/10.1016/j.fochms.2021.100067
12. Bakina OV, Glazkova EA, Lozhkomoev AS et al (2018) Cellulose acetate fibres surface modified with AlOOH/Cu particles: synthesis, characterization and antimicrobial activity. Cellulose 25:4487–4497. https://doi.org/10.1007/s10570-018-1895-z
13. Basu A, Celma G, Strømme M, Ferraz N (2018) In vitro and in vivo evaluation of the wound healing properties of nanofibrillated cellulose hydrogels. ACS Appl Bio Mater 1:1853–1863. https://doi.org/10.1021/acsabm.8b00370
14. Basu P, Narendrakumar U, Arunachalam R et al (2018) Characterization and evaluation of carboxymethyl cellulose-based films for healing of full-thickness wounds in normal and diabetic rats. ACS Omega 3:12622–12632. https://doi.org/10.1021/acsomega.8b02015
15. Bianchi SE, Machado BEK, da Silva MGC et al (2018) Coumestrol/hydroxypropyl-β-cyclodextrin association incorporated in hydroxypropyl methylcellulose hydrogel exhibits wound healing effect: in vitro and in vivo study. Eur J Pharm Sci 119:179–188. https://doi.org/10.1016/j.ejps.2018.04.019
16. Bocheva G, Slominski RM, Slominski AT (2019) Neuroendocrine aspects of skin aging. Int J Mol Sci 20:2798. https://doi.org/10.3390/IJMS20112798
17. Capanema NSV, Mansur AAP, de Jesus AC et al (2018) Superabsorbent crosslinked carboxymethyl cellulose-PEG hydrogels for potential wound dressing applications. Int J Biol Macromol 106:1218–1234. https://doi.org/10.1016/j.ijbiomac.2017.08.124
18. Carolin CF, Kamalesh T, Kumar PS et al (2023) A critical review on sustainable cellulose materials and its multifaceted applications. Ind Crops Prod 203:117221. https://doi.org/10.1016/j.indcrop.2023.117221
19. Chopra L, Manikanika (2022) Extraction of cellulosic fibers from the natural resources: a short review. Mater Today Proc 48:1265–1270. https://doi.org/10.1016/j.matpr.2021.08.267
20. Collazo-Bigliardi S, Ortega-Toro R, Chiralt A (2019) Improving properties of thermoplastic starch films by incorporating active extracts and cellulose fibres isolated from rice or coffee husk. Food Packag Shelf Life 22:100383. https://doi.org/10.1016/j.fpsl.2019.100383
21. Collazo-Bigliardi S, Ortega-Toro R, Chiralt Boix A (2018) Isolation and characterisation of microcrystalline cellulose and cellulose nanocrystals from coffee husk and comparative study with rice husk. Carbohydr Polym 191:205–215. https://doi.org/10.1016/j.carbpol.2018.03.022
22. Croitoru A-M, Ayran M, Altan E et al (2023) Development of gallic acid-loaded ethylcellulose fibers as a potential wound dressing material. Int J Biol Macromol 253:126996. https://doi.org/10.1016/j.ijbiomac.2023.126996
23. Dai J, Chae M, Beyene D et al (2018) Co-production of cellulose nanocrystals and fermentable sugars assisted by endoglucanase treatment of wood pulp. Materials 11:1645. https://doi.org/10.3390/ma11091645
24. Dairi N, Ferfera-Harrar H, Ramos M, Garrigós MC (2019) Cellulose acetate/AgNPs-organoclay and/or thymol nano-biocomposite films with combined antimicrobial/antioxidant properties for active food packaging use. Int J Biol Macromol 121:508–523. https://doi.org/10.1016/j.ijbiomac.2018.10.042
25. Das M, Zandraa O, Mudenur C et al (2022) Composite scaffolds based on bacterial cellulose for wound dressing application. ACS Appl Bio Mater 5:3722–3733. https://doi.org/10.1021/acsabm.2c00226
26. Derakhshani A, Hesaraki S, Nezafati N, Azami M (2022) Wound closure, angiogenesis and antibacterial behaviors of tetracalcium phosphate/hydroxyethyl cellulose/hyaluronic acid/gelatin composite dermal scaffolds. J Biomater Sci Polym Ed 33:605–626. https://doi.org/10.1080/09205063.2021.2008786
27. Devesa-Rey R, Vecino X, Varela-Alende JL et al (2011) Valorization of winery waste vs. the costs of not recycling. Waste Manage 31:2327–2335. https://doi.org/10.1016/j.wasman.2011.06.001
28. Dharmalingam K, Anandalakshmi R (2019) Fabrication, characterization and drug loading efficiency of citric acid crosslinked NaCMC-HPMC hydrogel films for wound healing drug

delivery applications. Int J Biol Macromol 134:815–829. https://doi.org/10.1016/j.ijbiomac.2019.05.027
29. Dixit K, Bora H, Chakrabarti R et al (2024) Thermoresponsive keratin-methylcellulose self-healing injectable hydrogel accelerating full-thickness wound healing by promoting rapid epithelialization. Int J Biol Macromol 263:130073. https://doi.org/10.1016/j.ijbiomac.2024.130073
30. Do MH, Van Thi KK, Huynh PT et al (2023) Synthesis, characterization, and antibacterial activity of amino-functionalized microcrystalline cellulose derivatives from cotton fibers. Biomass Convers Biorefin 13:10595–10603. https://doi.org/10.1007/s13399-022-02391-7
31. Domingues L, Hurbain I, Gilles-Marsens F et al (2020) Coupling of melanocyte signaling and mechanics by caveolae is required for human skin pigmentation. Nat Commun 11:1–14. https://doi.org/10.1038/s41467-020-16738-z
32. Domínguez-Perles R, Guedes A, Queiroz M et al (2016) Oxidative stress prevention and anti-apoptosis activity of grape (*Vitis vinifera* L.) stems in human keratinocytes. Food Res Int 87:92–102. https://doi.org/10.1016/j.foodres.2016.06.030
33. Duan H, Shao Z, Zhao M, Zhou Z (2016) Preparation and properties of environmental-friendly coatings based on carboxymethyl cellulose nitrate ester & modified alkyd. Carbohydr Polym 137:92–99. https://doi.org/10.1016/j.carbpol.2015.10.067
34. Feng Y-H, Cheng T-Y, Yang W-G et al (2018) Characteristics and environmentally friendly extraction of cellulose nanofibrils from sugarcane bagasse. Ind Crops Prod 111:285–291. https://doi.org/10.1016/j.indcrop.2017.10.041
35. Ferrer-Gallego R, Silva P (2022) The wine industry by-products: applications for food industry and health benefits. Antioxidants 11:2025. https://doi.org/10.3390/antiox11102025
36. García-Mesa Y, García-Piqueras J, García B et al (2017) Merkel cells and Meissner's corpuscles in human digital skin display Piezo2 immunoreactivity. J Anat 231:978–989. https://doi.org/10.1111/joa.12688
37. Garraud O, Hozzein WN, Badr G (2017) Wound healing: time to look for intelligent, 'natural' immunological approaches? BMC Immunol 18:23. https://doi.org/10.1186/s12865-017-0207-y
38. Godakanda VU, Li H, Alquezar L et al (2019) Tunable drug release from blend poly(vinyl pyrrolidone)-ethyl cellulose nanofibers. Int J Pharm 562:172–179. https://doi.org/10.1016/j.ijpharm.2019.03.035
39. Gouvinhas I, Pinto R, Santos R et al (2020) Enhanced phytochemical composition and biological activities of grape (*Vitis vinifera* L.) Stems growing in low altitude regions. Sci Hortic 265:109248. https://doi.org/10.1016/j.scienta.2020.109248
40. Grzybek P, Dudek G, van der Bruggen B (2024) Cellulose-based films and membranes: a comprehensive review on preparation and applications. Chem Eng J 495:153500. https://doi.org/10.1016/j.cej.2024.153500
41. Guan J, Wu C, He Y, Lu F (2023) Skin-associated adipocytes in skin barrier immunity: a mini-review. Front Immunol 14. https://doi.org/10.3389/fimmu.2023.1116548
42. Gurtner GC, Werner S, Barrandon Y, Longaker MT (2008) Wound repair and regeneration. Nature 453:314–321. https://doi.org/10.1038/nature07039
43. Hakkarainen T, Koivuniemi R, Kosonen M et al (2016) Nanofibrillar cellulose wound dressing in skin graft donor site treatment. J Control Release 244:292–301. https://doi.org/10.1016/j.jconrel.2016.07.053
44. Hoffman BU, Baba Y, Griffith TN et al (2018) Merkel cells activate sensory neural pathways through adrenergic synapses. Neuron 100:1401-1413.e6. https://doi.org/10.1016/j.neuron.2018.10.034
45. Ibrahim NA, Hasanin MS, Kamel S (2023) A new approach for improving the antimicrobial activity of cellulose pulp. Inorg Chem Commun 155:111009. https://doi.org/10.1016/j.inoche.2023.111009
46. Ioniţă M, Crică LE, Voicu SI et al (2018) Synergistic effect of carbon nanotubes and graphene for high performance cellulose acetate membranes in biomedical applications. Carbohydr Polym 183:50–61. https://doi.org/10.1016/j.carbpol.2017.10.095

47. Iqbal HMN, Kyazze G, Locke IC et al (2015) Poly(3-hydroxybutyrate)-ethyl cellulose based bio-composites with novel characteristics for infection free wound healing application. Int J Biol Macromol 81:552–559. https://doi.org/10.1016/j.ijbiomac.2015.08.040
48. Isopencu G, Deleanu I, Busuioc C et al (2023) Bacterial cellulose—carboxymethylcellulose composite loaded with turmeric extract for antimicrobial wound dressing applications. Int J Mol Sci 24:1719. https://doi.org/10.3390/ijms24021719
49. Iversen L, Johansen C, Kragballe K (2005) Signal transduction pathways in human epidermis. Eur J Dermatol 15:4–12
50. Karavasili C, Tsongas K, Andreadis II et al (2020) Physico-mechanical and finite element analysis evaluation of 3D printable alginate-methylcellulose inks for wound healing applications. Carbohydr Polym 247:116666. https://doi.org/10.1016/j.carbpol.2020.116666
51. Khoshnevisan K, Maleki H, Samadian H et al (2018) Cellulose acetate electrospun nanofibers for drug delivery systems: applications and recent advances. Carbohydr Polym 198:131–141. https://doi.org/10.1016/j.carbpol.2018.06.072
52. Kim MH, Park H, Nam HC et al (2018) Injectable methylcellulose hydrogel containing silver oxide nanoparticles for burn wound healing. Carbohydr Polym 181:579–586. https://doi.org/10.1016/j.carbpol.2017.11.109
53. Koike S, Yamasaki K (2020) Melanogenesis connection with innate immunity and toll-like receptors. Int J Mol Sci 21:9769. https://doi.org/10.3390/ijms21249769
54. Komi DEA, Khomtchouk K, Santa Maria PL (2020) A review of the contribution of mast cells in wound healing: involved molecular and cellular mechanisms. Clin Rev Allergy Immunol 58:298–312. https://doi.org/10.1007/s12016-019-08729-w
55. Kong P, Rosnan SM, Enomae T (2024) Carboxymethyl cellulose–chitosan edible films for food packaging: a review of recent advances. Carbohydr Polym 346:122612. https://doi.org/10.1016/j.carbpol.2024.122612
56. Kozlowska J, Stachowiak N, Sionkowska A (2018) Collagen/gelatin/hydroxyethyl cellulose composites containing microspheres based on collagen and gelatin: design and evaluation. Polymers (Basel) 10:456. https://doi.org/10.3390/polym10040456
57. Krasian T, Daranarong D, Punyodom W et al (2025) Electrospun composite membranes of ethyl cellulose and MXene (Ti_3C_2Tx): biocompatible platforms for enhanced drug delivery and antibacterial wound healing. Int J Biol Macromol 287:138596. https://doi.org/10.1016/j.ijbiomac.2024.138596
58. Krüger-Genge A, Blocki A, Franke R-P, Jung F (2019) Vascular endothelial cell biology: an update. Int J Mol Sci 20:4411. https://doi.org/10.3390/ijms20184411
59. Leal C, Gouvinhas I, Santos RA et al (2020a) Potential application of grape (*Vitis vinifera* L.) stem extracts in the cosmetic and pharmaceutical industries: valorization of a by-product. Ind Crops Prod 154. https://doi.org/10.1016/j.indcrop.2020.112675
60. Leal C, Santos RA, Pinto R et al (2020) Recovery of bioactive compounds from white grape (*Vitis vinifera* L.) stems as potential antimicrobial agents for human health. Saudi J Biol Sci 27:1009–1015. https://doi.org/10.1016/j.sjbs.2020.02.013
61. Liakos IL, Iordache F, Carzino R et al (2018) Cellulose acetate—essential oil nanocapsules with antimicrobial activity for biomedical applications. Colloids Surf B Biointerfaces 172:471–479. https://doi.org/10.1016/j.colsurfb.2018.08.069
62. Li H, Zhang Z, Godakanda VU et al (2019) The effect of collection substrate on electrospun ciprofloxacin-loaded poly(vinylpyrrolidone) and ethyl cellulose nanofibers as potential wound dressing materials. Mater Sci Eng, C 104:109917. https://doi.org/10.1016/j.msec.2019.109917
63. Li J, Zhang J, Sun H et al (2019) An optical fiber relative humidity sensor based on hollow-core fiber and hydroxypropyl methylcellulose hydrogel film. Optik (Stuttg) 195:163172. https://doi.org/10.1016/j.ijleo.2019.163172
64. Lin W-C, Lien C-C, Yeh H-J et al (2013) Bacterial cellulose and bacterial cellulose–chitosan membranes for wound dressing applications. Carbohydr Polym 94:603–611. https://doi.org/10.1016/j.carbpol.2013.01.076

65. López de Dicastillo C, Bustos F, Guarda A, Galotto MJ (2016) Cross-linked methyl cellulose films with murta fruit extract for antioxidant and antimicrobial active food packaging. Food Hydrocoll 60:335–344. https://doi.org/10.1016/j.foodhyd.2016.03.020
66. Lu H, Li X, Zhang M et al (2022) Antibacterial cellulose nanocrystal-incorporated hydrogels with satisfactory vascularization for enhancing skin regeneration. Front Bioeng Biotechnol 10. https://doi.org/10.3389/fbioe.2022.876936
67. Luo P, Liu L, Xu W et al (2018) Preparation and characterization of aminated hyaluronic acid/oxidized hydroxyethyl cellulose hydrogel. Carbohydr Polym 199:170–177. https://doi.org/10.1016/j.carbpol.2018.06.065
68. Luz EPCG, da Silva TF, Marques LSM et al (2024) Bacteria-derived cellulose membranes modified with graphene oxide-silver nanoparticles for accelerating wound healing. ACS Appl Bio Mater 7:5530–5540. https://doi.org/10.1021/acsabm.4c00650
69. Makrantonaki E, Zouboulis CC (2007) The skin as a mirror of the aging process in the human organism—state of the art and results of the aging research in the German National Genome Research Network 2 (NGFN-2). Exp Gerontol 42:879–886
70. Mangione R, Simões R, Pereira H et al (2022) Potential use of grape stems and pomaces from two red grapevine cultivars as source of oligosaccharides. Processes 10:1896. https://doi.org/10.3390/pr10091896
71. Maté I, Vargas M, Atarés L, Chiralt A (2024) Fractionation of winemaking grape stalks by subcritical water extraction to obtain added-value products. Foods 13:3566. https://doi.org/10.3390/foods13223566
72. M.H. Periayah ASHAZMS (2017) Mechanism action of platelets and crucial blood coagulation pathways in Hemostasis. Int J Hematol Oncol Stem Cell Res 319–327
73. Mirmoeini SS, Moradi M, Tajik H et al (2023) Cellulose/Salep-based intelligent aerogel with red grape anthocyanins: preparation, characterization and application in beef packaging. Food Chem 425:136493. https://doi.org/10.1016/j.foodchem.2023.136493
74. Misery L (1997) Skin, immunity and the nervous system. Bras J Dermatol 137:843–850
75. Mojahedi M, Heydari P, Kharazi AZ (2024) Preparation and characterization of an antibacterial CMC/PCL hydrogel films containing CIP/Cur: in vitro and in vivo evaluation of wound healing activity. Int J Biol Macromol 282:136570. https://doi.org/10.1016/j.ijbiomac.2024.136570
76. Moll R, Divo M, Langbein L (2008) The human keratins: biology and pathology. Histochem Cell Biol 129:705. https://doi.org/10.1007/s00418-008-0435-6
77. Mondal MdIH, MstS Y, Rahman MdS (2015) Preparation of food grade carboxymethyl cellulose from corn husk agrowaste. Int J Biol Macromol 79:144–150. https://doi.org/10.1016/j.ijbiomac.2015.04.061
78. Nair SS, Zhu J, Deng Y, Ragauskas AJ (2014) High performance green barriers based on nanocellulose. Sustainable Chem Processes 2:23. https://doi.org/10.1186/s40508-014-0023-0
79. Naomi R, Bt Hj Idrus R, Fauzi MB (2020) Plant-vs. Bacterial-derived cellulose for wound healing: a review. Int J Environ Res Public Health 17:6803. https://doi.org/10.3390/ijerph17186803
80. Naserian F, Mesgar AS (2022) Development of antibacterial and superabsorbent wound composite sponges containing carboxymethyl cellulose/gelatin/Cu-doped ZnO nanoparticles. Colloids Surf B Biointerfaces 218:112729. https://doi.org/10.1016/j.colsurfb.2022.112729
81. Negut I, Grumezescu V, Grumezescu AM (2018) Treatment strategies for infected wounds. Molecules 23:2392. https://doi.org/10.3390/molecules23092392
82. Niculescu V-C, Ionete R-E (2023) An overview on management and valorisation of winery wastes. Appl Sci 13:5063. https://doi.org/10.3390/app13085063
83. Noda T, Hatakeyama M, Kitaoka T (2022) Combination of polysaccharide nanofibers derived from cellulose and chitin promotes the adhesion, migration and proliferation of mouse fibroblast cells. Nanomaterials 12:402. https://doi.org/10.3390/nano12030402
84. Nogueira DP, Jiménez-Moreno N, Esparza I et al (2023) Evaluation of grape stems and grape stem extracts for sulfur dioxide replacement during grape wine production. Curr Res Food Sci 6:100453. https://doi.org/10.1016/j.crfs.2023.100453

85. Ogushi Y, Sakai S, Kawakami K (2007) Synthesis of enzymatically-gellable carboxymethylcellulose for biomedical applications. J Biosci Bioeng 104:30–33. https://doi.org/10.1263/jbb.104.30
86. Oosterhoff LA, Kruitwagen HS, van Wolferen ME et al (2019) Characterization of endothelial and smooth muscle cells from different canine vessels. Front Physiol 10. https://doi.org/10.3389/fphys.2019.00101
87. Palaninathan V, Raveendran S, Rochani AK et al (2018) Bioactive bacterial cellulose sulfate electrospun nanofibers for tissue engineering applications. J Tissue Eng Regen Med 12:1634–1645. https://doi.org/10.1002/term.2689
88. Pal N, Dubey P, Gopinath P, Pal K (2017) Combined effect of cellulose nanocrystal and reduced graphene oxide into poly-lactic acid matrix nanocomposite as a scaffold and its antibacterial activity. Int J Biol Macromol 95:94–105. https://doi.org/10.1016/j.ijbiomac.2016.11.041
89. Pandey M, Mohamad N, Low W-L et al (2017) Microwaved bacterial cellulose-based hydrogel microparticles for the healing of partial thickness burn wounds. Drug Deliv Transl Res 7:89–99. https://doi.org/10.1007/s13346-016-0341-8
90. Panthi KP, Gyawali A, Pandeya S et al (2022) The encapsulation of bioactive plant extracts into the cellulose microfiber isolated from G. optiva species for biomedical applications. Membranes (Basel) 12:1089. https://doi.org/10.3390/membranes12111089
91. Pant N, Wairkar S (2024) Mupirocin-doped α-cellulose nanopaper for wound dressing: development, in vitro characterization and antimicrobial studies. AAPS PharmSciTech 26:11. https://doi.org/10.1208/s12249-024-03013-3
92. Pappu A, Saxena M, Thakur VK et al (2016) Facile extraction, processing and characterization of biorenewable sisal fibers for multifunctional applications. J Macromol Sci Part A 53:424–432. https://doi.org/10.1080/10601325.2016.1176443
93. Prasathkumar M, George A, Sadhasivam S (2024) Influence of chitosan and hydroxyethyl cellulose modifications towards the design of cross-linked double networks hydrogel for diabetic wound healing. Int J Biol Macromol 265:130851. https://doi.org/10.1016/j.ijbiomac.2024.130851
94. Qiu Y, Qiu L, Cui J, Wei Q (2016) Bacterial cellulose and bacterial cellulose-vaccarin membranes for wound healing. Mater Sci Eng, C 59:303–309. https://doi.org/10.1016/j.msec.2015.10.016
95. Quijano L, Rodrigues R, Fischer D et al (2024) Bacterial cellulose cookbook: a systematic review on sustainable and cost-effective substrates. J Bioresour Bioprod 9:379–409. https://doi.org/10.1016/j.jobab.2024.05.003
96. Rodrigues M, Kosaric N, Bonham CA, Gurtner GC (2019) Wound healing: a cellular perspective. Physiol Rev 99:665–706. https://doi.org/10.1152/physrev.00067.2017
97. Rognoni E, Watt FM (2018) Skin cell heterogeneity in development, wound healing, and cancer. Trends Cell Biol 28:709–722. https://doi.org/10.1016/j.tcb.2018.05.002
98. Sader HS, Farrell DJ, Flamm RK, Jones RN (2014) Antimicrobial susceptibility of Gram-negative organisms isolated from patients hospitalized in intensive care units in United States and European hospitals (2009–2011). Diagn Microbiol Infect Dis 78:443–448. https://doi.org/10.1016/j.diagmicrobio.2013.11.025
99. Sahoo PK, Jena DK (2018) Synthesis and study of mechanical and fire retardant properties of (carboxymethyl cellulose -g-polyacrylonitrile)/montmorillonite biodegradable nanocomposite. J Polym Res 25:260. https://doi.org/10.1007/s10965-018-1659-3
100. Samadian H, Salehi M, Farzamfar S et al (2018) *In vitro* and *in vivo* evaluation of electrospun cellulose acetate/gelatin/hydroxyapatite nanocomposite mats for wound dressing applications. Artif Cells Nanomed Biotechnol 46:964–974. https://doi.org/10.1080/21691401.2018.1439842
101. Sarangi PK, Srivastava RK, Sahoo UK et al (2024) Biotechnological innovations in nanocellulose production from waste biomass with a focus on pineapple waste. Chemosphere 349:140833. https://doi.org/10.1016/j.chemosphere.2023.140833

102. Sardana S, Singh Z, Sharma AK et al (2022) Self-powered biocompatible humidity sensor based on an electrospun anisotropic triboelectric nanogenerator for non-invasive diagnostic applications. Sens Actuators B Chem 371:132507. https://doi.org/10.1016/j.snb.2022.132507
103. Satoh T, Nakagawa K, Sugihara F et al (2017) Identification of an atypical monocyte and committed progenitor involved in fibrosis. Nature 541:96–101. https://doi.org/10.1038/nature20611
104. Schuhladen K, Mukoo P, Liverani L et al (2020) Manuka honey and bioactive glass impart methylcellulose foams with antibacterial effects for wound-healing applications. Biomed Mater 15:065002. https://doi.org/10.1088/1748-605X/ab87e5
105. Schuhladen K, Raghu SNV, Liverani L et al (2021) Production of a novel poly(ε-caprolactone)-methylcellulose electrospun wound dressing by incorporating bioactive glass and Manuka honey. J Biomed Mater Res B Appl Biomater 109:180–192. https://doi.org/10.1002/jbm.b.34690
106. Schütz K, Placht A-M, Paul B et al (2017) Three-dimensional plotting of a cell-laden alginate/methylcellulose blend: towards biofabrication of tissue engineering constructs with clinically relevant dimensions. J Tissue Eng Regen Med 11:1574–1587. https://doi.org/10.1002/term.2058
107. Seddiqi H, Oliaei E, Honarkar H et al (2021) Cellulose and its derivatives: towards biomedical applications. Cellulose 28:1893–1931. https://doi.org/10.1007/s10570-020-03674-w
108. Seneschal J, Clark RA, Gehad A et al (2012) Human epidermal langerhans cells maintain immune homeostasis in skin by activating skin resident regulatory T cells. Immunity 36:873–884. https://doi.org/10.1016/j.immuni.2012.03.018
109. Shah SA, Sohail M, Karperien M et al (2023) Chitosan and carboxymethyl cellulose-based 3D multifunctional bioactive hydrogels loaded with nano-curcumin for synergistic diabetic wound repair. Int J Biol Macromol 227:1203–1220. https://doi.org/10.1016/j.ijbiomac.2022.11.307
110. Shao W, Liu H, Liu X et al (2015) Development of silver sulfadiazine loaded bacterial cellulose/sodium alginate composite films with enhanced antibacterial property. Carbohydr Polym 132:351–358. https://doi.org/10.1016/j.carbpol.2015.06.057
111. Shpichka A, Butnaru D, Bezrukov EA et al (2019) Skin tissue regeneration for burn injury. Stem Cell Res Ther 10:94. https://doi.org/10.1186/s13287-019-1203-3
112. Simões D, Miguel SP, Ribeiro MP et al (2018) Recent advances on antimicrobial wound dressing: a review. Eur J Pharm Biopharm 127:130–141. https://doi.org/10.1016/j.ejpb.2018.02.022
113. Sismaet HJ, Banerjee A, McNish S et al (2016) Electrochemical detection of *Pseudomonas* in wound exudate samples from patients with chronic wounds. Wound Repair and Regeneration 24:366–372. https://doi.org/10.1111/wrr.12414
114. Spigno G, Pizzorno T, De Faveri DM (2008) Cellulose and hemicelluloses recovery from grape stalks. Bioresour Technol 99:4329–4337. https://doi.org/10.1016/j.biortech.2007.08.044
115. Stuart Hogg (2013) Essential Microbiology, Second
116. Suhas GVK, Carrott PJM et al (2016) Cellulose: a review as natural, modified and activated carbon adsorbent. Bioresour Technol 216:1066–1076. https://doi.org/10.1016/j.biortech.2016.05.106
117. Su T, Wu Q-X, Chen Y et al (2019) Fabrication of the polyphosphates patched cellulose sulfate-chitosan hydrochloride microcapsules and as vehicles for sustained drug release. Int J Pharm 555:291–302. https://doi.org/10.1016/j.ijpharm.2018.11.058
118. Tabassum M, Shahruzzaman Md, Mukta NA et al (2024) Hydroxyapatite and cellulose nanocrystals loaded gelatin-chitosan based electrospun nanofibrous mats for rapid wound healing. Heliyon 10:e25871. https://doi.org/10.1016/j.heliyon.2024.e25871
119. Talagas M, Lebonvallet N, Leschiera R et al (2018) What about physical contacts between epidermal keratinocytes and sensory neurons? Exp Dermatol 27:9–13. https://doi.org/10.1111/exd.13411
120. Trompette A, Ubags ND (2023) Skin barrier immunology from early life to adulthood. Mucosal Immunol 16:194–207. https://doi.org/10.1016/j.mucimm.2023.02.005

121. Tudoroiu E-E, Dinu-Pîrvu C-E, Albu Kaya MG et al (2021) An overview of cellulose derivatives-based dressings for wound-healing management. Pharmaceuticals 14:1215. https://doi.org/10.3390/ph14121215
122. Vallejo M, Cordeiro R, Dias PAN et al (2021) Recovery and evaluation of cellulose from agroindustrial residues of corn, grape, pomegranate, strawberry-tree fruit and fava. Bioresour Bioprocess 8:25. https://doi.org/10.1186/s40643-021-00377-3
123. Vatanpour V, Pasaoglu ME, Barzegar H et al (2022) Cellulose acetate in fabrication of polymeric membranes: a review. Chemosphere 295:133914. https://doi.org/10.1016/j.chemosphere.2022.133914
124. Velazquez-Meza ME, Galarde-López M, Carrillo-Quiróz B, Alpuche-Aranda CM (2022) Antimicrobial resistance: one health approach. Vet World 743–749. https://doi.org/10.14202/vetworld.2022.743-749
125. Verma C, Singh V, AlFantazi A (2024) Cellulose, cellulose derivatives and cellulose composites in sustainable corrosion protection: challenges and opportunities. Phys Chem Chem Phys 26:11217–11242. https://doi.org/10.1039/D3CP06057H
126. Vidal Yucha SE, Tamamoto KA, Kaplan DL (2019) The importance of the neuro-immuno-cutaneous system on human skin equivalent design. Cell Prolif 52
127. Voss M, Kotrba J, Gaffal E et al (2021) Mast cells in the skin: defenders of integrity or offenders in inflammation? Int J Mol Sci 22:4589. https://doi.org/10.3390/ijms22094589
128. Wang C, Niu H, Ma X et al (2019) Bioinspired, injectable, quaternized hydroxyethyl cellulose composite hydrogel coordinated by mesocellular silica foam for rapid, noncompressible hemostasis and wound healing. ACS Appl Mater Interfaces 11:34595–34608. https://doi.org/10.1021/acsami.9b08799
129. Wang M, Li W, You C et al (2017) Triboelectric nanogenerator based on 317L stainless steel and ethyl cellulose for biomedical applications. RSC Adv 7:6772–6779. https://doi.org/10.1039/C6RA28252K
130. Wong TW, Ramli NA (2014) Carboxymethylcellulose film for bacterial wound infection control and healing. Carbohydr Polym 112:367–375. https://doi.org/10.1016/j.carbpol.2014.06.002
131. Yadav BK, Solanki N, Patel G (2024) Electrospun nanofibers as advanced wound dressing materials: comparative analysis of single-layered and multilayered nanofibers containing polycaprolactone, methylcellulose, and polyvinyl alcohol. J Biomater Sci Polym Ed 35:869–879. https://doi.org/10.1080/09205063.2024.2311448
132. Yang M, Ward J, Choy K (2020) Nature-inspired bacterial cellulose/methylglyoxal (BC/MGO) nanocomposite for broad-spectrum antimicrobial wound dressing. Macromol Biosci 20. https://doi.org/10.1002/mabi.202000070
133. Yavuzturk Gul B, Pekgenc E, Vatanpour V, Koyuncu I (2023) A review of cellulose-based derivatives polymers in fabrication of gas separation membranes: recent developments and challenges. Carbohydr Polym 321:121296. https://doi.org/10.1016/j.carbpol.2023.121296
134. Yu J, Wang L, Zhao Y, Zhou C (2021) Preparation, characterization, and antibacterial property of carboxymethyl cellulose derivatives bearing tetrabutylammonium salt. Int J Biol Macromol 176:72–77. https://doi.org/10.1016/j.ijbiomac.2021.02.063
135. Zhang J, Xu Y, Liu T et al (2019) In vivo construction of lymphoid node by implantation of adipose-derived stromal cells with hydroxypropyl methyl cellulose hydrogel in BALB/c nude mice. Organogenesis 15:85–99. https://doi.org/10.1080/15476278.2019.1656994
136. Zhong W, Hu R, Zhou S et al (2023) Spatiotemporally responsive hydrogel dressing with self-adaptive antibacterial activity and cell compatibility for wound sealing and healing. Adv Healthc Mater 12. https://doi.org/10.1002/adhm.202203241
137. Zmejkoski D, Spasojević D, Orlovska I et al (2018) Bacterial cellulose-lignin composite hydrogel as a promising agent in chronic wound healing. Int J Biol Macromol 118:494–503. https://doi.org/10.1016/j.ijbiomac.2018.06.067
138. The Burden of Wounds on EU Healthcare Systems
139. United States Census Bureau
140. Ageing Europe—statistics on population developments

141. Topical silver nanoparticles reduced with ethylcellulose enhance skin wound healing
142. FAO (2017)

Raquel Fernandes (R.F.) is a Ph.D. Student at CITAB-UTAD (FCT scholarship), on by-product research for natural-based targeted therapies for prostate cancer. She holds a B.Sc. in Biochemistry (2018) and a M.Sc. in Health Sciences (2021) at ICVS, University of Minho, on the role of glycogen synthase kinase-3 (GSK-3) inhibition in lung injury. From 2021–2022, she was focused on immunobiological and regenerative lung mechanisms under pathological conditions. From 2022–2023, she was as a researcher in the I&D Department of Mesosystem S.A, focusing on the identification and extraction of natural functional molecules and their implementation for cosmetics. She published 8 scientific articles, authored 1 book chapter, co-edited a book, she participated in 2 international award-winning projects, she was an internship tutor of two master students, and she presented more than 10 oral communications in international conferences.

Cláudia Botelho (C.B.) earned her degree in Biological Engineering from the University of Minho (UM) in 2000, beginning her research career in 1998 as an undergraduate. Her early work on bioceramics—exploring how microelements enhance bone cell metabolic activity and tissue regeneration—saw her collaborate with esteemed institutions such as Queen Mary University of London, the University of Cambridge, Addenbrooke's Hospital, and the Nara Institute of Science and Technology in Japan. Following her Ph.D., which earned her multiple awards including the European Biomaterials and Tissue Engineering Doctoral Award, Claudia secured patents for a novel biomaterial for bone regeneration and an innovative nanocarrier able to cross the Brain-Blood-Barrier. Her academic achievements were recently recognized with the title of Habilitation at ICBAS, University of Porto. She now leads research into innovative therapies using natural molecules and industrial by-products, while supervising Ph.D. students and advocating for stable scientific careers.

Ana Novo Barros (A.N.B.) is an Associate Professor with Habilitation in Chemistry at the University of Trás-os-Montes and Alto Douro. Her research focuses on the identification, separation, and recovery of bioactive molecules from various natural sources, with the aim of repurposing them for applications in the food, cosmetic, and pharmaceutical industries. Her ultimate goal is to extract high-value functional compounds and integrate them into sustainable solutions, ensuring their reutilization across the entire value chain—from agriculture to the final consumer. She has published more than 130 scientific documents, including research articles, reviews and editorials, within the highest impact factor journals in the Food Science and Technology field (35 h-index, SCOPUS, January 2025). She is the coordinator from the European project WASTELESS|TOPIC ID: HORIZON-CL6-2022-FARM2FORK-01–08.

Cellulose Nanofibers from Plant Sources and Their Use in Biomedical Applications

Sagnite Ventura-Cruz, Cristina Arely De León Condes, and Alberto Tecante

Abstract Nanocellulose is a purified cellulose derivative conventionally obtained from sources such as wood and cotton. However, alternative sources for its extraction are currently being studied, and lignocellulosic residues represent a viable option for obtaining this natural polymer. This material is classified into cellulose micro-, nanofibers, and nanocrystals and is a promising component in biomedical applications. Cellulose nanofibers are biocompatible and durable, enabling their use in developing advanced dressings. These fibers have been integrated with antibacterial agents and drugs to enhance tissue regeneration and combat infections, thereby improving the efficacy of wound treatment. Furthermore, cellulose nanofibers have been combined with metal oxide nanoparticles to create hybrid materials, enhancing antibacterial activity, magnetic capabilities, and adsorption properties. This combination makes them suitable for magnetically controlled drug release, biosensors, and substances that enhance magnetic resonance imaging (MRI) contrast agents. They also have the potential for bone disease treatments and advanced therapeutic approaches. In addition, using plant extracts for synthesizing metal nanoparticles and technological advancements, such as 3D printing, are driving significant advancements in biomedical applications. These innovations are improving the biocompatibility of composite materials, opening new possibilities for clinical use.

S. Ventura-Cruz
División de Ingeniería Ambiental, Tecnológico Nacional de México/Tecnológico de Estudios Superiores de Tianguistenco, Santiago Tilapa, Mexico
e-mail: sagnicruz@yahoo.com

C. A. De León Condes
División de Ingeniería Industrial, Tecnológico Nacional de México/Tecnológico de Estudios Superiores de Tianguistenco, Santiago Tilapa, Mexico
e-mail: cristina_arely@test.edu.mx

A. Tecante (✉)
Departamento de Alimentos y Biotecnología, Facultad de Química, Universidad Nacional Autónoma de México, Cd. Universitaria, Mexico City, Mexico
e-mail: tecante@unam.mx

R. Malviya and S. Sundram (eds.), *Sustainable Cellulose Materials for Biomedical Applications*, Sustainable Materials and Technology,
https://doi.org/10.1007/978-981-96-9904-9_8

Keywords Nanocellulose · Nanofiber · Drug delivery · Tissue engineering · Wound healing · Biomedical applications

1 Introduction

Nanocellulose (NC) is a term that includes extracts obtained from different natural or bacterial cellulose sources. Cotton and wood have been the traditional cellulose supplies, but many other plant substrates have also been used. In recent years, agro-industrial waste has gained importance as a source of cellulose recovery and obtaining nanocellulose as a derivative with added value. This situation has expanded the availability of cellulose stocks beyond the traditional ones and has given impetus to the search for solutions to the environmental pollution caused by the disposal of these wastes. Consequently, the greater availability of cellulose sources to obtain nanocellulose ensures good future sustainability support. This situation has, in turn, promoted research on obtaining nanocellulose and the diversification of its applications.

The production methods of nanocellulose fibers (CNFs) and crystals (CNCs) have been widely documented and include chemical and mechanical methods, as well as combinations of both treatments for cellulose nanofibers and the subsequent acid hydrolysis treatment for cellulose nanocrystals. On the other hand, bacterial nanocellulose (BNC) is produced mainly in small-scale submerged culture media or bioreactors by the growth of Gram-negative species of the genera Acetobacter, Pseudomonas, and Alcaligenes. However, *Acetobacter xylinum* is the most efficient and has the highest cellulose production capacity [46].

Each nanocellulose form has distinctive physicochemical properties that make it attractive in various applications in several areas, such as biomedical. The main applications include skin lesion dressings, implants, drug delivery, and tissue scaffolds due to low cytotoxicity, mechanical strength, biodegradability, and low cost [46].

The physicochemical properties are common, but each form has different behaviors or responses governed by its characteristics. This peculiarity means that although different forms can show mechanical strength, that does not imply that they have the same level of strength. For example, cellulose nanofibers have a higher mechanical strength than cellulose nanocrystals [38]. This difference is attributed to fibers' more significant aspect ratio over crystals. Compressive strength and flexibility are crucial in some applications, such as manufacturing wound dressings, supporting and transporting therapeutic agents, and forming mechanically stable matrices in tissue regeneration. CNCs also serve as drug carriers and reinforcing agents in manufacturing membranes and fibers [46].

In parallel to the current and potential nanocellulose application research, there is intense activity in studying low-cost agricultural and industrial biomass through increasingly refined and environmentally friendly methods. The commitment and, simultaneously, the challenge are to make the methods increasingly more advanced, refined, and ecological, but still simple, to develop scalable processes and explore new application areas.

One of the main problems in extracting nanocellulose from plant matrices is the removal of the amorphous regions of hemicellulose and lignin. The purification of extracts involves several stages of chemical treatments that generate effluents that would have to be treated for the process to be circular. Paradoxically, using waste to solve an environmental pollution problem generates residues that contaminate the environment. Bacterial nanocellulose does not have this problem because the substrates involved in its production are sources of carbon and nitrogen that do not contain hemicellulose and lignin. This aspect is not negligible; on the contrary, it makes the production of this form of nanocellulose attractive. However, the problems are different and have to do with the specificity of the microorganism, the control of its growth and productivity, the appropriate operating conditions in the bioreactor of any scale, and the recovery, refining, and conservation of the final product. All these factors are inherent to producing any metabolite by microbial means, and the production of bacterial nanocellulose is not exempt. Another advantage of bacterial nanocellulose production is that its final properties are more homogeneous, which has been exploited in producing matrices used in tissue regeneration and wound healing, where their cytotoxicity, biocompatibility, and biodegradability are crucial [38].

In summary, CNCs, CNFs, and BNC share characteristics, but they are not identical, representing an advantage and, simultaneously, a versatility that can be exploited to obtain materials with different individualities based on modifying their properties to create innovative and intelligent materials. This feature is the main factor driving and continues to drive nanocellulose research. Studying the behavior of nanocellulose forms in biological media will allow the development of improved materials.

This chapter highlights the main characteristics of the three mentioned forms of nanocellulose. The structure, properties relevant to the applications of its various forms, and general aspects are presented first to put the subject in perspective and move on to the examination of applications, where more detailed discussions identify the relevant role of nanocellulose in different biomedical employment. Delivery technologies, drugs and active substances, tissue engineering, injury bandaging, membranes to repair tissues, and the combination with metal oxide nanoparticles are discussed. Finally, the perspectives focus on progress, current knowledge, and future needs.

2 Cellulose Nanofibers

Cellulose is one of the most abundant materials on the planet; it is renewable and has low toxicity. It is hydrophilic, has a crystalline structure, and is thermally stable, porous, durable, with acceptable tensile strength, and biodegradable. It is present in plants, tunicates, algae, and fungi. It has a structure consisting of a long chain with repeating units of simple sugar D-glucose, thus maintaining a linear arrangement; this configuration makes it a stable natural polymer [54, 71]. The parameter indicates that the number of glucose units comprising the native cellulose chains ranges from 10,000 to 15,000.

Cellulose comprises fibrils representing regions with highly ordered crystalline structures, the crystalline zone, and disordered regions with a lower orientation, the amorphous zone. Purified cellulose from different sources can be transformed into micro and nanoscale materials, from which microcrystals, nanofibers, or nanocrystals of cellulose and bacterial cellulose can be obtained [2, 71]. This polymer, formed by glucose-repeating units linked via β-1-4-glucosidic bonds, is obtained from various animal, vegetable, and bacterial sources. Bacterial cellulose is the most extensively used polymer for biomedical applications because of its high chemical purity. It does not contain hemicellulose or lignin, which is non-toxic and biocompatible. In addition, it has a high mechanical resistance to tension, crystallinity, swelling capacity, and moldability. BC is an exopolysaccharide produced either by surface culture or submerged agitated batch fermentation in a medium rich in glucose. This biopolymer is obtained from aerobic bacteria of the *Acetobacter* and *Gluconacetobacter* genera. The substrate can impact the cost of BC industrial production. The strategies include brewing industrial and agricultural by-products, such as sugar cane and malt bagasse [19, 54]. These different types of cellulose are based on their dimension and morphology. The interest in obtaining them is related to their characteristics, such as better thermal stability, mechanical resistance, flexibility, and a high degree of crystallinity, mainly on the nanometric scale. This versatility allows applications in the pharmaceutical, food, wastewater treatment, cosmetics, construction, packaging, automobile, and biomedical materials industries [7, 10, 31, 77].

Cellulose nanofibers (CNFs) are composed of crystalline and amorphous domains. CNFs with lengths between 100 and 1800 nm are thin fibrils intertwined in the form of a network, with a diameter in the nanometric range, elongated, and with a surface area more extensive than microfibers. CNFs contain more amorphous regions than nanocrystals, so these fibers are more flexible. The extraction of cellulose nanofibers from lignocellulosic sources can also be achieved through chemical ionic liquid, alkali hydrolysis, and organosolv treatment, in addition to mechanical processes grinding, steam explosion, or phase separation [41, 77].

3 Structure and Properties

CNFs have interesting physical properties, such as non-toxicity, lightweight, high elastic modulus, crystallinity, and a higher aspect ratio than cellulose nanocrystals (CNCs). These features make CNFs applicable in potential areas such as biomedicine for developing scaffolds and tissue engineering, in the electrical and energy sectors, and for preparing membranes and hydrogels [4, 63].

Raw materials that have been recently used as a source for producing CNFs include banana peel [25], *Stipa obtusa* fibers [16], *Musa textileis* fibers [58], bagasse pulps

[40], pullulan [12], jute fibers [8], rice straw [34], and sugarcane bagasse residues [29]. The final properties of CNFs depend on the origin and type of treatment for their extraction, as well as the intensive mechanical treatment, fibrillation, to separate the nanofibrils.

The crystallinity of CNFs originates from the chemical structure of cellulose and depends on the extraction method and the production source. The degree of crystallinity of CNFs is variable but is usually between 40 and 85% lower than that of nanocrystalline cellulose due to the presence of amorphous and crystalline regions. CNFs have a more prominent length-to-diameter aspect ratio ($\geq$ 52), making them desirable as reinforcing materials with nanostructures with fiber diameters from 50 to 500 nm [4, 9]. CNFs have many hydroxyl groups forming strong hydrogen bonds, constituting a highly cohesive network. They exhibit gel-like behavior with a low solids content and can form transparent films with improved tensile strength, elongation at break, and Young's modulus [24].

CNFs have been obtained from different sources, such as homogeneous nanofibrous structures containing pullulan prepared by electrospinning. Different concentrations (g/100 g) of CNFs were used: 1, 3, and 5, respectively. Ultrafine, continuous, and microsphere-free electrospun nanofibers were obtained. Mechanical tests show that as the concentration of CNFs increased, the tensile strength and elastic modulus increased compared to the sample containing only pullulan. This phenomenon is a feature that can be attributed to the self-aggregation of CNFs within the electrospun fibers. On the other hand, nanofibers were used as carriers of salicylic acid. Their presence in the structure induced an enhanced delayed drug release at higher doses than pure pullulan nanofibers, suggesting that dressing systems with controlled release of salicylic acid can be developed [12].

On the other hand, CNFs were prepared from purified α and β cellulose from six different species of seaweed collected in Abu Qir, a city on the Mediterranean coast of Egypt. CNFs were isolated by an acid hydrolisis treatment with sulfuric acid (H_2SO_4) at 56 °C; the yield of CNFs and the density of the algae are directly proportional. The species *U. fasciata* and *C. sinuosa* contain the highest percentage of cellulose of 10 and 11 g/100 g, respectively. The results showed that CNFs can be an alternative to conventional polymers [57].

Likewise, cellulose nanofibers from *Stipa obtusa* fibers were isolated and characterized using a combined mechanical-chemical treatment in low reagent concentrations-steam explosion. The optimal conditions for obtaining CNFs were determined using an 8% concentration of oxalic acid ($C_2H_4O_2$). CNFs diameters of 19 and 27 nm and a crystallinity index of 52.92% were obtained. In addition, the results showed a progressive partial elimination of lignin and hemicellulose at each isolation stage [16].

Agricultural straw residues comprise cellulose, hemicellulose, and lignin, considered high-value raw materials for obtaining nanofibers using electrospinning technology. These electrospun nanofibers from straw residues (ESNF) have been applied to environmental adsorption processes, energy storage, and biomedical materials. Membranes developed from ESNF are promising for tissue engineering because they are expected to adhere better to an affected surface and have advantages such as good biocompatibility, degradability, low cost, and antimicrobial properties [64].

4 Preparation

In the biomedical field, cellulose nanofibers are used as carriers to administer and release drugs and some health supplements in a controlled manner without modifying them, as well as scaffolds to support tissues, wound dressings, implants, and medical devices [4, 28]. CNFs are produced by chemical, mechanical, or biological treatments such as enzymatic hydrolysis (Fig. 1). The separation of CNFs by chemical processes from purified cellulosic fibers includes several stages. The first is conditioning, which includes reduction and separation to obtain the same particle size. The second stage includes alkaline and acid hydrolysis under controlled conditions to eliminate the amorphous regions of the cellulose polymer and preserve the compact crystalline regions. The third step is a repetition of washings and centrifugation to neutralize the solution. Finally, the suspension is dried to produce solid CNFs [28, 35, 44].

The first step in the conventional methodology for isolating CNFs with chemical treatments is conditioning the extraction source. The biomass is ground and sieved to reduce the particle size before or after the chemical treatment to ensure good extraction efficiency. A homogeneous particle size increases the contact surface and improves chemical and biomass interaction. Subsequently, the fibers are extracted with benzene-ethanol and toluene-ethanol to eliminate waxes, pigments, and soluble sugars. In the second stage, alkaline hydrolysis with sodium hydroxide (NaOH) or potassium hydroxide (KOH) swells the fibers, solubilizes hemicellulose partially, and

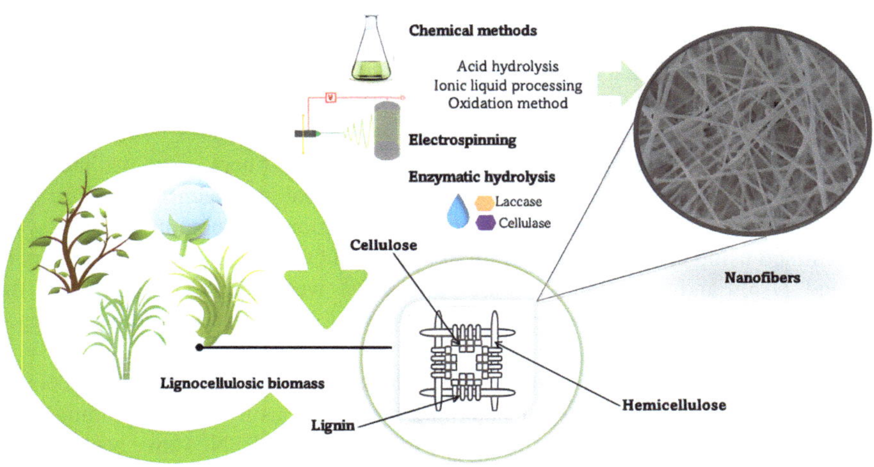

Fig. 1 Fabricating methods of nanofibers from lignocellulosic biomass

eliminates pectin and natural fats. Subsequently, the fibers are consecutively bleached with sodium chlorite ($NaClO_2$) and, in recent years, hydrogen peroxide (H_2O_2) [69]. Currently, the priority of research on chemical treatments for obtaining nanomaterials from lignocellulosic is to use mild chemical concentrations and process conditions to reduce the environmental impact and operating costs.

Electrospinning is one of the mechanical methods to synthesize cellulose nanofibers, which uses electrostatic forces to form and collect micro- and nanometer-sized fibers. Electrospun CNFs have a three-dimensionally interconnected porous structure that offers a higher area-to-volume ratio and is a potential biomaterial for making artificial skin, muscle tissue, vascular implants, cartilage, tendons, ligaments, and central and peripheral nervous system components. CNFs can be used as scaffolds for tissue engineering because they adopt the appearance of extracellular matrix functions [4, 77].

Electrospinning requires a high-voltage power supply, a needle-pump injection system, and a collector. The injection system pulls the polymer solution and delivers it between two opposite polar electrodes. The polymer solution layers surmount the surface tension and stretch uniaxially, sliding over each other. This extensional flow makes the polymer solution droplet thinner, creating a Taylor cone [12, 26]. Finally, the CNFs come from the syringe tip following the electric field, dry along the path, and are deposited onto the collector surface. The solution conductivity, surface tension, viscosity, voltage, needle-collector distance, flow rate, air temperature, and relative humidity around the electrospinning system influence the formation of CNFs. All these parameters should be adjusted to an adequate level to obtain bead-free CNFs [12, 26].

The enzymes laccase and cellulase participate in enzymatic hydrolysis. Laccase degrades lignin and hemicellulose, and cellulase hydrolyzes additional cellulose components. Compared to other enzymatic treatments, this one has not been extensively explored, but its ecological advantages can be highlighted, such as lower energy expenditure and zero emissions of hazardous substances [4, 53, 63]. Enzymatic hydrolysis depends on temperature, time, and concentration of cellulases relative to the mass of the cellulosic raw material. However, its disadvantages involve a long time and a high cost of enzymes. This process is suitable for reinforcing high-performance polymeric compounds and in biomedical applications.

Cellulose nanofibers are obtained from cellulose microfibers (CMFs) by mechanical methods. The isolated cellulose is disintegrated mechanically by any of the following operations: homogenization, refining, ultrasonication, and milling. High-pressure homogenization and microfluidization are the most common because of their high defibrillation efficiency and scalability. The high-shear force in high-pressure homogenization (HPH) produces cellulose fibers with crystalline and amorphous domains [29].

Nanometer-sized fibers are obtained by homogenizing a cellulosic pulp suspension at 50–2000 MPa. This suspension flows out a narrow die; consequently, cellulose reduction depends on the pressure, shear force, and flow conditions. Unlike chemical treatments, HPH does not produce chemical waste. However, its main drawbacks include high operating costs because of its high energy consumption, potential damage to the crystalline structure of CNFs, and wide particle size distribution. CNFs have been isolated from sugarcane bagasse waste using steam explosion, HPH, alkaline treatment, and bleaching. The CNFs had a crystallinity degree of 65.13–72.76%, 3–7 nm diameter, and thermal degradation temperatures of 296.62–305.83 °C [48, 65].

5 Applications

Nanomaterials obtained from lignocellulosic renewable sources have increased in recent years as an alternative to reducing the use of conventional synthetic materials. The applications of cellulose and its derivatives continue to be investigated because of their biodegradability, non-toxicity, and broad accessibility. Cellulose nanofibers in suspensions, powders, films, hydrogels, and aerogels have been widely applied in wound healing and dressings. These materials promote effective drug delivery, enhance wound healing, exhibit antibacterial activity, and support biosensing applications.

5.1 Drug Delivery

The skin protects the body as a barrier against microorganisms and external factors like thermal aggressions. This organ's structural and functional characteristics can be affected by injuries such as cuts, burns, physical forces, damage caused by chemical or cosmetic products, and metabolic disorders such as diabetes [63]. A skin wound is an alteration of the skin that occurs because of a simple or severe break, affecting normal functions. Wound healing is a fundamental process for the body to repair itself. This repair consists of a series of phases known as hemostasis, inflammation, proliferation, and remodeling [23]. Wound healing is a biological process influenced by direct and indirect factors. Hydration, temperature, oxygenation, and pressure are direct factors. For example, the synthesis of the extracellular medium and the formation of new blood vessels are strongly determined by adequate oxygenation. Indirect factors are related to systemic components such as nutrition, chronic diseases, and age; psychological factors may also influence [20].

Ointments, gels, and dressings deliver substances to skin wounds. However, they can be cumbersome or difficult to apply consistently compared to new forms such as drug delivery systems and innovative wound dressings that seek to maximize therapeutic efficiency and deliver exclusively to the wound site [66].

Drug delivery systems must guarantee the delivery of therapeutic substances directly to the wound site. These systems comprise technologies and materials to improve the wound healing process, such as regulating these therapeutic substances at the wound site, a complex microenvironment that includes changes such as tissue architecture and reduced blood flow [56].

The goal of localized drug delivery is to restore tissue integrity. Wound healing physiology must be understood to achieve this goal. Electrospun cellulose nanofibers can be applied to wound healing. These nanomaterials should be permeable and have morphological characteristics that favor the biological process that allows the restoration or stimulation of a damaged tissue's capacity; they can also release antimicrobial agents or bioactive molecules to the wound site, inhibit pathogen infiltration, lessen scar formation, and structurally resemble the extracellular matrix [20].

Innovative nanofiber-based wound dressings and drug delivery forms are promising because they accelerate healing and improve therapeutic efficiency. The controlled drug release offered by these systems provides continuous and exclusive delivery to the wound site. The porous nature of electrospun nanofibers and their large surface area prevent tissue dehydration, maintaining moisture at the wound site [21, 50]. Therefore, it is viable to develop self-bioactive nanofibers or nanofibers containing active pharmaceutics or natural products, e.g., plant extracts, nanoparticles, and antimicrobial peptides with potential wound healing [20, 42].

In the pharmaceutical industry, cellulose is an excipient because of its excellent compaction properties, which are essential for tableting. The application of nanocellulose to improve tablet fragmentation rates or extend drug release is currently being studied in formulations. Drug molecules adhere to nanocellulose crystals, and their slow release is determined by external stimuli, such as those imposed by human physiological conditions. The nanofiber must be compatible with the degradation of the fiber system for high release efficiency and percentage to be achieved [4].

5.2 *Tissue Engineering*

Bone is a rigid structural organ responsible for locomotion, mineral storage, soft tissue protection, and creating a microenvironment for bone marrow. Bone has the inherent ability to regenerate from minor failings. However, significant damage requires invasive interventions to restore bone losses resulting from trauma, accidents, surgeries, congenital malformations, and tumor extraction. Innovative nanomaterials have been developed with cellulose nanocrystals and nanofibers that promote the healing of bone, soft tissue, connective tissue, cartilage, ligaments, cardiac, vascular, and neuronal tissue [30].

Tissue engineering is a biomedical science discipline aiming to develop functional biomaterials to repair or regenerate damaged tissues or organs. Cells are implanted in an adequate environment and transformed into new tissues or organ parts that maintain the same structure and functional properties of the tissues to be regenerated [39]. This task combines scaffolds, cells, and biologically active molecules to create functional tissues. The materials, either synthetic, natural, or a mixture of both, must be biocompatible, biodegradable, and with active inorganic components. The biomaterials must be like the natural extracellular matrix so that cell transport, adhesion, growth, and supply of the required biomacromolecules can regenerate new tissues [4].

Various types of synthetic polymers, such as polycaprolactone (PCL), polyhydroxybutyrate (PHB), and polyvinyl alcohol (PVA), are used in bone tissue engineering because of their outstanding tunable mechanical properties and in vivo stability. However, they have disadvantages, such as insufficient bioactivity and cell adhesion for bone tissue engineering. On the other hand, hydrocolloids like chitosan, alginate, and starch are biodegradable and biocompatible but fragile. Collagen and gelatin are valuable proteins offering advantages of hemostatic qualities, low immunogenicity, biocompatibility, and enhanced cellular responses, but with disadvantages including poor mechanical strength, enzymatic breakdown, and stability under physiological conditions [20, 30].

Different cellulose types and derivatives can be chemically or mechanically modified to improve their specific mechanical properties, non-immunogenicity, non-toxicity, biodegradability, and availability of additional sources. For this reason, they are the most common materials to manufacture bone substitutes. In recent years, cellulose nanocrystals (CNCs) and CNFs have been studied to develop scaffolds that can function and regenerate like an original tissue or organ, especially in bone biomineralization, and to improve mechanical reinforcement properties [21, 30].

CNFs exhibit promising properties such as biocompatibility, biodegradability, porosity, improved mechanical properties, and crystallinity. This type of nanocellulose can produce new soft tissues because its chemical groups can be functionalized and accommodate various molecules or cells. The high-water absorption capacity conserves the injured skin, moisturizes it, and simultaneously absorbs wound exudate. CNFs are semi-crystalline units that resemble the nanoscale structure of the native extracellular medium, and hence, they can be used as an appropriate scaffold for skin cells to grow [5].

Different polymeric matrices such as hydrogels, foams or sponges, membranes, and electrospun nanofibers have been used to culture biological species. Electrospun nanofibers are cytocompatible, with controlled degradation and promising reinforcing materials in polylactic acid (PLA) mixtures. In vitro tests on a medium of CNFs and collagen have also shown biocompatibility and positive response of human endothelial and ligament cells regarding adhesion and proliferation in ligament repair. CNF-based aerogels have high porosity and are suitable for good gas permeability and adequate exchange of metabolic requirements. Consequently, CNFs are very useful in improving cell growth and proliferation. Aerogel microspheres filled with CNFs can promote fibroblast growth [1]. A suitable microenvironment for tissue regeneration can be created by incorporating CNFs in PLA and polybutylene succinate (PBS) structures. CNFs showed an excellent contribution to tissue regeneration and improved the proliferation of human fibroblast cells [3].

5.3 *Wound Dressing and Healing*

A wound covering is a sterile substance applied directly to an injury, providing a temporary physical barrier of protection, moisturization, infection prevention, and gas exchange facilitation. The production of wound dressings has increased due to the growing prevalence of chronic wound-causing diseases. These dressings accelerate healing, primarily restoring normal tissue regeneration and advancing healing progression [22]. Conventional wound dressings, such as gauze and cotton wool, primarily prevent contamination but lack bioactive components, and therefore, they are less effective for treating complex conditions like chronic, infected, or disease-related wounds [72]. Consequently, it is imperative to develop dressings that regulate moisture levels and facilitate an optimal environment for tissue regeneration. They should also be biocompatible, biodegradable, non-toxic, and hypoallergenic. Parallelly, they should facilitate gas exchange, tissue granulation, and re-epithelialization.

Additionally, they can be enriched with active components like antibiotics, pain relievers, or natural extracts possessing anti-inflammatory, epithelializing, antioxidant, and antimicrobial effects. Modern dressings include hydrocolloids, hydrogels, foams, and films, providing enhanced biocompatibility, analgesic effects, effective moisture maintenance, and degradability [74]. These modifications have shifted the research interest to composite materials reinforced with natural fibers possessing outstanding mechanical properties and low density, making these materials well-suited for technical applications.

The use of cellulose-based plant fibers as reinforcement is becoming increasingly widespread across various sectors and has long been employed to enhance the strength of plastics [68]. Cellulose-derived wound dressings have adequate applicability to improve wound healing and reduce patient discomfort because of substituting hydroxyl groups in the cellulose molecule with alkyl groups. Other appealing attributes include biocompatibility, biodegradability, cost-effectiveness, sufficient absorption of wound exudates, thermally induced gelation capacity, and antibacterial activity. Cellulose is the primary material for developing wound dressings. Cellulose nanofibers, or microfibrillated cellulose, nanofibrils, or microfibrils, are derived from cellulose fibers and typically exhibit sizes varying from 1 to 100 nm. The process of CNFs extraction is of considerable interest because it contributes to reducing agricultural waste by converting CNFs into value-added composite material [43]. Cellulose nanofibers consist of long, smooth chains formed via cross-linking of cellulose molecules, thus imparting flexibility and enabling fibers to intertwine with other cellulose chains [59]. Cellulose nanofibers exhibit mechanical properties similar to or superior to those of advanced synthetic fibers, such as ultra-high modulus polyethylene [75].

Consequently, cellulose can replace glass or carbon fibers due to its high mechanical strength, elasticity modulus, and tensile resistance. This replacement would offer an alternative to preparing fiber-reinforced composite materials. An additional advantage is that a low-cost biodegradable material would be available [11].

Cellulose nanofibers have attracted significant interest in research for developing wound dressings, as they enhance the mechanical strength of the materials to which they are incorporated. For this reason, they are considered promising for biomedical applications. For example, a study explored using silver sulfadiazine (SSD) combined with cellulose acetate (CA) to create wound dressings with enhanced antibacterial properties. SSD, a well-known topical antibacterial agent, was embedded in nanofibers for the first time. The resulting electrospun nanofibers showed uniform morphology and an average 292 $\pm$ 2 nm diameter. This compound with nanofibers exhibited significant antimicrobial effectiveness against the *Bacillus subtilis* and the *Escherichia coli* with the capacity for repeated use [36].

Yao et al. [73] used tetracycline hydrochloride (TCH) in combination with peanut protein (PPI) and poly-L-lactic acid (PLLA) to obtain membranes made of nanofibers. The aim was to develop potential wound dressings. Nanofibers of PPI-PLLA containing 5, 10, and 15% TCH were synthesized and analyzed using electron microscopy, FTIR spectroscopy, thermal analysis, and X-ray diffraction. These materials' surface wettability, mechanical characteristics, and drug release efficacy were examined. Their antimicrobial effectiveness against *Staphylococcus aureus* was assessed using a mouse wound model to evaluate healing outcomes. Increases in TCH concentration reduced fiber diameter and enhanced mechanical strength and wettability. Initially, TCH-PPI-PLLA nanofibers rapidly release TCH following Fickian diffusion. Hemolysis rates were below 5%, and TCH retained its anticoagulant activity with low toxicity. The study demonstrated the optimal antibacterial activity of the nanofibers against antibiotic-resistant bacteria, which was applied in wound healing experiments conducted on mice. These results indicate that nanofiber membranes have a strong potential as innovative materials for advanced wound dressing applications [67].

In 2020, Bie et al. [11] introduced a dual-functional wound dressing specific for diabetic patients. This innovative material was fabricated using a gentamicin-cellulose acetate (CA) composite through electrospinning. The nanofibers were further enhanced by incorporating silver nanoparticles (AgNPs) onto their surface via a straightforward wetting method. Gentamicin (Ge) addressed inflammation and infection, while the AgNPs provided robust antibacterial activity. Comprehensive characterization of the CA-Ge-Ag nanofibers was conducted, focusing on their morphology, chemical interactions, release dynamics, structural attributes, and antibacterial performance. The results showed that all CA-Ge-Ag nanofiber samples exhibited antibacterial properties and a significant drug release profile.

Carboxymethylcellulose-polyvinyl alcohol nanofibers were developed utilizing citric acid-based quantum dots because of their ability to be innovative and biocompatible cross-linking agents. Colistin was used as an antibacterial agent during electrospinning to enhance antimicrobial properties. The nanofibers (NFs) showed over 80% cytocompatibility with HFF-1 cells and demonstrated strong antibacterial effects against several strains, including Gram-positive Staphylococcus aureus, with inhibition zones of 0.8–1.5 cm. These results highlight their potential as effective wound healing scaffolds [45].

Tripathi et al. [67] employed a sodium alginate-chitosan (AC) matrix to incorporate oxidized cellulose nanofibers, enhancing the physicochemical properties of TOCNFs. They modified the composite by introducing a cell-free extracellular medium derived from skin. A distinctive AC-TOCNFs aerogel was successfully developed and confirmed to be non-toxic using mouse fibroblast L929 cells. In vitro hemolysis tests demonstrated the aerogel's significant platelet adhesion, fibrin network formation capabilities, and a rapid clotting time under 60 s. In vivo skin regeneration studies comparing AC-TOCNFs1-E0 and AC-TOCNFs1-E10 groups showed that AC-TOCNFs1-E10 significantly enhances wound recovery, characterized by improved re-epithelialization, enhanced collagen accumulation, and effective extracellular medium remodeling. These results indicate that AC-TOCNFs1-E10 has significant potential as an aerogel for regenerating skin defects due to its superior wound healing performance [6].

A pH- and near-infrared-sensitive dressing indicator was developed from nanofibers. This material is well-suited for photothermal therapies and can be used in wound healing caused by tumors and infections. The interlocking three-dimensional network structure imparts a high drug-loading capacity to the dressing, with 400 mg/g for doxorubicin and 25 mg/g for indocyanine green. The bionic switches enable controlled drug release directly at the injury location. The sensitivity indicator dressing exhibited outstanding antimicrobial properties against bacteria commonly associated with wounds. Its shape adaptability allows it to effectively treat irregular tumor wounds and drug-resistant infections through synergistic photothermal, photodynamic, and chemotherapy therapies [47].

Several bioprinting ink formulations containing alginate-cellulose nanofibers were developed for 3D printing applications, with their printing efficiency evaluated through rheological analysis. The freeze-dried scaffolds prepared with increasing CNFs concentration show reduced porosity, higher compressive modulus, and lower swelling capacity. Furthermore, curcumin was added to the selected bioinks to evaluate their potential as drug delivery systems. The lyophilized 3D-printed patches show that scaffold disintegration-controlled curcumin release, modulated by CNFs concentration. Increasing CNFs concentration slows down drug release, allowing delivery systems to be customized to patient needs. Thus, these bioinks are suitable for creating customized drug delivery systems using 3D printing [52].

Cellulose-derived nanofibers have been applied in devices that deliver therapeutic amounts of drugs in active molecules to the wound site only when the dressing is infected has been considerably studied. This system typically operates around 37 °C, and the wound's pH simultaneously triggers drug release and nanofibers degradation. The degradation rate of a nanofiber scaffold is crucial to achieve proper healing. The goal is for the cellulose to degrade simultaneously with the healing process [17, 73].

Materials such as cellulose nanofibers are drug carriers due to their excellent biocompatibility and biodegradability. Captopril, a vasodilation promotor; tetracycline hydrochloride, an antibiotic; curcumin with anti-infective, antioxidant, anti-inflammatory, and antimicrobial properties; titanium dioxide and activated carbon nanoparticles are among the substances used in wound dressing and healing [6].

CNFs are a novel material with extensive uses in wound healing due to properties like surface area, biological compatibility, mechanical strength, and moisture retention. Depending on their form—suspensions, powders, films, hydrogels, or aerogels—CNFs contribute to developing wound dressing by enhancing drug delivery, providing antibacterial effects, and enabling biosensing capabilities. For example, electrospun CNFs-based dressings integrated with antibacterial agents like silver sulfadiazine or gentamicin have demonstrated significant efficacy against bacteria. Among the most complex lesions to treat are drug-resistant and tumor-associated infections, but functionalized nanofibers allow controlled drug delivery and photothermal treatment. Additionally, CNF-based bioinks infused with curcumin facilitate customized drug delivery via 3D printing, enabling tailored release profiles to meet specific patient requirements. Hydrogels and aerogels incorporating TEMPO-oxidized CNFs with alginate-chitosan media or extracellular skin matrices enhance hemostasis, support collagen deposition, and promote tissue regeneration. These diverse applications position CNFs as groundbreaking material in advanced wound care, capable of accelerating healing and addressing the challenges of chronic and complex wounds.

6 Applications of Cellulose Nanofibers and Metal Oxide Nanoparticles

Nanocellulose, derived from plants or bacteria, is a sustainable material with remarkable properties owing to its durability, lightweight nature, biocompatibility, and biodegradability. It finds applications in multiple areas, including biosensors, energy storage, and water treatment, with notable advances in biomedical technologies. However, its lack of intrinsic antibacterial, magnetic, or electrical properties has led to the inclusion of metal oxide nanoparticles (MONPs), like ZnO, TiO_2, CuO, MgO, and Fe_3O_4, to enhance its functionality. The resultant hybrid materials exhibit enhanced antibacterial, magnetic, and sensing properties, contributing to drug delivery, wound healing, and medical imaging applications. These nanoparticles are particularly effective in antimicrobial applications, as they release reactive oxygen species to get rid of microorganisms and reduce the risk of microbial resistance often associated with antibiotics.

Iron oxide nanoparticles with superparamagnetic properties have been used in magnetic resonance imaging, sustained drug release, and biological sensing. Nanocellulose acts as a stabilizing matrix for nanoparticles, enhancing dispersion and preventing aggregation. These composites can be created by incorporating pre-synthesized (*ex-situ*) or synthesizing iron oxide nanoparticles directly within the nanocellulose matrix (in situ). The synergy between nanocellulose and iron oxide nanoparticles positions these hybrids as transformative materials for advanced healthcare applications, offering improved efficacy, versatility, and safety [27].

Nanotechnology has gained prominence due to its application in various products and its implementation in fields such as electronics, energy, and healthcare, where metallic nanoparticles (NPs) are valued for their antibacterial properties and ability to optimize structures' characteristics, tensile strength and the regulated delivery of bioactive substances. Conventional nanoparticle synthesis methods, such as physical and chemical methods, are costly and potentially toxic.

Therefore, ecological alternatives are preferred, which use natural sources with a high antioxidant content and can act as reducing agents, which can be found in fruits, barks, leaves, or plant roots. Green synthesis employs biomass sources, such as plant extracts, roots, or fungi, as reducing and coating agents, which reduces environmental impact, improves antimicrobial properties, and optimizes nanoparticle dispersion. This eco-friendly approach is at a lower cost and achieves synergistic effects through the participation of biomolecules. The synthesis of nanoparticles is complex, and this complexity increases due to key factors such as acidity or basicity, the number of reagents, and the time during which the reaction takes place, which influence the size of the nanoparticles, shape, and stability, which favor further research to address challenges such as nanoparticle aggregation, morphology control, and purification. Green-synthesized nanoparticles have several biomedical applications, including regenerative medicine, cancer treatment, drug release, and bioimaging. Plant extracts offer advantages due to their faster reduction of metal ions and higher stability than other organisms, making them ideal for scalable and sustainable nanoparticle production. Nanoparticles obtained by green synthesis show enhanced antibacterial activity compared to those obtained by physical or chemical methods due to the encapsulation of bioactive compounds [49]. The widespread use of metallic nanoparticles as essential components in various techniques and advances in nanotechnology have opened new possibilities in medicine, particularly in diagnosing and treating emerging diseases. The widespread use of metallic nanoparticles in biomedical applications has prompted researchers to explore greener synthesis methods that are simpler, safer, faster, and cost-effective than conventional approaches. Plant-derived cellulose is increasingly favored for synthesizing and stabilizing various nanoparticles of oxides such as zinc, magnetic, silver, manganese, copper, titanium, and gold. Among natural resources, cellulose is an effective matrix for forming and stabilizing various metallic nanoparticles [76].

Haider et al. [27] synthesized copper oxide nanoparticles embedded in nanocellulose by using alkaline lignin as a reducing agent. Characterization FTIR-ATR spectra revealed deacetalization during treatment with alkaline lignin. The cellulose nanofibers with copper nanoparticles were evaluated for permeability, maintaining values between 2100 and 1900 g/m^2, an appropriate range for air and moisture transport in wounds. Additionally, copper release remained stable at 80% daily for five days, demonstrating antibacterial properties. Based on these results, this material is proposed as an ideal candidate for dressings designed to accelerate wound healing [27].

Silver and copper nanoparticles incorporated into cellulose scaffolds demonstrate antimicrobial properties. The study highlights the significant advantages of these scaffolds as promotors of uniform nanoparticle distribution and agents enabling the regulated release of silver or copper at the contact surfaces. These characteristics make the material highly suitable for therapeutic agent delivery, regenerative medicine frameworks, and wound dressings. Due to nanometric size, silver nanoparticles have a larger surface area, which enhances their antibacterial activity compared to their pure metallic form, as they exhibit greater capacity to infiltrate the intricate cell wall and ensure controlled release dynamics [49]. Composite films incorporating commercial ZnO nanoparticles into cellulose nanofiber suspensions were produced. The antibacterial action against *Staphylococcus aureus* was evaluated, yielding favorable inhibition, which suggests the films' potential for specific biomedical applications. Rheological tests demonstrated that the nanoparticles did not affect the reduction in thinning caused by the shearing of the cellulose nanofibers. Thermogravimetric analysis revealed that the nanoparticles did not impact the thermal characteristics of the composite. The mechanical strength and strain rate decrease with the increasing ZnO concentration. The films obtained showed enhanced antimicrobial properties against bacteria commonly found on the skin, such as Gram-positive bacteria, compared to pure nanofibers, suggesting their potential as an alternative for specific biomedical applications [76].

Nanometric cellulose exhibits various morphologies, including cellulose nanocrystals, cellulose nanowhiskers, nanocrystalline cellulose, and cellulose nanofibers. These forms possess superior mechanical strength and a larger surface area than bulk cellulose. Numerous polar hydroxyl groups (-OH) allow cellulose to be functionalized, creating magnetic nanocellulose. When cellulose nanofibers contain both amorphous and crystalline fibrils, they display enhanced flexibility and uniformity close to that of a gel. This combination imparts unique properties, positioning it as a potential material for various applications, particularly in gene delivery, regenerative medicine, protein separation and purification, drug release, wound healing, and cancer treatment [37]. The information presented in Table 1 shows different applications of composite materials with cellulose fiber and metallic nanoparticles, including silver, cesium oxide, titanium oxide, and ferrite. These compounds demonstrate synergistic functionality with cellulose fibers, mainly enhancing antibacterial properties. These material combinations are highly versatile for biomedical applications, as shown in the Table 1, where various authors have reported their use in this field.

The integration of metal oxide nanoparticles with cellulose nanofibers has shown high promise in various biomedical applications thanks to the combination of structural integrity, functionality, and biocompatibility. Some nanoparticles, such as silver nanoparticles, have proven effective antibacterial activity. Due to their compatibility with biological systems, magnetite nanoparticles have been successfully applied in biomedicine. Incorporating Fe_3O_4 nanoparticles into cellulose nanofibrils enhances the dressing's breaking resistance and ease of use while providing a magnetic response. An iron oxide-based nanohybrid system has been explored as a drug-carrying platform for ampicillin, allowing controlled release of the antibacterial drug and improving efficacy in treating infected wounds.

Furthermore, Fe_3O_4 nanoparticles have been used in antigen–antibody interaction models, where TEMPO-oxidized cellulose nanofibers immobilize antibodies, facilitating magnetic capture and retrieval of antigens, which may be particularly useful in diagnostic applications. On the other hand, CeO_2 nanoparticles, with their antioxidant capacity, enhance tissue regeneration by mitigating the role of reactive oxygen species, which are involved in the inflammatory process during tissue repair.

Table 1 Utility of cellulose fiber with nanoparticles and various fields of application

Metal nanoparticle	Application	Components	Properties	Reference
Fe_3O_4	Wound dressing	Composited with cellulose nanofibrils poly (vinyl alcohol). Carbon quantum dots combined with Fe_3O_4 were presented as an innovative antibacterial agent	Antibacterial properties	[32]
Au	Healthcare and biomedical applications	CNFs impregnated with gold nanoparticles (AuNPs) show potential for antibacterial and industrial uses	Antibacterial properties	[60]
Ag	Tissue engineering	Cellulose nanofibers combined with nano-hydroxyapatite as nucleating agents and surface-grafted silver nanoparticles form scaffolds with antibacterial properties	Antibacterial properties	[61]
CeO_2	Wound dressing	Chitosan-containing nanoparticles encapsulated CeO_2NPs, integrated on polycaprolactone/ cellulose acetate-based nanofiber	Antibacterial and wound healing effects	[33]
FeO	Drug delivery	A carrier platform for ampicillin was developed based on a hybrid compound of cellulose nanofibers with iron nanoparticles	Controlled delivery system for an antibacterial drug	[14]

(continued)

Table 1 (continued)

Metal nanoparticle	Application	Components	Properties	Reference
Fe_3O_4	Antigen–antibody interaction model	(TEMPO)-oxidized cellulose nanofiber particles and Fe_3O_4 NPs function as immobilizers of the antibody onto the particles to capture the antigens present in the solution	Magnetic recovery of the particles	[55]
TiO_2	Wound dressing and healing	Cellulose acetate nanofibers incorporated with graphene oxide and TiO_2 were obtained by adding curcumin and released sustainably, making them suitable for wound dressing applications	Curcumin delivery	[51]

This dual functionality, antioxidant and antibacterial, makes cellulose nanofibers loaded with CeO_2 nanoparticles an option for treating chronic wounds. Incorporating metal oxide nanoparticles into cellulose-based materials provides benefits, including the ability to customize their properties through controlled nanoparticle synthesis, creating new opportunities for developing multifunctional biomaterials capable of addressing complex medical challenges.

The use of magnetic nanoparticles in various biomedical applications has been approved by the US Food and Drug Administration [15, 18].

Iron oxide nanostructures are materials with significant potential in the biomedical field, as they can positively influence cells by controlling the dosage and selectively targeting tumor cells. They can also deliver drugs, enabling the development of innovative medical treatments. These nanoparticles activate therapeutic molecules, including RNA and proteins, to promote cell repair, demonstrating efficacy in the in vitro and in vivo treatment of various diseases. This action opens new possibilities and innovations in biomedical applications [73].

Silver, copper, and zinc nanoparticles exhibit antimicrobial and antibacterial activity against various microorganisms. Antibacterial mechanism occurs because nanoparticles, due to their size, penetrate and damage the microbial cell membrane, inhibiting DNA replication. This effect, combined with other internal biological processes induced by nanoparticles, ultimately leads to cell death. In cancer treatment, nanoparticles activate the Caspase 3 pathway, triggering various reactions, including generating reactive oxygen species, ultimately leading to cell collapse and death [52].

Nanoparticles in medicine represent an expanding field of interest. However, the in vivo application remains limited due to a significant lack of research. Enhanced collaboration between physicians, biologists, and materials scientists is crucial for gaining a deeper understanding of cancer biology, designing intelligent nanoparticles, and determining the most effective path to clinical application. Such interdisciplinary efforts could develop advanced nanoparticles with superior precision in targeting cancer cells while sparing healthy ones. The favorable properties of ZnO nanoparticles, including their inherent selectivity and cytotoxicity toward cancer cells, hold immense potential as a cornerstone for next-generation cancer treatment [13, 62].

Vidakis et al. [70] investigated a photosensitive biomedical resin incorporating nanoparticles, which was used to prepare multifunctional 3D objects through photopolymerization. Cellulose nanofibers were employed as mechanical reinforcers, while copper nanoparticles, copper oxide nanoparticles, and a commercial antimicrobial powder served as antimicrobial agents. The particles were kept at 1/100 g to evaluate synergistic effects. They were characterized by improving chemical, thermal, and mechanical properties. This study paves the way for new multifunctional materials for biomedical applications by photopolymerization 3D printing in medical applications requiring enhanced mechanical performance with antibacterial properties [70].

Further structure modification of cellulose nanofibers with organic or inorganic components enables the customization of a wide range of functionalities. Figure 2 shows the process applied by the authors cited in this section, where cellulose nanofibers are functionalized with metallic nanoparticles for biomedical applications. The nanoparticles were synthesized using green methods, which are considered more effective, biocompatible, and less toxic than chemical and physical approaches. This condition is due to biogenic reducing and coating agents that create a biologically compatible surface around the nanoparticles. This application enhances properties that contribute to developing advanced structures obtained through various manufacturing processes. These advancements will further expand the applications of cellulose nanofibers combined with metallic nanoparticles, particularly in biomedical uses such as magnetic nanoparticles for cancer therapy, regenerative tissue repair, drug delivery, medical biosensors, and wound dressings.

Significant advances have been made in utilizing metallic nanoparticles for biomedical applications. These compounds have different modifications and specific characteristics, enabling their application in imaging and cancer therapies, considering their inherent properties, including plasmonic surface interactions, optical resonance, physicochemical characteristics, and electronic attributes. These advances will enable the development of cellulose nanofiber applications integrated with metallic nanoparticles for various promising biomedical applications.

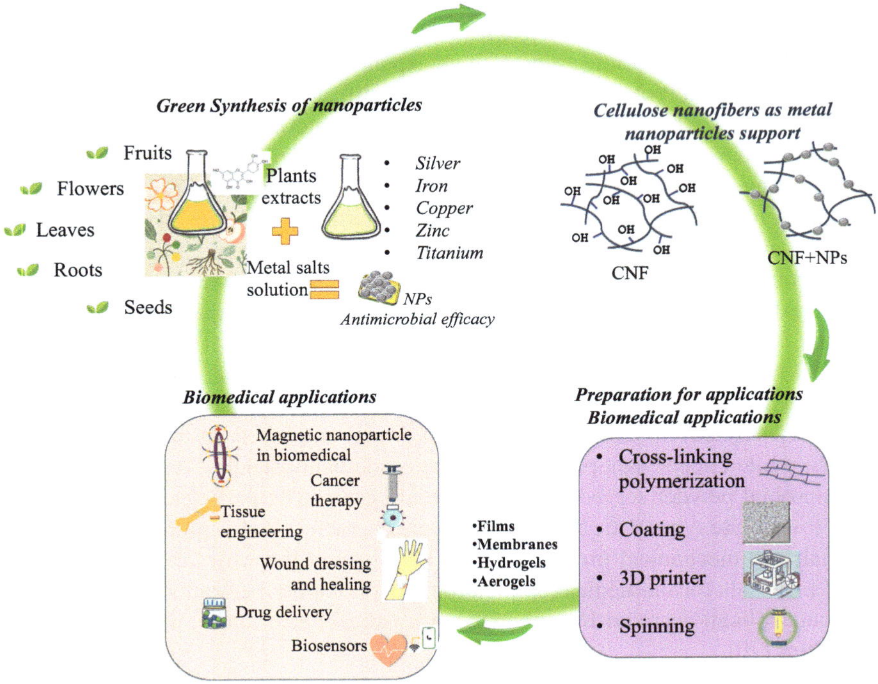

Fig. 2 Schematic representation of the green synthesis of metal nanoparticles, cellulose nanofiber functionalization, preparation processes, and their biomedical applications

7 Perspectives

Cellulose nanofiber, bacterial cellulose, and nanocrystals can act as polymeric support matrices to improve the stabilization and dispersion of metal nanoparticles. Due to their distinctive optical, magnetic, and electronic properties, the composites are innovative for biomedical applications. Metal oxide nanoparticle-nanofiber hybrids are preferred over antibiotics because they can avoid microbial resistance. Nanoparticles obtained by green synthesis show improved antibacterial activity compared to those obtained by physical or chemical methods due to the coating of organic compounds such as polysaccharides, quinones, amides, ketones, and aldehydes, which have significant antibacterial effects.

The development of cellulose-based micro- and nanofibrous materials holds great promise for innovative wound dressings due to their similarity to the extracellular matrix, structural versatility, and diverse functionalities. Innovations include stimuli-responsive, shape-memory, and conductive fibers, with designs ranging from two-dimensional fibrous membranes to emerging three-dimensional fibrous hydrogels and sponges. Applications span drug delivery, biomechanical and electrical stimulation, and wound treatment. Three-dimensional printing of materials containing silver nanoparticles with the best antibacterial properties will help find advanced wound healing solutions. Future wound dressings are anticipated to integrate intelligent drug delivery systems capable of reacting to physiological fluctuations and external triggers and providing genetic or cellular material, thereby improving cellular compatibility and adhesion.

Cellulose nanofibers have a broad spectrum of biomedical applications; however, they can acquire new properties when modified to form a support material for metal nanoparticles. Metal oxide nanoparticles (NPs) tend to aggregate in aqueous media or polymer matrices because of their high surface energy. Hybrid materials of cellulose nanofibers and metal oxides, including superparamagnetic iron oxide nanoparticles smaller than 15 nm, are being studied for their compatibility with materials designed for drug delivery controlled by their magnetic properties. These nanoparticles possess antibacterial, magnetic, and adsorption properties, contributing to rapid wound healing and acting as a contrast agent to differentiate tissues in magnetic resonance imaging and controlled drug release technologies, bioseparation, and microbial resistance.

In MRI applications, nanocellulose and cellulose nanofibers offer an excellent biocompatible matrix for superparamagnetic iron oxide nanoparticles.

Cellulose nanofibers are being explored for advanced biomedical therapies, focusing on maximizing their clinical effectiveness. Ongoing studies aim to enhance their adsorption capacity, enabling their application to deliver therapeutic substances.

Using cellulose nanofibers from diverse natural origins is increasingly being investigated for various applications. However, the challenge lies in improving their production while optimizing extraction methods and promoting scalable alternatives that must be addressed in an interdisciplinary manner. In the biomedical sector, applications are explored where long-term functionalization constitutes a significant challenge.

Nanocellulose, which is used as a scaffold, base, or transporter, could be used in advanced therapeutic approaches to investigate bone disease treatments. Its high mechanical strength and biocompatibility make nanocellulose attractive as a scaffold for bone regeneration and the development of biomaterials. Drugs and nanoparticles can be immobilized and transported in a nanocellulose base, and this preparation is used for the regulated delivery of innovative therapeutic agents and other nanostructured materials. Administration of pharmaceutical agents triggers an immune response and undesirable elimination from the body before their beneficial effect can be expressed. Encapsulation of the immobilized drugs to protect them from the immune reaction and their transference to a localized site is another desirable function of a nanomaterial. Nanocellulose could satisfy this function in bone disease treatment, but more research is still required.

8 Conclusions

Cellulose nanofibers are biocompatible and durable, ideal for advanced wound dressings. Their integration with antibacterial agents and drugs enhances tissue regeneration and combats infections. Electrospun cellulose nanofibers have excellent uses in regenerative medicine and pharmaceutical delivery. In dressings, cellulose nanofibers enable controlled drug release, photothermal therapy, and 3D-printed drug delivery systems, crucial in treating chronic and complex wounds. Nanocomposites of cellulose and metal oxides excel in biomedicine, facilitating drug delivery, wound healing, bioimaging, and targeted therapies. Green synthesis of nanoparticles enhances biocompatibility while emerging technologies like 3D printing drive advancements in biomedical applications.

References

1. Abdul Khalil HPS, Adnan AS, Yahya EB, Olaiya NG, Safrida S, Hossain MS, Balakrishnan V, Gopakumar CKA, Oyekanmi AA, Pasquini D (2020) A review on plant cellulose nanofibre-based aerogels for biomedical applications. Polymers 12(8):1759. https://doi.org/10.3390/polym12081759
2. Abolore RS, Jaiswal S, Jaiswal AK (2024) Green and sustainable pretreatment methods for cellulose extraction from lignocellulosic biomass and its applications: a review. Carbohydr Polym Technol Appl 7:100396. https://doi.org/10.1016/j.carpta.2023.100396
3. Abudula T, Saeed U, Memic A, Gauthaman K, Hussain MA, Al-Turaif H (2019) Electrospun cellulose nanofibril reinforced PLA/PBS composite scaffold for vascular tissue engineering. J Polym Res 26:1–15. https://doi.org/10.1007/s10965-022-03234-9
4. Ahmed S, Khan RA, Rashid TU (2025) Cellulose nanocrystal based electrospun nanofiber for biomedical applications—a review. Carbohydr Polym 348:122838. https://doi.org/10.1016/j.carbpol.2024.122838
5. Amaral HR, Wilson JA, do Amaral RJ, Pasçu I, de Oliveira FC, Kearney CJ, Freitas JCC, Heise A (2021) Synthesis of bilayer films from regenerated cellulose nanofibers and poly (globalide) for skin tissue engineering applications. Carbohydr Polym 252:117201. https://doi.org/10.1016/j.carbpol.2020.117201
6. Ambekar RS, Kandasubramanian B (2019) Advancements in nanofibers for wound dressing: a review. Eur Polym J 117:304–336. https://doi.org/10.1016/j.eurpolymj.2019.05.020
7. Anusiya G, Jaiganesh R (2022) A review on fabrication methods of nanofibers and a special focus on application of cellulose nanofibers. Carbohydr Polym Technol Appl 4:100262. https://doi.org/10.1016/j.carpta.2022.100262
8. Ara ZA, Wang L, Padhye R, Wang X (2021) Electrospinning of lignocellulose nanofibers from dissolved jute. J Nat Fibers 19:6469–6481. https://doi.org/10.1080/15440478.2021.1921662
9. Arfelis S, Aguado RJ, Civancik D, Fullana-i-Palmer P, Pèlach MÀ, Tarrés Q, Delgado-Aguilar M (2023) Sustainability of cellulose micro-/nanofibers: a comparative life cycle assessment of pathway technologies. Sci Total Environ 874:162482. https://doi.org/10.1016/j.scitotenv.2023.162482
10. Bangar SP, Esua OJ, Nickhil C, Whiteside WS (2023) Microcrystalline cellulose for active food packaging applications: a review. Food Packag Shelf Life 36:101048. https://doi.org/10.1016/j.fpsl.2023.101048
11. Bie X, Khan MQ, Ullah A, Ullah S, Kharaghani D, Phan DN, Tamada Y, Kim IS (2020) Fabrication and characterization of wound dressings containing gentamicin/silver for wounds in diabetes mellitus patients. Mater Res Express 7. https://doi.org/10.1088/2053-1591/ab8337

12. Biliuta G, Zhang S, Gradinaru LM, Bercea M, Baron RI, Bejan D, Coseri S (2024) Advanced pullulan nanofibers reinforced by cellulose fibrils as drug carriers for salicylic acid. Int J Biol Macromol 283:137624. https://doi.org/10.1016/j.ijbiomac.2024.137624
13. Bisht G, Rayamajhi S (2016) ZnO nanoparticles: a promising anticancer agent. Nanobiomedicine 3:1–11. https://doi.org/10.5772/63437
14. Borah N, Rather MA, Bhar B, Mandal BB, Mandalb M, Karak N (2023) A robust epoxy nanocomposite with iron oxide decorated cellulose nanofiber as a sustained drug delivery vehicle for antibacterial drugs. New J Chem 1–4. https://doi.org/10.1039/D3NJ03412G
15. Cai N, Li C, Han C, Luo X, Shen L, Xue Y, Yu F (2016) Tailoring mechanical and antibacterial properties of chitosan/gelatin nanofiber membranes with Fe_3O_4 nanoparticles for potential wound dressing application. Appl Surf Sci 369:492–500. https://doi.org/10.1016/j.apsusc.2016.02.053
16. Chavez BK, Garces-Porras K, Parada DC, Flores E (2023) Thermochemical isolation and characterization of nanofibrillated cellulose from Stipa obtusa fibers. Carbohydr Polym Tech App 6:100344. https://doi.org/10.1016/j.carpta.2023.100344
17. Chen R, Zhao C, Chen Z, Shi X, Zhu H, Bu Q, Wang L, Wang C, He H (2022) A bionic cellulose nanofiber-based nanocage wound dressing for NIR-triggered multiple synergistic therapy of tumors and infected wounds. Biomaterials 281:121330. https://doi.org/10.1016/j.biomaterials.2021.121330
18. Chen Y, Hou S (2023) Recent progress in the effect of magnetic iron oxide nanoparticles on cells and extracellular vesicles. Cell Death Discov 9:1–7. https://doi.org/10.1038/s41420-023-01490-2
19. Emenike EC, Iwuozor KO, Saliu OD, Ramontja J, Adeniyi AG (2023) Advances in the extraction, classification, modification, emerging and advanced applications of crystalline cellulose: a review. Carbohydr Polym Technol Appl 6:100337. https://doi.org/10.1016/j.carpta.2023.100337
20. Ejiohuo O, Folami SO, Edi D, Issac J (2024) Polyphenol encapsulated nanofibers in wound healing and drug delivery. EJMCR 12:100184. https://doi.org/10.1016/j.ejmcr.2024.100184
21. Ezike TC, Okpala US, Onoja UL, Nwike CP, Ezeako EC, Okpara OJ, Okoroafor CC, Eze SC, Kalu OL, Odoh EC, Nwadike UG, Ogbodo JO, Umeh BU, Ossai EC, Nwanguma BC (2023) Advances in drug delivery systems, challenges, and future directions. Heliyon 9:e17488. https://doi.org/10.1016/j.heliyon.2023.e17488
22. Farahani M, Shafiee A (2021) Wound healing: from passive to smart dressings. Adv Health Mater 10:1–32. https://doi.org/10.1002/adhm.202100477
23. Fernández-Guarino M, Hernández-Bule ML, Bacci S (2023) Cellular and molecular processes in wound healing. Biomedicines 11:2526. https://doi.org/10.3390/biomedicines11092526
24. Freville E, Pescheux-Sergienko J, Mujica RRC, Bras J (2024) Novel technologies for producing tridimensional cellulosic materials for packaging: a review. Carbohydr Polym 342:122413. https://doi.org/10.1016/j.carbpol.2024.122413
25. Gan PG, Sam ST, bin Abdullah MF, Omar MF (2020) Thermal properties of nanocellulose-reinforced composites: a review. J Appl Polym 137:48544. https://doi.org/10.1002/app.48544
26. Haider A, Haider S, Kang IK (2018) A comprehensive review summarizing the effect of electrospinning parameters and potential applications of nanofibers in biomedical and biotechnology. Arab J Chem 11(8):1165–1188. https://doi.org/10.1016/j.arabjc.2015.11.015
27. Haider MK, Ullah A, Sarwar MN, Saito Y, Sun L, Park S, Kim IS (2021) Lignin-mediated in-situ synthesis of CuO nanoparticles on cellulose nanofibers: a potential wound dressing material. Int J Biol Macromol 173:315–326. https://doi.org/10.1016/j.ijbiomac.2021.01.050
28. Hafid HS, Omar FN, Zhu J, Wakisaka M (2021) Enhanced crystallinity and thermal properties of cellulose from rice husk using acid hydrolysis treatment. Carbohydr Polym 260:117789. https://doi.org/10.1016/j.carbpol.2021.117789
29. Hongrattanavichit I, Aht-Ong D (2020) Nanofibrillation and characterization of sugarcane bagasse agro-waste using water-based steam explosion and high-pressure homogenization. J Clean Prod 277:123471. https://doi.org/10.1016/j.jclepro.2020.123471

30. Janmohammadi M, Nazemi Z, Salehi AOM, Seyfoori A, John JV, Nourbakhsh MS, Akbari M (2023) Cellulose-based composite scaffolds for bone tissue engineering and localized drug delivery. Bioact Mat 20:137–163. https://doi.org/10.1016/j.bioactmat.2022.05.018
31. Joshi NC, Joshi A, Mitra D, Gururani P, Kumar N, Joshi HK (2024) Removal of heavy metals using cellulose-based materials: a mini-review. Environ Nanotechnol Monit Manage 21:100942. https://doi.org/10.1016/j.enmm.2024.100942
32. Kamalipooya S, Fahimirad S, Abtahi H, Golmohammadi M, Satari M, Dadashpour M, Nasrabadi D (2023) Optimization of electrospun CQDs-Fe3O4-RE loaded PVA-cellulose nanofibrils via central composite design for wound dressing applications: Kinetics and in vitro release study. Int J Biol Macromol 237:3–4. https://doi.org/10.1016/j.ijbiomac.2023.124067
33. Kamalipooya S, Fahimirad S, Abtahi H, Golmohammadi M, Satari M, Dadashpour M, Nasrabadi D (2024) Diabetic wound healing function of PCL/cellulose acetate nanofiber engineered with chitosan/cerium oxide nanoparticles. Int J Pharm 653:123880. https://doi.org/10.1016/j.ijpharm.2024.123880
34. Kamel R, El-Wakil NA, Dufresne A, Elkasabgy NA (2020) Nanocellulose: from an agricultural waste to a valuable pharmaceutical ingredient. Int J Biol Macromol 163:1579–1590. https://doi.org/10.1016/j.ijbiomac.2020.07.242
35. Kassie BB, Daget TM, Tassew DF (2024) Synthesis, functionalization, and commercial application of cellulose-based nanomaterials: a review. Int J Biol Macromol 278:34990. https://doi.org/10.1016/j.ijbiomac.2024.134990
36. Kazeminava F, Javanbakht S, Nouri M, Adibkia K, Ganbarov K, Yousefi M, Ahmadi M, Gholizadeh P, Kafil HS (2022) Electrospun nanofibers based on carboxymethyl cellulose/polyvinyl alcohol as a potential antimicrobial wound dressing. Int J Biol Macromol 214:111–119. https://doi.org/10.1016/j.ijbiomac.2022.05.175
37. Kumar S, Ngasainao MR, Sharma D, Sengar M, Gahlot APS, Shukla S, Kumari P (2022) Contemporary nanocellulose-composites: a new paradigm for sensing applications. Carbohydr Polym 298:120052. https://doi.org/10.1016/j.carbpol.2022.120052
38. Leong MY, Kong YL, Harun MY, Looi CY, Wong WF (2023) Current advances of nanocellulose applications in biomedical field. Carbohydr Res 532:108899. https://doi.org/10.1016/j.carres.2023.108899
39. Liu D, Sui G, Dong Y (2021) Electrospun polymeric composite nanofibers with cellulose nanocrystals and their applications. In: Dong Y, Baji A, Ramakrishna S (eds) Electrospun polymers and composites. Woodhead Publishing, pp 147–178. https://doi.org/10.1016/B978-0-12-819611-3.00005-4
40. Lu Y, Tao P, Zhang N, Nie S (2020) Preparation and thermal stability evaluation of cellulose nanofibrils from bagasse pulp with differing hemicelluloses contents. Carbohydr Polym 245:116463. https://doi.org/10.1016/j.carbpol.2020.116463
41. Mahmud MA, Anannya FR (2021) Sugarcane bagasse—a source of cellulosic fiber for diverse applications. Heliyon 7(8):e07771. https://doi.org/10.1016/j.heliyon.2021.e07771
42. Miguel SP, Figueira DR, Sim˜oes D, Ribeiro MP, Coutinho P, Ferreira P (2018) Electrospun polymeric nanofibres as wound dressings: a review. Colloids Surf B: Biointerfaces 169: 60–71. https://doi.org/10.1016/j.colsurfb.2018.05.011
43. Mohamed MA, Abd Mutalib M, Mohd Hir ZA, Zain MFM, Mohamad AB, Jeffery Minggu L, Awang NA, W. Salleh WN (2017) An overview on cellulose-based material in tailoring biohybrid nanostructured photocatalysts for water treatment and renewable energy applications. Int J Biol Macromol 103: 1232–1256. https://doi.org/10.1016/j.ijbiomac.2017.05.181
44. Mokhena TC, John MJ (2020) Cellulose nanomaterials: new generation materials for solving global issues. Cellulose 27:1149–1194. https://doi.org/10.1007/s10570-019-02889-w
45. Olmos-Juste R, Alonso-Lerma B, Pérez-Jiménez R, Gabilondo N, Eceiza A (2021) 3D printed alginate-cellulose nanofibers based patches for local curcumin administration. Carbohydr Polym 264. https://doi.org/10.1016/j.carbpol.2021.118026
46. Ong XR, Chen AX, Li N, Yang YY, Luo HK (2023) Nanocellulose: recent advances toward biomedical applications. Small Sci 3:2200076. https://doi.org/10.1002/smsc.202200076

47. Oprea M (2020) Nanocellulose hybrids with metal oxides nanoparticles for biomedical applications. Molecules 25:4–6. https://doi.org/10.3390/molecules25184045
48. Otoni CG, Carvalho AS, Cardoso MV, Bernardinelli OD, Lorevice MV, Colnago LA, Watson L, Mattoso LH (2018) High-pressure microfluidization as a green tool for optimizing the mechanical performance of all-cellulose composites. ACS SCE 6(10):12727–12735. https://doi.org/10.1021/acssuschemeng.8b01855
49. Phan DN, Dorjjugder N, Khan MQ, Saito Y, Taguchi G, Lee H, Mukai Y, Kim IS (2019) Synthesis and attachment of silver and copper nanoparticles on cellulose nanofibers and comparative antibacterial study. Cellulose 26:6629–6640. https://doi.org/10.1007/s10570-019-02542-6
50. Prabhu ATA, Baliga V, Bhat S, Thenkondar ST, Nayak Y, Nayak UY (2023) Transforming wound management: nanomaterials and their clinical impact. Pharmaceutics 15:1560. https://doi.org/10.3390/pharmaceutics15051560
51. Prakash J, Venkataprasanna KS, Bharath G, Banat F, Niranjan R, Venkatasubbu GD (2021) In-vitro evaluation of electrospun cellulose acetate nanofiber containing Graphene oxide/TiO_2/Curcumin for wound healing application. Coll Surf A Physicochem Eng Asp 627:127166. https://doi.org/10.1016/j.colsurfa.2021.127166
52. Radulescu DM, Surdu VA, Ficai A, Ficai D, Grumezescu AM, Andronescu E (2023) Green synthesis of metal and metal oxide nanoparticles: a review of the principles and biomedical applications. Int J Mol Sci 24:1–39. https://doi.org/10.3390/ijms242015397
53. Ribeiro RSA, Pohlmann BC, Calado V, Bojorge N, Pereira N (2019) Production of nanocellulose by enzymatic hydrolysis: trends and challenges. Eng Life Sci 19(4):279–291. https://doi.org/10.1002/elsc.201800158
54. Riseh RS, Vazvani MG, Hassanisaadi M, Thakur VK (2024) Agricultural wastes: A practical and potential source for the isolation and preparation of cellulose and application in agriculture and different industries. Ind Crops Prod 208:117904. https://doi.org/10.1016/j.indcrop.2023.117904
55. Saipul Bahri NSN, Nguyen TT, Matsumoto K, Watanabe M, Morita Y, Septiani EL, Cao KLA, Hirano T, Ogi T (2024) Controlling the magnetic responsiveness of cellulose nanofiber particles embedded with iron oxide nanoparticles. ACS Appl Bio Mater 7:3227–3237. https://doi.org/10.1021/acsabm.4c00213
56. Saghazadeh S, Rinoldi C, Schot M, Kashaf SS, Sharifi F, Jalilian E, Nuutila K, Giatsidis G, Mostafalu P, Derakhshandeh H, Yue K, Swieszkowski W, Memic A, Tamayo A, Khademhosseini A (2018) Drug delivery systems and materials for wound healing applications. Adv Drug Deliv Rev 127:138–166. https://doi.org/10.1016/j.addr.2018.04.008
57. Salem DM, Ismail MM (2022) Characterization of cellulose and cellulose nanofibers isolated from various seaweed species. Egypt J Aquat Res 48(4):307–313. https://doi.org/10.1016/j.ejar.2021.11.001
58. Saragih S, Wirjosentono WB, Meliana Y (2020) Influence of cross-linking agent on the morphology, chemical, crystallinity and thermal properties of cellulose nanofiber using steam explosion. Case Stud Therm Eng 22:100740. https://doi.org/10.1016/j.csite.2020.100740
59. Sharma M, Aguado R, Murtinho D, Valente AJM, Mendes De Sousa AP, Ferreira PJT (2020) A review on cationic starch and nanocellulose as paper coating components. Int J Biol Macromol 162:578–598. https://doi.org/10.1016/j.ijbiomac.2020.06.131
60. Silva LPC, de Oliveira NA, Valotto RS, Monteiro FC, Alvarez LAC, Cesário LM, Zanardo TÉC, Nilo APMC, Schuenck RP, de Oliveira JP, Campos FV, Guimarães MCC (2016) Cellulose nanofibers infused with pomegranate gold nanoparticles display antibacterial activity. Nano-Struct Nano-Objects 36:1–23. https://doi.org/10.1016/j.nanoso.2023.101045
61. Sofi HS, Akram T, Shabir N, Vasita R, Jadhav AH, Sheikh FA (2021) Regenerated cellulose nanofibers from cellulose acetate: Incorporating hydroxyapatite (HAp) and silver (Ag) nanoparticles (NPs), as a scaffold for tissue engineering applications. Mater Sci Eng C 118:111547. https://doi.org/10.1016/j.msec.2020.111547
62. Stepankova H, Swiatkowski M, Kruszynski R, Svec P, Michalkova H, Smolikova V, Ridoskova A, Splichal Z, Michalek P, Richtera L, Kopel P, Adam V, Heger Z, Rex S (2021) The anti-proliferative activity of coordination compound-based ZnO nanoparticles as a promising agent

against triple negative breast cancer cells. Int J Nanomed 16:4431–4449. https://doi.org/10.2147/IJN.S304902
63. Surendran G, Sherje AP (2022) Cellulose nanofibers and composites: an insight into basics and biomedical applications. J Drug Deliv Sci Technol 75:03601. https://doi.org/10.1016/j.jddst.2022.103601
64. Tang B, Xiong Z, Zhou Z, Lu D, Sun Y, Ding P, Wang C (2024) Transforming agricultural straw waste into functional nanofibers via electrospinning: a sustainable approach. J Agric Food Res 18:101490. https://doi.org/10.1016/j.jafr.2024.101490
65. Teo HL, Wahab RA (2020) Towards an eco-friendly deconstruction of agro-industrial biomass and preparation of renewable cellulose nanomaterials: a review. Int J Biol Macromol 161:1414–1430. https://doi.org/10.1016/j.ijbiomac.2020.08.076
66. Tiwari R, Pathak K (2023) Local drug delivery strategies towards wound healing. Pharmaceutics 15(2):634. https://doi.org/10.3390/pharmaceutics15020634
67. Tripathi G, Park M, Lim H, Lee BT (2023) Natural TEMPO oxidized cellulose nano fiber/alginate/dSECM hybrid aerogel with improved wound healing and hemostatic ability. Int J Biol Macromol 243:125226. https://doi.org/10.1016/j.ijbiomac.2023.125226
68. Tudoroiu EE, Dinu-Pîrvu CE, Kaya MGA, Popa L, Anuţa V, Prisada RM, Ghica MV (2021) An overview of cellulose derivatives-based dressings for wound-healing management. Pharmaceuticals 14. https://doi.org/10.3390/ph14121215
69. Ventura-Cruz S, Tecante A (2019) Extraction and characterization of cellulose nanofibers from rose stems (Rosa spp.). Carbohydr Polym 220:53–59. https://doi.org/10.1016/j.carbpol.2019.05.053
70. Vidakis N, Petousis M, Michailidis N, Papadakis V, Korlos A, Mountakis N, Argyros A (2022) Multifunctional 3D-printed vat photopolymerization biomedical-grade resin reinforced with binary nano inclusions: the effect of cellulose nanofibers and antimicrobial nanoparticle agents. Polymers (Basel) 14:1–24. https://doi.org/10.3390/polym14091903
71. Vu AN, Nguyen LH, Yoshimura K, Tran TD, Van LH (2024) Cellulose nanocrystals isolated from sugarcane bagasse using the formic/peroxyformic acid process: structural, chemical, and thermal properties. Arab J Chem 17:105841. https://doi.org/10.1016/j.arabjc.2024.105841
72. Yang C, Zhu Y, Tian Z, Zhang C, Han X, Jiang S, Liu K, Duan G (2024) Preparation of nanocellulose and its applications in wound dressing: a review. Int J Biol Macromol 254:127997. https://doi.org/10.1016/j.ijbiomac.2023.127997
73. Yao F, Zheng Y, Gao Y, Du Y, Chen F (2023) Electrospinning of peanut protein isolate/poly-L-lactic acid nanofibers containing tetracycline hydrochloride for wound healing. Ind Crops Prod 194:116262. https://doi.org/10.1016/j.indcrop.2023.116262
74. Yazdanbakhsh MF, Rashidi A, Rahimi MK, Khajavi R, Shafaroodi H (2018) The effect of impregnated alpha-cellulose nanofibers with ciprofloxacin hydrochloride on staphylococcus aureus in vitro and healing process of wound in rat. Regen Eng Transl Med 4:247–256. https://doi.org/10.1007/s40883-018-0066-y
75. Zhang K, Barhoum A, Xiaoqing C, Li H (2019) Cellulose nanofibers: fabrication and surface functionalization techniques. In: Handbook of nanofibers. Springer, Cham, pp 409–449. https://doi.org/10.1007/978-3-319-53655-2_58
76. Zhang X, Ren S, He X, Dong L, Bai W, Lei T, Chen H (2019) Preparation and characterization of cellulose nanofiber/zinc oxide composite films. J Biobased Mater Bioenergy 14:203–208. https://doi.org/10.1166/jbmb.2020.1945
77. Zinge C, Kandasubramanian B (2020) Nanocellulose based biodegradable polymers. Eur Polym J 133:109758. https://doi.org/10.1016/j.eurpolymj.2020.109758

Sagnite Ventura-Cruz Ph.D. in Water Sciences from Universidad Autónoma del Estado de México (UAEMex), specializing in water quality. She was a postdoctoral fellow at the Faculty of Chemistry of the Universidad Nacional Autónoma de México (UNAM). Dr. Ventura's scientific contribution is aimed at using residual biomass and giving it added value by obtaining natural polymers of interest both scientifically and industrially for different applications. Dr. Ventura's work incorporates the philosophy of the circular economy to make the most of natural resources and raw materials considered waste in developing new biodegradable materials, mainly cellulosic ingredients.

Cristina Arely De León Condes Ph.D. in Materials Science from Universidad Autónoma del Estado de México (UAEMex), specializing in composite materials with metallic nanoparticles for environmental applications and system optimization. She has completed research stays in Spain and the USA, published articles in indexed journals, and presented at international conferences. A full-time professor with expertise in materials development, waste utilization, and optimization technologies, she serves as a technical lead on funded projects and is a Level 1 member of the Sistema Nacional de Investigadores (National System of Researchers), Mexico.

Alberto Tecante is a full professor in the Faculty of Chemistry at the Universidad Nacional Autónoma de México (UNAM). He holds a Ph.D. in Chemical Engineering from Université Laval, Quebec, QC, Canada. His expertise includes transport phenomena, the rheology of macromolecules of natural origin, and the physical chemistry of dispersed systems, and he has an interest in biopolymers in aqueous media. He is the author and co-author of numerous papers published in indexed peer-reviewed international journals, books, and book chapters. He is a member of the Editorial Board of Food Hydrocolloids and Food Hydrocolloids for Health, published by Elsevier.

Modified Cellulose-Supported Catalysts and Reagents in Three-Component Reactions for Sustainable Medicinal Chemistry

Vladimir V. Kouznetsov

Abstract Cellulose is a crucial renewable polymer widely utilized in the biomedical and pharmaceutical industries because of its natural abundance, biocompatibility, biodegradability, non-toxicity, hydrophilicity, antimicrobial properties, and ease of fabrication. Despite its potential, the application of microcrystalline cellulose and its modified derivatives as supports for catalysts and reagents in three-component condensation reactions (3C-CR) in pharmaceutical research, particularly in medicinal chemistry and drug manufacturing, is still underexplored and inadequately categorized. This chapter reviews key 3C-CR reactions, including the Betti, Biginelli, Kabachnik-Fields, Hantzsch, Passerini, Povarov, and Strecker reactions. Recognizing the significance of green chemistry in sustainable medicinal chemistry, the discussion underscores the role of cellulose-based catalysts and modified cellulose-based reagents in these reactions. Additionally, the chapter examines their structural classification, physicochemical properties, and related chemical processes in synthesizing drug-like molecules and innovative cellulose-based biomaterials.

Keywords Cellulose · Catalysis · Three-component reactions · Dialdehyde cellulose · Cellulose acetoacetate · Carboxymethylcellulose · Cellulose functionalization · Nanocellulose · Microcellulose · Biocatalytic systems

List of Abbreviations

Acet	Acetone
AGU	β-D-anhydrogluco-pyranose units
AI	1-Adamantyl isocyanide
AIBN	Azodiisobutyronitrile
AMCell-SO_3H	Amorphous milled sulfonated cellulose

V. V. Kouznetsov (✉)
Laboratorio de Química Orgánica y Biomolecular, Escuela de Química, Universidad Industrial de Santander, Bucaramanga, Colombia
e-mail: kouznet@uis.edu.co

R. Malviya and S. Sundram (eds.), *Sustainable Cellulose Materials for Biomedical Applications*, Sustainable Materials and Technology,
https://doi.org/10.1007/978-981-96-9904-9_9

ATP-FTIR	Attenuated total reflectance fourier-transform infrared
t-BAA	*tert*-Butyl acetoacetate
BCD	β-Cyclodextrin
CAA	Cellulose acetoacetate
CAA-FA	Cellulose acetoacetate-based fluorescent paper
Cell	Cellulose
CMC	Carboxymethylcellulose
CMC-MA	Methacrylated carboxymethyl cellulose
CN	Cellulose nanofibrils
CP-MAS	Cross-polarization magic angle spinning
CSA	Cellulose sulfuric acid
Cy	Cyclohexyl
3C-C	Three-component condensation
DA	Dodecyl aldehyde
DAC	Dialdehyde cellulose
DAR	Diels-Alder reaction
DBU	1,8-Diazabicyclo[5.4.0]undec-7-ene
DEGMA	Diethylene glycol methyl ether methacrylate
DHP	Dihydropyridine
DHPM	3,4-Dihydropyrimidin-2(1H)-ones/thione
DI	Deionized
DMSO	Dimethyl sulfoxide
DNA	Deoxyribonucleic acid
DO	Degree of oxidation
DS	Degree of substitution
DSC	Differential scanning calorimetry
DSSCs	Dye-sensitized solar cells
EDX	Energy-dispersive X-ray
ESBL	Extended-spectrum β-lactamases
EtBrPyr	Ethyl bromopyruvate
EtF_3AcAc	Ethyl 4,4,4-trifluoro acetoacetate
EtF_3Pyr	Ethyl 3,3,3-trifluoropyruvate
EtPyr	Ethyl pyruvate
5-FAM	5-Carboxyfluorescein
HA	1-Hexanoic acid
IL	Ionic liquid
IPA	Isopropanol
KF-3C-C	Kabachnik–Fields three-component condensation
MA	Methacrylic acid
MC	Microcellulose
MCR	Multicomponent reaction
MFC	Cellulose microfibrils
MPC	3-Mercaptopropylcellulose
MV	Methacrylated vanillin
NaAcs	Sodium salt of ascorbic acid

NaCMC	Sodium carboxymethylcellulose
NC	Nanocellulose
p-NIPAm	Poly(N-isopropyl acrylamide)
NMMC	Nano-γ-Fe_2O_3@microcellulose
NMNC	Nano-γ-Fe_2O_3@nanocellulose
NMR	Nuclear magnetic resonance
ODA	Octadecyl amine
OEGMA	Oligoethylene glycol methyl ether methacrylate
PA	4-Pentynoic acid
PAP	*p*-Anisidine phosphite
g-PMV	Graft-poly(methacrylated vanillin)
ProAm	Propargylamine
SA	Succinic anhydride
SEM-EDX	Scanning electron microscopy-energy dispersive X-ray
SFP	Succinylated filter paper
TEMPO	2,2,6,6-Tetramethylpiperidine 1-oxyl
TGA	Thermogravimetric analysis
THQ	Tetrahydroquinoline
TMSCN	Trimethylsilyl cyanide
Top2	Topoisomerase II
UV	Ultraviolet
XPS	X-ray photoelectron spectroscopy

1 Introduction: Cellulose Is an Abundant Macromolecule in the World

Global efforts to enhance sustainability drive the growing need for bio-based renewable natural resources. Plants stand out as plentiful sources of lignocellulosic biomass, which is composed of principal polymers such as cellulose (40–50 wt%), hemicellulose (20–40 wt%), and lignin (20–30 wt%). That makes plants one of the richest natural resources available on Earth. Cellulose, the main component of plants, is a linear syndiotactic homopolymer that provides essential structural elements and strength in trees or wood [15, 69]. Moreover, cellulose exhibits some valuable properties, such as excellent biocompatibility, non-toxicity, and biodegradability.

Historically, cellulose was isolated from plants by the French chemist Anselme Payen in 1839. He also gave this name. Hermann Staudinger, Nobel Prize winner in chemistry, later ascertained its polymer structure in 1920 [9]. Chemically, its trivial name is (1 → 4)-β-D-glucopyranose based on the β-D-anhydrogluco-pyranose units (AGU) and linked by β-(1–4)-glycosidic bonds. The AGU has hydroxyl groups, including the primary hydroxyl group at C6 and the secondary hydroxyl groups at C2 and C3 (Fig. 1). There are around 1500 AGU rings in the native form of this natural homopolymer.

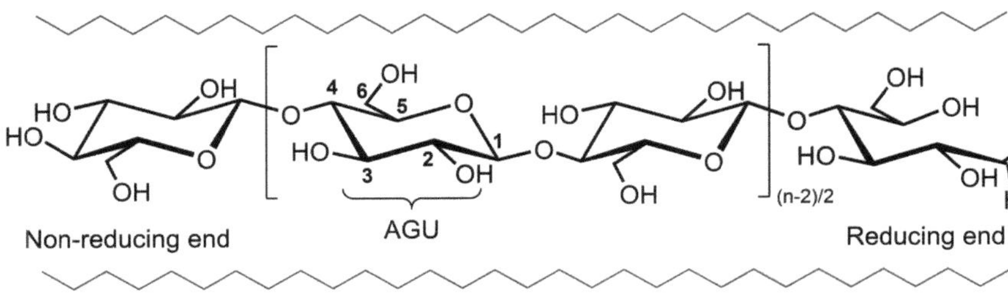

Fig. 1 Structural details of a polymeric cellulose chain include the AGU unit, chair conformation, atom numbering, glycosidic link, and reducing and non-reducing ends

Cellulose, with three hydroxyl groups per anhydroglucose unit (AGU) within its chains, serves as an exceptional podium for diverse chemical modifications to create looked-for functional groups. These hydroxyl groups, behaving as typical alcoholic hydroxyls, enable cellulose to participate in various reactions, including esterification, etherification, oxidation, silylation, and sulfonation. As a result, diverse modified celluloses can be prepared through these reactions. Thus, through additional mild post-hydrolysis reactions, aldehyde (–CHO), amino (–NH), or thiol (–SH) groups can be incorporated onto the cellulose surface. Moreover, aldehyde-functionalized cellulose can also participate in reactions with specific nucleophilic reagents. In contrast, amine-functionalized cellulose can also act as the nucleophilic reagent in reactions with electrophilic reagents.

Accordingly, cellulose and its modified forms can be used as a support material for active catalytic sites or modified to act as a catalyst. Indeed, cellulose's high surface area and hydroxyl groups make it an ideal material for anchoring metal or metal oxide nanoparticles or other active species. Thus, many metal catalysts are absorbed into the cellulose superficies [22]. Cellulose can also be impregnated with two or more metals to create a bimetallic catalyst, where the metals work synergistically to enhance catalytic activity [36]. Besides metals, non-metallic catalytic components such as transition metal oxides, solid acids, or zeolites may also be combined with cellulose to form hybrid catalysts for specific applications [70, 72, 73].

This heterogeneous support presents several gains over other types, including enhanced absorption capacity and the presence of hydroxyl groups that facilitate the anchoring of various functional groups. These catalysts also have advantages in renewability and sustainability. Cellulose is abundant, biodegradable, and comes from renewable sources, making these catalysts eco-friendlier than traditional materials.

They are also low-cost catalysts because cellulose is relatively inexpensive as a natural polymer than typical bases like alumina, silica, or activated carbon. Finally, cellulose can offer a high surface area after modification, enhancing its effectiveness as a catalyst or a catalyst support. More commonly, cellulose supports catalytic species or is part of an environmentally friendly, sustainable catalytic system, where the cellulose itself is not the active reagent but helps facilitate the reaction.

Conversely, cellulose can be used directly in organic synthesis. However, its role typically depends on its chemical modifications and the specific reaction type. While native cellulose is usually not reactive enough to participate directly in multicomponent reactions, modified cellulose can be a highly effective reagent or catalyst, depending on the specific reaction conditions and functional groups present. Usually, the organic chemistry of cellulose can produce a varied range of new, valuable products for multiple purposes.

Therefore, the present chapter's goal is to emphasize particular current progress in the chemical application of cellulose or its modified forms as cellulose-supported catalysts or chemical reagents for synthesizing new products/materials with promising properties through three-component reactions.

2 Three-Component Reactions for Synthesizing New and Multifaceted Heterocyclic Skeletons, Focusing on Applications in Medicinal Chemistry and the Pharmaceutical Industry

Three-component condensation reactions (3C-C reactions), a multicomponent reaction (MCR), are highly efficient one-pot processes where three reactants combine to form target molecules, effectively incorporating most of the starting materials. These reactions proceed in a single step and exhibit excellent atom economy [6, 7, 52, 74]. Key advantages of 3C-C reactions include their efficiency, minimal waste generation, and alignment with green chemistry principles.

A principal goal of the current research in this area is to substitute waste-producing and hazardous reagents with sustainable and environmentally friendly feedstock [6]. Consequently, sustainable cellulose-based catalysts and reagents significantly impact sustainable medicinal chemistry development.

In modern organic synthesis, several named 3C-C reactions, such as Strecker, Biginelli, Passerini, Betti, Gewald, Groebke–Blackburn–Bienayme, Mannich, Ugi, and Petasis, as well as classified newly described 3C-C reactions, have been broadly used in the preparation of numerous biologically active small heterocyclic molecules. However, a few were selected in this chapter. Table 1 captures the selected 3C-C reactions in which cellulose-based catalysts and cellulose reagents participate efficiently.

They are potent tools in medicinal chemistry for drug discovery and development. They are also time- and cost-effective green methods for identifying and optimizing lead bioactive compounds [7, 74]. In this context, MCR reactions leading to heterocycle-based agents or drugs play a particular role in producing active pharmaceutical ingredients.

Table 1 Representative examples of three-component condensation reactions involving cellulose and its modified forms participation

Reaction*	Scheme
Betti	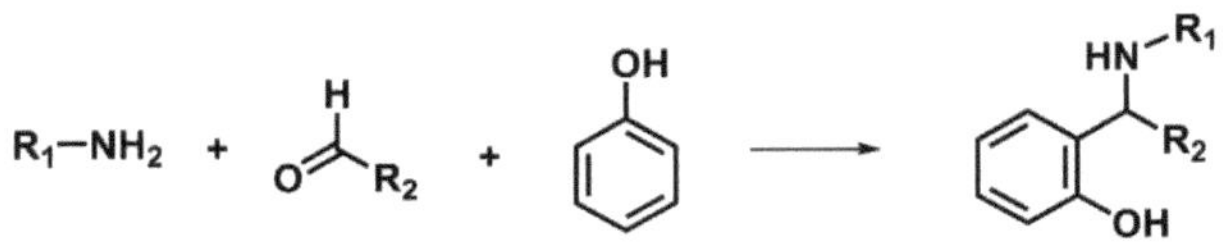
Biginelli	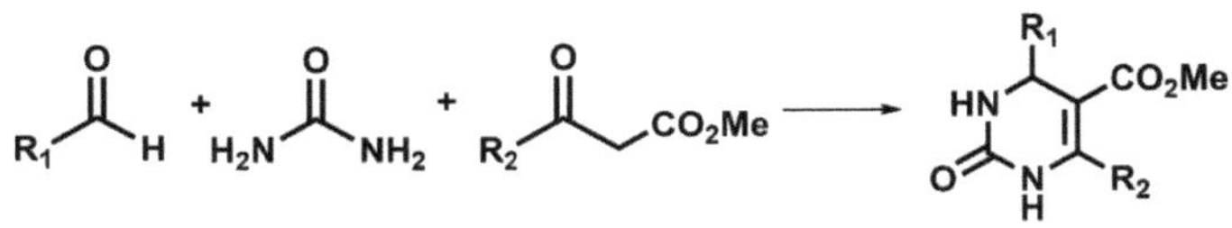
Hantzsch	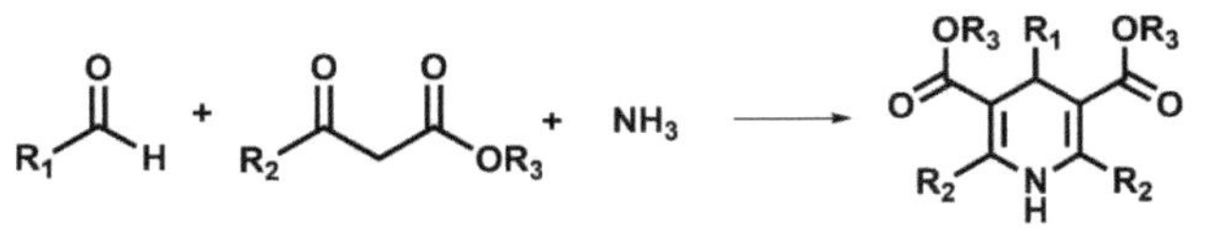
Kabachnik-Fields	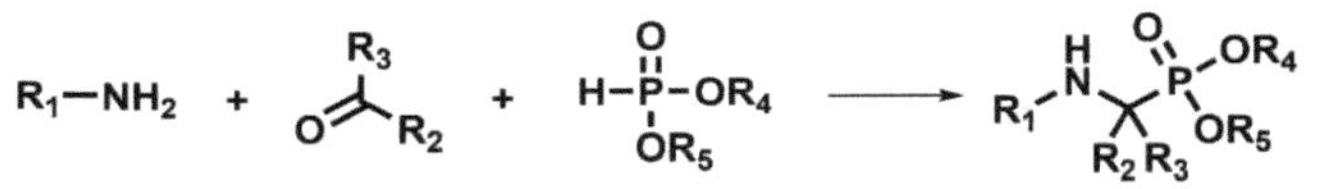
Passerini	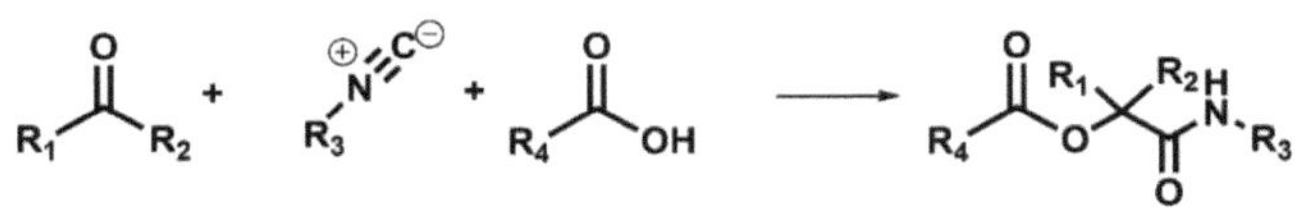
Povarov	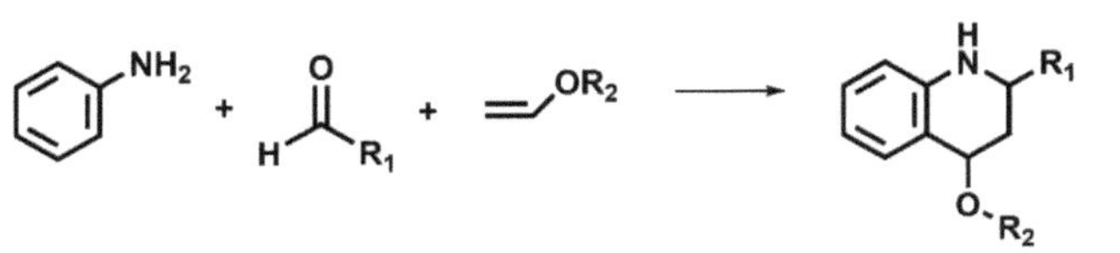
Strecker	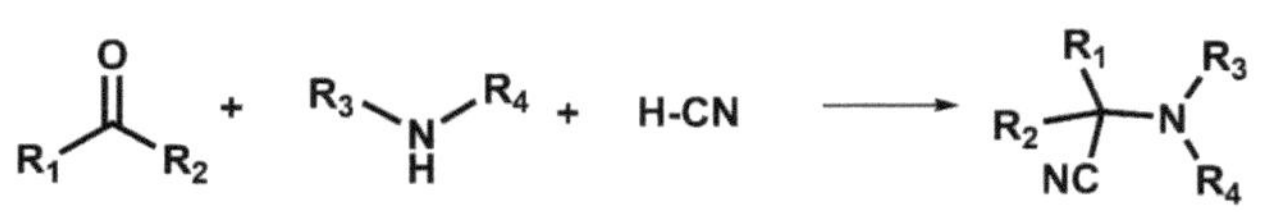

* They are presented alphabetically and will be discussed in this order

3 Functionalized Cellulose Supports in the Catalyzed Three-Component Reactions and Functionalized Cellulose Materials Through the Three-Component Reactions to Develop Sustainable Medicinal Chemistry

Cellulose, nanocellulose, and modified celluloses can assume a dual role in three-component condensation reactions. They are first used in various catalytic applications, especially in organic synthesis [1, 7, 8, 26, 52, 66, 71]. In contrast, studies on cellulose functionalization with 3C-C reactions are less developed; however, in recent years, the renaissance of cellulose functionalization/modification with 3C-C reactions could be noted [2, 3, 75].

Our discussion on the uses of cellulose derivatives in 3C-C reactions is constructed according to the order defined in Table 1. Thus, the Betti reaction gives the beginning.

3.1 *Betti Reaction*

Notably, the Betti reaction is a specific variation of the Mannich reaction, yielding products known as Betti bases, which are typically α-aminobenzyl phenols [50].

However, these products can be converted into 1,3-oxazine derivatives under specific reaction conditions. For example, mixing 2-naphthol **1** (1 equiv.), formalin **2** (2 equiv.), and primary amine **3** (1 equiv.), Azad and Mirjalili [4] obtained naphtho-oxazine derivatives **4**, an essential class of heterocycles, presenting many biological and pharmacological activities [78] (Fig. 2).

The final products were efficiently obtained using a Lewis acid catalyst, Fe_3O_4@nanocellulose/TiCl, under solvent-free conditions via a manual grinding method ("mortar and pestle" technique). This was prepared from cotton fibers through a two-step procedure, which included the direct co-precipitation of the ferric and ferrous ions using ammonium hydroxide in the nanocellulose solution. This was followed by the treatment of magnetite (Fe_3O_4), having a nanocellulose as the base, and a $TiCl_4$ mixture.

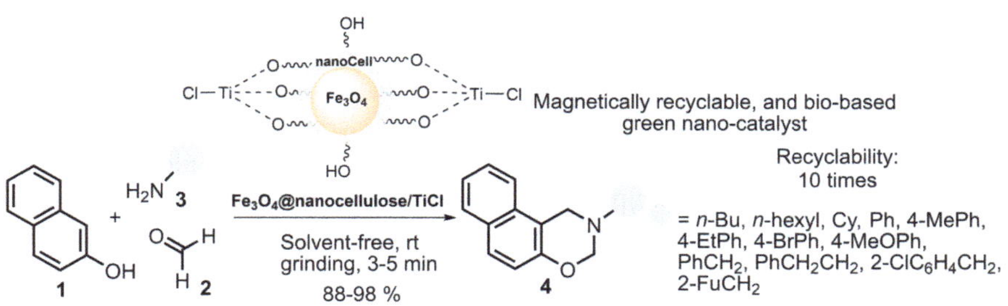

Fig. 2 Nanocellulose-supported Lewis acid-catalyzed Betti reaction

This magnetic nanocellulose-based catalyst was easily recoverable using an exterior magnetic body, followed by washing with ethanol and drying at 80 °C. The recovered catalyst exhibited outstanding reusability, maintaining its catalytic activity across 10 successive cycles with negligible performance degradation. This procedure is keyed by an environmentally benign technique, simple work-up procedure, and nanocatalyst reusability. Nanocellulose can be classified into two primary types, cellulose nanocrystals and cellulose nanofibrils, based on their size, source, and processing methods. This study utilized cellulose nanocrystals, which can be obtained through acid hydrolysis of cellulosic materials like plants, tunicates, and agricultural biomass. In addition, this Fe_3O_4@nanocellulose/TiCl nanocomposite is also a promising nanofiller for new dye-sensitized solar cells (DSSCs) design [44].

Cellulose functionalization through the 3C-C reactions is an excellent choice for discovering new biomaterials or heterogeneous biocatalytic systems. Therefore, Shaabani and colleagues obtained a new heterogeneous catalyst, Cu(II)@DAC-Betti, via functionalization using the Betti reaction involving dialdehyde cellulose (DAC), 2-naphthol **1**, and benzylamine **5** [5] (Fig. 3). The essential DAC, an appropriate reagent for the 3C-C reactions, could be easily prepared via oxidizing nanocellulose by potassium periodate (KIO_4) in deionized water. Its functionalization through the Betti reaction was achieved in acetic acid (AcOH) at 22 °C for 24 h. As a result, a solid yellowish DAC-Betti product was obtained, in which Cu^{2+} ions were immobilized, presenting a Cu(II)@DAC-Betti biocatalyst. The chemical behavior of the obtained catalyst was assessed in the preparation of 2-aryl-5-pyridinyl-disubstituted oxadiazoles **7** from the respective N-arylidenenicotinohydrazides **6** through an intramolecular C–H functionalization cyclization (Fig. 3). The best amount of catalyst was 0.30 mmol%. However, its recyclability creates doubts because the yield decreases (from 75 to 60%) after five-cycle runs. It should be noted that Betti-functionalized DAC could be an attractive bioactive platform for creating novel antibacterial materials.

3.2 *Biginelli Reaction*

This includes the condensation of commercial aldehydes, (thio)urea, and β-ketoesters, which is catalyzed by a strong acid or base to directly afford 3,4-dihydropyrimidin-2(1*H*)-ones/thiones (DHPMs). DHPM scaffolds with unique pharmacological properties are promising nuclei for various therapeutic targets [25]. Due to the medicinal significance of these heterocycles arising from the Biginelli reaction, catalytic methods have been developed. The adaptability of cellulose-based acid catalysts, coupled with the biological relevance of these structures, offers a strong incentive to optimize the reaction conditions of the Biginelli reaction within the framework of green pharmaceutical chemistry. Despite this, it is surprising that the first protocol in the reaction using the Lewis acid-supported nanocellulose catalyst was proposed by Iranian scientists only in 2022 [25]. This work continues to be the only one on this topic. The developed procedure involves the synthesis of 3,4-DHPMs

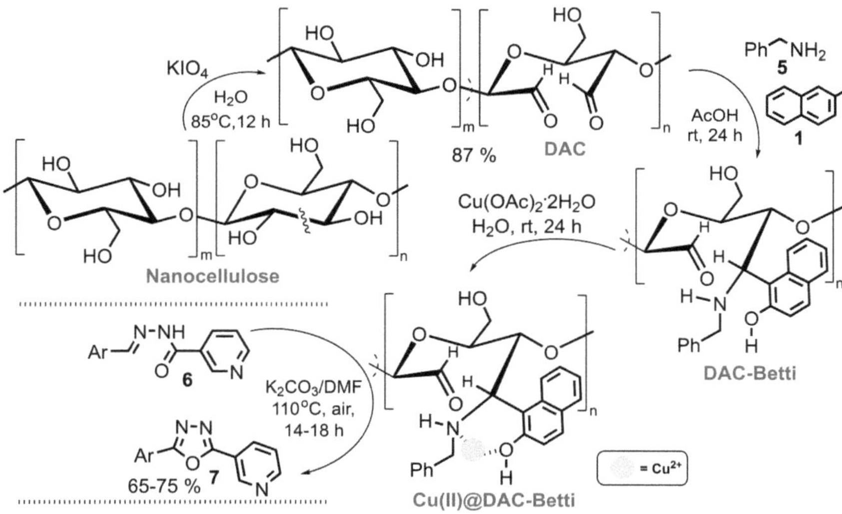

Fig. 3 Route for the Cu(II)@DAC-Betti biocatalyst preparation and its application in bioactive oxadiazole synthesis

11 mixing ethyl acetoacetate **8** (3.5 equiv.), aromatic aldehyde **9** (1 equiv.), and urea **10** (3.5 equiv.) in the presence of nanocellulose/BF_3/Fe_3O_4 catalyst in ethanol at 22 °C for 5–12 min (Fig. 4).

The targeted DHPM compounds **11** were rapidly prepared with good to excellent isolated yields. After condensation, the catalyst is separated using an external magnet, cleaned with ethanol, and dehydrated at 50 °C. After the condensation, the catalyst is removed with an external bar magnet, washed with ethanol, and dried at 50 °C. The test of its reusability showed excellent recyclability without significant loss after five sequential runs (from 98 to 90%). Thus, this procedure could be recommended for synthesizing monastrol-based anticancer agents due to the fast and efficient reaction performance and easy reusability process.

Similarly, Sui et al. s' study on transforming cellulose acetoacetate (CAA) through the Biginelli reaction remains the sole research on this topic to date [10]. Nonetheless, chemists have greatly appreciated the "click chemistry" attributes of the Biginelli reaction in the development of functional polymers [11, 12]. Thus, CAA was used as the substrate with *p*-substituted benzaldehydes **9** and urea **10** to amend cellulose via the Biginelli reaction (Fig. 5).

Cellulose acetoacetate (CAA), incorporated within the anhydroglucose unit (AGU) ring, is derived from wood pulp containing 99% cellulose through a trans-esterification reaction of glucosidic HO- groups with *tert*-butyl acetoacetate (*t*-BAA) or diketene. For catalyzing the Biginelli reaction, a Lewis acid $MgCl_2{\cdot}6H_2O$ was utilized in AcOH at 70 °C. That results in CAA-DHPM polymeric products **12**. The products were precipitated in 50% aqueous ethanol to provide excellent 87–98% yields. Analysis of the obtained CAA-DHPM products using standard spectroscopic

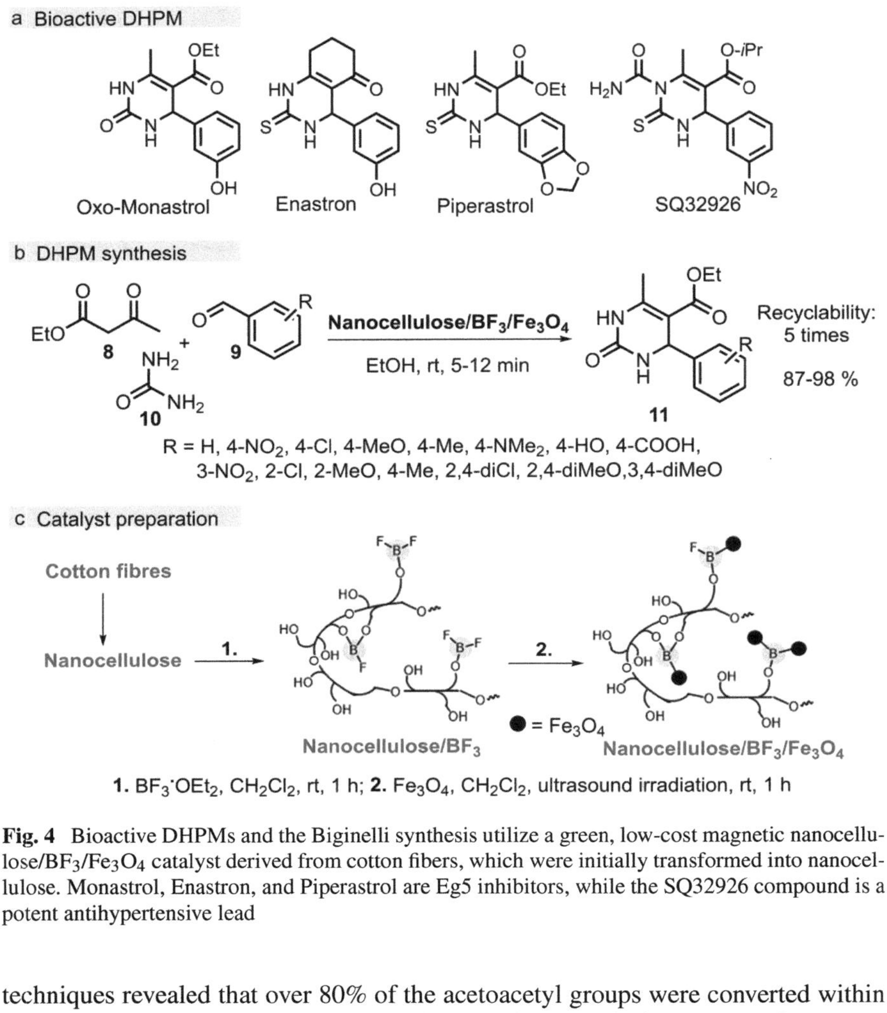

Fig. 4 Bioactive DHPMs and the Biginelli synthesis utilize a green, low-cost magnetic nanocellulose/BF_3/Fe_3O_4 catalyst derived from cotton fibers, which were initially transformed into nanocellulose. Monastrol, Enastron, and Piperastrol are Eg5 inhibitors, while the SQ32926 compound is a potent antihypertensive lead

techniques revealed that over 80% of the acetoacetyl groups were converted within 4 h. They presented a good solubility in chosen solvents and advantageous thermostability with the start of structural deterioration at circa 260 °C. Thus, the Biginelli reaction was successfully exploited as a modular cellulose modification method, and a diverse set of promising cellulose-based biomaterials can be attained.

The synthesis of condensed pyrimidine derivatives through 3C-CR of β-ketoesters **8**, benzaldehydes **9**, and 2-amino heterocycles like benzothiazoles or tetrazoles **13–14** could be considered the Biginelli-type condensation. The respective pyrimidobenzothiazole derivatives **15** and 7-dihydrotetrazolo[1,5-*a*]pyrimidine molecules **16** were straightforwardly assembled employing diverse Lewis acid-supported nanocellulose catalysts [13–17] (Fig. 6).

Magnetic nanocellulose-based catalysts, nanocellulose/BF_3/Fe_3O_4 (Fig. 4c), Fe_3O_4@nanocellulose/Sb(V) (Fig. 6b), Fe_3O_4@nanocellulose/TiCl (Fig. 2),

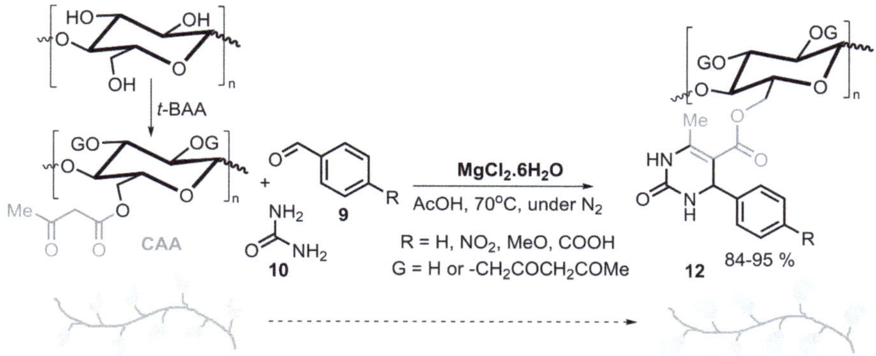

Fig. 5 Preparation of new CAA derivatives using the $MgCl_2$-catalyzed Biginelli reaction. The functionalized CAA-DHPM polymers possess various functional groups in their side chains, showing improved thermostability than CAA and good solubility in the tested solvent (AcOH, DMF, and DMSO) at diverse temperatures (40 and 60 °C)

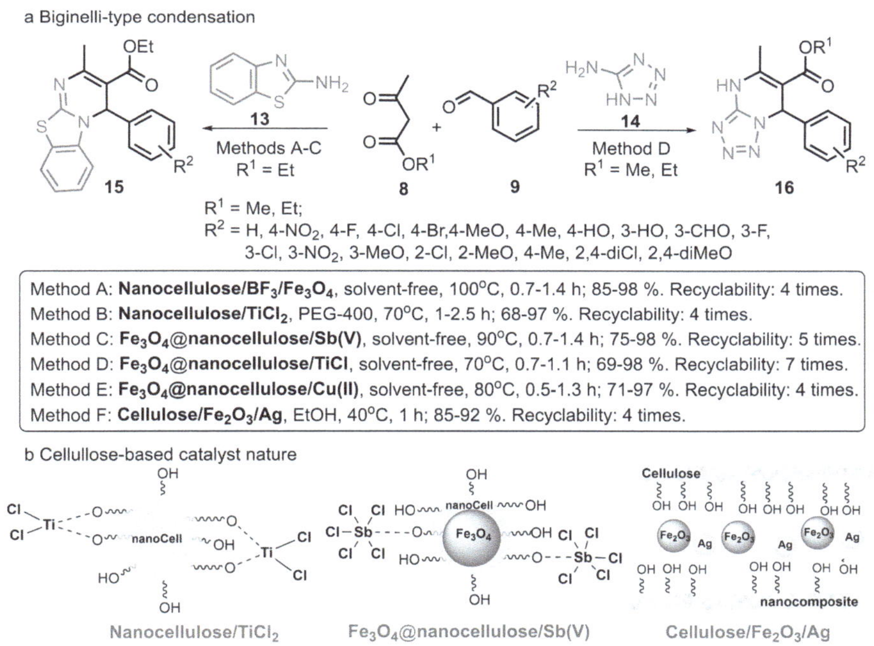

Fig. 6 Diverse nanocellulose catalysts supported by Lewis acids for synthesizing condensed pyrimidines via Biginelli-type condensation reactions

Fe_3O_4@nanocellulose/Cu(II), and cellulose/γ-Fe_2O_3/Ag (Fig. 6b), easily detachable, were very efficient in the synthesis of pyrimidine derivatives **15** and **16**, respectively (methods A [13], C [15], D [16], E [17], and F [18]). Their good thermal stability allows appropriate recoverability and reusability. They are suitable catalysts for the solvent-free synthesis of the organic molecules and materials. Not being magnetic, nanocellulose/$TiCl_2$ catalyst equals them in efficiency and reusability (method B) [14]. However, a metal-free Lewis acid-supported nanocellulose catalyst (nanocellulose/BF_3/Fe_3O_4) would be more suitable for environmentally benign conditions. Interestingly, cellulose/γ-Fe_2O_3/Ag nanocomposite exhibited antimicrobial activity against Gram-positive bacteria, *Staphylococcus aureus*, demonstrating an ability to be applied in the therapeutic and healthcare fields [19]. Indeed, besides the catalyst role, magnetic nanocellulose composites can be used in several fields, such as microcontroller devices, photovoltaic cells, or drug design [20].

In 2022, Jahanbakhshi and Farahi [21] conducted an interesting study. The researchers described the synthesis and characterization of a new Fe_3O_4/MPC-[IL] catalyst, incorporating the Brønsted acidic ionic liquid 1-(propyl-3-sulfonate) vinyl imidazolium hydrogen sulfate [IL], supported on modified magnetic cellulose. This catalyst was evaluated in a condensation reaction involving 3-amino-1*H*-1,2,4-triazole **17**, malononitrile (or ethyl cyanoacetate) **18**, and benzaldehydes **9** synthesizing triazolopyrimidine derivatives **19** (Fig. 7).

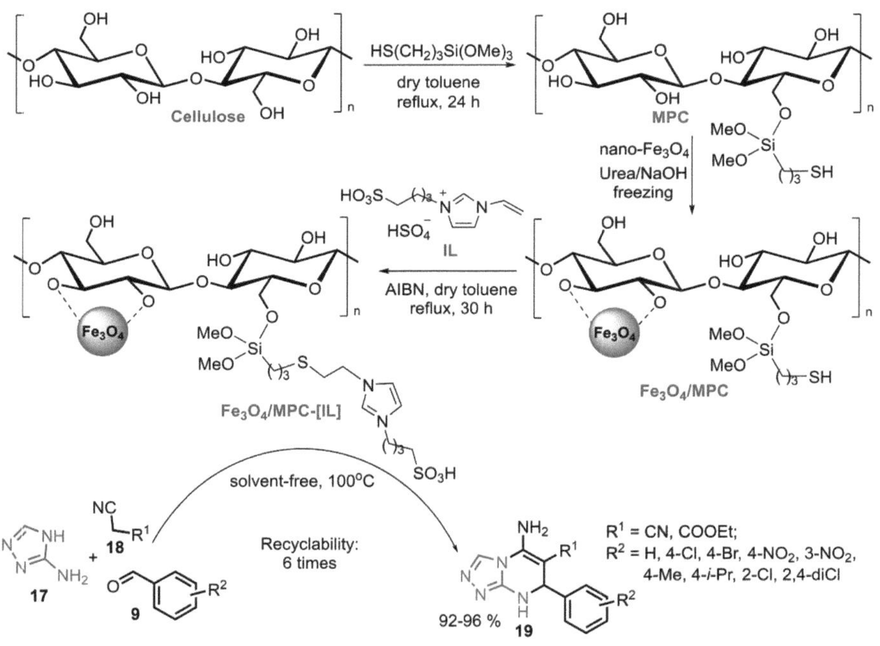

Fig. 7 Preparation of Fe_3O_4/MPC-[IL] catalyst and application in the Biginelli-type condensation for triazolo[4,3-*a*]pyrimidines

The magnetic Fe_3O_4-cellulose-supported ionic liquid catalyst preparation is based on the convergent route: (a) Cellulose and 3-mercaptopropyltrimethoxysilane were mixed in anhydrous toluene and refluxed for 24 h to obtain 3-mercaptopropylcellulose (MPC). The latter was added to Fe_3O_4, dispersed in an aqueous solution with NaOH (7 wt%) and urea (12 wt%) to furnish Fe_3O_4/MPC composite; (b) 1,4-Butanesultone reacts with 1-vinyl imidazole in Et_2O at 0 °C for 24 h to give a solid salt, which was dissolved in water, dropping equal molar sulfuric acid to form acidic Brønsted ionic liquid [IL]. Finally, both reagents from routes a–b were combined and refluxed in the presence of 5 mol% azodiisobutyronitrile (AIBN) and anhydrous toluene under atmospheric nitrogen for 30 h, proving the desired heterogeneous acidic nanocatalysts.

In their quest to optimize the reaction conditions for the Biginelli-type reaction, the authors identified that using 0.003 g of Fe_3O_4/MPC-[IL] catalyst under solvent-free conditions at 100 °C (for 1 mmol of each reagent) was ideal. This setup allowed the catalyst to be reused across six consecutive cycles without markedly dropping its catalytic activity.

3.3 Hantzsch Reaction

The classic condensation reaction permits the preparation of NH-dihydropyridine (DHP) compounds from aldehydes with two equivalents of β-keto-esters and ammonium acetate (or ammonia) using a broad range of catalysts, including nanomaterials [22]. An aromatic/aliphatic amine can replace the latter nitrogen source, providing N-substituted dihydropyridines (DHPs) [23, 24]. The 1,4-dihydropyridines are widely known for their role in calcium channel modulation (Fig. 8a). These compounds are highly significant due to their broad spectrum of pharmacological and biological activities, particularly in treating inflammation, hypertension, and cardiovascular illnesses.

A cellulose-based bimetallic nanocomposite (γ-Fe_2O_3/Cu@cellulose) can catalyze the classic Hantzsch (1,4-dihydropyridine) reaction, as Maleki and coworkers described [25]. An aqueous solution mixture containing ferric and ferrous ions (1:2 ratio) was added to a homogeneous mixture of cellulose, sodium hydroxide, and urea (81:7:21 ratio) to synthesize γ-Fe_2O_3-implanted cellulose, which was subsequently treated with a $CuSO_4 \cdot 4H_2O$ solution. This allows the desired magnetic nanocatalyst to hold C, O, Cu, and Fe elements according to the energy-dispersive X-ray (EDX) data. Thus, ethyl acetoacetate **8** (2 equiv.), benzaldehydes **9** (1 equiv.), and ammonium acetate **20** (2 equiv.) were mixed and vigorously stirred using γ-Fe_2O_3/Cu@cellulose as a catalyst at 22 °C for 15–35 min and without any solvents to afford Hantzsch dihydropyridines **21** in high yields (80–93%) (Fig. 8b, method A).

Notably, the developed catalyst is significantly more efficient than the cellulose sulfuric acid (cellulose-SO_3H, CSA) catalyst used earlier in the same solvent-free condensation (100 °C, 2–5 h) for DHPs preparation (method B) [26]. These

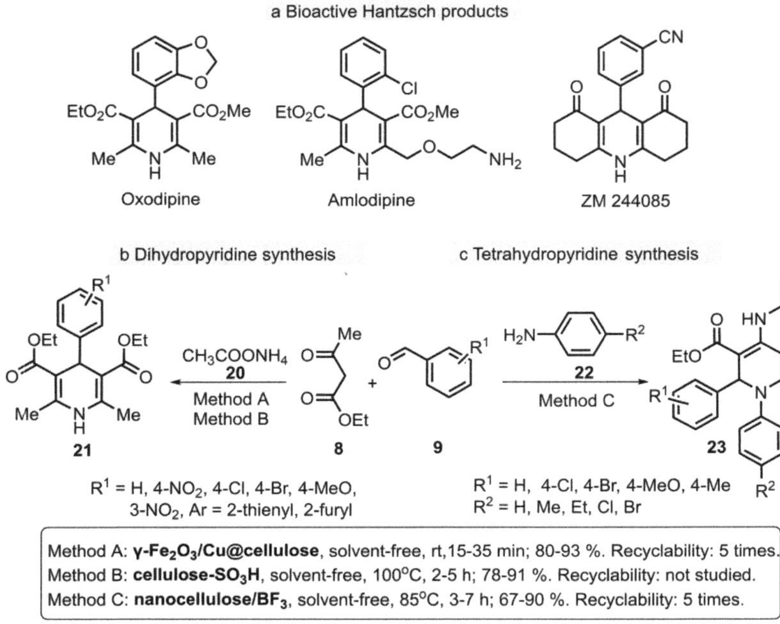

Fig. 8 Bioactive Hantzsch products utilize cellulose-based catalysts to synthesize dihydro(tetrahydro)-pyridines through a three-component Hantzsch condensation. Oxodipine and amlodipine, both DHP derivatives, are recognized antihypertensive agents for cardiovascular diseases. A tricyclic dihydropyridine agent, ZM244085, can activate K^+ channels with bladder selectivity without impacting Ca^{2+} channels. This feature may help in treating bladder control loss, particularly urinary incontinence

compounds possess good antibacterial activity in twenty-seven ESBL (extended-spectrum β-lactamases) isolates of *Klebsiella pneumoniae* and *Escherichia coli.*

On the other hand, Mirjalili and coworkers applied Lewis acid nanocellulose/ BF_3 catalyst to prepare 1,2,5,6-tetrahydropyridine-3-carboxylates **23** using ethyl acetoacetate **8** (1 equiv.), benzaldehydes **9** (2 equiv.), and *para*-substituted anilines **22** (2 equiv.) in formally the three-component Hantzsch-type condensation under mild reaction conditions (85 °C) without solvents [26] (Fig. 8c, method C). This biodegradable catalyst generates highly functionalized tetrahydropyridines in moderate to good yields (67–90%) for 3–7 h.

Hantzsch dihydropyridines **21** were proposed to be obtained in the presence of complex magnetic, catalytic systems comprising cellulose, β-cyclodextrin (BCD), and graphene oxide (GO) like $GO/Cu_2O/Fe_3O_4$ nanocomposite embedded into the cellulose-BCD hydrogel, which acts as recoverable Lewis acid [27]. Despite the high yields of these simple heterocycles, one wonders why one should complicate the system when the simple ones work pretty well.

Other cellulose-based nanocomposites were employed as low-cost, simple, and environment-benign catalysts for the straightforward 3C-C reaction synthesis of polycyclic dihydropyridine derivatives starting from structurally diverse 1,3-diketones, benzaldehydes, and aniline derivatives. For example, a magnetic cellulose nanocomposite with the embedded SO_3H groups, Fe_3O_4@cellulose-SO_3H, was successfully used in the construction of tri- and tetracyclic dihydropyridine compounds **27** and **28**, respectively [28] (Fig. 9). The condensation reactions between arylaldehydes **9**, 6-amino-1,3-dimethyluracil **24**, and 1,3-cyclohexanediones **25** (or 1,3-indanedione **26**) smoothly occurred in water at 80 °C in short reaction times. The corresponding heterocyclic products **27–28** were synthesized in excellent yields.

Upon completion of the reaction, the catalyst was effortlessly removed employing an external bar magnet, treated with ethanol, driėd at 70 °C, and reused for subsequent cycles. This straightforward recovery process, combined with the catalyst's chemical stability, enabled the condensation reactions to be performed in barely five consecutive cycles, lacking a noticeable decline in catalytic efficiency.

The same heterogeneous nano-biocomposite catalyst was used to synthesize indenoquinolinone derivatives **30** via the three-component Hantzsch-type reaction using benzaldehydes **9**, indanedione **26**, and 1-naphthylamine **29** [29] (Fig. 10a). The transformation proceeded rapidly under mild reaction conditions (40 °C) in a solvent-free environment, yielding excellent results with efficiencies ranging from 79 to 98%.

On the other hand, the Fe_3O_4@nanocellulose/Cu(II) catalyst mentioned above was found to catalyze similar condensation reactions between benzaldehydes **9**, indanedione **26**, and amino-substituted pyrimidinone **31** in ethanol at 60 °C to provide tetracyclic fused pyrimidines with an indenopyrido skeleton **32** with good effectiveness ranging from 79 to 97% [33] (Fig. 10b).

Both protocols offer significant advantages, including easy-going reaction conditions, rapid reaction times, straightforward work-up processes, high product purity, excellent yields, and the nanocatalyst's recycling.

In 2020, Tamaddon et al. conducted an interesting study comparing the catalytic activities of magnet-based micro-DAC and nano-DAC immobilized urease enzymes in similar Hantzsch and Biginelli condensations [32]. In their research, microcellulose, -MC and nanocellulose, -NC, derived from cotton, were converted into the respective nanomagnetic microcellulose (nano-γ-Fe_2O_3@microcellulose (NMMC) and nanomagnetic nanocellulose (nano-γ-Fe_2O_3@nanocellulose (NMNC), which

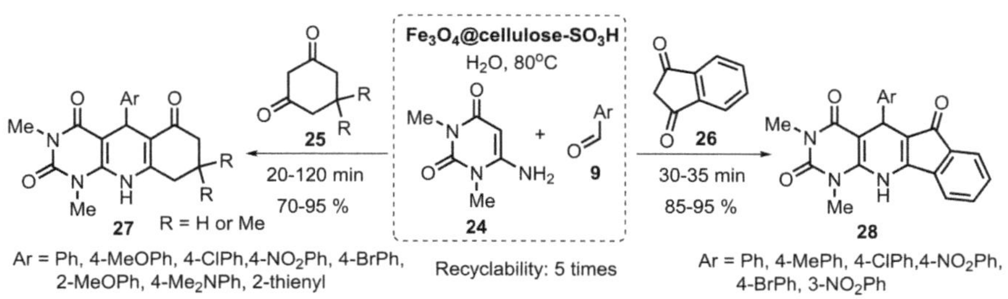

Fig. 9 Recyclable and economical Fe_3O_4@cellulose-SO_3H catalyst for straightforward water synthesis of functionalized polycyclic dihydropyridine

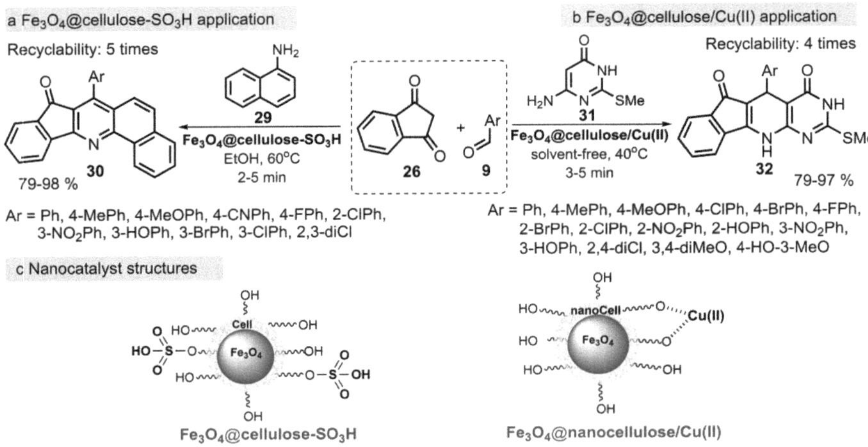

Fig. 10 Three-component Hantzsch-type condensation reactions using magnetic, cost-effective sulfonic cellulose-coated Fe_3O_4 and Fe_3O_4@nanocellulose/Cu(II) nanocatalysts

were then transformed to the corresponding dialdehyde derivatives NMMDAC and NMNDAC through oxidation reaction by periodate under controlled conditions. Their aldehyde functions were reacted with enzyme urease in phosphate-buffered saline at 4 °C for 24 h to afford NMMDAC/urease or NMNDAC/urease via the Schiff-base covalent immobilization process (Fig. 11).

Having the immobilized NMMDAC/urase and NMMDAC/urase in their hands, the authors associated the catalytic activity of these immobilized ureases, nanomagnetic micro/nanocellulose materials, and free enzyme in the condensation between ethyl acetoacetate **8** (2 equiv.), benzaldehyde **9** (1 equiv.), and urea **10** (1equiv.) in water that may conduce to the formation of the Biginelli product (DHPM-1) and Hantzsch product (1,4-DHP-1) (Fig. 12). It was found that the exclusive Hantzsch product was observed when the free urease was employed. In comparison, the isolated Biginelli cycloadduct yields were achieved with 92% and 68% using immobilized urease on magnetized NMMDAC and NMNDAC, respectively. This cycloadduct was also found in the presence of NMMC and NMNC polymers [32] (Table 2).

The results from Table 2 indicate that the selectivity for immobilized urease enzymes arises from their ability to mediate urea condensation over urea hydrolysis, resulting in the formation of the Biginelli cycloadduct DHPM-1, which acts as an enzyme inhibitor.

Assembling long-lasting compact luminous polysaccharide materials built using the Hantzsch condensation is a promising topic because these materials can be employed in inhibiting bacterial growth or intelligent recognition as fluorescent probes and portable test strips [31, 30]. The classic Hantzsch reaction, involving formaldehyde ($H_2C{=}O$), ammonium acetate ($AcONH_4$), and cellulose acetoacetate (CAA), derived from cotton fabrics, directly led to a hydrogel CAA-DHP with the 1,4-DHP rings [34] (Fig. 13). The obtained cellulosic material showed an apparent

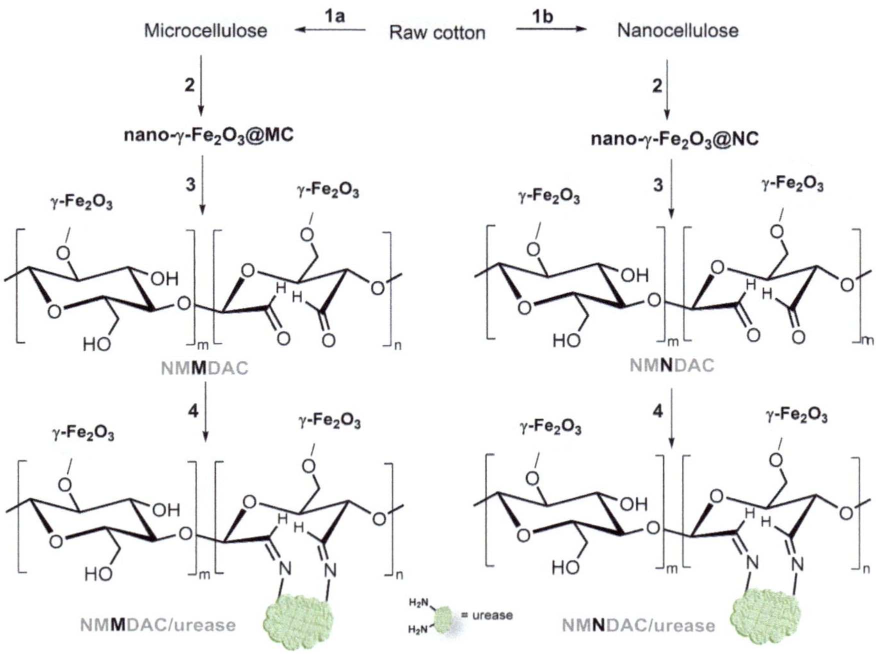

Fig. 11 Simplified schematic preparation of NMNDAC/urease and NMMDAC/urease from raw cotton via magnetization, oxidation, and covalent immobilization of urease processes: **1a** 18% NaOH, reflux; **1b** 35% H_2SO_4, 35 °C, 3 h; **2** H_2O, urea, magnetization; **3** KIO_3/dark; **4** Urease, phosphate buffer, 4 °C, 24 h

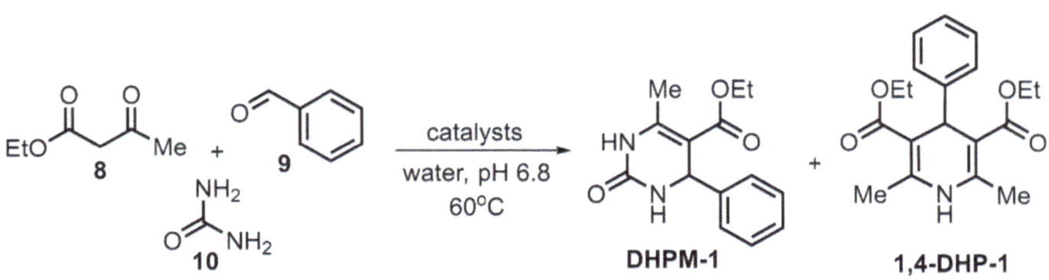

Fig. 12 Formation of the Biginelli product (DHPM-1) and Hantzsch product (1,4-DHP-1) through enzymatic three-component Hantzsch-type condensation reaction

fluorescent emission around 460 nm, reached in 10 min at room temperature in water. The incorporation of 1,4-DHP rings on the cotton fabric surface was verified through XPS (X-ray photoelectron spectroscopy), fluorescence spectroscopy, and elemental analysis. The CAA-DHP hydrogel exhibited a faint yellow under visible light and fluoresced bright blue when illuminated with UV light at 365 nm.

Table 2 Catalytic evaluation of immobilized ureases, nanomagnetic micro/nanocellulose materials, and free urease components

Catalysts	DHPM-1 (%)	1,4-DHP-1 (%)	Selectively (%)	Time (min)
Pure urease	0	100	100	180
NMMC	28	0	100	180
NMNC	20	0	100	180
NMMDAC/urease	92	0	100	90
NMNDAC/urease	68	0	100	130

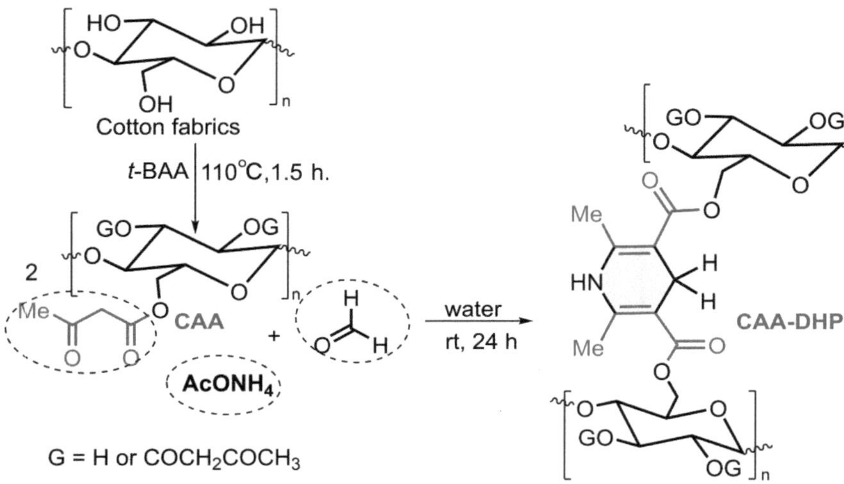

Fig. 13 Functionalizing cotton fabric to provide a fluorescent hydrogel CAA-DHP through the Hantzsch reaction of cellulose acetoacetate (CAA), $H_2C{=}O$, and $AcONH_4$ in water

Notably, the proposed method of simple surface esterification is highly suitable for producing cellulose acetoacetate-based fluorescent paper (CAA-FA). This material is fabricated from CAA-paper (structurally analogous to CAA-DHP) through the Hantzsch reaction, starting with bagasse cellulose fibers [35]. Substituting formaldehyde and ammonium acetate with benzaldehydes and amines in the Hantzsch condensation could further modify cotton fibers with luminous and hydrophobic functioning. In 2022, Qi et al. described a protocol for the fast building of biodegradable luminous paper-based materials in which DHP rings contain diverse arylaldehydes such as 2-phenylpropionaldehyde, 3,4-dimethoxybenzaldehyde, vanillic aldehyde, and benzaldehyde (Fig. 14a). These luminescent materials hold significant potential for various applications, particularly in anti-counterfeiting measures [35].

The same year, Liu et al. prepared a series of similar fluorescent cellulose materials with similar groups in the DHP rings (Fig. 14b) using the abovementioned Hantzsch reaction scheme. The obtained transparent cellulose-based films exhibit prospective

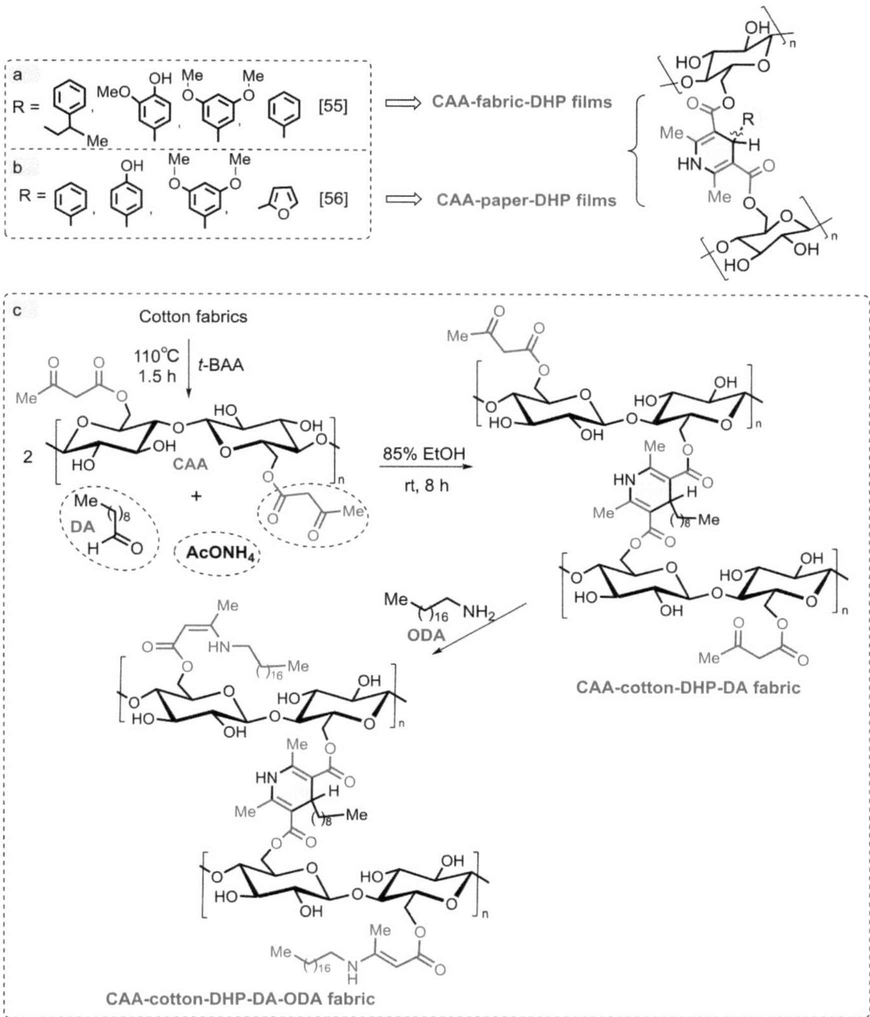

Fig. 14 Functionalization of CAA-fabric (**a**) and CAA-paper with aromatic aldehydes: simplified structure of obtained corresponding CAA-fabric-DHP and CAA-paper-DHP films. The Hantzsch reaction of cellulose acetoacetate fabric, dodecyl aldehyde (DA) (**b**), and ammonium acetate to afford CAA-cotton-DHP-DA fabric and its transformation in CAA-cotton-DHP-DA-ODA fabric via an amidation reaction (**c**)

applications in ultraviolet shielding and anti-falsification processes, as well as smart remote light-emitting diode devices [38].

Finally, the Qi team fabricated a long-lasting luminescent and water-repellent cotton material through surface amendment [37]. They used the Hantzsch reaction of CAA fiber, dodecyl aldehyde (DA), and $AcONH_4$ in EtOH at room temperature to obtain CAA-DHP fabric. Moreover, the hydrophobic octadecylamine (ODA) moiety was incorporated into the surface of the CAA-DA material by an amidation process, providing a fluorescent and hydrophobic CAA-DA-ODA fabric (Fig. 14c).

Therefore, the fluorescent properties of the DHP rings contribute to the modified cellulose surface. At the same time, diverse aldehyde functions donate intense and high-stability fluorescence and/or more hydrophobic properties of cellulose-based biomaterials.

3.4 Kabachnik-Fields Reaction

This is the primary method for preparing α-amino phosphonates. The latter can be achieved by irradiating the reactants with microwave radiation without any catalyst or solvent. However, the conventional approach involves the condensation of equimolar amounts of the three components (a carbonyl compound, an amine, and dialkyl phosphite) in organic solvents, typically employing various catalysts, most commonly Lewis or Brønsted acids [37]. Significantly, α-amino phosphonates, structure equivalents of α-amino acids, have garnered considerable interest for their diverse uses in biological and medicinal chemistry.

The Kabachnik-Fields reaction can also synthesize functional polymers. Developing novel, fully functioning synthetic polymers built on the remarkable physicochemical characteristics of α-aminophosphonate structures (such as metal chelation, antibacterial, antifungal, and anticancer activities) presents a promising avenue for cellulose exploration [39, 43]. However, the surface Kabachnik–Fields three-component condensation reactions (KF-3C-C reactions) are still less underdeveloped [40–42]. In one of the few studies on cellulose modification via KF-3C-C reactions, Hamada and coworkers performed a solid KF-3C-C reaction between *p*-anisidine **33**, diisopropyl phosphite **34**, and aldehyde groups from the cellulose-based methacrylate polymer, called a cellulose-graft-poly(methacrylated vanillin) aldehyde (Cell-*g*-PMV) (Fig. 15) [42].

The latter was synthesized with an inserting polymerization yield of 71% using pre-irradiated cellulose fabric (50 kGy) and methacrylated vanillin (MV). The KF-3C-C reaction was conducted at 60 °C for 5 h in MeOH. In the ATP-FTIR spectrum, the distinctive ester and CH=O bands of this Cell-*g*-PMV precursor were detected at 1753 and 1692 cm^{-1}, respectively.

Thus, the KF-3C-C reaction of these three components was performed in refluxing 1,4-dioxane for 24 h, forming a fiber structure that includes 4-metoxyaniline (*p*-anisidine) and diisopropyl phosphite moieties, i.e., Cell-*g*-PAP biomaterial with improved thermal resistance compared with that of the initial cellulose fabrics. This is due to the nitrogen and phosphorus atoms resulting from α-amino phosphonates. Solid-state ^{31}P NMR confirmed the biding of phosphorus through surface KF-3C-C reaction with Cell-*g*-PAP. Additionally, nitrogen presence was detected using scanning electron microscopy-energy-dispersive X-ray (SEM–EDX) spectroscopy. These findings enable the development of novel functioning fiber materials derived from

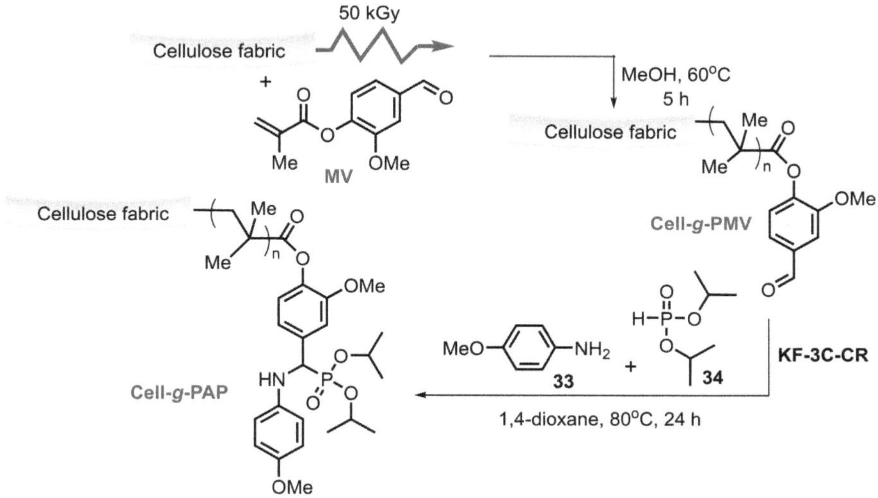

Fig. 15 Functionalization of cellulose fabrics with KF-3C-C reaction: Combining irradiated cellulose fabrics with methacrylated vanillin (MV) to obtain poly(methacrylated vanillin)-grafted cellulose and its KF-3C-C reaction with amine and phosphite, resulting in the formation of a high-thermal biomaterial Cell-g-PAP

wood biomass, particularly the versatile Cell-PMV aldehyde, which can be applied in other 3C-C reactions.

The cellulose sulfuric acid (cellulose-SO_3H, CSA), the recyclable solid acid-based catalyst for acid-promoted transformations, could also be helpful in the KF-3C-C reaction. One of the examples is the CSA-catalyzed condensation reaction between benzaldehydes **35**, anilines **36**, and dimethyl phosphonate **37** in water at 22 °C for three hours that provides α-aminophosphonates **38** in excellent yields (94–95%) [44] (Fig. 16).

The synthetic procedure is simple and efficient. The reaction mixture was treated with 5% (wt%) HCl, and the target product was purified through recrystallization

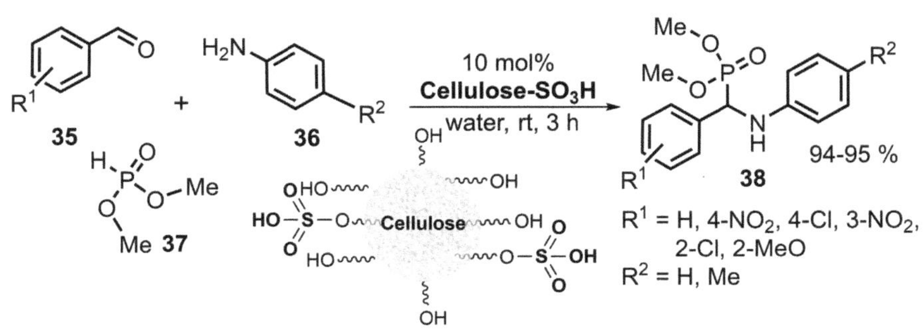

Fig. 16 CSA-catalyzed KF-3C-C reaction yielding α-amino phosphonates

using hexane. Furthermore, the filtered catalyst was reprocessed in four consecutive cycles without a notable reduction in catalytic performance.

3.5 Passerini Reaction

This is the earliest known isocyanide-based multicomponent reaction [45], where carboxylic acids, aldehydes or ketones, and isocyanides combine to afford drug-like α-acyloxyamide compounds, also referred to as acyclic depsipeptides [46]. These pseudo-peptidic molecules are the typical Passerini products. Some of them are Hydrastine alkaloid, an herbal supplement to treat symptoms of flu, cold, and gastrointestinal problems, and Mandipropamid, a fungicide to control oomycete pathogens (Fig. 17).

This 3C-C reaction is one of the most versatile in heterocyclic, macrocyclic, and organic chemistry, including life science, polymeric research, and the medicinal industry [47–50]. This condensation can be accomplished in conventional solvents and aqueous mediums using ambient temperature and rapid reaction times, and it easily tolerates diverse chemical functions.

The Passerini reaction has also emerged as a pivotal method for the design and synthesis of materials derived from polysaccharides. The modified cellulose biopolymer can be easily functionalized because cellulose derivatives containing carboxylic acid and aldehyde functions can be used in this 3C-C reaction. The most utilized derivatives are carboxymethylcellulose (CMC), known as cellulose gum or carboxymethylcellulose sodium, a polar cellulose derivative (–CH_2COOH or as Na^+ salt, carboxymethylcellulose sodium salt), dialdehyde cellulose (DAC), and Cell-PMV aldehyde.

The earliest reports of CMC modification using the Passerini reaction date back to 1999, marking the development of novel chemically cross-linked hydrogels composed of carboxymethylcellulose and various macromolecular carbohydrate biopolymers [51, 52]. After 25 years, this topic is still actual and vital [53–61]. Among published works, particular ones deserve to be discussed here. CMC can be prepared from cellulose via an O-alkylation process using chloroacetic acid and

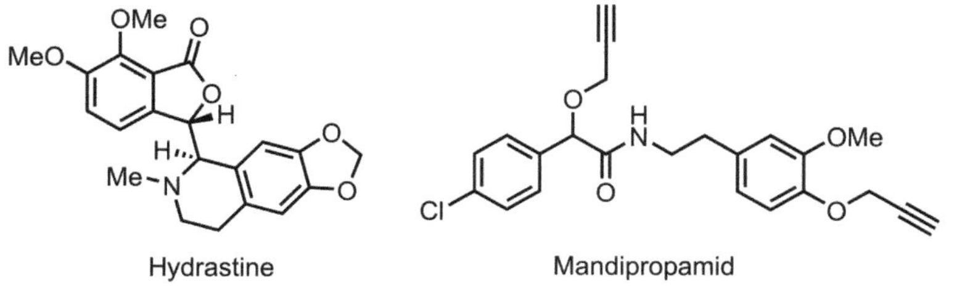

Fig. 17 Bioactive α-acyloxyamide compounds synthesized by Passerini protocol

sodium hydroxide in anhydrous ethanol at 65 °C for three hours, resulting in the in situ formation of sodium carboxymethylcellulose (NaCMC) [62] (Fig. 18).

Commercially available NaCMC has a degree of substitution (DS) of 1.2, indicating the typical amount of $-OH$ groups replaced by $-CH_2COOH$ groups per anhydroglucose unit and a molecular weight of 250 kDa. At pH 3.7 and 3.2, 58 and 74% of the initial COO^- groups are protonated, matching 0.7–0.9 acidic functional groups for each sugar unit. Fleury et al. [58] chemically modified this product by reacting the carboxyl groups of CMC with various aldehydes and isocyanides. This environmentally friendly one-pot process yielded a diverse collection of Passerini-altered CMC products (CMC-Ald-Isocyanides).

After the reaction ended, the formed CMC derivatives were crystallized in isopropanol (IPA) and recuperated in good mass yields (59−85%). These modified CMC products were systematically examined to corroborate the amendment and establish the DS.

Examining chemical reactivity for the selected reagents, it was found that among two tested isocyanides, cyclohexyl (Cy) isocyanide and *tert*-butyl isocyanide, the latter was more reactive in the Passerini reaction. Regarding aliphatic aldehydes, such as acetaldehyde and formaldehyde, demonstrated high reactivity, achieving a DS value of 0.41–0.43, corresponding to a grafting efficiency of 58–61%. In contrast, the reactivity of aromatic aldehydes (furfural and benzaldehyde) was much lower than aliphatic aldehydes [58]. Nonetheless, the obtained Passerini-modified CMC products showed a promising adsorption ability to initiate a new, diverse route toward the straightforward modification of CMC surface properties.

Various structurally diverse α-substituted ketones ($R^1R^2C=O$) and cyclohexyl isocyanide (Cy–NC) were also used to modify CMC via a commercial NaCMC

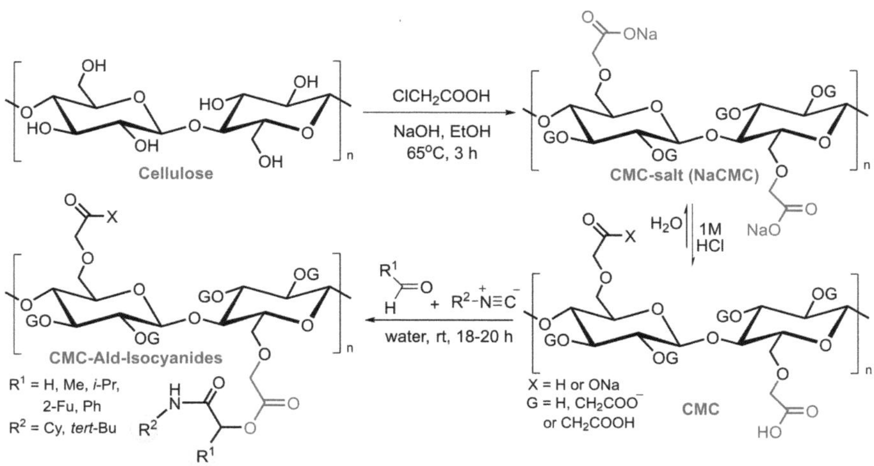

Fig. 18 Synthesis of the NaCMC and CMC functionalization using the Passerini reaction with aldehydes and isocyanides to the Passerini cellulose adducts, CMC-Ald-Isocyanides

product using the Passerini reaction [59] (Fig. 18). Followed the abovementioned protocol, the obtained CMC product was subjected to the three-component condensation reaction in water at 22 °C for 18 h to get a crude mixture reaction solution, which was decanted into cold ISP, obtaining white powders, CMM-Ket-Cy-NC products, in good mass yields (72–98%). Nuclear magnetic resonance (^{1}H- and ^{13}C-NMR) and Infrared (IR) spectroscopy proved the structure of these materials.

A series of the α-substituted ketones, depicted in Fig. 19, from acetone (Acet) to ethyl 4,4,4-trifluoro acetoacetate (EtF_3AcAc), was used in the Passerini CMC modification. Assessing the influence of ketone structure on reactivity, the authors noted that no grafting occurred when small quantities of ketone and isocyanide were used. However, the reaction proceeded when a higher amount was applied (with a ketone/isocyanide ratio of 1.14:1), yielding a DS of 0.038. Among the pyruvate-containing ketones, EtPyr exhibited higher reactivity (DS = 0.116), while the reactivity of EtBrPyr was significantly enhanced, achieving a DS of 0.231. In contrast, the EtF_3Pyr ketone just feebly reacted in the Passerini-3C-C reaction, i.e., there was no quantifiable modification.

Notably, the CMM-McEtAcAc-Cy-NC product, which contains methacrylate reactive sites (referred to as methacrylated carboxymethyl cellulose or CMC-MA),

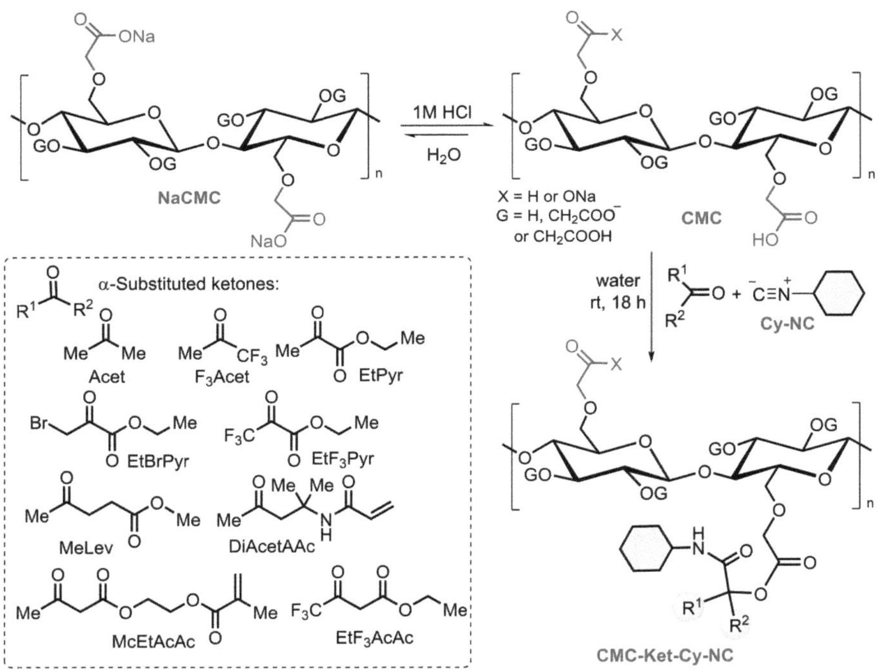

Fig. 19 Reactivation of CMC through an acidification process and its functionalization by the Passerini-3C-C reaction with cyclohexyl isocyanide and various α-substituted compounds to provide Passerini-modified CMC products

can form chemical hydrogels through thermal radical reactions with oligoethylene glycol methyl ether methacrylate (OEGMA) and diethylene glycol methyl ether methacrylate (DEGMA) co-monomers. The copolymerization of CMC-MA with 50 equivalents of OEGMA and 120 equivalents of DEGMA was successfully carried out in water using the VA-044 initiator at 50 °C for five hours, resulting in a macroscopic hydrogel, denoted as CMC-MA-co-OEGMA(50)-co-DEGMA(120). This densely cross-linked hydrogel demonstrated a promising thermal response, exhibiting modular contraction in reaction to temperature change [59].

An alternative method for producing a cellulose backbone containing carbonate anions, CMM-CO_2^- involves activating microcrystalline cellulose with CO_2 (2–5 bar) in DMSO at 50 °C for 30 min and using DBU (1,8-diazabicyclo[5.4.0]undec-7-ene) as a base organocatalyst. This approach, proposed by Meier and coworkers in 2018, allowed an easy in situ succinylation reaction of cellulose filters with succinic anhydride (SA), leading to white succinylated filter paper, denoted as SFP-Cell, in 86% yield with DS of 2.64 [60] (Fig. 20).

Then, the SFP-Cell product was suspended in DMSO and stirred at 50 °C, achieving a clear solution. Subsequently, the reaction mixture gradually introduced the corresponding aldehydes (R^1–CH=O) and isocyanides (R^2–$^+$N≡C^-). Then, it was stirred at 60 °C for 48 h to obtain white Passerini products, SFP-Cell-Ald-Isocyanides, in good mass yields (87–92%). Their glass transition temperatures

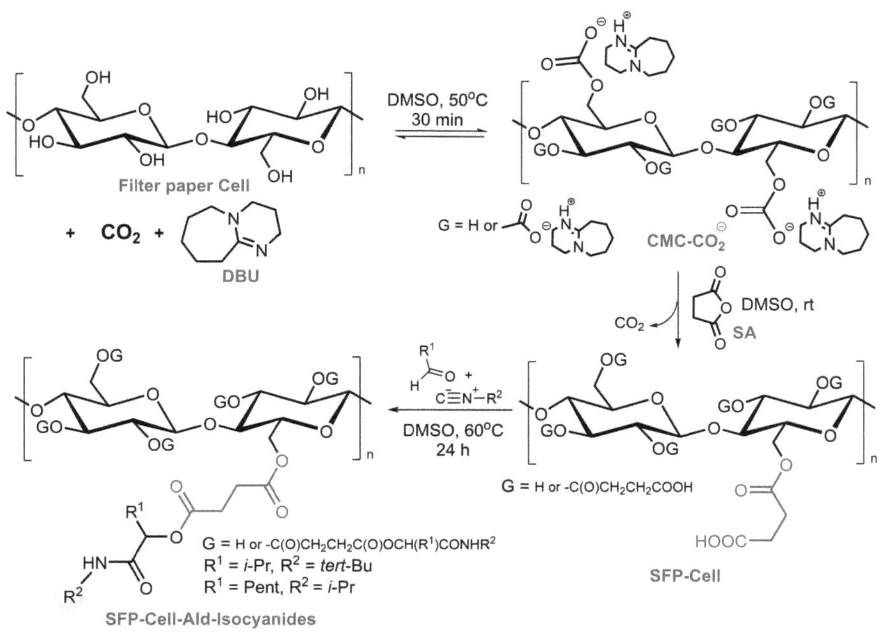

Fig. 20 A CO_2/DBU-based procedure for easily functionalizing filter paper cellulose via the Passerini reaction to create cellulose-containing materials with adjustable thermodynamic properties

(T_g) were found to be 99–116 °C, and their TGA measurements showed high thermostability, with decomposition between 263 and 290 °C, upper than that of the SFP-Cell precursor (202 °C). The developed solvent DMSO-DBU system is efficient, environmentally, and inexpensive. DBU is less toxic and more straightforward to recover from. Furthermore, this solvent system did not require any supplementary catalyst, as DBU, part of the solvent, functioned as the catalyst. Additionally, the recovery of DMSO and DBU was efficiently achieved through a straightforward work-up process [60]. A similar Passerini-modified CMC product (R^1 = Ph, R^2 = Cy) is a base for preparing hydrogels in cross-linking as an oral gentamicin drug administration system. Such hydrogels, loaded with gentamicin, serve as effective antibacterial agents against the Gram-positive bacterium *Staphylococcus aureus* and the Gram-negative bacterium *Escherichia coli* [54].

Similar cellulose-CO_2 products can be synthesized through TEMPO (2,2,6,6-tetramethylpiperidine-1-oxyl radical)-mediated oxidation of the cellulose pulp derived from the Kraft pulping process [63] or microcrystalline cellulose (MCC) [61]. In the latter case, oxidized cellulose nanofibrils (CN) through TEMPO radical (TOCN) with 6-carboxylic acid functions (TOCN-COOH) are a suitable reagent for the Passerini-3C-C reaction (Fig. 21). Therefore, the condensation of TOCN-COOH, poly(N-isopropyl acrylamide) polymer with the terminal –CH=O function (pNIPAm-CH=O), and cyclohexyl isocyanide (Cy–NC) was easily achieved in water at room temperature for 30 min to afford thermoresponsive pNIPAm-grafted cellulose fibrils, pNIPAm-*g*-TOCN-Cy-NC, in 36% of grafting efficiency [61]. NMR and IR spectroscopy supported the structure of the obtained material. Additionally, the XRD technique established the structural integrity of unchanged and altered nanocellulose (TOCN-COOH and pNIPAm-*g*-TOCN-Cy-NC, respectively). The latter, containing thermoresponsive polymers (pNIPAm), showed improved aqueous dispersibility and structural integrity. However, the TEMPO-oxidation process directs to a degree of oxidation (DO = 0.3–0.4).

In comparison, periodate oxidation allows for a high degree of oxidation (DO), ranging from 0 to 2. This process forms 2,3-dialdehyde cellulose (DAC) (or periodate-oxidized cellulose, POC). In 2020, Esen and Meier described their study on the green functionalizing of 2,3-dialdehyde cellulose via the Passerini-3C-C reaction [64]. The starting block, DAC, was prepared from commercial microcrystalline cellulose (MCC) using the oxidation reaction in DI (demineralized) water at 55 °C for 15 h without direct contact with light (Fig. 22). Its ATR-IR spectrum showed the presence of two distinctive bands at 1730 and 881 cm^{-1}, the elongating vibration of –CH=O function, and the C–O–C vibration of hemiacetal, respectively. The degree of oxidation of this dialdehyde was 1.23.

In the subsequent step, dissolved DAC (hot water, 90–95 °C) was subjected to the Passerini-3-C reaction with water-soluble carboxylic acids (R^1–COOH) and isocyanides (R^2–NC) at room temperature for 24 h to furnish white products, MCC/DAC-Acid-Isocyanides, in good mass yields (67–91%) with varied but elevated DS of 0.68–0.94 (Table 3), signifying that the reaction proceeded efficiently.

According to the thermal characterization (DSC and TGA), the obtained products revealed increased thermal stability ($T_{d,5\%}$ = 190–217 °C), i.e., degradation (T_d)

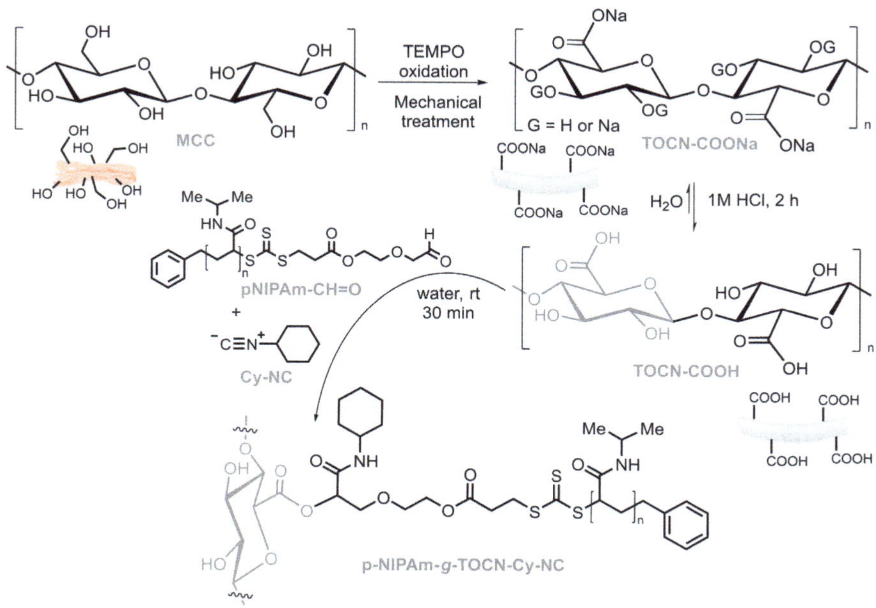

Fig. 21 Graphic preparation of polymer-grafted cellulose fibrils, pNIPAm-*g*-TOCN-Cy-NC, through the Passerini-3C-C reaction. The functionalization of cellulose is performed via a TEMPO-oxidation process using TEMPO reagent, sodium bromide, and sodium hypochlorite in an aqueous environment at ambient temperature and pH 10

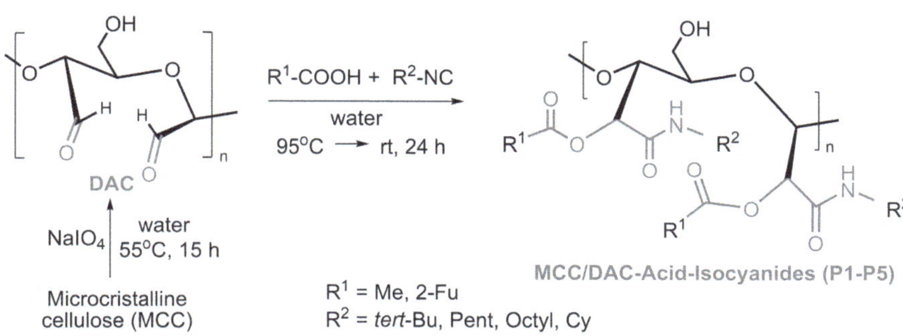

Fig. 22 Preparation of modified cellulose, MCC/DAC-Acid-Isocyanide products through the Passerini 3C-CR reaction

than that of DAC ($T_{d,5\%} = 156$ °C) and their glass transition temperature (T_g) ranges from 121 to 166 °C (Table 3). Thus, these new cellulose-containing products with diverse structural and thermodynamic characteristics can be helpful as models for the further development of cellulose chemistry.

Table 3 Thermal characterization data of the obtained MCC/DAC-Acid-Isocyanide products (P1–P5)

Sample	R^1-Acid	R^2-Isocyanide	Yield (%)	DS	T_g (°C)	$T_{d,5\%}$ (°C)
P1	Me	*tert*-Bu	73	0.94	166	217
P2	Me	Cyclohexyl	88	0.77	159	201
P3	Me	*n*-Pentyl	91	0.68	121	190
P4	Me	*n*-Octyl	72	0.75	133	198
P5	2-Furyl	*tert*-Bu	67	0.82	154	211

Fleury and coworkers also published their interesting study on functionalizing cellulose microfibrils (MFC) using the 2,3-DAC chemistry and the Passerini-3C-C reaction in 2020 [65]. Following the reported work, the prepared dialdehyde cellulose from MFC has a DO of 0.77 at a 9.7 h time reaction. Having an aldehyde component, they chose a hydrophobic isocyanide (*tert*-Bu–NC), merging it with carboxylic acids, 4-pentynoic acid (PA), or methacrylic acid (MA) for the Passerini reaction performance. Note that these acids are non-toxic molecules that are highly reactive in a subsequent modification stage. The Passerini-3-C reaction was completed by introducing the chosen acid followed by *tert*-Bu–NC into a DAC dispersion in purified water and mixing the obtained product at room temperature for 18 h (Fig. 23).

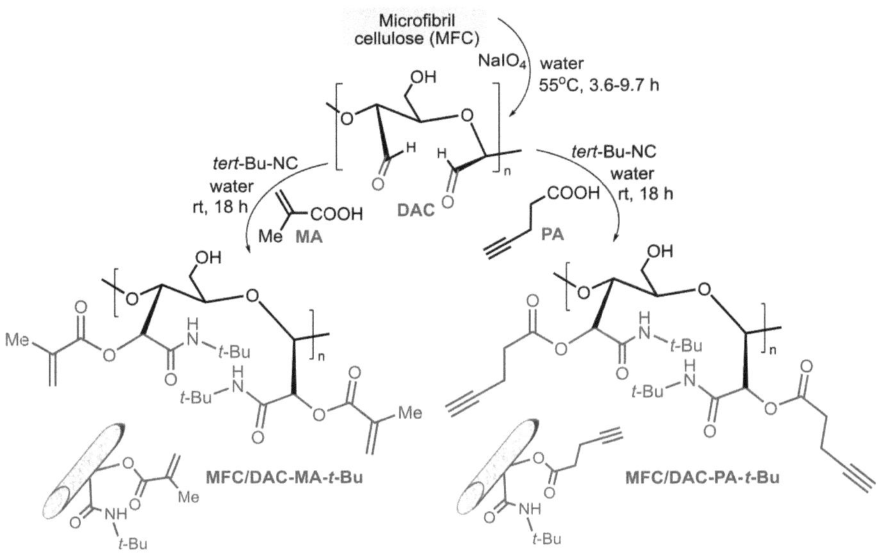

Fig. 23 Functionalization of MFC via DAC employing the Passerini 3C-C reaction with *tert*-Bu–NC and MA or PA reagents. Jean and coworkers obtained and characterized analogous modified MFC products derived from butanoic (heptanoic, undecanoic) acids and *tert*-butyl or cyclohexyl isocyanides in 2024 [66]

The obtained modified MFC products, MFC/DAC-MA-*t*-Bu and MFC/DAC-PA-*t*-Bu, were characterized by the combination of Raman and Fourier Transform Infrared attenuated total reflection (FTIR-ATR) techniques as well as ^{13}C CP-MAS NMR spectroscopy. Regarding their thermogravimetric analysis (TGA), the products exhibited an evident influence of the Passerini modification on their thermophysical properties, with decomposition temperatures of 345–350 °C compared to 330 °C for the unmodified MFC/DAC.

However, one of these products' most remarkable properties is their high chemical reactivity, which allowed for the following post-modifications. Initially, the reactivity of the MFC/DAC-MA-*t*-Bu product was evaluated through its interaction (aza-Michael addition) with 2-propynylamine (propargylamine, ProAm), yielding an MFC/DAC-ProAm-*t*-Bu adduct featuring a terminal alkyne group (Fig. 24a).

In contrast, alkyne-functionalized MFC/DAC-PA-*t*-Bu was exposed to the copper-catalyzed Huisgen cycloaddition reaction with a fluorescent dye, namely, 5-fluorescein azide (5-FAM-Azide), that allows efficient labeling of biomolecules by click reactions (Fig. 24b). The observable fluorescence response confirmed the formation of the anticipated MFC/DAC-FAM-*t*-Bu cycloadduct, indicating the successful click ligation of the cellulose surface with the fluorescent tag.

Thus, the obtained Passerini-modified MFC/DAC, reactive derivatives decorated with methacrylate and terminal alkyne functions, could be conveniently post-derivatized by engaging aqueous coupling ligations providing new, attractive

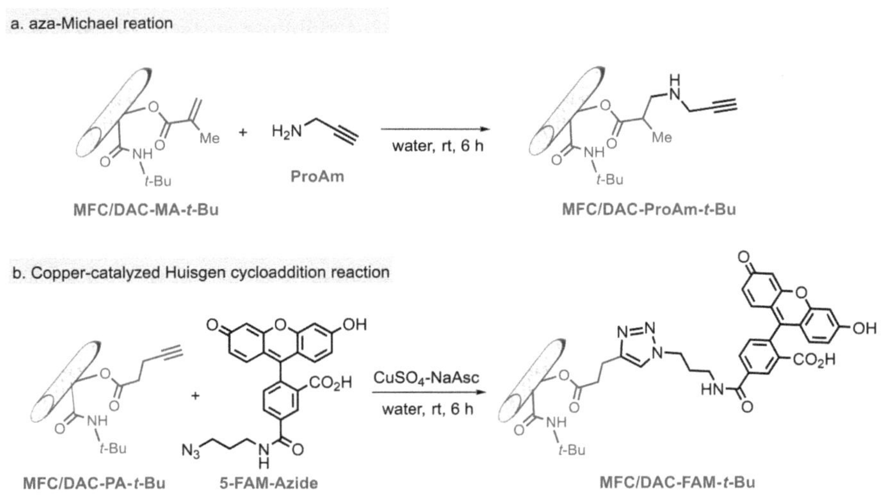

Fig. 24 Heterogeneous later alteration of Passerini MFC/DAC materials: **a** aza-Michael addition with 2-propynylamine (ProAm) and MFC/DAC-MA-*t*-Bu derivative to obtain MFC/DAC-ProAm-*t*-Bu adduct; **b** Cu-catalyzed Huisgen 1,3-dipolar cycloaddition using 5-FAM azide and MFC/DAC-PA-*t*-Bu derivative for preparing MFC/DAC-FAM-*t*-Bu cycloadduct. NaAcs (sodium salt of ascorbic acid) reduces copper(II) sources, like $CuSO_4$, in the cooper-catalyzed azide-alkyne reaction, i.e., the Cu-catalyzed Huisgen cycloaddition reaction

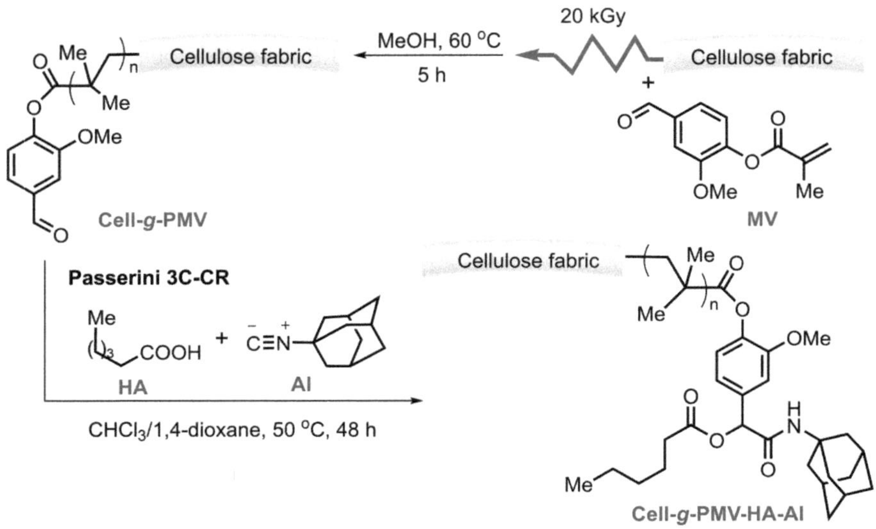

Fig. 25 Surface Passerini-3C-C reaction of Cell-*g*-PMV with HA and (AI) that conduces to forming Cell-*g*-PMV-HA-AI materials with the α-acyloxyamide function decoration

functional cellulosic materials, so-called tailor-made dually functionalized MFC/DAC-ProAm(FAM)-*t*-Bu products.

Another aldehyde function incorporated in the cellulose surface, which can be employed in the Passerini-3C-C reaction, is the mentioned above Cell-*g*-PMV, i.e., cellulose-embedded polymethacrylate vanillin polymer, prepared from pre-irradiated cellulose fabric at γ-irradiation of 20 kGy and methacrylated vanillin (MV). Kakuchi and coworkers explored the reactivity of this aldehyde cellulose derivative in the three-component condensation reaction. Mainly, the Passerini 3C-CR between 1-hexanoic acid (HA) and 1-adamantyl isocyanide (AI) resulted in the formation of the Cell-*g*-PMV-HA-AI product (Fig. 25) [67].

The reaction was conducted using the following reagent ratios: Cell-g-PMV ([CHO] = 1.0 equiv.), HA (200 equiv.), and AI (100 equiv.). The researchers observed that the PMV aldehyde conversion rapidly reached approximately 80% within two hours of reaction initiation, progressively increasing to accomplishment after three hours (83–87% conversion). The reaction medium influenced the yield of the final products; the surface Passerini-3C-C reaction in $CHCl_3$ proceeded inefficiently, while using 1,4-dioxane slightly enhanced the process efficiency. The relatively small reactivity of Cell-*g*-PMV-HA-AI materials in $CHCl_3$ was attributed to the mismatch of the polar and aprotic chloroform with the water-affine cellulose outers. The yield of the final products depends on the reaction medium: the surface Passerini-3C-CR reaction in $CHCl_3$ did proceed awkwardly, and the use of 1,4-dioxane slightly improved process efficiency. The relative low reactivity for Cell-*g*-PMV-HA-AI in $CHCl_3$ was endorsed to the incompatibility of apolar chloroform with the hydrophilic cellulose surfaces. At the same time, 1,4-dioxane is compatible with water but exhibits an elevated ability to dissolve polar substances and enhances the solubility of Cell-*g*-PMV. Consequently, cellulose-embedded polymethacrylate vanillin polymer (Cell-*g*-PMV) was exposed to the Passerini 3C-C reaction in a $CHCl_3$/dioxane mixture (1/1, v/v).

It is interesting to note that more simple "blocks" from biopolymers, like D-glucopyranose aldehydes and various saccharide molecules, were employed in the Passerini reaction with the respective classic reagents (carboxylic acids and isocyanides) to create innovative materials based on biomass [68, 69].

3.6 Povarov Reaction

This is an inverse electron-demand acid-catalyzed reaction of the 2-azadiene moieties of electron-poor N-aryl-methanimines derived from primary aryl amines and usually aromatic aldehydes and electron-rich olefins. During this reaction, two carbon–carbon and carbon–nitrogen σ-bonds are formed. At the same time, carbon–carbon and carbon–nitrogen π-bonds allow the easy construction of nitrogen-containing six-membered rings, especially tetrahydroquinoline and quinoline skeletons [70, 71]. Since then, this type of cycloaddition reaction, i.e., the imino-Diels–Alder (imino-DAR) or Povarov reactions, has become a popular, potent synthetic tool to generate numerous quinoline molecules, including alkaloids, agrochemicals, pharmaceuticals, and organic materials. Povarov reaction's popularity can be attributed to the reaction's simplicity, fast speed, extensive applicability, and mild reaction conditions.

However, such imino-DAR, involving cellulose derivatives, is still less developed. In contrast, the typical Diels–Alder reaction involving cellulose-based reagents is helpful in synthetic organic chemistry [72, 73, 75]. Kumar and colleagues described the first example of using modified cellulose as a biocatalyst for the Povarov-3C-C reaction in 2011.

The authors described the synthesis of 2-aryl-tetrahydroquinoline (THQ) libraries **42–43** using reaction between aldimines **39** (in situ produced from anilines **36** and benzaldehydes **35**) and cyclic dienophiles like pyran (**40**) or furan (**41**) derivatives in the presence of the mentioned above cellulose-SO_3H catalyst [74] in acetonitrile or water as solvents (Fig. 26) [76]. *Trans*- and *cis*-THQ molecules (**42** and **43**) were synthesized under similar conditions, with *trans*- being the predominant isomer and *cis*- the minor isomer when acetonitrile was used as the solvent.

Interestingly, using water as the reaction medium inverted the stereochemistry relationship of the products, favoring the *cis*-isomer. However, the diastereoselective formation in both solvents was neither highly efficient nor strongly pronounced. Notably, some resulting isomeric THQ compounds exhibited significant activity against both the attenuated (H37Ra) and the virulent (H37Rv) strains of the bacterium *Mycobacterium tuberculosis.*

In 2014, Merchán Arenas and Kouznetsov developed a green protocol for a solvent-free cycloaddition-cyclization cascade process involving Povarov-3C-C

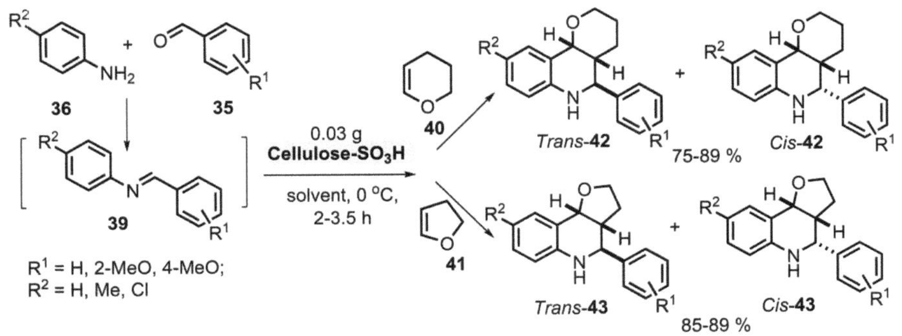

Fig. 26 Simple cellulose-SO_3H catalyst facilitates the three-component Povarov reaction in acetonitrile or water

reaction and intramolecular cyclization with lactam ring formation. The cascade reaction, promoted by acid AMCell-SO_3H biopolymer (an amorphous milled sulfonated cellulose), could provide diverse isoindoloquinolinone molecules, which display compelling activity against various types of infection and cancer cell lines, strongly inhibiting enzymes that regulate the structure of deoxyribonucleic acid such as DNA gyrase and Top2 (topoisomerase II) (Fig. 27) [77].

The desired densely diverse tetracyclic compounds **47** were *trans*-selectively prepared with good performances through solvent-free condensation reaction between various anilines **44**, 2-formyl benzoic acid **45**, and *trans*-isoeugenol **46a** or *trans*-anethole **46**b using an amorphous milled cellulose (AMCell) containing sulfonic functions as a green promoter at 90 °C for 4–8 h. The recyclability of this new catalyst was evaluated throughout five consecutive cycles.

Notably, the sulfonated AMCell promoter displayed outstanding reusability, as it was readily recuperated through filtration and reprocessed multiple times with minimal loss of catalytic activity. The resulting products retained biologically significant structural frameworks. The AMCell support was obtained from a commercial microcrystalline cellulose product employing a high-speed vibratory ball mill set at

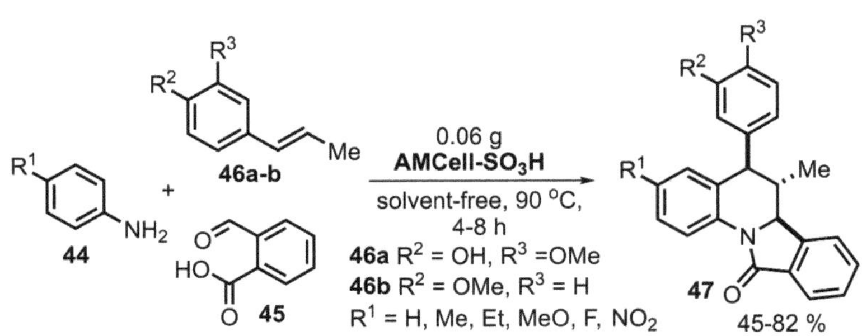

Fig. 27 Solvent-free Povarov-3C-C reaction followed intramolecular cyclization with lactam ring formation promoted by the acid AMCell-SO_3H catalyst

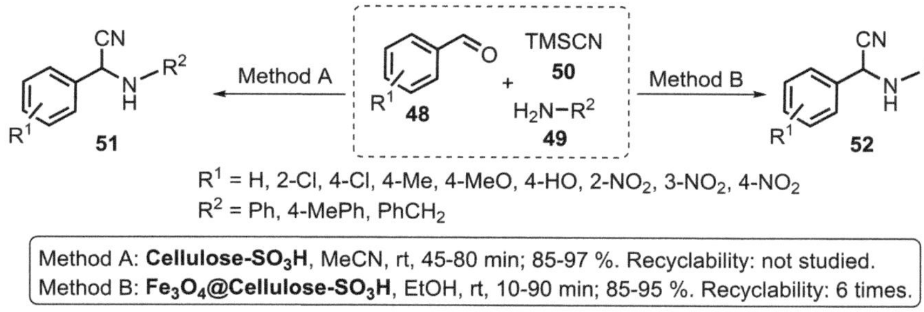

Fig. 28 Sulfonated cellulose-catalyzed Strecker reactions

60 Hz. The milling process involved a single agate ball (diameter: 51.6 mm, weight: 186.25 g) through impact within an agate dish [77].

3.6.1 Strecker Reaction

This provides α-amino nitrile compounds, significant intermediate products in the synthetic organic chemistry of crucially vital carbocyclic and heterocyclic derivatives. These can assist as suitable prototypes in medicinal chemistry and pharmacological exploration [78]. The Strecker reaction is the crucial 3C-C reaction, reported by Adolph Strecker for the first time in 1850, and it is the oldest example of this type of reaction. Nevertheless, like the Povarov reaction, the Strecker condensation reaction, comprising cellulose derivatives, has not yet progressed.

Only two similar procedures describe the preparation of α-amino nitrile compounds **51–52** employing the Strecker-3C-C reaction between benzaldehydes **48**, arylamines **49**, and trimethylsilyl cyanide (TMSCN) **50** in the presence of cellulose-based catalysts: cellulose-SO_3H was used as a bio-supported catalyst in this condensation (Fig. 28, method A) [79], while Fe_3O_4@cellulose-SO_3H was utilized as a green nanocatalyst in the same reaction (Fig. 28, method B) [80].

The straightforward reaction protocols offer excellent performance (high yields and remarkably short reaction times). The magnetic Fe_3O_4@cellulose-SO_3H nanocatalyst can be effortlessly separated from the reaction mixture and reprocessed in later reactions with minimal reduction of catalytic efficiency. Employing economical, innocuous, and noncorrosive solid acid catalysts, such as those described, is crucial for developing new methods to synthesize α-amino nitriles.

4 Conclusion

Cellulose is an essential renewable polymer material for use in the biomedical and pharmaceutical industries because it is widely available in nature, biocompatible, biodegradable, non-toxic, hydrophilic, antimicrobial, and easy to fabricate. Cellulose's versatility, sustainability, and cost-effectiveness make it attractive for catalysis and organic and medicinal chemistry in laboratory and industrial settings.

Cellulose-based catalysts are prepared using various strategies, either by modifying cellulose to introduce active catalytic sites or by using it as a support for metal-based or other catalysts. These catalysts are utilized in a wide range of reactions, including biomass conversion, green chemistry processes, and sustainable catalytic applications. The molecular design for functional materials and fine chemicals based on cellulose skeleton has also garnered general attention. On the other hand, in contemporary synthetic organic chemistry, numerous three-component condensation reactions, 3C-C reactions, named Betti, Biginelli, Kabachnik-Fields, Hantzsch, Passerini, Povarov, and Strecker reactions, play a significant role in developing medicinal chemistry.

This review highlights the use of 3C-C reactions for the amendment and functioning of cellulose and their applications in cellulose-based catalysis. Reactions such as the Hantzsch, Biginelli, and Passerini condensations are frequently employed to modify cellulose derivatives like dialdehyde cellulose, cellulose acetoacetate, and carboxymethylcellulose, leading to the development of innovative biomaterials. Among these, the Passerini reaction has emerged as a pivotal method for designing and synthesizing cross-linked hydrogels capable of absorbing and retaining substantial amounts of water without dissolving. Functionalization of cellulose acetoacetate fabric through the Hantzsch reaction enables the creation of fluorescent hydrogels-containing cellulose with stable fluorescence and/or enhanced hydrophobicity. Similarly, cellulose acetoacetate derivatives modified via the Biginelli reaction exhibit improved thermostability and solubility. In contrast, the Povarov, Betti, and Strecker reactions remain underutilized for surface cellulose modification. These functionalized cellulose materials hold promise for various applications, including meniscus implants, transparent films, antibacterial agents, hydrogels, aerogels, drug delivery systems, tissue engineering, and more. Among cellulose-base catalysts for 3C-C reactions, magnetic sulfated nanocellulose-based nanocomposites and simpler cellulose sulfonic acids are more employed as biocatalysts for synthesizing drug-like molecules valuable in sustainable medicinal chemistry. These catalysts stand out as more suitable for environmentally benign conditions.

Therefore, a clever combination of cellulose's versatility and the chemical possibilities of 3C-C reactions could open novel opportunities for cellulose assessment to generate diverse drug-like compounds and more efficient biomaterials sustainably.

References

1. Abdelhamid HN, Mathew AP (2022) Cellulose–metal organic frameworks (CelloMOFs) hybrid materials and their multifaceted applications: a review. Coord Chem Rev 451:214263
2. Afshari R, Shaabani A (2018) Materials functionalization with multicomponent reactions: state of the art. ACS Comb Sci 20(9):499–528
3. Alizadeh MH, Pooresmaeil M, Namazi H (2023) Carboxymethyl cellulose@ multi wall carbon nanotubes functionalized with Ugi reaction as a new curcumin carrier. Int J Biol Macromol 234:123778
4. Azad S, Mirjalili BBF (2019) One-pot solvent-free synthesis of 2,3-dihydro-2-substituted-1H-naphtho[1,2-e][1,3]oxazine derivatives using Fe_3O_4@nano-cellulose/TiCl as a bio-based and recyclable magnetic nano-catalyst. Mol Divers 23:413–420
5. Bahri F, Shadi M, Mohammadian R et al (2021) Cu-decorated cellulose through a three-component Betti reaction: an efficient catalytic system for the synthesis of 1,3,4-oxadiazoles via imine CH functionalization of N-acylhydrazones. Carbohydr Polym 265:118067
6. Cioc RC, Ruijter E, Orru RV (2014) Multicomponent reactions: advanced tools for sustainable organic synthesis. Green Chem 16(6):2958–2975
7. Cores Á, Clerigué J, Orocio-Rodríguez E et al (2022) Multicomponent reactions for the synthesis of active pharmaceutical ingredients. Pharmaceuticals 15(8):1009
8. Ellebracht NC, Jones CW (2019) Optimized cellulose nanocrystal organocatalysts outperform silica-supported analogues: cooperativity, selectivity, and bifunctionality in acid–base aldol condensation reactions. ACS Catal 9(4):3266–3277
9. Eyley S, Thielemans W (2014) Surface modification of cellulose nanocrystals. Nanoscale 6:7764–7779
10. Fazeli-Attar SA, Mirjalili BBF (2019) Nano-BF_3/cellulose as a biodegradable novel catalyst for synthesis of highly functionalized tetrahydropyridines. Iran J Catal 9(4):321–328
11. Guo L, Liu H, Peng F et al (2021) Efficient and portable cellulose-based colorimetric test paper for metal ion detection. Carbohydr Polym 274:118635
12. Guo L, Liu H, Peng F et al (2022) High stability and recyclable cellulose-based fluorescent paper derived from waste bagasse for anti-counterfeiting. Cellulose 29(10):5765–5778
13. Hamada T, Yamashita S, Omichi M et al (2019) Multicomponent-reaction-ready biomass-sourced organic hybrids fabricated via the surface immobilization of polymers with lignin-based compounds. ACS Sustain Chem Engin 7(8):7795–7803
14. Hamzah O, Vandenbrouck T, Heux L et al (2024) Insight into the hydrophobic functionalization of cellulose microfibrils using the Passerini three-component reaction. Carbohydr Polym 341:122323
15. Heinze T (2016) Cellulose: structure and properties. In: Rojas O (ed) Cellulose chemistry and properties: fibers, nanocelluloses and advanced materials, vol 271. Springer, Cham, pp 152
16. Hosseinikhah SS, Mirjalili BBF (2022) Fe_3O_4@ NCs/Sb (V): as a cellulose based nano-catalyst for the synthesis of 4H-Pyrimido[2,1-b]benzothiazoles. Polycycl Aromat Compd 42(4):1013–1022
17. Jahanbakhshi A, Farahi M (2022) Immobilized sulfonic acid functionalized ionic liquid on magnetic cellulose as a novel catalyst for the synthesis of triazolo [4, 3-a] pyrimidines. Arab J Chem 15(12):104311
18. Javanbakht S, Nazeri MT, Shaabani A et al (2020) Green one-pot synthesis of multicomponent-crosslinked carboxymethyl cellulose as a safe carrier for the gentamicin oral delivery. Int J Biol Macromol 164:2873–2880
19. Javanbakht S, Shaabani A (2019) Encapsulation of graphene quantum dot-crosslinked chitosan by carboxymethylcellulose hydrogel beads as a pH-responsive bio-nanocomposite for the oral delivery agent. Int J Biol Macromol 123:389–397
20. Kakuchi R (2019) The dawn of polymer chemistry based on multicomponent reactions. Polym J 51(10):945–953
21. Kakuchi R, Tsuji R, Fukasawa K et al (2021) Polymers of lignin-sourced components as a facile chemical integrant for the Passerini three-component reaction. Polym J 53(4):523–531

22. Kamel S, Khattab TA (2021) Recent advances in cellulose supported metal nanoparticles as green and sustainable catalysis for organic synthesis. Cellulose 28(8):4545–4574
23. Kargar PG, Noorian M, Chamani E et al (2021) Synthesis, characterization and cytotoxicity evaluation of a novel magnetic nanocomposite with iron oxide deposited on cellulose nanofibers with nickel (Fe_3O_4@ NFC@ ONSM-Ni). RSC Adv 11(28):17413–17430
24. Keglevich G, Bálint E (2012) The Kabachnik-Fields reaction: mechanism and synthetic use. Molecules 17(11):12821–12835
25. Khasimbi S, Ali F, Manda K et al (2021) Dihydropyrimidinones scaffold as a promising nucleus for synthetic profile and various therapeutic targets: a review. Curr Org Synth 18(3):270–293
26. Kheilkordi Z, Ziarani GM, Badiei A et al (2022). Recent advances in the application of magnetic bio-polymers as catalysts in multicomponent reactions. RSC Adv 12(20):12672–12701
27. Khine YY, Ganda S, Stenzel MH (2018) Covalent tethering of temperature responsive pNIPAm onto TEMPO-oxidized cellulose nanofibrils via three-component Passerini reaction. ACS Macro Lett 7(4):412–418
28. Kouznetsov VV (2009) Recent synthetic developments in a powerful imino Diels-Alder reaction (Povarov reaction): application to the synthesis of N-polyheterocycles and related alkaloids. Tetrahedron 65(14):2721–2750
29. Kouznetsov VV, Puerto Galvis CE (2018) Strecker reaction and α-amino nitriles: recent advances in their chemistry, synthesis, and biological properties. Tetrahedron 74(8):773–810
30. Kumar S, Arora A, Kumar S et al (2023) Passerini reaction: synthesis and applications in polymer chemistry. Eur Polym J 190:112004
31. Kumar B, Maity J, Shankar B et al (2021) Synthesis of D-glycopyranosyl depsipeptides using Passerini reaction. Carbohydr Res 500:108236
32. Kumar A, Srivastava S, Gupta G et al (2011) Natural product inspired diversity oriented synthesis of tetrahydroquinoline scaffolds as antitubercular agent. ACS Comb Sci 13(1):65–71
33. Kumar A, Srivastava S, Gupta G (2010) Supramolecular carbohydrate scaffold-catalyzed synthesis of tetrahydroquinolines. Tetrahedron Lett 51(3):517–520
34. Kurhade RR, Shaikh MS, Nagulwar V et al (2024) Advancements in carboxymethyl cellulose (CMC) modifications and their diverse biomedical applications: a comprehensive review. Int J Polym Mater 125
35. Lavanya G, Jariya AKI, SB, Sasikala M, et al (2024) Recent advances in the synthesis of dihydropyridine and their corresponding fused systems via multi-component Hantzsch reaction using catalytic nanomaterials. ChemistrySelect 9(43):e202403664
36. Liu X, Liu F (2023) Bimetallic (AuAg, AuPd and AgPd) nanoparticles supported on cellulose-based hydrogel for reusable catalysis. Carbohydr Polym 310:120726
37. Liu Y, Olsén P, Qi H (2023) Passerini three-component reaction for the synthesis of saccharide branched cellulose. Int J Biol Macromol 253:127367
38. Liu H, Yin Y, Zhou J et al (2022) Fabrication of durable fluorescent and hydrophobic cotton fabrics by multiple surface modifications. Ind Crops Prod 175:114238
39. Ma Z, Wang B, Tao L (2022) Stepping further from coupling tools: development of functional polymers via the Biginelli reaction. Molecules 27(22):7886
40. Maleki A, Akhlaghi E, Paydar R (2016) Design, synthesis, characterization and catalytic performance of a new cellulose-based magnetic nanocomposite in the one-pot three-component synthesis of α-aminonitriles. Appl Organomet Chem 30(6):382–386
41. Maleki A, Eskandarpour V, Rahimi J et al (2019) Cellulose matrix embedded copper decorated magnetic bionanocomposite as a green catalyst in the synthesis of dihydropyridines and polyhydroquinolines. Carbohydr Polym 208:251–260
42. Maleki A, Movahed H, Paydar R (2016) Design and development of a novel cellulose/γ-Fe_2O_3/Ag nanocomposite: a potential green catalyst and antibacterial agent. RSC Adv 6(17):13657–13665
43. Maleki A, Nooraie Yeganeh N (2017) Facile one-pot synthesis of a series of 7-aryl-8H-benzo[h]indeno[1,2-b]quinoline-8-one derivatives catalyzed by cellulose-based magnetic nanocomposite. Appl Organomet Chem 31(12):e3814

44. Maleki A, Ravaghi P, Aghaei M et al (2017) A novel magnetically recyclable silver-loaded cellulose-based bionanocomposite catalyst for green synthesis of tetrazolo[1,5-a]pyrimidines. Res Chem Intermed 43:5485–5494
45. Mazloum Tabaei MA, Bamoniri A, Mirjalili BBF (2022) One-pot Biginelli synthesis of 3,4-dihydropyrimidin-2(1H)-ones using nano-cellulose/BF3/Fe_3O_4. J Iran Chem Soc 19(7):2679–2691
46. Mazloum-Ardakani M, Arazi R, Mirjalili BBF et al (2019) Synthesis and application of Fe_3O_4@ nanocellulose/TiCl as a nanofiller for high performance of quasisolid-based dye-sensitized solar cells. Int J Energy Res 43(9):4483–4494
47. Merchan Arenas DR, Kouznetsov VV (2014) Diastereoselective synthesis of dihydroisoindolo[2,1-a]quinolin-11-ones by solvent-free AMCell-SO3H-catalyzed imino Diels–Alder/intramolecular amide cyclization cascade reactions. J Org Chem 79(11):5327–5333
48. Mirjalili BBF, Aref F (2018) Nano-cellulose/BF3/Fe_3O_4: a magnetic bio-based nano-catalyst for the synthesis of pyrimido[2,1-b]benzothiazoles under solvent-free conditions. Res Chem Intermed 44:4519–4531
49. Murthy YLN, Rajack A, Ramjid MT et al (2012) Design, solvent free synthesis, and antimicrobial evaluation of 1,4 dihydropyridines. Bioorg Med Chem Lett 22(18):6016–6023
50. Nazeri MT, Nasiriani T, Farhid H et al (2022) Sustainable synthesis of pseudopeptides via isocyanide-based multicomponent reactions in water. ACS Sustain Chem Eng 10(25):8115–8134
51. Novo DC, Edgar KJ (2024) Smart fluorescent polysaccharides: recent developments and applications. Carbohydr Polym 324:121471
52. Olyaei A, Sadeghpour M (2024) Recent advances in the transformation reactions of the Betti base derivatives. RSC Adv 14(17):11811–11848
53. Osanlou F, Nemati F, Sabaqian S (2017) An eco-friendly and magnetized biopolymer cellulose-based heterogeneous acid catalyst for facile synthesis of functionalized pyrimido[4,5-b]quinolines and indeno fused pyrido[2,3-d]pyrimidines in water. Res Chem Intermed 43:2159–2174
54. Patel P, Vaghani H, Kumbhani J et al (2024) Efficiency and diversity in chemical synthesis: exploring one-pot multicomponent reactions. Russ J Org Chem 60(9):1752–1760
55. Peng F, Liu H, Xiao D et al (2022) Green fabrication of high strength, transparent cellulose-based films with durable fluorescence and UV-blocking performance. J Mater Chem A 10(14):7811–7817
56. Pettignano A, Daunay A, Moreau C et al (2019) Sustainable modification of carboxymethyl cellulose by Passerini three-component reaction and subsequent adsorption onto cellulosic substrates. ACS Sustain Chem Eng 7(17):14685–14696
57. Pettignano A, Leguy J, Heux L et al (2020) Multifunctionalization of cellulose microfibrils through a cascade pathway entailing the sustainable Passerini multi-component reaction. Green Chem 22(20):7059–7069
58. Remy L, Sudre G, Charlot A et al (2023) α-Substituted ketones as reagent for Passerini modification of carboxymethyl cellulose: toward dually functionalized derivatives and thermo-sensitive chemical hydrogels. Carbohydr Polym 320:121228
59. Rong L, Zeng M, Liu H et al (2019) Biginelli reaction on cellulose acetoacetate: a new approach for versatile cellulose derivatives. Carbohydr Polym 209:223–229
60. Rostamnia S, Hassankhani A (2014) Application of biodegradable supramolecular polymer-supported catalyst for multicomponent synthesis of α-aminophosphonates Kabachnik–Fields reaction. Supramol Chem 26(1012):736–739
61. Safajoo N, Mirjalili BBF, Bamoniri A (2021) A facile and clean synthesis of indenopyrido[2,3-d]pyrimidines in the presence of Fe_3O_4@ NCs/Cu(II) as bio-based magnetic nano-catalyst. Polycycl Aromat Compd 41(6):1241–1248
62. Safajoo N, Mirjalili BBF, Bamoniri A (2019) Fe_3O_4@ nano-cellulose/Cu (II): a bio-based and magnetically recoverable nano-catalyst for the synthesis of 4H-pyrimido[2,1-b]benzothiazole derivatives. RSC Adv 9(3):1278–1283

63. Shaabani A, Maleki A (2007) Cellulose sulfuric acid as a bio-supported and recyclable solid acid catalyst for the one-pot three-component synthesis of α-amino nitriles. Appl Catal A General 331:149–151
64. Söyler Z, Onwukamike KN, Grelier S et al (2018) Sustainable succinylation of cellulose in a CO_2-based switchable solvent and subsequent Passerini 3-CR and Ugi 4-CR modification. Green Chem 20(1):214–224
65. Tamaddon F, Arab D, Ahmadi-Ahmad Abadi E (2020) Urease immobilization on magnetic micro/nano-cellulose dialdehydes: urease inhibitory of Biginelli product in Hantzsch reaction by urea. Carbohydr Polym 229:115471
66. Vala RM, Patel HM (2023) Recent developments in the Hantzsch synthesis of dihydropyridines. Adv Heterocycl Chem 141:179–208
67. Varga PR, Keglevich G (2021) Synthesis of α-aminophosphonates and related derivatives; the last decade of the Kabachnik-Fields reaction. Molecules 26(9):2511
68. Vekariya RH, Patel HD (2015) Cellulose sulfuric acid (CSA) and starch sulfuric acid (SSA) as solid and heterogeneous catalysts in green organic synthesis: recent advances. Arkivoc 1:136159
69. Vuillet C, Le Mouël C, Albertini D et al (2025) Dual functionalization of carboxymethyl cellulose and alginate via Passerini three-component reaction to graft two hydrophobic moieties: toward modular thin films. Carbohydr Polym 351:123066
70. Wang R, Sun S, Wang B et al (2021) Robust fabrication of fluorescent cellulosic materials via Hantzsch reaction. Macromol Rapid Commun 42(6):2000496
71. Wang Y, Wang X, Xie Y et al (2018) Functional nanomaterials through esterification of cellulose: a review of chemistry and application. Cellulose 25:3703–3731
72. Xie X, Shi J, Pu Y et al (2020) Cellulose derived nitrogen and phosphorus co-doped carbon-based catalysts for catalytic reduction of p-nitrophenol. J Colloid Interface Sci 571:100–108
73. Xin C, Chunxia C, Jinsong P (2021) Research progress of cellulose and its derivatives supported copper catalyst catalyzed organic reactions. Chin J Org Chem 41(4):1319–1336
74. Xiong L, Yu W, Lu X et al (2025) A novel dual-function antiscalant carboxymethyl cellulose-graft-protocatechuic acid for efficient scaling and biofouling prevention. J Membr Sci 713:123357
75. Xiong C, Zheng C, Zhang Z et al (2025) Polyaniline@cellulose nanofibers multifunctional composite material for supercapacitors, electromagnetic interference shielding and sensing. J Materiomics 11(1):100841
76. Yazdani H, Hooshmand SE, Stenzel MH (2022) Fusion of cellulose and multicomponent reactions: benign by design. ACS Sustain Chem Eng. 10(14):4359–4373
77. Younus HA, Al-Rashida M, Hameed A et al (2021) Multicomponent reactions (MCR) in medicinal chemistry: a patent review (2010–2020). Expert Opin Ther Pat 31(3):267–289
78. Yuan R, He X, Zhu C et al (2024) Recent developments in functional polymers via the Kabachnik-Fields reaction: the state of the art. Molecules 29(3):727
79. Zhu C, Yang B, Zhao Y et al (2013) A new insight into the Biginelli reaction: the dawn of multicomponent click chemistry? Polym Chem 4(21):5395–5400
80. Zinad DS, Mahal A, Mohapatra RK et al (2020) Medicinal chemistry of oxazines as promising agents in drug discovery. Chem Biol Drug Des 95(1):1647

Vladimir V. Kouznetsov earned his M.Sc. (1981) and Ph.D. (1986) in organic chemistry from Peoples' Friendship University in Moscow, USSR/Russia. From 1990 to 1992, he conducted postdoctoral research at the Enrique Moles Institute of Organometallic Chemistry at the University of Oviedo, Spain. In 1994, he completed his post-habilitation, receiving a Doctor of Chemical Sciences (D.Sc.) degree and was appointed Full Professor at Peoples' Friendship University. That same year, he joined the Industrial University of Santander in Bucaramanga, Colombia, where he is currently Laureate Professor and Director of the Laboratory of Organic and Biomolecular Chemistry within the School of Chemistry. Professor Kouznetsov has received numerous awards for his contributions to pharmacology, oncology, and phytopharmaceutical sciences. In 2012, he was honored with the National Science Award from the Colombian Association for the Advancement of Science. In 2014, he was elected a corresponding member of the Colombian Academy of Exact, Physical, and Natural Sciences. His current research focuses on heterocyclic diversity-oriented synthesis, green synthetic methodologies, medicinal chemistry, chemical biology, and organic materials chemistry.

Cellulose-Derived Matrices for Enhanced Healing and Regeneration in Plastic Surgery

Giovanni Miotti, Davide Quaglia, Pier Camillo Parodi, Mutali Musa, Fabiana D'Esposito, Caterina Gagliano, and Marco Zeppieri

Abstract Novel cellulose-based matrix products in plastic surgery promote tissue repair, wound healing, and soft tissue regeneration. Natural polysaccharide cellulose is non-toxic, biodegradable, and versatile, making it perfect for medical usage. This paper examines advances in cellulose-based matrix technology and its benefits and limitations. Due to its purity, structural stability, and moisture retention, bacterial cellulose (BC) seems promising for medicine. Plant-derived cellulose must be processed extensively to be sterile and compatible with human tissue. Cellulose bioengineering allows surgical matrices with regulated porosity, tensile strength, and

Giovanni Miotti and Davide Quaglia contributed equally and shared the first authorship.

G. Miotti · D. Quaglia · P. C. Parodi
Department of Plastic Surgery, University Hospital of Udine, Udine, Italy
e-mail: quaglia.davide.dott@icloud.com

P. C. Parodi
e-mail: piercamillo.parodi@uniud.it

M. Musa
Department of Optometry, University of Benin, Benin City, Nigeria
e-mail: mutali.musa@uniben.edu

F. D'Esposito
Imperial College Ophthalmic Research Group (ICORG) Unit, Imperial College, London, UK
e-mail: f.desposito@imperial.ac.uk

F. D'Esposito · C. Gagliano
Department of Medicine and Surgery, University of Enna "Kore", Enna, EN, Italy
e-mail: caterina.gagliano@unikore.it

C. Gagliano
Eye Center "G. B. Morgagni-DSV", Catania, Italy

M. Zeppieri (✉)
Department of Ophthalmology, University Hospital of Udine, Udine, Italy
e-mail: markzeppieri@hotmail.com

Department of Medicine, Surgery and Health Sciences, University of Trieste, Trieste, Italy

R. Malviya and S. Sundram (eds.), *Sustainable Cellulose Materials for Biomedical Applications*, Sustainable Materials and Technology,
https://doi.org/10.1007/978-981-96-9904-9_10

degradation rates. In reconstructive and cosmetic plastic surgery, cellulose-based polymers are useful. Cellulose matrices moisturize wounds, re-epithelialize, and reduce scars, improving aesthetics. Temporary wound coverings that encourage cell proliferation and angiogenesis assist in treating chronic wounds, burns, and donor sites in skin graft surgeries. Antibacterial cellulose formulations reduce surgical and traumatic wound infection risks. Cellulose scaffolds aid cell adhesion, migration, and differentiation, making them suitable for breast and face reconstruction. Recent research has incorporated growth factors, antibiotics, and nanoparticles into cellulose matrices to increase their therapeutic benefits. Advanced wound healing, inflammation, and vascularization can maximize complex wound treatment. Crosslinking agents have also made cellulose matrices more durable for load-bearing or high-movement areas like plastic surgery face and joints. Cellulose-based products offer potential, but degradation rates and biocompatibility are problems. In vitro investigations show promise, but in vivo and clinical trials are needed to standardize their use in various surgical settings. High-quality medical-grade cellulose matrices must be affordable and scalable for clinical application. Finally, cellulose matrix products are a breakthrough in plastic surgery biomaterials, providing wound dressings and tissue scaffolds for regeneration and cosmetic enhancement. Optimizing their properties, mixing bioactive chemicals, and conducting extensive clinical trials to show their usefulness and safety in reconstructive and aesthetic procedures are underway. As technology advances, cellulose-based matrices could replace synthetic materials in plastic surgery and enhance outcomes and patient satisfaction.

Keywords Cellulose-based matrices · Tissue regeneration · Wound healing · Diabetes · Burns · Biocompatibility · Scalability · Plastic surgery

List of Abbreviations

BC	Bacterial cellulose
BNC	Bacterial nanocellulose
CNCs	Cellulose nanocrystals
CNFs	Cellulose nanofibers
CMC	Carboxymethyl cellulose
HPMC	Hydroxypropyl methylcellulose
HEC	Hydroxyethylcellulose
PLA	Poly(lactic acid)
PEG	Polyethylene glycol
SA	Sodium alginate
NC	Nanocellulose
TEMPO	2,2,6,6-Tetramethylpiperidine-1-oxyl
MSCs	Mesenchymal stem cells
BC-PVA	BC-polyvinyl alcohol
MC	Methylcellulose

HPMC	Hydroxypropyl methylcellulose
ORC	Oxidized regenerated cellulose
3D	Three-dimensional
NPs	Nanoparticles
GF	Growth factors
ROS	Reactive oxygen species

1 Introduction

Cellulose, an abundant natural polymer, has garnered significant attention for its diverse applications in biomedicine due to its biocompatibility, biodegradability, and exceptional mechanical properties. Derived from plant cell walls, algae, fungi, and bacteria, cellulose exists in various forms, including cellulose nanocrystals (CNCs), cellulose nanofibers (CNFs), and bacterial cellulose (BC), each offering unique properties suited for specific applications. CNCs and CNFs are chemically and mechanically extracted from plants, while BC, the purest form, is synthesized by bacteria such as "*Gluconacetobacter*" and "*Agrobacterium.*" CNCs exhibit superior thermal stability, tensile strength, and crystallinity compared to CNFs, while BC offers exceptional purity and mechanical toughness. Cellulose and its derivatives, including carboxymethyl cellulose and hydroxypropyl methylcellulose, are extensively used in water treatment, biomedical applications, pharmaceuticals, and food technology [38].

Cellulose is a polysaccharide composed of β-D-glucose units linked by β-1,4-glycosidic bonds. This natural polymer exists in multiple forms, each distinguished by its source and structural characteristics:

1. Plant-Derived Cellulose: extracted from wood, cotton, and other plant materials, this cellulose often contains impurities such as lignin and hemicellulose. Processing methods like delignification and hydrolysis are required to enhance its purity and biocompatibility for medical applications.
2. Bacterial Cellulose (BC): synthesized by bacteria like "*Gluconacetobacter xylinus,*" BC is highly pure and possesses a unique nanofibrous structure. Its superior mechanical strength, high crystallinity, and biocompatibility make it suitable for advanced biomedical uses.
3. Cellulose Nanocrystals (CNCs): obtained through acid hydrolysis of cellulose fibers, CNCs are rod-like particles with high tensile strength and crystallinity. They are utilized to reinforce materials and create drug delivery systems.
4. Cellulose Nanofibers (CNFs): produced by mechanical fibrillation, CNFs exhibit both crystalline and amorphous regions, offering flexibility and strength for diverse applications
5. Cellulose Derivatives: modified forms like carboxymethyl cellulose (CMC) and hydroxypropyl methylcellulose (HPMC) expand the versatility of cellulose by improving solubility, gelation, and functionalization [21, 31, 37, 38, 56].

1.1 Methods of Cellulose Production

The production of cellulose can follow two primary routes: natural extraction or biosynthesis. Natural extraction involves isolating cellulose from plant fibers through processes such as alkaline or acidic treatments, which remove lignin, hemicellulose, and other impurities. Advanced enzymatic methods are increasingly employed to achieve high purity while reducing environmental impact. On the other hand, bacterial cellulose is synthesized by certain strains of bacteria, such as "*Komagataeibacter xylinus,*" in nutrient-rich media. The biosynthetic process offers advantages, including the absence of lignin, higher crystallinity, and the ability to tailor structural properties through variations in culture conditions. Despite these benefits, the scalability of bacterial cellulose remains a challenge due to its slow production rate and high costs. Innovations in bioreactor design and genetic engineering are actively pursued to address these limitations and enhance yield efficiency. Moreover, advancements in chemical functionalization have allowed cellulose derivatives to be tailored for specific biomedical applications. For example, the oxidation of cellulose can introduce carboxyl groups, enhancing its compatibility with hydrogel matrices or enabling conjugation with therapeutic agents. These methods, combined with ongoing research into sustainable production practices, position cellulose as a key material in advanced medical technologies. The development of hybrid production methods that combine synthetic and biosynthetic processes also holds the potential to improve efficiency and scalability.

1.2 Emerging Trends in Cellulose-Based Material Development

The development of cellulose-based materials has expanded significantly in recent years, with a focus on hybrid materials that combine cellulose with other natural or synthetic polymers. This approach aims to enhance the mechanical, chemical, and biological properties of cellulose for various advanced applications. For instance, the integration of cellulose with poly(lactic acid) (PLA) or polyethylene glycol (PEG) has resulted in composites with improved flexibility and degradability, catering to needs in tissue engineering and drug delivery. Additionally, the use of nanotechnology to develop cellulose-based nanoparticles and nanogels has broadened its applicability, especially in areas such as precision medicine, where controlled release and targeted drug delivery are critical.

1.3 Advances in Cellulose-Based Hydrogels

1.3.1 Properties and Synthesis

Hydrogels are three-dimensional, hydrophilic polymeric materials with a remarkable capacity to absorb and retain water due to functional groups like amine, hydroxyl, carboxyl, or sulphonic groups and osmotic pressure. Their structural integrity during swelling is ensured through physical or chemical crosslinking, with the latter being permanent. Hydrogels have versatile applications, including drug delivery, wound healing, water purification, and biocompatible coatings. They are classified as natural, synthetic, or semi-synthetic, each offering distinct advantages in terms of biocompatibility, durability, and biodegradability. Polysaccharide-based hydrogels, particularly cellulose and alginate, are widely studied for their biomedical and water treatment applications, with combined cellulose/alginate hydrogels showing promising potential. The structural and mechanical adaptability of cellulose-based hydrogels has also been a focus of recent research. By modulating the degree of crosslinking, researchers have been able to fine-tune the elasticity and strength of these materials for diverse biomedical applications. For example, hydrogels designed for cartilage repair require high mechanical integrity to withstand joint stresses, whereas those for wound healing prioritize flexibility and moisture retention [21, 31, 38, 56] (Donnaloja et al. 2022).

In wound healing, cellulose enhances fibroblast activity in the proliferative phase by releasing essential growth factors, expediting progression to the remodeling phase. Cellulose hydrogels promote tissue repair by boosting collagen deposition, angiogenesis, cell adhesion, and granulation tissue production. Their superior hemostatic and healing capabilities make cellulose hydrogels a promising medical dressing for wound treatment [4, 14, 28].

The development of hydrogels involves four principal techniques that enhance their mechanical and viscoelastic properties for biomedical use. Chemical crosslinking, a widely used method, consists of linking polymer chains or incorporating monomers using crosslinking agents like aldehydes. This approach is highly effective in creating robust networks by exploiting the reactive functional groups in synthetic and biopolymers, such as carboxylic, amine, and alcohol groups. Physical crosslinking, by contrast, has garnered significant attention for its simplicity and avoidance of chemical agents; by carefully adjusting parameters like pH, polymer concentration, and hydrocolloid type, a range of gel textures can be achieved through mechanisms such as ionic interactions, hydrogen bonding, and freeze-thawing cycles.

Radiation crosslinking offers another versatile approach, utilizing high-energy sources like gamma rays or electron beams to generate free radicals and induce polymer crosslinking. This method not only preserves the biocompatibility of natural polymers but also combines sterilization and modification into a single efficient process. The polymer environment—whether solid-state or in solution—plays a crucial role in determining the efficacy of this technique. Finally, graft polymerization involves covalently bonding monomers as side chains onto a primary polymer backbone, creating graft copolymers with enhanced functionality. These polymers, comprising structurally distinct grafted chains and main chains, allow for significant diversity in their applications. Despite these advanced methods and their resulting innovations, hydrogels remain constrained by challenges such as difficulty in securing dressings and reduced effectiveness in managing wounds with moderate-to-high exudate levels. These factors underscore the ongoing need to refine hydrogel technologies to fully harness their potential in wound care and other biomedical applications [39]. Bacterial cellulose (BC) and nanocellulose, derived from cellulosic biomass, are versatile materials for hydrogel development due to their unique structural and functional properties. BC, composed of β-1–4-linked glucoside chains with a crystalline arrangement, forms a highly hydrated, nonwoven gel with exceptional tensile strength and elasticity. Nanocellulose hydrogels, produced via chemical or mechanical methods, benefit from nanofiber entanglement, electrostatic stabilization, and chemical crosslinking, enhancing their stability and mechanical properties. Modified cellulose derivatives like hydroxyethyl cellulose (HEC) and hydroxypropyl methylcellulose (HPMC) improve hydrophilicity, strength, and antibacterial properties. These hydrogels exhibit superior tissue adhesion and antimicrobial performance compared to commercial alternatives.

In a rat tail amputation study, hydrogel outperformed gauze in hemostasis and proved more effective than Tegaderm™ in promoting wound healing [27].

Hydroxyethyl cellulose (HEC), derived from cellulose, enhances hydrogels' biocompatibility, hydrophilicity, and tunable mechanical properties owing to its water solubility and uniform solution-forming ability [44].

Wang et al. developed an injectable quaternized hydroxyethyl cellulose hydrogel coordinated with mesocellular silica foam, enabling rapid hemostasis and promoting effective wound healing [53].

1.3.2 Composite Hydrogels of Cellulose

Composite hydrogels of cellulose are prepared by blending with polymers or materials like graphene oxide to enhance functionality and versatility. Sodium alginate is a hydrophilic polysaccharide derived from brown algae, classified as an anionic polyelectrolyte composed of β-d-mannuronic acid (M units) and α-L-guluronic acid (G units) linked by 1–4-glycosidic bonds, forming distinct MG, MM, and GG block arrangements. Its structure, rich in free hydroxyl (–OH) and carboxyl (–COOH) groups, enables facile modification through techniques such as esterification, oxidation, amidation, sulfation, and graft copolymerization. These functionalization processes enhance its hydrophobicity, solubility, and biological properties, tailoring the material for a wide range of specialized applications [38].

Alginate-based hydrogels are extensively employed across diverse fields, including water treatment, agriculture, biomedical applications, and food packaging. However, their limitations, such as low thermal strength, poor mechanical stability, and restricted cell infiltration, necessitate enhancement. Incorporating cellulose or nanocellulose (NC) into alginate hydrogels significantly improves mechanical toughness, syneresis resistance, and water absorption and release properties, expanding their utility. NC addition enhances attributes like bead diameter, surface morphology, and structural compactness while also altering coloration. Various cellulose derivatives, including TEMPO (2,2,6,6-tetramethylpiperidine-1-oxyl)-oxidized nanocellulose, reinforce alginate hydrogels, yielding superior porosity, mechanical stability, and cell viability. Hydrogels are synthesized through chemical or physical crosslinking, leveraging the functional groups of cellulose and alginate to tailor network structure, porosity, and mechanical properties, optimizing their suitability for specific applications.

Cellulose-alginate-based hydrogels have gained widespread application in wound healing due to their exceptional biocompatibility, ability to maintain a moist physiological environment, efficient liquid absorption, and notable antibacterial properties. Moreover, these hydrogels exhibit a remarkable capacity for the controlled, sustained release of therapeutic agents, including antibiotics and analgesics, while concurrently promoting wound healing. To further enhance their efficacy, these hydrogels have been integrated with various inorganic materials, such as clays, iron oxide, and silver chloride, as well as organic substances like casein, chitosan, gelatin, aloe vera, and honey. In vitro and in vivo evaluations of inorganic fillers-loaded cellulose-alginate hydrogels, pure CA/alginate hydrogels, and organic fillers-loaded cellulose-alginate hydrogels have demonstrated their considerable potential in facilitating wound healing [53].

Tissue engineering utilizes advanced technologies to develop biomaterials, such as scaffolds or implants, to enhance tissue functionality and promote regeneration. Cellulose/alginate-based hydrogels are particularly effective in mimicking the extracellular matrix and facilitating tissue growth by supporting cell adhesion, differentiation, and migration more efficiently than synthetic polymers. These biopolymers have garnered significant interest in biomedical applications due to their hydrophilicity, water absorption, porous structure, mechanical strength, biocompatibility, and eco-friendliness. Furthermore, their ability to encapsulate cells and aid in the regeneration or repair of damaged tissue makes them promising materials for therapeutic purposes such as bone tissue engineering [50, 55], cardiac tissue engineering [40], human meniscus engineering [23], oral tissue regeneration[3], skin and soft tissue engineering [9, 46]. Figure 1 summarizes various applications of cellulose-alginate hydrogels.

1.3.3 Cellulose-Based Scaffolds for Mesenchymal Stem Cells (MSCs)

Mesenchymal stem cells (MSCs), valued for their regenerative potential and multilineage differentiation, are widely applied in biomedicine, either in their native form or integrated with biomaterials such as nanoparticles or polymer scaffolds. These materials—ranging from hydrogels like alginate and chitosan to polymers like PCL,

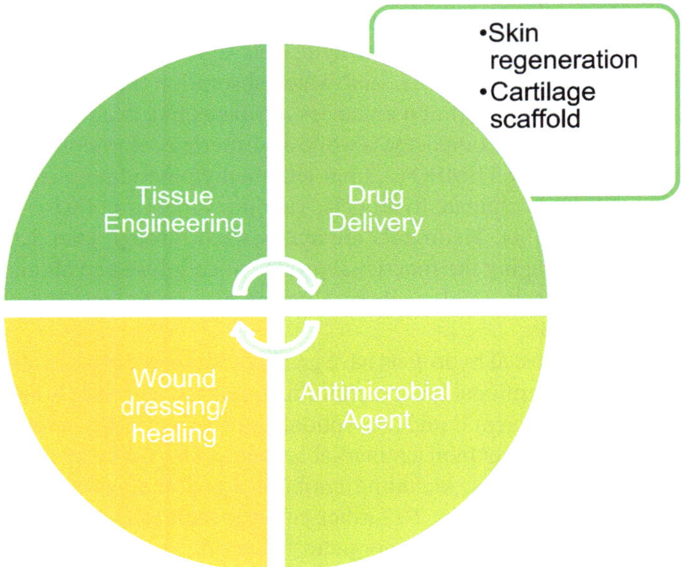

Fig. 1 Applications of cellulose-alginate hydrogels in plastic surgery and regenerative medicine

PLGA, and collagen—encapsulate or support MSCs, enabling advanced nanotherapies for wound healing, tissue engineering, and treating cardiac and neurological disorders.

Effective delivery of mesenchymal stem cells (MSCs) is essential, as evidence reveals their limited migratory efficiency, frequent entrapment in capillary networks, and short lifespan within the hematopoietic system. To overcome these limitations and ensure precise targeting, advanced MSC-based therapies utilize encapsulation within hydrogels or scaffolds or co-delivery with drug-loaded nanoparticles (NPs), thereby enhancing their therapeutic efficacy.

Among all scaffolds described, cellulose-based ones demonstrate exceptional potential in enhancing bone regeneration and promoting osteogenic differentiation. The structural alignment of these scaffolds can be precisely modulated using advanced fabrication techniques [7, 37]. For instance, cellulose-chitin nanofibrils can be engineered through reverse templating of hydrogen scaffolds, yielding a robust three-dimensional architecture that supports MSC adhesion and osteogenic differentiation. Furthermore, composite nanofiber mats composed of cellulose nanocrystals in combination with polylactic acid (PLA) or silk fibroin significantly improve the scaffold's biocompatibility, tensile strength, and elastic modulus, thereby advancing their utility in regenerative applications [7, 31].

2 Cellulose Types and Their Applications

2.1 *Bacterial Cellulose-Generalities*

Bacterial cellulose (BC), also known as microbial cellulose, is a high purity, gelatinous polysaccharide produced by certain bacteria, particularly "*Gluconacetobacter xylinum.*" This bacterium synthesizes cellulose at the liquid–air interface under aerobic conditions, where it forms flat, twisting ribbons. BC shares the same molecular formula $(C_6H_{10}O_5)_n$ as plant cellulose but differs significantly in its physical and chemical properties. The degree of polymerization in BC is much lower, typically ranging from 2000 to 6000, compared to plant cellulose's 13,000–140,000. BC is composed of twisted ribbon-shaped microfibrils with diameters of 50–100 nm and lengths of 1–9 μm, stabilized by extensive hydrogen bonding, which contributes to its robust structural integrity [1].

Bacterial cellulose (BC) is a homopolysaccharide synthesized through the synthase pathway, beginning with the precursor UDP-glucose. BC forms a three-dimensional nanostructured material characterized by high crystallinity and an extensive surface area, contributing to its functional properties and versatility in biomedical applications. It is widely used in wound dressings due to its biocompatibility, flexibility, superior water-holding capacity, vapor permeability, elasticity, and non-toxic nature.

The production of BC has seen exponential growth over recent decades. In 2016, the BC market was valued at USD 207.36 million and is projected to exceed USD 700 million by 2026. However, BC remains an expensive exopolysaccharide, partly due to the use of carbon sources such as glucose, fructose, and sucrose, combined with the low productivity of strains typically yielding less than 10 g/L. This economic factor poses challenges for scaling BC production for broader applications [39]. Bacterial cellulose (BC) holds immense potential in the biomedical sector, supporting advanced applications from diagnostics to regenerative tissue engineering. However, challenges in production remain significant. Variability in BC yields across Acetic Acid Bacteria (AAB) strains and dependence on costly carbon sources like glucose and fructose contribute to high production costs, accounting for ~ 30% of expenses. Industrial by-products, such as corn steep liquor (CSL), date syrup, and molasses, offer cost-effective alternatives that rival traditional media like Hestrin–Schramm. Additionally, agricultural and industrial wastes, including fruit juices, cotton textile waste, and dairy by-products, present untapped opportunities to enhance BC production efficiency.

Further innovation is needed to transition these alternative methods toward scalable biomedical applications. Strategies include integrating biopolymers or particulates into BC during synthesis and circumventing limitations of post-synthetic modifications like oxidation and grafting. These advancements could streamline BC production, reduce costs, and facilitate the development of high-value medical products [11]. Recent reviews have highlighted the advancements in functional BC-based materials for biomedical applications, including wound healing, drug delivery, tissue engineering, and cancer therapy. Specific BC structures like films, membranes, and fibers, as well as their composites, have been extensively researched for their potential to enhance therapeutic outcomes in various medical fields [39].

The biosynthesis of BC involves a multi-step biochemical pathway starting from glucose, catalyzed by enzymes including glucokinase, phosphoglucomutase, UDP-glucose pyrophosphorylase, and cellulose synthase. This pathway results in the production of cellulose through UDP-glucose as an intermediate. BC exhibits distinctive crystallinity, with a higher content of cellulose Iα (60%) compared to plant cellulose, which predominantly contains cellulose Iβ. The physical properties of BC, such as crystallinity, crystallite size, and cellulose content, are influenced by the cultivation conditions. Static cultures produce dense, leather-like pellicles at the liquid–air interface, while agitated cultures result in more branched, fibrous forms.

Bacterial cellulose (BC) has emerged as a promising material for wound healing due to its exceptional properties, such as high water absorption, moisture retention, and biocompatibility. Unlike conventional wound dressings like gauze, which can cause trauma upon removal, BC offers superior performance in promoting wound healing by maintaining optimal moisture levels, preventing infection, and accelerating tissue regeneration. BC membranes, produced by "*Gluconacetobacter xylinum*," can be directly applied as wound dressings or further processed to meet specific healing requirements. Early applications of BC as temporary skin substitutes, such as BioFill® (now Dermafill™), demonstrated its effectiveness in treating burns, cuts, and ulcers.

BC's ability to retain exudate and maintain moisture conditions conducive to healing is crucial for effective wound care. The nanofiber network of BC prevents bacterial infiltration, providing a physical barrier against infection. Furthermore, BC's high water vapor permeability facilitates the necessary moisture balance for efficient re-epithelialization and reduced healing time. Experimental studies have shown that BC dressings with optimal water vapor transmission rates significantly enhance wound healing by maintaining ideal moisture levels.

In addition to its moisture retention properties, BC exhibits impressive mechanical strength and flexibility, with a tensile strength of up to 260 MPa, making it suitable for use on dynamic wound sites such as joints. Its cytocompatibility, demonstrated by high fibroblast viability and enhanced cell attachment, promotes accelerated healing by supporting cell growth and tissue regeneration. BC's biodegradability allows for controlled degradation, making it adaptable to different wound environments [1].

2.1.1 Bacterial Cellulose (BC) and Plant Cellulose (PC)

Bacterial cellulose (BC) and plant cellulose (PC) represent two distinct forms of cellulose with unique properties that influence their clinical applications, particularly in wound healing. BC, synthesized by bacteria such as *Gluconacetobacter*, is inherently pure and highly crystalline. Its nanoscale fibers form a three-dimensional network resembling the extracellular matrix (ECM), making it highly suitable for biomedical uses. Conversely, PC, derived from sources like cotton or wood, requires extensive purification to remove lignin and other impurities, and its crystallinity is generally lower, with more structural variability.

BC's key advantages lie in its remarkable biocompatibility, which supports cell adhesion and proliferation, and its hydrophilic nature, which ensures excellent moisture retention. These qualities create a conducive environment for angiogenesis and tissue regeneration. Furthermore, BC's porosity and uniform pore structure facilitate oxygen exchange, which is crucial for wound healing. Its potential for functionalization, such as incorporating antimicrobial nanoparticles, enhances its clinical versatility. However, BC's production is resource-intensive, and its mechanical properties, including tensile strength, are generally inferior to PC, potentially limiting its utility in load-bearing applications. Additionally, BC lacks intrinsic antimicrobial properties, requiring chemical modifications to address this shortcoming.

PC offers robust mechanical strength and is readily available, making it more cost-effective. It can be chemically modified to enhance its hemostatic and therapeutic properties, such as in the production of sodium carboxymethyl cellulose. However, PC's lower hydrophilicity compared to BC can result in less effective moisture retention, a critical factor in wound care. Moreover, its initial composition often includes impurities, and its degradation in vivo is slower, potentially delaying integration in biomedical applications.

Clinically, BC demonstrates exceptional efficacy in wound healing, particularly for chronic wounds, where its ECM-like structure promotes cellular growth and expedites recovery. Its ability to maintain a moist environment while facilitating oxygen permeability makes it a favored choice in advanced wound care. On the other hand, PC's mechanical resilience and scalability lend it to applications such as tissue scaffolding and hemostasis. Its derivatives, such as oxidized regenerated cellulose, have shown promise in controlling bleeding and supporting tissue regeneration.

In summary, the choice between bacterial and plant-derived cellulose in clinical settings depends on the specific application. BC excels in biocompatibility and functional adaptability, while PC's advantages include greater mechanical strength and economic viability. Each offers unique contributions to advancing therapeutic materials and improving patient outcomes [32].

2.1.2 BC Applications in Regenerative Medicine and Plastic Surgery

While plant cellulose exhibits strong inter- and intramolecular interactions, its poor aqueous solubility necessitates modification into derivatives like carboxymethyl cellulose and hydroxypropyl cellulose for applications such as hydrogel-based wound dressings. Bacterial cellulose shares a similar chemical structure with plant cellulose but offers superior crystallinity, mechanical strength, hydrophilicity, and therapeutic properties. Despite its advantages, BC faces challenges, including susceptibility to microbial contamination and low productivity in culture. Pristine BC lacks inherent antimicrobial properties and adequate cell adhesion sites, necessitating its functionalization with antimicrobials, nanoparticles, and cationic peptides to broaden its applicability. Its structural complexity complicates 3D printing; thus, composites with polysaccharides like alginate or gelatin have been developed to facilitate bioprinting and create functional constructs.

Emerging approaches integrate next-generation antimicrobials that suppress virulent mechanisms and broaden BC's therapeutic potential. The incorporation of smart materials and sensors in printed BC constructs enables real-time wound monitoring and enhances clinical precision. Advances in BC composite development leverage external agents and cost-effective substrates to optimize its intrinsic properties for biomedical applications [24]. BC also offers versatility in pharmaceutical applications, enabling the incorporation of drugs or antibiotics for controlled release. Its unique, moldable nature allows BC to conform to various wound shapes, providing a customizable and comfortable solution not achievable with traditional dressings. These characteristics position BC as a highly effective and biocompatible material for advanced wound care [1].

Several notable applications of bacterial cellulose (BC) and its composites include their use in the development of artificial skin, wound dressings, tissue regeneration, and the production of facial masks for acne and pimple treatment. Millon et al. demonstrated that BC-polyvinyl alcohol (BC-PVA) composites could serve as potential substitute materials for articular cartilage [16, 30]. The primary medical application of bacterial cellulose (BC) is as a support material for skin wound dressings. Its absorbent properties are highly effective in absorbing wound exudates and gradually releasing moisture, thereby preventing the wound from drying out too rapidly. Kwak et al. showed that bacterial cellulose membrane may enhance the wound healing process in burn-injured skin of Sprague–Dawley rats by modulating angiogenesis and promoting the formation of connective tissue while not causing any discernible toxicity to the liver or kidneys [22].

In addition to pure bacterial cellulose, its composites with other polymers, nanomaterials, and biomaterials have demonstrated significant potential in wound healing applications. Composites incorporating zinc oxide (ZnO) and titanium dioxide (TiO_2) have exhibited enhanced biocompatibility, thereby showing promise in medical applications [6, 43]. Moreover, BC-based nanocomposites have also gained attention for their ability to address various tissue regeneration challenges. Specifically, the limited regenerative capacity of cartilage has led to the exploration of BC as a potential material for cartilage replacement. Ideal materials for cartilage substitutes must possess toughness, elasticity, non-toxicity, and biodegradability, which makes BC an excellent candidate. Studies have shown that chondrocytes cultured on BC exhibit high proliferation rates and collagen type II production. However, BC's inherent high crystallinity and poor biodegradability limit its applications. To overcome these challenges, modifications have been made to BC, such as the development of biodegradable and lysozyme-susceptible versions through metabolic engineering. These modified BCs, when used with human mesenchymal stem cells, successfully produced cartilage-specific extracellular matrix markers, suggesting their potential for cartilage regeneration [20].

BC has also demonstrated promise in the development of artificial blood vessels. The creation of BC tubes with precise dimensions immersed in silicone for enhanced mechanical strength highlights their potential for use as microvessels. Additionally, BC-polyvinyl alcohol (BC-PVA) composites have shown mechanical properties comparable to native porcine heart valves. Tubular hydrogels derived from BC-producing strains have successfully been used to replace carotid arteries in sheep for up to three months without signs of thrombogenic or inflammatory responses.

Furthermore, BC has been explored for the development of 3D organ-like hydrogels, showing potential for use in implantable tissue scaffolds for organs such as the kidney and liver.

Combining BC with alginate or other polymers improves its printability and bioactivity, with techniques like electrospinning enhancing fiber orientation. Innovations include incorporating antimicrobial agents like Cu nanoparticles and gold nanoparticles (GNPs) to enhance mechanical properties, biocompatibility, and osteogenic differentiation. BC composites demonstrate promise in bone regeneration, flexible sensors, and thermal insulation, leveraging BC's biodegradability, mechanical strength, and self-healing capabilities for diverse applications [16]. The commercialization of bacterial cellulose (BC) production faces significant economic hurdles due to its low yield, substantial capital investment needs, and high operational expenses. Reducing energy consumption and material preparation costs remains a critical challenge. The potential of BC in advanced material applications is further limited by insufficient understanding of optimal production conditions, difficulties in scaling up processes, and inefficiencies in separation and purification methods. As a result, BC-based wound dressings have yet to achieve widespread clinical adoption.

Exploring the use of low-cost substrates, such as agricultural and industrial waste from breweries, food production, and municipal sources, offers a promising path toward economically viable BC production. However, the high selling price of BC restricts its use to niche, high-value markets. To make BC-based wound dressings more accessible, development efforts should focus on creating cost-effective and straightforward production processes. Emphasis should also be placed on environmentally friendly methods, ensuring safety, eliminating potential toxicity, and achieving superior physical and chemical performance while incorporating sustainable sterilization techniques [54]. Bacterial Cellulose Matrices characteristics for wound healing are shown in Fig. 2.

2.2 *Nanocelluloses—Further Insights*

Nanocellulose, derived from natural sources like wood or bacteria, is ideal for biomedical applications, especially in advanced wound healing and dressings. There are three primary types of nanocellulose: cellulose nanocrystals (CNCs), cellulose nanofibrils (CNFs), and bacterial nanocellulose (BNC). BNC is particularly promising due to its high purity, mechanical strength, and biocompatibility, making it suitable for tissue engineering, wound healing, and infection prevention. BNC-based dressings also facilitate moisture retention, promote tissue regeneration, and can be functionalized with bioactive molecules for enhanced therapeutic properties.

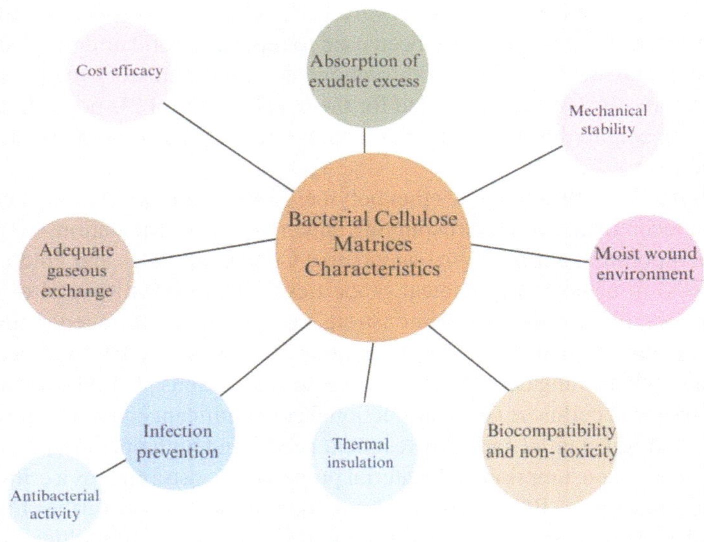

Fig. 2 Bacterial cellulose matrices characteristics for wound healing

BNC's unique properties include high water retention, flexibility, and ability to promote cell migration and angiogenesis. Its similarity to collagen enhances tissue regeneration, making it effective for managing various wound types, including chronic and traumatic wounds. BNC's antimicrobial properties can also prevent infections, and it can be functionalized with growth factors, antibiotics, or peptides to speed up healing. However, the high production cost of BNC remains a challenge, requiring cost-reduction strategies to make it more accessible for commercial use.

BNC's ideal characteristics for wound dressings include bacterial protection, oxygen circulation, moisture retention, and biocompatibility. Produced by bacteria like "*Komagataeibacter xylinus*," BNC varies in size and mechanical properties depending on the bacterial strain used. BNC's hydrophilicity, flexibility, and ability to adapt to wound shapes make it highly effective in wound care. It also promotes faster healing by reducing pain, accelerating epithelialization, and maintaining moisture balance. Despite these advantages, BNC lacks inherent antibacterial properties, limiting its standalone use in wound healing. Therefore, it must be combined with antimicrobial agents for optimal performance. Research into optimizing BNC production and incorporating advanced monitoring technologies further enhances its potential in wound healing applications. Bacterial infections in wounds such as burn scars, surgical sites, and trauma can delay healing, cause pain, and reduce quality of life. To address this, BNC (bacterial nanocellulose)-based wound dressings have been developed with antimicrobial agents to prevent infection and promote healing.

The addition of antibiotics like vancomycin and ciprofloxacin, or antibacterial materials such as chitosan, silver nanoparticles (AgNPs), metal oxides (ZnO, TiO_2), and other compounds, has proven effective in reducing bacterial growth in wounds.

In one study, BNC-based dressings were fabricated using electrohydrodynamic printing to load antibiotics such as amoxicillin, ampicillin, and kanamycin. The presence of these antibiotics increased the porosity, water absorption, and swelling ratios of the mats, resulting in controlled drug release over 14 days and effective antibacterial action against "*Staphylococcus aureus*" and "*Escherichia coli*." Silver nanoparticles (AgNPs) were also used, where they interacted with bacterial membranes and exhibited strong antimicrobial and antioxidant activities, enhancing cell proliferation and promoting vascular network formation. Additionally, zinc oxide (ZnO) nanoparticles integrated into BNC dressings showed broad-spectrum antibacterial effects and promoted wound healing by stimulating fibroblast proliferation and collagen deposition. Titanium dioxide (TiO_2) nanoparticles also demonstrated antibacterial action and enhanced healing by promoting re-epithelialization and fibroblast migration.

The incorporation of plant-derived materials into BNC-based dressings, such as poly-L-lysine, polyvinyl alcohol (PVA), and polyacrylamide, further improved antibacterial properties and facilitated wound healing. Mussel-inspired hydrogels decorated with polydopamine (PDA) and loaded with doxycycline exhibited excellent antibacterial effects and promoted healing by reducing inflammation and accelerating tissue regeneration. The addition of plant extracts like *Boswellia serrata*, Curcumin, and *Teucrium polium* enhanced antimicrobial, anti-inflammatory, and collagen production activities, providing additional support for wound healing. BNC dressings can also influence the wound microenvironment to enhance healing. For example, the use of siRNA against matrix metalloproteinase 9 (MMP-9) in a BNC-based dressing accelerated diabetic wound healing by reducing inflammation. Smart wound dressings, incorporating materials like silver nanowires (AgNWs) and Ti_3C_2Tx MXene, offer real-time monitoring and responsive release of therapeutic agents. These materials promote healing by stimulating angiogenesis, collagen deposition, and epithelialization, as well as providing antibacterial protection through ROS production [52].

The integration of bacterial cellulose (BC) and MXene into composite materials is advancing the field of multifunctional nanocomposites for biomedical and engineering applications. MXene, a two-dimensional material characterized by high electrical conductivity, large surface area, and excellent electromagnetic interference (EMI) shielding, enhances the functional properties of composites. When combined, BC serves as a scaffold, while MXene imparts electrical and mechanical enhancements. BC/MXene composites exhibit a synergistic relationship where BC's porous network stabilizes MXene layers, improving film strength, flexibility, and conductivity. These composites find applications in wound dressings, wearable sensors, and EMI shielding. In wound care, BC/MXene hydrogels provide moist environments conducive to healing, while MXene's conductive properties enable electrical stimulation to accelerate tissue regeneration. For sensors, BC/MXene films demonstrate exceptional sensitivity, durability, and real-time response, making them suitable for monitoring biological activities like heartbeats and gestures. Additionally, their high EMI shielding efficiency makes them valuable in telecommunications and aerospace.

The synthesis of BC/MXene composites involves integrating MXene nanosheets into BC matrices through chemical and physical crosslinking. Advanced methods, such as vacuum filtration and low-temperature solvent treatments, ensure homogeneous dispersion and robust interfaces. Optimized composites maintain lightweight structures with enhanced mechanical and functional properties. The versatility of BC/MXene nanocomposites underpins their potential in high-performance biomedical devices and sustainable engineering applications, marking a significant step forward in material innovation [42].

A typical plastic surgery ward issue is represented by skin graft donor site healing: CNF dressings are a possibility when treating split-thickness skin graft donor sites. The NFC dressing was applied directly to the donor site and adhered gradually as it dehydrated during the initial days of treatment. In an RCT by Hakkarainen et al., donor site epithelialization occurred more rapidly with the NFC dressing than with Suprathel®. Healthy, epithelialized skin was consistently revealed beneath the NFC dressing, which self-detached 11–21 days post-application, in contrast to the 16–28 days observed with Suprathel®. On average, NFC dressings detached four days earlier. This accelerated detachment correlated with improved pliability, attributed to adjustments in the NFC material composition, which allowed for better attachment to the wound bed. Importantly, no allergic reactions or inflammatory responses to NFC were observed, and patients reported no increase in pain compared to traditional treatments. These findings suggest that NFC dressings are a promising option for skin graft donor site treatment. Their biocompatibility, ease of application, secure adherence, and natural detachment once healing is complete make them a valuable addition to the arsenal of wound care technologies [13].

2.3 Oxidized Regenerated Cellulose

Oxidized regenerated cellulose (ORC)/collagen and ORC/collagen/silver-ORC play significant roles in wound healing, impacting three of the four phases: hemostasis, inflammation, and repair. These materials exhibit hemostatic properties, as evidenced by Cheng et al., who highlighted their ability to control bleeding. This characteristic is particularly useful in achieving hemostasis following injury. During the inflammatory phase, elevated levels of matrix metalloproteinases (MMPs) and reactive oxygen species can impede healing, especially in chronic wounds. ORC/collagen helps mitigate these effects by absorbing wound exudate, creating a moist environment conducive to healing. Studies demonstrate ORC/collagen's ability to promote granulation tissue formation and re-epithelialization. Hart et al. reported improved fibroblast migration, proliferation, and accelerated repair in a diabetic mouse model. At the same time, a clinical trial observed greater wound size reduction and favorable wound fluid biochemistry within eight days of ORC/collagen application compared to standard care.

ORC/collagen/silver-ORC further enhances healing by reducing bacterial bioburden, thereby minimizing inflammation and infection risk. Gottrup et al. found fewer infections in diabetic foot ulcers (DFUs) treated with ORC/collagen/silver-ORC than those receiving standard care, underscoring its antimicrobial efficacy. In the repair phase, the collagen component supports fibroblast migration, collagen synthesis, angiogenesis, and granulation tissue formation. In a rat model, ORC/collagen combined with a hydrocolloid dressing enhanced re-epithelialization and increased growth factor concentrations compared to control wounds. This highlights its potential to improve dermal and epidermal healing.

Overall, ORC/collagen and its silver variant contribute significantly to wound healing by managing inflammation, reducing bacterial load, and facilitating tissue repair, making them valuable in chronic and acute wound management [47]. A meta-analysis of 1521 participants with chronic skin wounds assessed the efficacy of oxidized regenerated cellulose (ORC)/collagen dressings compared to controls. Among the participants, 763 used ORC/collagen, while 758 received standard care. ORC/collagen dressings showed significantly higher rates of complete wound healing, greater wound size reduction, and fewer adverse events. However, the analysis highlighted limitations, including small sample sizes in most studies and limited comparisons.

The protective effects of ORC/collagen are attributed to its ability to bind excess iron, absorb oxygen free radicals, and protect growth factors, helping to regulate the wound environment. Chronic wounds often exhibit prolonged inflammation and high protease activity, which impair healing by breaking down the extracellular matrix. ORC/collagen functions as a protease inhibitor, reducing the activity of proteases such as elastase, plasmin, and gelatinase, thus supporting tissue repair. Studies also demonstrated that ORC/collagen encourages fibroblast migration and proliferation, which facilitates epithelialization and accelerates wound closure. In diabetic mice, ORC/collagen improved wound closure and histological outcomes. Despite these advantages, there was no significant difference in the overall time to complete wound closure when compared with similar dressings.

While ORC/collagen dressings are effective in promoting healing and reducing protease activity, they require removal during re-epithelialization or when exudate ceases. The findings underscore the potential of ORC/collagen in managing chronic wounds but emphasize the need for further high-quality randomized controlled trials. These should explore variations in outcomes based on age, gender, and ethnicity to validate and expand the evidence [2]. Overview of different types of cellulose is given Table 1.

Table 1 Comparison of different types of cellulose

Type of cellulose	Source	Key properties	Applications in plastic surgery
Plant-derived cellulose	Wood, cotton, plant fibers	It requires purification to remove lignin and hemicellulose, as well as moderate mechanical strength and biocompatibility	Temporary wound coverings and scaffold for tissue regeneration
Bacterial cellulose (BC)	*Gluconacetobacter xylinus*	High purity, biocompatibility, superior mechanical strength, and hydrophilicity	Wound healing, skin grafts, burn treatment, scaffolds for tissue engineering
Cellulose nanocrystals (CNCs)	Acid hydrolysis of plant cellulose	High tensile strength, crystallinity, and thermal stability	Drug delivery systems, tissue reinforcement
Cellulose nanofibers (CNFs)	Mechanical fibrillation	Combination of crystalline and amorphous regions; flexible and strong	Advanced wound dressings and flexible scaffolds for reconstruction
Modified cellulose derivatives	Chemical modifications	Enhanced solubility, gelation, and functionalization. Examples: carboxymethyl cellulose and hydroxypropyl cellulose	Hydrogels, injectable wound fillers, and bioactive dressing materials

3 Cellulose 3D Bioprinting

In recent years, 3D bioprinting has emerged as a transformative technology in regenerative medicine, enabling the creation of tailored wound dressings with unparalleled precision and efficacy. Among the materials at the forefront of this revolution, cellulose, and its derivatives have proven invaluable due to their remarkable biocompatibility, tunable properties, and versatility. The application of cellulose in 3D bioprinting for wound care highlights its potential to meet the complex demands of personalized medicine [12]. 3D bioprinting is an advanced manufacturing technique that creates biologically functional structures by depositing layers of bioinks that include cells, biomaterials, and growth factors. This method is used to produce customized wound dressings, skin substitutes, and scaffolds for tissue engineering.

The primary approaches in 3D bioprinting involve extrusion-based, inkjet-based, laser-assisted, and stereolithography printing. Extrusion bioprinting uses a pressurized system to deposit thick bioinks, allowing for the inclusion of living cells and various biomaterials, though it faces challenges with resolution and complexity.

Inkjet bioprinting creates high-resolution patterns through droplet deposition but can be limited by nozzle clogging and compatibility with biomaterials. Laser-assisted bioprinting provides precision and maintains cell viability, making it suitable for viscous materials and in situ applications, although it remains expensive. Stereolithography employs photocrosslinkable bioinks to produce highly detailed constructs, although it is constrained by material compatibility and the effects of UV exposure.

These bioprinting techniques enable the production of structures that mimic the extracellular matrix (ECM), enhance wound healing, and support cellular functions. Cellulose-based bioinks play a crucial role in these innovations, offering customizable mechanical properties and biocompatibility, thereby improving the effectiveness of therapeutic solutions for complex biomedical needs [41]. Cellulose's compatibility with 3D bioprinting lies in its ability to form bioinks with desirable rheological and mechanical properties. Bioinks derived from cellulose exhibit shear-thinning behavior, enabling smooth extrusion through print nozzles and rapid gelation post-printing, which ensures structural fidelity. Furthermore, cellulose-based bioinks can be engineered to maintain optimal moisture levels, enhance cellular adhesion, and support the controlled release of therapeutic agents, all of which are essential for wound healing. 3D bioprinting allows the precise deposition of cellulose-based bioinks to fabricate scaffolds that mimic the microenvironment of healthy tissue. By tailoring pore sizes, geometries, and material compositions, these printed constructs can optimize cell proliferation, vascularization, and wound closure [12].

4 Factors of Cellulose

4.1 *Nanocellulose and Bioinks*

Nanocellulose, encompassing cellulose nanofibers (CNFs) and nanocrystals (CNCs), has gained prominence as a bioink component. CNFs are highly versatile, providing scaffolds with mechanical strength, shape fidelity, and moisture retention. They also facilitate cellular proliferation due to their anisotropic porous structures. Studies have demonstrated that CNFs can be combined with polymers like alginate or gelatin methacryloyl to create hybrid inks. These inks enhance printability and provide tunable mechanical properties, which are crucial for applications in tissue regeneration. CNFs improve scaffold printability, mechanical strength, and cell adhesion. For instance, CNF-based dressings exhibit anisotropic porous structures promoting cell proliferation and controlled drug release, aiding wound healing. CNCs, known for their rod-like structure and strength, enable the incorporation of bioactive agents, offering antibacterial and regenerative properties. CNC-based hydrogels enhance swelling, targeted therapeutic delivery, and structural integrity, demonstrating efficacy in in vivo wound healing studies.

BC, synthesized by bacteria, offers a pure nanofiber network resembling an extracellular matrix (ECM). BC-based inks combined with polymers like alginate or thermoplastics yield dressings with superior hydrophilicity, mechanical stability, and controlled drug release. Modifications such as carboxymethylation improve printability and bioactivity, fostering angiogenesis and re-epithelialization in chronic wounds. 3D bioprinted cellulose-based dressings represent a paradigm shift in wound management, merging material innovation with precise customization for enhanced healing outcomes [35]. CNCs, on the other hand, act as structural reinforcements within bioinks. When integrated into bioprinted scaffolds, CNCs improve the stability of porous structures and enable controlled swelling, which is essential for maintaining a conducive healing environment. CNC-based bioinks also show great promise in delivering antibacterial agents and bioactive molecules, making them particularly effective in addressing chronic and infected wounds.

4.2 Bacterial Cellulose: Mimicking Extracellular Matrices

Bacterial cellulose (BC) offers a nanofibrous network that resembles extracellular matrices, fostering cell adhesion and supporting tissue growth. BC's inherent properties, such as high purity, crystallinity, and water retention, make it a prime candidate for wound healing. BC-based bioinks have been used to develop 3D-printed dressings with enhanced mechanical properties and drug delivery capabilities. For example, combining BC with thermoplastics like polycaprolactone has resulted in scaffolds that provide sustained antibiotic release and exhibit tailored porosity for wound-specific needs.

4.3 Modified Cellulose Derivatives: Expanding Functionalities

Chemical modifications of cellulose, such as carboxymethylation and methylation, have expanded its application in 3D bioprinting. Carboxymethyl cellulose (CMC) is extensively utilized due to its excellent water absorbency, biocompatibility, and ability to form hydrogels. In 3D-printed dressings, CMC bioinks can be functionalized with growth factors or antimicrobial agents to enhance tissue regeneration and prevent infections. Recent advances include PRP-loaded CMC scaffolds that accelerate angiogenesis and epithelialization in diabetic wounds.

Methylcellulose (MC) and hydroxypropyl methylcellulose (HPMC) have also been integrated into bioinks to provide temperature-sensitive gelation and tunable degradation rates. These derivatives are often blended with other polymers, such as alginate or gelatin, to improve mechanical stability and enhance bioactivity.

Such composite bioinks are particularly advantageous in creating dynamic wound dressings that adapt to the healing process.

4.4 Tailored Wound Dressings Through 3D Bioprinting

The precision of 3D bioprinting enables the fabrication of dressings tailored to the specific dimensions and requirements of individual wounds. This customization includes the ability to control scaffold porosity, enhance oxygen and nutrient transport, and regulate the release of incorporated therapeutic agents. For instance, bioinks incorporating cellulose derivatives have been used to print scaffolds that deliver antibiotics and anti-inflammatory drugs directly to the wound site, minimizing systemic side effects.

Additionally, the incorporation of bioactive molecules, such as growth factors and cytokines, into cellulose-based bioinks has advanced the development of "smart" wound dressings. These dressings not only provide structural support but actively modulate the wound environment to promote faster and more effective healing.

4.5 Future Perspectives

While cellulose-based bioinks have unlocked significant potential in 3D bioprinting, challenges remain. Optimizing the rheological properties of bioinks to balance printability and biological performance is an ongoing endeavor. Moreover, integrating multiple functionalities—such as antimicrobial activity, controlled degradation, and enhanced cellular interactions—within a single scaffold demands further innovation.

Despite these challenges, the synergy between 3D bioprinting and cellulose materials represents a promising frontier in wound care. By harnessing the unique properties of cellulose and its derivatives, researchers are paving the way for highly effective, patient-specific solutions in regenerative medicine [2]. The integration of three-dimensional (3D) and four-dimensional (4D) printing represents an innovative frontier in manufacturing technologies, blending engineering, materials science, and biology. 3D printing enables precise fabrication of complex structures via layer-by-layer material deposition, while 4D printing introduces dynamic transformations triggered by stimuli such as temperature or water. These advancements are pivotal for tissue engineering, facilitating patient-specific scaffolds that promote functional tissue regeneration.

3D bioprinting technologies are categorized based on their ability to incorporate living cells, using techniques like fused deposition modeling (FDM), stereolithography (SLA), and selective laser sintering (SLS). Bioinks, comprising natural polymers such as alginate, cellulose, and chitosan, and synthetic polymers like polylactic acid (PLA) and polycaprolactone (PCL), allow the creation of mechanically stable, bioactive constructs. Cellulose-based bioinks, particularly those derived from cellulose nanofibrils (CNFs), exhibit exceptional printability and tunable properties, enhanced through crosslinking with cations like Ca^{2+} and Mg^{2+}.

4D printing extends the functionality of 3D structures by incorporating time as a dimension, enabling shape memory and self-actuation. This technology employs responsive materials that adapt to external stimuli, offering temporal and spatial control over transformations. Recent breakthroughs include hydroxybutyl-methacrylated chitosan hydrogels for dynamic mobility and renewable soybean oil-based scaffolds for biomedical applications. Overall, 3D and 4D printing technologies herald transformative advancements in wound care and tissue engineering, delivering customized, multifunctional scaffolds with applications spanning from skin regeneration to antimicrobial treatments. These methodologies exemplify the potential of additive manufacturing in driving innovation across biomedical sciences [15].

5 Specific Applications of Cellulose Matrices in Regenerative Medicine and Plastic Surgery

5.1 Hemostatic Cellulose Materials in Clinical Practice

Cellulose-based materials, derived from delignified wood fibers, represent a versatile linear biopolymer with significant biomedical applications. Nanocellulose and its derivatives have garnered increasing attention due to their biocompatibility, high surface area, negative surface charge, non-toxicity, and cost-effectiveness. Among cellulose derivatives, oxidized cellulose stands out as a pivotal hemostatic agent, first introduced in 1942. The inaugural product based on regenerated oxidized cellulose, Surgicel®, debuted in 1960. Notably, oxidized non-regenerated cellulose demonstrates superior hemostatic performance compared to its regenerated counterpart, attributed to its frayed fiber structures, which enhance the available surface area for clot formation and hemostatic action; Develle et al. compared the efficacy and safety of two cellulose-based hemostatic agents, HemoStyp (a neutralized oxidized regenerated cellulose, NORC) and Surgicel (an oxidized regenerated cellulose, ORC), in managing surgical wound bleeding during abdominal, thoracic, and vascular procedures. Conducted as a prospective, multicenter, randomized, open-label trial, the research demonstrated that haemostasis was achieved significantly faster with NORC than ORC ($p < 0.0001$). All patients in the NORC group achieved haemostasis within 2 min, compared to 81% in the ORC group. For grade 1 bleeds, median bleeding control times were 24 s for NORC versus 51 s for ORC. For grade 2 bleeds, the times were 76 s and 116 s, respectively.

The findings highlight that NORC was more effective in achieving rapid haemostasis compared to ORC across surgical contexts involving mild to moderate bleeding. This supports its potential as a preferred choice for managing surgical wound bleeding efficiently [5, 25, 51]. Aerogels have garnered significant attention due to their exceptional properties, including ultralow density, extensive surface area, remarkable mechanical strength, and high porosity. Notably, aerogels exhibit a porosity exceeding 95%, surpassing the ideal threshold of 90% for effective hemostatic materials. This characteristic positions nanocellulose fiber-based composite aerogels as highly promising candidates for applications as hemostatic sponges and scaffolds in tissue engineering [26].

5.2 *Burns*

Bacterial cellulose (BC) has established itself as a pivotal material in the treatment of burn wounds, primarily due to its ability to modulate wound healing and mitigate scarring. Produced by bacteria such as **Acetobacter xylinum**, BC exhibits a nanofibrous structure akin to the extracellular matrix (ECM), which facilitates cellular adhesion, proliferation, and tissue remodeling. BC's unique physicochemical characteristics, including high water retention capacity, exceptional gas permeability, and superior mechanical strength, render it an ideal candidate for burn wound management. Its inherent biocompatibility ensures no adverse reactions even during prolonged applications. Furthermore, BC can be synthesized in tailored dimensions and conformations, allowing precise adaptation to the irregular geometries of burn injuries.

One of BC's most significant roles lies in its regulation of fibroblast activity, directly influencing collagen deposition and alignment. Excessive fibroblast proliferation and disorganized collagen secretion are key contributors to hypertrophic scars and keloids, often associated with severe burns. BC mitigates these processes, promoting a more organized tissue architecture and reducing the visibility of scars. The healing of deep burns is frequently complicated by dysregulated fibroblast activity, leading to the excessive accumulation of extracellular matrix components such as collagen. BC has demonstrated an ability to downregulate macrophage markers, such as scavenger receptor-A (SR-A), associated with inflammation. This action shifts the macrophage phenotype from pro-inflammatory (M1) to pro-healing (M2), reducing inflammation and fostering tissue regeneration.

Histological studies of BC-treated wounds reveal improved collagen organization, reduced granulocyte infiltration, and a more structured wound matrix. These outcomes are critical for minimizing the development of hypertrophic scars, which are characterized by thickened, disordered connective tissue. BC membranes with engineered surface patterns, such as microgrooves or stripes, have proven effective in further controlling fibroblast activity. These designs limit cell adhesion and modulate collagen deposition, resulting in reduced scar size and a more organized fibrotic tissue. Scar regions treated with micropatterned BC membranes show angular, less extensive scars with reduced fibrous density, as evidenced by histopathological analyses.

The therapeutic potential of BC has been significantly enhanced through its integration with bioactive agents, stem cells, and growth factors. For instance, BC composites seeded with adipose-derived stem cells (ASCs) accelerate epithelialization and enhance collagen type I deposition. These composites also promote angiogenesis, a critical process for effective wound healing, while minimizing inflammatory responses. BC infused with anti-inflammatory agents, such as curcumin, has been shown to expedite wound closure while dampening inflammatory pathways. This dual action not only promotes faster healing but also reduces the likelihood of fibrosis and scar formation. Similarly, BC-based constructs augmented with angiogenic growth factors like vascular endothelial growth factor (VEGF) enhance vascularization in the wound bed, ensuring improved nutrient and oxygen delivery to the regenerating tissue.

BC has demonstrated exceptional efficacy in clinical scenarios, including pediatric burn patients. The material's ability to maintain a moist wound environment and provide a cooling effect alleviates pain while supporting optimal healing conditions. Moreover, its non-adherent properties allow for painless removal of dressings, significantly improving patient comfort and reducing the risk of secondary infections. BC plays a critical role in regulating collagen synthesis and alignment within the wound site. BC-treated wounds show enhanced deposition of well-aligned collagen fibrils, which contribute to the development of a more functional and aesthetically favorable regenerated tissue. The material also reduces collagen density within scars, thereby mitigating the formation of hypertrophic and keloid scars.

Histological evaluations confirm that BC accelerates epithelialization, decreases inflammatory cell infiltration, and facilitates dermal remodeling. These effects contribute to the formation of a resilient and elastic tissue matrix, effectively preventing the fibrotic outcomes commonly observed in severe burn injuries [17]. A recent meta-analysis by Wu et al. reveals that Aquacel-Ag dressings (primarily consisting of sodium carboxymethylcellulose and silver) significantly outperform silver sulfadiazine in burn treatment, offering faster wound healing, higher healing rates, reduced inflammatory markers (e.g., TNF-α), and fewer dressing changes. These dressings maintain a moist wound environment, inhibit eschar formation, and enhance epithelial cell proliferation, granulation tissue development, and autolytic debridement, accelerating recovery. Burn-induced inflammatory responses, such as elevated interleukin-6 and TNF-α levels, often result in immune impairment, pain sensitivity, and endothelial damage, increasing infection risks. Aquacel-Ag mitigates these effects by sustainably releasing active silver ions, ensuring prolonged bactericidal and anti-inflammatory efficacy.

Clinical evidence underscores its effectiveness across diverse contexts: Aquacel-Ag's ability to shorten wound healing in diabetic foot ulcers, and its efficacy is confirmed in pediatric head and facial burns, improving healing time and quality while reducing scar formation. Safety data from eight studies showed no significant adverse reactions, though further research is needed to conclusively determine its safety superiority. Overall, Aquacel-Ag emerges as a potent, innovative dressing for enhanced wound management in burn patients, combining effective healing support with robust infection prevention; it conforms closely to the wound bed, minimizing dead space and creating a soft, moist environment that optimally supports the wound healing process [48].

In a randomized clinical trial by Karlsson compare the effectiveness of two dressing treatments for partial-thickness burns: a biosynthetic cellulose dressing (BsC) and a porcine xenograft. The findings revealed no significant differences between the two groups in terms of primary or secondary outcomes during the initial follow-up. Both dressings demonstrated comparable performance in healing time, pain management, infection rates, and the overall impact on daily activities. However, at the 12-month follow-up, patients treated with the porcine xenograft rated their scars as thicker and more pronounced in relief on the POSAS scale. Interestingly, these patient-reported differences were not corroborated by the observer evaluations. The study concludes that both the biosynthetic cellulose dressing and the porcine xenograft are effective for treating partial-thickness burns, with no significant distinctions in most outcomes [18].

5.3 Diabetic Wounds Treatment

The "TIME" framework underpins conventional chronic wound management, focusing on tissue debridement, infection control, moisture regulation, and wound edge assessment. Wound bed preparation is pivotal for facilitating healing, with debridement removing necrotic tissue to promote recovery. Infection management is critical, as pathogens amplify inflammation and impair tissue repair. Maintaining a warm, moist microenvironment fosters cellular activity, while wound edge evaluation monitors healing progress. However, these strategies are insufficient for diabetic wounds, with nearly 25% of patients requiring amputations. Advanced tissue engineering, including hydrogels and bioactive scaffolds, offers transformative potential for enhancing skin regeneration and improving patient quality of life [33, 34].

Unlike sponges and conventional bandages, natural polymeric hydrogels are ideal for diabetic wound dressings due to their biocompatibility, antibacterial properties, and ability to maintain a moist, microporous environment. Mimicking the extracellular matrix, they promote cell proliferation, migration, and gas exchange. Hydrogels can also deliver encapsulated drugs, enhancing wound healing and stimulating tissue regeneration[33]. Cellulose's high absorbent capacity makes it effective for intermediate to severe exudate wounds. Carboxymethyl cellulose (CMC), a hydrophilic, biocompatible polysaccharide, promotes fibroblast proliferation and migration, aiding diabetic wound healing. Bacterial cellulose (BC), produced by various bacterial species (e.g., "*Komagataeibacter xylinus*," offers excellent properties for wound dressings, including high water content, tensile strength, elasticity, and biocompatibility. Its nanofiber network and crystallinity support moist environments and exudate control. However, BC-based dressings often require antimicrobial treatments like fusidic acid, risking allergic reactions. Notably, "*K. intermedius*" produces BC with distinct properties, yet its unmodified use in diabetic wound healing remains underexplored [33].

Cellulose-based hydrogels, often blended with polymers and chemically modified to enhance mechanical strength and wound healing properties, have transformed antibacterial dressing design. However, further research is essential to develop advanced cellulose hydrogels capable of combating antibiotic-resistant strains and supporting biofilm-mediated wound regeneration. Hydrofiber dressings from CMC plant fibers form a soft gel upon contact with wound exudates. Combining CMC and carbomer 940 with "*Periplaneta americana*" extract effectively polarized M2 macrophages, enhanced collagen deposition, and promoted angiogenesis in diabetic wounds. CMC/sericin hydrogels, with antioxidant, anti-inflammatory, and antibacterial properties, accelerated re-epithelialization in diabetic rats. Cellulose-derived cotton gauze dressings facilitated wound healing by releasing growth factors (FGF, EGF, PDEGF) to stimulate fibroblast migration, proliferation, and bacterial inhibition [8, 45].

Chronic wounds, particularly diabetic wounds, exhibit prolonged inflammation due to elevated reactive oxygen species (ROS) levels and poor vascularization. A novel sustained oxygenation system, comprising MnO_2 nanosheets embedded in a dual-network hydrogel of silk fibroin (SF) and carboxymethyl cellulose (CMC), addresses these challenges. The hydrogel, strengthened by enzymatic and UV crosslinking, shows a 116.2% increase in compression modulus and shear-thinning properties, enabling injectable and shape-customizable applications for irregular diabetic wounds. MnO_2 nanosheets catalyze ROS into oxygen, reducing oxidative stress (threefold) and increasing oxygen levels (17-fold) in vitro, facilitating angiogenesis, inflammation reduction, and extracellular matrix remodeling. Additionally, the hydrogel mitigates excessive matrix metalloproteinase (MMP) activity ($> 80\%$) in diabetic wounds. In vivo, the system achieves rapid wound closure ($> 76\%$ in 7 days, 100% in 14 days) compared to commercial dressings ($< 30\%$ and $< 80\%$, respectively). This bioengineered hydrogel offers a versatile and effective solution for personalized chronic wound management [36]. Bacterial cellulose (BC) reinforced with multi-walled carbon nanotubes (MWCNT) forms a robust composite for diabetic wound healing. The BC-MWCNT film exhibits strong antibacterial activity and accelerates wound closure (99% in 21 days vs. 77% in controls). Molecular studies reveal reduced pro-inflammatory cytokines (IL-1α, TNF-α) and enhanced VEGF expression, promoting faster healing, re-epithelialization, and healthy granulation tissue formation [19].

Adhesive hydrogels have emerged as a promising solution for chronic diabetic wound repair, though challenges remain, including inadequate wet adhesion, poor adaptability to irregular wounds, and oxidative stress. A mussel-inspired hydrogel dressing, comprising catechol-modified carboxymethyl cellulose (CMC-DA) and tannic acid, was developed to address these issues. This hydrogel exhibits injectability, self-healing, and rapid shape adaptability, with robust tissue adhesion (45.9 kPa) and sustained wet adhesion (14.9 kPa), suitable for irregular and mobile wound sites. Detachable by ice-cooling to avoid secondary damage, it offers antioxidant, antibacterial, and hemostatic properties with excellent biocompatibility. In vivo, the hydrogel effectively reduced inflammation, alleviated oxidative stress, and promoted neovascularization, significantly improving diabetic wound healing. This biosafe adhesive hydrogel holds substantial potential for advanced diabetic wound management [49]. In cases of chronic wounds, such as diabetic ulcers, prolonged healing times often lead to complications. Advanced formulations incorporating growth factors, like vascular endothelial growth factor (VEGF) or platelet-derived growth factor (PDGF), have shown promising results in enhancing tissue regeneration and reducing healing durations.

Bacterial cellulose-based wound dressings impregnated with "*Zanthoxylum acanthopodium*" microencapsulants have been successfully developed, showcasing significant promise in diabetic wound management. Adjustments in gelatin composition within the microencapsulants reduced their size. Swelling tests revealed that increasing gelatin content diminished swelling capacity in simulated exudate media. The dressings exhibited potent antibacterial activity, effectively inhibiting "*Escherichia coli*" and "*Staphylococcus aureus*." Moreover, they accelerated wound closure by a remarkable eightfold on day 17 compared to untreated wounds. While these findings highlight the therapeutic potential of these wound dressings, further research is required to substantiate their application. Key areas include TEM-based microencapsulation analysis, surface contact angle evaluation, drug release kinetics, and mechanical property characterization, ensuring robust and reliable clinical utility [10].

5.4 Aesthetic Medicine

BC properties make this compound ideal for applications such as facial masks, scrubs, and cleansing formulations. BC masks deliver active ingredients deeply into fine lines and wrinkles, surpassing conventional sheet masks. Its nanostructure and hydrogen bonding capabilities enhance the release and penetration of infused substances, such as essential oils and plant extracts, offering prolonged skin hydration and comfort. BC masks improve skin moisturization and are well-tolerated, with no significant irritation reported. Glycerin addition enhances moisturizing effects, while BC fragments improve transdermal permeation, exfoliation, and sebum absorption. Innovative BC-based products combine moisturization, whitening, anti-aging, and cleansing functions. BC masks exemplify advanced skincare technology, providing sterile, toxin-free solutions for enhanced hydration and active ingredient delivery. This innovation underscores BC's transformative role in cosmetology [29].

Comparing the properties of cellulose-based matrices with traditional materials is given in Table 2.

Table 2 Properties of cellulose-based matrices compared to traditional materials

Material	Biocompatibility	Hydration capability	Mechanical strength	Degradability	Cost	Applications in plastic surgery
Cellulose-based matrices	High	Excellent moisture retention	Moderate to high (customizable)	Controlled, tunable	Moderate	Wound healing, tissue scaffolds, burn treatment, skin regeneration
Collagen	High	Good	Moderate	Biodegradable	High	Skin grafts, wound healing, and injectable fillers
Synthetic polymers	Variable	Low to moderate	High	Slow, less tunable	Low to moderate	Implants, load-bearing applications, and temporary wound dressings
Hydrogels	High	Excellent	Moderate (requires crosslinking)	Biodegradable, tunable	Moderate	Injectable fillers, moist wound dressings, and drug delivery
Traditional dressings	Low to moderate	Low	Low	Non-biodegradable	Low	Basic wound coverage, protection from contamination

6 Conclusion

Cellulose-based materials represent a frontier in regenerative medicine, aesthetic, and plastic surgery and biomedical engineering. Their intrinsic properties, combined with advancements in nanotechnology and 3D printing, position them as pivotal in future healthcare innovations. Their application in wound healing, burn treatment, and reconstructive surgery demonstrates not only their therapeutic potential but also their capacity to improve cosmetic outcomes, significantly enhancing patient quality of life. Despite these advantages, challenges such as scalability and cost-effective production remain critical barriers. Overcoming these will require interdisciplinary collaboration between material scientists, clinicians, and industry stakeholders. The integration of cellulose with emerging technologies, such as bioelectronics and AI-driven manufacturing systems, could open new pathways for innovation, transforming personalized medicine and creating smarter, more responsive biomedical devices. Looking ahead, the adaptability of cellulose-based platforms positions them to address pressing healthcare challenges, from chronic wound management to organ regeneration. With sustained investment in research and the translation of laboratory findings to clinical applications, cellulose-derived materials are poised to play a transformative role in the future of medicine.

Financial Disclosures/Support and Conflict of Interests None.

References

1. Ahmed J, Gultekinoglu M, Edirisinghe M (2020) Bacterial cellulose micro-nano fibres for wound healing applications. Biotechnol Adv 41:107549. https://doi.org/10.1016/j.biotechadv.2020.107549. Epub 2020 Apr 14. PMID: 32302653
2. Chen Y, Du P, Lv G (2023) A meta-analysis examined the effect of oxidised regenerated cellulose/collagen dressing on the management of chronic skin wounds. Int Wound J 20(5):1544–1551. https://doi.org/10.1111/iwj.14009. Epub 2022 Dec 8. PMID: 36480562; PMCID: PMC10088825
3. Chiaoprakobkij N, Sanchavanakit N, Subbalekha K, Pavasant P, Phisalaphong M (2011) Characterization and biocompatibility of bacterial cellulose/alginate composite sponges with human keratinocytes and gingival fibroblasts. Carbohydr Polym 85(3):548–553. https://doi.org/10.1016/j.carbpol.2011.03.011
4. Deng Y, Yang X, Zhang X, Cao H, Mao L, Yuan M, Liao W (2020) Novel fenugreek gum-cellulose composite hydrogel with wound healing synergism: facile preparation, characterization and wound healing activity evaluation. Int J Biol Macromol 1(160):1242–1251. https://doi.org/10.1016/j.ijbiomac.2020.05.220. Epub 2020 May 29. PMID: 32479939
5. Develle R, Schaerf R, Najibi S, Conrad J, Abate G (2020) Efficacy and safety of regenerated cellulose topical gauze haemostats in managing secondary haemostasis: a randomised control trial. J Wound Care 29(11):670–677. https://doi.org/10.12968/jowc.2020.29.11.670. PMID: 33175627

6. Dieter K, Dieter S, Ulrike U, Silvia M (2001) Bacterial synthesized cellulose—artificial blood vessels for microsurgery. Prog Polymer Sci 26(9):1561–1603. https://doi.org/10.1016/S0079-6700(01)00021-1
7. Donnaloja F, Jacchetti E, Soncini M, Raimondi MT (2020) Natural and synthetic polymers for bone scaffolds optimization. Polymers (Basel) 12(4):905. https://doi.org/10.3390/polym12040905. PMID: 32295115; PMCID: PMC7240703
8. El-Samad LM, Hassan MA, Basha AA, El-Ashram S, Radwan EH, Abdul Aziz KK, Tamer TM, Augustyniak M, El Wakil A (2022) Carboxymethyl cellulose/sericin-based hydrogels with intrinsic antibacterial, antioxidant, and anti-inflammatory properties promote re-epithelization of diabetic wounds in rats. Int J Pharm 629:122328. https://doi.org/10.1016/j.ijpharm.2022.122328. Epub 2022 Oct 22. PMID: 36280221
9. Erkoc P, Uvak I, Nazeer MA, Batool SR, Odeh YN, Akdogan O, Kizilel S (2020) 3D printing of cytocompatible gelatin-cellulose-alginate blend hydrogels. Macromol Biosci 20(10):e2000106. https://doi.org/10.1002/mabi.202000106. Epub 2020 Aug 13. PMID: 32790232
10. Gea S, Putra IB, Lindarto D, Pasaribu KM, Saraswati Y, Karina M, Goei R, Tok AIY (2023) Bacterial cellulose impregnated with andaliman (*Zanthoxylum acanthopodium*) microencapsulation as diabetic wound dressing. Int J Biol Macromol 253(Pt 1):126572. https://doi.org/10.1016/j.ijbiomac.2023.126572. Epub 2023 Aug 28. PMID: 37648123
11. Gorgieva S, Trček J (2019) Bacterial cellulose: production, modification and perspectives in biomedical applications. Nanomaterials (Basel) 9(10):1352. https://doi.org/10.3390/nano9101352. PMID: 31547134; PMCID: PMC6835293
12. Gouripriya DA, Adhikari J, Debnath P, Ghosh S, Ghosh P, Thomas S, Ghandilyan E, Gorbatov P, Kuchukyan E, Gasparyan S, Saha P (2024) 3D printing of bacterial cellulose for potential wound healing applications: Current trends and prospects. Int J Biol Macromol 279(Pt 2):135213. https://doi.org/10.1016/j.ijbiomac.2024.135213. Epub 2024 Aug 30. PMID: 39216564
13. Hakkarainen T, Koivuniemi R, Kosonen M, Escobedo-Lucea C, Sanz-Garcia A, Vuola J, Valtonen J, Tammela P, Mäkitie A, Luukko K, Yliperttula M, Kavola H (2016) Nanofibrillar cellulose wound dressing in skin graft donor site treatment. J Control Release 244(Pt B):292–301. https://doi.org/10.1016/j.jconrel.2016.07.053. Epub 2016 Aug 1; PMID: 27491880
14. He W, Wu J, Xu J, Mosselhy DA, Zheng Y, Yang S (2021) Bacterial cellulose: functional modification and wound healing applications. Adv Wound Care (New Rochelle) 10(11):623–640. https://doi.org/10.1089/wound.2020.1219. Epub 2020 Sep 28. PMID: 32870775; PMCID: PMC8392072
15. Hema S, Unni VV, Gayathri, Niranjan B, Chandran S, Sambhudevan S (2023) Bio-based polymers containing traditional medicinal fillers for wound healing applications—an evaluation of neoteric development and future perspectives. Biotechnol J 18(8):e2300006. https://doi.org/10.1002/biot.202300006. Epub 2023 May 17. PMID: 37170732.
16. Islam SU, Ul-Islam M, Ahsan H, Ahmed MB, Shehzad A, Fatima A, Sonn JK, Lee YS (2021) Potential applications of bacterial cellulose and its composites for cancer treatment. Int J Biol Macromol 31(168):301–309. https://doi.org/10.1016/j.ijbiomac.2020.12.042. Epub 2020 Dec 11; PMID: 33316340
17. Jabbari F, Babaeipour V (2024) Bacterial cellulose as an ideal potential treatment for burn wounds: a comprehensive review. Wound Repair Regen 32(3):323–339. https://doi.org/10.1111/wrr.13163. Epub 2024 Mar 6; PMID: 38445725
18. Karlsson M, Elmasry M, Steinvall I, Huss F, Olofsson P, Elawa S, Larsson A, Sjöberg F (2022) Biosynthetic cellulose compared to porcine xenograft in the treatment of partial-thickness burns: a randomised clinical trial. Burns 48(5):1236–1245. https://doi.org/10.1016/j.burns.2021.09.017. Epub 2021 Sep 24; PMID: 34629186
19. Khalid A, Madni A, Raza B, Islam MU, Hassan A, Ahmad F, Ali H, Khan T, Wahid F (2022) Multiwalled carbon nanotubes functionalized bacterial cellulose as an efficient healing material for diabetic wounds. Int J Biol Macromol 1(203):256–267. https://doi.org/10.1016/j.ijbiomac.2022.01.146. Epub 2022 Jan 29; PMID: 35093443

20. Khan S, Ul-Islam M, Khattak WA et al (2015) Bacterial cellulose-titanium dioxide nanocomposites: nanostructural characteristics, antibacterial mechanism, and biocompatibility. Cellulose 22:565–579. https://doi.org/10.1007/s10570-014-0528-4
21. Kumar A, Lee Y, Kim D, Rao KM, Kim J, Park S, Haider A, Lee DH, Han SS (2017) Effect of crosslinking functionality on microstructure, mechanical properties, and in vitro cytocompatibility of cellulose nanocrystals reinforced poly (vinyl alcohol)/sodium alginate hybrid scaffolds. Int J Biol Macromol 95:962–973. https://doi.org/10.1016/j.ijbiomac.2016.10.085. Epub 2016 Oct 25; PMID: 27793679
22. Kwak MH, Kim JE, Go J, Koh EK, Song SH, Son HJ, Kim HS, Yun YH, Jung YJ, Hwang DY (2015) Bacterial cellulose membrane produced by *Acetobacter* sp. A10 for burn wound dressing applications. Carbohydr Polym 122:387–398. https://doi.org/10.1016/j.carbpol.2014.10.049. Epub 2014 Nov 13; PMID: 25817683
23. Lan X, Ma Z, Szojka ARA, Kunze M, Mulet-Sierra A, Vyhlidal MJ, Boluk Y, Adesida AB (2021) TEMPO-oxidized cellulose nanofiber-alginate hydrogel as a bioink for human meniscus tissue engineering. Front Bioeng Biotechnol 5(9):766399. https://doi.org/10.3389/fbioe.2021.766399. PMID: 34805119; PMCID: PMC8602093
24. Lan X, Du T, Zhuo J, Wang T, Shu R, Li Y, Zhang W, Ji Y, Wang Y, Yue X, Wang J (2024) Advances of biomacromolecule-based antibacterial hydrogels and their performance evaluation for wound healing: a review. Int J Biol Macromol 279(Pt 4):135577. https://doi.org/10.1016/j.ijbiomac.2024.135577. Epub 2024 Sep 11; PMID: 39270907
25. Lewis KM, Spazierer D, Urban MD, Lin L, Redl H, Goppelt A (2013) Comparison of regenerated and non-regenerated oxidized cellulose hemostatic agents. Eur Surg 45(4):213–220. https://doi.org/10.1007/s10353-013-0222-z. Epub 2013 Jul 4. PMID: 23950762; PMCID: PMC3739866
26. Lu T, Li Q, Chen W, Yu H (2014) Composite aerogels based on dialdehyde nanocellulose and collagen for potential applications as wound dressing and tissue engineering scaffold. Compos Sci Technol 94:132–138. https://doi.org/10.1016/j.compscitech.2014.01.020
27. Lu S, Zhang X, Tang Z, Xiao H, Zhang M, Liu K, Chen L, Huang L, Ni Y, Wu H (2021) Mussel-inspired blue-light-activated cellulose-based adhesive hydrogel with fast gelation, rapid haemostasis and antibacterial property for wound healing. Chem Eng J 417:129329. https://doi.org/10.1016/j.cej.2021.129329
28. Ma C, Du L, Guo Y, Yang X (2024) A review of polysaccharide hydrogels as materials for skin repair and wound dressing: construction, functionalization and challenges. Int J Biol Macromol 280(Pt 3):135838. https://doi.org/10.1016/j.ijbiomac.2024.135838. Epub ahead of print. PMID: 39317293
29. Mbituyimana B, Liu L, Ye W, Ode Boni BO, Zhang K, Chen J, Thomas S, Vasilievich RV, Shi Z, Yang G (2021) Bacterial cellulose-based composites for biomedical and cosmetic applications: research progress and existing products. Carbohydr Polym 273:118565. https://doi.org/10.1016/j.carbpol.2021.118565. Epub 2021 Aug 27; PMID: 34560976
30. Millon LE, Wan WK (2006) The polyvinyl alcohol-bacterial cellulose system as a new nanocomposite for biomedical applications. J Biomed Mater Res B Appl Biomater 79(2):245–253. https://doi.org/10.1002/jbm.b.30535. PMID: 16680717
31. Mohan T, Hribernik S, Kargl R, Stana-Kleinschek K (2015) Nanocellulosic materials in tissue engineering applications. Cellulose-fundamental aspects and current trends, pp 251–273. Textbook of pathology, 7/e ed. ISBN 9789351523697 https://doi.org/10.5005/jp/books/12412_10
32. Naomi R, Bt Hj Idrus R, Fauzi MB (2020) Plant- vs. bacterial-derived cellulose for wound healing: a review. Int J Environ Res Public Health 17(18):6803. https://doi.org/10.3390/ijerph17186803. PMID: 32961877; PMCID: PMC7559319
33. Oprita EI, Iosageanu A, Craciunescu O (2023) Natural polymeric hydrogels encapsulating small molecules for diabetic wound healing. Gels 9(11):867. https://doi.org/10.3390/gels9110867. PMID: 37998956; PMCID: PMC10671021
34. Oyebode OA, Jere SW, Houreld NN (2023) Current therapeutic modalities for the management of chronic diabetic wounds of the foot. J Diabetes Res 10(2023):1359537. https://doi.org/10.1155/2023/1359537. PMID: 36818748; PMCID: PMC9937766

35. Pita-Vilar M, Concheiro A, Alvarez-Lorenzo C, Diaz-Gomez L (2023) Recent advances in 3D printed cellulose-based wound dressings: a review on in vitro and in vivo achievements. Carbohydr Polym 321:121298. https://doi.org/10.1016/j.carbpol.2023.121298. Epub 2023 Aug 15. PMID: 37739531
36. Pu Y, Wang P, Yang R, Tan X, Shi T, Ma J, Xue W, Chi B (2022) Bio-fabricated nanocomposite hydrogel with ROS scavenging and local oxygenation accelerates diabetic wound healing. J Mater Chem B 10(21):4083–4095. https://doi.org/10.1039/d2tb00343k
37. Raghav PK, Mann Z, Ahlawat S, Mohanty S (2022) Mesenchymal stem cell-based nanoparticles and scaffolds in regenerative medicine. Eur J Pharmacol 918:174657. https://doi.org/10.1016/j.ejphar.2021.174657. Epub 2021 Dec 3; PMID: 34871557
38. Rana AK, Gupta VK, Hart P, Thakur VK (2024) Cellulose-alginate hydrogels and their nanocomposites for water remediation and biomedical applications. Environ Res 243:117889. https://doi.org/10.1016/j.envres.2023.117889. Epub 2023 Dec 10; PMID: 38086501
39. Revin VV, Liyaskina EV, Parchaykina MV, Kurgaeva IV, Efremova KV, Novokuptsev NV (2023) Production of bacterial exopolysaccharides: xanthan and bacterial cellulose. Int J Mol Sci 24(19):14608. https://doi.org/10.3390/ijms241914608. PMID: 37834056; PMCID: PMC10572569
40. Sedighim S, Chen Y, Xu C, Mohindra R, Liu H, Agrawal DK, Thankam FG (2023) Carboxymethyl cellulose-alginate interpenetrating hydroxy ethyl methacrylate crosslinked polyvinyl alcohol reinforced hybrid hydrogel templates with improved biological performance for cardiac tissue engineering. Biotechnol Bioeng 120(3):819–835. https://doi.org/10.1002/bit.28291. Epub 2022 Nov 26. PMID: 36412070; PMCID: PMC9931685
41. Tabatabaei Hosseini BS, Meadows K, Gabriel V, Hu J, Kim K (2024) Biofabrication of cellulose-based hydrogels for advanced wound healing: a special emphasis on 3D bioprinting. Macromol Biosci 24(5):e2300376. https://doi.org/10.1002/mabi.202300376. Epub 2023 Dec 20. PMID: 38031512
42. Talipova AB, Buranych VV, Savitskaya IS, Bondar OV, Turlybekuly A, Pogrebnjak AD (2023) Synthesis, properties, and applications of nanocomposite materials based on bacterial cellulose and MXene. Polymers (Basel) 15(20):4067. https://doi.org/10.3390/polym15204067. PMID: 37896311; PMCID: PMC10610809
43. Ul-Islam M, Khattak WA, Ullah MW et al (2014) Synthesis of regenerated bacterial cellulose-zinc oxide nanocomposite films for biomedical applications. Cellulose 21:433–447. https://doi.org/10.1007/s10570-013-0109-y
44. Wang C, Niu H, Ma X, Hong H, Yuan Y, Liu C (2019) Bioinspired, injectable, quaternized hydroxyethyl cellulose composite hydrogel coordinated by mesocellular silica foam for rapid, noncompressible hemostasis and wound healing. ACS Appl Mater Interfaces 11(38):34595–34608. https://doi.org/10.1021/acsami.9b08799. Epub 2019 Sep 10. PMID: 31464418
45. Wang T, Liao Q, Wu Y, Wang X, Fu C, Geng F, Qu Y, Zhang J (2020) A composite hydrogel loading natural polysaccharides derived from *Periplaneta americana* herbal residue for diabetic wound healing. Int J Biol Macromol 1(164):3846–3857. https://doi.org/10.1016/j.ijbiomac.2020.08.156. Epub 2020 Aug 22. PMID: 32841667
46. Wei J, Wang B, Li Z, Wu Z, Zhang M, Sheng N, Liang Q, Wang H, Chen S (2020) A 3D-printable TEMPO-oxidized bacterial cellulose/alginate hydrogel with enhanced stability via nanoclay incorporation. Carbohydr Polym 238:116207. https://doi.org/10.1016/j.carbpol.2020.116207. Epub 2020 Apr 8. PMID: 32299554
47. Wu S, Applewhite AJ, Niezgoda J, Snyder R, Shah J, Cullen B, Schultz G, Harrison J, Hill R, Howell M, Speyrer M, Utra H, de Leon J, Lee W, Treadwell T (2017) Oxidized regenerated cellulose/collagen dressings: review of evidence and recommendations. Adv Skin Wound Care 30(11S Suppl 1):S1–S18. https://doi.org/10.1097/01.ASW.0000525951.20270.6c. PMID: 29049055; PMCID: PMC5704727
48. Wu JJ, Zhang F, Liu J, Yao HJ, Wang Y (2023) Effect of silver-containing hydrofiber dressing on burn wound healing: a meta-analysis and systematic review. J Cosmet Dermatol 22(5):1685–1691. https://doi.org/10.1111/jocd.15639. Epub 2023 Jan 22. PMID: 36683314

49. Xie H, Shi G, Wang R, Jiang X, Chen Q, Yu A, Lu A (2024) Bioinspired wet adhesive carboxymethyl cellulose-based hydrogel with rapid shape adaptability and antioxidant activity for diabetic wound repair. Carbohydr Polym 334:122014. https://doi.org/10.1016/j.carbpol.2024.122014
50. Yan M, Chen T, Zhang S, Lu T, Sun X (2021) A core-shell structured alginate hydrogel beads with tunable thickness of carboxymethyl cellulose coating for pH responsive drug delivery. J Biomater Sci Polym Ed 32(6):763–778. https://doi.org/10.1080/09205063.2020.1866350
51. Yu P, Zhong W (2021) Hemostatic materials in wound care. Burns Trauma 9:tkab019. https://doi.org/10.1093/burnst/tkab019. PMID: 34541007; PMCID: PMC8445204
52. Zarepour A, Gok B, Budama-Kilinc Y, Khosravi A, Iravani S, Zarrabi A (2024) Bacterial nanocelluloses as sustainable biomaterials for advanced wound healing and dressings. J Mater Chem B 12(48):12489–12507. https://doi.org/10.1039/d4tb01024h. PMID: 39533945
53. Zhang R, Tian Y, Pang L, Xu T, Yu B, Cong H, Shen Y (2022) Wound microenvironment-responsive protein hydrogel drug-loaded system with accelerating healing and antibacterial property. ACS Appl Mater Interfaces 14(8):10187–10199. https://doi.org/10.1021/acsami.2c00373. Epub 2022 Feb 17. PMID: 35172579
54. Zheng L, Li S, Luo J, Wang X (2020) Latest advances on bacterial cellulose-based antibacterial materials as wound dressings. Front Bioeng Biotechnol 23(8):593768. https://doi.org/10.3389/fbioe.2020.593768. PMID: 33330424; PMCID: PMC7732461
55. Zheng Y, Wang L, Bai X, Xiao Y, Che J (2022) Bio-inspired composite by hydroxyapatite mineralization on (bis)phosphonate-modified cellulose-alginate scaffold for bone tissue engineering. Colloids Surf A Physicochem Eng Aspects 635:127958. https://doi.org/10.1016/j.colsurfa.2021.127958
56. Zou P, Yao J, Cui YN, Zhao T, Che J, Yang M, Li Z, Gao C (2022) Advances in cellulose-based hydrogels for biomedical engineering: a review summary. Gels 8(6):364. https://doi.org/10.3390/gels8060364. PMID: 35735708; PMCID: PMC9222388

Dr. Giovanni Miotti graduated in 2016 at the University of Udine and specialized with top marks in 2023 in Plastic, Reconstructive and Aesthetic surgery at the University of Trieste. He currently works at the plastic surgery department of the academical hospital of Udine and as a freelancer. His fields of interest are in particular facial and breast surgery and regenerative medicine. He has completed multiple publications (most of all about facial and periorbital surgery) and various work and study experiences from leading specialists in the field. He strongly believes in serious and independent research, as an essential growth factor for the medical profession and the training of new doctors.

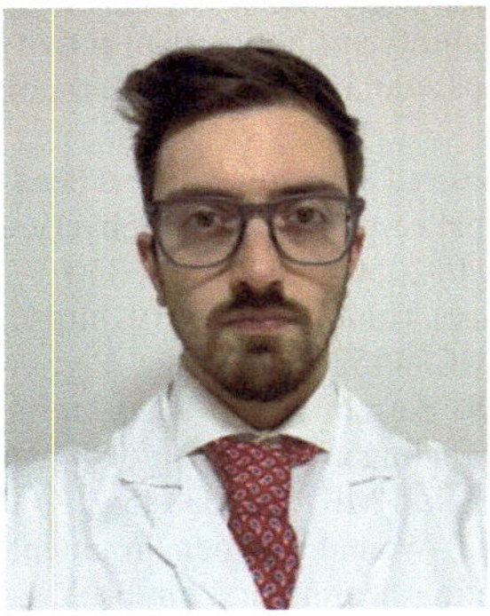

Dr. Davide Quaglia is currently a third year Plastic, Reconstructive and Aesthetic Surgery Resident at University of Trieste. He is attending as a trainee the Plastic Surgery Department of Ospedale Santa Maria della Misericordia in Udine. Devoted researcher, his interests embrace all fields of Plastic Surgery sub-specialties with distinguished commitment to oculo-plastics, that represents his main research publication domain. Novel materials and high-tech applications in Plastic Surgery are the milestone of his research work and the major uncharted territories to explore in his creed as researcher.

Prof. Pier Camillo Parodi attended the degree course in Medicine at the University of Parma, where he was graduated in 1985 and, subsequently, attended the School of Specialization in Plastic, Reconstructive and Aesthetic Surgery, obtaining the Specialization in 1990. In 1995, he also specialized in Maxillofacial Surgery at the University of Milan. In 2007, he became Associate Professor for the Plastic Surgery Disciplinary Sector at the Faculty of Medicine of Udine and Director of the SOC Clinic for Reconstructive Plastic Surgery of the "Santa Maria della Misericordia di Udine" University Hospital. He is author of over one hundred publications and multiple international book chapters, he boasts great national and international professional experience.

Dr. Mutali Musa is a consultant clinical Optometrist with 15 years of experience, specializing in both Academic and Primary eye care. He has played a crucial role in establishing and nurturing two of the busiest eye hospitals in southern Nigeria. He has experience in providing end-of-care Low Vision Care and managing a dedicated Glaucoma clinic. His Education has taken him from Doctor of Optometry, Masters in Vision Science, and Ongoing Ph.D. at the University of Benin. He also holds a Fellowship in Rehabilitative Optometry and Low Vision Care from the Postgraduate College of Optometrists. He is widely published and cited in top international journals and serves as a guest editor to many. His research interests include: clinical examination and counseling of glaucoma and cataract patients; visual fields, tonometry, and data analysis related to glaucoma disease progression; and, measurement of light properties as they interact with the human eye. He is a member of several Research Laboratories and academic societies including the Low Vision Assessment, Spectrophotometry, and Quality testing of Ophthalmic lenses unit at the University of Benin and the ocular Biometrics unit at Africa Eye Laser Center Ltd. He has received multiple scholarships and awards and has presented papers in local and international scientific sessions. He is currently a Lecturer at the Department of Optometry, University of Benin and is also an Eye Care Program Consultant to Vision Spring. He is the Head of Operations, and Senior Consultant Optometrist at Africa Eye Laser Center Ltd and Center for Sight Africa Ltd.

Fabiana D'Esposito is an Ophthalmologist and a Ph.D. in Neuro-ophthalmology. She qualified in Naples University "Federico II" and trained in London at the Institute of Ophthalmology, Department of Molecular Genetics (UCL). At present she is visiting professor of Ophthalmic Genetics at Naples University "Federico II", and Honorary Senior Lecturer at Imperial College. Member of ESASO Faculty, board member of the Italian Society of Pediatric Ophthalmology and Strabismus (SIOPS). Associate Editor of Scientific Reports and European Journal of Ophthalmology. Her current main activity consists in the clinical and molecular diagnosis of genetic eye diseases, and in genotype–phenotype correlations.

Caterina Gagliano is a Full Professor in Ophthalmology at the Kore University Enna (Italy) and is the Director of the Ophthalmology Department at the "G.B. Morgagni" Foundation, Catania (Italy). Head of the Ocular Immunology and Rare Diseases Unit AOU Policlinico—San Marco Hospital, Catania from 2012 to 2024. She worked as Evaluator for the European Commission "Horizon 2020." She was the President and Scientific Director of NEST (Neurovisual Science Technology srl), a transfer society that specializes in clinical and preclinical research design. Furthermore, she is the founder of GRIDO (Group of Research in Innovative Discovery in Ophthalmology) with main fields of interest including rare ocular diseases, severe inflammatory eye conditions, and neuro-ophthalmological diseases. Her work is conducted in a multidisciplinary environment, promoting continuous dialogues among colleagues and encouraging the participation of younger coworkers. She works in an environment surrounded by one of the largest gardens in Europe, the Villa Bellini in Catania. This, along with the sunshine of the beautiful Sicily, instills an overwhelming and engaging sense of optimism in her daily life.

Marco Zeppieri was born in Ontario, Canada and currently lives in Udine, Italy. He graduated with a Degree in Biology from the University of Toronto, Canada in1992. He obtained his Degree in Medicine and Surgery at the University of Milan-Bicocca, Italy in 2003. He then obtained his Specialization in Ophthalmology at the University of Udine in 2007. In 2012, he completed his Ph.D. from the University of Udine. He also completed a Post-Doctoral Glaucoma Research Fellowship at Discoveries In Sight, Devers Eye Institute in Portland, Oregon, USA in 2006. He obtained his National Qualifications as a 2° level University Professor in 2018. Since January 2008 to date, he has worked as an Ophthalmologist at the Department of Ophthalmology at Santa Maria della Misericordia Hospital in Udine, Italy. He has won several awards such as the Best Video SOI (2004), AIGS Travel Grant (2007), and ARVO GB Bietti Foundation for Ophthalmology Travel Grant (2007). His scientific interests include new techniques for the morphological and functional diagnosis of ocular hypertension and glaucoma, the use of stem cells for corneal wound lesions, ocular surface and lid diagnostics and treatments and community based screening procedures. He has participated as invited speaker at numerous national and international ophthalmology conferences. He is a Reviewer for numerous international peer reviewed journals. He is author of over 200 indexed scientific publications in international scientific journals and book chapters cited on PubMed.

Development of Dosage Form Using Cellulose-Based Materials

Rajendra Chauhan, Ramakant Joshi, Mandeep Kumar Singh, Deepesh Parashar, Wasim Akram, and Navneet Garud

Abstract Polymers represent a critical category of excipients employed in contemporary pharmaceutical technology, significantly contributing to the formulation of therapeutic dosage forms. Cellulose derivatives exhibit distinct physicochemical properties, including swellability, viscosity, biodegradability, pH sensitivity, and mucoadhesion, which influence their application in the formulation of diverse dosage forms. Biocompatible, biodegradable, and bioavailable excipients, notably cellulose and its derivatives, are crucial for drug delivery systems. These materials have been extensively researched for their distinctive encapsulating and binding capabilities, along with their ecologically sustainable attributes. They are often employed in continuous or regulated medication delivery systems. Cellulose and its derivatives typically modify the solubility and gelling properties of pharmaceuticals, resulting in diverse approaches to control drug release profiles. This chapter offers a brief summary of what is currently understood about the chemistry and structure of common cellulose derivatives and their use in drug delivery systems. The development of new cellulose-derived materials for sustained drug delivery, such as micro-cellulose (MC) and nano-cellulose (NC), is also addressed.

Keywords Cellulose · Cellulose derivative · Drug delivery systems · Biocompatible polymer · Dosage form development

R. Chauhan · N. Garud
School of Studies in Pharmaceutical Sciences, Jiwaji University, Gwalior, M.P, India

R. Joshi (✉) · D. Parashar · W. Akram
Amity Institute of Pharmacy, Amity University Madhya Pradesh, Gwalior, M.P, India
e-mail: joshiram78@gmail.com

M. K. Singh
Amity Institute of Pharmacy, Amity University Rajasthan, Jaipur, Rajasthan, India

R. Malviya and S. Sundram (eds.), *Sustainable Cellulose Materials for Biomedical Applications*, Sustainable Materials and Technology,
https://doi.org/10.1007/978-981-96-9904-9_11

List of Abbreviations

API	Active pharmaceutical ingredients
CMC	Carboxymethyl cellulose
CAB	Cellulose acetate butyrate
DSE	Degradative solvent extraction
DS	Degree of substitution
EC	Ethylcellulose
HPMC	Hydroxypropyl methylcellulose
HEC	Hydroxyethyl cellulose
HPC	Hydroxypropyl cellulose
IPCs	Inter polymer complexes
MC	Micro-cellulose
MLX	Meloxicam
MC	Methyl cellulose
NC	Nano-cellulose
MCC	Microcrystalline cellulose
NaCMC	Sodium carboxymethylcellulose
TDDS	Transdermal drug delivery systems

1 Introduction

In 1838, the French scientist Anselme Payen identified a substance during his investigation of several wood species, which he recognized was not starch (the stored form of glucose or sugar), yet could be decomposed into its constituent glucose molecules similarly to starch. After isolating this new substance from plant cell walls, he designated it as "cellulose." Payen identified its chemical formula after separating it from plant debris [63]. Cellulose is the predominant biopolymer in nature [17, 41]. The main ingredient in many natural fibers, such as cotton and higher plants, is cellulose [36, 70]. Except for the terminal ends, each cellulose molecule has three hydroxyl groups per anhydro-D-glucopyranose unit (AGU), constituting its extensive chains. The robust intramolecular and intermolecular hydrogen bonding among cellulose chains essentially explains its low solubility, making it insoluble in water and most common solvents [17]. Cellulose is employed in various goods, including composites, netting, upholstery, coatings, packaging, and paper, despite its limited solubility. Cellulose undergoes chemical modification to enhance processing capabilities and generate cellulose derivatives (cellulosics) designed for particular industrial applications [5]. Cellulosics are utilized in many biological applications, including blood purification membranes and similar ones, since they are generally robust, repeatable, recyclable, and biocompatible [24]. Therefore, cellulosics have expanded their application and created a window of opportunity through derivatization.

The most prevalent organic substance on Earth is probably cellulose, which is mostly produced by plants. It is the primary structural component of herbal tissues and cells. One long-chain polymer that occurs naturally and indirectly affects the human food chain is cellulose. This polymer has several uses in a variety of industries, including textiles, veterinary nutrition, wood and paper production, and as an excipient in the cosmetic and pharmaceutical industries. Numerous semi-synthetic derivatives of cellulose are widely utilized in the cosmetics and medicinal sectors. The two primary types of cellulose derivatives are cellulose ethers and cellulose esters, each of which has distinct mechanical and physicochemical characteristics. These polymers are frequently utilized in the manufacturing of pharmaceuticals and dosage forms. These substances are necessary for many pharmaceutical applications, including osmotic drug delivery systems, extended and controlled-release matrices, extended and delayed release coated dosage forms, bioadhesive and mucoadhesive, compression tablets as enhancers of compressibility, liquid dosage forms as stabilizers and thickeners, granules and tablets as binders, and semisolid preparations as gelling agents [85].

The pharmaceutical sector gradually adopted distinct characteristics. Cellulose is characterized by its stability, versatility, non-toxicity, and the ability to become water-soluble or swellable through straightforward chemical modifications. Cellulose derivatives, referred to as cellulosics, are commonly utilized as excipients in various pharmacological formulations; however, in specific contexts, they are regarded as the active ingredient of the medicinal product [19]. Currently, cellulose and cellulose-based polymers have gained significant importance in the pharmaceutical sector, driven by the development of new derivatives and the exploration of novel applications for existing compounds by researchers.

2 Sources of Cellulose and Their Chemistry

Plant cell walls include cellulose, a complex carbohydrate, as a structural element. Comprising approximately 33% of all plant matter, it is the most widespread organic polymer on Earth. Another valuable resource is cellulose, which is used in the manufacturing of paper, textiles, biofuels, and numerous other goods. Cellulose comes from a variety of sources, each having special qualities and possible applications [52].

One of the most popular sources of cellulose is wood pulp. Wood fibers are broken down mechanically or chemically to create wood pulp. The pulping process is the first step in the production of paper and synthetic cellulosic fibers like viscose rayon. Cotton fibers are found in bolls that are roughly 6 cm across. Many seeds are linked to multiple fibers in each boll; before using, the fibers are ginned away from the seeds. Historically, cotton and linen (a fabric derived from flax fiber) were used to make paper, and today, premium writing and money papers employ cotton linters, which are fibers too short to be spun into skeins. Dialysis membranes now contain liners. Recycling cellulose is still important, particularly for paper goods. The hairs on seeds, such as those of cotton, kapok, and milkweed, provide some cellulose. Plants including hemp, kenaf, flax (linen), ramie (a perennial Asian nettle), and jute produce what are collectively referred to as bast fibers from their stems. To release the fibers, these plants' stems are retted. In addition to "soft" bast fibers, "hard" cellulose fibers can be found in the leaves of plants like pineapple, bananas, and agave, particularly sisal. *Dictyostellium discoideum* is a slime mold that is neither plant nor animal. The fruiting bodies are supported by cellulose stalks. The significant amounts of cellulose found in some materials, such corn stover (the stalks and leaves of maize), wheat, or rice straw, are starting to be widely used for the generation of ethanol in addition to more conventional applications like direct burning. We can keep innovating and producing sustainable products that benefit people and the environment by investigating and creating new cellulose sources [31, 33].

A d-glucose unit and the non-reducing C_4–OH group make up one end of cellulose, while the reducing C_1–OH group with the aldehyde structure is the terminating group. Certain technological celluloses, such as bleached wood pulp, have additional carboxy and carbonyl groups. It's important characteristics hydrophilicity, degradability, chirality, and chemical unpredictability because of the donor group's strong reactivity are caused by the molecular structure. Cellulose gains crystalline fiber architectures from the superior hydrogen bonding. Cellulose production from plants is the most well-known and often utilized method. It is well-known that cotton seed hairs contain the purest form of cellulose. Large-scale chemical pulping and purifying procedures, on the other hand, further isolate the wood cellulose from the composite it forms with lignin and other polysaccharides. Apart from the majority of plants, cellulose can also be derived from fungi, algae, and certain types of bacteria. Research on cellulose architectures, reactivity, and crystallinity is conducted using supramolecular structures, with additional attention paid to the creation of novel chemicals and biomaterials. For almost 3.5 billion years, cyanobacteria have been known to biosynthesize cellulose. The synthesis of cellulose by cellucellulase, catalyzed by cellobiosyl fluoride and processed through the ring-opening polymerization of d-glucose moieties, is recognized as the first in vitro cellulose synthesis. Over the past ten years, cellulose has been the focus of wide-ranging examination. For cellulose to undergo changes and processing overall, its structure and characteristics are important [87]. Structure of cellulose is given in Fig. 1.

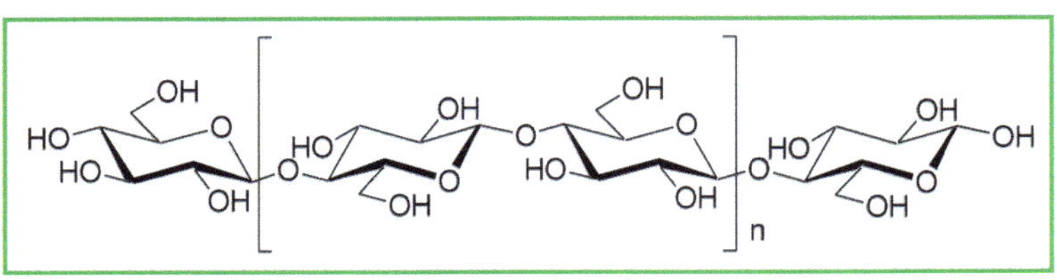

Fig. 1 Basic structure of cellulose "reused from [87] under CC By 3.0 creative common license"

3 Methods of Extraction for Cellulose Materials

One of the most prevalent organic substances on Earth is cellulose, a biopolymer that is present in plant cell walls. Paper, textiles, and pharmaceuticals are just a few of the industries that use it extensively. A vital stage in the creation of these goods is the extraction of cellulose from plant sources. Cellulose can be extracted from plants using a variety of techniques, each with unique benefits and drawbacks [30].

3.1 *Chemical Method*

A popular chemical technique for separating cellulose from lignocellulosic feedstocks for the production of biofuels and other products with added value is acid pretreatment. By treating the biomass with acid, the hemicellulose and lignin components are broken down, leaving only the cellulose, which can subsequently be fermented into biofuels or modified or enzymatically converted into sugars. Concentrated or diluted mineral acids like hydrochloric acid, phosphoric acid, sulfuric acid, or organic acids like acetic acid, citric acid, and formic acid are frequently employed in the processing of lignocellulosic biomass in order to isolate cellulose [2].

3.2 *Enzymatic Method*

In the initial phase of the process, fermentations were conducted in 500 mL conical flasks with 5 g of solid substrate. After that, the moisture content was adjusted to a predetermined amount and the solid substrate was supplemented. After covering each flask with hydrophobic cotton, they were autoclaved for 20 min at 121 °C. According to preliminary research, there were no discernible changes in the substrate's moisture content following autoclaving. Following cooling, each flask was infected in accordance with the previously mentioned guidelines and incubated for ninety-six hours in a temperature- and humidity-controlled chamber [35].

3.3 *Mechanical Method*

The fibers are extracted by blasting dried sugarcane bagasse with steam at 195 °C and 13 bar of pressure. 20 μg of xylanase (1:10 fiber to liquid ratio) is added to these fibers and they are continuously agitated for one hour at 50 °C. The dried bagasse was then bleached for an hour at 70 °C using acetic acid and 0.7% sodium chlorite ($NaClO_2$) to maintain a pH of 4. Until the fibers turn white, acetic acid and sodium chlorite are added hourly. The product was filtered and cleaned to a neutral pH before being dried at 55 °C for a day [49].

3.4 Solvent Method

The lignocellulosic biomass wastes have been upgraded and deoxygenated using the degradative solvent extraction (DSE) process, which yields high-quality extracts with a range of value-added applications. The primary constituents of lignocellulosic biomass are cellulose, hemicellulose, and lignin; the way in which these three substances interact throughout the DSE process is yet unknown and requires clarification [100].

3.5 Steam Explosion

One technique for achieving the steam exploded treatment—the approach of interest in this study—is high pressure steaming, which entails heating the materials at high temperatures and pressures before forcefully releasing the processed material into a collection tank. Steam treatment is an effective pre-treatment technique for wheat straw fibers that removes the bundle of rice straw fibers and eliminates the lignin percentage. A longer retention period and a higher steam temperature produced more uniform fiber with enhanced surface wettability. The steam explosion treatment increased the treated fiber's tensile strength while decreasing the amount of ash and extractives (like silicon) [18].

3.6 Ultrasound-Assisted Extraction

By applying high mechanical shear forces to the plant matrix, ultrasound-assisted extraction is a novel, eco-friendly method that disassembles a plant's fundamental structure and extracts certain components. The high shear rates are caused by a physical phenomenon called acoustic cavitation, which is characterized by the collapse of air or water vapor bubbles produced by the compression and rarefaction of ultrasonic waves traveling in the solvent [32].

3.7 Hydrothermal Treatment

Hydrothermal treatment, which involves exposing the plant material to high pressure and hot water or steam, is an additional method for cellulose extraction. This method can be energy-intensive and requires specialized equipment, but it is effective at extracting cellulose from a range of plant sources [34].

3.8 Hydrolysis Method

Cotton fibers were converted into cellulose nanocrystals using a two-step process that involved hydrolysis in mildly acidic conditions after an ionic liquid swelling treatment. In 1-butyl-3-methylimidazolium chloride at 80 °C, cellulosic fibers swelled under control although crystalline portions did not dissolve completely. Even in mild conditions, the hydrolysis took place because the substrate was far more accessible, and it required up to 60 times less sulfuric acid than the conventional extraction techniques that relied on concentrated sulfuric acid. Investigations were conducted into how process parameters affected the stability, composition, and shape of nanoparticles [53].

4 Cellulose and Their Derivatives-Based Materials

Cellulose is a repeating D-glucose homo-biopolymer characterized by β 1 $\rightarrow$ 4-glycosidic connections. It has a high molecular weight and exhibits a linear structure. A cellulose terminus converts into a hemiacetal. The packing and aggregation of cellulose chains differ among cellulose-producing species, thereby complicating its structure. Cellulose contains three hydroxyl groups per glucose unit and can polymerize into chains ranging from 1,000 to 15,000 glucose units, influenced by its source and processing techniques. Unbranched cellulose chains result from the intra-chain hydrogen bonding between hydroxyl groups and the oxygen of nearby cellulose rings, which stabilizes the connection [15, 82].

Cellulose is a partly crystalline substance, with its crystallinity influenced by the source, extraction procedures, and pre-treatment processes. In cellulose obtained from wood and plants, crystallinity typically ranges from 40 to 60%. Conversely, cellulose derived from sources such as bacteria and tunicin has markedly increased crystallinity, varying from 80 to 100%. Cellulose exists in several allomorphic crystalline forms, namely cellulose I, II, III, and IV [15].

The chemical treatment or functionalization of cellulose derivatives enhances their utility and adaptability in biomedicine. The kind and degree of substitution, along with polymer chain functionalization, dictate the properties of cellulose derivatives. The regioselective synthesis of cellulose derivatives presents challenges owing to the rigidity and bulkiness of their main chain, as well as their restricted solubility in organic solvents. The hydroxyl group is the most targeted reactive site on cellulose. Hydroxyl groups in cellulose are weak nucleophiles that require vigorous reactions, complicating the utilization of the minor reactivity variations among the 2–, 3–, and 6–OH groups. Consequently, the synthesis of cellulose derivatives continues to face challenges in region-selective substitution. The commercial manufacture of cellulose derivatives necessitates comprehension of their regiochemical structure and characteristics [28]. The derivatization of cellulose's hydroxyl groups often entails etherification and esterification processes but they differ in chemical composition, moisture absorption, water interaction, surface activity, and solubility.

4.1 Modified Cellulose Ethers

Cellulose ethers constitute a significant category of economically vital water-soluble polymers utilized in the construction and building sectors. Cellulose ethers can enhance the viscosity of aqueous solutions. The viscosifying capacity of a cellulose ether is predominantly determined according to its molecular weight, the chemical substituents involved, and the structural features of the polymer chain [65].

In building materials, cellulose ethers are used as water-retention agents, film formers, thickeners, and binders. They also function as lubricants, protective colloids, emulsifiers, suspension agents, and surfactants. Certain cellulose ethers exhibit heat gelation in aqueous solutions, a unique characteristic that is essential for a number of applications [47].

Cellulose ethers play a noteworthy part in drug delivery systems through various applications. They are employed in microencapsulation, acting as carriers or wall materials to encapsulate drugs, enabling controlled, sustained, or targeted delivery. Carboxymethyl cellulose and hydroxypropyl methylcellulose are two types of cellulose that are widely recognized for their function in the manufacture of microencapsulated medicines that have a long duration of action. Cellulose ethers serve a dual purpose within transdermal systems. They can be used as matrices or matrix additives, which ensures efficient drug delivery and prevents crystal formation. Additionally, they can be used as skin adhesives or gelation agents. In addition, interpolymer complexes (IPCs) that are based on cellulose ether are utilized for the controlled release of encapsulated pharmaceuticals. These complexes also have uses in the fields of coatings, adhesives, and food packaging. In addition, the ability of cellulose ethers to absorb water also makes it possible to create hydrogels, which are particularly helpful in the treatment of inflammation as well as a variety of skin conditions [22, 97].

4.1.1 Hydroxypropyl Methyl Cellulose

Hydroxypropyl methylcellulose, often known as HPMC, is a type of cellulose derivative that is distinguished by the presence of methoxyl and hydroxypropyl substituents that are connected to a cellulose backbone by means of ether bonds. One of the characteristics of HPMC is that it is an ether polymer that is nonionic, water-soluble, enzyme-resistant, and exhibits chemical stability across a pH range of 3.0–11.0. The hydrophobic interactions that occur between the polymer chains are responsible for the gel-forming ability of high-performance polymer blends (HPMC). There is a clear correlation between this capacity and the molecular weight as well as the degree of backbone replacement [38, 46].

The production of bioadhesive hydrogels for topical administration is one of the most significant applications of high-performance microcapsules (HPMC) in new drug delivery systems. These hydrogels offer continuous drug release while also retaining the drug for longer periods of time on the skin. When it comes to the delivery of topical medications, the HPMC hydrogel that was previously synthesized functions as an effective mechanism for prolonged release. Meloxicam (MLX) is a potent analgesic that is used in the management of pain and inflammation related to colorectal cancer. One more noteworthy use is the formulation of microparticles based on HPMC for the transport of Meloxicam (MLX) to the colon. Utilizing crosslinked high-performance microcapsules (HPMC) and gold nanoparticles, Raghunath Das and colleagues developed a nanocomposite consisting of ornidazole and 5-ASA for the purpose of colon-targeted drug delivery. The c-HPMC/Au nanocomposite is a viable option for the regulated release of colonic medications, as this work show [73, 94].

HPMC is used in a variety of medication delivery methods. It plays a crucial part in the administration of medication at the oral and oromucosal sites by creating a thick, viscous gel layer that regulates the drug's release. When it comes to delivering drugs to the eye, HPMC acts as a polymeric matrix in films, filaments, and inserts, a gelling agent in injections, and a viscosity modifier in eye drops. Tablets, capsule shells, and 3D printed pharmaceutical products all make use of it as a hydrophilic matrix for controlled release. Tablets make use of it. Moreover, nanofiber-based drug delivery systems make use of HPMC as a potential component. Significant swellability is exhibited by HPMC, which is characterized by its hydrophilicity, biocompatibility, and water solubility. This swellability is critically important for the kinetics of drug release. The structure of the delivery system, including its composition and shape, has a substantial impact on the mechanism by which the medicine is released into the body [40, 62].

4.1.2 Methyl Cellulose

About 27–32% of the hydroxyl groups in the cellulose ether derivative are converted to the methyl ether (CH_3O) form. White to milky-white in color, methyl cellulose (MC) is derived from natural cellulose. The powder is insipid. It dissolves in water and exhibits thermal gelation properties in its aqueous solution. Commercially accessible viscosity grades vary from 5 to 75,000 cP at 2%, which corresponds to an average molecular weight of 10,000–220,000 Da. It has several functions, such as film formation, emulsifiability, adhesive characteristics, and thickening abilities, and is often employed as a culinary component in baked products, fried dishes, sweets, confections, and soups. Methylcellulose is frequently utilized as a binder, coating agent, and tablet disintegrant in oral solid pharmaceutical formulations. It is also utilized in matrix tablets for prolonged release [29, 40].

MC is a multifaceted molecule with extensive applications in numerous industries, such as pharmaceuticals, food, and cosmetics. Methyl cellulose is utilized in drug delivery systems for controlled-release formulations, including tablets, capsules, and oral solutions, due to its capacity to create hydrophilic gels that modulate the release of active components. In the food industry, methyl cellulose acts as a thickening agent, stabilizer, and emulsifier in products such as salad dressings, sauces, and ice cream, thereby enhancing their texture and consistency. Similar roles, such as increasing texture and stability, are performed by it in cosmetics and personal care products, such as creams, lotions, shampoos, and conditioners. In addition to this, it is utilized in pharmaceutical coatings with the purpose of enhancing mechanical characteristics and protecting drugs from moisture and degradation. Methyl cellulose is used in the oil and gas industry as a viscosifier in drilling fluids. This helps to improve the flow of fluids and the efficiency of drilling. During the building process, it is put into cement and mortar mixtures in order to enhance the workability of the mixture and to retain water. In addition, methyl cellulose is utilized in the production of films and fibers, such as material for biodegradable packaging and fibers for medicinal applications. Its biocompatibility and the fact that it can be formed into hydrogels are two further reasons why it is valuable in the field of biotechnology for use as cell culture media and tissue engineering. The capacity of methyl cellulose to form gels, to hold moisture, and to maintain a stable, non-toxic structure in a variety of applications is the source of its flexibility [43, 61].

4.1.3 Carboxymethyl Cellulose (CMC)

Made from cellulose, carboxymethyl cellulose (CMC) is an anionic polymer that dissolves in water. It is created when cellulose reacts with a strong base, such sodium hydroxide and chloroacetic acid. The incorporation of carboxymethyl groups (–CH_2COOH) into the backbone of cellulose results in the production of a polymer that is hydrophilic at water and negatively charged. In a manner similar to that of cellulose, CMC is characterized by its linear configuration, where glucose units are linked together through β-1,4-glycosidic linkages. The average number of hydroxyl groups that are replaced by another substance is known as the degree of substitution (DS) carboxymethyl groups on each glucose unit, has an effect on the anionic properties that are imparted to the molecule by carbamethyl groups. The characteristics of CMC's viscosity and solubility are affected by this replacement. The anionic properties of the polymer make it compatible with a wide range of ions and chemicals, as well as water soluble [50].

When it comes to water, CMC is extremely soluble, and depending on the molecular weight and substitution, it can produce a solution that is rather concentrated. Because of its pseudoplasticity, it is able to reduce viscosity when subjected to shear stress, which enables it to function as a good thickener and stabilizer in a variety of compositions. The production of gels that play significant roles in the fields of medicine, food, and cosmetics is possible with carboxymethyl cellulose (CMC) under specific conditions. In wound dressings and moisturizers, this polymer is particularly beneficial since it can absorb a significant amount of water and expand as a result of this absorbency. CMC is stable across a large pH range, often from 4 to 10, despite the fact that its solubility and viscosity fluctuate according on the pH. Its carboxymethyl groups have negative charges, which allow it to interact with cations and form complexes. These complexes are especially important in drug delivery systems because they allow active chemicals to be attached to the molecule [67].

Several industries, such as the food industry, the pharmaceutical industry, and the cosmetics industry, make use of CMC as a thickening, a stabilizing agent, and an emulsifier. This is due to the unique chemical properties that it possesses. Controlled-release medication formulations are a particularly good application for this substance because of its ability to hold water and to create gels. In addition, the water-retention characteristics of CMC are advantageous in the production of personal care products, in agricultural settings as a dispersion agent, and in biomedical applications such as dressing wounds. When it comes to a wide variety of industrial applications, the solubility, viscosity control, and stability of the chemical are all quite important [60].

CMC is a versatile molecule that has various applications across a variety of industries. This is due to its remarkable features, which include the ability to thicken, gel, and emulsify, as well as its water solubility. In tablet formulations used in the pharmaceutical business, CMC functions as a binder, disintegrant, and controlled-release agent. Additionally, it acts as a suspending agent in oral liquids and topical gels for the purpose of medicine administration. Sauces, dressings, ice cream, and baked goods are just some of the products that can benefit from its thickening, stabilizing, and emulsifying properties in the food sector. These properties allow it to improve the texture, consistency, and shelf life of these products. Increasing the texture and effectiveness of formulations like shampoos, conditioners, lotions, and creams is one of the most important functions that CMC performs in the cosmetics and personal care items products industry. In addition, CMC is utilized in detergents to prevent the separation of ingredients, in agriculture as a dispersion agent for pesticides and fertilizers, and in biomedical applications for the purpose of wound healing through the utilization of hydrogels and dressings [66, 75].

4.1.4 Ethyl Cellulose

Ethyl cellulose is a modified derivative of cellulose, a natural biopolymer derived from plant cell walls. Synthesis occurs through the ethylation of cellulose, involving the interaction of ethyl alcohol (ethanol) with the hydroxyl groups of the cellulose molecule. This chemical modification alters the physical and chemical properties of cellulose, making it soluble in organic solvents such as ethanol, chloroform, and acetone, while it remains insoluble in water. The degree of substitution (DS) in ethyl cellulose reflects the quantity of hydroxyl groups in the cellulose molecule that have been replaced by ethyl groups, thereby influencing its solubility and applications. The ethyl groups are attached to the cellulose structure. Ethyl cellulose is chemically inert, non-toxic, and stable under typical environmental circumstances [58].

Ethyl cellulose is characterized by its white, odorless, free-flowing powder or granules, with a molecular weight ranging from 50,000 to several hundred thousand, influencing its solubility and viscosity. It possesses exceptional film-forming properties, rendering it a significant element in coatings and as a binder in diverse formulations. The viscosity escalates with increased concentrations and higher degrees of substitution. Ethyl cellulose possesses strong thermal stability, enabling it to endure processing temperatures of approximately 180 °C. Significantly, it is non-hygroscopic, indicating it does not collect moisture from the atmosphere [96].

Due to the fact that it possesses specific chemical and physical properties, ethyl cellulose is a polymer has many different uses across a variety of sector industries. It is largely used as a binder in tablet formulations in the pharmaceutical sector. This ensures that the tablets maintain their structural integrity and the requisite smoothness to be effective. For controlled-release formulations, it is an excellent choice because of its film-forming properties, which make it easier to manage the release of the active component over a period of time. In addition, ethyl cellulose is utilized in the production of coatings, such as enteric coatings, which shield pharmaceuticals from the acidic effects of the stomach and ensure that they are broken down in the right portion of the digestive pipeline [3, 68].

In addition to its use as an emulsifying agent, thickening agent, and stabilizing agent, ethyl cellulose is also utilized in the food sector for the production of food coatings and packaging materials. Since it is not regarded to be harmful, it is frequently utilized in edible films for packaging purposes. In the cosmetics sector, ethyl cellulose is used as a binder and emulsifies lotions, creams, and ointments to improve the quality and texture of these products. The fact that ethyl cellulose is soluble, stable, and capable of forming films is the reason why it is extremely significant in a wide variety of sectors [77].

4.1.5 Hydroxyethyl Cellulose

Hydroxyethyl cellulose, often known as HEC, is a cellulosic product that is produced through the reaction of cellulose with ethylene oxide, which is water soluble. Through this process, hydroxyethyl ($-CH_2CH_2OH$) is added to the ether bond, which ultimately results in an alteration of the physical properties of cellulose as well as it's the ability to dissolve in water. Natural sources of cellulose that contain plant fibers, such as cotton and wood, are the most abundant sources for the manufacture of HEC resources. Additionally, HEC is a material that is both biodegradable and favorable to the environment [98].

HEC's cellulose backbone is altered using hydroxyethyl groups. In water, solubility and viscosity change with increasing substitution; greater substitution increases solubility. While most organic solvents, such alcohols and acetone, HEC is insoluble, it is soluble in both cold and hot water. Under normal conditions, it is chemically inert and stable; it resists interactions with most compounds and degradation [8].

HEC is a free-flowing, white, odorless powder or granules. The viscosity of the material is directly impacted by the molecular weight, which can vary from 50,000 to several million. It is commonly known that HEC has superior thickening properties in water, which can be explained by its increased viscosity, which is associated with its higher molecular weight and substitution levels. It is appropriate for a variety of industrial uses since it is pH-stable and works well in a pH range of 3–10 [43]. HEC is a crucial ingredient in many different medicinal dosage forms due to its exceptional thickening, binding, and stabilizing qualities. It is frequently employed as a binder in tablet formulations to assist guarantee that the tablets keep their integrity during handling and digestion. Additionally, it facilitates the elements' cohesion. HEC thickens oral solutions by increasing their viscosity and ensuring that the active components are dispersed uniformly throughout the suspension [42].

Topical dose forms like gels, creams, and lotions use HEC to improve texture and stability and give the skin a smooth, non-greasy feel. In therapies like wound healing gels, it acts as a film-forming agent, allowing the skin to develop protective layers. Additionally, by creating a gel matrix, HEC is used in oral controlled-release formulations to adjust the pace at which the active medicinal ingredient is released. Because of its aqueous solubility, non-toxicity, and biocompatibility, HEC is a good excipient for a variety of pharmaceutical formulations [12].

4.1.6 Hydroxypropyl Cellulose

A flexible, water-soluble cellulose derivative, hydroxypropyl cellulose (HPC) is commonly used in medical contexts because to its distinct chemical and physical characteristics. It is a nonionic cellulose ether that is distinguished by the presence of hydroxypropyl groups in place of hydroxyl groups on the cellulose backbone. This substitution imparts solubility to HPC in both aqueous and organic solvents, a trait crucial for its many uses. The degree of substitution and molecular weight influence solubility, viscosity, and film-forming characteristics. HPC demonstrates thermoplastic characteristics and is recognized for its superior adhesive, binding, and thickening capabilities. It produces transparent, pliable films, exhibits resistance to microbial proliferation, and possesses low toxicity, rendering it an appropriate excipient in medicinal compositions [13, 25].

In pharmaceutical applications, HPC fulfills various functions owing to its multifunctional characteristics. It is frequently utilized as a binder in tablet formulations, augmenting mechanical strength and guaranteeing consistent medication distribution. It functions as a film-forming agent in controlled-release coatings, safeguarding pharmaceuticals from external influences while modulating their release. HPC serves as a thickening agent in ocular solutions and topical gels, enhancing viscosity and stability. Moreover, its solubility in organic solvents renders it an optimal selection for solvent-based coatings. The biocompatibility and nonionic characteristics of HPC guarantee minimal interaction with active pharmaceutical ingredients (APIs), preserving their stability and efficacy. Their qualities, along with regulatory recognition, highlight the crucial significance of HPC in contemporary pharmaceutical development and drug delivery systems [26, 80].

4.2 Cellulose Ester

Cellulose esters are chemically altered derivatives of cellulose, a natural polymer prevalent in plant cell walls. These chemicals are synthesized by replacing the hydroxyl groups (–OH) in the cellulose framework with organic acid moieties, such as acetate, nitrate, propionate, or butyrate, via reactions with acids or acid anhydrides. This alteration endows cellulose with distinctive characteristics, rendering it appropriate for many industrial applications. Prevalent cellulose esters comprise cellulose acetate, cellulose nitrate (nitrocellulose), cellulose propionate, and cellulose butyrate. Each variety differs in its chemical composition, physical qualities, and areas of application [10, 44].

Cellulose esters are esteemed for their thermal stability, mechanical strength, clarity, and solubility in organic solvents. Cellulose acetate is extensively utilized in textiles, photography, and film production because of its transparency and pliability. Nitrocellulose, recognized for its combustibility, is utilized in lacquers, varnishes, and explosives. Mixed esters, including Cellulose Acetate Butyrate (CAB), are employed in automotive coatings and packaging due to their durability and moisture resistance [20].

The biodegradability and renewable supply of cellulose esters are substantial benefits that correspond with sustainability objectives. Nonetheless, constraints such as humidity sensitivity and flammability, particularly in nitrocellulose, provide issues in specific applications. Nonetheless, cellulose esters are indispensable in sectors including pharmaceuticals, medical products, and paints and coatings. Their plasticity and adaptability highlight their significance in contemporary materials science, guaranteeing their ongoing relevance across a wide range of applications [78, 89].

4.2.1 Cellulose Acetate

Cellulose acetate is a significant derived of cellulose that finds extensive application in pharmaceutical formulations, especially in dosage forms, owing to its biocompatibility, stability, and adjustable characteristics. Its capacity to create films and matrices renders it an ideal candidate for systems designed for controlled and sustained drug release. The preparation of cellulose acetate involves the acetylation of cellulose using acetic acid or acetic anhydride, with the degree of acetylation playing a crucial role in determining its solubility and permeability, thus enabling tailored drug delivery solutions [22].

Osmotic drug delivery systems, such as osmotic pumps, frequently use cellulose acetate. The semi-permeable membrane of cellulose acetate used in these devices allows water to enter the drug core, dissolve the medication, and allow for controlled release through a small hole. This process improves patient adherence to treatment by guaranteeing a consistent and long-lasting therapeutic effect [7].

Furthermore, cellulose acetate is crucial for enteric coatings, which prevent drugs from degrading in the stomach's acidic environment and guarantee that they are released in the intestines in the appropriate way. This substance's ability to form films is also used in the microencapsulation procedure, which entails encasing active pharmaceutical ingredients in a protective layer. This layer hides any undesirable flavors and improves the contents' stability [101].

Cellulose acetate is a matrix-forming component used in controlled-release tablets. This component allows to regulate drug diffusion and release rates. Additionally, it is non-toxic and biodegradable is another factor that makes it suitable for use in medical applications. In the field of contemporary pharmaceutical science, cellulose acetate is an essential excipient, despite the fact that it has a number of significant limitations, such as its susceptibility to moisture and problems during its processing. The development of more advanced and efficient methods of administering medication is made possible as a result of this [48].

4.2.2 Nitrocellulose

Nitrocellulose, which is also known as cellulose nitrate, is a type of cellulose that has been subjected to a chemical modification in which nitrate groups have been substituted for hydroxyl groups. One of the most common applications for cellulose nitrate is in the coatings, film, and explosives sectors; nevertheless, it is also utilized in the development of pharmaceutical products, particularly in dosage forms. Several drug delivery applications can benefit from its controlled permeability, chemical stability, and film-forming capabilities. These characteristics make it an attractive candidate [64].

Because of its hydrophobic and semi-permeable qualities, cellulose nitrate is utilized in dosage forms for sustained-release formulations. These properties allow for regulated drug diffusion, contributing to the formulation's effectiveness. In the case of osmotic pump systems, for instance, it is able to generate coatings or membranes that regulate the amount of water that enters the drug core and allow for the consistent release of the drug. Microencapsulation methods, which use it to form a barrier of protection around active pharmaceutical substances, are particularly valuable for its adaptability, which makes it particularly useful for microencapsulation operations. Due to the presence of this barrier, the stability of pharmaceuticals is enhanced, unpleasant flavors are concealed, and they are protected from light and moisture [7].

Although cellulose nitrate has several advantages, its usage in the pharmaceutical industry should be approached with caution because to its flammability, sensitivity to solvents, and temperature sensitivity. In most cases, it is only utilized in formulations in specific circumstances where the peculiar features it possesses offer evident advantages. When creating systems based on cellulose nitrate, it is imperative that stringent safety rules be adhered to in order to ensure that the use of these systems in the administration of therapeutic agents is not only safe but also effective. Despite the fact that it is utilized less frequently than other cellulose derivatives, cellulose nitrate possesses the potential to be utilized in specialized pharmaceutical applications that require regulated drug release and stability [85].

4.2.3 Cellulose Sulfate

A chemically altered derivative of cellulose, cellulose sulfate is created when cellulose is sulfated, which substitutes sulfate groups (–OSO_3H) for some of the hydroxyl groups (–OH). This alteration gives cellulose sulfate special qualities like enhanced solubility, biodegradability, and anionic charge, which makes it a material with great potential for use in pharmaceuticals, especially in dosage forms. It has drawn a lot of interest in drug delivery systems due to its biocompatibility and capacity to interact with biological molecules [55].

Cellulose sulfate is utilized as a matrix-forming agent in pharmaceutical formulations to create controlled drug release systems. Because of its anionic properties, it can combine with positively charged medications or molecules to form ionic complexes, which improves drug stability and permits sustained release. These complexes can enable targeted delivery to particular sites and shield active pharmaceutical ingredients (APIs) from gastrointestinal tract degradation [57].

Hydrogels and microparticle formulations also use cellulose sulfate. It can encapsulate medications as a hydrogel, providing a stable and biocompatible delivery system for proteins, peptides, and other medicinal substances. It allows for controlled drug diffusion over long periods of time due to its capacity to swell in aqueous environments. Additionally, because of its potential as a bioadhesive agent, drug formulations have a longer residence time at mucosal sites, which improves drug efficacy and absorption [22]. The sensitivity of cellulose sulfate to enzymatic degradation and processing conditions necessitates careful formulation design, despite its many benefits, which include ease of chemical modification and compatibility with a variety of APIs. However, because of its special qualities, it is a useful excipient in sophisticated drug delivery systems, guaranteeing effective and focused therapeutic results [9]. Various structure of cellulose derivative is given in Fig. 2.

5 Dosage Form Development from Cellulose-Based Material and Application in Drug Delivery System

With the emergence of novel cellulose-based materials, dosage forms and drug delivery systems have advanced due to their biodegradability, biocompatibility, and multifunctionality. These materials are sourced from renewable resources. In this regard, cellulose and its derivatives, namely hydroxypropyl methylcellulose (HPMC), microcrystalline cellulose (MCC), and carboxymethyl cellulose (CMC) serve as the building blocks of varied drug delivery systems. The role of cellulose-based materials in dosage form development, focusing on their application in oral solid dosage form delivery, controlled-release systems, mucoadhesive drug delivery, osmotic drug delivery, gelling agent in semisolid drug delivery, nano-particulate drug delivery, transdermal drug delivery, ocular drug delivery, protein/peptide drug delivery, and nasal drug delivery.

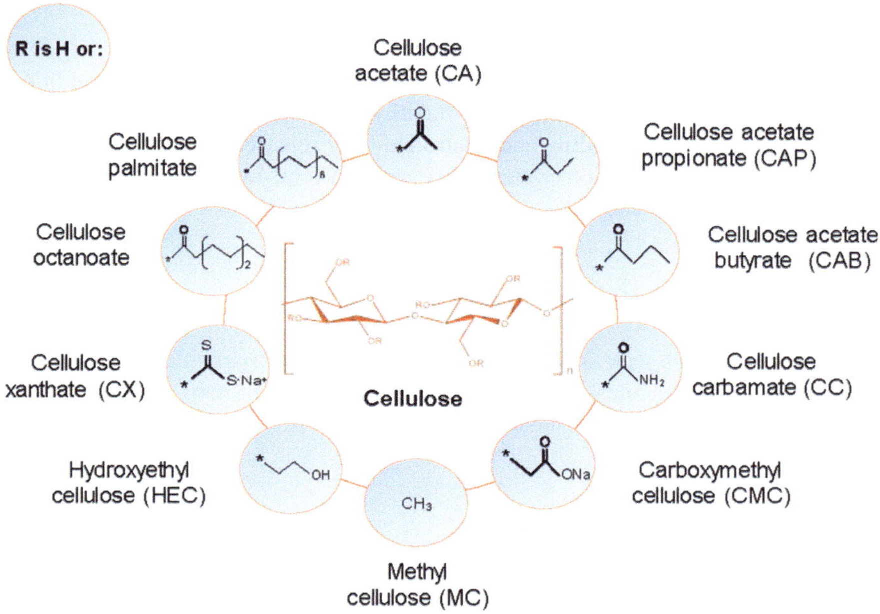

Fig. 2 Cellulose derivatives structure (where the R can be a hydrogen (H) atom or one of the groups presented in the figure) "reused from [27] under CC By 4.0 creative common license"

5.1 Oral Solid Dosage Form Delivery

Oral solid dosage forms such as tablets and capsules are among the furthermost widely used methods for administering drugs due to ease of use, stability of active ingredients, and adherence to therapy. Many cellulose-based materials perform a number of functions including excipients, binders, disintegrants, and coating agents.

5.1.1 Role as Binders and Disintegrants

Microcrystalline cellulose is one of the widely used binder materials in the tablets dispensing. Its great adaptability to/lots of compressibility and binding properties provide good tablet integrity for compactions, even during high-speed compression processes. Hence, microcrystalline cellulose serves as a disintegrant aiding disintegration upon swallowing by absorbing water and swelling [45]. Carboxymethylcellulose, sodium carboxymethylcellulose crosslinked, widely known as croscarmellose sodium, are examples of superdisintegrants. Superdisintegrants enhance drug dissolution through rapid disintegration, usually in aqueous surroundings, which can be critical for drugs with low solubilities [79].

5.1.2 Coating and Film Formation

The HPMC is favorable in these tablet coatings against environmental conditions such as moisture and light, facilitating stability in the drug. The HPMC films are preferably used for taste hiding and controlled drug release. Its good film-forming capacity and solubility in both water and organic solvents make it a handy polymer for coating applications [37].

5.1.3 Extended Applications

Cellulose derivatives such as ethylcellulose (EC) use multi-particulate dosage forms such as pellets or granules that can later be encapsulated in hard gelatin capsules. These kinds of formulations enhance the drug release profile and reduce variability in absorption [23].

5.2 Controlled-Release Dosage Forms

Controlled-release dosage forms comprise systems that were specially developed for the preparation of drugs, ensuring a controlled and predetermined rate of release of drugs; thus, helping pharmaceutical effectiveness and compliance while minimizing adverse effects. A great deal of development of such systems has relied on cellulose-based materials.

5.2.1 Matrix Systems

HPMC and MCC are widely used in hydrophilic matrix systems for controlled drug release. Upon contact with intestinal fluid, these polymers swell to form a sort of gel-like barrier that regulates the release of the drug from the matrix. The release rate of the drug can be controlled by adjusting the polymer ratio and molecular weight [54]. Ethyl cellulose (EC) is an insoluble cellulose derivative used in lipid-based matrix systems. EC matrices obey a diffusion mechanism of drug release; therefore, the formulation is helpful for hydrophobic drugs or drugs that require an extended-release profile [14].

5.2.2 Coating for Sustained Release

Cellulose acetate and EC are some of the traditional compounds used as coating materials in sustained-release systems. Those polymers form semi-permeable membranes that regulate drug release through osmotic or diffusive mechanisms. For optimizing drug release kinetics, it can be modified with multilayer coatings with cellulose derivatives [39].

5.2.3 Drug Layering

Drug layering in the multi-particulate systems, cellulose polymers are used for drug layering. The term "drug layering" indicates that the active pharmaceutical ingredient (API) is deposited onto the core substrate. This technique ensures a uniform drug dose administered and provides a customized release profile based on polymer composition [69].

5.3 Mucoadhesive Drug Delivery

Mucoadhesive drug delivery is thus designed to hold onto surfaces alongside enhancement of some drug properties. Such materials should highlight hydrophilic properties, especially cellulose derivatives.

5.3.1 Mechanism of Mucoadhesion

Cellulose derivatives, including HPMC and sodium carboxymethylcellulose (NaCMC), demonstrate mucoadhesive properties through the formation of hydrogen bonds with mucins. The created interface offers strong adhesion, thereby improving drug retention at the application site [93].

5.3.2 Applications in Buccal and Nasal Delivery

HPMC-based films and gels in the buccal delivery systems enhance bioavailability of certain drugs especially that of peptides and proteins subjected to degradation by digestive juices. The incorporation of NaCMC in nasal delivery systems has also shown bear promises for enhancing the delivery of polar drugs across the nasal mucosa [91].

5.3.3 Vaginal and Rectal Delivery

These products are also employed in vaginal and rectal formulations using cellulose derivatives. The hydrogels based on HPMC and CMC provide sustained drug release and encourage biocompatibility with sensitive mucosal tissues. These systems are useful in hormone administration or treatment of local infections [95].

5.4 *Osmotic Drug Delivery*

The osmotic agent employed in such drug delivery systems is osmotic pressure that enables a very well-controlled and precise drug release. The cellulose-based materials, specifically cellulose acetate, are the frontiers in this particular technology.

5.4.1 Semi-permeable Membranes

The semi-permeable polymer most commonly used in an osmotic drug delivery system is cellulose acetate. Upon exposure to aqueous fluids, the cellulose acetate makes use of osmotic pressure to allow water to enter the dosage form through a delivery orifice and thus act as a driving force for the release of the drug. The medication's release can be managed by altering the thickness and/or porosity of the cellulose acetate membrane [51].

5.4.2 Push–Pull Systems

Push–Pull Systems In a push–pull osmotic system, one possible mechanism is that, after the polymer layer swells following hydration, pushes the drug-containing layer through the orifice. HPMC and CMC are often used as the swelling agents in the push compartment for a steady release of the drug over time [59]. An osmotic system can be ideal because, in a therapeutic sense, the drug delivery achieves steady drug release and provides a more constant plasma concentration level for those drugs with narrow therapeutic windows like antihypertensive drugs, antidiabetics, and gastrointestinal drugs [86].

5.5 *Gelling Agent in Semisolid Drug Delivery*

Gels are semisolid systems that are dispersions of large molecules or small particles in an aqueous vehicle, which on addition of a gelling agent develop a jelly-like consistency. Two important gelling agents used in pharmaceutical formulations are hydroxypropyl methylcellulose (HPMC) and carboxymethyl cellulose (CMC). These substances are less susceptible to microbial contamination than natural agents such tragacanth, acacia, sodium alginate, agar, pectin, and gelatin. With the exception of MC grades, cellulose derivatives often show higher solubility in hot water, and heat induction causes them to gel. To make the gel, the polymer powder is first mixed with cold water using a mechanical mixer. With the exception of MC grades, the dispersion is then heated to between 60 and 80 °C and then allowed to cool to room temperature in order to promote gel formation [90].

5.6 Nano-Particulate Drug Delivery

The application of cellulose nanoparticles, i.e., microfibrillated cellulose, as strengthening in nanocomposites is a relatively new research area. Other than being a raw material expense, the utilization of cellulosic particles as a reinforcement phase in nanocomposites offers the advantages of low density, renewability, global availability of fillers, energy conservation, improved specific properties, low abrasiveness during processing, biodegradability, a moderately reactive surface for grafting desired functional groups, and an unlimited supply. However, cellulose nanoparticles also have limitations for reinforcement uses like high moisture absorbency, low wettability, incompatibility with most polymer matrices, and restrictions upon processing temperatures. The degradation of lignocellulosic substrates begins at around 220 °C, which restricts the variety of matrices to be blended with natural fillers [11].

5.7 Transdermal Drug Delivery

By avoiding gastrointestinal and hepatic pre-systemic metabolism, controlling blood drug concentrations to improve the efficacy-to-tolerability ratio, and permitting prolonged drug administration, transdermal drug delivery seeks to overcome the limitations of oral and parenteral methods. Nanocellulose's beneficial qualities make it a promising material for transdermal drug delivery systems. Nanocellulose-based transdermal medication delivery devices have been the subject of recent research. Cellulose derivatives like ethyl cellulose (EC) and hydroxypropyl methylcellulose (HPMC) are essential components of transdermal drug delivery systems (TDDS). They are used as film-forming agents in patches for the controlled release of active pharmaceutical ingredients through the epidermal barrier [6].

5.8 Ocular Drug Delivery

The ocular administration of pharmaceuticals is challenging due to its unique anatomy, which features various defense mechanisms that impede the ingress of foreign particles or drug molecules, including systemic absorption, enzymatic breakdown, nasolacrimal drainage, protein binding, tear turnover, and intricate penetration barriers [16].

Polymers such as cellulose, alginate, pectin, and xanthan gum are extensively employed in the preparation of topical ophthalmic formulations [74]. Cellulose derivatives are commonly utilized to enhance the viscosity of medicinal solutions and dispersed systems, including suspensions and emulsions. The polymers can also augment the viscosity of non-aqueous pharmaceutical solutions, such as organic-based coating formulations. The increase in viscosity of pharmaceutical solutions provides several advantages, such as enhanced dosage control and prolonged retention of medications in mucosal and topical applications, thereby improving the bioavailability of ocular, nasal, or topical formulations [4].

5.9 Protein/Peptide Drug Delivery

The development of an oral dosage form that guarantees therapeutic peptides or proteins have adequate bioavailability might revolutionize the way some illnesses are treated. As a result, this objective has been the focus of several efforts. One example is the use of cellulose derivatives in the creation of oral dosage forms. The principal rationale for utilizing these molecules is their potential ability to safeguard against gastrointestinal barriers, including acidity and proteolytic enzymes. They also confer bioadhesive and mucoadhesive characteristics to the formulations. The predominant bioadhesive polymers are derivatives of cellulose and polyacrylic acid [83].

5.10 Nasal Drug Delivery

A range of pharmaceutical-grade cellulose derivatives is utilized worldwide for all methods of administration. Both insoluble cellulose derivatives like ethylcellulose and microcrystalline cellulose, as well as soluble cellulose derivatives like hydroxypropyl methylcellulose, hydroxypropyl cellulose, methylcellulose, and carboxymethyl cellulose, have shown promise in improving intranasal drug absorption. Because of their remarkable mucoadhesive qualities, cellulose derivatives significantly lengthen the time that medications remain in the nasal cavity. Medication release is further prolonged by their increased viscosity in the nasal cavity after hydration. Therefore, using cellulose derivatives as absorption enhancers can boost intranasal absorption and bioavailability [21]. Overview of dosage forms developed using cellulose-based materials is given in Table 1.

6 Conclusion

Cellulose is a multifaceted polymer employed in medical applications, sourced from several origins. Chemical modification of cellulose yields cellulose derivatives that are typically resilient, cost-effective, repeatable, recyclable, and biocompatible, making them suitable for medical applications. Cellulose derivatives are frequently employed to alter drug release in tablet and capsule formulations, and serve as binding agents, thickeners, rheological modifiers, film formers, water retainers, and enhancers of adhesive strength. MCC serves as a diluent and disintegrating agent to enhance

Table 1 List of dosage forms developed using cellulose-based materials

S. No.	Dosage form	Cellulose material	Year	References
1.	Alginate-cellulose nano crystal	Cellulose nanocrystals	2018	[92]
2.	Nanopaper and nanofoam of cellulose	Cellulose nanocrystals	2017	[56]
3.	Mucoadhesive drug carrier	Microcrystalline cellulose	2015	[88]
4.	Membrane composed of cellulose acetate butyrate for osmotic drug administration	Cellulose acetate butyrate	2007	[84]
5.	In situ gel	Cellulose nanocrystals	2019	[71]
6.	Nanocomposite hydrogels	Cellulose nanocrystals	2017	[76]
7.	Cellulose nanocrystals nanocomposites	Cellulose nanocrystals	2018	[1]
8.	Transdermal drug delivery system utilising cellulose nanofibrils and chitosan	Cellulose nanofibrils	2017	[81]
9.	Ophthalmic formulations based on cellulose nanofibrils grafted with nanocollagen	Cellulose nanofibrils	2019	[72]
10.	Microsphere of ethyl cellulose for nasal delivery	Ethylcellulose	2021	[99]

the release of oral solid dosage forms. HEC and HPC are employed in hydrophilic matrix systems, while EC is utilized in hydrophobic matrix systems. Their main roles as binders (in direct compression or wet granulation), disintegrants, viscosity modifiers, and components of coatings or matrices that enable modified release are evident from the presence of cellulosics in the materials gathered. Two of the most often used cellulose derivatives in the pharmaceutical sector are HPMC and MCC. In several dosage forms, including topical, parenteral, transdermal, and oral treatments, cellulose and its derivatives are necessary ingredients. Nevertheless, tablet and capsule versions remain the primary means of administration. Cellulosics, recognized for their established safety and inertness, remain prevalent in contemporary formulations, sustaining their status as versatile excipients.

References

1. Abo-Elseoud WS, Hassan ML, Sabaa MW, Basha M, Hassan EA, Fadel SM (2018) Chitosan nanoparticles/cellulose nanocrystals nanocomposites as a carrier system for the controlled release of repaglinide. Int J Bio Macro 111:604–613
2. Abolore RS, Jaiswal S, Jaiswal AK (2024) Green and sustainable pretreatment methods for cellulose extraction from lignocellulosic biomass and its applications: a review. Carb Poly Technol App 7:100396

3. Adeleke OA (2019) Premium ethylcellulose polymer based architectures at work in drug delivery. Int J Pharm: X 1:100023
4. Adibkia K, Shadbad MR, Nokhodchi A, Javadzedeh A, Barzegar-Jalali M, Barar J, Mohammadi G, Omidi Y (2007) Piroxicam nanoparticles for ocular delivery: physicochemical characterization and implementation in endotoxin-induced uveitis. J Drug Targ 15(6):407–416
5. Akira I (2001) Chemical modification of cellulose. In: Wood and cellulosic chemistry. Hon DNS, Shiraishi N (eds). Marcel Dekker, New York
6. Alexander A, Dwivedi S, Giri TK, Saraf S, Saraf S, Tripathi DK (2012) Approaches for breaking the barriers of drug permeation through transdermal drug delivery. J Cont Rel 164(1):26–40
7. Almoshari Y (2022) Osmotic pump drug delivery systems—a comprehensive review. Pharmaceut 15(11):1430
8. Arai K, Shikata T (2020) Hydration/dehydration behavior of hydroxyethyl cellulose ether in aqueous solution. Molecules 25(20):4726
9. Ashique S, Mishra N, Mohanto S, Gowda BJ, Kumar S, Raikar AS, Masand P, Garg A, Goswami P, Kahwa I (2024) Overview of processed excipients in ocular drug delivery: opportunities so far and bottlenecks. Heliyon 15;10(1)
10. Aziz T, Farid A, Haq F, Kiran M, Ullah A, Zhang K, Li C, Ghazanfar S, Sun H, Ullah R, Ali A (2022) A review on the modification of cellulose and its applications. Polymers 14(15):3206
11. Azizi Samir MA, Alloin F, Dufresne A (2005) Review of recent research into cellulosic whiskers, their properties and their application in nanocomposite field. Biomacromol 6(2):612–626
12. Barnes TM, Mijaljica D, Townley JP, Spada F, Harrison IP (2021) Vehicles for drug delivery and cosmetic moisturizers: review and comparison. Pharmaceutics 13(12):2012
13. Baron RI, Biliuta G, Macsim AM, Dinu MV, Coseri S (2023) Chemistry of hydroxypropyl cellulose oxidized by two selective oxidants. Polymers 15(19):3930
14. Bechard SR, Leroux JC (1992) Coating technologies for controlled drug delivery. Drug Dev Ind Pharm 18(9):1063–1089
15. Benalaya I, Alves G, Lopes J, Silva LR (2024) A review of natural polysaccharides: sources, characteristics, properties, food, and pharmaceutical applications. Int J Mol Sci 25(2):1322
16. Bisht R, Mandal A, Jaiswal JK, Rupenthal ID (2018) Nanocarrier mediated retinal drug delivery: overcoming ocular barriers to treat posterior eye diseases. Wiley Interd Rev Nanomed Nanobiotech 10(2):e1473
17. Bochek AM (2003) Effect of hydrogen bonding on cellulose solubility in aqueous and nonaqueous solvents. Russ JApp chem 76:1711–1719
18. Boonterm M, Sunyadeth S, Dedpakdee S, Athichalinthorn P, Patcharaphun S, Mungkung R, Techapiesancharoenkij R (2016) Characterization and comparison of cellulose fiber extraction from rice straw by chemical treatment and thermal steam explosion. J Cle Prod 134:592–599
19. Carpita N, Vergara C (1998) A recipe for cellulose. Science 279(5351):672–673
20. Carter EA, Swarbrick B, Harrison TM, Ronai L (2020) Rapid identification of cellulose nitrate and cellulose acetate film in historic photograph collections. Her Sci 8:1–3
21. Chaturvedi M, Kumar M, Pathak K (2011) A review on mucoadhesive polymer used in nasal drug delivery system. J Adv Pharm Technol Res 2(4):215–222
22. Ciolacu DE, Nicu R, Ciolacu F (2020) Cellulose-based hydrogels as sustained drug-delivery systems. Materials 13(22):5270
23. Colombo P, Bettini R, Santi P, Peppas NA (2000) Swellable matrices for controlled drug delivery: gel-layer behaviour, mechanisms and optimal performance. Pharma Sci Technol Tod 3(6):198–204
24. Conner AH (1995) Size exclusion chromatography of cellulose and cellulose derivatives. Chrom Sci Ser 69:331–352
25. Cremer G, Danthine S, Van Hoed V, Dombree A, Laveaux AS, Damblon C, Karoui R, Blecker C (2023) Variability in the substitution pattern of hydroxypropyl cellulose affects its physicochemical properties. Heliyon 1;9(2)

26. David J (2004) Pharmaceutical applications of polymers for drug delivery. Rapra Technol Limit 50(15):124
27. Erdal NB, Hakkarainen M (2022) Degradation of cellulose derivatives in laboratory, man-made, and natural environments. Biomacromol 23(7):2713–2729
28. Etale A, Onyianta AJ, Turner SR, Eichhorn SJ (2023) Cellulose: a review of water interactions, applications in composites, and water treatment. Chem Rev 123(5):2016–2048
29. Feller RL, Wilt MH (1991) Evaluation of cellulose ethers for conservation. Getty Pub 21
30. Femina KS, Asokan A (2024) Extraction of cellulose. In: Handbook of Biomass, Springer Nature Singapore. Singapore
31. Figueiredo JA, Ismael MI, Anjo CM, Duarte AP (2010) Cellulose and derivatives from wood and fibers as renewable sources of raw-materials. CarbSus Dev 117–128
32. Freitas PA, González-Martínez C, Chiralt A (2022) Applying ultrasound-assisted processing to obtain cellulose fibres from rice straw to be used as reinforcing agents. Innov Food Sci Emer Technol 76:102932
33. French AD, Pérez S, Bulone V, Rosenau T, Gray D (2018) Encyclopedia of polymer science and technology. Cellulose 1838:1
34. Gao Y, Wang XH, Yang HP, Chen HP (2012) Characterization of products from hydrothermal treatments of cellulose. Energy 42(1):457–465
35. Gasparotto JM, Werle LB, Foletto EL, Kuhn RC, Jahn SL, Mazutti MA (2015) Production of cellulolytic enzymes and application of crude enzymatic extract for saccharification of lignocellulosic biomass. App Biochem Biotechnol 175:560–572
36. Gross RA, Scholz C (2001) Biopolymers from polysaccharides and agroproteins. American Chemical Society, Washington, DC
37. Gullapalli RP (2010) Manufacturing techniques of multiparticulate formulations for controlled release. J Pharm Sci 99(2):501–517
38. Gupta B, Mishra V, Gharat S, Momin M, Omri A (2021) Cellulosic polymers for enhancing drug bioavailability in ocular drug delivery systems. Pharmaceuticals 14(11):1201
39. Hagerstrom H, Edsman K (2004) Limitations of the rheological characterization of mucoadhesive gels. Eur J Pharm Sci 22(6):469–477
40. Harika K, Sunitha K, Maheshwar K, Rao M (2012) Basic concepts of cellulose polymers-a comprehensive review. Arch Pharm Prac 3(3–2012):202–216
41. Hinterstoisser B, Salmén L (2000) Application of dynamic 2D FTIR to cellulose. Vibrat Spect 22(1–2):111–118
42. Ho HN, Le HH, Le TG, Duong TH, Ngo VQ, Dang CT, Tran TH, Nguyen CN (2022) Formulation and characterization of hydroxyethyl cellulose-based gel containing metronidazole-loaded solid lipid nanoparticles for buccal mucosal drug delivery. Int J Bio Macromol 194:1010–1018
43. Hosny KM, Alkhalidi HM, Alharbi WS, Md S, Sindi AM, Ali SA, Bakhaidar RB, Almehmady AM, Alfayez E, Kurakula M (2021) Recent trends in assessment of cellulose derivatives in designing novel and nanoparticulate-based drug delivery systems for improvement of oral health. Polymers 14(1):92
44. Jiang Z, Ngai T (2022) Recent advances in chemically modified cellulose and its derivatives for food packaging applications: a review. Polymers 14(8):1533
45. Jivraj M, Martini LG, Thomson CM (2000) An overview of the different excipients useful for the direct compression of tablets. Pharm Sci Tech Tod 3(2):58–63
46. Joshi SC (2011) Sol-gel behavior of hydroxypropyl methylcellulose (HPMC) in ionic media including drug release. Materials 4(10):1861–1905
47. Jurczyková T, Kmeťová E, Kačík F, Lexa M, Dědič D (2024) Evaluating the effectiveness of cellulose-based surfactants in expandable graphite wood coatings. Polymers 16(19):2832
48. Khaing EM, Lertsuphotvanit N, Thammasut W, Rojviriya C, Chansatidkosol S, Phattarateera S, Pichayakorn W, Phaechamud T (2024) Cellulose acetate butyrate-based in situ gel comprising doxycycline hyclate and metronidazole. Polymers 16(24):3477
49. Khan R, Jolly R, Fatima T, Shakir M (2022) Extraction processes for deriving cellulose: a comprehensive review on green approaches. Poly Adv Technol 33(7):2069–2090

50. Kriplani P, Guarve K (2021) Chitosan-a potential polymer to prepare anticancer drug delivery systems: patent review. Rec Pat Anti-Cancer Drug Dis 16(2):249–257
51. Kumar R, Patil MB, Patil SR, Paschapur MS (2010) Polysaccharides based on mucoadhesive systems for drug delivery: a review. Int J Pharm Pharm Sci 2(4):39–44
52. Lavanya DK, Kulkarni PK, Dixit M, Raavi PK, Krishna LN (2011) Sources of cellulose and their applications—a review. Int J Drug Form Res 2(6):19–38
53. Lazko J, Sénéchal T, Landercy N, Dangreau L, Raquez JM, Dubois P (2014) Well defined thermostable cellulose nanocrystals via two-step ionic liquid swelling-hydrolysis extraction. Cellulose 21:4195–4207
54. Lecomte F, Siepmann J, Walther M, MacRae RJ, Bodmeier R (2004) Polymer blends used for the coating of multiparticulates: comparison of aqueous and organic coating techniques. Pharma Res 21:882–890
55. Li W, Xue Y, He M, Yan J, Lucia LA, Chen J, Yu J, Yang G (2021) Facile preparation and characteristic analysis of sulfated cellulose nanofibril via the pretreatment of sulfamic acid-glycerol based deep eutectic solvents. Nanomaterials 11(11):2778
56. Löbmann K, Wohlert J, Müllertz A, Wågberg L, Svagan AJ (2017) Cellulose nanopaper and nanofoam for patient-tailored drug delivery. Adv Mater Interfaces 4(9):1600655
57. Lukova P, Katsarov P, Pilicheva B (2023) Application of starch, cellulose, and their derivatives in the development of microparticle drug-delivery systems. Polymers 15(17):3615
58. Magalhães S, Fernandes C, Pedrosa JF, Alves L, Medronho B, Ferreira PJ, Rasteiro MD (2023) Eco-friendly methods for extraction and modification of cellulose: an overview. Polymers 15(14):3138
59. Maroni A, Zema L, Del Curto MD, Foppoli A, Gazzaniga A (2013) Film coatings for oral pulsatile release. Int J Pharm 457(1):362–371
60. Martins D, Rocha C, Dourado F, Gama M (2021) Bacterial cellulose-carboxymethyl cellulose (BC: CMC) dry formulation as stabilizer and texturizing agent for surfactant-free cosmetic formulations. Colloids Surf A Physicochem Eng Asp 617:126380
61. Maryam S, Krukiewicz K (2024) Sweeten the pill: multi-faceted polysaccharide-based carriers for colorectal cancer treatment. Int J Bio Macromol 20:136696
62. Mašková E, Kubová K, Raimi-Abraham BT, Vllasaliu D, Vohlídalová E, Turánek J, Mašek J (2020) Hypromellose–a traditional pharmaceutical excipient with modern applications in oral and oromucosal drug delivery. J Cont Rel 324:695–727
63. Mastropietro DJ, Omidian H (2013) Prevalence and trends of cellulosics in pharmaceutical dosage forms. Drug Dev Ind Pharm 39(2):382–392
64. Mattar H, Baz Z, Saleh A, Shalaby AS, Azzazy AE, Salah H, Ismail I (2020) Nitrocellulose: structure, synthesis, characterization, and applications. Water Energy Food Environ J 3:1–5
65. McMullen RL, Ozkan S, Gillece T (2022) Physicochemical properties of cellulose ethers. Cosmetics 9(3):52
66. Mehrabi A, Jalise SZ, Hivechi A, Habibi S, Kebria MM, Haramshahi MA, Latifi N, Karimi A, Milan PB (2024) Evaluation of inherent properties of the carboxymethyl cellulose (CMC) for potential application in tissue engineering focusing on bone regeneration. Poly Adv Technol 35(1):e6258
67. Michaelis JU, Kiese S, Amann T, Folland C, Asam T, Eisner P (2023) Thickening properties of carboxymethyl cellulose in aqueous lubrication. Lubricants 11(3):112
68. Mittal V, Kriplani P, Guarve K (2023) Ethylcellulose-an amazing polymer for anticancer formulations. Cur App Poly Sci 6(2):61–75
69. Morales JO, McConville JT (2011) Manufacture and characterization of mucoadhesive buccal films. Euro J Pharm Biopharm 77(2):187–199
70. Myasoedova VV (2000) Physical chemistry of non-aqueous solutions of cellulose and its derivatives
71. Orasugh JT, Dutta S, Das D, Pal C, Zaman A, Das S, Dutta K, Banerjee R, Ghosh SK, Chattopadhyay D (2019) Sustained release of ketorolac tromethamine from poloxamer 407/cellulose nanofibrils graft nanocollagen based ophthalmic formulations. Int J Bio Macromol 140:441–453

72. Orasugh JT, Sarkar G, Saha NR, Das B, Bhattacharyya A, Das S, Mishra R, Roy I, Chattoapadhyay A, Ghosh SK, Chattopadhyay D (2019) Effect of cellulose nanocrystals on the performance of drug loaded in situ gelling thermo-responsive ophthalmic formulations. Int J Bio Macromol 124:235–245
73. Pan P, Svirskis D, Waterhouse GI, Wu Z (2023) Hydroxypropyl methylcellulose bioadhesive hydrogels for topical application and sustained drug release: the effect of polyvinylpyrrolidone on the physicomechanical properties of hydrogel. Pharmaceutics 15(9):2360
74. Patchan M, Graham JL, Xia Z, Maranchi JP, McCally R, Schein O, Elisseeff JH, Trexler MM (2013) Synthesis and properties of regenerated cellulose-based hydrogels with high strength and transparency for potential use as an ocular bandage. Mat Sci Eng C 33(5):3069–3076
75. Rahman MS, Hasan MS, Nitai AS, Nam S, Karmakar AK, Ahsan MS, Shiddiky MJ, Ahmed MB (2021) Recent developments of carboxymethyl cellulose. Polymers 13(8):1345
76. Rao KM, Kumar A, Han SS (2017) Poly (acrylamidoglycolic acid) nanocomposite hydrogels reinforced with cellulose nanocrystals for pH-sensitive controlled release of diclofenac sodium. Pol Test 64:175–182
77. Rogers TL, Wallick D (2012) Reviewing the use of ethylcellulose, methylcellulose and hypromellose in microencapsulation. Part 3: applications for microcapsules. Drug Dev Ind Pharm 38(5):521–539
78. Romão S, Bettencourt A, Ribeiro IA (2022) Novel features of cellulose-based films as sustainable alternatives for food packaging. Polymers 14(22):4968
79. Rowe RC, Sheskey PJ, Weller PJ (2006) Handbook of pharmaceutical excipients, 4th edn. Pharmaceutical Press, London
80. Salawi A (2022) Pharmaceutical coating and its different approaches, a review. Polymers 14(16):3318
81. Sarkar G, Orasugh JT, Saha NR, Roy I, Bhattacharyya A, Chattopadhyay AK, Rana D, Chattopadhyay D (2017) Cellulose nanofibrils/chitosan based transdermal drug delivery vehicle for controlled release of ketorolac tromethamine. New J Chem 41(24):15312–15319
82. Seddiqi H, Oliaei E, Honarkar H, Jin J, Geonzon LC, Bacabac RG, Klein-Nulend J (2021) Cellulose and its derivatives: towards biomedical applications. Cellulose 28(4):1893–1931
83. Shaji J, Patole V (2008) Protein and peptide drug delivery: oral approaches. Ind J Pharm Sci 70(3):269
84. Shanbhag A, Barclay B, Koziara J, Shivanand P (2007) Application of cellulose acetate butyrate-based membrane for osmotic drug delivery. Cellulose 14:65–71
85. Shokri J, Adibkia K (2013) Application of cellulose and cellulose derivatives in pharmaceutical industries. In: Cellulose-medical, pharmaceutical and electronic applications, IntechOpen
86. Singh BN, Kim KH (2000) Floating drug delivery systems: an approach to oral controlled drug delivery via gastric retention. J Cont Rel 63(3):235–259
87. Siqueira G, Bras J, Dufresne A (2010) Cellulosic bionanocomposites: a review of preparation, properties and applications. Polymers 2(4):728–765
88. Songsurang K, Siraleartmukul K, Muangsin N (2015) Mucoadhesive drug carrier based on functional-modified cellulose as poorly water-soluble drug delivery system. J Microencap 32(5):450–459
89. Szabó L, Milotskyi R, Sharma G, Takahashi K (2023) Cellulose processing in ionic liquids from a materials science perspective: turning a versatile biopolymer into the cornerstone of our sustainable future. Green Chem 25(14):5338–5389
90. Tas C, Özkan Y, Savaser A, Baykara T (2003) In vitro release studies of chlorpheniramine maleate from gels prepared by different cellulose derivatives. Il Farmaco 58(8):605–611
91. Theeuwes F (1975) Elementary osmotic pump. J Pharm Sci 64(12):1987–1991
92. Thomas D, Latha MS, Thomas KK (2018) Synthesis and in vitro evaluation of alginate-cellulose nanocrystal hybrid nanoparticles for the controlled oral delivery of rifampicin. J Drug Del Sci Technol 46:392–399
93. Valenta C (2005) The use of mucoadhesive polymers in vaginal delivery. Adv Drug Del Rev 57(11):1692–1712

94. Vanti G, Wang M, Bergonzi MC, Zhidong L, Bilia AR (2020) Hydroxypropyl methylcellulose hydrogel of berberine chloride-loaded escinosomes: dermal absorption and biocompatibility. Int J Bio Macromol 164:232–241
95. Verma RK, Krishna DM, Garg S (2002) Formulation aspects in the development of osmotically controlled oral drug delivery systems. J Cont Rel 79(1–3):7–27
96. Wasilewska K, Winnicka K (2019) Ethylcellulose–a pharmaceutical excipient with multidirectional application in drug dosage forms development. Materials 12(20):3386
97. Xu Y, Yan X, Zheng H, Li J, Wu X, Xu J, Zhen Z, Du C (2024) The application of encapsulation technology in the food industry: classifications, recent advances, and perspectives. Food Chem X 19:101240
98. Yoshida M, Iwase H, Shikata T (2022) Conformation and structure of hydroxyethyl cellulose ether with a wide range of average molar masses in aqueous solutions. Polymers 14:4532
99. Zafar A, Afzal M, Quazi AM, Yasir M, Kazmi I, Al-Abaasi FA, Alruwaili NK, Alharbi KS, Alzarea SI, Sharma S, Kaur R (2021) Chitosan-ethyl cellulose microspheres of domperidone for nasal delivery: preparation, in-vitro characterization, in-vivo study for pharmacokinetic evaluation and bioavailability enhancement. J Drug Del Sci Technol 63:102471
100. Zhu X, Liu M, Sun Q, Ma J, Xia A, Huang Y, Zhu X, Liao Q (2022) Elucidation of the interaction effects of cellulose, hemicellulose and lignin during degradative solvent extraction of lignocellulosic biomass. Fuel 1(327):125141
101. Zhu Y, Wang Z, Bai L, Deng J, Zhou Q (2021) Biomaterial-based encapsulated probiotics for biomedical applications: current status and future perspectives. Mater Des 210:110018

Rajendra Chauhan is a young and dynamic professional working as Assistant Professor in the S.O.S in Pharmaceutical Sciences, Jiwaji University, Gwalior. His specialization in Pharmaceutics. He has 11 years of teaching and research experience. His various research articles have been published in various scientific journals. He has attended and presented a paper in various International/National conference/Seminars. He is a life member of the Indian Pharma Educational Society [IPES]. His area of specialization in research is Pharmaceutics with transdermal drug delivery system. He has also post graduate diploma in computer application.

Ramakant Joshi is working as an Associate Professor at the Amity Institute of Pharmacy, Amity University Madhya Pradesh, Gwalior. He has more than 10 years of teaching and research experience. He has more than 20 publications in various International/national journals of repute. His area of specialization is transdermal drug delivery systems and novel drug delivery systems. He is the author/co-author of 6 books and 12 book chapters for an International/national publisher and filed 03 patents. He is GPAT qualified. He is a life member of the Association of Pharmaceutical Teachers in India (APTI) and the Indian Pharma Educational Society (IPES). He has attended and presented a paper at various International/National conferences/Seminars. He is a reviewer in the peer-reviewed journal Indian Journal of Pharmaceutical Sciences (0250-474X), which is the official scientific publication of the Indian Pharmaceutical Association.

Mandeep Kumar Singh a distinguished academician and researcher, has made remarkable contributions to the field of pharmaceutical sciences. With a career spanning over a decade, his expertise in pharmacology, drug delivery systems, and metabolic disorders has earned him recognition in both academia and research. Holding a Ph.D. from Dr. Harisingh Gour University, Sagar, and an M. Pharm. in Pharmacology, Dr. Singh's academic journey has been marked by excellence. His dedication to research is evident in his numerous publications in high-impact journals, focusing on immunomodulatory agents, nanogel drug delivery systems, and neurobiology. Dr. Singh has contributed significantly as an educator, serving as an Assistant Professor at Amity Institute of Pharmacy, Amity University Rajasthan, India. His commitment to shaping young minds is reflected in his extensive teaching and mentoring experience. Beyond academia, his achievements include an awarded UK patent for medical and laboratory equipment, highlighting his inventive mindset. Recognized through various accolades, including the CSIR Direct SRF Award, Dr. Singh's contributions to pharmaceutical sciences continue to inspire. With an unwavering passion for research and teaching, Dr. Singh stands as a beacon of excellence in the field of pharmaceutical sciences. His journey is a testament to dedication, innovation, and the pursuit of knowledge, making him a role model for aspiring researchers and academicians.

Deepesh Parashar is working as an Associate Professor at the Amity Institute of Pharmacy, Amity University Madhya Pradesh, Gwalior. He has more than 12 years of teaching and research experience. He has more than 10 publications in various International/national journals of repute. His area of specialization is HPLC, HPTLC, and GC method development, standardization of herbal drugs, cytoprotection against cancer chemotherapy. He is the author/co-author of 2 books and 5 book chapters for an International/national publisher. He is a life member of the Association of Pharmaceutical Teachers in India (APTI). Dr. Parashar is CCSEA nominee and Got Selected in UGC-NRC Scheme and worked at UIPS, Panjab University for research.

Wasim Akram is working as Assistant Professor at Amity Institute of Pharmacy, Gwalior. He has a total experience of 08 years in teaching, research and the pharmaceutical industry. His research interests include designing and developing biopolymers-based novel drug delivery systems. He is actively involved in formulating nanoparticulate drug delivery systems for treating Inflammatory bowel disorders and cancer. Dr. Akram has been bestowed with several honors; a few of them worth mentioning include the Young Talent Award by the Indian Pharmaceutical Education Society; Young Researcher Award by; Arjyopa during World Congress on Drug Development; Awarded JRF & SRF By University Grant Commission (UGC), New Delhi; in addition, several awards for best paper presentations at conferences; Indian Pharmacy Graduate Association, and Controlled Release Society, Indian Chapter. Dr. Akram shared his research and teaching experience in several conferences, seminars, symposiums and workshops as a representative or invited speaker. He is also actively engaged in various committees of the department, Institute and University. He has published 12 peer-reviewed papers in the field of pharmaceutical sciences in national and international journals, four book chapters and four co-authored books. He has also been granted 1 Indian patent. He is also active as a reviewer for several international scientific journals. He is a member/Life member of several Indian scientific organizations.

Navneet Garud is Ph.D. in Pharmaceutics currently working as Professor at School of Studies in Pharmaceutical Sciences, Jiwaji University, Gwalior. He has more than 22 years of teaching and research experience. He has more than 70 publications in various International/national journals of repute. His area of specialization is a controlled drug delivery system. He is author of 03 books. He is a life member of various professional bodies like IPA, IPS, APTI, IPGA.

Sustainable Cellulose Materials for Biomedical Application

Fatma Akpınar, Kübra Gençdağ Şensoy, Merve Esen Keçeci, and Mihrican Muti

Abstract Cellulose is a biopolymer that is abundant, biodegradable, and renewable. Cellulose is a preferred material in biosensor development due to its properties such as biocompatibility, chemical structure versatility, and mechanical durability. When cellulose and nanocellulose are included in sensor modification, the sensitivity, selectivity, and environmental friendliness of the sensor are enhanced, thereby improving its overall performance. Cellulose-modified sensors are widely used in many areas such as health, environmental monitoring, and food safety. Cellulose-modified biosensors benefit from the material's capacity to immobilize biomolecules, including enzymes, antibodies, and nucleic acids, thus enhancing biosensor performance through improved molecular recognition, catalytic activity, and signal transduction. The two basic forms of nanocellulose, cellulose nanofibrils (long and thin cellulose derivatives) and nanocrystals (shorter and more regular cellulose particles), have unique properties such as larger surface area, greater possibilities for chemical and physical interactions, higher aspect ratio, and improved mechanical and optical properties. Cellulose-modified sensors are effective in detecting substances such as pathogens, toxins, and environmental contaminants. Furthermore, cellulose's inherent biodegradability and low toxicity enable the design of disposable and environmentally friendly sensor devices, addressing the growing demand for sustainable solutions in biosensing. This paper synthesizes the latest progress in cellulose-modified biosensors, discussing fabrication strategies, sensing mechanisms, and integration with nanomaterials, as well as emerging applications in point-of-care diagnostics and wearable sensors. The exploration of cellulose-based materials in biosensing not only contributes to the development of sensitive and selective detection tools but also promotes sustainable and scalable approaches for rapid, portable biomolecular analysis in real-world settings.

F. Akpınar · M. E. Keçeci · M. Muti (✉)
Department of Chemistry, Faculty of Science, Aydın Adnan Menderes University, Aydın, Türkiye
e-mail: mihricanmuti@adu.edu.tr

K. G. Şensoy
Department of Food Processing, Köşk Vocational High School, Aydın Adnan Menderes University, Aydın, Türkiye

R. Malviya and S. Sundram (eds.), *Sustainable Cellulose Materials for Biomedical Applications*, Sustainable Materials and Technology,
https://doi.org/10.1007/978-981-96-9904-9_12

1 Introduction

1.1 Cellulose Structure

Cellulose is the primary and most widespread biopolymer found on Earth [79, 177]. It can be produced naturally by plants (such as wood, cotton, and straws), animals (certain marine species) and microorganisms (like fungi) with an estimated global production of around 10^{11}–10^{12} tons per year [55, 177] Cellulose obtained from plants and secreted by a microorganism can be transformed into variety of nanostructures [94]. The hierarchical arrangement of cellulose in plant-derived biomass is depicted in Fig. 1 [159].

Plant fibers act as natural cellulose reservoirs, forming intricate biocomposites in their unprocessed state. These fibers, which consist of single cells, exhibit lengths ranging from 1 to 50 mm and diameters spanning 10 to 50 µm [15]. The typical cell wall of a plant fiber measures 4–6 µm in thickness. The part of the nanofibril bundles that forms irregular networks, approximately 100 nm thick, is called the primary wall. This part is rich in cellulose microfibrils, which give the structure rigidity and flexibility. The secondary wall (S wall) has a thickness of 3–5 µm and is situated between the primary cell wall and the plasma membrane. It is stiffer and supports water transport due to the lignin it contains. Cellulose microfibrils are characterized using microfibril angle (MFA) θ. The measurement of MFA is particularly useful in understanding the strength, flexibility, and other physical properties of plant fibers. The hollow part located at the center of plant fibers is called the lumen (L). The lumen influences the structural properties of the fiber and plays an important role in

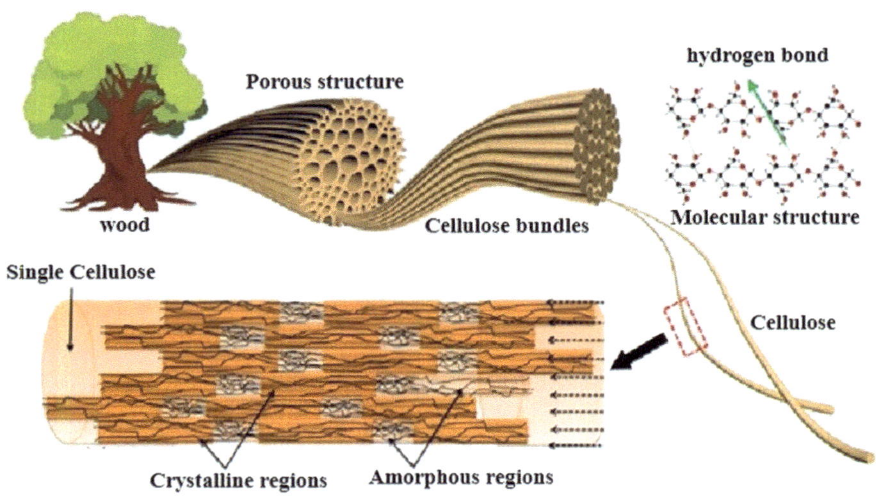

Fig. 1 Hierarchical arrangement of cellulose in plant-derived biomass. Adapted from Nanotechnology Reviews, 2024, 13, 20240006, CC BY 4.0

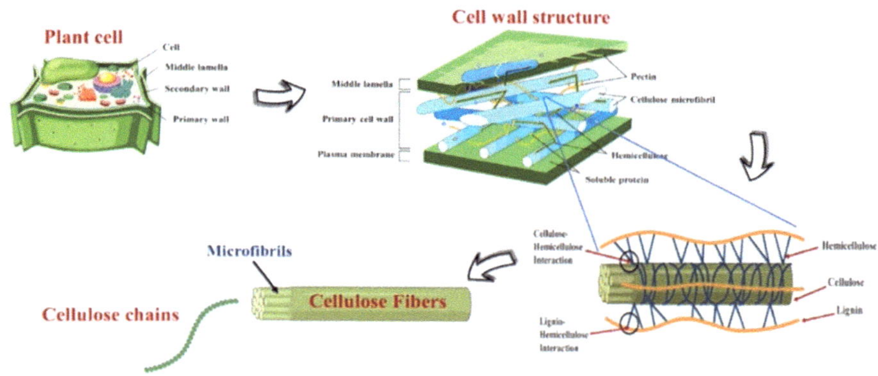

Fig. 2 Hierarchical structure of natural fiber. Adapted with permission from Channab et al. [15]

functions such as water transport in the plant. It can also contribute to the mechanical properties of the fiber. The total width of plant fibers, including the lumen, ranges between 15 and 30 μm. S2 layer, is the most prominent layer and has 2–4 μm in thickness. 85–95% of the fiber's mass is composed from S2 layer. This layer contains nanofibril bundles and lamellae aligned parallel to each other, oriented at a narrow angle relative to the fiber axis. This precise arrangement enhances the mechanical strength of cellulose fibers [26, 61]. The arrangements of natural fibers, showcasing the composition of the plant cell wall, including cellulose microfibrils is illustrated in Fig. 2.

(a) ***Crystal Structure of Cellulose***

There are different crystalline polymorphic forms of cellulose. Cellulose I is known as "native" cellulose and is found in many organisms. It is thermodynamically metastable and can be converted to cellulose II or cellulose III. Cellulose II is the most stable and technologically important form and is usually produced by regeneration (dissolution and recrystallization) and mercerization (treatment with sodium hydroxide). Cellulose II has a monoclinic crystal structure. Cellulose I and cellulose II are converted to cellulose III_1 and cellulose III_2 when treated with ammonia; these forms share the same unit cell structure and are reversible. Cellulose III can also be converted to cellulose IV_1 and IV_2 by treatment with glycerol at high temperatures, but this conversion is incomplete. Cellulose is a polymer composed of β-D-glucopyranose units, containing glucose chains linked together by β-1,4-glycosidic bonds, and this structure provides the crystalline and fibrous properties of cellulose enhanced by strong hydrogen bonds [91].

(b) ***Hydrogen Bondings***

Cellulose consists of a straightforward primary structure, composed of a linear sequence of β-glucose molecules which covalently linked through 1,4′ glycosidic bonds (C–O–C). These cellulose chains are organized into partially crystalline fibers

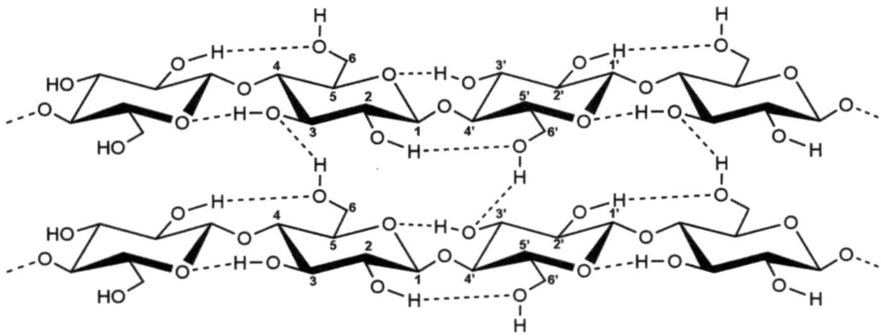

Fig. 3 Primary structure of cellulose and the hydrogen bonds formed between its chains Reproduced with permission from Chen et al. [19]

known as microfibrils, typically about 3 nm in diameter. Within each microfibril, the glucan chains are organized in layers, with hydrogen bonds forming both between adjacent chains and between monomers within the same chain [48]. The two hydrogen bonds involving the hydroxyl groups of C6 and C2, as well as the oxygen of C5 and the hydroxyl group of C3, stabilize the glycosidic linkage and contribute to the structural rigidity [19, 31]. The primary and secondary structures of cellulose are depicted in Figs. 3 and 4, respectively. Natural microfibrils are further stabilized by intra- and inter-chain OH–O hydrogen bonds, unconventional CH–O hydrogen bonds, and van der Waals forces between glucan chains in adjacent layers. This complex interaction network prevents cellulose from dissolving in most solvents [5].

Each repeating unit of cellulose has three hydroxyl groups, and the overall structure of cellulose is mainly shaped by van der Waals forces and hydrogen bonds. The hydrogen bonds between adjacent cellulose chains are important for maintaining the regularity of the chains and give cellulose fibers their strength and thermal stability.

1.2 Cellulose Properties

Cellulose is unique due to its linear and ordered polysaccharide structure, along with its strong intermolecular hydrogen bonds. Cellulose is a renewable material that can be abundantly and easily obtained from wood and other plant fiber sources, and does not melt at high temperatures, and is not soluble in conventional organic solvents [93]. Its non-toxic properties for the human body, biological compatibility, and the potential to be converted into nanoparticles with unique properties make it highly interesting, as it offers potential for various applications [111]. In addition to these features, its low production costs and its ability to be modified for different uses increase its area of use over time [91]. The strong intra and intermolecular hydrogen bonds in cellulose give hardness to the material [134].

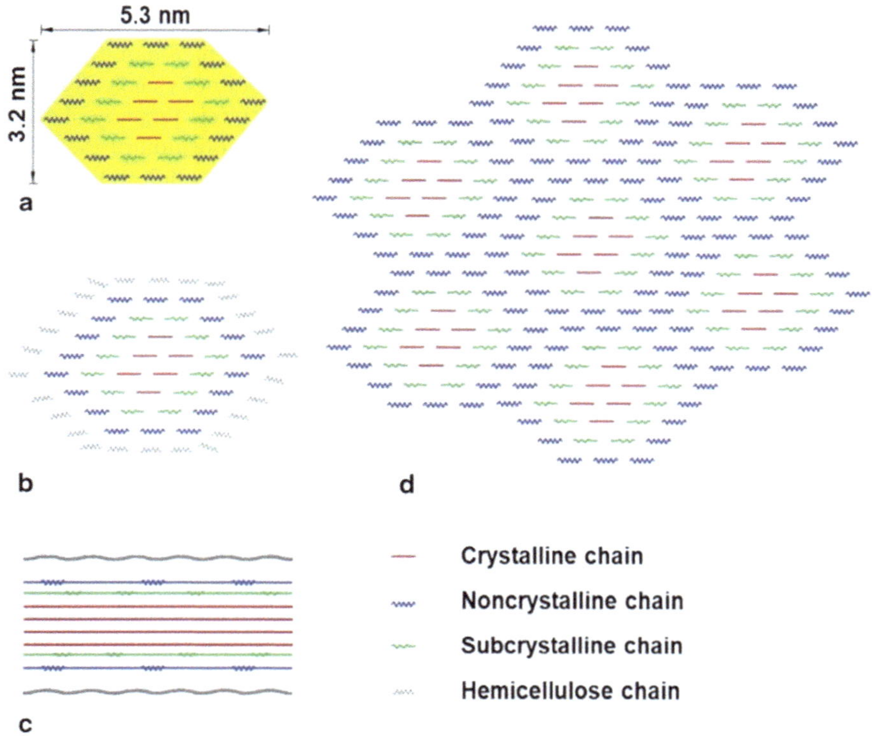

Fig. 4 Secondary structure of cellulose, **a** a cross-sectional view of the elementary fibril, which contains 36 glucan chains, **b** the cross-section of the microfibril, where hemicellulose covers the surface of the core fibril, **c** a longitudinal section of the microfibril, and **d** a cross-sectional view of the macrofibril, which is formed by the aggregation of elementary fibrils. Reproduced with permission from Chen et al. [19]

Cellulose attracts water due to the hydroxyl (–OH) groups in its structure, that is, it is hydrophilic. However, it may be difficult to dissolve in some organic solvents. The most important feature of cellulose is that it has a strong affinity for substances containing –OH groups, such as water [148]. This causes it to attract water very well. In addition, the hydroxyl group on carbon 6 is more reactive than the hydroxyl groups on other carbons, that is, this group interacts with water more easily [4]. Cellulose reacts quickly with water and accelerates this interaction by forming –OH bonds under the influence of water. In addition, cellulose can turn into glucose when treated at high temperatures and with strong acids [148].

The intra and intermolecular hydrogen bonds are responsible for making natural cellulose insoluble in water. When cellulose is modified and the hydrogen bonds are disrupted, causing the crystalline structure of cellulose to be lost, the resulting cellulose derivative becomes soluble in water [39].

Cellulose has highly stable supramolecular structures made up of microfibrils composed of several dozen macromolecules. The parallel orientation of the macromolecules forming the microfibrils makes the structure thermodynamically more stable. Cellulose exhibits amphiphilic properties. The hydrophilic property of cellulose is due to the hydroxyl (–OH) groups in the equatorial position of the anhydroglucose unit. These groups tend to attract water and allow cellulose to interact with water. On the other hand, the hydrophobic property arises due to the methyl groups and carbon–carbon bonds of the anhydroglucose ring. These structures have a water-repellent effect and prevent cellulose from interacting with water. These two properties together prevent cellulose from dissolving in most conventional solvents. In other words, cellulose both attracts water and tends to repel water due to some of its building blocks, which makes it insoluble in most solvents [16].

(a) ***Toxicity***

Cellulose is a natural and non-toxic biopolymer as it is obtained from wood, plant cell walls, certain algae and bacteria [16]. Cellulose-based materials are generally considered biocompatible (compatible with the body) and non-toxic. Because they are derived from cellulose, a naturally occurring substance in nature. However, cellulose-derived nanostructures in nanometer sizes can become more reactive when obtained through various processes, which can lead to toxic substances. In recent years, cellulose-derived nanomaterials (CNMs) have been investigated for their toxicity, as they promise a wide range of applications in industry and biomedical fields. Most studies have examined exposure to these nanomaterials through inhalation. In other words, although cellulose-derived nanomaterials appear potentially safe, they may cause toxic effects because the nanostructures may be more reactive, and this is still under investigation.

Upon reviewing the literature, it has been observed that cellulose nanomaterials with different physicochemical properties can cause different toxicological effects. CNMs in gel or powder form are likely to trigger various immunological responses associated with their propensity to aggregate when distributed in body fluids (in vivo) or culture media (in vitro). In general, it has been understood that the toxicity of CNMs is significantly governed by their physicochemical attributes and it is believed to be strongly related to the cellulose source from which the CNMs are produced, the preparation procedure, or sterilization. Continued research on the toxicity of various nanomaterials derived from cellulose sources is necessary to ensure a clearer understanding [162, 164].

(b) ***Mechanical Properties***

The mechanical properties of cellulose arise from the interactions between these regions. The Young modulus, which indicates the elasticity of plant cellulose fibers, is generally between 20 and 30 GPa. However, for cellulose with a high degree of crystallization, values can reach up to 138 GPa. Studies have shown that under high temperatures (355–400 °C) and moderate pressures (25–33 MPa), natural cellulose is soluble in supercritical water and reprecipitate as cellulose II [13].

The elasticity of cellulose's crystalline regions have been extensively characterized using X-ray diffraction, Raman spectroscopy, and Atomic Force Microscopy (AFM). Molecular dynamics simulations determined the structure and Young's modulus of cellulose I and II, reporting a modulus of 134–135 GPa for cellulose I [36]. In 2006, Tanaka et al. reported a longitudinal modulus of 155 GPa for regenerated cellulose at room temperature and an elastic modulus of 124–155 GPa for the crystalline regions of cellulose Iβ [152]. In the same year, Eichhorn and colleagues used molecular mechanics modeling and reported elastic modulus values of 136–155 GPa for cellulose Iα, 116–149 GPa for cellulose Iβ, and 109–166 GPa for cellulose II [35]. In a different study by Bergenstråhle et al., the thermal response of the cellulose Iβ structure was examined at 300 K and 500 K, with reported longitudinal Young's moduli of 156 GPa and 117 GPa, respectively [10].

(c) ***Solubility***

Although cellulose contains numerous hydrogen bonds within its chains, it is insoluble in water and most organic solvents, as its molecules form hydrogen bonds with each other rather than with solvent molecules [110].

Cellulose can dissolve in alkaline urea solutions and fibers, membranes, microspheres, hydrogels and various cellulose derivatives can be obtained from these solutions [87].

Ionic liquids are effective solvents for cellulose because of their unique structure, which is highly amphiphilic. This means they have both hydrophilic (water-attracting) and hydrophobic (water-repelling) properties within the same molecule. The hydrophilic component interacts with the polar cellulose molecules, while the hydrophobic component helps to disrupt the strong hydrogen bonds between cellulose chains. This dual characteristic allows ionic liquids to dissolve cellulose effectively, overcoming the resistance that cellulose typically has to dissolving in most solvents. The same applies to other solvents, such as N-methylmorpholine N-oxide or aqueous alkylammonium hydroxides [84]. Cellulose is also soluble in other ionic solvents such as N-methylmorpholine N-oxide (NMMO) and 1-butyl-3-methylimidazolium acetate. These solvents are also called "green solvents" because they can be purified and reused in cellulose processes [142].

Despite their advantages, many ionic liquids are expensive and difficult to recover. The ionic solvent N-methylmorpholine-N-oxide (NMMO), mentioned above, is a suitable ionic liquid for cellulose dissolution due to its non-toxicity and the ability to recover up to 90% of it [14].

(d) ***Hygroscopic Properties/Water Interactions***

Hygroscopy is a characteristic of cellulose fibers. This property is affected by environmental factors like humidity and temperature. As the material absorbs moisture, water molecules can become trapped between its molecules during processing. These changes can affect the material's physical properties, leading to changes in its boiling point, as well as increases in volume and viscosity. Cellulose fibers are hygroscopic, meaning they absorb moisture from the environment. When cellulose comes into contact with water, it swells and disperses. This swelling is believed to occur because water molecules are "packed" into the amorphous regions of its semi-crystalline structure [96]. Cellulose interacts well with water through the hydroxyl groups found in its structure. In its natural state, cellulose always maintains a unique relationship with water. This relationship is beneficial for many industrial processes, such as paper production and fiber spinning. However, for highly crystalline, insoluble celluloses, water acts as an anti-solvent, and it has been observed that higher dissolution is achieved in the absence of water. In contemporary applications, excess water is considered undesirable, requiring active drying processes to effectively eliminate moisture from cellulose [77, 157].

1.3 Cellulose Sources

Cellulose can be sourced from a variety of materials. This material is found in plants, some bacteria, algae, fungi, and tunicates [3, 91].

Plant fibers can be considered a source of cellulose because they contain cellulose and hemicellulose structures. However, the proportion of cellulose, hemicellulose, lignin, pectin, and wax contained in plants varies depending on the climate and soil in which they are grown [181].

Lignocellulosic biomass can also be referred to as a source of cellulose. These include wood, non-wood plants, agricultural waste, and industrial waste. Other sources of cellulose include by-products of various agricultural and industrial processes, such as empty fruit bunches of palm trees, residues from the beer and sugar industries, and by-products of the paper industry [3].

(a) ***Wood Biomass***

Wood biomass represents the most abundant and accessible source of biomass [182]. Approximately 40–50% of the weight of wood biomass is composed of cellulose. Wood biomass can be processed to yield high-purity cellulose for the paper industry via sulfite, kraft, or soda pulping techniques [144].

Wood biomass includes trees, leaves, and other woody plants found in forests, as well as residues from wood processing. It is typically divided into two categories: softwood and hardwood. Softwood examples are pine, spruce, and fir trees, while hardwood examples include birch, trembling aspen, eucalyptus, and acacia trees. Hardwood fibers are shorter and finer than softwood fibers. Hardwood generally are richer in cellulose and hemicellulose and poorer in lignin. Softwood contains higher resin content [3].

(b) ***Plant***

Similar to woody biomass, plants can also undergo processes that allow the extraction of cellulose from their structures. Cotton, flax, wheat straw, can be used as attractive sources of cellulose [97].

(c) ***Tunicate***

Tunicates, which are the only animals capable of producing cellulose microfibrils, are part of the marine animal family. In many studies, the term "sea squirts" (Ascidiacea), referring to a class of tunicates with over 2300 species, has been commonly used. Marine organisms known as tunicates produce extremely high-purity cellulose with high crystallinity in their outer tissues. The cellulose structures possess a significantly large aspect ratio and a surface area between 150 and 170 m^2/g^{-1} [187].

(d) ***Algae***

Due to their high abundance and availability, marine macroalgae (seaweeds) present a promising renewable source of cellulose. These organisms nourish by atmospheric and water-soluble carbon dioxide, light, and nutrients; their growth conditions are quite simple, and they reproduce rapidly. By these positive attributes, they form an interesting and alternative source of cellulose [89]. Based on their pigmentation, marine macroalgae can be divided into three groups: green (Chlorophyta), red (Rhodophyta), and brown (Ochrophyta-Phaeophyceae) [113]. Most researchers who study cellulose use green pigmentation from these groups. The species most frequently studied include *Micrasterias denticulata*, *Micrasterias rotata*, *Valonia*, and *Caldophora* [61].

(e) ***Bacteria***

Bacterial cellulose (BC) has attracted considerable attention owing to its remarkable properties, such as a high degree of polymerization, biodegradability, superior crystallinity, and outstanding mechanical characteristics. Bacterial cellulose (BC) features an exceptionally high surface area per unit mass and demonstrates strong hydrophilic properties, resulting in an outstanding water retention capacity [43].

Various bacterial species are capable of producing cellulose through biosynthesis from carbon sources available in the culture medium. Glucose, fructose, and glycerol are traditionally used as cellulose sources for production. In an effort to lower the production costs of bacterial cellulose synthesized by bacteria, numerous studies have explored alternative and varied sources such as cotton-based textile waste, fruit by-products, dairy industry waste, and wine fermentation residues [43, 129].

2 Cellulose Derivatives

Cellulose is a regenerative, sustainable, hydrophilic, biodegradable, inexpensive, and abundant polymer found in nature. It comprises large crystalline regions held together by numerous hydrogen bonds. Because of the strong hydrogen bonds that are present in its structure, cellulose can be modified chemically. But these linkages also make cellulose less chemically reactive and less soluble in water and most organic solvents [54, 68]. Derivatization of cellulose involves chemically modifying it to enhance properties such as solubility, thermal stability, and biocompatibility, thereby expanding its industrial and biomedical applications. This process, which addresses the drawbacks of raw cellulose and introduces new functional properties, is typically achieved through etherification and esterification [33, 45, 66].

2.1 Etherification of Cellulose

Etherification is the process by which the alkyl groups of organic compounds, especially alkyl halides, combine with the hydroxyl groups (–OH) of cellulose. These changes improve cellulose's solubility, thermal stability, lifetime, and functional qualities, which makes it appropriate for a range of uses, including biomedical ones [54, 161].

Industrial cellulose ethers are prepared by slurring cellulose in different concentrations of NaOH solution and an organic solvent, followed by swelling (Fig. 5). By converting cellulose into alkali form the activity of cellulose can be increased. For this, alkaline treatment is applied with sodium hydroxide. The interaction between the organic solvent and the hydroxyl groups of cellulose allows the formation of the ether bond, efficient polymer dispersion, heat transfer during the reaction, and recovery of the resulting product [72, 176].

Although many cellulose ether compounds have been synthesized, hydroxyethyl cellulose (HEC), carboxymethyl cellulose (CMC), and methyl cellulose (MC) are among the most widely used in industrial and biomedical product formulations [183]. The properties, chemical modifications and biomedical applications of cellulose ether derivatives with widespread industrial use are summarized in Table 1.

(a) ***Methyl Cellulose (MC)***

Methyl cellulose is an alkyl ether synthesized by methylating alkali cellulose with dimethyl sulfate or methyl chloride. It is an important commercial cellulose ether

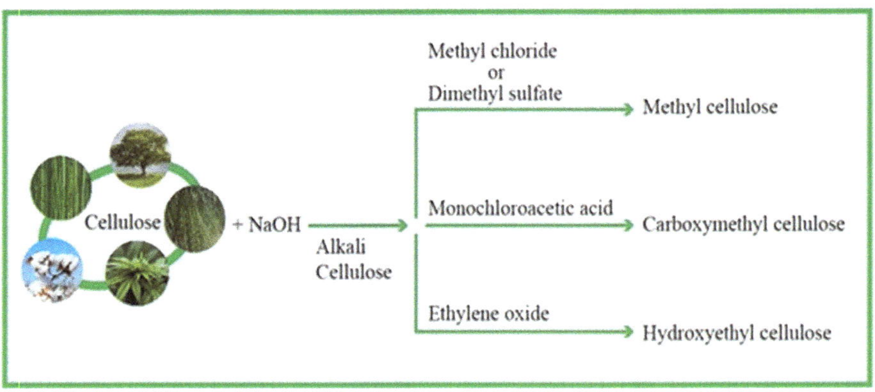

Fig. 5 Preparation procedures of the most commonly used cellulose ethers

Table 1 Key properties chemical modification patterns and biomedical applications of some cellulose ester derivatives

Type of cellulose ether	Chemical structure/ modification	Key properties	Biomedical applications
Hydroxyethylcellulose (HEC)	Hydroxyethyl (–CH_2CH_2OH) groups are substituted for hydroxyl groups	• Hydrophilic and water-soluble • Non-ionic thickening agent	• **Bioadhesives** for wound closure • **Thickener** in topical gels and creams
Carboxymethylcellulose (CMC)	Carboxymethyl (–CH_2COOH) groups are substituted for hydroxyl groups	• Water-soluble, forming anionic hydrogels • High biocompatibility and biodegradability	• **Wound dressings** for maintaining moist healing environments • Stabilizer in protein drug formulations
Methylcellulose (MC)	Methyl (–CH_3) groups are substituted for hydroxyl ones	• Thermoreversible gelation • Forms gels at elevated temperatures • High viscosity	• **Hydrogels that can be injected** to deliver drugs • Tissue engineering **scaffolds**
Ethylcellulose (EC)	Substitution of hydroxyl groups with ethyl (–C_2H_5) groups	• Hydrophobic and water-insoluble • Forms stable films and coatings	• **Microcapsules** for drug delivery • **Film coatings** for sustained-release drugs
Hydroxypropylcellulose (HPC)	Hydroxypropyl (–$CH_2CHOHCH_3$) groups are substituted for hydroxyl groups	• Thermoplastic and hydrophilic • Both water and some organic solvents can dissolve it	• **Drug carriers** for controlled release • Coating agent for pharmaceutical tablets
Hydroxypropyl methylcellulose (HPMC)	Substitution with both hydroxypropyl (–$CH_2CHOHCH_3$) and methyl groups	• Excellent film-forming ability • pH-independent swelling • Sustained drug release	• **Sustained-release drug formulations** • **Ophthalmic gels** for dry eye treatment
Sodium carboxymethylcellulose (NaCMC)	Sodium salt of carboxymethylcellulose	• Strong water-binding capacity • Highly anionic polymer	• **Wound-healing hydrogels** • **Thickener** in injectable drug formulations
Methyl Hydroxyethylcellulose (MHEC)	Substitution with methyl and hydroxyethyl groups	• Improves water retention and adhesion • Moderate gelation properties	• **Controlled-release drug systems** • **Adhesives** for biomedical applications

that is water-soluble and forms viscous solutions that gel at high temperatures. Its non-ionic nature allows it to dissolve in a wide range of organic liquids [24, 59]. Because it is non-reactive and biocompatible, methyl cellulose is frequently used in biomedical applications such as pharmaceutical formulations [18], wound healing [32], and tissue engineering [107]. Furthermore, its emulsifying properties enhance its role in drug delivery systems [22, 23].

(b) ***Carboxymethyl Cellulose (CMC)***

The synthesis of carboxymethyl cellulose consists of two main steps: mercerization and etherification. The first stage involves dispersing cellulose in a solution of water, alcohol, and NaOH. The NaOH solution renders the cellulose alkaline, dissolves hydroxyl groups, and prepares it for etherification. In the final step, alkali cellulose reacts with monochloroacetic acid or sodium monochloroacetic acid to form carboxymethyl cellulose ether [121, 133].

A significant cellulose derivative, carboxymethyl cellulose is distinguished by its biodegradability, non-toxicity, and exceptional chemical stability. It dissolves easily in water but remains insoluble in solvents like methanol, ethanol, chloroform, and benzene. Its hydrophilic nature contributes to its high viscosity [119, 125]. In biomedical applications, it serves as a viscosity enhancer, binder, and stabilizer. Thanks to its anionic structure, attributed to the carboxymethyl groups, it can interact with positively charged molecules such as proteins and drugs, making it valuable in drug transport systems [62, 143] and tissue engineering [95, 163]. Moreover, hydrogels based on carboxymethyl cellulose are frequently utilized in wound care and healing [73, 132].

(c) ***Hydroxyethyl Cellulose (HEC)***

When alkali cellulose and ethylene oxide combine, hydroxyethyl groups are attached to the cellulose backbone, resulting in the creation of hydroxyethyl cellulose. For a number of applications, this alteration increases cellulose's stability and solubility. A non-ionic cellulose derivative called hydroxyethyl cellulose forms viscous solutions when dissolved in water and a variety of organic solvents. It is non-toxic, biodegradable, and well-suited for biomedical applications [92, 109, 189].

The ability of hydroxyethyl cellulose to create films, retain moisture, and interact with bioactive chemicals makes it a common ingredient in wound care treatments [37, 83]. In ophthalmic formulations, it serves as a viscosity enhancer, improving retention time on the ocular surface and enhancing effectiveness [28, 51]. It is appropriate for tissue engineering applications due to its mechanical qualities and biocompatibility [29, 190]. Additionally, it is utilized in drug delivery systems for its ability to increase drug bioavailability and control drug release [44, 53].

2.2 *Esterification of Cellulose*

When cellulose reacts with acid anhydrides, acid chlorides, or organic acids when a catalyst is present, the hydroxyl groups (–H) in the cellulose structure are replaced with ester groups (–COOR), a process known as cellulose esterification (Fig. 6). In comparison with raw cellulose, this alteration improves the resulting cellulose ester derivatives' mechanical strength, solubility in organic solvents, and thermal stability. Cellulose esters are now more useful and adaptable for usage in industrial, medicinal, and biological applications thanks to these enhanced qualities. The three cellulose ester derivatives that are most commonly used are cellulose sulfate (CS), cellulose nitrate (CN), and cellulose acetate (CA) [45, 121]. The properties, chemical modifications and biomedical applications of cellulose ester derivatives with widespread industrial use are summarized in Table 2.

(a) ***Cellulose Acetate (CA)***

Acetic anhydride and cellulose's hydroxyl groups react to generate acetyl groups (–$COCH_3$) in the presence of a catalyst, such as sulfuric or perchloric acid, to produce cellulose acetate. Acetic acid is used as the solvent. This reaction produces cellulose acetate polymers with varying degrees of substitution. Cellulose that has been partially acetylated dissolves in water. On the other hand, fully acetylated cellulose dissolves in organic solvents including ethanol, dichloromethane, and acetone but is insoluble in water. Cellulose acetate polymers decompose thermally above 230 °C, making them suitable for medium-temperature applications [21, 112]. Key characteristics, chemical alteration strategies, and biomedical uses of selected cellulose ester derivatives is given in Table 2.

Cellulose acetate is among the key cellulose esters, offering versatility and sustainability with widespread applications in biomedical, environmental, and industrial

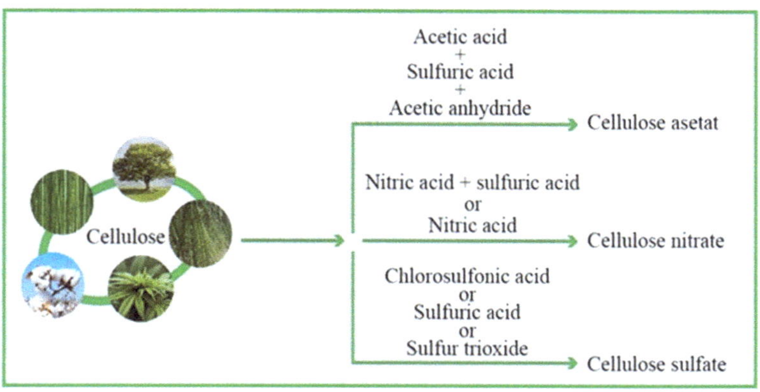

Fig. 6 Preparation procedures of the most commonly used cellulose esters

Table 2 Key properties chemical modification patterns and biomedical applications of some cellulose ester derivatives

Type of cellulose ester	Chemical structure/ modification	Key properties	Biomedical applications
Cellulose sulfate (CS)	Sulfation of cellulose by introducing sulfate groups at hydroxyl positions	Water solubility, polyelectrolyte behavior, thermal stability, biocompatibility, bioactivity	• **Wound healing** materials • Scaffolds for **tissue engineering** • **Anticoagulant** coatings • **Drug delivery** systems
Cellulose nitrate (CN)	Substitution with nitrate (–NO_3) groups	• High flammability • Excellent film-forming ability	• Previously used in **wound care films** (now limited)
Cellulose acetate (CA)	Substitution of hydroxyl groups with acetyl (–$COCH_3$) groups	• Biodegradable and thermoplastic • Transparent, strong, and flexible • Moisture-resistant	• **Drug delivery membranes** for controlled release • **Wound dressings** with breathability
Cellulose propionate (CP)	Substitution with propionyl (–COC_2H_5) groups	• Hydrophobic • High durability and flexibility	• **Biofilms** for drug encapsulation • **Scaffold materials** for tissue engineering
Cellulose acetate phthalate (CAPh)	Substitution with acetyl (–$COCH_3$) and phthaloyl (–COC_6H_4COOH) groups	• pH-sensitive, dissolving in alkaline environments • Acid-resistant	• **Enteric coatings** for protecting drugs from gastric acid • Oral sustained-release tablets
Cellulose acetate butyrate (CAB)	Substitution with acetyl and butyryl (-COC_3H_7) groups	• Hydrophobic with enhanced impact resistance • Low water absorption	• **Drug release systems** for hydrophobic drugs • **Protective coatings** for biomedical devices
Cellulose acetate propionate (CAP)	Substitution with both acetyl and propionyl (–COC_2H_5) groups	• Improved toughness and moisture resistance • Hydrophobic	• **Film coatings** for prolonged drug release • **Scaffold materials** in tissue engineering
Cellulose succinate (CS)	Substitution with succinyl (–$COCH_2CH_2COOH$) groups	• pH-responsive (dissolves in basic pH) • Biodegradable	• **Oral drug transport systems** for controlled release in the intestines • **pH-responsive hydrogels**
Cellulose esters with fatty acids	Substitution with long-chain fatty acids (e.g., lauryl, stearyl)	• Amphiphilic, providing hydrophilic-lipophilic balance • Biocompatible	• **Surfactants** in biomedical formulations • **Drug carriers** for poorly water-soluble drugs

fields. It is a great option for overcoming obstacles in drug administration, tissue engineering, and sustainability because of its remarkable qualities, which include non-toxicity, biocompatibility, biodegradability, and customisable functionality. Thanks to its high strength, selectivity, and flexibility, cellulose acetate can form films, making it widely used in coatings and hemodialysis membranes for biomedical applications. These membranes selectively remove toxins and small molecules while preserving essential proteins [8]. In osmotic drug delivery systems, cellulose acetate membranes regulate water flow and drug release [116]. The extracellular matrix (ECM) is mimicked by porous cellulose acetate scaffolds, which promote cell adhesion, proliferation, and differentiation. In bone and cartilage engineering, cellulose acetate is combined with bioactive agents, such as hydroxyapatite, to promote bone regeneration [74]. Additionally, cellulose acetate films and hydrogels are applied as wound dressings due to their breathability, moisture retention, and ability to incorporate antimicrobial agents [138].

(b) ***Cellulose Nitrate (CN)***

After cellulose's hydroxyl groups are swapped out for nitro groups during the nitration process, cellulose nitrate is produced. Cellulose nitrate is more commonly used as nitrocellulose.

A solution of sulfuric and concentrated nitric acids is used to nitrate cellulose. In this process, the sulfuric acid acts as both a catalyst and a dehydrating agent to increase the efficiency. The degree of nitration significantly impacts the material's properties, with higher degrees leading to increased flammability and solubility in organic solvents. Research has examined the synthesis of nitrocellulose from plant and bacterial celluloses, highlighting the ways in which various cellulose sources affect the final nitrocellulose's characteristics [12, 99].

Applications for nitrocellulose include coatings and films since it dissolves in organic solvents like ethanol and acetone. Due to its high nitrogen content, it is highly flammable and has historically been used as a propellant and explosive. It is capable of forming flexible and durable films in industrial applications [64, 99].

Although nitrocellulose is primarily used in industrial applications, recent research has explored its potential in biomedical fields. Nitrocellulose membranes are widely used in diagnostic assays, such as lateral flow tests, due to their excellent protein-binding capabilities [41]. Its biocompatibility and ability to form porous structures make it a potential candidate for supporting cell growth, though concerns about its biodegradability and potential cytotoxicity require further study [78]. The development of wound dressings has made use of nitrocellulose's ability to produce films. Its ability to form a protective barrier, combined with antibacterial properties and a porous structure, may aid in wound healing [34, 122].

(c) ***Cellulose Sulfate (CS)***

Cellulose sulfate is formed as a result of the esterification of cellulose with sulfating agents, and this process is carried out through several methods. In the first method, cellulose is treated with chlorosulfonic acid or another sulfonating agent in an organic solvent such as pyridine or dimethylformamide. This reaction adds sulfate groups to the hydroxyl positions of the cellulose backbone. The second method uses sulfur trioxide in combination with pyridine to achieve controlled sulfation, enabling the production of cellulose sulfate with a defined degree of substitution [17, 101, 180].

The degree of substitution (DS), which expresses the number of sulfate groups replaced by hydroxyl groups in the cellulose molecule, significantly affects the solubility, charge density, and biological activity of cellulose sulfate. Sulfation increases the hydrophilicity of cellulose, making it highly soluble in water. Cellulose sulfate acts as a polyelectrolyte due to its charged sulfate groups, which facilitate interactions with oppositely charged molecules. Although sulfation can reduce the thermal stability of cellulose, cellulose sulfate remains stable within the temperature ranges required for biomedical applications [44, 82, 180].

Cellulose sulfate is a promising material for biomedical uses, owing to its distinctive characteristics such as biocompatibility, bioactivity, minimal cytotoxicity, and water solubility. Cellulose sulfate hydrogels and nanoparticles enable the controlled and sustained release of drugs, particularly ionic and hydrophilic molecules. It has shown efficacy in delivering antiviral agents through electrostatic binding with viral particles [128, 150]. Scaffolds based on cellulose sulfate replicate the natural extracellular matrix (ECM), facilitating cell proliferation, differentiation, and tissue repair [178]. The incorporation of sulfate groups enhances scaffold interactions with growth factors, improving osteogenic differentiation [115]. Additionally, cellulose sulfate mimics heparin, a natural anticoagulant, and is used in biomedical coatings to prevent thrombosis. Cellulose sulfate hydrogels provide a moist environment and interact with growth factors to accelerate the wound healing process [49, 128, 160].

3 Cellulose Particles

In our developing world, products that are made from low-cost, sustainable, renewable, non-petroleum-based, and ecologically safe are in greater demand. Humans have been using natural cellulose-based products for thousands of years for a variety of purposes. Yet, traditional cellulose-based materials often fail to meet the requirements of today's advancing world. To address this need, many of the drawbacks of conventional cellulose have been addressed by the development of cellulose particles while incorporating features needed in industrial applications [7, 97]. Cellulose particles are a multipurpose nanomaterial that have attracted a lot of interest lately because of its special structural, mechanical, and functional characteristics. These particles exhibit remarkable characteristics, including a high surface area, tunable surface chemistry, biocompatibility, and exceptional mechanical strength. Numerous uses are made possible by these qualities, especially in the biomedical industry. Recent advancements position cellulose particles as an environmentally friendly substitute for artificial polymers, showcasing their potential in tissue engineering, drug delivery, and biosensor technologies [7, 47, 97, 147].

3.1 Classification of Cellulose Particles

Linear chains of β-D-glucopyranose units joined by β-1,4-glycosidic linkages make up cellulose molecules. The hydroxyl groups at positions C2, C3, and C6 facilitate the formation of a dense hydrogen bond network. This network stabilizes the crystalline regions of cellulose while imparting flexibility to the amorphous regions. The hierarchical structure of cellulose extends from microfibrils to fibril bundles and eventually to macroscopic fibers. Each particle type exhibits distinct characteristics, including crystal structure, morphology, aspect ratio, size, and physicochemical properties. Generally, cellulose particles are divided into groups according to their size, shape, and manner of manufacture. These groups include Cellulose Nanofibrils (CNFs), Cellulose Nanocrystals (CNCs), Microcrystalline Cellulose (MCC), and Bacterial Nanocellulose (BC) (Table 3) [45, 97, 126].

(a) ***Microcrystalline Cellulose (MCC)***

Wood pulp or cotton is hydrolyzed with acid and neutralized with alkali to produce microcrystalline cellulose. The amorphous portions of cellulose are eliminated during hydrolysis, resulting in crystalline particles that range in diameter from 10 to 50 μm. Functionalization strategies, such as polymer grafting, surface oxidation, and nanoparticle incorporation, are used to enhance MCC's mechanical and biological properties. For instance, TEMPO-mediated oxidation introduces carboxy groups onto MCC surfaces, improving dispersibility in aqueous media and enabling its use in hydrogels. Because of its qualities like mechanical resistance, MCC, a biocompatible cellulose derivative, finds extensive use in biological applications including food and pharmaceuticals [52, 97, 137]. Its crystallinity index generally ranges between 70 and 90%, based on the processing conditions and the source. MCC particles are typically spherical or rod-like with a high surface area, which allows for various chemical modifications. MCC maintains structural integrity up to 300 °C, making it suitable for applications requiring high thermal stability [70, 97, 127]. MCC is

Table 3 Properties of cellulose particle types [7, 47, 52, 97, 126, 137, 147]

Property	Microcrystalline cellulose (MCC)	Cellulose nanocrystals (CNCs)	Cellulose nanofibrils (CNFs)	Bacterial nanocellulose (BNC)
Crystallinity	~ 70–90%	> 80%	~ 50–70%	> 60%
Diameter	10–50 μm	5–20 nm	10–100 nm	20–100 nm
Length	50–200 μm	100–500 nm	Several micrometers	Variable; can be several micrometers
Morphology	Irregular, rod-like	Rod-like, needle-shaped	Long, flexible fibrils	3D nanofiber network
Production methods	Mild acid hydrolysis	Strong acid hydrolysis	Mechanical fibrillation	Biosynthesis by bacteria

a naturally occurring polymer that is safe for human tissues, biodegradable, and non-toxic. It serves as an excipient in tablet formulations, acting as a binder and disintegrant. MCC's high compressibility ensures uniform tablet strength, while its porosity allows for controlled drug release [185]. Additionally, MCC composites are used in hydroxyapatite-based scaffolds to enhance osteoblast adhesion and proliferation, aiding bone regeneration [103]. Because of its capacity to absorb exudates and offer a barrier against microbial contamination, MCC is also used in wound dressings [30, 174].

(b) ***Cellulose Nanocrystal (CNC)***

The stiff, rod-shaped nanostructures called cellulose nanocrystals (CNCs), sometimes referred to as whiskers or nanocrystalline cellulose, are formed from the crystalline portions of cellulose fibers. CNC is typically produced by removing amorphous regions through controlled acid hydrolysis. The size, crystallinity, and surface chemistry of the CNC are affected by parameters such as the concentration of the acid used during hydrolysis, reaction time, and temperature. For instance, sulfuric acid hydrolysis produces negatively charged CNC with sulfate ester groups, which enhance water dispersibility but reduce thermal stability. In contrast, hydrochloric acid hydrolysis yields CNC without functional groups, resulting in neutral surfaces and improved thermal properties [7, 126]. CNC particles have a high degree of crystallinity (usually above 80%) and range in length from 100 to 500 nm and diameter from 5 to 20 nm [75, 97, 126]. Excellent mechanical qualities, such as strong tensile strength and a nanoscale modulus equivalent to steel, are conferred by this high crystallinity. The abundant hydroxyl groups on CNC surfaces make CNC highly amenable to modifications, such as esterification, oxidation, and grafting. Additionally, CNC's surface charges enable effective dispersion in polar solvents like water. CNC generally exhibits thermal decomposition temperatures around 200 °C, although functionalization can influence its thermal behavior [75, 126]. Due to its large and functionalizable surface, CNC can be converted into hydrogels and used as carriers for the controlled release of various drugs such as antibiotics and anticancer drugs [7, 75, 126]. CNC enhances the mechanical characteristics of hydrogels and scaffolds, mimicking the stiffness of native tissues. It has been shown to improve chondrocyte adhesion and proliferation when integrated into cartilage regeneration scaffolds [126]. The functionalized surface of CNC also facilitates biomolecule conjugation, making it suitable for use in biosensors for detecting glucose, DNA, and other biomarkers [71, 158].

(c) ***Cellulose Nanofibril (CNF)***

Long, flexible fibrils with both crystalline and amorphous sections make up cellulose nanofibrils (CNFs), also referred to as nanofibrillated cellulose. CNF is primarily produced through mechanical methods, although chemical or enzymatic processes are often combined with mechanical techniques to reduce energy consumption [97, 141]. CNF typically has diameters of 10–100 nm and a high aspect ratio due to its micron-scale lengths [7, 126]. The long and flexible structure of CNF enables the formation of interconnected networks in hydrogels and composites. Because of its strong water retention capacity, CNF is a great option for scaffolds and wound dressings. Although less rigid than cellulose nanocrystals (CNCs), CNF demonstrates impressive tensile strength and flexibility, which make it suitable for load-bearing applications. The abundant hydroxyl groups on CNF surfaces allow for chemical modifications, enabling properties to be tailored for specific applications [7, 75, 126]. By preserving a wet environment, CNF-based hydrogels encourage cell proliferation and hasten wound healing [86]. CNF scaffolds mimic the extracellular matrix, supporting the growth of osteoblasts and chondrocytes [27]. Additionally, CNF hydrogels are used for the sustained release of hydrophilic drugs. The entangled network structure of CNF slows drug diffusion, providing long-term therapeutic effects [141]. CNF has also been incorporated into biomedical films for controlled permeability, such as oxygen barrier coatings in wound care and encapsulation materials for implantable devices [33, 139].

(d) ***Bacterial Nanocellulose (BNC)***

Acetobacter xylinum and other bacteria produce bacterial nanocellulose (BNC), an extremely pure extracellular polysaccharide. It is produced using a static culture medium that contains carbon sources such as glucose. Parameters such as pH, temperature, and oxygen availability significantly influence the yield and morphology of BNC. BNC has a crystallinity of 60–80% since it is free of hemicellulose, lignin, and other contaminants, in contrast to plant cellulose. With sizes ranging from 20 to 100 nm, BNC nanofibrils create a porous structure that is perfect for tissue regeneration and cell infiltration [75, 97, 126]. Due to its nanofibrillar structure, BNC can absorb water up to 200 times its dry weight and maintain a moist condition, which is critical for injury healing and tissue engineering. BNC exhibits excellent tensile strength and flexibility, even when moistened, allowing it to maintain these properties. These characteristics make BNC suitable for biomedical applications involving load-bearing processes. The high durability and flexibility of BNC support its use as a material with load-bearing capacity in medical devices, tissue engineering, and other biomedical fields. BNC-based dressings, like Biofill® and XCell®, are widely used because they promote cell proliferation, maintain a moist environment, and reduce infection risks [105]. Experimental studies demonstrate that BNC enhances granulation tissue formation and accelerates wound closure [147]. BNC's biocompatibility, high mechanical strength, ease of modification, cellular communication, and growth support make it suitable for use as a scaffold in cartilage, bone, and tissue engineering. Studies have shown the incorporation of growth factors into BNC scaffolds to support chondrocyte growth for cartilage regeneration [75, 155]. The nanoporosity and biocompatibility of BNC make it an excellent carrier for sustained drug release. Antimicrobial agents incorporated into BNC films effectively reduce bacterial colonization at wound sites [47]. Furthermore, other nanoparticles can be embedded into the BNC structure or functionalized with active agents via surface modifications, enhancing its properties for specific applications such as scaffolds for targeted drug delivery or antimicrobial wound dressings [2, 137].

3.2 Cellulose Particle Synthesis and Preparation Methods

The synthesis of nanocellulose involves breaking down bulk cellulose into nanoscale structures using mechanical, chemical, or enzymatic methods (Fig. 7). Every technique affects the final nanocellulose's composition, characteristics, and possible uses [165].

(a) ***Mechanical Treatment***

High-pressure homogenization, grinding, microfluidization, cryocrushing, and ultrasonication are examples of mechanical treatment methods. These methods rely on physical forces to separate cellulose fibers into nanoscale fibrils. Mechanical methods are highly effective in producing cellulose nanofibers (CNFs) with high aspect ratios while preserving both crystalline and amorphous regions [42, 58].

High-Pressure Homogenization: Cellulose fibers are suspended in water and repeatedly passed through a homogenizer at pressures of up to 2000 bar. Processes that use high shear rate, impact and cavitation forces are used to separate cellulose fibers, especially cellulose nanofibrils. High shear forces physically separate the fibers, while impact and cavitation effects support this separation process. As a result, these forces transform the cellulose fibers into smaller, thinner and nanometric CNFs. The number of homogenization cycles gradually reduces fiber size, influencing the degree of fibrillation and the quality of the resulting CNFs. Cellulose nanofibrils with diameters varying from 10 to 100 nm and lengths extending to several microns are produced. These CNFs exhibit high aspect ratios and outstanding mechanical properties [135].

Grinding: Cellulose fibers are processed by applying grinding, shearing, and friction forces through a disk mill. Factors such as disk gap size, rotation speed, and cellulose pretreatment significantly influence the morphology of the resulting cellulose nanofibrils (CNFs). Grinding generates a wider range of particle sizes and uses less energy than high-pressure homogenization (HPH) or microfluidization [120].

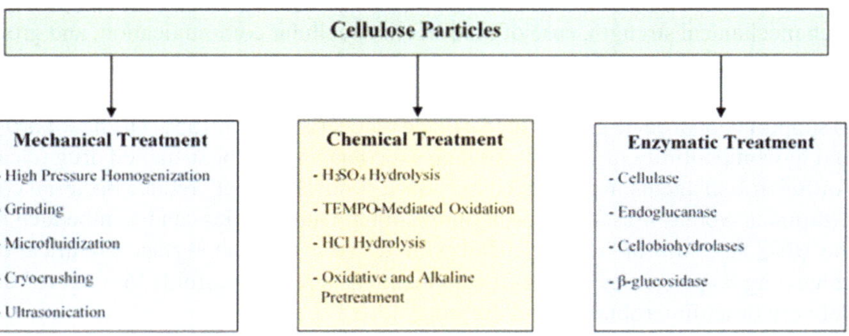

Fig. 7 Cellulose particles preparation methods

Microfluidization: Cellulose suspensions are forced through narrow microchannels at high pressures (500–1500 bar). The intense shear and impact forces generated result in highly uniform CNFs with fine particle distribution and controllable dimensions [123].

Cryocrushing: Cryocrushing is a technique used to create cellulose nanostructures. It is the process of freezing cellulose fibers at very low temperatures using liquid nitrogen and then mechanically crushing them. This process causes the cellulose fibers to break and separate into nanometers. The presence of liquid nitrogen makes the cellulose fibers more brittle and therefore easier to crush [67, 114].

Ultrasonication: Ultrasonic waves generate cavitation forces that break cellulose fibers into nanofibrils. This method is particularly effective for producing short fibrils and can be combined with chemical pretreatments to enhance fibrillation efficiency [67, 114].

(b) ***Chemical Treatment***

Chemical processes involve the selective removal of amorphous regions of cellulose through the use of acids, oxidants or other reagents to obtain highly crystalline cellulose nanocrystals (CNCs). These methods are essential for applications that demand high crystallinity and precise particle dimensions.

Sulfuric Acid Hydrolysis is used to separate crystalline nanocellulose by selectively removing amorphous regions of cellulose. The resulting CNCs typically exhibit a crystallinity of 70–90% and rod-like dimensions (5–20 nm in diameter and 100–500 nm in length). The process involves treating cellulose with a 64% (w/w) sulfuric acid solution at temperatures between 45 and 50 °C for 30–60 min. Using this technique, the CNC surface is exposed to negatively charged sulfate ester groups, improving dispersibility in aqueous systems but reducing thermal stability. Post-treatment steps, including washing and dialysis, are required to remove residual acid and neutralize the CNC suspension [7, 67].

TEMPO-Mediated Oxidation: This technique converts the main hydroxyl groups on cellulose into carboxyl groups by using 2,2,6,6-tetramethylpiperidin-1-oxyl (TEMPO) as a catalyst. TEMPO oxidation is performed at ambient temperature under mild circumstances (pH 10–11), providing excellent control over surface functionality. The resulting carboxylated cellulose nanofibers exhibit enhanced water dispersibility and compatibility with cationic species [7, 97, 126].

Hydrochloric Acid Hydrolysis: Sometimes, hydrochloric acid is utilized instead of sulfuric acid to create CNCs. This method results in CNCs with neutral surfaces, making them suitable for applications where surface charges are undesirable. The neutral surface enhances compatibility with hydrophobic polymers, which is advantageous for nanocomposite applications [137].

Oxidative and Alkaline Pretreatment: Oxidative treatments using hydrogen peroxide or alkaline pretreatment with sodium hydroxide improve the compatibility of nanocellulose with synthetic polymers in composite applications. These treatments modify the surface chemistry of cellulose, enhancing its integration into polymer matrices [145].

(c) ***Enzymatic Treatment***

Enzymatic methods utilize enzymes to selectively cleave glycosidic bonds in the amorphous regions of cellulose while preserving its crystalline structure. This environmentally friendly approach minimizes damage to the natural properties of nanocellulose. Enzymatic treatments are often combined with mechanical methods to improve yield and reduce energy consumption. For instance, enzymatic hydrolysis of pretreated fibers can significantly lower the energy required for mechanical fibrillation. This method is particularly suitable for applications demanding high purity and biocompatibility [7, 11, 130].

4 Usage of Cellulose in Biomedical Applications

Their biocompatibility, biodegradability, and condition-specific properties allow cellulose and its derivatives to find applications in sensor technology. Cellulose-based sensors have found applications in diagnostics, monitoring of health conditions, and environmental sensing.

4.1 Drug Delivery System

Drug delivery systems are being developed to make treatment processes more efficient and safer. Nanometals used for therapeutic purposes pose significant risks due to cytotoxicity and adverse effects on some organs (kidney, spleen, brain, liver, and cardiovascular system) [1]. Cellulose-supported biomaterials have a very important place in minimizing the adverse effects of basic drugs and increasing their bioavailability. These biomaterials place active ingredients in the cellulose chain, enabling the regulation of harmful effects and ensuring controlled release. This mechanism facilitates a controlled release over a prolonged duration, ensuring an effective dosage [140]. Consequently, the 3D network within cellulose's intermolecular structure captures the molecules, preserving their activity and mitigating toxicity [179]. Doxorubicin, an anticancer drug used in the treatment of breast cancer, has the potential to cause significant invasive side effects during treatment [38]. However, cellulose loaded with doxorubicin exhibited enhanced therapeutic efficiency and fewer side effects at reduced dosages compared to its unbound counterpart [65].

Moreover, drug conditions have been optimized to account for variables like temperature, pH, chemical interactions, light, and magnetic or electric fields. Pure cellulose has long been widely employed in drug delivery applications, serving roles such as fillers, suspension agents, and stabilizers [131]. In modern applications, cellulose and cellulose derivatives are identified as vital components in drug development and production of medical agents. Cellulose is widely used in drug delivery systems for antimicrobial, anticancer, antioxidant, antiviral, and antidiabetic therapies; hydrogel forms are preferred for skin or external body surface applications, and solution forms are preferred for liquid drugs such as syrups and drops. Ethylcellulose (EC), bacterial cellulose, carboxymethyl cellulose (CMC), microcrystalline cellulose, hydroxypropyl methylcellulose (HPMC), methylcellulose (MC), and cellulose nanofibers are used either alone or in combination with other polysaccharides (such as alginate or chitosan) in drug delivery system. These systems are often designed to perform functions such as increasing the bioavailability of drugs, providing controlled release, improving stability, and targeted delivery. These materials enhance drug targeting, stability, and bioavailability while also strengthening the effects of therapeutic regulators. Another notable advantages of these systems lies in smart drug carrier technologies, which are specifically engineered to transport drugs directly to targeted organs or tissues. These systems can also release the drug in response to specific conditions at the target site, such as temperature or pH levels. This ensures that the drug acts precisely where needed, enhancing effectiveness and reducing side effects. These systems are commonly developed using nano-sized cellulose fibers, HPMC, and CMC [124].

4.2 Tissue Engineering

Cellulose and its derivatives provide new opportunities for tissue culture engineering by supporting cell growth in culture without aggregation, media smoothly, and with greater variability compared to traditional conjugation methods. Cellulose micro/nanofibers are modified to create films integrated with surgical sutures, aiming to control cell response following surgery [102]. These films are also utilized both in vitro and in vivo. For surface modification of nanofiber scaffolds in tissue engineering, as well as supporting cell culture and facilitating tissue regeneration [88].

MC is utilized in tissue engineering within various biomaterials, particularly as a thermoresponsive hydrogel system, providing adjustable crosslinking kinetics, scaffolds, and hydrogel ink for supporting 3D bioprinting [118]. EC and CMC are used in fibroblast cell tissue engineering as hydrogels or biocomposites, where they control stimulation, spreading, adhesion, and migration. Additionally, in bone tissue engineering, they are used as hydrogels combined with hydroxyapatite to reduce osteoclastogenesis and accelerate regeneration and repetition by murine bone marrow progenitors [136]. Cellulose acetate (CA) is a biocompatible material. It is utilized as a non-toxic biopolymer in artificial heart valves, demonstrates the ability to reduce lipid accumulation, and proves effective in repairing other damaged valves [117].

4.3 *Wound Dressings*

Cellulose has long been utilized in wound dressings, and it serves as a physical barrier, reducing the wound's exposure to harmful environmental factors and lowering the likelihood of secondary bacterial infections. By creating a dry, oxygen-rich environment around the wound, cotton-based dressings effectively inhibit the growth and reproduction of most pathogenic anaerobic bacteria. Modern dressings also incorporate advanced features such as antimicrobial and anti-inflammatory effects, improved exudate management, stimulation of cell proliferation, and even bacterial diagnostic functions [46]. Cellulose-based materials offer a combination of mechanical support, fluid management, and therapeutic delivery that is highly effective for wound healing applications.

Preventing inflammation resulting from bacterial infections is a widely researched topic in the development of dressings. One approach to imparting antimicrobial properties to cellulose is through the modification of its surface groups, enabling the material to exhibit inherent antimicrobial activity [104]. The incorporation of chemical groups like carboxymethyl groups has enabled CNF films to influence monocyte and macrophage activity [57]. Furthermore, the oxidizing groups in cellulose enhance its antibacterial properties by adding negative charges and promoting electrostatic interactions. This modification also facilitates the dispersion of cellulose suspensions [56]. Research indicates that 2,3-dialdehyde nanocellulose, synthesized via sodium iodate oxidation, exhibits enhanced antibacterial effectiveness against *S. aureus* and MRSA, primarily due to pH alterations resulting from the presence of two aldehyde groups [100, 168].

In the design of cellulose-based dressings, it is often prioritized combining cellulose materials with diverse antibacterial agents, such as metallic nanoparticles, metal ions, biological extracts, and antimicrobial medications [9]. The versatility of cellulose, ranging from its use as a dry fabric to an injectable gel, coupled with its customizable surface modifications, makes it an ideal material for practical and effective antimicrobial loading applications. An example is the development of the antimicrobial gel BC-Au-DAPT, which contains Au particles and allows for tunable antimicrobial effectiveness by adjusting the concentration of these particles. When stabilized on a solid carrier, nisin, an antimicrobial peptide, achieves an extended shelf life. CNC, modified with aldehyde groups (–CHO) and sterically stabilized, offers an exceptionally large specific surface area, significantly improving its binding capacity with other substances. This characteristic facilitates the incorporation of a higher concentration of antimicrobial agents into wound dressings, improving the attachment of Nisin [153].

Cellulose aerogels, with their exceptionally high specific surface area, excel at binding antimicrobial drugs while also supporting hemostasis, and wound healing. An antimicrobial composite aerogel, created using CNC and a biological peptide obtained from the skin of *Andrias davidianus*, rapidly absorbed fluids and expanded upon being injected into a wound, exerting pressure to stop the bleeding [188]. By efficiently collecting erythrocytes and thrombocytes, the aerogel demonstrated remarkable hemostatic properties, speeding up the clotting process and greatly minimizing blood loss.

Electrotherapy's capacity to promote cell regeneration has been utilized in numerous clinical areas, such as nerve repair, in vivo cellular growth, and the enhancement of wound healing [90, 108, 167]. As a biocompatible, non-toxic material that is not metabolized in the body, cellulose holds significant potential for the development of implantable agents. An injectable nanofiber hydrogel designed for cell infiltration is highly durable and capable of quickly regaining its shape after being injected into the body. Even after repeated usage cycles, the cellulose hydrogel maintained a high retention rate, significantly prolonging the controlled drug release period. Beyond supporting the healing of skin wounds, cellulose-based materials also hold significant potential for the healing of internal organ wounds (e.g., heart, liver) and the prevention of post-operative adhesions [6, 60, 173].

4.4 Flexible Biosensor Devices

Biosensors, which include flexible electronics that can be ingested, worn, or implanted, are capable of monitoring various signals such as joint movements, pressure, temperature, and neural activity. These devices are essential for diagnosing medical conditions and evaluating the effectiveness of specific treatments [174]. Cellulose, known for its affordability, biodegradability, and flexibility, is an ideal substrate for creating lightweight and adaptable biosensor devices. By integrating cellulose and its derivatives with materials such as metal nanoparticles or nanowires [20], graphene, conductive polymers [169], inorganic oxides, carbon nanotubes [166], and MXene [149], their electron and ion conductivity can be tailored for sensor applications. This capability enables it to respond to external stimuli, paving the way for creating functional biosensors like electronic skins (e-skins).

4.5 Biocompatible Cellulosic E-Skins for Motion Detection

The performance of stress and strain sensors is primarily influenced by changes in resistance or capacitance, which result from variations in the spacing between the measurement points. For instance, stress, and strain sensing materials were created using CNF/carbon nanotube filaments, which were produced by wet-spinning durable nanocellulose networks onto carbon nanotubes [76]. Apart from incorporating conductive materials, carbonization is a widely used technique to enhance the electrical conductivity of cellulose. Liu and their team produced sensors from cellulose fabrics modified with reduced graphene oxide (rGO). These sensors' wide pressure range sensitivity was primarily attributed to the fabric's curved surface structure [171]. Moreover, it was reported by the group that, hydrogels made of Ag/tannic acid @CNCs, evenly distributed throughout the nanocomposite, demonstrated improved electrical conductivity (4.61 S m^{-1}), and remarkable stretchability (> 4000%), showing exceptional sensitivity to various stress and deformation conditions [25, 81].

Researchers developed an e-skin with enhanced sensitivity to internal stimuli using a cellulosic ionic gel. The researchers were reported that the e-skin demonstrated outstanding ionic conductivity, and flexibility effortlessly adapting to human skin and delivering precise touch feedback [184]. Furthermore, a multifunctional smart fabric was engineered by coating cellulose fiber nonwovens with 2D $Ti_3C_2T_x$ nanosheets, resulting in an MXene-based material. This innovative cellulose-based fabric demonstrated a reversible and highly sensitive reaction to moisture, making it an ideal candidate for wearable respiratory monitoring applications [186].

4.6 Cellulosic Biomarker Monitoring Systems

User-friendly medical testing devices play a crucial role in promoting the widespread adoption of scientific testing. Thanks to its amphiphilic nature, flexibility, affordability, and portability, cellulose is an excellent functional carrier material, making it highly suitable for use in portable, disposable, and integrated testing devices. These devices can monitor a variety of biomarkers essential for assessing human health, including blood glucose, uric acid, proteins, and viruses [175].

Using lateral flow analysis, Kim et al. designed a cellulose-based device for the diagnosis and monitoring of diabetes, enabling the simultaneous measurement of the glycation ratio of human serum albumin and glucose levels [69]. This portable health device demonstrated an average glucose detection recovery rate exceeding 85% and achieved a glycosylation ratio of 98%. In many applications, cellulose, which naturally lacks detection activity, primarily functions as a substrate responsible for carrying active substances. As an example, Yan's team developed a biochemical sweat monitor using a pencil-paper method [172], where graphite patterns drawn with a pencil created conductive pathways and sensing electrodes, enabling accurate and sensitive detection of biomarkers like uric acid, glucose, and pH.

4.7 Cellulose-Based Sensors for Environmental Monitoring

Detecting trace levels of heavy metals remains a critical challenge in the analysis of environmental and medical samples [106]. Functionalized with activated carbon, cellulose-based materials demonstrate high efficiency in adsorbing various heavy metals. Incorporating fluorescent molecules into these materials enables visual and dependable monitoring of heavy metal contamination.

A detection system for copper ions was created by integrating cyanobacterial C-phycocyanin into a cellulose-based substrate [170]. As the copper ion concentration increased, the fluorescence intensity decreased, making the system ideal for analyzing intricate medical samples, including human serum filtrate. A similar approach applies to a study where the luminescent bacterium *Vibrio fischeri* is used [85]. The luminescence of the system diminished in the presence of higher concentrations of toxic substances, providing a quick method for assessing contaminant toxicity. This detection approach was recognized for its efficiency, affordability, speed, and user-friendly nature.

Cellulose has been employed in disposable test strips designed to detect pesticide residues [50, 98]. Zhang's team introduced a cellulose-based test strip that responds to ammonia, relying on a unique mechanism of color differentiation [63]. Exhibiting clear color changes, this innovative test strip offered a quick, highly sensitive, and linear response across ammonia concentration levels ranging from 5.0 to 2.5×10^4 ppm.

4.8 Health Products

Historically, cellulose has been primarily used in medical applications as an excipient in tablet formulations [154]. For instance, Li and their team modified the CNF surface by immobilizing quercetin (QT), a natural flavonoid known for its antioxidant properties, to assess its potential as a material for nutraceutical delivery [80].

The consumption of high-fat diets poses a growing threat to the health of elderly, and middle-aged individuals, contributing to "tri-hyper" conditions such as hyperglycemia, hypertension, and hyperlipidemia. Studies indicate that cellulose has potential as a fat substitute and, in combination with soy protein, can be utilized in developing meal-replacement foods. For instance, Sun and their team incorporated CNF as an edible ingredient in ice cream, effectively lowering calorie absorption and increasing satiety [151]. The cellulose-infused ice cream showed exceptional thermal stability, retaining its structure and shape at 30 °C, unlike conventional ice cream, which melted within an hour.

Carboxymethyl cellulose and methylcellulose have been extensively utilized as excipients in pharmaceutical formulations for decades. Recent studies have exploited methylcellulose's amphiphilic and thermoresponsive properties to integrate it with hydrophobic active pharmaceutical ingredients. This method resulted in the creation of a novel thermo-gel nano-emulsion through a flexible thermal processing technique [18]. This nano-emulsion not only helped lower calorie consumption but also increased the feeling of satiety.

4.9 Others

Beyond the examples mentioned, cellulosic materials exhibit significant potential in diverse fields, including fingerprint recognition, virus isolation, DNA extraction, protein recovery, biometric imaging, as well as data and image encryption [40].

In the fight against cancer, alongside traditional methods like tumor surgery, chemotherapy using cancer-suppressing radiation therapy, drugs, and gene therapy has surfaced as a powerful and rapidly advancing treatment option. Researchers have achieved substantial inhibition of hepatocellular carcinoma mutation and growth by targeting maternally-expressed gene-3, a tumor-suppressor lncRNA. To facilitate gene delivery, nanocellulose serves as an excellent nanocarrier, combining the characteristics of nanoparticles and polysaccharides. These supramolecular assemblies provide reliable platforms for cell regeneration, and gene delivery demonstrating their potential in advanced medical treatments [146].

Carbon dots derived from cellulose offer great potential as biocompatible materials for bioimaging applications. Cellulose acts as an ideal functional carrier for quantum dots with high fluorescence. Fluorescent nanoparticles [156] have the ability to integrate quantum dots or be fabricated into fluorescent films, making them particularly advantageous for bioimaging and the detection of toxic ions.

Fingerprints are distinguished by their uniqueness and permanence. As mobile telecommunication technology has advanced, fingerprint-based unlocking and check-in systems have become widely adopted. High-quality fingerprint capture technology is essential for precise fingerprint identification.

5 Conclusion

Cellulose is a biocompatible natural polysaccharide and due to this feature it is used in many different biomedical applications. Due to its biocompatibility, and biodegradability, it can be utilized as a compatible material for the body, as it exhibits no toxic effects. Its high water retention capacity makes it effective in wound dressings and other applications where biological fluids must be absorbed. Cellulose acts as a skeleton material that provides a suitable microenvironment for the growth and proliferation of cells, and due to this feature, cellulose is used in the renewal of some tissues such as cartilage, bone, skin, and veins. In addition to all these properties, its antimicrobial feature makes cellulose a safe material in terms of infection in applications such as wound healing. Thanks to its flexibility, and durability, it has a field of use in biomedical devices, especially implants or prostheses. Cellulose derivatives can serve as carriers for the controlled release of drugs. These systems enhance treatment efficacy by facilitating the release of therapeutic agents at a precise and controlled rate. As a result, cellulose serves as an ideal material for various applications in the biomedical field, particularly in tissue engineering, wound care products, drug delivery systems, and implants. Being natural, biocompatible and processable makes it a preferred material in biomedical applications.

References

1. Abbasian M, Mahmoodzadeh F, Khalili A et al (2019) Chemotherapy of breast cancer cells using novel pH-responsive cellulose-based nanocomposites. Adv Pharm Bull 9:122–131
2. Abdelhamid HN, Mathew AP (2022) Cellulose-based nanomaterials advance biomedicine: a review. Int J Mol Sci 23(10):5405
3. Abolore RS, Jaiswal S, Jaiswal AK (2024) Green and sustainable pretreatment methods for cellulose extraction from lignocellulosic biomass and its applications: a review. Carbohydr Polym Technol Appl 7:100396
4. Allaoui BE, Benzeid H, Zari N et al (2023) Functional cellulose-based beads for drug delivery: preparation, functionalization, and applications. J Drug Deliv Sci Technol 88:104899
5. Altaner CM, Thomas LH, Fernandes AN et al (2014) How cellulose stretches: synergism between covalent and hydrogen bonding. Biomacromol 15:791–798
6. An JM, Shahriar SMS, Hasan MN et al (2021) Carboxymethyl cellulose, pluronic, and pullulan-based compositions efficiently enhance antiadhesion and tissue regeneration properties without using any drug molecules. ACS Appl Mater Interfaces 13:15992–16006
7. Ansari MM, Heo Y, Do K et al (2024) Nanocellulose derived from agricultural biowaste by-products—sustainable synthesis, biocompatibility, biomedical applications, and future perspectives: a review. Carbohydr Polym Technol Appl 8:100529
8. Azhar O, Jahan Z, Sher F et al (2021) Cellulose acetate-polyvinyl alcohol blend hemodialysis membranes integrated with dialysis performance and high biocompatibility. Mater Sci Eng C 126:112127
9. Baker S, Volova T, Prudnikova SV et al (2018) Bio-hybridization of nanobactericides with cellulose films for effective treatment against members of ESKAPE multi-drug-resistant pathogens. Appl Nanosci 8:1101–1110
10. Bergenstråhle M, Berglund LA, Mazeau K (2007) Thermal response in crystalline Iβ cellulose: a molecular dynamics study. J Phys Chem B 111:9138–9145
11. Bondancia TJ, Florencio C, Baccarin GS et al (2022) Cellulose nanostructures obtained using enzymatic cocktails with different compositions. Int J Biol Macromol 207:299–307
12. Budaeva VV, Gismatulina YA, Mironova GF et al (2019) Bacterial nanocellulose nitrates. Nanomaterials 9(12):1694
13. Cabrera RQ, Meersman F, McMillan PF et al (2011) Nanomechanical and structural properties of native cellulose under compressive stress. Biomacromol 12(6):2178–2183
14. Carrillo-Varela I, Mendonc RT, Pereira M et al (2022) Methylene blue adsorption onto hydrogels made from different Eucalyptus dissolving pulps. Cellulose 29:445–468
15. Channab BE, Idrissi AE, Essamlali Y et al (2024) Nanocellulose: structure, modification, biodegradation and applications in agriculture as slow/controlled release fertilizer, superabsorbent, and crop protection: a review. J Environ Manag 352:119928
16. Chashchilov DV (2024) Regenerated cellulose review part 1. Cellulose structure, solvent systems, and dissolution mechanisms. Polym Sci Ser D 17(2):466–473
17. Chaudhary P, Deepa Meena DR, Aalam MJ et al (2021) Cellulose sulfate: an efficient heterogeneous catalyst for the ring-opening of epoxides with alcohols and anilines. Synth Commun 51(12):1834–1846
18. Chen LH, Doyle PS (2021) Design and use of a thermogelling methylcellulose nanoemulsion to formulate nanocrystalline oral dosage forms. Adv Mater 33(29):2008618
19. Chen YL, Zhang X, You TT et al (2019) Deep eutectic solvents (DESs) for cellulose dissolution: a mini-review. Cellulose 26:205–213

20. Cheng R, Zeng J, Wang B et al (2021) Ultralight, flexible and conductive silver nanowire/nanofibrillated cellulose aerogel for multifunctional strain sensor. Chem Eng J 424:130565
21. Cheng W, Zhu Y, Jiang G et al (2023) Sustainable cellulose and its derivatives for promising biomedical applications. Prog Mater Sci 138:101152
22. Cheng Y, Qin H, Acevedo NC et al (2021) Development of methylcellulose-based sustained-release dosage by semisolid extrusion additive manufacturing in drug delivery system. J Biomed Mater Res Part B 109(2):257–268
23. Chung JY, Ko JH, Lee YJ et al (2018) Surfactant-free solubilization and systemic delivery of anti-cancer drug using low molecular weight methylcellulose. J Control Release 276:42–49
24. Coughlin ML, Liberman L, Ertem SP et al (2021) Methyl cellulose solutions and gels: fibril formation and gelation properties. Prog Polym Sci 112:101324
25. Cui C, Fu Q, Meng L et al (2021) Recent progress in natural biopolymers conductive hydrogels for flexible wearable sensors and energy devices: materials, structures, and performance. ACS Appl Bio Mater 4:85–121
26. Czibula C, Seidlhofer T, Ganser C et al (2021) Longitudinal and transverse low-frequency viscoelastic characterization of wood pulp fibers at different relative humidity. Materialia 16:101094
27. Deng J, Song Q, Liu S et al (2022) Advanced applications of cellulose-based composites in fighting bone diseases. Compos Part B Eng 245:110221
28. Destruel PL, Zeng N, Seguin J et al (2020) Novel in situ gelling ophthalmic drug delivery system based on gellan gum and hydroxyethylcellulose: innovative rheological characterization, in vitro and in vivo evidence of a sustained precorneal retention time. Int J Pharm 574:118734
29. Dey K, Agnelli S, Serzanti M et al (2019) Preparation and properties of high-performance gelatin-based hydrogels with chitosan or hydroxyethyl cellulose for tissue engineering applications. Int J Polym Mater Polym Biomater 68(4):183–192
30. Ding M, Wang X, Man J et al (2023) Antibacterial and hemostatic polyvinyl alcohol/microcrystalline cellulose reinforced sodium alginate breathable dressing containing *Euphorbia humifusa* extract based on microfluidic spinning technology. Int J Biol Macromol 239:124167
31. Ding SY, Himmel ME (2006) The maize primary cell wall microfibril: a new model derived from direct visualization. J Agric Food Chem 54(3):597–606
32. Dixit K, Bora H, Chakrabarti R et al (2024) Thermoresponsive keratin-methylcellulose self-healing injectable hydrogel accelerating full-thickness wound healing by promoting rapid epithelialization. Int J Biol Macromol 263:130073
33. Du H, Liu W, Zhang M et al (2019) Cellulose nanocrystals and cellulose nanofibrils based hydrogels for biomedical applications. Carbohydr Polym 209:130–144
34. Du S, Liu B, Li Z et al (2021) A nanoporous graphene/nitrocellulose membrane beneficial to wound healing. ACS Appl Bio Mater 4(5):4522–4531
35. Eichhorn SJ, Davies GR (2006) Modelling the crystalline deformation of native and regenerated cellulose. Cellulose 13:291–307
36. Eichhorn SJ, Young RJ (2001) The Young's modulus of a microcrystalline cellulose. Cellulose 8(3):197–207
37. El Fawal GF, Abu-Serie MM, Hassan MA et al (2018) Hydroxyethyl cellulose hydrogel for wound dressing: fabrication, characterization and in vitro evaluation. Int J Biol Macromol 11:649–659
38. El-Naggar ME, Hasanin M, Hashem AH (2021) Eco-friendly synthesis of superhydrophobic antimicrobial film based on cellulose acetate/polycaprolactone loaded with the green biosynthesized copper nanoparticles for food packaging application. J Polym Environ 30:1820
39. Etale A, Onyianta AJ, Turner SR et al (2023) Cellulose: a review of water interactions, applications in composites, and water treatment. Chem Rev 123(5):2016–2048
40. Fan J, Bao B, Wang Z et al (2021) Flexible, switchable and wearable image storage device based on light responsive textiles. Chem Eng J 404:126488

41. Fan M, Li Y, Chen J et al (2023) Plasmonic internal standard-decorated nitrocellulose membranes for duplex detection of circulating tumor biomarkers. Sens Actuators B Chem 395:134508
42. Fernandes A, Cruz-Lopes L, Esteves B et al (2023) Nanotechnology applied to cellulosic materials. Materials 16(8):3104
43. Fernandes IAA, Pedro AC, Ribeiro VR et al (2020) Bacterial cellulose: from production optimization to new applications. Int J Biol Macromol 164:2598–2611
44. Franca JR, Foureaux G, Fuscaldi LL et al (2019) Chitosan/hydroxyethyl cellulose inserts for sustained-release of dorzolamide for glaucoma treatment: in vitro and in vivo evaluation. Int J Pharm 570:118662
45. Gabrielli V, Frasconi M (2022) Cellulose-based functional materials for sensing. Chemosensors 10(9):352
46. Gao Y, Qiu Z, Liu L et al (2022) Multifunctional fibrous wound dressings for refractory wound healing. J Polym Sci 60:2191–2212
47. Ghilan A, Nicu R, Ciolacu DE et al (2023) Insight into the latest medical applications of nanocellulose. Materials 16(12):4447
48. Gopi S, Balakrishnan P, Chandradhara D et al (2019) General scenarios of cellulose and its use in the biomedical field. Mater Today Chem 13:59–78
49. Groth T, Willems C, Zhang K et al (2020) Development of bioactive cellulose sulfates for biomedical applications. Acta Med Medianae 59(3)
50. Guo L, Wang T, Wu Z et al (2020) Portable food-freshness prediction platform based on colorimetric barcode combinatorics and deep convolutional neural networks. Adv Mater 32(45):2004805
51. Gupta B, Mishra V, Gharat S et al (2021) Cellulosic polymers for enhancing drug bioavailability in ocular drug delivery systems. Pharmaceuticals 14(11):120
52. Halib N, Perrone F, Cemazar M et al (2017) Potential applications of nanocellulose-containing materials in the biomedical field. Materials 10(8):977
53. Hartzke D, Pössl A, Schlupp P et al (2022) Evaluation of hydroxyethyl cellulose grades as the main matrix former to produce 3D-printed controlled-release dosage forms. Pharmaceutics 14(10):2103
54. Heinze T, El Seoud OA, Koschella A (2018) Cellulose derivatives: synthesis, structure, and properties. Springer, Cham
55. Hernández-Becerra E, Osorio M, Marín D et al (2023) Isolation of cellulose microfibers and nanofibers by mechanical fibrillation in a water-free solvent. Cellulose 30:4905–4923
56. Hu J, Quan Y, Lai Y et al (2017) A smart aminoglycoside hydrogel with tunable gel degradation, on-demand drug release, and high antibacterial activity. J Controlled Release 247:145–152
57. Hua K, Alander E, Lindstrom T et al (2015) Surface chemistry of nanocellulose fibers directs monocyte/macrophage response. Biomacromol 16:2787–2795
58. Huang P, Wang C, Huang Y et al (2019) Structure and properties of cellulose nanofibrils. In: Huang J, Dufresne A, Lin N (eds) Nanocellulose from fundamentals to advanced materials, 1st edn. Wiley, Weinheim
59. Hynninen V, Mohammadi P, Wagermaier W et al (2019) Methyl cellulose/cellulose nanocrystal nanocomposite fibers with high ductility. Eur Polym J 112:334–345
60. Imani SM, Ladouceur L, Marshall T et al (2020) Antimicrobial nanomaterials and coatings: current mechanisms and future perspectives to control the spread of viruses including SARS-CoV-2. ACS Nano 14:12341–12369
61. Ioelovich M (2008) Cellulose as a nanostructured polymer: a short review. BioResources 3(4):1403–1418
62. Javanbakht S, Shaabani A (2019) Carboxymethyl cellulose-based oral delivery systems. Int J Biol Macromol 133:21–29
63. Jia R, Tian W, Bai H et al (2019) Amine-responsive cellulose-based ratiometric fluorescent materials for real-time and visual detection of shrimp and crab freshness. Nat Commun 10(1):795

64. John J, Archana KS, Thomas AM et al (2024) Nitrocellulose unveiled: a brief exploration of recent research progress. Sustain Chem Eng 15:146–167
65. Jorfi M, Foster EJ (2015) Recent advances in nanocellulose for biomedical applications. J Appl Polym Sci 132:41719
66. Joseph B, Sagarika VK, Sabu C et al (2020) Cellulose nanocomposites: fabrication and biomedical applications. J Bioresour Bioproducts 5(4):223–237
67. Joshi G, Shukla SR, Chauhan SS (2024) Nanocellulose extraction from lignocellulosic materials and its potential applications: a review. J Indian Acad Wood Sci 21(1):1–23
68. Kamel SA, Khattab T (2020) Recent advances in cellulose-based biosensors for medical diagnosis. Biosensors 10(6):67
69. Ki H, Jang H, Oh J et al (2020) Simultaneous detection of serum glucose and glycated albumin on a paper-based sensor for acute hyperglycemia and diabetes mellitus. Anal Chem 92:11530–11534
70. Kian LK, Saba N, Jawaid M et al (2020) Characterization of microcrystalline cellulose extracted from olive fiber. Int J Biol Macromol 156:347–353
71. Kim H, Han S, Choi K et al (2023) Self-assembled nanocomplex using cellulose nanocrystal based on zinc/DNA nanocluster for gene delivery. Biotechnol Bioprocess Eng 28(1):83–90
72. Kim J, Rackstraw NB, Weinstein TJ et al (2024) Cellulose etherification with glycidol for aqueous rheology modification. ACS Appl Polym Mater 6(11):6714–6725
73. Koneru A, Dharmalingam K, Anandalakshmi R (2020) Cellulose-based nanocomposite hydrogel films consisting of sodium carboxymethylcellulose–grapefruit seed extract nanoparticles for potential wound healing applications. Int J Biol Macromol 148:833–842
74. Laboy-López S, Méndez Fernández PO, Padilla-Zayas JG et al (2022) Bioactive cellulose acetate electrospun mats as scaffolds for bone tissue regeneration. Int J Biomater 1:3255039
75. Leong MY, Kong YL, Harun MY et al (2023) Current advances of nanocellulose application in biomedical field. Carbohydr Res 17:108899
76. Leppiniemi J, Lahtinen P, Paajanen A et al (2017) 3D-printable bioactivated nanocellulose–alginate hydrogels. ACS Appl Mater Interfaces 9:21959–21970
77. Leskinen M, Kilpeläinen I, Hietala S (2024) Interaction of cellulose and water upon drying and swelling by 13C CP/MAS NMR. Cellulose 31:10745–10769
78. Li A, Wang Y, Deng L et al (2013) Use of nitrocellulose membranes as a scaffold in cell culture. Cytotechnology 65:71–81
79. Li T, Chen C, Brozena AH et al (2021) Developing fibrillated cellulose as a sustainable technological material. Nature 590:47–56
80. Li X, Liu Y, Yu Y et al (2019) Nanoformulations of quercetin and cellulose nanofibers as healthcare supplements with sustained antioxidant activity. Carbohydr Polym 207:160–168
81. Lin F, Wang Z, Shen Y et al (2019) Natural skin-inspired versatile cellulose biomimetic hydrogels. J Mater Chem A 7:26442–26455
82. Lin N, Dufresne A (2014) Surface chemistry, morphological analysis and properties of cellulose nanocrystals with gradiented sulfation degrees. Nanoscale 6(10):5384–5393
83. Lin P, Liu L, He G et al (2020) Preparation and properties of carboxymethyl chitosan/oxidized hydroxyethyl cellulose hydrogel. Int J Biol Macromol 162:1692–1698
84. Lindman B, Medronho B, Alves L et al (2017) The relevance of structural features of cellulose and its interactions to dissolution, regeneration, gelation and plasticization phenomena. Phys Chem Chem Phys 19:23704–23718
85. Liu J, Morales-Narváez E, Orozco J et al (2017) Bioluminescent nanopaper for rapid screening of toxic substances. Nano Res 11:114–125
86. Liu J, Shi Y, Cheng L et al (2021) Growth factor functionalized biodegradable nanocellulose scaffolds for potential wound healing application. Cellulose 28(9):5643–5656
87. Luo X, Zhang L (2013) New solvents and functional materials prepared from cellulose solutions in alkali/urea aqueous system. Food Res Int 52:387–400
88. Lovett ML, Nieland TJF, Dingle YTL et al (2020) Innovations in 3-dimensional tissue models of human brain physiology and diseases. Adv Funct Mater 30:1909146

89. Machado B, Costa SM, Costa I et al (2024) The potential of algae as a source of cellulose and its derivatives for biomedical applications. Cellulose 31(6):3353–3376
90. Mao L, Hu S, Gao Y et al (2020) Biodegradable and electroactive regenerated bacterial cellulose/MXene ($Ti_3C_2T_x$) composite hydrogel as wound dressing for accelerating skin wound healing under electrical stimulation. Adv Healthc Mater 9(19):2000872
91. Mbisana M, Keroletswe N, Nareetsile F et al (2024) Nanocellulose composites: synthesis, properties, and applications to wastewater treatment. Cellulose 31:10651–10678
92. McMullen RL, Ozkan S, Gillece T (2022) Physicochemical properties of cellulose ethers. Cosmetics 9(3):52
93. Medronho B, Lindman B (2015) Brief overview on cellulose dissolution/regeneration interactions and mechanisms. Adv Colloid Interface Sci 222:502–508
94. Menon MP, Selvakumar R, Kumar PS et al (2017) Extraction and modification of cellulose nanofibers derived from biomass for environmental application. RSC Adv 7:42750–42773
95. Mohan T, Dobaj Štiglic A, Beaumont M et al (2020) Generic method for designing self-standing and dual porous 3D bioscaffolds from cellulosic nanomaterials for tissue engineering applications. ACS Appl Bio Mater 3(2):1197–1209
96. Mokhothua TH, John MJ (2015) Review on hygroscopic aging of cellulose fibres and their biocomposites. Carbohydr Polym 131:337–354
97. Moon RJ, Martini A, Nairn J et al (2011) Cellulose nanomaterials review: structure, properties and nanocomposites. Chem Soc Rev 40:3941–3994
98. Moradi M, Tajik H, Almasi H et al (2019) A novel pH-sensing indicator based on bacterial cellulose nanofibers and black carrot anthocyanins for monitoring fish freshness. Carbohydr Polym 222:115030
99. Morris E, Pulham CR, Morrison CA (2023) Structure and properties of nitrocellulose: approaching 200 years of research. RSC Adv 13(46):32321–32333
100. Mou K, Li J, Wang Y et al (2017) 2,3-Dialdehyde nanofibrillated cellulose as a potential material for the treatment of MRSA infection. J Mater Chem B 5:7876–7884
101. Muhitdinov B, Heinze T, Turaev A et al (2019) Homogenous synthesis of sodium cellulose sulfates with regulable low and high degree of substitutions with SO_3/Py in N,N-dimethylacetamide/LiCl. Eur Polym J 119:181–188
102. Müller WEG, Tolba E, Schröder HC et al (2015) A new printable and durable N, O-carboxymethyl chitosan–Ca^{2+}–polyphosphate complex with morphogenetic activity. J Mater Chem B 3:1722–1730
103. Murizan NI, Mustafa NS, Ngadiman NH et al (2020) Review on nanocrystalline cellulose in bone tissue engineering applications. Polymers 12(12):2818
104. Narh C, Badoe W, Howard EK et al (2020) Synthesized OH-radical rich bacteria cellulosic pockets with photodynamic bacteria inactivation properties against *S. aureus* and *E. coli*. Mater Sci Eng C Mater Biol Appl 116:111230
105. Naomi R, Bt Hj Idrus R, Fauzi MB (2020) Plant-vs. bacterial-derived cellulose for wound healing: a review. Int J Environ Res Public Health 17(18):6803
106. Nawaz H, Tian W, Zhang J et al (2019) Visual and precise detection of pH values under extreme acidic and strong basic environments by cellulose-based superior sensor. Anal Chem 91:3085–3092
107. Niemczyk-Soczynska B, Gradys A, Kolbuk D et al (2022) A methylcellulose/agarose hydrogel as an innovative scaffold for tissue engineering. RSC Adv 12(41):26882–26894
108. Nishiguchi A, Taguchi T (2020) A thixotropic, cell-infiltrative nanocellulose hydrogel that promotes in vivo tissue remodeling. ACS Biomater Sci Eng 6:946–958
109. Noreen A, Zia KM, Tabasum S et al (2020) A review on grafting of hydroxyethylcellulose for versatile applications. Int J Biol Macromol 150:289–303
110. O'Brien CT, Virtanen T, Donets S et al (2021) Control of the aqueous solubility of cellulose by hydroxyl group substitution and its effect on processing. Polymer 223:123681
111. Oguzlu H, Danumah C, Boluk Y (2016) The role of dilute and semi-dilute cellulose nanocrystal (CNC) suspensions on the rheology of carboxymethyl cellulose (CMC) solutions. Can J Chem Eng 94:1841–1847

112. Oprea M, Voicu SI (2020) Recent advances in composites based on cellulose derivatives for biomedical applications. Carbohydr Polym 247:116683
113. Otero P, Carpena M, Garcia-Oliveira P et al (2023) Seaweed polysaccharides: emerging extraction technologies, chemical modifications, and bioactive properties. Crit Rev Food Sci Nutr 63(13):1901–1929
114. Padhi S, Singh A, Routray W (2023) Nanocellulose from agro-waste: a comprehensive review of extraction methods and applications. Rev Environ Sci Bio/Technol 22(1):1–27
115. Palaninathan V, Raveendran S, Rochani AK et al (2018) Bioactive bacterial cellulose sulfate electrospun nanofibers for tissue engineering applications. J Tissue Eng Regen Med 12(7):1634–1645
116. Pak HT, Choe JA, Ju KS et al (2024) Novel method for the prediction of distribution and elution characteristic of several drugs in cellulose diacetate-water system: interaction between mixed solvent and cellulose diacetate. Chem Phy Lett 834:140979
117. Pasqui D, Torricelli P, De Cagna M et al (2014) Carboxymethyl cellulose-hydroxyapatite hybrid hydrogel as a composite material for bone tissue engineering applications. J Biomed Mater Res Part A 102(5):1568–1579
118. Petreus T, Stoica BA, Petreus O et al (2014) Preparation and cytocompatibility evaluation of cellulose acetate–hydroxyapatite composite membranes. J Mater Sci Mater Med 25:1115–1127
119. Pettignano A, Charlot A, Fleury E (2019) Carboxyl-functionalized derivatives of carboxymethyl cellulose: towards advanced biomedical applications. Polym Rev 59(3):510–560
120. Phanthong P, Reubroycharoen P, Hao X et al (2018) Nanocellulose: extraction and application. Carbon Resour Convers 1(1):32–43
121. Pourmadadi M, Rahmani E, Shamsabadipour A et al (2023) Novel carboxymethyl cellulose-based nanocomposite: a promising biomaterial for biomedical applications. Process Biochem 130:211–226
122. Pudžiuvelytė L, Drulytė E, Bernatonienė J (2023) Nitrocellulose-based film-forming gels with cinnamon essential oil for covering surface wounds. Polymers 15(4):1057
123. Punyanunt S, Nanta P, Puttipan R (2025) Green-chemical nanofibrillated cellulose from pineapple cores using microfluidization. Trends Sci 22(2):9085
124. Qiu X, Hu S (2013) Smart materials based on cellulose: a review of the preparations, properties, and applications. Materials 6(3):738–781
125. Rahman MS, Hasan MS, Nitai AS et al (2021) Recent developments of carboxymethyl cellulose. Polymers 13(8):1345
126. Randhawa A, Dutta SD, Ganguly K et al (2022) A review of properties of nanocellulose, its synthesis, and potential in biomedical applications. Appl Sci 12(14):7090
127. Rasheed M, Jawaid M, Karim Z et al (2020) Morphological, physiochemical and thermal properties of microcrystalline cellulose (MCC) extracted from bamboo fiber. Molecules 25(12):2824
128. Ray B, Ali I, Jana S et al (2021) Antiviral strategies using natural source-derived sulfated polysaccharides in the light of the COVID-19 pandemic and major human pathogenic viruses. Viruses 14(1):35
129. Revin V, Liyaskina E, Nazarkina M et al (2018) Cost-effective production of bacterial cellulose using acidic food industry by-products. Braz J Microbiol 49(1):151–159
130. Ribeiro RS, Pohlmann BC, Calado V et al (2019) Production of nanocellulose by enzymatic hydrolysis: trends and challenges. Eng Life Sci 19(4):279–291
131. Saigal N, Baboota S, Ahuja A et al (2009) Microcrystalline cellulose as a versatile excipient in drug research. J Young Pharm 1(1):6–12
132. Sadeghi S, Nourmohammadi J, Ghaee A et al (2020) Carboxymethyl cellulose-human hair keratin hydrogel with controlled clindamycin release as antibacterial wound dressing. Int J Biol Macromol 147:1239–1247
133. Sahoo PK, Jena DK (2018) Synthesis and study of mechanical and fire retardant properties of (carboxymethyl cellulose-g-polyacrylonitrile)/montmorillonite biodegradable nanocomposite. J Polym Res 25(12):260

134. Salimi M, Idrissi AE, Channab B et al (2024) Recent advances in cellulose-based hydrogels for wastewater treatment: preparation, properties, and applications. Cellulose 31:10679–10726
135. Samsalee N, Meerasri J, Sothornvit R (2023) Rice husk nanocellulose: extraction by high-pressure homogenization, chemical treatments and characterization. Carbohydr Polym Technol Appl 6:100353
136. Seddiqi H, Oliaei E, Honarkar H et al (2021) Cellulose and its derivatives: towards biomedical applications. Cellulose 28:1893–1931
137. Shabbir M (ed) (2023) Regenerated cellulose and composites: morphology-property relationship. Springer, Singapore
138. Sharaf SS, El-Shafei AM, Refaie R et al (2022) Antibacterial and wound healing properties of cellulose acetate electrospun nanofibers loaded with bioactive glass nanoparticles; in-vivo study. Cellulose 29(8):4565–4577
139. Sharip NS, Ariffin H (2019) Cellulose nanofibrils for biomaterial applications. Mater Today Proc 16:1959–1968
140. Shehabeldine A, Hasanin M (2019) Green synthesis of hydrolyzed starch–chitosan nano-composite as drug delivery system to gram negative bacteria. Environ Nanotechnol Monit Manag 12:100252
141. Sheikhi A (2019) Emerging cellulose-based nanomaterials and nanocomposites. In: Karak N (ed) Nanomaterials and polymer nanocomposites, 1st edn. Elsevier, Amsterdam
142. Shanks RA, Pardo IRM (2018) Cellulose solubility, gelation, and absorbency compared with designed synthetic polymers. In: Mondal IH (ed) Cellulose-based superabsorbent hydrogels. Springer, Cham
143. Sheng Y, Gao J, Yin ZZ et al (2021) Dual-drug delivery system based on the hydrogels of alginate and sodium carboxymethyl cellulose for colorectal cancer treatment. Carbohydr Polym 269:118325
144. Si X, Lu F, Chen J (2017) A strategy for generating high-quality cellulose and lignin simultaneously from woody biomass. Green Chem 19:4849–4857
145. Soleimani S, Heydari A, Fattahi M (2022) Isolation and characterization of cellulose nanocrystals from waste cotton fibers using sulfuric acid hydrolysis. Starch-Stärke 74(11–12):2200159
146. Song HQ, Pan W, Li RQ et al (2018) Rodlike supramolecular nanoassemblies of degradable poly(aspartic acid) derivatives and hydroxyl-rich polycations for effective delivery of versatile tumor-suppressive ncRNAs. Small 14:1703152
147. Subhedar A, Bhadauria S, Ahankari S et al (2021) Nanocellulose in biomedical and biosensing applications: a review. Int J Biol Macromol 166:587–600
148. Sudhaik A, Raizada P, Ahamad T et al (2023) Recent advances in nanocomposites for integrated photocatalytic techniques in water pollution remediation. Int J Biol Macromol 226:1284–1308
149. Su T, Liu N, Lei D et al (2022) Flexible MXene/bacterial cellulose film sound detector based on piezoresistive sensing mechanism. ACS Nano 16:8461–8471
150. Su T, Wu QX, Chen Y et al (2019) Fabrication of the polyphosphates patched cellulose sulfate-chitosan hydrochloride microcapsules and as vehicles for sustained drug release. Int J Pharm 555:291–302
151. Sun L, Chen W, Liu Y et al (2015) Soy protein isolate/cellulose nanofiber complex gels as fat substitutes: rheological and textural properties and extent of cream imitation. Cellulose 22:2619–2627
152. Tanaka F, Iwata T (2006) Estimation of the elastic modulus of cellulose crystal by molecular mechanics simulation. Cellulose 13(5):509–517
153. Tavakolian M, Okshevsky M, van de Ven TGM et al (2018) Developing antibacterial nanocrystalline cellulose using natural antibacterial agents. ACS Appl Mater Interfaces 10:33827–33838
154. Than A, Duong PK, Zan P et al (2020) Lancing drug reservoirs into subcutaneous fat to combat obesity and associated metabolic diseases. Small 16(31):2002872
155. Thomas P, Duolikun T, Rumjit NP et al (2020) Comprehensive review on nanocellulose: recent developments, challenges and future prospects. J Mech Behav Biomed Mater 110:103884

156. Tian W, Zhang J, Yu J et al (2018) Phototunable full-color emission of cellulose-based dynamic fluorescent materials. Adv Funct Mater 28:1703548
157. Timofeev O, Torvinen K, Sievänen J et al (2014) Drying of pigment-cellulose nanofibril substrates. Materials 7:6893–6907
158. Tracey CT, Torlopov MA, Martakov IS et al (2020) Hybrid cellulose nanocrystal/magnetite glucose biosensors. Carbohydr Polym 247:116704
159. Tu W, Wang S, Deng Q et al (2024) Review on nanocellulose composites and CNFs assembled microfiber toward automotive applications. Nanotechnol Rev 13(1):20240006
160. Tudoroiu EE, Dinu-Pîrvu CE, Albu Kaya MG et al (2021) An overview of cellulose derivatives-based dressings for wound-healing management. Pharmaceuticals 14(12):1215
161. Umoren SA, Solomon MM, Saji VS (2022) Polymeric materials in corrosion inhibition: fundamentals and applications. Elsevier, Amsterdam
162. Ventura C, Pinto F, Lourenc AF (2020) On the toxicity of cellulose nanocrystals and nanofibrils in animal and cellular models. Cellulose 27(5):5509–5544
163. Verma N, Pramanik K, Singh AK et al (2022) Design of magnesium oxide nanoparticle incorporated carboxymethyl cellulose/poly vinyl alcohol composite film with novel composition for skin tissue engineering. Mater Technol 37(8):706–716
164. Vital N, Ventura C, Kranendonk M et al (2022) Toxicological assessment of cellulose nanomaterials: oral exposure. Nanomaterials 12(19):3375
165. Vishnu R, Revathi R (2019) Methods of isolation of nanocellulose from wood: a review. J Pharmacogn Phytochem 8(5):2386–2392
166. Wan Z, Chen C, Meng T et al (2019) Multifunctional wet-spun filaments through robust nanocellulose networks wrapping to single-walled carbon nanotubes. ACS Appl Mater Interfaces 11:42808–42817
167. Wang L, Hu S, Ullah MW et al (2020) Enhanced cell proliferation by electrical stimulation based on electroactive regenerated bacterial cellulose hydrogels. Carbohydr Polym 249:116829
168. Wang X, Chang CH, Jiang J et al (2019) The crystallinity and aspect ratio of cellulose nanomaterials determine their pro-inflammatory and immune adjuvant effects in vitro and in vivo. Small 15(42):1901642
169. Wang X, Zhou J, Zhu Y et al (2020) Assembly of silver nanowires and PEDOT:PSS with hydrocellulose toward highly flexible, transparent and conductivity-stable conductors. Chem Eng J 392:123644
170. Weishaupt R, Siqueira G, Schubert M et al (2017) A protein-nanocellulose paper for sensing copper ions at the nano- to micromolar level. Adv Funct Mater 27:1604291
171. Wu R, Ma L, Patil A et al (2020) Graphene decorated carbonized cellulose fabric for physiological signal monitoring and energy harvesting. J Mater Chem A 8:12665–12673
172. Xu Y, Zhao G, Zhu L et al (2020) Pencil-paper on-skin electronics. Proc Natl Acad Sci 117:18292–18301
173. Yang Y, Liu X, Li Y et al (2017) A postoperative anti-adhesion barrier based on photoinduced imine-crosslinking hydrogel with tissue-adhesive ability. Acta Biomater 62:199–209
174. Yin H, Song P, Chen X et al (2022) A self-healing hydrogel based on oxidized microcrystalline cellulose and carboxymethyl chitosan as wound dressing material. Int J Biol Macromol 221:1606–1617
175. Yin M, Alexander Kim Z, Xu B (2022) Micro/nanofluidic-enabled biomedical devices: integration of structural design and manufacturing. Adv NanoBiomed Res 2:2100117
176. You J, Zhang X, Mi Q et al (2022) Mild, rapid and efficient etherification of cellulose. Cellulose 29(18):9583–9596
177. You L, Marcolini B, Bour J et al (2025) Physicochemical, morphological, and rheological properties of cellulose nanofibrils produced via ultra-high-pressure homogenization. Carbohydr Polym Technol Appl 9:100635
178. Zeng K, Groth T, Zhang K (2019) Recent advances in artificially sulfated polysaccharides for applications in cell growth and differentiation, drug delivery, and tissue engineering. ChemBioChem 20(6):737–746

179. Zhang N, Li S, Xiong L et al (2015) Cellulose-hemicellulose interaction in wood secondary cell-wall. Modell Simul Mater Sci Eng 23(8):085010
180. Zhang Q, Lin D, Yao S (2015) Review on biomedical and bioengineering applications of cellulose sulfate. Carbohydr Polym 132:311–322
181. Zhang R, Belwal T, Li L et al (2020) Recent advances in polysaccharides stabilized emulsions for encapsulation and delivery of bioactive food ingredients: a review. Carbohydr Polym 242:116388
182. Zhang X, Yang W, Blasiak W (2011) Modeling study of woody biomass: interactions of cellulose, hemicellulose, and lignin. Energy Fuels 25:4786–4795
183. Zhang Z, Cai S, Li Y et al (2020) High performances of plant fiber reinforced composites—a new insight from hierarchical microstructures. Compos Sci Technol 194:108151
184. Zhao D, Zhu Y, Cheng W et al (2020) A dynamic gel with reversible and tunable topological networks and performances. Matter 2:390–403
185. Zhao H, Zhao L, Lin X et al (2022) An update on microcrystalline cellulose in direct compression: functionality, critical material attributes, and co-processed excipients. Carbohydr Polym 278:118968
186. Zhao X, Wang LY, Tang CY et al (2020) Smart Ti_3C_2Tx MXene fabric with fast humidity response and joule heating for healthcare and medical therapy applications. ACS Nano 14:8793–8805
187. Zhao Y, Zhang Y, Lindström ME (2015) Tunicate cellulose nanocrystals: preparation, neat films and nanocomposite films with glucomannans. Carbohydr Polym 117:286–296
188. Zheng L, Wang Q, Zhang YS et al (2021) A hemostatic sponge derived from skin secretion of Andrias davidianus and nanocellulose. Chem Eng J 416:129136
189. Zia F, Zia KM, Aftab W et al (2021) Synthesis and characterization of hydroxyethyl cellulose copolymer modified polyurethane bionanocomposites. Int J Biol Macromol 179:345–352
190. Zulkifli FH, Hussain FS, Harun WS et al (2019) Highly porous hydroxyethyl cellulose biocomposite scaffolds for tissue engineering. Int J Biol Macromol 122:562–571

Fatma Akpınar is a researcher in the field of analytical chemistry and biosensors, with a focus on electrochemical nanobiosensors for biomolecular interactions. She earned her Ph.D. from Aydın Adnan Menderes University, specializing in the design and development of DNA-based electrochemical biosensors. Her research spans biosensor technology, nanomaterials, and their applications in medical diagnostics and environmental monitoring.

Kübra Gençdağ Şensoy completed her undergraduate education in Chemistry at Ege University in 2010, her master's degree in 2013, and her doctorate in 2017 at the Aydın Adnan Menderes University. She is currently serving as an academic staff member at Adnan Menderes University Her research focuses on the development of electrochemical sensors using nanomaterials and polymers, particularly molecularly imprinted polymers, for the analysis of drugs, pesticides, and nucleic acids.

Merve Esen Keçeci earned her bachelor's degree in Chemical Engineering from Istanbul University in 2015. She completed her master's degree in 2018 at the same university, in the Department of Process and Reactor Design, with a thesis titled "Production of Biofuel from Cellulose," under a TÜBİTAK Project scholarship. She is currently continuing her doctoral studies in Chemistry at Aydın Adnan Menderes University. In parallel, she works as a Chemical Engineer at a company in Aydın, where she focuses on production of nanocellulose or microfibrillated cellulose production and natural cellulose fiber manufacturing.

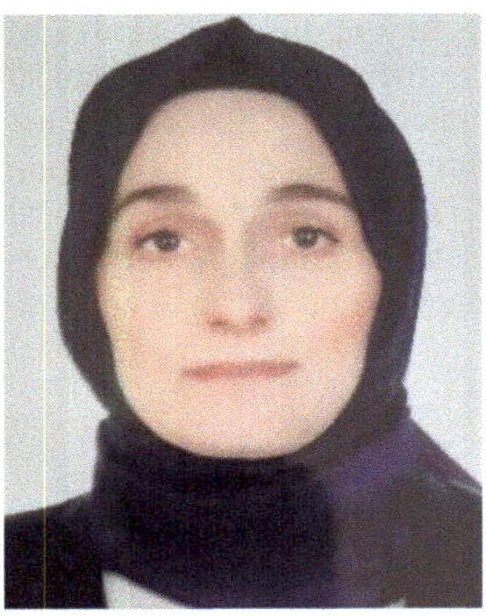

Mihrican Muti obtained her Ph.D. in Chemistry from Aydın Adnan Menderes University in 2010. She is currently a professor at the same university. Her research focuses on electrochemistry, particularly the development and characterization of sensors and biosensors based on nanomaterials, polymers, and molecularly imprinted polymers. Additionally, she conducts research on modified electrodes for the electrochemical detection of pesticides, toxins, biological samples, nucleic acid hybridization, and drug-DNA interactions.

Cellulose-Based Material as Antibacterial Agent

Bhanu Sharma, Preeti Maiti, Edwin Davidson, Cecilia Ezeanya, Jorge Pereira, and Swadeshmukul Santra

Abstract In recent years, there has been increased efforts on the development of innovative cellulose-based materials such as cellulose nanocrystals (CNCs), cellulose nanofibrils (CNFs), and bacterial cellulose (BC), among others, gaining much interest as potential sustainable platforms for the incorporation of antibacterial agents toward biomedical applications. These materials are obtained from the highly abundant natural cellulose and showcased enhanced biocompatibility, mechanical strength, and chemical surface functionalization. These properties facilitate the incorporation of antimicrobial active ingredients such as metal ions (e.g., copper, silver, and zinc), nanoparticles (e.g. silver and gold), or molecular compounds (e.g., antibiotics, antiseptics, and antifungals). This study investigates the recently reported applications of functionalized cellulose-based materials loaded with antimicrobial actives ingredients, as alternative strategies against bacterial infections. These cellulose materials are suitable for a broad scope of applications, ranging from biomedical relevant with wound dressings, medical coatings, all the way to healthcare with water purification and pathogen-resistant packaging. Furthermore, the versatile surface modification and green syntheses associated with cellulose-based materials as delivery systems support their appeal as sustainable antimicrobial strategy. Collectively, these wide range of cellulose materials showcased effective antimicrobial activity and increased stability, while preserving great biocompatibility with living tissues. Thus, corroborating the potential suitability and applicability of cellulose-based materials as effective antibacterial treatment for applications in healthcare and biomedical sectors. Herein, this review summarizes recently developed cellulose-based antibacterial materials, highlighting their synthesis methods, reported results, and their relevancy to multiple antimicrobial applications within the biomedical sector.

B. Sharma · P. Maiti · E. Davidson · C. Ezeanya · J. Pereira · S. Santra (✉)
NanoScience Technology Center, University of Central Florida, Orlando, FL, USA
e-mail: swadeshmukul.santra@ucf.edu

B. Sharma · P. Maiti · E. Davidson · S. Santra
Department of Chemistry, University of Central Florida, Orlando, FL, USA

C. Ezeanya · S. Santra
Burnett School of Biomedical Sciences, University of Central Florida, Orlando, FL, USA

R. Malviya and S. Sundram (eds.), *Sustainable Cellulose Materials for Biomedical Applications*, Sustainable Materials and Technology,
https://doi.org/10.1007/978-981-96-9904-9_13

Keywords Cellulose materials · Antimicrobial cellulose · Nanocellulose

Abbreviations

AgNPs	Silver nanoparticles
AI	Artificial intelligence
Alg	Sodium alginate
BC	Bacterial cellulose
BNC	Bacterial nanocellulose
BNC-Ag	Bacterial nanocellulose-silver composite
BSE	*Boswellia serrata* extract
CBA	Cinnamon bark oil
CBO	Clove bud oil
CEF	Cefoperazone
CEO	Cinnamon essential oil
Chi	Chitosan
CMC	Carboxymethyl cellulose
CMCCs	Carboxymethyl chitosan
CNCs	Cellulose nanocrystals
CNFs	Cellulose nanofibrils
DDS	Drug delivery systems
EDC	1-ethyl-3-(3-dimethylaminopropyl) carbodiimide
EDS	Energy dispersive spectroscopy
FDA	Food and Drug Administration
FTIR	Fourier transform infrared spectroscopy
GMP	Good manufacturing practice
HEC	Hydroxyethyl cellulose
HMP	High-methoxylated pectin
HSA	Human Serum Albumin
HTCMCh	N-2-hydroxylpropyl-3-trimethylammonium-O-carboxymethyl chitosan
ISO	International Organization for Standardization
LDH	Lactate dehydrogenase
Levo	Levofloxacin
LRV	Log retention value
MgO	Magnesium oxide
MOFs	Metal–organic frameworks
MTT	3-(4,5-dimethylthiazol-2-yl)-2,5-diphenyltetrazolium bromide (cell viability assay)
NC	Nanocellulose
NHS	N-hydroxysuccinimide
NIR	Near-infrared
PAA	Poly(acrylic acid)

PDA	Polydopamine
PDAc	Polydiacetylene
PEG	Polyethylene glycol
PLA	Polylactic acid
PVA	Polyvinyl alcohol
ROS	Reactive oxygen species
SEM	Scanning electron microscopy
SFNPs	Silk fibroin nanoparticles
SLNs	Solid lipid nanoparticles
TEMPO	(2,2,6,6-tetramethylpiperidin-1-yl)oxyl
TGA	Thermogravimetric analysis
UV	Ultraviolet
XPS	X-ray photoelectron spectroscopy
XRD	X-ray diffraction
ZIF-8	Zeolitic imidazolate framework-8
ZnO	Zinc oxide
ZnO-CNCs	Zinc oxide-cellulose nanocrystals composite

1 Introduction

The rising threat of antimicrobial resistance represents one of the most serious challenges facing global health today, with deaths from drug-resistant infections projected to exceed 39 million by 2050 [69]. This rising concern highlights an essential need for new, sustainable antibacterial agents. As an abundant, biocompatible and biodegradable natural biopolymer, cellulose has attracted great attention for the construction of effective antibacterial materials [16]. These characteristics, along with its plentiful availability in plants, bacteria, and algae, make cellulose a versatile substrate in the design of novel materials for biomedical applications. Various types of cellulose-based materials with different structural and functional properties (e.g. cellulose nanocrystals (CNCs), cellulose nanofibrils (CNFs), and bacterial cellulose (BC)) (Fig. 1). Among these BC is distinctive in its high purity, great ability for water retention and mechanical strength. These characteristics make BC a highly attractive candidate for biomedical uses, including wound dressings, tissue scaffolds and medical coatings [73, 124].

But cellulose is not inherently antibacterial despite its excellent biocompatibility and sustainability. This limitation can be solved by developing functionalization strategies to impart antimicrobial activity. Incorporation of metal nanoparticles (including silver, copper, and zinc oxide), polymer grafted onto the cotton surface and bioactive compounds embedded in addition to their significant antibacterial activity [124]. For instance, quaternization (functionalization of cellulose with quaternary ammonium groups) interacts with bacterial membranes and causes drying, ruptured membranes, and ultimately, cell death. This approach has been effective against a

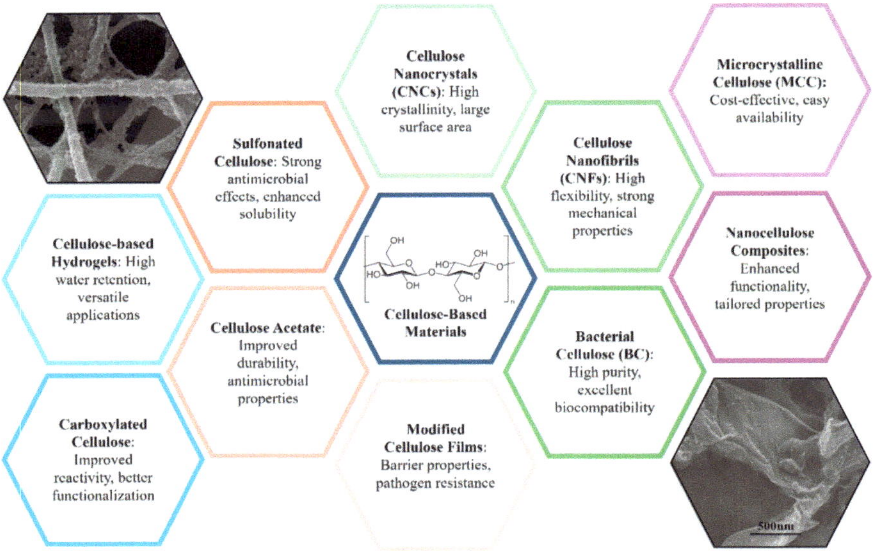

Fig. 1 Types of commonly used cellulose-based materials in antibacterial applications. Reproduced with permission from Avatefi et al. [15]

wide range of pathogens, from *Escherichia coli* and *Staphylococcus aureus*, with applications as antimicrobial coatings, wound dressings, or food packaging [2, 15].

In addition to chemical modifications, cellulose-based composites have been developed into efficient antimicrobial agents. Loading cellulose with metal, functionalized cellulose (silver or copper) can increase antibacterial activity via ROS production, disrupting the bacterial membrane and impaired enzymatic processes [5]. In such drug-loaded cellulose systems, antibiotics, essential oils, or plant polyphenols can be incorporated to sustain the antibacterial activity by controlled release. Moreover, electrostatic interaction of chitosan-cellulose composites or cellulose-polymer provides microbial adhesion and biofilm formation prevention characteristics to be used in medical textiles, surgical implants, and antimicrobial coatings [5, 83].

These antibacterial cellulose-based materials have been adopted in various fields. In healthcare, they are implemented in wound healing, drug delivery, and medical coatings to mitigate the risk of infection. Moreover, cellulose-based antimicrobial packaging has been widely studied to provide food preservation, limiting bacterial contamination and enhancing shelf life. Another avenue of interest is the functionalization of cellulose membranes to be utilized for water purification, effectively removing pathogens and heavy metals. Novel trends also show their capacity in tissue engineering and smart biomaterials, including self-cleaning and biofilm resistant surfaces [1, 111].

While there is great promise associated with these technologies, there are still hurdles to overcome, including scalability, cost-efficiency, and regulatory clearance for commercial use. Overcoming these challenges will be critical to translate cellulose-based antibacterial materials from lab studies to real-world applications. Herein, we explore different functionalization strategies, antibacterial mechanisms, and applications, providing a thorough analysis of recent advancements in cellulose-based nanoparticles. Furthermore, this work highlights the current challenges and future research directions. Particularly, providing a collection of recent research to guide researchers, materials scientists, and industrial scientists in developing next-generation cellulose-based materials as sustainable antimicrobial products.

2 Antimicrobial Applications of Cellulose-Based Materials

Cellulose-based materials showed high versatility that was exploited to overcome microbial challenges in a wide variety of applications. They can be used for wound healing to prevent infections and promote tissue healing, for the fabrication of drug delivery systems for controlled and prolonged release of bioactive molecules, and as antimicrobial coatings on medical devices to prevent infections (Fig. 2). Such materials are also essential in the development of pathogen-resistant food packaging, and in water purification applications, making food safer and free of dangerous pollutants [5]. In this work, we will discuss these various applications in the following sections, outlining how cellulose-based materials are being applied to solve some major healthcare and environmental issues in novel ways.

2.1 Wound Healing

Wound healing is a multistage process consisting of hemostasis, inflammation, proliferation, and remodeling [83]. The process of hemostasis (vasoconstriction and clot

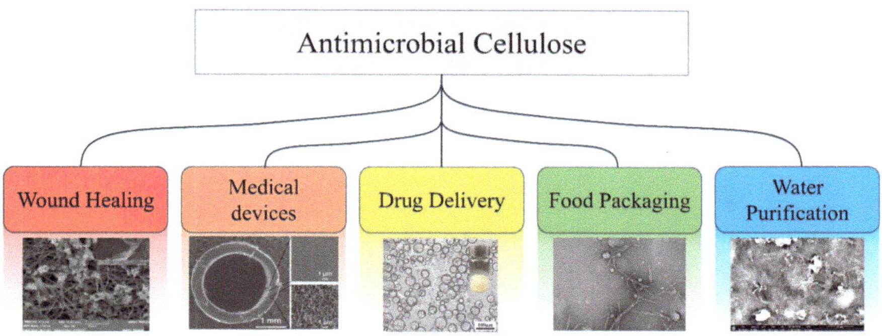

Fig. 2 Diagram representing the major applications of antimicrobial cellulose materials. The micrographs show examples of each of these applications. Reproduced with permission from Huang et al. [42], Tian et al. [107], Meng et al. [65] and Hassan et al. [36]

formation) prevents the loss of blood [77]. In the second stage of inflammation, white blood cells arrive at the injury site to clear out the wound, a mechanism that can also cause swelling, heat, and pain. In the proliferation phase, new tissue and blood vessels form, and epithelial cells cover the wound and form a new surface [35]. The last is remodeling, where collagen is restructured, and excess cells are removed, and scarring may occur [83]. Wound healing dressings can be grouped into passive, interactive, and active types. Passive dressings like gauze and adhesive bandages create an easy physical barrier but do not actively help air-tighten the body. Hydrocolloids, films, and hydrogels are examples of interactive dressings that not only create a moist environment but also interact with the wound bed, resulting in the formation of granulation tissue and reduced microbial colonization [100]. Active dressings go a step further and introduce bioactive agents (e.g., antimicrobial agents, growth factors, anti-inflammatory agents), stimulating the healing process [16]. Some common types of cellulose-based dressings are shown in Fig. 3.

After all, conventional polymeric wound dressings prepared from synthetic polymers suffered from a number of disadvantages, which, in turn, encouraged the need for discovery of novel wound therapies and care methods; cellulose-based dressings, bacteria cellulose (BC), cellulose acetate (CA), and carboxymethyl cellulose (CMC) have become the advanced solutions for wound care due to their tunable properties, biocompatibility and ability to release therapeutic agents [1, 50]. Cellulose acetate is a semi-synthetic derivative with remarkable mechanical strength, hydrolytic stability, and biocompatibility, which has been used alone or mixed with other polymers [1]. For instance, Ullah et al. investigated the addition of different amounts of Monuka

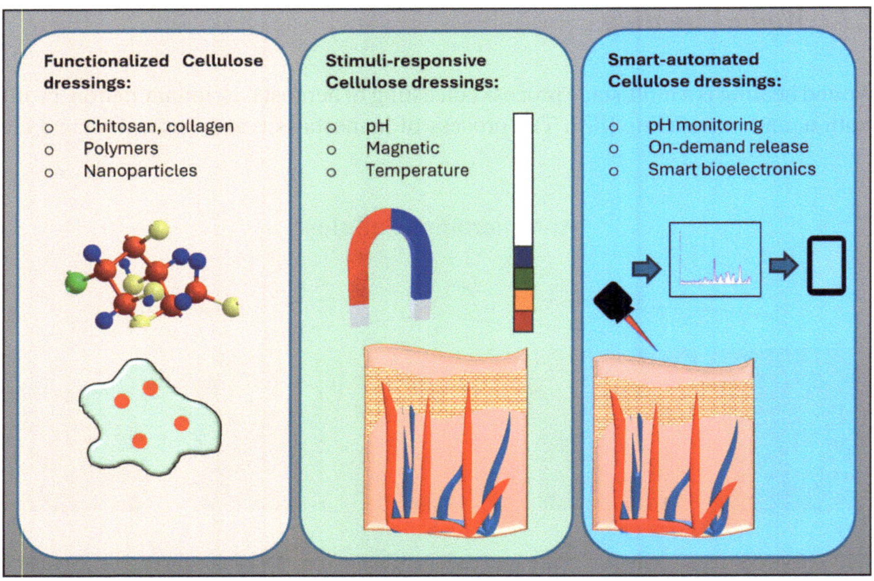

Fig. 3 Types of cellulose-based wound dressings

honey into CA nanofibers and found that the increase of Monuka honey increased the antibacterial and antioxidant activity of the nanofibers but also increased the diameter of the nanofibers [108]. This modification impaired the mechanical strength of the mats, affecting the osmotic absorption and vapor permeability. Nevertheless, the honey-loaded mats showed better cell growth and antibacterial activity in vitro than the neat mats. Similarly, Wang et al. developed fibrous tri-layer wound dressing consisting of PVP-berberine hydrochloride/PVP-CA-berberine hydrochloride/CA-aloin electrospinning [111]. Using this design, they obtained a dual-release profile: rapid release of antibacterial berberine hydrochloride and a prolonged, slow release of healing-promoting Aloin. Fibers manifested high water absorption and moisture retention for wound management and potent antibacterial performance against *S. aureus*. The material performed differently in terms of incomplete drug dissolution and reduced mechanical strength, and this composite showed a short release period for long-term wound care. Avateli et al. developed a self-disinfecting nanofibrous membrane using cellulose acetate electrospun fibers modified by a mussel-inspired polydopamine coating method with a graphene oxide-copper iodide nanocomposite [15]. The characterization revealed uniform distribution of the nanocomposite and high antibacterial (biofilm inhibition > 50%) and antioxidant properties. Yet, challenges such as optimizing synthesis to balance antimicrobial efficacy and cytotoxicity still persist. CA is further functionalized with antimicrobial agents (silver, copper, and zinc nanoparticles) to exhibit antibacterial, antifungal, antiparasitic, antiprotozoal, and anti-inflammatory activities. For instance, Aldossary et al. developed the CA films blended with gallium oxide, zirconium oxide, and graphene oxide by a casting technique [11]. The addition of nanoparticles resulted in increased hydrophilicity of the films and reduced their contact angle (from 40.9° to 27.1°), thus favoring their biodegradability and cell adhesion. These films were able to proliferate the cells as indicated by the cell viability tests with a maximum of 120% cell growth which suggests the ability of the films to promote healing. Nevertheless, limited nanoparticle dispersion and toxicity at high concentrations are key obstacles to this process.

A water-soluble derivative of cellulose, carboxymethyl cellulose (CMC), stands out among these polymers in its ability to form hydrogels and nanofibers characterized by high absorbency in fluids and moisture retention, both of which are necessary for the maintenance of proper conditions in the wound environment [50, 97]. It is highly effective against infections due to its biocompatibility and its ability to incorporate antimicrobial agents (like silver nanoparticles or bioactive medicines) [1, 97]. However, mechanical strength improvements of CMC-based materials could require the introduction of crosslinking procedure as well, which can complicate synthesis. Also, some formulations create a barrier for wet wounds, requiring fixation from the outside [1]. For example, Thakare et al. crosslinked the films based on CMC and polyvinyl alcohol with copper ions using a solution casting method [106]. Copper ion crosslinking was beneficial in terms of improving mechanical properties, swelling ability, hydrophobicity, and antibacterial effect against *S. aureus* and *E. coli*. Cell viability and proliferation were observed at lower copper concentrations, showing potential for wound healing (especially relevant to second-degree burns). But greater copper levels lowered cell adhesion and made the cell stiffer. On the contrary, hydrogels based on CMC-chitosan were designed by Li et al. for improvements in burn wound healing utilizing a Schiff-base reaction [52]. This hydrogel exhibited favorable tissue adhesion (27.3 kPa), self-healing, injectability, and painless on-demand detaching using glycine. It facilitated cell growth, leading to a 95.1% burn healing rate within 15 days. Nevertheless, the mechanical strength and gelation time were moderate so the enhancement in the properties such as the multi-crosslinked networks is required.

Bacterial cellulose (BC), another cellulose-based material, is a biopolymer with exceptional purity and mechanical strength, it can demonstrate a significantly larger water-holding capacity and porosity that can help with hydration and gas exchange at the area of injury [41]. This material has innate biocompatibility, minimalizing inflammatory responses and encouraging cell proliferation [101]. BC itself does not possess inherent antimicrobial properties but can be functionalized with antimicrobial agents or growth factors. Rapid biodegradation of BC material makes it more useful, but requires modification. A powerful alternative is represented by bacterial nanocellulose (BNC), a nanoscale derivative of BC possessing a greater surface area for functionalization and thus further improved antimicrobial and mechanical properties for advanced wound care applications [120]. As an example, Shaheen et al. added silver nanoparticles to the BNC produced with *Komagataeibacter hansenii* SA3.1 [90]. Green, solvent-free synthesis of silver (Ag) nanoparticles reduced in situ and stabilized on the BNC. The composite of such a kind on one hand manifested high antibacterial efficacy against both Gram-positive and Gram-negative bacteria along with fungi, and on the other hand, encouraged antioxidant response to alleviate oxidative stress and cellular proliferation. Nonetheless, the water absorption and holding capacity were comparatively lower for Ag added BNC than pure BNC, which may be attributed to the size of metallic silver nanoparticles disrupting the porous structure of the bacteria cell and thus affecting the porosity. Similarly, Khalid et al. integrated ZnO nanoparticles in BC sheets to develop BC–ZnO nanocomposites, and characterized them using techniques like SEM, XRD, and FTIR to confirm the developed burn wound dressings [49]. These nanocomposites have high antimicrobial activity against pathogens *E. coli* and *S. aureus* and enhanced wound phagocytosis, wound healing and tissue regeneration in a burn wound mice model compared to only BC. Compared to conventional drugs such as silver sulfadiazine, the healing efficacy was still relatively limited, and further optimization is necessary prior to clinical adoption. A different strategy to BC is employment of plant extracts, as reported by Karl et al. with unloaded BC and *Boswellia serrata* extract (BSE) in single and double nanoemulsions [47]. To overcome the incompatibility between lipophilic BSE and hydrophilic BNC, the water within the fiber network was substituted with nanoemulsions containing BSE that enabled efficient drug release with a biphasic profile, and good storage stability. This strategy has limitations, such as a reduction of the water absorption and retention capacity of BNC after drug loading and the need for advanced delivery systems to modulate drug release.

In the past decade another avenue of interest is the development of smart wound dressings as an emerging class of materials that are engineered to facilitate advanced wound management. They combine real-time biomarker sensing with a drug-delivery-on-demand system, thus providing therapeutic performance to locally feedback-controlled therapy within the wound microenvironment [74]. Recently, Liu et al. reported a novel bacterial cellulose-based dressings incorporating antibacterial agent and electric stimulation for advanced wound healing [58]. The authors reported the used vacuum filtration to synthesize conducting, antibacterial bandages made of ultrathin, breathable BC membranes that contained silver nanowires. This bandage could promote the secretion of vascular endothelial growth factor, thus leading to faster wound healing, when tested in a rat full skin thickness wound model. The bandage concept also aided wound healing phase classification using artificial intelligence (AI) that reached high accuracy (94.7%), superior mechanical properties, antibacterial efficacy, and biocompatibility were also evident. In 2024, Xu et al. took another approach for diabetic wounds with the synthesis of a cellulose-based Janus dressing with pH-responsive and unidirectional moisture drainage properties [115]. It is composed of a hydrophilic cellulose layer embedded with commercially available antioxidant anthocyanins for pH monitoring, and a second hydrophobic polycaprolactone layer which contains the antibacterial treatment chlorhexidine to prevent infection. This smart dressing could not only promote healing by dealing with wound exudates and lowering oxidative stress but also sensed the stages of wound healing through color shifts due to changes in pH that may be measured through smartphones. Nonetheless, the complexities of filtering devices, costs, and regulatory issues impose limits on smart wound bandages.

Altogether, these cellulose-based materials have applications for different kinds of wounds. For instance, hydrogels of CMC and CA nanofibers functionalized with antimicrobial agents such as chitosan or silver nanoparticles reduce the occurrence of diabetic wound infections and accelerate the production of granulated tissue [10]. BC-based hydrogels containing silver nanoparticles have been used to mode burn injuries due to their ability to maintain hydration, prevent bacterial colonization, and promote epithelialization. CA films and BC nanofibers also provide high mechanical stability and antimicrobial activity, reducing the risk of post-surgical wound infectious diseases [1]. Several other cellulose-based wound dressings are listed in Table 1.

2.2 *Drug Delivery*

The recent development of cellulose-based materials, particularly hydrogels and films used as hybrid systems, are gradually becoming a mainstay component in drug delivery systems (DDS) to enhance the controlled release of antimicrobial agents [24, 31, 72, 103, 116]. These systems could use endogenous stimuli (pH, temperature, ionic-strength) or exogenous stimuli (light, magnetic field, and electric field) [34], to regulate drug release, ensure targeted delivery at infection sites [48, 79], minimize off-target effects [57], and enable the sustained release of drug over an extended period [45]. For instance, Shahzadi et al. utilized magnesium oxide (MgO)

Table 1 Composition, advantages, and limitation of cellulose-based materials applied in wound healing

Type of structure	The test model	Advantages	Limitations	References
Xylan/chitosan-Ag-TiO_2 NPs nanocomposite scaffold	*E. coli* and *S. aureus*	• Strong antimicrobial activity • Biocompatibility • High porosity for effective wound healing	• Higher nanoparticle concentrations may cause cytotoxicity • Potential nanoparticle leaching	[12]
CA-graphene oxide, titanium dioxide, and curcumin nanofiber	*E. coli*, *S. aureus*, *Pseudomonas aeruginosa*, and *Enterococcus faecalis*	• Excellent antibacterial activity • Sustained drug release • Biocompatibility	• Reduced porosity • Slower degradation, which may limit certain applications	[78]
Polyvinyl alcohol/sodium alginate—TEMPO-oxidized cellulose nanofibrils-calcium phosphate hydrogels	Human skin fibroblast cells (HFB-4 and HSF)	• Excellent biocompatibility • Potential for bone and wound healing	• High concentrations of calcium phosphate caused cytotoxicity • Necessitates careful dosage optimization	[3]
Collagen (porcine, bovine, equine) and carboxymethyl cellulose blend	Artificial wound model	• Excellent flexibility • Mechanical durability • Acidic pH maintenance essential for wound healing	• Equine collagen exhibited slightly inferior pH stability compared to porcine and bovine variants	[105]
Biopaper made of hydroxyapatite nanowires and cellulose fibers	Human umbilical vein endothelial cells (HUVECs)	• Effectively promotes angiogenesis and wound healing • Excellent biocompatibility • Excellent flexibility	• Challenges in scalability	[29]

(continued)

Table 1 (continued)

Type of structure	The test model	Advantages	Limitations	References
Polyvinyl alcohol/cellulose hydrogel with curcumin	*S. aureus* and *P. aeruginosa*	• Excellent antibacterial and anticancer drug delivery • Hemocompatibility • Biodegradability • Moisture retention	• Reduced drug release efficacy after prolonged use beyond five days	[86]
Bacterial cellulose/collagen fabric membranes with additives like lactic acid, tannic acid, and Tween 80	*E. coli* and *S. aureus*	• Excellent antibacterial, hydrophilic and absorptive properties	• Challenges with dehydration processing • Difficulty maintaining structural integrity during functionalization	[76]
3D hydrogels composed of cellulose and dextran	Human fibroblasts and endothelial cells	• High biocompatibility • Controlled release of bioactive compounds • Potent anti-inflammatory effects	• Reduced mechanical strength when dextran content is high	[70]
2-hydroxyethyl cellulose-salicylic acid films	*S. aureus*	• Effective antibacterial properties against bacteria at higher salicylic acid compositions	• Reduced mechanical strength • Limited structural integrity	[7]
Chitosan/dialdehyde cellulose nanocrystals-silver nanoparticles and loaded with quercetin hydrogel films reinforced with oil-in-water emulsions	Full-thickness excisional skin wound model in mice	• Excellent biocompatibility • Enhanced wound healing through controlled quercetin release • Synergistic effects with silver nanoparticles	• Structural integrity may be compromised at higher emulsion or glycerol content	[93]

(continued)

Table 1 (continued)

Type of structure	The test model	Advantages	Limitations	References
Bacterial cellulose—silk fibroin and epigallocatechin-3-gallate-based composite membranes	*S. aureus*	• Excellent antibacterial and antioxidant properties • Biocompatibility	• High hydrophilicity and biodegradability may limit long-term structural stability	[122]
Electrically modified bacterial cellulose-allantoin-zein-terpinen-4-ol (antimicrobial agent) composite	*E. coli* and *S. aureus*	• Combines antimicrobial and biocompatible properties • Promotes rapid and complete wound healing	• Cytotoxicity of terpinen-4-ol slightly impacts cell viability	[4]

deposited onto CNC grafted with poly(acrylic acid) (PAA) to form a pH-responsive hydrogel for controlled doxorubicin release under acidic conditions (pH 5.8) over 24 h [91]. This nanocomposite hydrogel (MgO/CNC-g-PAA) inhibited *S. aureus* growth and showed potential for biomedical application such as the targeted hydrophilic drug delivery in cancer therapy. Few months after this development, Madivoli et al. developed a dual responsive hydrogel that utilized another metal oxide called zinc oxide (ZnO) which upon been crosslinked with polydiacetylene (PDAc) endowed carboxymethyl chitosan (CMCCs)-hydroxyethyl cellulose (HEC) hydrogels with pH and thermal-responsive capabilities [61]. The resulting PDAc-ZnO-CMCCs-HEC hydrogel displayed notable reduction in the growth of *E. coli* based off on the gradual and sustained release of zinc oxide nanoparticles, in contrast to the non-functionalized CMCs-HEC hydrogel [61]. Dai et al. also utilized a dual, yet synergistic pH- and magnetic-responsive hydrogel based on pineapple peel CMC, regenerated nanocellulose, and polyvinyl alcohol (PVA), and Fe_3O_4. These combinations served as a reliable cellulose-based hydrogel drug delivery vehicle for antimicrobial agent as demonstrated by the enhanced loading, entrapment efficiency and sustained release of naringin over 11 h [25].

Besides the above examples of single- or dual-stimuli-responsive hydrogels, hybrid systems integrating nanocellulose and other materials have proven effective in improving drug release profiles. In 2023, for instance, Karimi et al. developed a triple-network hydrogel composed of CMC, Sodium Carboxymethyl Cellulose, and Agarose [45]. These three components were further enhanced by incorporating ciprofloxacin-loaded silk fibroin nanoparticles (SFNPs) modified with polydopamine (PDA). The resulting hydrogel exhibited enhanced drug loading and pH-dependent release of approximately 75% ciprofloxacin for 504 h (about 3 weeks), in alkaline conditions. Moreover, the authors recorded a significant capacity of ciprofloxacin loaded into a triple network to inhibit the growth of *P. aeruginosa* [55]. Notably, hybrid systems combining organic polymers with Metal–Organic Frameworks (MOFs), also called porous coordination polymers [75] have been shown to mitigate the burst effect commonly observed in conventional drug delivery systems (DDS) [54, 68]. This improved performance for such DDS can be attributed to the intrinsic properties of each component in the hybrid system. MOF has porous structure that allows for gradual drug release [60], while also offering superb resistance to elevated temperature and sensitivity to pH changes in acidic environments [63]. Meanwhile, other components such as organic polymer can contribute to enhanced biocompatibility and targeted delivery [54, 117]. For example, Karimzadeh et al. developed graphene oxide (GO) modified with carboxymethyl cellulose (CMC), and Zn-based MOF-5, which effectively loaded tetracycline and regulated its release in acidic conditions for 8 h [46]. This combination also demonstrated the ability to reduce *E. coli* growth upon contact with the released antibiotic and can potentially be used for gastrointestinal infections and oral delivery [46]. Liu et al. also developed a hydrogel using cellulose nanofibrils (CNF) crosslinked with PDA to load tetracycline, achieving sustained for 24 h under low pH or near-infrared (NIR) stimuli while effectively targeting pathogenic *E. coli* and *S. aureus* [59]. Overall, while various kinds of cellulose have been explored as drug delivery vehicles, multi-stimuli-responsive hydrogel depicts another hybrid system that could address burst release challenges.

Incorporating nanoparticles into MOFs and not just organic polymers into MOFs is also emerging as a promising strategy for drug delivery [9]. Nano-MOFs, when integrated into hydrogels, offer sustained release and tissue-specific targeting [37]. Research on such nanoparticle-hydrogel platforms indicates that they improve drug efficacy by enhancing the release profiles and reducing off-target effects [6]. This is achieved by loading nanoparticles, which encapsulate drugs, into three-dimensional (3D) matrices such as hydrogels [70]. For example, the cellulose derivative hydroxyethyl cellulose was used to create a 3D hydrogel for the controlled release of metronidazole-loaded solid lipid nanoparticles (SLNs) in periodontal therapy. The incorporation of SLNs enhanced the sustained release of the antibiotic due to its ability to restrict drug movement, with continuous release observed over 24 h. This release profile demonstrated effectiveness against *S. aureus* but showed limited activity against the gram-negative strain, *E. coli* [38].

Moreover, biopolymeric blends incorporating antibacterial agents have been explored for their dual functionality in exerting antibacterial action and controlling antibiotic release. For instance, Sultan et al. recently formed hydrogel films made up of cellulose and Arabic gum which in turn demonstrated effective antimicrobial properties through controlled drug release. This biopolymeric blend prepared as hydrogel film was used to load sulfadiazine, which allowed for its sustained delivery for over 12 h under physiological pH conditions and reduced growth of pathogens, including pathogenic yeast [102]. Further research has shown that biopolymeric blends, particularly film-based scaffolds, can enhance drug delivery systems when Human Serum Albumin (HSA), a protein known to regulate release kinetics, is incorporated [19]. Cacicedo et al. demonstrated this by modifying BC in situ with high-methoxylated pectin (HMP), a biopolymer, to control the release of levofloxacin (Levo), an antibiotic [19]. The resulting BC-HMP-HSA film exhibited sustained release of Levo for approximately 12 h, regardless of the pH conditions. However, the antibacterial efficacy of the BC-HMP-HSA film against *S. aureus* was comparable to that of BC-HMP film without HSA.

In addition to the above-mentioned strategies, BC, has also served as an exceptional 3D hydrogel matrix [39], expanding the scope of drug delivery systems (DDS) [81, 113]. Its unique nanofibrous structure and hydrophilicity enables functionalization with antimicrobial agents like zinc oxide (ZnO) and layering with Zn^{2+}-crosslinked sodium alginate (Alg) to form dual-layer BC/ZnO/CAlg composite. This composite which was developed by Song et al. sustained Zn^{2+} release, effectively inhibiting 99.9% of *S. aureus* and 85% of *E. coli* [99]. Additionally, chitosan (Chi), another biopolymer [58] has been incorporated into BC films, which enabled a hyperbolic release of ciprofloxacin (Cip), for over six hours than utilizing only BC. The BC-Chi film or membrane had strong and sustained antibacterial activity against the biofilms of *P. aeruginosa* and *S. aureus* [20]. Since BC lacks inherent antimicrobial properties [84], it has also been functionalized with cefoperazone (CEF) to prepare antibacterial membranes (BC-CEF), which also functioned as a DDS. The study revealed that BC been able to utilize the carboxyl groups in the CEF facilitated drug loading and extended drug release, for over 48 h (about 4 days). BC-CEF exhibited strong antibacterial activity against *S. aureus* and *E. coli* [118].

Collectively, most of the delivery vehicles discussed in this section demonstrated slightly greater effectiveness against gram-positive strains. This could be due to the absence of an outer membrane in gram-positive bacteria, unlike gram-negative strains such as *E. coli*, where the outer membrane reduces the susceptibility to the released antimicrobial agent. A few systems were also effective against other microbes such as yeast. Further research is required to either modify these existing systems' surfaces or develop novel systems to enhance the penetration of antimicrobial agents through the cell walls of these organisms. Hybrid systems can also be harnessed even more effectively, especially those that tweak fibrillated cellulose-based delivery vehicles. This includes cellulose nanofibers, cellulose nanocrystals, or bacterial cellulose. These systems can be tailored to incorporate MOFs or utilize three-dimensional structure or employ multiple stimuli to trigger antimicrobial agent release for diverse therapeutic needs.

2.3 *Medical Devices*

As discussed in the previous sections, a large portion of the literature on the biomedical applications of antimicrobial cellulose-based materials has centered on their roles in wound healing and drug delivery. Nevertheless, the superior functionality, biocompatibility, and biodegradability of cellulose, as a natural polymer, has enabled an additional research avenue focusing on the development of innovative strategies for medical devices (e.g., prosthetic grafts, catheters, and coating, among others). Regarding synthetic grafts decades ago, Charpentier et al. reported the cellulose-based surface modification of medical grade polyesters to produce vascular prosthetic devices [22]. In addition, to the antimicrobial properties, the incorporation of cellulose aimed to prevent thrombosis and immune responses. Since then, tissue engineered vascular grafts based on cellulose have gained great interest.

Recently, Fusco et al. reported a small-diameter cellulose graft for coronary artery bypass in a pig model [28]. In this study, the grafts were implanted as an aortocoronary bypass in the left anterior descending occluded, showing a complete patency at 4 weeks. Then, the grafts were explanted and showcased functional remodeling with the colonization of smooth muscle cells and a luminal layer of endothelial cells. Moreover, the work from Li et al. further this research with the development of small-diameter cellulose and polyurethane conduit with enhanced mechanical matching and strength, burst pressure, and suture strength [51]. This graft achieved a long-term patency of 9 months in rat models with unobstructed blood flow and reduced inflammatory response. The graft also prevented microbial infection with no evidence of sepsis or bacterial biofilm formation. Other groups have also reported enhanced benefits of the incorporation or modification of cellulose-based for vascular prosthetic grafts/conduits. In particular, the work from Tian et al. describing the development of a biomimetic artificial vascular material with dual network entanglement [56, 57, 88, 107]. This work reports the fabrication of a cellulose-based small-diameter artificial blood vessels, with a stiff regenerative cellulose hydrogel as matrix. This artificial blood vessel exhibited great cell and blood compatibility in vitro, corroborated by the long-term stability, patency, low fibrinogen adsorption, and platelet adhesion. Nevertheless, this study would benefit from the exploration of the well-known antimicrobial properties of cellulose-based materials to further support their claim of cellulose superiority within the medical devices field. Collectively, these reports highlight the promising potential of cellulose as a synthetic vascular graft, with exceptional antimicrobial properties, angiogenesis stimulation, and biocompatibility for longer patency. It is important to note that in most studies with innovative vascular prosthetics the antimicrobial inhibition is often overlooked, focusing mainly on the performance evaluation of grafts. For this reason, future work should provide more insight into the bactericidal properties to address the emerging concern of bacterial infection during surgical procedures with prosthetics or medical devices.

On the other hand, cellulose-based materials have also shown promise in the development of catheters to reduce urinary tract bacterial infections. In 2024, Zhao et al. reported an amphiphilic nanofibrillated cellulose-polyurethane composite with outstanding antimicrobial and antifouling properties to serve as a catheter with a subcutaneous encapsulation model [123]. In this study, the cellulose-based catheter exhibited enhanced mechanical strength with self-healing ability under urine conditions, with good biocompatibility demonstrating minimal inflammatory response and absence of bacterial infections. Furthermore, cellulose is an exceptionally versatile material that could be used as a substrate for hydrogel formation. Hence, the recently reported by Li et al. a carboxymethyl cellulose-based hydrogel composite to serve as coating of urinary catheters to prevent bacterial infections [53]. This composite incorporated multiple bactericidal agents to foster a synergistic antibacterial mechanism, measured with inhibition zone tests. Furthermore, this coating was fully characterized with multiple techniques and demonstrated the advantage of including cellulose into the composite to provide great in vitro biocompatibility with fibroblast and hemolysis test.

The cellulose versatility was also harnessed by Archana et al. with the development of an anti-infective 3D printed urinary catheter composed with methylcellulose and other ingredients (Fig. 4a, b) [14]. In this work, the 5% cellulose incorporation in the substrate influenced the water retention and drug loading. The drug-loaded cellulose matrix showcased an initial 79% drug release followed by a sustained release described by quasi-Fickian diffusion model. The catheter showed disruption of biofilm formation and suppressed virulence factors of relevant pathogens associated to urinary tract infections. Other previously reported studies have shown the significant impact of cellulose incorporation to catheter materials with antimicrobial properties [13, 21, 40, 85]. Altogether these reports corroborate the remarkable suitability of cellulose-based materials in the development of innovative catheters in combination with other biocidal ingredients to foster a new generation of multifunctional implantable technologies.

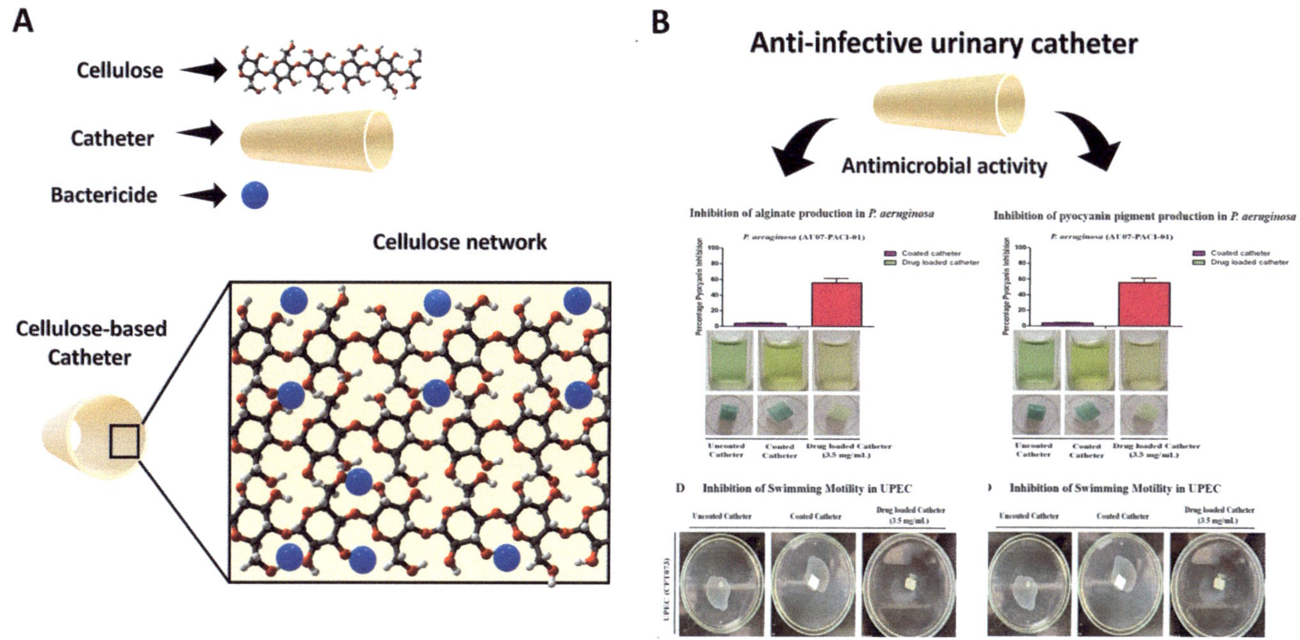

Fig. 4 **a** Schematic of a cellulose-based catheter loaded with antimicrobial active ingredient. **b** Adapted and modified with permission from Archana et al. [14]

2.4 Food Packaging

Moreover, cellulose-based materials have been extensively studied as substitutes for plastic packaging. Although this application is not directly related to healthcare it indirectly contributes to a decreased transmission of food-borne pathogens and food spoilage, which benefits public health. The main benefit of using cellulose materials for packing is their biodegradability and raw material availability. Even though cellulose does not possess intrinsic antimicrobial properties it can be blended with other polymers and antimicrobial components to alter packaging's physical properties and prevent microbial growth. For example, Wang et al. utilized cellulose extracted from bamboo leaf sheath to develop a Cinnamaldehyde Essential Oil (CEO)-PLA/Cellulose packaging film [112]. For this, the extracted cellulose was modified using a silane coupling agent, vacuum dried and pressed into cellulose sheets. Then a CEO-PLA solution was dried onto the cellulose sheets to create the final packing films. The silane modification was confirmed through FTIR by identifying the Si–O stretches on the material. SEM and XRD studies demonstrate that this modification did not substantially alter the thickness of cellulose fibers or their crystallinity, respectively. Furthermore, the addition of CEO-PLA into the cellulose films produced a smoother surface, increased water's contact angle, and lowered water absorption, without compromising the film's mechanical properties. The addition of CEO bestowed the material with antimicrobial properties, as shown by disc diffusion assays against *Aspergillus niger* and *Trichoderma harzianum.* Similar cellulose-polymer-essential oil blends have also been utilized to fabricate antimicrobial food packing. Sayed et al. proposed a PVA-Cellulose film infused with Clove Essential Oil (CEO) for antimicrobial packaging [87]. The films were generated by suspending alkaline treated cellulose in a PVA solution containing CEO, irradiated, and cast on a container. It was found that irradiation is key for optimizing the physical properties of the film, such as tensile strength. As in the previous study, the incorporation of CEO imparted antimicrobial properties to the material, demonstrating effectiveness against *Candida albicans, E. coli, S. aureus* and *P. aeruginosa.* Generally, these studies exemplify large body of work utilizing cellulose-based materials blended into polymeric matrixes containing essential oils for antimicrobial packaging.

However, cellulose materials on their own, such as Kraft paper, have also been functionalized with essential oil components by using 1-(3-dimethylaminopropyl)-3-ethylcarbodiimide/N-hydroxysuccinimide (EDS/NHS) chemistry to bind active components on the surface of γ-irradiated cellulose [44]. The successful binding of active ingredients was corroborated through FTIR and DSC, which increased the water's contact angle with the paper. Furthermore, this also proved to be effective against *Listeria monocytogenes*, *Salmonella enteritidis*, and *E. coli* in vitro, as well as in fresh food. Likewise, Sethunga et al. developed cellulose acetate-essential oils nanofibrous membranes for food packaging using electrospinning [89]. To generate the membranes, cellulose acetate was dissolved in organic solvents containing cinnamon bark oil (CBA) and clove bud oil (CBO), and subsequently electrospun into mats using an administration rate of 1.5 mL/h and a bias voltage of 16 kV. SEM studies confirmed a reduction in fiber thickness from an average of ~ 250 to ~ 80 nm when the material was loaded with 30% w/w of CBA and CBO. GC–MS revealed that the materials entrapped almost 100% of the essential oils and that it released them in a time dependent manner over a period of 65 days. These mats were found to be effective against *S. aureus* and *E. coli* in disc diffusion assays and demonstrated substantial radical scavenging activity. Moreover, the packaging was found to extend the shelf life of fresh tomatoes and grapes. Although many different essential oils have been infused into cellulose-based packaging, it has been proposed that their antimicrobial mode of action is comparable, due to cellular ATP depletion, lipid peroxidation, increased cellular permeability and/or microbial membrane disruption [27]. These works showcased how cellulose materials can be modified and used as packing materials without the need for other polymers.

Additionally, it is possible to create antimicrobial cellulose composites by using other organic active ingredients. Dang et al. functionalized Microcrystalline Cellulose (MCC) with indole acetic acid using EDC to generate, alongside polycaprolactone, cellulose-based nonionic biopolymeric films [26]. The functionalization was corroborated through FTIR, NMR, XPS, and TGA. The incorporation of the indole modified MCC reduced slightly the tensile strength of the polycaprolactone films but increased the water vapor transmittance in a weight dependent manner. It was found that the materials did not leach IAA into making them more environmentally friendly than IAA loaded films. Furthermore, disc diffusion assays of the materials showed that this material is antimicrobial against *E. coli* and *S. aureus*. On the other hand, Zhang et al. blended *N*-2-hydroxylpropyl-3-trimethylammonium-*O*-carboxymethyl chitosan (HTCMCh), a type of zwitterionic chitosan derivative into carboxymethyl cellulose films to enhance their antimicrobial potential [121]. In this regard, HTCMCh increased the zone of inhibition on disc diffusion assays in a dose dependent manner against both *E. coli* and *S. aureus* Cytotoxicity studies on HFF-1 cells fibroblast demonstrate low cytotoxicity, with approximately 100% cell viability when the material was supplied at 2.0 mg/mL. Moreover, the films prevented bacterial growth on pork meat by 48 h compared to carboxymethyl cellulose films. This report shows the advantages of chitosan derivatives, which possess low cytotoxicity while providing high antimicrobial potential. More novel organic biocides such as antimicrobial peptides have also been incorporated successfully into cellulose-based films. Si et al. created a bacterial cellulose—PVA hydrogel loaded with antimicrobial peptide FengycinA-M3 to enhance the preservation of refrigerated chicken [95]. The hydrogel was synthesized by suspending the bacterial cellulose and FengycinA-M3 in a PVA solution and then reacting to the suspension with a solution of borax to form a three-dimensional matrix. The composite hydrogels exhibited significantly less hardness and improved swellability. This allowed the composites to progressively release 90% of the FengycinA-M3 over 120 h. The composites significantly inhibited *E. coli* and *S. aureus* at 256 μg/mL as compared to unloaded hydrogels. The materials also demonstrated to be non-cytotoxic against RAW264.7 macrophages as they did not reduce cell viability even when supplied at 1024 μg/mL. The total viable count of microorganisms in raw chicken stored in the hydrogels was also statistically lower than those found the control after 3 days and maintained the difference up to 12 days of storage. These results show the benefits of incorporating highly specific antimicrobial compounds into cellulose-based hydrogels for meat storage. Even though novel antimicrobials can be incorporated into cellulose packing films, more ubiquitous small molecules have also been shown to be effective. To this regard, PVA-carboxymethyl cellulose films loaded with ascorbic acid has been studied as a biodegradable packing alternative [43]. The films were generated by drying an aqueous PVA-carboxymethyl cellulose solution containing ascorbic acid. It was found that including higher percentages of carboxymethyl cellulose in the films increased water contact angle, reduced the water vapor transmission rate, and significantly enhanced the tensile strength of the material. The film-forming solutions incorporating ascorbic acid exhibited a significant inhibition zone in disc diffusion assays against *E. coli* and *S. aureus.* This inhibition was likely due to a localized pH alteration. This study shows how cellulose-based films can be loaded with green small molecules to prevent the growth of bacterial pathogens.

Besides functionalizing or infusing cellulose-based packaging with organic compounds it is also possible to use inorganic biocides. Usually, these active ingredients are micron/nanoparticles containing transition metals, which account for their antimicrobial activity. In this regard, Marrez et al. functionalized cellulose acetate films with silver nanoparticles synthesized through polyphenol reduction [62]. The polyphenolic capping agent allowed for homogeneous dispersion of the silver nanoparticles in the cellulose acetate film. Furthermore, the incorporation of the particles allowed the composites to swell significantly more than pure cellulose acetate films. The antimicrobial properties of the composites were tested on *E. coli, S. aureus, Salmonella typhi, P. aeruginosa,* among others, all of which were inhibited by the cellulose nanocomposite. These findings demonstrate the potential of inorganic biocides, such as silver nanoparticles, as they usually possess broader antimicrobial potential and can protect food from a variety of food-borne pathogens even at low concentrations. Although silver nanoparticles have been heavily reported as an effective antimicrobial to incorporate into packaging, there are several concerns regarding their safety, cost, and sustainability. Therefore, several other alternatives have been explored in recent years. Huang et al. reported using ZnO nanoparticles to functionalize CNCs and incorporate them into a PLA film for food packaging [42]. The CNCs were synthesized from textile waste using sulfuric acid hydrolysis and functionalized by precipitating the nanoparticles on the CNCs using NaOH. The functionalization was confirmed using XRD and SEM, which showed the characteristic crystal pattern of ZnO and distinct nanoparticles on the CNCs, respectively. Subsequently, this composite was sprayed and dried to functionalize the surface PLA films. Incorporating the ZnO-CNCs did not change the water contact angle significantly but reduced the water vapor transmission through the films. This composite was then tested against *E. coli* and *S. aureus* through several different antimicrobial assays and was successful at inhibiting these pathogens. When used for packing, the composites also significantly reduced the total viable count of microorganisms of refrigerated pork meat after 3 days of storage. Moreover, this cellulose functionalization strategy also proved to be viable using N-halamine and quaternary ammonium salt, instead of ZnO nanoparticles. However, it is also possible to incorporate both inorganic and organic antimicrobials into cellulose materials. Motelica et al. incorporated ZnO nanoparticles and CEO loaded mesoporous silica into hydroxyethyl-cellulose-based films [67]. The nanoparticles were prepared separately and added to the hydroxyethyl-cellulose solutions before casting it into films, which afforded good distribution of both Zn and Si according to the SEM-EDS of the composites. It was found that loading greater amount of ZnO nanoparticles into the composites reduced the water vapor permeability in a concentration dependent manner. Upon immersion, the material released by the CEO gradually in the first 5 h, after which the release diminished. Afterward, the composite inhibited both *S. aureus* and *E. coli* in vitro but displayed additive antimicrobial activity compared to films containing only one active ingredient. The cytotoxicity and HaCaT keratinocytes viability of the composite were measured using the MTT and LDH assays. The results show that the composites did not produce any cytotoxic effect but slightly increased cell viability. Overall, this work shows how incorporating two different active ingredients can enhance the antimicrobial activity of cellulose-based materials against food-borne pathogens.

2.5 Water Purification

The water crisis is one of the major pressing issues faced by large number of populations which are very evident due to the unavailability of safe drinking water, imposing of health and socio-economic structure. According to the data provided by the World Economic Forum, approximately one billion people currently lack access to safe drinking water, which makes it one of the most significant societal risks globally [8]. Water scarcity can pose severe risks and can damage public health and socio-economic stability. Recent studies show that in the United States, per capita water consumption exceeds 500 L per day, raising concerns about the long-term availability of purified water [23]. To supply enough water to meet the demand sustainable water management practices have been proposed, which include effective recycling. However, effective recycling necessitates advanced filtration and purification technologies that are sustainable, reliable, and economical [125].

Both household and industrial wastewater undergo treatment before reusing. As it often contains a diverse range of contaminants, including dyes, pharmaceuticals, fertilizers, pesticides, heavy metals, and other toxic chemicals. Additionally, because of pollutants, it usually harbors lots of microorganisms (e.g., bacteria, viruses, fungi), which are often pathogenic making it unfit for any use or consumption [33]. In recent years, more advanced methods are being used for water purification, such as ozone treatment, ultraviolet (UV) disinfection, biological treatment, microfiltration, reverse osmosis, and electro-assisted methods have been employed to enhance water purification efficiency [114]. Although these methods are very efficient, they are not cost-effective and can be cumbersome to set in place. To address this problem, more practical and sustainable purification methods must be developed with a focus on renewable resources. Innovative, cost-effective, and environmentally friendly filtration technologies are essential for ensuring clean water availability around the world.

Recent advances have led to the use of polymers for various purposes and water filtration being the majorly used processes. Employing solution diffusion method for water filtration with efficient porous filters, it can remove various pollutants, including microorganisms [64]. However, many conventional polymer-based membranes are non-biodegradable, raising concerns about their long-term environmental impact [17]. Cellulose, on the other hand, is a biodegradable polymer that presents a sustainable alternative for membrane fabrication. Its inherent structural properties make it suitable for developing membranous filaments that can be effectively utilized in filtration systems.

Recently nanocellulose (NC) has garnered significant attention due to its advanced filtration capabilities over conventional or regular cellulose, which makes it a better alternative of any water purification methods. The superiority of NC is proven by its exceptional absorbent properties, due to its high surface area to volume ratio which enhances its ability to capture contaminants and show better incorporation into composites. It is better alternative too because of its sustainable synthesis and low production cost making it a promising candidate for large-scale water filtration systems [64]. Besides high surface area, NC also demonstrates excellent thermal stability, strong interactions with metals ions and no adverse effects, enables them to be incorporated into different filtration processes. For instance, their metal ion binding affinity enables them to effectively absorb heavy metals for wastewater treatment for its effluents [92]. In recent years NC has shown potential in removing up to 90% of Cu(II) and Fe(II) like heavy and toxic metals which makes it an incredibly useful for water filtration [55, 94].

Another advantage of NC is their flexibility by functionalizing their surface which enhances their physicochemical properties for specific filtration applications [18]. This tunable surface chemistry allows for efficient incorporation of antimicrobial agents, such as including silver nanoparticles, or chemical modifications (e.g., carboxylation or amination), enhancing its ability to adsorb heavy metals, organic pollutants, and biological material like microorganisms [98]. Carboxylation and amination modifications have shown significant efficacy in capturing toxic metal ions such as Ni(II), Cu(II), and Cd(II) using microfabricated cellulose. This efficient formation of microfabrics is efficient in screening out microbes and metal ions [119]. The adsorption capacities for Ni(II), Cu(II), and Cd(II) were reported to be 2.734, 3.150, and 4.195 mmol/g, respectively [104].

NC is considered a better alternative as further work has shown NC can be fabricated into various forms with different sizes range and enhanced porosities, making it a perfect material for microbial filtration of water. Optimum pore size can be achieved to tailor for specific filtration requirements, including the removal of viruses and bacteria through filtration processes, as illustrated in Fig. 5 [92]. Moreover, NC-based membranes can be composed into nanofibrils, to create an efficient platform for water purification given their controlled permeability and enhanced contaminant removal [71]. This multifunctionality makes NC a promising material not only for

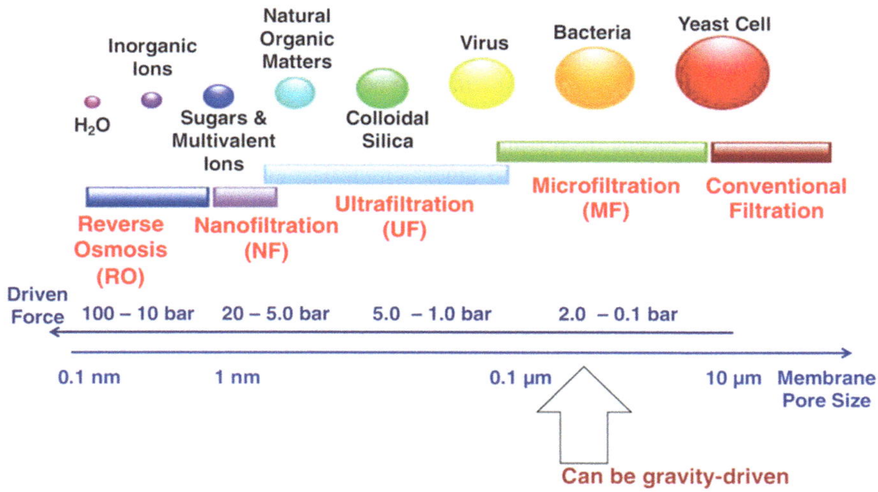

Fig. 5 Comparison of nanocellulose size with biological material. Adapted from [92]

heavy metal adsorption but also for microbial filtration. Their use as effective microbial filtration systems has been extensively studied for their ability to remove human pathogens.

For instance, Metreveli et al. utilized NC embedded in filter paper to enhance viral filtration efficiency [66]. This new technique decreased the pore size while simultaneously increasing the available surface area for adsorption. The nonwoven structure of these NC-based filters features controlled pore sizes, enabling the effective removal of microorganisms, including viruses. Quellmalz et al. functionalized NC through crosslinking with citric acid to enhance its filtration capabilities. The crosslinking process done by him increased the absorbent properties of NC, improving its efficiency in capturing microorganisms [80]. These simple chemical modifications significantly improved the filtering performance for particulates larger than 20 nm in size, demonstrating potential for microbial filtration.

Like earlier work Hassain et al. developed a bilayer membrane composed of CNFs combined with activated charcoal and TEMPO-oxidized CNF for the removal of bacteria like *E. coli* from contaminated water [14]. This bilayer membrane, with a thickness of 0.75 µm, demonstrated remarkable efficiency in bacterial filtration. The effectiveness of the system was confirmed through SEM, where *E. coli* cells were visibly trapped on the surface of the CNF bilayer membrane. This imaging gave a substantial proof of how NC can be an effective screening material for microbes.

Additionally, Goetz et al. employed an electro spun cellulose acetate embedding chitin nanocrystals [30]. This formulation enhanced the hydrophilicity, antifouling ability, and water filtration efficiency of cellulose acetate membranes and showed 48% decreased biofilm growth. Gouda et al. incorporated silver nanoparticles to enhance not only filtration but also impart antimicrobial properties, further enhancing the material's ability to eliminate pathogenic microbes [32]. The synergy between nanocellulose and silver nanoparticles underscores the potential of NC-based materials as advanced filtration platforms for water purification.

One of the advantages among the plethora of benefits is that NC can be sustainably synthesized through plant waste. For example, Sijabat et al. and Rini et al. synthesized BNC and NC fibrils from banana and pineapple peel respectively for filtering out microbes [82, 96]. The resulting fibril, which was engineered by them into a membranous structure like a woven mat which can be utilized for filtering water to remove microbes like *E. coli* and *Brevundimonas diminuta.* Similar strategies have been tested to purify water effluents from viruses [109]. For instance, Wang et al. developed a two-layered nanoscale polyacrylonitrile mat interwoven with NC fibers whose diameter was 5 nm whereas average scaffold diameter is 202–240 nm in size. This mat like design demonstrated the ability of removing viruses, bacteria and heavy metal particles. It was tested against *E. coli* and it showed complete removal of the bacteria and reached log retention value (LRV) of 4 for MS2 virus removal, which depicted that the retention time of the virus was good [110].

3 Challenges and Future Directions

As shown throughout this review cellulose-based materials have recognizable potential as a sustainable and versatile alternative strategy for applications regarding wound healing, medical device coatings, drug delivery systems, food and healthcare packaging, and water purification. Based on the vast number of reports documenting the outstanding biocompatibility, biodegradability, and functionalization to enhance the antimicrobial potency of cellulose-based materials, lead us to believe that they may have a great impact in both healthcare and biomedical fields in the near future. However, in the biomedical sector there are some considerations to address regarding their translation to market and industrial implementation. For instance, potential limiting factors can be associated to production scalability and cost, synthesis reproducibility, and environmental risks. In particular, scalable engineering processes and thorough long-term studies with animals and humans are necessary to establish clinical viability. This is a main challenge for wound dressings, stimuli-responsive drug release devices, prosthetics, and medical device coatings applications. In the future, an alternative to mitigate the scalable production cost, could be led by the recent innovation in 3D bioprinting of polymers such as cellulose. This technique offers improved automatization of material synthesis with lower costs and a personalized medical approach based on individual. Future studies are necessary to provide insight into the human hazards associated with nanoscale cellulose materials, understanding their mode of action, transport, and translocation upon interactions with biological systems.

On the other hand, in the healthcare and food packaging sectors the technologies encounter price dynamics that challenge cellulose-based substitutes, which need to rival the low cost associated with conventional plastics. Further work can be made in refining production techniques and implementing multiple stimuli-responsive antimicrobial films to balance output with resource efficiency. Finally, water purification is another frontier with cellulose-based materials showing excellent pathogen removal but problems such as heavy metal contamination and high production costs remain to be considerable challenge. Thus, providing an important avenue to explore and develop innovative cellulose-based material for safer and sustainable antimicrobial systems. Overall, these challenges are opportunities to advance interdisciplinary research collaboration, characterization techniques, and developments in cutting-edge technology to further 3D bioprinting or nanotechnology to overcome these major limitations. Once these challenges are addressed, we envision cellulose-based materials to gain widespread incorporation across multiple industries, fostering sustainable solutions to worldwide pressing issues.

4 Conclusion

In summary, the recently developed cellulose-derived materials have shown considerable promise as sustainable antibacterial platforms with enhanced biocompatibility, biodegradability, and surface functionalization. In addition, these cellulose materials exhibited high compatibility with other active ingredients such as metal nanoparticles, antibiotics, bioactive compounds, and polymeric composites, among others. Their versatile range of applications has been corroborated with the vast amount of recently reported literature discussed in this review. These materials have showcased significant improvement on current bactericidal strategies for wound therapy, medical devices coating, controlled drug delivery, food packaging and water purification applications. Hence, the interest within the scientific community to further study and innovate cellulose-based materials as a suitable alternative to conventional antimicrobial strategies. The technological advancements in nanoscale materials, smart biomaterials, and biofabrication, are contributing to the development of novel cellulose-based materials to foster multifunctional systems with higher antimicrobial potency and controlled release of active ingredients. Future studies may continue to explore the implementation of artificial intelligence technologies to develop smart dressings with precise controlled release and targeted delivery of the bactericidal bioactive.

As the world pivots toward sustainability, the role of cellulose-based materials will become more prevalent and widely incorporated. These materials can significantly mitigate environmental polymeric plastic waste and provide suitable replacements for conventional antibacterial treatments. Furthermore, research endeavors should progress to better understand the specific functionality and properties of each cellulose raw material, leading to an informed decision-making process to determine the materials suitable applications across multiple disciplines. We believe that the next generation of antimicrobial technologies should incorporate biodegradability, biocompatibility, and readily abundance, which are some of the main advantages gained while utilizing cellulose-based materials.

References

1. Abazari MF, Gholizadeh S, Karizi SZ, Birgani NH, Abazari D, Paknia S et al (2021) Recent advances in cellulose-based structures as the wound-healing biomaterials: a clinically oriented review. Appl Sci 11(17):7769
2. Abdelhamid HN, Mathew AP (2022) Cellulose-based nanomaterials advance biomedicine: a review. Int J Mol Sci 23(10):5405
3. Abouzeid RE, Salama A, El-Fakharany EM, Guarino V (2022) Mineralized polyvinyl alcohol/sodium alginate hydrogels incorporating cellulose nanofibrils for bone and wound healing. Molecules 27(3):697
4. Adhikari M, Bakadia BM, Wang L, Li Y, Shi Z, Yang G (2025) Electrically modified bacterial cellulose tailored with plant based green materials for infected wound healing applications. Biomater Adv 167:214087
5. Ahmad H (2021) Celluloses as support materials for antibacterial agents: a review. Cellulose 28(5):2715–2761
6. Ahmad N, Bukhari SNA, Hussain MA, Ejaz H, Munir MU, Amjad MW (2024) Nanoparticles incorporated hydrogels for delivery of antimicrobial agents: developments and trends. RSC Adv 14(19):13535–13564
7. Ahmad Sohaimy MIH, Ismail MAA, Ahmad Khiar AS, Mhd Sarbon N, Md Salim NS, Yahya HN et al (2024) Development of cellulose-based films containing salicylic acid for wound dressing applications: fabrication, properties and in vitro assessment. J Bioact Compat Polym 08839115241253560
8. Ahuja S (2016) Chemistry and water: the science behind sustaining the world's most crucial resource. Elsevier
9. Alavi SE, Alavi SF, Koohi M, Raza A, Ebrahimi Shahmabadi H (2024) Nanoparticle-integrated metal–organic frameworks: a revolution in next-generation drug delivery systems. J Pharm Investig 54(6):751–783. https://doi.org/10.1007/s40005-024-00691-w
10. Alavi M (2019) Modifications of microcrystalline cellulose (MCC), nanofibrillated cellulose (NFC), and nanocrystalline cellulose (NCC) for antimicrobial and wound healing applications. e-Polymers 19(1):103–119
11. Aldossary HA, Khalaf MM, Gouda M, Elmushyakhi A, Abou Taleb MF, Abd El-Lateef HM (2023) Wound dressing candidate materials based on casted films of cellulose acetate modified with zirconium oxide (ZrO_2), and gallium oxide (Ga_2O_3). Mater Today Commun 34:105299
12. Ali A, Bano S, Poojary SS, Priyadarshi R, Choudhary A, Kumar D et al (2022) Comparative analysis of TiO_2 and Ag nanoparticles on xylan/chitosan conjugate matrix for wound healing application. Int J Polym Mater Polym Biomater 71(5):376–385
13. Alshehri SM, Aldalbahi A, Al-Hajji AB, Chaudhary AA, in het Panhuis M, Alhokbany N et al (2016) Development of carboxymethyl cellulose-based hydrogel and nanosilver composite as antimicrobial agents for UTI pathogens. Carbohydr Polym 138:229–236
14. Archana M, Ali A, Rubini D, Dharshini KP, Hari BNV, Jayasankari S, Ramyadevi D et al (2023) Development of an anti-infective urinary catheter composed of polyvinyl alcohol/sodium alginate/methylcellulose/polyethylene glycol by using a pressure-assisted 3D-printing technique. Int J Biol Macromol 249:126029. https://doi.org/10.1016/j.ijbiomac.2023.126029
15. Avatefi M, Moghadam NB, Asghari S, Bakhshesh M, Kruppke B, Khonakdar HA et al (2024) Cellulose acetate-based electrospun nanocomposites improved by mussel-inspired polydopamine coatings and copper iodide-decorated graphene oxide: a self-disinfecting nanofibrous membrane with potential biomedical applications. Inorg Chem Commun 170:113181
16. Bao Y, He J, Song K, Guo J, Zhou X, Liu S (2022) Functionalization and antibacterial applications of cellulose-based composite hydrogels. Polymers 14(4):769

17. Bellanthudawa B, Nawalage N, Handapangoda H, Suvendran S, Wijayasenarathne K, Rathnasuriya M et al (2023) A perspective on biodegradable and non-biodegradable nanoparticles in industrial sectors: applications, challenges, and future prospects. Nanotechnol Environ Eng 8(4):975–1013
18. Bethke K, Palantöken S, Andrei V, Ross M, Raghuwanshi VS, Kettemann F et al (2018) Functionalized cellulose for water purification, antimicrobial applications, and sensors. Adv Funct Mater 28(23). https://doi.org/10.1002/adfm.201800409
19. Cacicedo ML, Islan GA, Drachemberg MF, Alvarez VA, Bartel LC, Bolzán AD et al (2018) Hybrid bacterial cellulose–pectin films for delivery of bioactive molecules. New J Chem 42(9):7457–7467
20. Cacicedo ML, Pacheco G, Islan GA, Alvarez VA, Barud HS, Castro GR (2020) Chitosan-bacterial cellulose patch of ciprofloxacin for wound dressing: preparation and characterization studies. Int J Biol Macromol 147:1136–1145. https://doi.org/10.1016/j.ijbiomac.2019.10.082
21. Cai Y, Gu R, Dong Y, Zhao Q, Zhang K, Cheng C et al (2023) Fabrication of antibacterial polydopamine-carboxymethyl cellulose-Ag nanoparticle hydrogel coating for urinary catheters. J Biomater Appl 38(1):73–84
22. Charpentier PA, Maguire A, Wan W-k (2006) Surface modification of polyester to produce a bacterial cellulose-based vascular prosthetic device. Appl Surf Sci 252(18):6360–6367. https://doi.org/10.1016/j.apsusc.2005.09.064
23. Chartres C, Varma S (2010) Out of water: from abundance to scarcity and how to solve the world's water problems. FT Press
24. Ciolacu DE, Nicu R, Ciolacu F (2020) Cellulose-based hydrogels as sustained drug-delivery systems. Materials 13(22):5270
25. Dai H, Zhang H, Ma L, Zhou H, Yu Y, Guo T et al (2019) Green pH/magnetic sensitive hydrogels based on pineapple peel cellulose and polyvinyl alcohol: synthesis, characterization and naringin prolonged release. Carbohydr Polym 209:51–61
26. Dang XG, Yu ZF, Wang XC, Du YM, Wang CH (2023) Sustainable one-pot synthesis of novel soluble cellulose-based nonionic biopolymers for natural antimicrobial materials. Chem Eng J 468. https://doi.org/10.1016/j.cej.2023.143810
27. Firmanda A, Fahma F, Warsiki E, Syamsu K, Arnata IW, Sartika D et al (2023) Antimicrobial mechanism of nanocellulose composite packaging incorporated with essential oils. Food Control 147. https://doi.org/10.1016/j.foodcont.2023.109617
28. Fusco D, Meissner F, Podesser BK, Marsano A, Grapow M, Eckstein F et al (2022) Small-diameter bacterial cellulose-based vascular grafts for coronary artery bypass grafting in a pig model. Front Cardiovasc Med 9:881557
29. Gao J, Hao L-S, Ning B-B, Zhu Y-K, Guan J-B, Ren H-W et al (2022) Biopaper based on ultralong hydroxyapatite nanowires and cellulose fibers promotes skin wound healing by inducing angiogenesis. Coatings 12(4):479
30. Goetz LA, Jalvo B, Rosal R, Mathew AP (2016) Superhydrophilic anti-fouling electrospun cellulose acetate membranes coated with chitin nanocrystals for water filtration. J Membr Sci 510:238–248
31. Gong J, Hou L, Ching YC, Ching KY, Hai ND, Chuah CH (2024) A review of recent advances of cellulose-based intelligent-responsive hydrogels as vehicles for controllable drug delivery system. Int J Biol Macromol 264:130525. https://doi.org/10.1016/j.ijbiomac.2024.130525
32. Gouda M, Hebeish A, Al-Omair M (2014) Development of silver-containing nanocellulosics for effective water disinfection. Cellulose 21:1965–1974
33. Grimm J, Bessarabov D, Sanderson R (1998) Review of electro-assisted methods for water purification. Desalination 115(3):285–294
34. Gunathilake TMSU, Ching YC, Chuah CH, Rahman NA, Liou N-S (2020) Recent advances in celluloses and their hybrids for stimuli-responsive drug delivery. Int J Biol Macromol 158:670–688. https://doi.org/10.1016/j.ijbiomac.2020.05.010
35. Gurtner GC, Werner S, Barrandon Y, Longaker MT (2008) Wound repair and regeneration. Nature 453(7193):314–321

36. Hassan M, Abou-Zeid R, Hassan E, Berglund L, Aitomäki Y, Oksman K (2017) Membranes based on cellulose nanofibers and activated carbon for removal of *Escherichia coli* bacteria from water. Polymers 9(8):335
37. Hezari S, Olad A, Dilmaghani A (2022) Modified gelatin/iron-based metal–organic framework nanocomposite hydrogel as wound dressing: synthesis, antibacterial activity, and *Camellia sinensis* release. Int J Biol Macromol 218:488–505. https://doi.org/10.1016/j.ijbiomac.2022.07.150
38. Ho HN, Le HH, Le TG, Duong THA, Ngo VQT, Dang CT et al (2022) Formulation and characterization of hydroxyethyl cellulose-based gel containing metronidazole-loaded solid lipid nanoparticles for buccal mucosal drug delivery. Int J Biol Macromol 194:1010–1018. https://doi.org/10.1016/j.ijbiomac.2021.11.161
39. Hu W, Ouyang Q, Jiang C, Huang S, Alireza N-E, Guo D et al (2024) Biomedical metal–organic framework materials on antimicrobial therapy: perspectives and challenges. Mater Today Chem 41:102300. https://doi.org/10.1016/j.mtchem.2024.102300
40. Hu J, Yu Q, Wang L, Shi H, Luan S (2025) Recent progress in antibacterial surfaces for implant catheters. BME Front 6:0063
41. Huang Y, Zhu C, Yang J, Nie Y, Chen C, Sun D (2014) Recent advances in bacterial cellulose. Cellulose 21:1–30
42. Huang ST, Zou S, Wang YX (2023) Construction of compostable packaging with antibacterial property and improved performance using sprayed coatings of modified cellulose nanocrystals. Carbohydr Polym 305. https://doi.org/10.1016/j.carbpol.2023.120539
43. Hussain R, Batool SA, Aizaz A, Abbas M, Rehman MAU (2023) Biodegradable packaging based on poly(vinyl alcohol) and carboxymethyl cellulose films incorporated with ascorbic acid for food packaging applications. ACS Omega 8(45):42301–42310. https://doi.org/10.1021/acsomega.3c04397
44. Irimia A, Grigoras VC, Popescu CM (2024) Active cellulose-based food packaging and its use on foodstuff. Polymers 16(3). https://doi.org/10.3390/polym16030389
45. Karimi T, Mottaghitalab F, Keshvari H, Farokhi M (2023) Carboxymethyl chitosan/sodium carboxymethyl cellulose/agarose hydrogel dressings containing silk fibroin/polydopamine nanoparticles for antibiotic delivery. J Drug Deliv Sci Technol 80:104134. https://doi.org/10.1016/j.jddst.2022.104134
46. Karimzadeh Z, Javanbakht S, Namazi H (2019) Carboxymethylcellulose/MOF-5/graphene oxide bio-nanocomposite as antibacterial drug nanocarrier agent. Bioimpacts 9(1):5–13. https://doi.org/10.15171/bi.2019.02
47. Karl B, Alkhatib Y, Beekmann U, Bellmann T, Blume G, Steiniger F et al (2020) Development and characterization of bacterial nanocellulose loaded with *Boswellia serrata* extract containing nanoemulsions as natural dressing for skin diseases. Int J Pharm 587:119635
48. Kaur H, Gogoi B, Sharma I, Das DK, Azad MA, Pramanik DD et al (2024) Hydrogels as a potential biomaterial for multimodal therapeutic applications. Mol Pharm 21(10):4827–4848. https://doi.org/10.1021/acs.molpharmaceut.4c00595
49. Khalid A, Khan R, Ul-Islam M, Khan T, Wahid F (2017) Bacterial cellulose-zinc oxide nanocomposites as a novel dressing system for burn wounds. Carbohydr Polym 164:214–221
50. Lee JH, Lim S-J, Oh DH, Ku SK, Li DX, Yong CS et al (2010) Wound healing evaluation of sodium fucidate-loaded polyvinylalcohol/sodium carboxymethylcellulose-based wound dressing. Arch Pharmacal Res 33:1083–1089
51. Li G, Bao L, Hu G, Chen L, Zhou X, Hong FF (2024) Development and performance evaluation of a novel elastic bacterial nanocellulose/polyurethane small caliber artificial blood vessels. Int J Biol Macromol 268:131685. https://doi.org/10.1016/j.ijbiomac.2024.131685
52. Li M, Qu H, Li Q, Lu S, Wu Y, Tang Z et al (2024) A carboxymethyl cellulose/chitosan-based hydrogel harvests robust adhesive, on-demand detachment and self-healing performances for deep burn healing. Chem Eng J 498:155552
53. Li J-X, Yang H, Gu R-H, Dong Y-H, Cai Y-W, Zhao Q et al (2023) Fabrication of antibacterial Fe^{3+}-carboxymethylcellulose-polyacrylamide-Ag nanoparticles hydrogel coating for urinary catheters. Colloids Surf A Physicochem Eng Asp 672:131680. https://doi.org/10.1016/j.colsurfa.2023.131680

54. Li R, Wang X, Lin C, Zhang X, Yuan Z (2025) The synergistic effect of MOF and biomass achieving dual breakthroughs in material structure and performance. J Mater Chem A
55. Liu P, Borrell PF, Božič M, Kokol V, Oksman K, Mathew AP (2015) Nanocelluloses and their phosphorylated derivatives for selective adsorption of Ag^{+}, Cu^{2+} and Fe^{3+} from industrial effluents. J Hazard Mater 294:177–185
56. Liu L, Ji X, Mao L, Wang L, Chen K, Shi Z et al (2022) Hierarchical-structured bacterial cellulose/potato starch tubes as potential small-diameter vascular grafts. Carbohydr Polym 281:119034
57. Liu D, Meng Q, Hu J (2023) Bacterial nanocellulose hydrogel: a promising alternative material for the fabrication of engineered vascular grafts. Polymers 15(18):3812
58. Liu K, Wang Y, Liu W, Zheng C, Xu T, Du H et al (2024) Bacterial cellulose/chitosan composite materials for biomedical applications. Chem Eng J 494:153014. https://doi.org/10.1016/j.cej.2024.153014
59. Liu Y, Sui Y, Liu C, Liu C, Wu M, Li B (2018) A physically crosslinked polydopamine/nanocellulose hydrogel as potential versatile vehicles for drug delivery and wound healing. Carbohydr Polym 188:27–36. https://doi.org/10.1016/j.carbpol.2018.01.093
60. Livesey TC, Mahmoud LAM, Katsikogianni MG, Nayak S (2023) Metal–organic frameworks and their biodegradable composites for controlled delivery of antimicrobial drugs. Pharmaceutics 15(1):274
61. Madivoli ES, Schwarte JV, Kareru PG, Gachanja AN, Fromm KM (2023) Stimuli-responsive and antibacterial cellulose–chitosan hydrogels containing polydiacetylene nanosheets. Polymers 15(5):1062
62. Marrez DA, Abdelhamid AE, Darwesh OM (2019) Eco-friendly cellulose acetate green synthesized silver nano-composite as antibacterial packaging system for food safety. Food Packag Shelf Life 20. https://doi.org/10.1016/j.fpsl.2019.100302
63. Di Matteo V, Di Filippo MF, Ballarin B, Gentilomi GA, Bonvicini F, Panzavolta S et al (2023) Cellulose/zeolitic imidazolate framework (ZIF-8) composites with antibacterial properties for the management of wound infections. J Funct Biomater 14(9):472
64. Mautner A (2020) Nanocellulose water treatment membranes and filters: a review. Polym Int 69(9):741–751
65. Meng Q, Xue Z, Chen S, Wu M, Lu P (2023) Smart antimicrobial Pickering emulsion stabilized by pH-responsive cellulose-based nanoparticles. Int J Biol Macromol 233:123516. https://doi.org/10.1016/j.ijbiomac.2023.123516
66. Metreveli G, Wågberg L, Emmoth E, Belák S, Strømme M, Mihranyan A (2014) A size-exclusion nanocellulose filter paper for virus removal. Adv Healthc Mater 3(10):1546–1550
67. Motelica L, Ficai D, Petrisor G, Oprea OC, Trusca RD, Ficai A et al (2024) Antimicrobial hydroxyethyl-cellulose-based composite films with zinc oxide and mesoporous silica loaded with cinnamon essential oil. Pharmaceutics 16(9). https://doi.org/10.3390/pharmaceutics16091225
68. Nadar SS, Vaidya L, Maurya S, Rathod VK (2019) Polysaccharide based metal organic frameworks (polysaccharide–MOF): a review. Coord Chem Rev 396:1–21. https://doi.org/10.1016/j.ccr.2019.05.011
69. Naghavi M, Vollset SE, Ikuta KS, Swetschinski LR, Gray AP, Wool EE et al (2024) Global burden of bacterial antimicrobial resistance 1990–2021: a systematic analysis with forecasts to 2050. Lancet 404(10459):1199–1226. https://doi.org/10.1016/S0140-6736(24)01867-1
70. Nicu R, Ciolacu DE, Petrovici A-R, Rusu D, Avadanei M, Mihaila AC et al (2023) 3D matrices for enhanced encapsulation and controlled release of anti-inflammatory bioactive compounds in wound healing. Int J Mol Sci 24(4):4213
71. Norrrahim MNF, Mohd Kasim NA, Knight VF, Ong KK, Mohd Noor SA, Abdul Halim N et al (2021) Emerging developments regarding nanocellulose-based membrane filtration material against microbes. Polymers 13(19):3249
72. Ortega MA, De Leon-Oliva D, Boaru DL, Fraile-Martinez O, García-Montero C, Diaz R et al (2023) Unraveling the new perspectives on antimicrobial hydrogels: state-of-the-art and translational applications. Gels 9(8). https://doi.org/10.3390/gels9080617

73. Pandit A, Kumar R (2024) Bacterial cellulose-based materials for biomedical applications. In: Encyclopedia of green materials. Springer, pp 105–112
74. Pang Q, Yang F, Jiang Z, Wu K, Hou R, Zhu Y (2023) Smart wound dressing for advanced wound management: real-time monitoring and on-demand treatment. Mater Des 229:111917
75. Pettinari C, Pettinari R, Di Nicola C, Tombesi A, Scuri S, Marchetti F (2021) Antimicrobial MOFs. Coord Chem Rev 446:214121. https://doi.org/10.1016/j.ccr.2021.214121
76. Phan HN, Bui HM, Vu NK, Trinh HTK (2023) Fabrication of fabric-like Bacterial Cellulose/ Collagen membranes by applying textile padding method for wound dressing applications. Cellulose 30(4):2289–2321
77. Pool J (1977) Normal hemostatic mechanisms: a review. Am J Med Technol 43(8):776–780
78. Prakash J, Venkataprasanna K, Bharath G, Banat F, Niranjan R, Venkatasubbu GD (2021) In-vitro evaluation of electrospun cellulose acetate nanofiber containing graphene oxide/TiO_2/ curcumin for wound healing application. Colloids Surf A Physicochem Eng Asp 627:127166
79. Qin S, Niu Y, Zhang Y, Wang W, Zhou J, Bai Y et al (2024) Metal Ion-containing hydrogels: synthesis, properties, and applications in bone tissue engineering. Biomacromolecules 25(6):3217–3248. https://doi.org/10.1021/acs.biomac.3c01072
80. Quellmalz A, Mihranyan A (2015) Citric acid cross-linked nanocellulose-based paper for size-exclusion nanofiltration. ACS Biomater Sci Eng 1(4):271–276
81. Ramezani G, Stiharu I, van de Ven TGM, Nerguizian V (2024) Advancements in hybrid cellulose-based films: innovations and applications in 2D nano-delivery systems. J Funct Biomater 15(4):93
82. Rini, Suryanto H, Hari PD, Syukri D, Jaswandi, Kurniawan F et al (2023) Application of nanocellulose biofilter from pineapple peel waste for water microbes removal. J Environ Public Health 2023(1):5823207
83. Rodrigues M, Kosaric N, Bonham CA, Gurtner GC (2019) Wound healing: a cellular perspective. Physiol Rev 99(1):665–706. https://doi.org/10.1152/physrev.00067.2017
84. Roque-Borda CA, Carnero Canales CS, Primo LMDG, Colturato VMM, Polinário G, Di Filippo LD et al (2024) Cellulose from bacteria as a delivery system for improved treatment of infectious diseases: a review of updates and prospects. Int J Biol Macromol 277:133831. https://doi.org/10.1016/j.ijbiomac.2024.133831
85. Rugaie OA, Abdellatif AA, El-Mokhtar MA, Sabet MA, Abdelfattah A, Alsharidah M et al (2022) Retardation of bacterial biofilm formation by coating urinary catheters with metal nanoparticle-stabilized polymers. Microorganisms 10(7):1297
110. Sankarganesh P, Parthasarathy V, Ganesh Kumar A, Ragu S, Saraniya M, Udayakumari N, Anbarasan RJ (2024) Preparation of cellulose-PVA blended hydrogels for wound healing applications with controlled release of the antibacterial drug: an in vitro anticancer activity. Biomass Convers Biorefin 14(3):3385–3395. https://doi.org/10.1007/s13399-022-02586-y
87. Sayed A, Safwat G, Abdel-raouf M, Mahmoud GA (2023). Alkali-cellulose/ Polyvinyl alcohol biofilms fabricated with essential clove oil as a novel scented antimicrobial packaging material. Carbohydr Polym Technol Appl 5. https://doi.org/10.1016/j.carpta.2022.100273.
88. Scherner M, Reutter S, Klemm D, Sterner-Kock A, Guschlbauer M, Richter T (2014) In vivo application of tissue-engineered blood vessels of bacterial cellulose as small arterial substitutes: proof of concept? J Surg Res 189(2):340–347
89. Sethunga M, Gunathilake K, Ranaweera K, Munaweera I (2024) Antimicrobial and antioxidative electrospun cellulose acetate-essential oils nanofibrous membranes for active food packaging to extend the shelf life of perishable fruits. Innov Food Sci Emerg Technol 97. https://doi.org/10.1016/j.ifset.2024.103802
90. Shaheen TI, El-Gamal MS, Desouky SE, Hassan SED, Alemam AM (2021) Benign production of AgNPs/bacterial nanocellulose for wound healing dress: antioxidant, cytotoxicity and in vitro studies. J Cluster Sci 1–17
91. Shahzadi I, Islam M, Saeed H, Haider A, Shahzadi A, Haider J et al (2022) Formation of biocompatible MgO/cellulose grafted hydrogel for efficient bactericidal and controlled release of doxorubicin. Int J Biol Macromol 220:1277–1286. https://doi.org/10.1016/j.ijbiomac.2022.08.142.

92. Sharma PR, Sharma SK, Lindström T, Hsiao BS (2020). Nanocellulose-enabled membranes for water purification: perspectives. Adv Sustain Sys 4(5):1900114
93. Sharma G, George Joy J, Sharma AR, Kim JC (2024) Accelerated full-thickness skin wound tissue regeneration by self-crosslinked chitosan hydrogel films reinforced by oxidized CNC-AgNPs stabilized Pickering emulsion for quercetin delivery. J Nanobiotechnol 22(1):323
94. Sheikhi A, Safari S, Yang H, Van De Ven TG (2015) Copper removal using electrosterically stabilized nanocrystalline cellulose. ACS Appl Mater Interfaces 7(21):11301–11308
95. Si S, Huang XX, Wang Q, Manickam S, Zhao D, Liu YN (2024) Enhancing refrigerated chicken breasts preservation: Novel composite hydrogels incorporated with antimicrobial peptides, bacterial cellulose, and polyvinyl alcohol. Int. J. Biolog. Macromol 281. https://doi.org/10.1016/j.ijbiomac.2024.136505
96. Sijabat EK, Nuruddin A, Aditiawati P, Purwasasmita BS (2019) Synthesis and characterization of bacterial nanocellulose from banana peel for water filtration membrane application. In: Journal of physics: conference series, vol 1230, pp 012085. IOP Publishing
97. Singh B, Panda N, Mund R, Pramanik K (2016) Carboxymethyl cellulose enables silk fibroin nanofibrous scaffold with enhanced biomimetic potential for bone tissue engineering application. Carbohydr Polym 151:335–347
98. Sirviö JA, Hasa T, Leiviskä T, Liimatainen H, Hormi O (2016) Bisphosphonate nanocellulose in the removal of vanadium (V) from water. Cellulose 23:689–697
99. Song B, Zhang T, Yang K, Tian G, Dang Y, Ma J (2023) Zn^{2+} ions cross-linking sodium alginate encapsulated bacterial cellulose/ZnO composite film for antibacterial application. Cellulose 30(12):7853–7864. https://doi.org/10.1007/s10570-023-05371-w
100. Stoica AE, Chircov C, Grumezescu AM (2020) Nanomaterials for wound dressings: an up-to-date overview. Molecules 25(11):2699
101. Sulaeva I, Henniges U, Rosenau T, Potthast A (2015) Bacterial cellulose as a material for wound treatment: properties and modifications. A review. Biotechnol Adv 33(8):1547–1571
102. Sultan M, Mohamed OA, El-Masry HM, Taha G (2023) Fabrication and evaluation of antimicrobial cellulose/Arabic gum hydrogels as potential drug delivery vehicle. Int J Biol Macromol 242:125083. https://doi.org/10.1016/j.ijbiomac.2023.125083
103. Sun B, Zhang M, Shen J, He Z, Fatehi P, Ni Y (2019) Applications of cellulose-based materials in sustained drug delivery systems. Curr Med Chem 26(14):2485–2501
104. Suopajärvi T, Liimatainen H, Hormi O, Niinimäki J (2013) Coagulation–flocculation treatment of municipal wastewater based on anionized nanocelluloses. Chem Eng J 231:59–67
105. Tenorová K, Masteiková R, Pavloková S, Kostelanská K, Bernatonienė J, Vetchý D (2022) Formulation and evaluation of novel film wound dressing based on collagen/microfibrillated carboxymethylcellulose blend. Pharmaceutics 14(4):782
106. Thakare NR, Gogoi P, Bharali P, Hazarika S (2024) Influence of copper ion cross-linked CMC-PVA film on cell viability and cell proliferation study. Int J Biol Macromol 282:136645
107. Tian M, Shuai J, Bishop BA, Zhang W, Chen J, Wang X (2023) Plant cellulose-based biomimetic artificial small-diameter vascular materials enabled by gradient dual-network entanglement. Chem Eng J 476:146751. https://doi.org/10.1016/j.cej.2023.146751
108. Ullah A, Ullah S, Khan MQ, Hashmi M, Nam PD, Kato Y et al (2020) Manuka honey incorporated cellulose acetate nanofibrous mats: fabrication and in vitro evaluation as a potential wound dressing. Int J Biol Macromol 155:479–489
109. Voisin H, Bergström L, Liu P, Mathew AP (2017) Nanocellulose-based materials for water purification. Nanomaterials 7(3):57
110. Wang R, Guan S, Sato A, Wang X, Wang Z, Yang R et al (2013) Nanofibrous microfiltration membranes capable of removing bacteria, viruses and heavy metal ions. J Membr Sci 446:376–382
111. Wang M, Huang C, Chen Y, Ji Y, Yu D-G, Bligh SWA (2025) Medicated tri-layer fibers based on cellulose acetate and polyvinylpyrrolidone for enhanced antibacterial and wound healing properties. Carbohydr Polym 348:122856
112. Wang WT, Niu B, Liu RL, Chen HJ, Fang XJ, Wu WJ et al (2023) Development of bio-based PLA/cellulose antibacterial packaging and its application for the storage of shiitake mushroom. Food Chem 429. https://doi.org/10.1016/j.foodchem.2023.136905

113. Weyell P, Beekmann U, Küpper C, Dederichs M, Thamm J, Fischer D et al (2019) Tailor-made material characteristics of bacterial cellulose for drug delivery applications in dentistry. Carbohydr Polym 207:1–10
114. Wijmans JG, Baker RW (1995) The solution–diffusion model: a review. J Membr Sci 107(1–2):1–21
115. Xu Z, Fan J, Tian W, Ji X, Cui Y, Nan Q et al (2024) Cellulose-based pH–responsive Janus dressing with unidirectional moisture drainage for exudate management and diabetic wounds healing. Adv Funct Mater 34(3):2307449
116. Yang M, Zhang Y, Liu Z, Liu L, Wang X, Qian L (2023) Advances in research on cellulose-based drug carriers. Paper Biomater 8(4):55–68. https://doi.org/10.26599/PBM.2023.9260026
117. Yao T, Zeng X, Tao X, Xu H (2024) Recent progress of MOF-based antibacterial hydrogels. Chem Eng J 487:150641. https://doi.org/10.1016/j.cej.2024.150641
118. Ye J, Li J, Wang X, Wang Q, Wang S, Wang H et al (2024) Preparation of bacterial cellulose-based antibacterial membranes with prolonged release of drugs: emphasis on the chemical structure of drugs. Carbohydr Polym 323:121379. https://doi.org/10.1016/j.carbpol.2023.121379
119. Yu X, Tong S, Ge M, Wu L, Zuo J, Cao C et al (2013) Adsorption of heavy metal ions from aqueous solution by carboxylated cellulose nanocrystals
120. Zarepour A, Gok B, Kılıç YB, Khosravi A, Iravani S, Zarrabi A (2024) Bacterial nanocelluloses as sustainable biomaterials for advanced wound healing and dressings
121. Zhang CH, Yang XD, Li Y, Qiao CD, Wang SJ, Wang XJ et al (2020) Enhancement of a zwitterionic chitosan derivative on mechanical properties and antibacterial activity of carboxymethyl cellulose-based films. Int J Biol Macromol 159:1197–1205. https://doi.org/10.1016/j.ijbiomac.2020.05.080
122. Zhao X, Shi Y, Yang M, Sun L, Fu Y, Gao G et al (2024) Barnacle-inspired and polyphenol-assisted modification of bacterial cellulose-based wound dressings for promoting infectious wound healing. Int J Biol Macromol 279:135291
123. Zhao X, Yang K, Song B, Qiu H, Zhao J, Liu H et al (2024) Amphiphilic nanofibrillated cellulose/polyurethane composites with antibacterial, antifouling and self-healing properties for potential catheter applications. Int J Biol Macromol 263:130407. https://doi.org/10.1016/j.ijbiomac.2024.130407
124. Zheng L, Li S, Luo J, Wang X (2020) Latest advances on bacterial cellulose-based antibacterial materials as wound dressings. Front Bioeng Biotechnol 8:593768
125. Zhu R, Chen Q, Zhou Q, Xi Y, Zhu J, He H (2016) Adsorbents based on montmorillonite for contaminant removal from water: a review

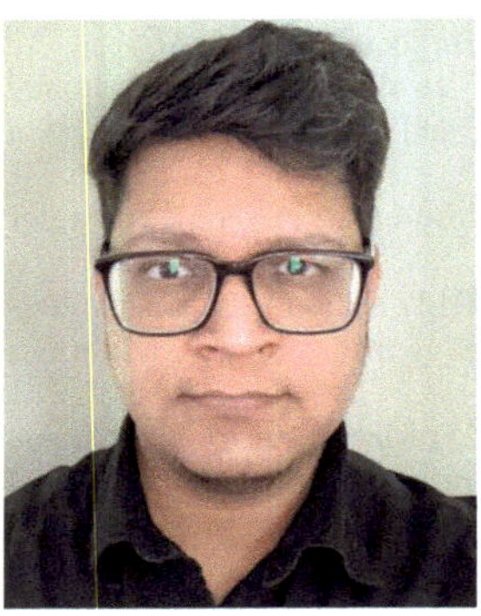

Bhanu Sharma earned a B.Sc. in Chemistry from the University of Delhi in 2016 before completing a Master's in Organic Chemistry at Jamia Hamdard in 2018. He then joined the Chemistry Ph.D. program in 2019 at the University of Central Florida. His research focuses on the synthesis and functionalization of cellulose-based materials for antimicrobial applications in agriculture and biomedical fields, as well as the synthesis and characterization of metal nanoparticles.

Preeti Maiti completed her B.Sc. from Banaras Hindu University at Institute of Agricultural Science. She is currently enrolled in chemistry Ph.D. program at University of Central Florida. Also, part of the NanoScience and Technology Center, her work mainly focuses on the developing nanodelivery systems for agricultural and biomedical applications.

Dr. Edwin Davidson is a chemist with expertise in nanotechnology, agricultural chemistry, and analytical biochemistry. He earned his M.Ed. from the Universidad Tecnológica de Panamá in 2018, followed by an M.S. in Nanotechnology in 2020 and a Ph.D. in Chemistry in 2024 from the University of Central Florida. His research focused on developing nanodelivery systems and utilizing various analytical techniques to assess the active ingredients of interest. These contributions have led to several research articles and publications in high impact factor journals. Currently, he works as a senior scientist at the Gatorade Sport Science Institute—PepsiCo, fostering research and development on hydration products to optimize athletic performance.

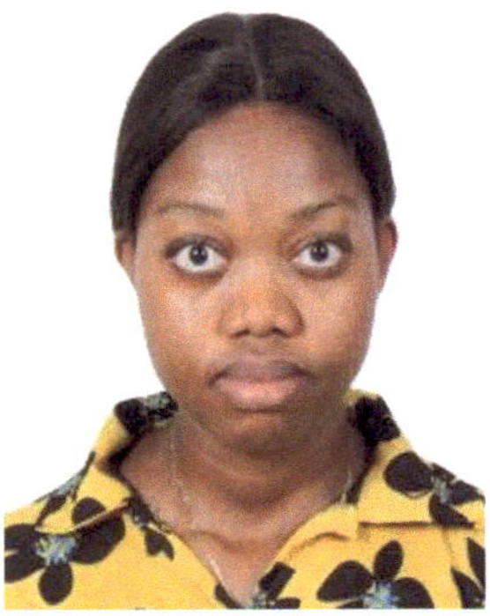

Ms. Cecilia Ezeanya holds a bachelor's degree in Botany from the University of Ibadan and is currently pursuing a master's in Biotechnology at the College of Medicine, University of Central Florida. Her research focuses on optimizing polyphenol-polymer nanoparticles for enhanced encapsulation and delivery of oxytetracycline in agricultural and biomedical applications. She has experience in synthesis formulation, antimicrobial assays, analytical techniques, and controlled environment studies, including screenhouse and controlled release experiments. Cecilia has authored a peer-reviewed article and has served as a reviewer for two scientific journals. She has presented her research at major scientific conferences, including the American Chemical Society, NanoFlorida, the Sustainable Nanotechnology Organization, and the Burnett School of Biomedical Sciences Symposium. Passionate about nanomedicine and drug delivery innovation, she aims to pursue a Ph.D. to further explore novel therapeutic strategies.

Dr. Jorge Pereira studied his Industrial Chemistry B.S. at Universidad Nacional de Honduras, then emigrated to Florida where he completed his Chemistry Ph.D. at the University of Central Florida. His research focuses on the development of multi-nutrient nanoparticle systems capable of targeted delivery and synergistic bactericidal activity for improving modern crop management practices. Additionally, he studies the use of natural-based excipients, such as polyphenols and cellulose, as sustainable components in drug delivery nanosystems. His work has led to several patent applications and has been published in different peer-reviewed scientific journals.

Dr. Swadeshmukul Santra is a professor at the University of Central Florida holding a joint appointment with the NanoScience Technology Center and the Department of Chemistry. Dr. Santra is the founding director of UCF Materials Innovation for Sustainable Agriculture (MISA) Center. He has been actively working in the field of nanoscience and nanotechnology for over 20 years focusing on nanoagriculture and nanomedicine areas. Under his leadership, the UCF MISA center has significantly advanced translational research to address some of the most pressing crop protection challenges. In partnership with industry, he has developed several agrochemical formulations for the immediate benefit of Florida citrus industry. He has published 110 peer-reviewed research articles, 9 review articles, 12 book chapters, and delivered 125 invited talks. He has been awarded 41 patents including 32 US patents.

Bacterial Nanocellulose-Based Biomaterial Composite for Tissue Engineering

Phasuwit P. Phatchayawat and Muenduen Phisalaphong

Abstract Bacterial nanocellulose (BNC) or bacterial cellulose (BC) is a precious biomaterial with some special characteristics, therefore extremely useful in a wide range of biomedical fields, exclusively in biomedical applications, especially tissue engineering. In this chapter, there is an extensive discussion of BNC-based composite starting from bacterial strains, biosynthesis processes, metabolic pathway, properties and BNC-based composites. Physical and chemical characteristics of BNC, including high purity, biocompatibility, mechanical properties and water adsorption, are addressed, which are useful characteristics for tissue regeneration. Modifications of BNC and its functionalities such as by chemical modification, 3D bioprinting and hybridization with other biomaterials have been approached for particular biomedical applications. These methods not only improve the mechanical strength and architecture of BNC composites but also affect their biodegradability and biocompatibility, which can be used for tissue regeneration. In addition, this chapter reviews the use of BNC in fields of tissue engineering, including its uses in bone, cartilage, skin, vascular and urethral tissue engineering. Finally, this chapter is discussed on an overview of future trends and challenges in BNC-based biomaterials in tissue engineering application.

Keywords Bacterial nanocellulose · Tissue engineering · Biomaterial · Composites · Biomedical applications

---

P. P. Phatchayawat
Biomedical Engineering Program, Faculty of Engineering, Chulalongkorn University, Bangkok, Thailand

M. Phisalaphong (✉)
Bio-Circular-Green-Economy Technology and Engineering Center, BCGeTEC, Department of Chemical Engineering, Faculty of Engineering, Chulalongkorn University, Bangkok, Thailand
e-mail: muenduen.p@chula.ac.th

R. Malviya and S. Sundram (eds.), *Sustainable Cellulose Materials for Biomedical Applications*, Sustainable Materials and Technology,
https://doi.org/10.1007/978-981-96-9904-9_14

List of Abbreviations

3D	3-Dimensional
4D	4-Dimensional
BC	Bacterial cellulose
BMP	Bone morphogenetic protein
BNC	Bacterial nanocellulose
ECM	Extracellular matrix
EGF	Epidermal growth factor
FDA	Food and drug administration
FGF	Fibroblast growth factor
HAp	Hydroxyapatite
HGF	Hepatocyte growth factor
iPSCs	Induced pluripotent stem cells
LbL	Layer-by-layer
MSCs	Mesenchymal stem cells
TGF-β	Mesenchymal stem cells
UDP	Uridine Diphosphate
VEGF	Vascular endothelial g

1 Introduction

Bacterial nanocellulose (BNC) or bacterial cellulose (BC) is a biomaterial produced by some bacteria [8]. BNC differs from the plat-based cellulose in its nanostructured characteristics [102, 125]. The high crystallinity, purity and biocompatibility of biomaterials, as well as properties such as excellent water adsorption, biodegradability and high mechanical strength, establish it as an important biomaterial to field of tissue engineering for wound dressing, vascular grafting, bone and cartilage tissue regeneration [101, 109, 127, 141].

BNC is synthesized through a complex biosynthetic pathway, mainly by *Acetobacter* or *Gluconacetobacter* genus, whereby carbon sources are converted into cellulose [8, 88]. Thereafter, many fabrication techniques, such as in situ, ex situ, electrospinning, freeze-drying, layer-by-layer, cross-linking, solvent casting, particulate-leaching, injectable, 3D bioprinting and chemical, physical and biological modifications of BNC, have been developed to enhance its mechanical strength and adaptability for specific applications and developments [46, 109]. Studies show that BNC can retain an impressive amount of water—up to 700 times its dry weight, thus increasing its value in medical applications where moisture retention is critical [46, 109].

While the features are promising, there are still a few obstacles in the path of fabrication and modification of BNC for tissue engineering, mostly in mechanical property optimization and production scalability [123]. Although possessing good tensile strength, applications that require load-bearing ability will further demand its improvement. Consistence of BNC production by fermentation methods remains an area of research activity, as variability may affect reliability as a commercial biomaterial [41, 71, 109].

The development of BNC-based composite materials aims to improve physical and biological properties, overcome limitations, and expand applications. Continuing research in fabrication techniques, including high-end 3D bioprinting provides the possibility for the further expanding uses in tissue engineering. In addition, there remain some challenges of the improvement in mechanical performance and compatibility that need to be overcame for BNC composite to reach its full potential in mimicking the intricate structure of some natural tissues [15, 41]. The supplementation, modification methods, and production conditions of biomaterials have a significant impact on the characteristics and properties of BNC composite biomaterials in advanced biomedical applications and biomaterial development.

2 Biosynthesis and Properties of BNC

2.1 Biosynthesis of BNC

BNC biosynthesis is relevant for some bacteria as extracellular nanocellulose fiber production. The key actions which are part of BNC biosynthesis are briefly described below:

Bacterial Strains and Cultivation

BNC is synthesized extracellularly by gram-negative bacterial strains including *Acetobacter*, *Achromobacter*, *Aerobacter*, *Agrobacterium*, *Alcaligenes*, *Azotobacter*, *Gluconacetobacter*, *Komagataeibacter*, *Pseudomonas*, *Rhizobium*, *Salmonella* and *Sarcina*. Among these, *Gluconacetobacter* genus is recognized as the most efficient producer of BNC [8, 20, 87, 101]. These organisms utilize various carbon sources as raw biomaterials for the biosynthesis of cellulose fibrils. Bacteria incorporates cellulose fibrils forming a biofilm and then the self-assembly results of a cellulose pellicle component of a 3D matrix structural of nanocellulose fibers [43].

Metabolic Pathway

Construction of cellulose in bacteria involves a complex metabolism, as shown in Fig. 1. Glucose from the culture medium is utilized for UDP-glucose synthesis, which is a starting material for cellulose biosynthesis [99]. Enzymatically controlled reactions using cellulose synthesizing polymer enzymes breakdown UDP-glucose polymers into insoluble pieces [16, 130]. These chains are then extruded outside the bacterial cell forming nanofibrils that assemble into the final BNC structure [106].

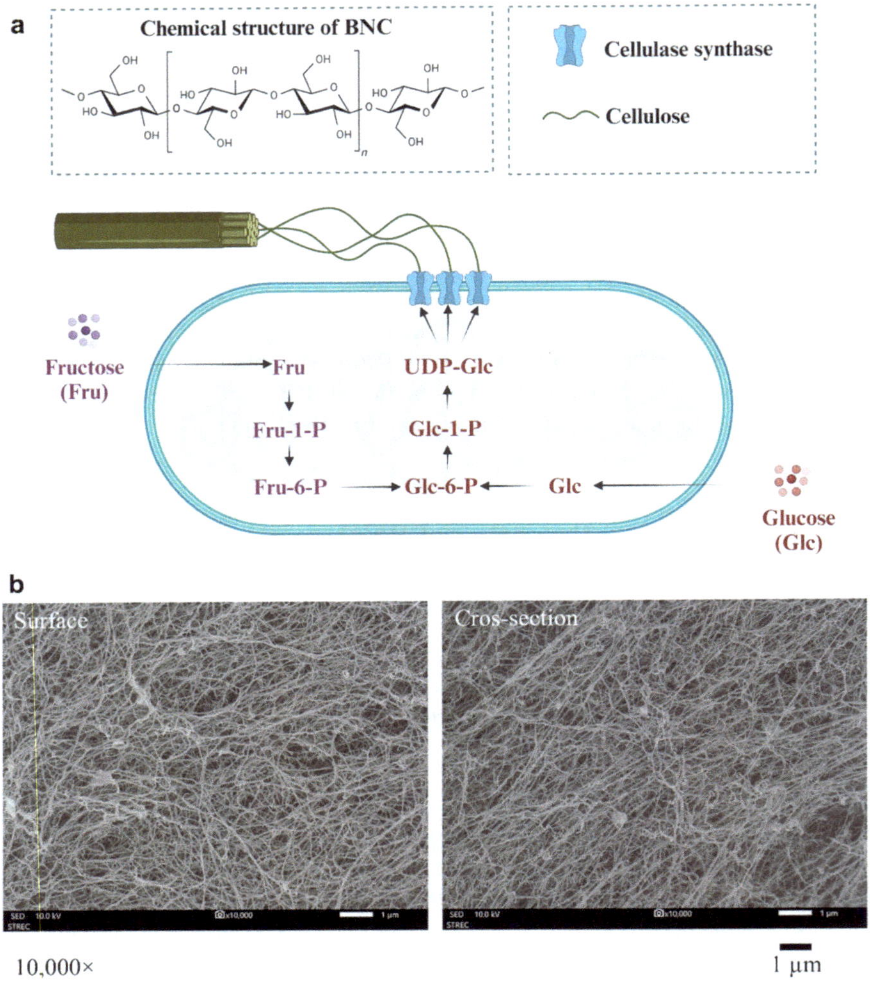

Fig. 1 Biosynthesis of BNC; **a** metabolic pathway and **b** morphology of structural of BNC; (created with BioRender.com; access date 1 December 2024)

Environmental Factors for BNC Production

Regarding the BNC production rate and BNC produced yield, pH, temperature, concentration of nutrients (primarily glucose) and aeration can be denoting environmental parameters affecting BNC production [99, 106]. Bacteria that can be used for BNC production should be grown in the growth temperature range of 28–30 °C, growth pH value of approximately 4.5–5.5, and an appropriate glucose concentration [13, 16, 128]. Under these conditions, a type of bacteria produces a thick cellulose which has better mechanical properties.

Purification of BNC

BNC requires purification once it has been synthesized as a way of eliminating remaining bacterial cells, culture media, and other contaminants [30, 110]. Purification can be achieved by treating the BNC film with an alkaline solution such as sodium hydroxide solution at room temperature or higher to kill bacterial cells and other contaminants and then washing with deionized water. Purified BNCs are typically available in gel or membrane form and can be further processed for use in medical devices such as wound dressings and tissue regeneration [37, 54]. Such treatment helps to establish biocompatibility of the material and improve the characteristics necessary for its application [140].

2.2 Properties of BNC

BNC possesses various beneficial properties, which deliver potential biomaterial in numerous applications, particularly in tissue engineering. These properties are attributed to their distinctive structural and chemical properties [46].

Unique Structural

BNC consists of ultra-fine and highly pure fibrous network structures with high mechanical strength as well as excellent elasticity [36, 53, 127]. The microfibril size of BNC ranges from around 20–100 nm, significantly lower than that of plant cellulose. This makes it more surface area and extremely porous in nature, capable of retaining water. Such a feature is crucial in keeping tissue engineering constructs well-hydrated [46]. Additionally, highly crystalline BNC is chemically stable and resistant, thereby making it compatible with unfavorable conditions [79, 127].

Biocompatibility and Bioactivity

The biocompatibility of BNC with biological systems makes it ideal for several biomedical applications, particularly in tissue engineering scaffold production [36, 79, 150]. Its chemical purity at a high-level insurance that it will not cause undesirable biological responses, a critical component in assure the successful incorporation of implants into living tissue. The structural homology of BNC to component of the natural ECM may also facilitate cellular attachment and proliferation, thus triggering tissue regulation processes [36, 41].

Sustainability and Cost-Effectiveness

BNC can be produced by cheap raw materials from agricultural industrial waste, such as by-products of brewing, food, fruit juice, sugar and coconut milk industries [87]. This environmentally friendly production process not only cuts down on waste but also helps towards a circular economy, meets the modern-day environmental objectives [79, 111]. In addition, the utilization of natural biomaterials like BC can also reduce or no harm upon ecosystems or the environment [71].

Various Applicative Possibilities

The diverse physical features of BNC facilitate its modification and tailoring for specific applications in tissue engineering. Research shows that BNC can be engineered to have different degrees of porosity, mechanical properties and degradation rates make it appropriate for wide applications in biomedicine [46, 111]. The Food and Drug Administration (FDA) has approved the use of BNC for tissue regeneration, including wound healing, bone and cartilage regeneration. [68]. The capability to design three-dimensional architecture with contained properties open new avenues for the development of advanced scaffolds that biomimetic the native structure of tissues, enhancing the likelihood of successful tissue regeneration [46, 150].

3 BNC-Based Composites

BNC composites can be processed into various structural to meet specific needs in terms of tissue regeneration, wound healing and drug delivery applications [115, 132]. The structural organization of BNC is very important because it can affect various physical, mechanical and biological characteristics, including mechanical strength, cell attachment, nutrient transfer and degradation rate [122].

By changing the physical structure of BNC composites, researchers can attain certain character, such as porosity, specific surface area, and mechanical strength. Such modifications are necessary to optimize biological performance and ensure that the material is suitable for specific uses [92, 126]. The common configurations of BNC-based composites are in the structure of fibrous, porous, hydrogel and 3D bioprinting, as shown in Fig. 2. Each structure has its own advantages and is chosen according to the compatibility of the structure with the application. The ability to design BNC composites in different forms makes them highly effective in applications, i.e., scaffold development for tissue regeneration, regenerative medicine, and controlled drug delivery systems [115, 149]. Different types of scaffold BNC composites and applications are shown in Table 1.

3.1 Fibrous Composites

Fibrous composites based on bacterial nanocellulose (BNC) are created to biomimetic the fibrous structure of the extracellular matrix (ECM) [107]. The ECM is a consist of the human organism, providing both architecture and biochemical support to the cells surrounding it. These fibrous composites incorporate either natural or synthetic fibers into the BNC matrix, hence resulting in scaffolds characterized by enhanced surface area and mechanical durability [84]. Fibrous architecture facilitates cell attachment and proliferation, both of which are essential to the process of tissue regeneration. Specifically, fibrous BNC composites have been shown to

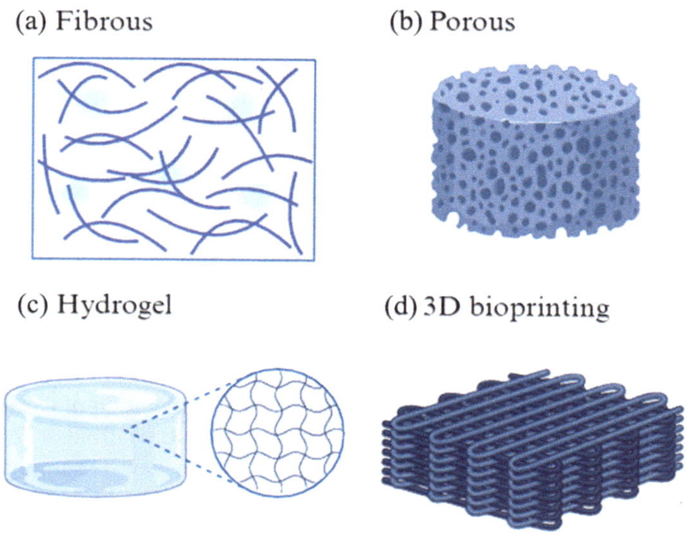

Fig. 2 Different structures of BNC composites, **a** fibrous, **b** porous, **c** hydrogel, and D) 3D bioprinting (created with BioRender.com; access date 1 December 2024)

Table 1 Different structures and applications of composite scaffolds

Type	Structure	Applications	Advantages	References
Fibrous	Interwoven fibers	Skin, cartilage, vascular grafts	High surface area, good mechanical strength	[35]
Porous	Interconnected pores	Bone, soft tissue	Tunable porosity, supports vascularization	[103]
Hydrogel	Hydrated matrix	Wound healing, cartilage repair	Soft tissue mimicry, bioactive molecule incorporation	[44]
3D Bioprinting	Precise, layered constructs	Complex tissues, organoids	Customizable, multi-cellular incorporation	[40, 90]

be highly promising for application in bone, cartilage and skin tissue engineering [4, 132]. As osteoblast cell adhesion and growth support biomaterials are needed for bone tissue engineering [28], cartilage regeneration is conducive to chondrocyte alignment and interaction with fibrous scaffolds [9]. Besides, healing of skin also depends on the communication of fibroblasts with fibrous scaffolds [55]. Fibrous composites are versatile and can be designed and developed to meet the specific demands on individual tissue types. This should in turn enable scaffolding to be endowed with sufficient strength, biomimetic native tissue microenvironment [4].

3.2 Porous Composites

Another highly studied arrangement is porous BNC composites, with an interconnected porous architecture [77]. The morphology is essential to allow for cell migration, nutrient transfers and waste exchange that are needed for tissue regeneration and growth. 3D porous scaffold feature similarly to a native ECM, providing an ideal setting to support cell activity [1]. The porosity of BNC can also be controlled by different fabrication methods, e.g., freeze-drying or lyophilization [78]. Using these processes, the size, distribution and total porosity of the product can be precisely controlled. Depending on the particular application,for example, fewer pores may be desirable where greater mechanical properties is needed, whereas larger pores offer excellent nutrients transfers and cell activity [76]. Substantial tissue engineering applications, researches, particularly for the regeneration of cartilage, bone and skin, have been performed on the use of porous BNC scaffolds [77, 78]. This is because pores with interconnections create space for new tissue regeneration, enabling cells within scaffolds to deposit growth elements of ECM [1]. For materials used in drug delivery, increased porosity increases their drug delivery capacities to transfer substances, such as bioactive agents, growth factors and drugs, to target specific areas [82].

3.3 Hydrogel Composites

Generally, hydrogel composites consist of 3D polymer networks, which swell in water. Their hydrophilic structures have the ability to hold enormous amounts of water in3D networks [139]. Hydrogel composites of BNC in tissue engineering are noteworthy because of their soft, wet environment that simulates the characteristics of native tissue [38]. BNC hydrogel composites have been designed for applications in soft tissues like skin, cartilage and blood vessels [23]. Hydrogel is a compliant but stabilizing matrix that is permissive to cell differentiation and growth. Its wet environment also facilitates the transfers of nutrients and waste, thus enhancing cell viability and function [66, 139]. Through functionalization, BNC hydrogel composite can be loaded with bioactive molecules, including growth factors, drugs or some proteins [22]. The target substances can be released in a controlled rate, thus enhancing tissue repair and regeneration. For example, the presence of growth factors in BNC hydrogel can be potentially use to induce proliferation and differentiation of cells, thus accelerating the healing process [73]. In addition, BNC hydrogel is very biocompatible and could be blended with other biomaterials to achieve necessary mechanical and biological properties [10, 48].

3.4 Three-Dimensional (3D) Bioprinting

3D bioprinting is a novel tissue engineering technique that enables the fabrication of highly precise and customized scaffolds [17]. It depends on the employment of bioinks, which could be made up from BNC blended with other biomaterials. Bioink is printed layer-by-layer to generate intricate structures that mimic nature tissue [72]. BNC bioink could be developed in order to exhibit substantial advantages, such as high mechanical strength, biocompatibility and porosity regulation [62]. The increased biocompatibility makes BNC bioink to promote cell viability, cells adhesion, migration, growth and differentiation [139]. The ability of bioink to control scaffold geometry and internal structure allows the development of constructs for given applications in tissue engineering [85], for instance, cartilage tissue engineering can incorporate the use of BNC bioink for the design of scaffold having gradient structure matching the nature cartilage [62]. In bone tissue engineering, bioink derived from BNC are combined with osteoinductivity factors for bone induction [55]. Vascular tissue engineering makes use of precision in 3D bioprinting to develop interconnected porous channels of the scaffold with blood vessels growth factors [139]. In addition to enabling the achievement of structural accuracy, 3D bioprinting enables the incorporation of various cell types, growth factors and gradient biomaterials in scaffolds [17, 62]. This aspect could enhance the functions of the scaffold itself, together with its integration with host tissues. In addition, the layer-by-layer fabrication facilitates hierarchical structure that can reproduce nature tissue complexity [45].

4 Modification and Fabrication of BNC-Based Biomaterials Composite

4.1 Modification of BNC

The modification of BNC composites for tissue engineering applications is one of interesting biomedical research topics. BNC is a versatile material that holds great potential for the modulation to meet the requirements in tissue engineering using various modification techniques, such as chemical modification, physical modifications and biological modification [96, 108]. The summary of modification of BNC composites is shown in Table 2.

Chemical Modification

The chemical modification of BNC is among the most versatile methods to enhance its properties [96]. Different techniques, such as oxidation, carboxylation, phosphorylation, or grafting of polymers, enable the introduction of functional groups that improve bioactivity, hydrophilicity, or mechanical properties [96, 116]. For example, carboxymethylation enhances water solubility, whereas phosphorylation

Table 2 Summary of modification of BNC composites

Technique	Features	Properties	Structure	Advantages	Limitations	Applications	References
Chemical modification	Cross-linking with agents (e.g., glutaraldehyde, genipin), grafting functional groups	Alters chemical properties, improves stability	Increased mechanical strength, reduced swelling, tailored biodegradability	Retains fibrillar structure with modified chemistry, enhanced stability and durability	Potential cytotoxicity of cross-linkers, limited biodegradability	Material for wound treatment, antibacterial hydrogel	[118, 131]
Physical modification	Freeze-drying, solvent casting, mechanical blending	Alters porosity, shape, and size	Controlled porosity, improved mechanical properties	Porous or sheet-like structures, Simple method, nontoxic chemicals	Less molecular-level changes	Wound dressing, scaffold for tissue engineering	[39, 67]
Biological modification	Functionalization with growth factors, proteins, or enzymes	Enhances bioactivity	Promotes cell adhesion, proliferation, and differentiation	ECM-like structure with bioactive surfaces, Biocompatible, tailored for specific biological functions	Limited mechanical enhancement	Scaffold for tissue engineering, regenerative medicine	[14, 29]

enhances mineralization and is thus suitable for bone tissue engineering [108]. While this significantly enlarges the scope of possible applications for BNC, chemical routes often utilize toxic chemicals and may involve stringent purification to achieve biocompatibility [116].

Physical Modification

Physical modification strategies include freeze-drying, electrospinning and thermal cross-linking, with the aim of changing the architecture and mechanical properties of BNC without changing the chemical composition [96]. For instance, freeze-drying generates a porous architecture ideal for cell infiltration, while thermal cross-linking enhances thermal and mechanical stabilities [108]. Physical methods have the advantage of involving minimum chemical intervention and hence preserving the biocompatibility of BNC [117]. However, these methods might suffer from scalability and reproducibility issues at large scales [89].

Biological Modification

BNC has been bioengineered with several bioactive molecules, such as Bone morphogenetic protein-2 (BMP-2) growth factor, for facilitating cell attachment, proliferation and differentiation [83]. BNC also acts as a scaffold with the capability to encapsulate other cell or bioactive molecules. Expanding its role in tissue engineering applications [69, 117].

4.2 Fabrication Techniques

BNC composites fabrication involves the blending of BNC with other components such as polymers, nanoparticles and bioactive molecules to increase its mechanical, biological and functional features for tissue regeneration and regenerative medicine. Listed below are some of the commonly used fabrication methods for BNC composites. The summary of fabrication techniques of BNC composites, as shown in Table 3.

In situ

In situ synthesis involves the introduction of nanoparticles, polymers, or bioactive molecules directly into the BNC matrix while it is produced [107]. In the process of bacterial synthesis, hydroxyapatite, protein, silk fibroin or metal nanoparticles can be introduced into the culture medium for integration into BNC [31]. This results in a homogeneous distribution of additives in a fibrous structure that enhances mechanical strength, biocompatibility, and bioactivity [60].

Ex situ

Ex situ, BNC is synthesized first, and the other components are integrated into it afterward. This method involves immersion or coating of pre-synthesized BNC with functional agents and subsequent drying or post-processing [52]. This technique

Table 3 Summary of fabrication techniques of BNC composites

Technique	Features	Properties	Structure	Advantages	Limitations	Applications	References
In situ	Materials embedded during BNC biosynthesis	Homogeneous integration, bioactive matrix	Dense or porous matrix	Simple process, bioactive material is incorporated uniformly	Limited to compatible materials with biosynthesis	Bone scaffolds, drug delivery	[31, 60]
Ex situ	Materials added after BNC synthesis	Customizable composite with varied material incorporation	Films, gels, or 3D scaffolds	Broad material compatibility, precise tailoring	May lacks uniformity; labor-intensive	Skin regeneration, wound healing	[19, 52]
Electrospinning	Creates nanofibrous scaffolds using electric fields	High porosity, high surface area, tunable fiber diameter	Nanofibrous mats	Mimics extracellular matrix (ECM), suitable for soft tissues	Requires polymer blending; low mechanical strength	Skin, cartilage, vascular tissue engineering	[35, 98]
Freeze drying	Water removal via sublimation for porosity	High porosity, low density, tunable pore size	Porous 3D scaffolds or sponges	Retains bioactivity, suitable for load-bearing tissues	Fragile structure, time-consuming process	Bone, cartilage scaffolds	[64, 85, 103]
Layer-by-layer (LbL)	Multilayered structure via alternating deposition	Controlled surface functionality, bioactivity, or stimuli-responsiveness	Multilayered films or coatings	Precise functionalization, suitable for drug delivery and surface modification	Labor-intensive for large-scale production	Drug delivery, wound dressings	[8, 14, 32]

(continued)

Table 3 (continued)

Technique	Features	Properties	Structure	Advantages	Limitations	Applications	References
Cross-linking	Chemical or physical bonding of BNC with polymers or nanoparticles	Improved mechanical properties, tunable degradation rates	Dense or porous 3D scaffolds	Enhanced stability and durability	Cross-linkers may affect biocompatibility	Bone, soft tissue engineering, wound healing	[70, 86]
Solvent casting	Dissolves BNC and components, casts into thin films or membranes	Flexible, homogenous thin structures	Films, membranes	Simple and cost-effective method	Limited to thin structures	Wound dressings, drug delivery patches	[33, 114]
Particulate leaching	Uses porogens (e.g., salt) for interconnected porosity	Highly porous, adjustable pore size	Porous 3D scaffolds	High porosity suitable for cell infiltration	Poor mechanical strength; porogens need removal	Bone and cartilage engineering	[51, 120, 142]
Injectable	Injectable hydrogels or suspensions for minimally invasive applications	Biocompatible, gel-like properties, in situ cross-linking	Hydrogels	Minimally invasive, adaptable for irregular tissue defects	Limited to soft tissues; requires sterile conditions	Cartilage repair, soft tissue engineering	[44, 119]
3D bioprinting	Layer-by-layer additive fabrication using bioinks	High precision, patient -specific designs	Complex 3D structures	Customization, integration with cells for bioactivity	Expensive equipment; slow process	Personalized tissue scaffolds	[40, 90]

mostly yields a layered or coated structure that provides flexibility in tailoring surface properties such as hydrophobicity or antibacterial activity [20].

Electrospinning

Electrospinning is a relatively versatile method for producing BNC-based nanofiber scaffolds by applying a high-voltage electric field to a polymer solution. [7]. As such, thin fibers in this instance are deposited onto a substrate to create an open-type scaffold with great surface areas and porosity; it thus finds great applications, for example, in drug delivery and cell adhesion [35, 98].

Freezing-drying

Freeze-drying is the extensively used technique to prepare porous BNC-based composites by freezing the material, followed by sublimating the water under vacuum [95]. The method yields scaffolds with interconnected porous networks critical for tissue engineering applications. The high porosity allows nutrient diffusion and infiltration of cells [49, 95].

Layer-by-Layer (LbL)

The Layer-by-Layer (LbL) assembling method provides a build-up of multilayered structures through alternating application of functional materials, such as polyelectrolytes or nanoparticles, on the BNC films [148]. Such a technique enables layer thickness and composition to be tuned with care, allowing films with tunable mechanical, optical, and bioactive properties [32].

Cross-linking

Cross-linking improvement the stability and mechanical properties of BNC composites by creating covalent or ionic bonds with different chemical agents, such as calcium ions or glutaraldehyde [70]. This process increases the internal cohesion of the composite, hence increasing its mechanical strength, elasticity, and durability [70, 86].

Solvent Casting

The solvent casting technique is an uncomplicated method for fabricating thin films or coatings by casting a solution containing BNC and other materials into molds and allowing them to dry. The obtained films are flat and homogeneous, which makes them suitable for flexible applications in wound dressings [114].

Particulate Leaching

Particulate leaching involves the creation of porous BNC scaffolds by embedding porogens into the BNC matrix, followed by their leaching out [142]. This technique gave rise to highly porous structures with interconnected networks that will allow infiltration of cells and transport of nutrients [120].

Injectable

Injectable systems based on BNC composites can provide a minimally invasive method for delivering biomaterials. Typically, these are hydrogel or nanoparticle-integrated formulations that undergo in situ solidification following injection and thereby allow the material to take the shape of tissue defects and release bioactive molecules or cells locally [44].

3D Bioprinting

3D bioprinting enables the fabrication of custom-shaped BNC-based scaffolds by layer-by-layer displacement of composite inks using a 3D printer. This way, it is possible to have accurate control over scaffold constructs and porosity to create application-specific designs [43] (Fig. 3).

5 Applications of BNC-Based Biomaterials Composite for Tissue Engineering

In recent years, BNC containing biomaterials composites have received much attention in tissue engineering [55]. Due to its essential properties, including high surface area, excellent biocompatibility and superior mechanical strength, BNC has been blended with other biomaterials to design composite scaffolds with superior functionalities [112]. BNC composites that can simulate the ECM are being developed in an attempt to provide the best conditions for cell proliferation and tissue regeneration [59]. The combination of BNC with other biomaterials such as collagen, gelatin, silk, alginate, chitosan, and hydroxyapatite creates a structure with engineering properties to meet the specific needs of various tissues, which can be applied to bone, cartilage, skin, blood vessels, and urethral tissues [55, 59, 112]. The ultimate goal of tissue engineering is to build scaffolding, which not only helps cells adhere, migrate, expand, and divide, but also enables cells to only act on tissues, improving the regeneration of damaged tissues [115]. BNC composites have been provide to possess excellent potential in this interest and their applications in different fields of tissue engineering are explicated in this section [113] and summarized as shown in Table 4.

5.1 Bone Tissue Engineering

Bone tissue engineering has been created to replace bone tissue with potential to treat defects caused by injuries, fractures, and other bone disorders such as osteoporosis [129]. The main goal is to establish scaffolds to support the adhesion, migration and proliferation of bone forming cells and improve the growth of new bone tissue [58]. BNC scaffolds for bone tissue engineering show distinct improvements, including

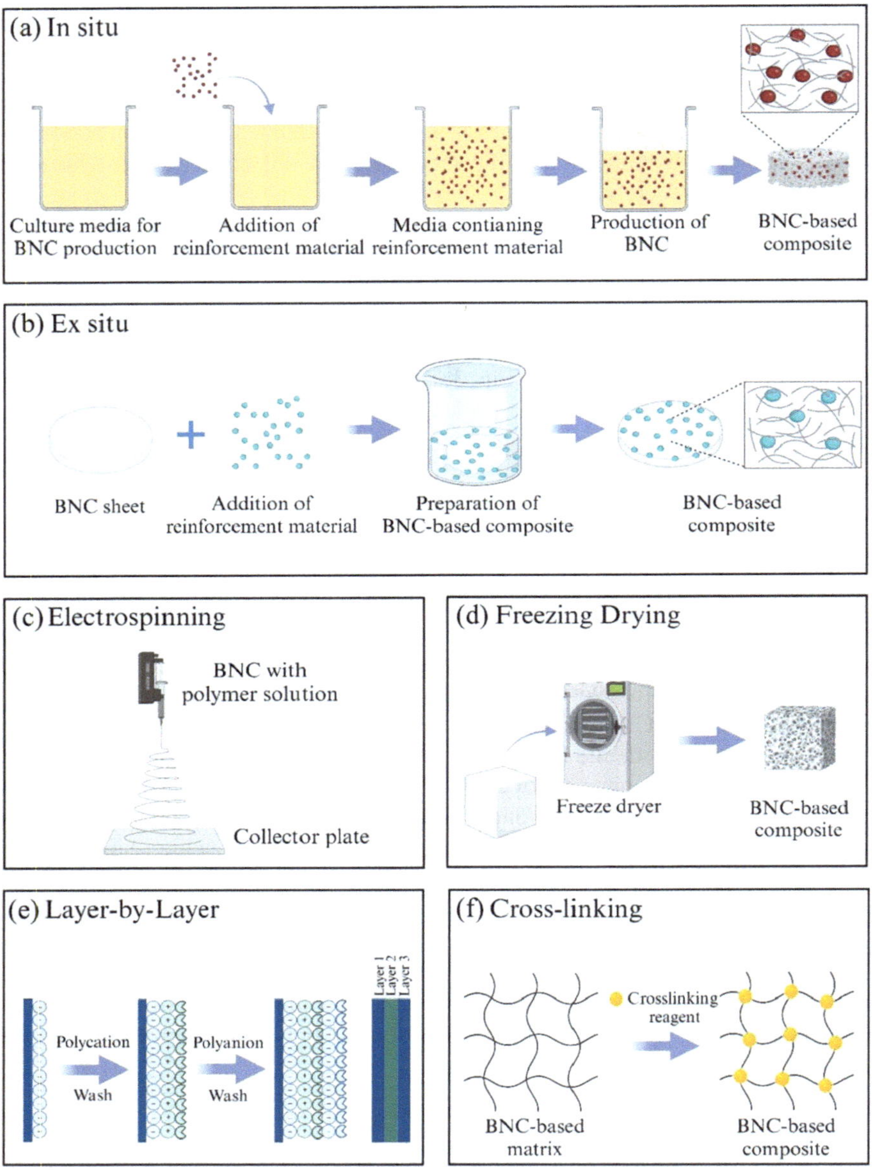

Fig. 3 Fabrication of BNC composites (created with BioRender.com; access date 1 December 2024)

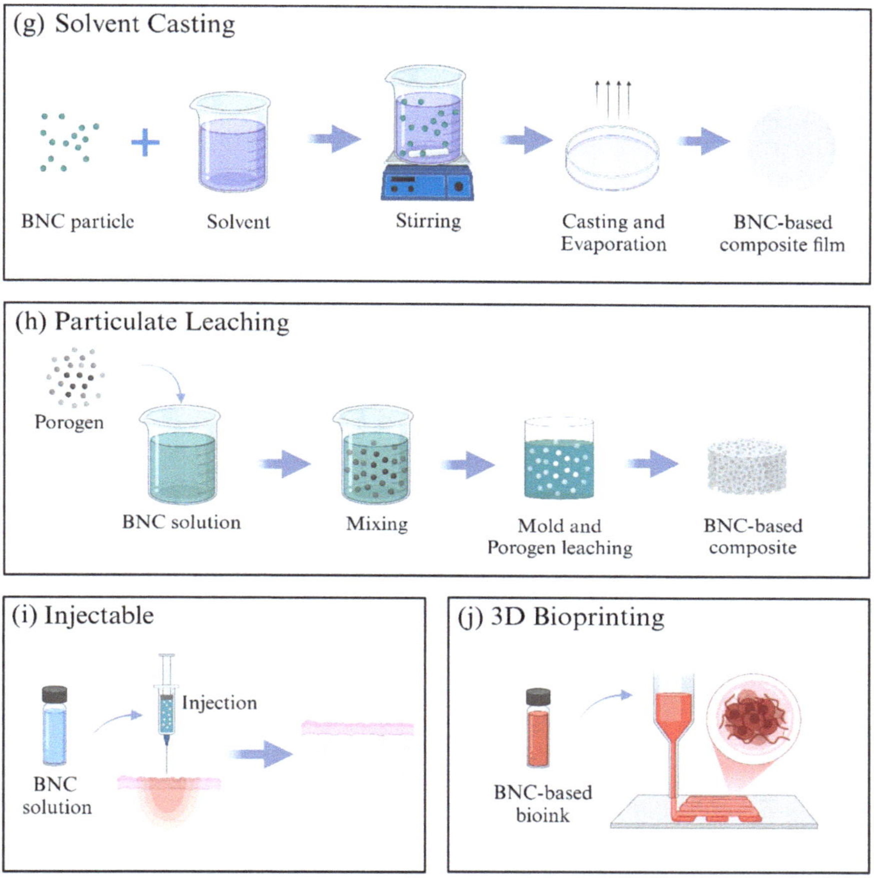

Fig. 3 (continued)

high surface area, porosity, and optimized mechanical properties, all of which are requirements for biomimetic architecture and natural bone function [34, 58]. Especially, BNC may be incorporated with inorganic materials, such as hydroxyapatite (HAp) or bioactive glass, to further improve the osteoconductive and osteoinductive ability of the scaffolds [151]. In bone tissue engineering, BNC scaffolds are commonly seeded with osteoblasts, osteocytes or mesenchymal stem cells (MSCs), which all possess the ability to differentiate into osteoblasts. MSCs may be derived from a number of sources including bone morrow, fat tissue or umbilical cord blood [146]. These cells are osteogenic by nature and can proliferate under the optimal culture conditions if supplemented with the right growth factor stimulation, such as bone morphogenetic protein-2 (BMP-2), a very potential osteogenic growth factor engineered to induce bone formation [83]. Also, BNC scaffolds could be modified by impregnation of osteoinductivity agents or chemical mineralization to enhance

Table 4 Applications of BNC composites for tissue engineering

Materials	Modification/fabrication techniques	Structure	Function/feature	Application	References
BNC with magnetic particles	In situ synthesis	Fibrous	High mechanical strength, biocompatibility	Tissue engineering	[124]
BC–silk	In situ synthesis	Film	Enhance mechanical properties, enhance cell adhesion	Wound healing	[18]
BC/HAp with gum tragacanth	In situ synthesis	Porous	High mechanical strength, and excellent osteoblast viability and proliferation	Tissue engineering	[104]
BC-Gelatin containing mangosteen extract	Solvent casting, evaporation and incorporation	Film	Good biocompatibility, improved antibacterial and anticancer	Wound healing and tissue engineering	[24]
BNC-chitosan-alginate-gelatin	Cross-linking	Hydrogel	Good compressive strength, Excellent biocompatibility, high potential for cartilage tissue engineering	Cartilage tissue engineering	[100]
BC/ethanolic extract of *Garcinia mangostana* peel	Ex situ, solvent casting and evaporation	Film	Enhanced biocompatibility, improved antimicrobial and anticancer properties	Wound healing and tissue engineering	[121]
BC/alginate/gelatin with curcumin	Solvent casting, evaporation and incorporation	Film	Antibacterial properties, biocompatibility, anticancer properties	Wound healing	[26]
BC	Immersion and freeze-drying	Porous	Promotes macrophage activity for tissue healing	Tissue engineering	[37]

(continued)

Table 4 (continued)

Materials	Modification/fabrication techniques	Structure	Function/feature	Application	References
BC conjugated with osteopontin	Freeze-drying	Film	Promotes calcification and periodontal ligament repair	Bone tissue engineering	[68]
Methacrylated gelatin/BNC	Immersion and cross-linking	Hydrogel	Maintains chondrocyte phenotype	Cartilage tissue engineering	[42]
BC/bladder acellular matrix	Chemical modification and Freeze-drying	Porous	Promotes angiogenesis and epithelialization	Urethral tissue engineering	[136]
HAp-BNC with cellulose nanocrystals	In situ synthesis	Porous	Promotes cell adhesion and osteogenic potential	Bone tissue engineering	[91]
BNC-modified lotus root starch and HAp	Particulate leaching	Porous	Articular cartilage engineering	Cartilage tissue engineering	[138]
BC reinforced calcium phosphate	Incorporation, blending, solvent casting and cross-linking	Porous	High mechanical strength for bone replacement	Bone tissue engineering	[146]
HAp–BC	In situ synthesis	Porous	Enhances osteoblast differentiation	Bone tissue engineering	[12]
BC-collagen	Freeze-drying	Porous	Promotes osteogenesis in mesenchymal stem cells	Bone tissue engineering	[93]
Inorganic calcium-filled BC	Solvent casting	Porous hydrogel	Promotes osteoblast differentiation	Bone tissue engineering	[11]
BC/polycaprolactone	3D bioprinting	Porous	Promotes cell proliferation (NIH/3T3 fibroblasts)	Tissue engineering	[5]

(continued)

Table 4 (continued)

Materials	Modification/fabrication techniques	Structure	Function/feature	Application	References
BC/keratin	Electrospinning	Fibrous	Supports wound healing and soft tissue repair	Wound dressing/tissue engineering	[7]
BC/alginate/gelatin	Blending, solvent casting and cross-linking	Film	Improved mechanical properties, enhanced biocompatibility	Wound healing and wound dressings	[25]
Biomimetic BC with HAp	Cross-linking	Hydrogel	Promotes osteochondral tissue repair	Bone tissue engineering	[151]
VEGF-loaded heparinized BC/gelatin	Freeze-drying	Porous	Supports endothelial cell growth	Vascular tissue engineering	[135]
Silk fibroin/BC	Freeze-drying	Fibrous	Supports keratinocytes and muscle cells	Urethral tissue engineering	[80]
Dialdehyde BNC with DL-allo-hydroxylysine and chitosan	Freeze-drying	Porous	Cartilage regeneration	Cartilage tissue engineering	[137]
BC loaded with fisetin	Freeze-drying	Porous	Promotes osteogenic differentiation	Bone tissue engineering	[129]
BC/silk fibroin/HAp	In situ synthesis	Porous	Supports osteoprogenitor cell growth	Bone tissue engineering	[58]
Silver nanoparticles/BC	In situ synthesis	Porous	Nontoxic and antimicrobial	Wound dressings	[56]
BC-based	Freeze-drying	Hydrogel	Supports keratinocyte and fibroblast proliferation	Wound healing	[75]
BC-poly(acrylic acid)	Immersion and freeze-drying	Porous	Antimicrobial properties for wound care	Wound dressings	[27]

(continued)

Table 4 (continued)

Materials	Modification/fabrication techniques	Structure	Function/feature	Application	References
BNC/gelatin loaded with VEGF-silk fibroin	Freeze-drying and immersion	Porous	Enhances angiogenesis for tissue regeneration	Vascular Tissue Engineering	[134]
Silk fibroin/BNC	Freeze-drying	Porous	Excellent mechanical properties and improve bone regeneration	Bone tissue engineering	[21]
BC/Gelatin/HAp	Freeze-drying	Porous	Enhances osteoinductive properties	Bone tissue engineering	[50]
BC/HAp	Immersion and freeze-drying	Porous	Enhance cell proliferation and differentiation	Bone tissue engineering	[105]
Gelatin-BC	Freeze-drying	Porous	Biocompatible and supports tissue growth	Tissue engineering	[65]
Polylactic acid (PLA)/ polyethylene glycol (PEG)/ bacterial cellulose nanofibers (BCNFs)	Blending, solvent casting and particulate leaching	Porous	Enhanced mechanical properties, biocompatibility, biodegradability	Tissue Engineering	[2]
BC/gelatin	Solvent casting, particulate leaching and freeze-drying	Porous	Supports cellular adhesion and proliferation, enhanced flexibility and moderate strength suitable	Tissue engineering and wound healing	[61]
BNC with ZnO	Solvent casting and ex situ	Film	Antibacterial properties, environmentally friendly	Wound healing	[57]
BC-alginate	Freeze-drying	Porous	Promotes cell proliferation	Wound healing and Tissue engineering	[63]

(continued)

Table 4 (continued)

Materials	Modification/fabrication techniques	Structure	Function/feature	Application	References
BC/silk fibroin	Immersion and freeze-drying	Porous	Enhance mechanical properties and cell viability	Wound healing and tissue engineering	[94]
HAp coated BC	Coating and freeze-drying	Porous	Osteoconductive; supports osteogenesis	Bone tissue engineering	[3]
Agarose particle in BNC templated	Particulate leaching	Porous	Promotes cartilage growth	Cartilage tissue engineering	[143]

further bone formation potential [83, 151]. Bone tissue engineering requires an environment maintaining appropriate mechanical load on the scaffold, mimicking the natural bone's load-bearing capacity. Besides, the porosity of the scaffold has to be designed in such a manner that it will allow the exchange of nutrients and vascularization for the maintenance of osteogenesis [12]. Vascularization becomes an important factor in larger bone defects because bone tissue requires a blood supply to deliver nutrients and remove waste products. Hence, an optimally designed microenvironment, which combines appropriate mechanical, chemical, and biological properties, is important for the successful regeneration of bone [104, 147].

5.2 *Cartilage Tissue Engineering*

Cartilage tissue engineering targets to replace damaged cartilage arising from osteoarthritis, injuries, or congenital anomalies. The challenge in cartilage repair is that the tissue has an inherently limited natural healing capacity [145]. Cartilage is avascular and therefore lacks significant cellularity [84]. BNC-based scaffolds have been shown to hold great promise in cartilage tissue engineering by providing a unique combination of structural support with the facilitation of chondrocytes-the cells responsible for cartilage formation-infiltration and proliferation [100]. Such hydrophilic characteristics of BNC provide excellent water retention, an important factor that cartilage tissue is made from a predominant composition of water and glycosaminoglycans [42, 47]. For cartilage tissue engineering, BNC scaffolds are generally produced with a fibrous porous matrix for cell attachment and migration [84]. BNC scaffolds could be functionalized with certain bioactive molecules, including bone morphogenetic proteins (BMPs) and transforming growth factor-beta (TGF-β), to induce osteoblast and chondrocyte differentiation and regulate ECM of cartilage [42, 138]. These bioactive molecules can control the biosynthesis of many ECMs, including proteoglycans and collagens, which are pivotal in cartilage structure integrity maintenance [74]. The cell source for cartilage tissue regeneration is generally derived from chondrocyte harvested through biopsies of healthy cartilage or chondrocytes derived from induced pluripotent stem cells (iPSCs) [97]. Induction of iPSCs is a promising substitute for autologous chondrocytes since it avoids donor site morbidity and provides an unlimited source of cells [97]. The culture medium for cartilage tissue engineering should be formulated to mimic the environment of native cartilage cells, with specific focus on low oxygen tension, high water content and bioactive molecules for cartilage cell activities [47].

5.3 *Skin Tissue Engineering*

Skin tissue engineering involves the restoration or replacement of burned, traumatized and chronically wounded skin through the utilization of scaffold biomaterials that will heal wounds and regenerate tissue [124]. BNC scaffolds have gained attention due to the outstanding properties of these biomaterials, such as good water retention capacity, good biocompatibility, and mechanical properties that mimic the dermis layer of dermis [25, 124]. BNC scaffolds present the formation of a perfect microenvironment for skin cell proliferation and migration, allowing for excellent regeneration in both the epidermis and dermis layers [37]. During wound healing, BNC-derived scaffolds confer a condition of retaining moisture, with resultants enhanced with minimal scar formation [144]. In addition, porous nature of BNC scaffolds offers the potential for transfers of nutrients and release of metabolic products, which facilitates cell proliferation and healing of tissue. Generally, skin cells such as keratinocytes and fibroblasts—the most common cell types involved in skin regeneration—are used for cell seeding onto biocompatible scaffolds [7]. The skin and stem cells are obtainable from patient's own tissue, hence being compatible and reducing immune rejection [37, 124]. The microenvironment of BNC scaffolds is highly crucial in repair and regenerating skin tissue [94]. Many growth factors, i.e., epidermal growth factor (EGF) and fibroblast growth factor (FGF), can be added to the scaffolds in order to encourage cell proliferation, migration and differentiation [124]. As dermis provides elasticity support to allow for skin regeneration, but in line with its comfort and flexibility [2, 61], the mechanical properties of scaffold, such as elasticity, should also be tuned to biomimetic properties of skin tissue.

5.4 *Vascular Tissue Engineering*

The field of vascular tissue engineering is focused on the evolution of blood vessels that are responsible for restoring blood flow to tissues that have been ischemic due to injury, disease, or surgery [6]. Restoration of the blood vessel network is critical to increasing tissue viability, mostly in large tissue-engineered constructions, such as bone, skin, and cartilage. Scaffolds of BNC origin have been used to create structures promoting angiogenesis-the evolution of new blood vessels from existing vessels [56]. High porosity and excellent biocompatibility of BNC enhance the infiltration of endothelial cells, which are instrumental in forming blood vessel networks [133]. The BNC scaffolds used in vascular tissue engineering can be combined with vascular endothelial cells, vascular smooth muscle cells and pericytes to promote the formation of vascular structures [6]. These cell components can be separated from human vascular endothelial cells, and are widely used in vascular tissue engineering because they can form capillary-like structures [10]. Moreover, stem cells or progenitor cells can further improve the process of vascular regeneration [134]. Various growth factors incorporated within the BNC scaffold, such as vascular endothelial growth factor (VEGF) and fibroblast growth factor (FGF), can enhance proliferation and migration of endothelial cells, thereby promoting the angiogenic process [56, 134]. In vascular tissue engineering, the environment has to support endothelial cell alignment, cell–cell junctions, and the construction of a feature lumen that allows blood to flow through [6]. The application of BNC scaffolds with controlled porosity and mechanical properties, together with growth factors and co-cultures of different cell types, creates an optimal microenvironment for the regeneration of vascular tissue [133]. 3D bioprinting techniques are also under investigation to fabricate BNC-based vascular scaffolds with a precise control over vascular network formation and spatial arrangement [133].

5.5 *Urethral Tissue Engineering*

Urothelial and urethral regeneration is critical to repair the defects resulting from trauma, congenital anomalies, or diseases of the urinary system. Urothelium is the epithelial lining of the urinary bladder and urethra, and its regeneration is important to maintain normal urinary function [80, 136]. BNC-based scaffolds have been explored in urothelial and urethral regeneration because of their biocompatibility, structural integrity, and capability to support the adhesion, migration and growth of urothelial cells [81].

Review of urethral and urothelial regeneration, BNC scaffolds can be seeded with urothelial cells or stem cells that can differentiate into urothelium [136]. These cell sources can be obtained from the patient's own tissues or from other reservoirs of stem cells. Use of BNC scaffolds with incorporation of growth factors, such as hepatocyte growth factor (HGF) and epidermal growth factor (EGF), which enhance the migration and growth of urothelial cells, secures achievement of the goal of tissue regeneration [81]. The environment of the scaffold should necessarily be moist, flexible, and mechanically supportive since the urothelium and urethra are subjected to dynamic mechanical forces in their physiological functioning [80, 136]. Moreover, BNC scaffolds can also be integrated with other biomaterials, such as elastomers, to enhance their flexibility and mechanical strength and therefore be applicable for urothelial regeneration [80, 81].

6 Future Perspectives and Challenges

BNC-based biomaterials have great potential in tissue engineering. However, there are many challenges to ensure that biomaterials can be more widely applied in the biomedical field. One of the main challenges is to expand the production scale of BNC. The fermentation methods developed until now often result in heterogeneous quality and quantity, hence restricted feasibility for industrial use. The development of new bioprocess optimization and use of low-cost substrates might feature to bypass these constraints. Moreover, while BNC has very good mechanical properties, its inherent brittleness and limited load-bearing capacity require enhancement through either composite fabrication or material functionalization. Biocompatibility is one of the essential properties of scaffolds, enhancing cell adhesion, migration, proliferation and differentiation. Application of advanced surface modification, incorporation of bioactive molecules and hybrid composite synthesis are a few optimistic approaches to enhanced cells interactions. Regulatory hurdles are especially important in creating a bottleneck in clinical translation of BNC-derived biomaterials. The establishment of standardized criteria regarding safety, efficacy and biodegradability entails large-scale preclinical and clinical investigations. Additionally, the high degree of economic investment needed for 3D bioprinting technologies, as well as the intricacies involved in generating constructs specific to individual patient requirements, represent additional challenges. Improvement in 4D bioprinting techniques and decreasing cost of equipment through technological advancements will be able to facilitate cheaper and more versatile scaffolds. Subsequently, multidisciplinary convergence is needed to address these challenges. Scientists, clinicians and industry collaborators will need to collaborate to optimize the manufacturing process, enhance biomaterials properties and develop regulatory routes for the general application of BNC in tissue engineering.

7 Conclusions

BNC composites are considered as an alternative revolution material in the field of tissue engineering because of their higher biocompatibility, superior mechanical properties and other customizable characteristics. The combination of the simulation ability of ECM with various manufacturing technologies such as biological 3D bioprinting makes it a great potential choice in various medical applications, such as wound healing, cartilage and skin tissue engineering. While these advantages are tremendously promising, there are several challenges that need to be overcome, such as manufacturability at large scale, improvement of mechanical properties and compliance with regulatory requirements. Advances in biomaterial engineering, process automation and interdisciplinary coordination should be helpful in order to overcome these challenges. In summary, BNC-based composite materials are considered innovative and environmentally friendly biomaterials for new strategies in tissue engineering. These composite materials have high potential to contribute to the transformation of tissue engineering and regenerative medicine to improve patient outcomes, especially when overcoming existing barriers and incorporating the latest technologies.

References

1. Abbasi N, Hamlet S, Love RM, Nguyen N-T (2020) Porous scaffolds for bone regeneration. J Sci Adv Mater Devices 5(1):1–9. https://doi.org/10.1016/j.jsamd.2020.01.007
2. Abu Ghalia M, Dahman Y (2017) Fabrication and enhanced mechanical properties of porous PLA/PEG copolymer reinforced with bacterial cellulose nanofibers for soft tissue engineering applications. Polym Testing 61:114–131. https://doi.org/10.1016/j.polymertesting.2017.05.016
3. Ahn S-J, Shin YM, Kim SE, Jeong SI, Jeong J-O, Park J-S, Gwon H-J, Seo DE, Nho Y-C, Kang SS (2015) Characterization of hydroxyapatite-coated bacterial cellulose scaffold for bone tissue engineering. Biotechnol Bioprocess Eng 20:948–955. https://doi.org/10.1007/s12257-015-0176-z
4. Akhlaghi MA, Bagherpour R, Kalhori H (2020) Application of bacterial nanocellulose fibers as reinforcement in cement composites. Constr Build Mater 241:118061. https://doi.org/10.1016/j.conbuildmat.2020.118061
5. Altun E, Ekren N, Kuruca SE, Gunduz O (2019) Cell studies on electrohydrodynamic (EHD)-3D-bioprinted bacterial cellulose\polycaprolactone scaffolds for tissue engineering. Mater Lett 234:163–167. https://doi.org/10.1016/j.matlet.2018.09.085
6. Arias SL, Shetty A, Devorkin J, Allain JP (2018) Magnetic targeting of smooth muscle cells in vitro using a magnetic bacterial cellulose to improve cell retention in tissue-engineering vascular grafts. Acta Biomater 77:172–181. https://doi.org/10.1016/j.actbio.2018.07.013
7. Azarniya A, Tamjid E, Eslahi N, Simchi A (2019) Modification of bacterial cellulose/keratin nanofibrous mats by a tragacanth gum-conjugated hydrogel for wound healing. Int J Biol Macromol 134:280–289. https://doi.org/10.1016/j.ijbiomac.2019.05.023
8. Bagewadi ZK, Dsouza V, Mulla SI, Deshpande SH, Muddapur UM, Yaraguppi DA, Reddy VD, Bhavikatti JS, More SS (2020) Structural and functional characterization of bacterial cellulose from Enterobacter hormaechei subsp. steigerwaltii strain ZKE7. Cellulose 27 (16):9181–9199. https://doi.org/10.1007/s10570-020-03412-2
9. Bao L, Li C, Tang M, Chen L, Hong FF (2022) Potential of a composite conduit with bacterial nanocellulose and fish gelatin for application as small-diameter artificial blood vessel. Polymers 14 (20). https://doi.org/10.3390/polym14204367
10. Bao L, Tang J, Hong FF, Lu X, Chen L (2020) Physicochemical properties and in vitro biocompatibility of three bacterial nanocellulose conduits for blood vessel applications. Carbohyd Polym 239:116246. https://doi.org/10.1016/j.carbpol.2020.116246
11. Basu P, Saha N, Saha P (2019) Inorganic calcium filled bacterial cellulose based hydrogel scaffold: novel biomaterial for bone tissue regeneration. Int J Polym Mater Polym Biomater 68(1–3):134–144. https://doi.org/10.1080/00914037.2018.1525733
12. Bayir E, Bilgi E, Hames EE, Sendemir A (2019) Production of hydroxyapatite–bacterial cellulose composite scaffolds with enhanced pore diameters for bone tissue engineering applications. Cellulose 26:9803–9817. https://doi.org/10.1007/s10570-019-02763-9
13. Blanco Parte FG, Santoso SP, Chou CC, Verma V, Wang HT, Ismadji S, Cheng KC (2020) Current progress on the production, modification, and applications of bacterial cellulose. Crit Rev Biotechnol 40(3):397–414. https://doi.org/10.1080/07388551.2020.1713721
14. Bodin A, Bharadwaj S, Wu S, Gatenholm P, Atala A, Zhang Y (2010) Tissue-engineered conduit using urine-derived stem cells seeded bacterial cellulose polymer in urinary reconstruction and diversion. Biomaterials 31(34):8889–8901. https://doi.org/10.1016/j.biomaterials.2010.07.108
15. Bouhlouli M, Pourhadi M, Karami F, Talebi Z, Ranjbari J, Khojasteh A (2021) Applications of bacterial cellulose as a natural polymer in tissue engineering. Asaio J 67(7):709–720. https://doi.org/10.1097/mat.0000000000001356
16. Buldum G, Mantalaris A (2021) Systematic understanding of recent developments in bacterial cellulose biosynthesis at genetic, bioprocess and product levels. Int J Mol Sci 22(13). https://doi.org/10.3390/ijms22137192

17. Cai Y, Chang SY, Gan SW, Ma S, Lu WF, Yen C-C (2022) Nanocomposite bioinks for 3D bioprinting. Acta Biomater 151:45–69. https://doi.org/10.1016/j.actbio.2022.08.014
18. Chaikhunsaeng J, Phatchayawat PP, Kirdponpattara S, Phisalaphong M (2024) Bacterial cellulose-silk hydrogel biosynthesized by using coconut skim milk as culture medium for biomedical applications. Gels 10(11):714. https://doi.org/10.3390/gels10110714
19. Chawla P, Bajaj IB, Survase SA, Singhal RS (2009) Microbial cellulose: fermentative production and applications. Food Technol Biotechnol 47:107–124
20. Chawla PR, Bajaj IB, Survase SA, Singhal RS (2009b) Microbial cellulose: fermentative production and applications. Food Technol Biotechnol 47(2)
21. Chen J, Zhuang A, Shao H, Hu X, Zhang Y (2017) Robust silk fibroin/bacterial cellulose nanoribbon composite scaffolds with radial lamellae and intercalation structure for bone regeneration. J Mater Chem B 5(20):3640–3650. https://doi.org/10.1039/C7TB00485K
22. Chen T, Yan Y, Liu W, Zhou X, Wei D, Feng Y, Zhang Y, Wang W, Zhang R, Qu D, Zhang H, Zhu Y, Cheng X, Wu G, Zhao Y (2025) An oxidative stress relieving hydrogel based on aminosilane-modified bacterial nanocellulose and Cordyceps militaris polysaccharides. Mater Today Commun 42:111289. https://doi.org/10.1016/j.mtcomm.2024.111289
23. Chen XY, Low HR, Loi XY, Merel L, Mohd Cairul Iqbal MA (2019) Fabrication and evaluation of bacterial nanocellulose/poly (acrylic acid)/graphene oxide composite hydrogel: characterizations and biocompatibility studies for wound dressing. J Biomed Mater Res B Appl Biomater 107(6):2140–2151. https://doi.org/10.1002/jbm.b.34309
24. Chiaoprakobkij N, Seetabhawang S, Okhawilai M, Uyama H, Phisalaphong M (2022) Multifunctional bacterial cellulose-gelatin containing mangosteen extract films with improved antibacterial and anticancer properties. Cellulose 29(12):6811–6830. https://doi.org/10.1007/s10570-022-04685-5
25. Chiaoprakobkij N, Seetabhawang S, Sanchavanakit N, Phisalaphong M (2019) Fabrication and characterization of novel bacterial cellulose/alginate/gelatin biocomposite film. J Biomater Sci Polym Ed 30(11):961–982. https://doi.org/10.1080/09205063.2019.1613292
26. Chiaoprakobkij N, Suwanmajo T, Sanchavanakit N, Phisalaphong M (2020) Curcumin-loaded bacterial cellulose/alginate/gelatin as a multifunctional biopolymer composite film. Molecules 25(17):3800. https://doi.org/10.3390/molecules25173800
27. Chuah C, Wang J, Tavakoli J, Tang Y (2018) Novel bacterial cellulose-poly (acrylic acid) hybrid hydrogels with controllable antimicrobial ability as dressings for chronic wounds. Polymers 10(12). https://doi.org/10.3390/polym10121323
28. Cielecka I, Szustak M, Kalinowska H, Gendaszewska-Darmach E, Ryngajłło M, Maniukiewicz W, Bielecki S (2019) Glycerol-plasticized bacterial nanocellulose-based composites with enhanced flexibility and liquid sorption capacity. Cellulose 26(9):5409–5426. https://doi.org/10.1007/s10570-019-02501-1
29. Cruz MA, Flor-Unda O, Avila A, Garcia MD, Cerda-Mejía L (2024) Advances in bacterial cellulose production: a scoping review. Coatings 14(11):1401. https://doi.org/10.3390/coatings14111401
30. Cytiva (2020) Use of sodium hydroxide for cleaning and sanitization of chromatography resins and systems. https://cdn.cytivalifesciences.com/api/public/content/digi-20986-original#:~:text=Sodium%20hydroxide%20has%20shown%20to,yeasts%2C%20fungi%2C%20and%20endotoxins. Accessed 6 July 2024
31. Czaja WK, Young DJ, Kawecki M, Brown RM Jr (2007) The future prospects of microbial cellulose in biomedical applications. Biomacromol 8(1):1–12. https://doi.org/10.1021/bm060620d
32. Decher G (1997) Fuzzy nanoassemblies: toward layered polymeric multicomposites. Science 277(5330):1232–1237
33. Dufresne A (2017) Nanocellulose: from nature to high performance tailored materials. Walter de Gruyter GmbH & Co KG. https://doi.org/10.1515/9783110254600
34. Dydak K, Junka A, Szymczyk P, Chodaczek G, Toporkiewicz M, Fijałkowski K, Dudek B, Bartoszewicz M (2018) Development and biological evaluation of Ti_6Al_7Nb scaffold implants coated with gentamycin-saturated bacterial cellulose biomaterial. PLoS ONE 13(10):e0205205. https://doi.org/10.1371/journal.pone.0205205

35. Flores-Rojas GG, Gómez-Lazaro B, López-Saucedo F, Vera-Graziano R, Bucio E, Mendizábal E (2023) Electrospun scaffolds for tissue engineering: a review. Macromol 3(3):524–553. https://doi.org/10.3390/macromol3030031
36. Fooladi S, Nematollahi MH, Rabiee N, Iravani S (2023) Bacterial cellulose-based materials: a perspective on cardiovascular tissue engineering applications. ACS Biomater Sci Eng 9(6):2949–2969. https://doi.org/10.1021/acsbiomaterials.3c00300
37. Frone AN, Panaitescu DM, Nicolae CA, Gabor AR, Trusca R, Casarica A, Stanescu PO, Baciu DD, Salageanu A (2020) Bacterial cellulose sponges obtained with green cross-linkers for tissue engineering. Mater Sci Eng, C 110:110740. https://doi.org/10.1016/j.msec.2020.110740
38. Gatenholm P, Klemm D (2010) Bacterial nanocellulose as a renewable material for biomedical applications. MRS Bull 35(3):208–213. https://doi.org/10.1557/mrs2010.653
39. Gorgieva S, Trček J (2019) Bacterial cellulose: production, modification and perspectives in biomedical applications. Nanomaterials (Basel) 9(10). https://doi.org/10.3390/nano9101352
40. Groll J, Boland T, Blunk T, Burdick JA, Cho D-W, Dalton PD, Derby B, Forgacs G, Li Q, Mironov VA (2016) Biofabrication: reappraising the definition of an evolving field. Biofabrication 8(1):013001. https://doi.org/10.1088/1758-5090/8/1/013001
41. Grunberg N, Mcmeeking A, Dieckmann E, Cheeseman C (2024) Development of printable bacterial nanocellulose bioinks for bioprinting applications. Biotechnol Sustain Mater 1(1):14. https://doi.org/10.1186/s44316-024-00015-w
42. Gu L, Li T, Song X, Yang X, Li S, Chen L, Liu P, Gong X, Chen C, Sun L (2020) Preparation and characterization of methacrylated gelatin/bacterial cellulose composite hydrogels for cartilage tissue engineering. Regen Biomater 7(2):195–202. https://doi.org/10.1093/rb/rbz050
43. Gullo M, La China S, Falcone PM, Giudici P (2018) Biotechnological production of cellulose by acetic acid bacteria: current state and perspectives. Appl Microbiol Biotechnol 102(16):6885–6898. https://doi.org/10.1007/s00253-018-9164-5
44. Guo WY, Wang WH, Xu PY, Kankala RK, Chen AZ (2024) Decellularised extracellular matrix-based injectable hydrogels for tissue engineering applications. Biomater Transl 5(2):114–128. https://doi.org/10.12336/biomatertransl.2024.02.003
45. Heidarian P, Kouzani AZ (2023) A self-healing nanocomposite double network bacterial nanocellulose/gelatin hydrogel for three dimensional printing. Carbohyd Polym 313:120879. https://doi.org/10.1016/j.carbpol.2023.120879
46. Hickey RJ, Pelling AE (2019) Cellulose biomaterials for tissue engineering. Front Bioeng Biotechnol 7:45. https://doi.org/10.3389/fbioe.2019.00045
47. Horbert V, Boettcher J, Foehr P, Kramer F, Udhardt U, Bungartz M, Brinkmann O, Burgkart RH, Klemm DO, Kinne RW (2019) Laser perforation and cell seeding improve bacterial nanocellulose as a potential cartilage implant in the in vitro cartilage punch model. Cellulose 26:647–664. https://doi.org/10.1007/s10570-019-02286-3
48. Hu G, Li G, Chen L, Hong FF (2023) Production of novel elastic bacterial nanocellulose/polyvinyl alcohol conduits via mercerization and phase separation for small-caliber vascular grafts application. Int J Biol Macromol 239:124221. https://doi.org/10.1016/j.ijbiomac.2023.124221
49. Huang Y, Wang J, Yang F, Shao Y, Zhang X, Dai K (2017) Modification and evaluation of micro-nano structured porous bacterial cellulose scaffold for bone tissue engineering. Mater Sci Eng C Mater Biol Appl 75:1034–1041. https://doi.org/10.1016/j.msec.2017.02.174
50. Huang Y, Wang J, Yang F, Shao Y, Zhang X, Dai K (2017) Modification and evaluation of micro-nano structured porous bacterial cellulose scaffold for bone tissue engineering. Mater Sci Eng, C 75:1034–1041. https://doi.org/10.1016/j.msec.2017.02.174
51. Hutmacher DW (2000) Scaffolds in tissue engineering bone and cartilage. Biomaterials 21(24):2529–2543. https://doi.org/10.1016/S0142-9612(00)00121-6
52. Iguchi M, Yamanaka S, Budhiono A (2000) Bacterial cellulose—a masterpiece of nature's arts. J Mater Sci 35(2):261–270. https://doi.org/10.1023/A:1004775229149
53. Iravani S, Varma RS (2022) Cellulose-based composites as scaffolds for tissue engineering: recent advances. Molecules 27(24). https://doi.org/10.3390/molecules27248830

54. Jagtap A, Krushnatray A, Dastager SG (2025) Bacterial nanocellulose (BNC) produced from sorbitol as a sustainable nano-filter for oil-water separation. Int J Biol Macromol 284:137953. https://doi.org/10.1016/j.ijbiomac.2024.137953
55. Jain P, Yu-Tong Lin R, Mishra K, Handral H, Dubey N (2024) Three-dimensional eco-friendly bacterial nanocellulose (BNC) scaffold for regenerative dentistry: characterization, cytocompatibility and differentiation potential. Dent Mater 40(1):151–157. https://doi.org/10.1016/j.dental.2023.11.001
56. Jalili Tabaii M, Emtiazi G (2018) Transparent nontoxic antibacterial wound dressing based on silver nano particle/bacterial cellulose nano composite synthesized in the presence of tripolyphosphate. J Drug Deliv Sci Technol 44:244–253. https://doi.org/10.1016/j.jddst.2017.12.019
57. Janpetch N, Saito N, Rujiravanit R (2016) Fabrication of bacterial cellulose-ZnO composite via solution plasma process for antibacterial applications. Carbohydr Polym 148:335–344. https://doi.org/10.1016/j.carbpol.2016.04.066
58. Jiang P, Ran J, Yan P, Zheng L, Shen X, Tong H (2018) Rational design of a high-strength bone scaffold platform based on in situ hybridization of bacterial cellulose/nano-hydroxyapatite framework and silk fibroin reinforcing phase. J Biomater Sci Polym Ed 29(2):107–124. https://doi.org/10.1080/09205063.2017.1403149
59. Jiji S, Maharajan K, Kadirvelu K (2022) CHAPTER 5 - Recent developments of bacterial nanocellulose porous scaffolds in biomedical applications. In: Oraon R, Rawtani D, Singh P, Hussain DCM (eds) Nanocellulose materials. Elsevier, pp 83–104. https://doi.org/10.1016/B978-0-12-823963-6.00002-8
60. Jonas R, Farah LF (1998) Production and application of microbial cellulose. Polym Degrad Stab 59(1):101–106. https://doi.org/10.1016/S0141-3910(97)00197-3
61. Khan S, Ul-Islam M, Ikram M, Ullah MW, Israr M, Subhan F, Kim Y, Jang JH, Yoon S, Park JK (2016) Three-dimensionally microporous and highly biocompatible bacterial cellulose–gelatin composite scaffolds for tissue engineering applications. RSC Adv 6(112):110840–110849. https://doi.org/10.1039/C6RA18847H
62. Kim H, Dutta SD, Randhawa A, Patil TV, Ganguly K, Acharya R, Lee J, Park H, Lim K-T (2024) Recent advances and biomedical application of 3D printed nanocellulose-based adhesive hydrogels: a review. Int J Biol Macromol 264:130732. https://doi.org/10.1016/j.ijbiomac.2024.130732
63. Kirdponpattara S, Khamkeaw A, Sanchavanakit N, Pavasant P, Phisalaphong M (2015) Structural modification and characterization of bacterial cellulose-alginate composite scaffolds for tissue engineering. Carbohydr Polym 132:146–155. https://doi.org/10.1016/j.carbpol.2015.06.059
64. Kirdponpattara S, Khamkeaw A, Sanchavanakit N, Pavasant P, Phisalaphong M (2015) Structural modification and characterization of bacterial cellulose–alginate composite scaffolds for tissue engineering. Carbohyd Polym 132:146–155. https://doi.org/10.1016/j.carbpol.2015.06.059
65. Kirdponpattara S, Phisalaphong M, Kongruang S (2017) Gelatin-bacterial cellulose composite sponges thermally cross-linked with glucose for tissue engineering applications. Carbohyd Polym 177:361–368. https://doi.org/10.1016/j.carbpol.2017.08.094
66. Klemm D, Heublein B, Fink HP, Bohn A (2005) Cellulose: fascinating biopolymer and sustainable raw material. Angew Chem Int Ed Engl 44(22):3358–3393. https://doi.org/10.1002/anie.200460587
67. Klemm D, Kramer F, Moritz S, Lindström T, Ankerfors M, Gray D, Dorris A (2011) Nanocelluloses: a new family of nature-based materials. Angew Chem Int Ed Engl 50(24):5438–5466. https://doi.org/10.1002/anie.201001273
68. Klinthoopthamrong N, Chaikiawkeaw D, Phoolcharoen W, Rattanapisit K, Kaewpungsup P, Pavasant P, Hoven VP (2020) Bacterial cellulose membrane conjugated with plant-derived osteopontin: preparation and its potential for bone tissue regeneration. Int J Biol Macromol 149:51–59. https://doi.org/10.1016/j.ijbiomac.2020.01.158

69. Kołaczkowska M, Siondalski P, Kowalik MM, Pęksa R, Długa A, Zając W, Dederko P, Kołodziejska I, Malinowska-Pańczyk E, Sinkiewicz I, Staroszczyk H, Śliwińska A, Stanisławska A, Szkodo M, Pałczyńska P, Jabłoński G, Borman A, Wilczek P (2019) Assessment of the usefulness of bacterial cellulose produced by Gluconacetobacter xylinus E25 as a new biological implant. Mater Sci Eng, C 97:302–312. https://doi.org/10.1016/j.msec.2018.12.016
70. Liang J, Wang R, Chen R (2019) The impact of cross-linking mode on the physical and antimicrobial properties of a chitosan/bacterial cellulose composite. Polymers 11(3). https://doi.org/10.3390/polym11030491
71. Liang S (2023) Advances in drug delivery applications of modified bacterial cellulose-based materials. Front Bioeng Biotechnol 11:1252706. https://doi.org/10.3389/fbioe.2023.1252706
72. Ligon SC, Liska R, Stampfl J, Gurr M, Mülhaupt R (2017) Polymers for 3D printing and customized additive manufacturing. Chem Rev 117(15):10212–10290. https://doi.org/10.1021/acs.chemrev.7b00074
73. Lin L, Chen L, Chen G, Lu C, Hong FF (2024) Effects of heterogeneous surface characteristics on hemocompatibility and cytocompatibility of bacterial nanocellulose. Carbohyd Polym 335:122063. https://doi.org/10.1016/j.carbpol.2024.122063
74. Lin L, Jiang S, Yang J, Qiu J, Jiao X, Yue X, Ke X, Yang G, Zhang L (2023) Application of 3D-bioprinted nanocellulose and cellulose derivative-based bio-inks in bone and cartilage tissue engineering. Int J Bioprint 9(1):637. https://doi.org/10.18063/ijb.v9i1.637
75. Loh EYX, Mohamad N, Fauzi MB, Ng MH, Ng SF, Mohd Amin MCI (2018) Development of a bacterial cellulose-based hydrogel cell carrier containing keratinocytes and fibroblasts for full-thickness wound healing. Sci Rep 8(1):2875. https://doi.org/10.1038/s41598-018-21174-7
76. Luo H, Li J, Zhou F (2018) Advances in hard tissue engineering materials—nanocellulose-based composites. Paper Biomater 3(4). https://doi.org/10.26599/PBM.2018.9260028
77. Luo L, Deng J, Zhang Q, Lan Y, Du G, Fan M, Zhao W (2025) Synergistically enhancing the efficient energy storage of B/N co-doped porous carbon with abundant micropores via microwave-induced strategy. J Alloy Compd 1010:177094. https://doi.org/10.1016/j.jallcom.2024.177094
78. Luo L, Zhou Y, Yan W, Wu X, Wang S, Zhao W (2020) Two-step synthesis of B and N co-doped porous carbon composites by microwave-assisted hydrothermal and pyrolysis process for supercapacitor application. Electrochim Acta 360:137010. https://doi.org/10.1016/j.electacta.2020.137010
79. Lv P, Yao Y, Li D, Zhou H, Naeem MA, Feng Q, Huang J, Cai Y, Wei Q (2017) Self-assembly of nitrogen-doped carbon dots anchored on bacterial cellulose and their application in iron ion detection. Carbohydr Polym 172:93–101. https://doi.org/10.1016/j.carbpol.2017.04.086
80. Lv X, Feng C, Liu Y, Peng X, Chen S, Xiao D, Wang H, Li Z, Xu Y, Lu M (2018) A smart bilayered scaffold supporting keratinocytes and muscle cells in micro/nano-scale for urethral reconstruction. Theranostics 8(11):3153–3163. https://doi.org/10.7150/thno.22080
81. Lv X, Yang J, Feng C, Li Z, Chen S, Xie M, Huang J, Li H, Wang H, Xu Y (2016) Bacterial cellulose-based biomimetic nanofibrous scaffold with muscle cells for hollow organ tissue engineering. ACS Biomater Sci Eng 2(1):19–29. https://doi.org/10.1021/acsbiomaterials.5b00259
82. Manzar R, Saeed M, Shahzad U, Al-Humaidi JY, ur Rehman S, Althomali RH, Rahman MM (2024) Retracted: recent advancements in boron carbon nitride (BNC) nanoscale materials for efficient supercapacitor performances. Prog Mater Sci 144:101286. https://doi.org/10.1016/j.pmatsci.2024.101286
83. Marshall KM, Wojciechowski JP, Echalier C, Callens SJ, Yang T, Øvrebø Ø, Zhou K, Jayawarna V, Kanczler JM, Stevens MM (2024) Using laponite to deliver BMP-2 for bone tissue engineering–in vitro, chorioallantoic membrane assay and murine subcutaneous model validation. Adv Mater Interfaces 11(29):2400332. https://doi.org/10.1002/admi.202400332
84. Martínez Ávila H, Feldmann E-M, Pleumeekers MM, Nimeskern L, Kuo W, de Jong WC, Schwarz S, Müller R, Hendriks J, Rotter N, van Osch GJVM, Stok KS, Gatenholm P (2015)

Novel bilayer bacterial nanocellulose scaffold supports neocartilage formation in vitro and in vivo. Biomaterials 44:122–133. https://doi.org/10.1016/j.biomaterials.2014.12.025
85. Matai I, Kaur G, Seyedsalehi A, McClinton A, Laurencin CT (2020) Progress in 3D bioprinting technology for tissue/organ regenerative engineering. Biomaterials 226:119536. https://doi.org/10.1016/j.biomaterials.2019.119536
86. Mbituyimana B, Liu L, Ye W, Ode Boni BO, Zhang K, Chen J, Thomas S, Vasilievich RV, Shi Z, Yang G (2021) Bacterial cellulose-based composites for biomedical and cosmetic applications: research progress and existing products. Carbohyd Polym 273:118565. https://doi.org/10.1016/j.carbpol.2021.118565
87. Mishra S, Singh PK, Pattnaik R, Kumar S, Ojha SK, Srichandan H, Parhi PK, Jyothi RK, Sarangi PK (2022) Biochemistry, synthesis, and applications of bacterial cellulose: a review. Front Bioeng Biotechnol 10:780409. https://doi.org/10.3389/fbioe.2022.780409
88. Mona S, Bajar S, Deepak B, Kiran B, Kaushik A (2019) Chapter 11 - Microbial cellulose: production and application. In: Grumezescu V, Grumezescu AM (eds) Materials for biomedical engineering. Elsevier, pp 309–322. https://doi.org/10.1016/B978-0-12-818415-8.00011-5
89. Moradi M, Jacek P, Farhangfar A, Guimarães JT, Forough M (2021) The role of genetic manipulation and in situ modifications on production of bacterial nanocellulose: a review. Int J Biol Macromol 183:635–650. https://doi.org/10.1016/j.ijbiomac.2021.04.173
90. Murphy SV, Atala A (2014) 3D bioprinting of tissues and organs. Nat Biotechnol 32(8):773–785. https://doi.org/10.1038/nbt.2958
91. Niamsap T, Lam NT, Sukyai P (2019) Production of hydroxyapatite-bacterial nanocellulose scaffold with assist of cellulose nanocrystals. Carbohyd Polym 205:159–166. https://doi.org/10.1016/j.carbpol.2018.10.034
92. Nicu R, Ciolacu F, Ciolacu DE (2021) Advanced Functional Materials Based on Nanocellulose for Pharmaceutical/Medical Applications. Pharmaceutics 13(8). https://doi.org/10.3390/pharmaceutics13081125
93. Noh YK, Dos Santos Da Costa A, Park YS, Du P, Kim IH, Park K (2019) Fabrication of bacterial cellulose-collagen composite scaffolds and their osteogenic effect on human mesenchymal stem cells. Carbohydr Polym 219:210–218. https://doi.org/10.1016/j.carbpol.2019.05.039
94. Oliveira Barud HG, Barud Hda S, Cavicchioli M, do Amaral TS, de Oliveira Junior OB, Santos DM, Petersen AL, Celes F, Borges VM, de Oliveira CI, de Oliveira PF, Furtado RA, Tavares DC, Ribeiro SJ (2015a) Preparation and characterization of a bacterial cellulose/silk fibroin sponge scaffold for tissue regeneration. Carbohydr Polym 128:41–51. https://doi.org/10.1016/j.carbpol.2015.04.007
95. Oliveira Barud HG, Barud HdS, Cavicchioli M, do Amaral TS, Junior OBdO, Santos DM, Petersen ALdOA, Celes F, Borges VM, de Oliveira CI, de Oliveira PF, Furtado RA, Tavares DC, Ribeiro SJL (2015b) Preparation and characterization of a bacterial cellulose/silk fibroin sponge scaffold for tissue regeneration. Carbohydrate Polymers 128:41–51. https://doi.org/10.1016/j.carbpol.2015.04.007
96. Osorio M, Ortiz I, Gañán P, Naranjo T, Zuluaga R, van Kooten TG, Castro C (2019) Novel surface modification of three-dimensional bacterial nanocellulose with cell-derived adhesion proteins for soft tissue engineering. Mater Sci Eng, C 100:697–705. https://doi.org/10.1016/j.msec.2019.03.045
97. Owaidah AY (2024) Induced pluripotent stem cells in cartilage tissue engineering: a literature review. Biosci Rep 44(5). https://doi.org/10.1042/BSR20232102
98. Parham S, Kharazi AZ, Bakhsheshi-Rad HR, Ghayour H, Ismail AF, Nur H, Berto F (2020) Electrospun nano-fibers for biomedical and tissue engineering applications: a comprehensive review. Materials (Basel) 13(9). https://doi.org/10.3390/ma13092153
99. Penttilä PA, Imai T, Capron M, Mizuno M, Amano Y, Schweins R, Sugiyama J (2018) Multimethod approach to understand the assembly of cellulose fibrils in the biosynthesis of bacterial cellulose. Cellulose 25(5):2771–2783. https://doi.org/10.1007/s10570-018-1755-x
100. Phatchayawat PP, Khamkeaw A, Yodmuang S, Phisalaphong M (2022) 3D bacterial cellulose-chitosan-alginate-gelatin hydrogel scaffold for cartilage tissue engineering. Biochem Eng J 184:108476. https://doi.org/10.1016/j.bej.2022.108476

101. Picheth GF, Pirich CL, Sierakowski MR, Woehl MA, Sakakibara CN, de Souza CF, Martin AA, da Silva R, de Freitas RA (2017) Bacterial cellulose in biomedical applications: a review. Int J Biol Macromol 104(Pt A):97–106. https://doi.org/10.1016/j.ijbiomac.2017.05.171
102. Popa L, Ghica MV, Tudoroiu E-E, Ionescu D-G, Dinu-Pîrvu C-E (2022) Bacterial cellulose—a remarkable polymer as a source for biomaterials tailoring. Materials 15(3). https://doi.org/10.3390/ma15031054
103. Portela R, Leal CR, Almeida PL, Sobral RG (2019) Bacterial cellulose: a versatile biopolymer for wound dressing applications. Microb Biotechnol 12(4):586–610. https://doi.org/10.1111/1751-7915.13392
104. Quan VM, Torgbo S, Kamonsutthipaijit N, Sukyai P (2024) In situ preparation of bacterial cellulose/hydroxyapatite scaffold in colloidal culture media containing gum tragacanth. Cellulose 31(3):1787–1800. https://doi.org/10.1007/s10570-023-05711-w
105. Ran J, Jiang P, Liu S, Sun G, Yan P, Shen X, Tong H (2017) Constructing multi-component organic/inorganic composite bacterial cellulose-gelatin/hydroxyapatite double-network scaffold platform for stem cell-mediated bone tissue engineering. Mater Sci Eng, C 78:130–140. https://doi.org/10.1016/j.msec.2017.04.062
106. Römling U, Galperin MY, Gomelsky M (2013) Cyclic di-GMP: the first 25 years of a universal bacterial second messenger. Microbiol Mol Biol Rev 77(1):1–52. https://doi.org/10.1128/MMBR.00043-12
107. Samyn P (2021) A platform for functionalization of cellulose, chitin/chitosan, alginate with polydopamine: a review on fundamentals and technical applications. Int J Biol Macromol 178:71–93. https://doi.org/10.1016/j.ijbiomac.2021.02.091
108. Samyn P, Meftahi A, Geravand SA, Heravi MEM, Najarzadeh H, Sabery MSK, Barhoum A (2023) Opportunities for bacterial nanocellulose in biomedical applications: review on biosynthesis, modification and challenges. Int J Biol Macromol 231:123316. https://doi.org/10.1016/j.ijbiomac.2023.123316
109. Sandhu S, Arpa A, Chen X, Kumar R, Jaworski J (2019) Engineering bacterial cellulose for diverse biomedical applications. Recent Prog Mater 01(04):006. https://doi.org/10.21926/rpm.1904006
110. Sandle T (2013) A comparative study of different methods for endotoxin destruction. https://www.americanpharmaceuticalreview.com/Featured-Articles/148858-A-Comparative-Study-of-Different-Methods-for-Endotoxin-Destruction/. Accessed 5 Aug 2024
111. Shah N, Ul-Islam M, Khattak WA, Park JK (2013) Overview of bacterial cellulose composites: a multipurpose advanced material. Carbohydr Polym 98(2):1585–1598. https://doi.org/10.1016/j.carbpol.2013.08.018
112. Sharma C, Bhardwaj NK (2019) Bacterial nanocellulose: present status, biomedical applications and future perspectives. Mater Sci Eng, C 104:109963. https://doi.org/10.1016/j.msec.2019.109963
113. Shen T, Dong H, Wang P (2024) Research progress on nanocellulose and its composite materials as orthopedic implant biomaterials. Alex Eng J 87:575–590. https://doi.org/10.1016/j.aej.2024.01.003
114. Shrivastav P, Pramanik S, Vaidya G, Abdelgawad MA, Ghoneim MM, Singh A, Abualsoud BM, Amaral LS, Abourehab MAS (2022) Bacterial cellulose as a potential biopolymer in biomedical applications: a state-of-the-art review. J Mater Chem B 10(17):3199–3241. https://doi.org/10.1039/d1tb02709c
115. Sousa RB, Dametto AC, de Mesquita GF, Ricardo PH, Helaehil JV, Medina-Carrasco S, Furtini JAO, Caetano GF, Barud HS, Silva-Filho EC (2024) Investigation of bacterial nanocellulose/calcium phosphates-based composite containing cerium for bone repair. Colloids Surf B Biointerfaces:114476. https://doi.org/10.1016/j.colsurfb.2024.114476
116. Stanisławska A, Szkodo M, Staroszczyk H, Dawidowska K, Kołaczkowska M, Siondalski P (2024) Effect of the ex situ physical and in situ chemical modification of bacterial nanocellulose on mechanical properties in the context of its potential applications in heart valve design. Int J Biol Macromol 269:131951. https://doi.org/10.1016/j.ijbiomac.2024.131951

117. Staňková L, Kutová A, Doubková M, Kvítek O, Vokatá B, Sedlář A, Idriss H, Slepička P, Švorčík V, Bačáková L (2024) Argon plasma-modified bacterial nanocellulose: cell-specific differences in the interaction with fibroblasts and endothelial cells. Carbohydr Polym Technol Appl 7:100470. https://doi.org/10.1016/j.carpta.2024.100470
118. Sulaeva I, Henniges U, Rosenau T, Potthast A (2015) Bacterial cellulose as a material for wound treatment: Properties and modifications: a review. Biotechnol Adv 33(8):1547–1571. https://doi.org/10.1016/j.biotechadv.2015.07.009
119. Sun Y, Nan D, Jin H, Qu X (2020) Recent advances of injectable hydrogels for drug delivery and tissue engineering applications. Polym Testing 81:106283. https://doi.org/10.1016/j.polymertesting.2019.106283
120. Tan Q, Li S, Ren J, Chen C (2011) Fabrication of porous scaffolds with a controllable microstructure and mechanical properties by porogen fusion technique. Int J Mol Sci 12(2):890–904. https://doi.org/10.3390/ijms12020890
121. Taokaew S, Chiaoprakobkij N, Siripong P, Sanchavanakit N, Pavasant P, Phisalaphong M (2021) Multifunctional cellulosic nanofiber film with enhanced antimicrobial and anticancer properties by incorporation of ethanolic extract of Garcinia mangostana peel. Mater Sci Eng, C 120:111783. https://doi.org/10.1016/j.msec.2020.111783
122. Thomas B, Raj MC, B AK, H RM, Joy J, Moores A, Drisko GL, Sanchez C (2018) Nanocellulose, a versatile green platform: from biosources to materials and their applications. Chem Rev 118(24):11575–11625. https://doi.org/10.1021/acs.chemrev.7b00627
123. Thongsomboon W, Serra DO, Possling A, Hadjineophytou C, Hengge R, Cegelski L (2018) Phosphoethanolamine cellulose: a naturally produced chemically modified cellulose. Science 359(6373):334–338. https://doi.org/10.1126/science.aao4096
124. Uchimura M, Tsumori F (2024) In-situ fabrication process of bacterial cellulose compositesfor soft robots. IEEJ Trans Electr Electron Eng 19(5):930–934. https://doi.org/10.1002/tee.24009
125. Ul-Islam M, Khan S, Ullah MW, Park JK (2019) Comparative study of plant and bacterial cellulose pellicles regenerated from dissolved states. Int J Biol Macromol 137:247–252. https://doi.org/10.1016/j.ijbiomac.2019.06.232
126. Ullah MW, Rojas OJ, McCarthy RR, Yang G (2021) Nanocellulose: a multipurpose advanced functional material. Front Bioeng Biotechnol 9:738779. https://doi.org/10.3389/978-2-88971-276-2
127. Utoiu E, Manoiu VS, Oprita EI, Craciunescu O (2024) Bacterial cellulose: a sustainable source for hydrogels and 3D-printed scaffolds for tissue engineering. Gels 10(6). https://doi.org/10.3390/gels10060387
128. Vadanan SV, Basu A, Lim S (2022) Bacterial cellulose production, functionalization, and development of hybrid materials using synthetic biology. Polym J 54(4):481–492. https://doi.org/10.1038/s41428-021-00606-8
129. Vadaye Kheiry E, Parivar K, Baharara J, Fazly Bazzaz BS, Iranbakhsh A (2018) The osteogenesis of bacterial cellulose scaffold loaded with fisetin. Iran J Basic Med Sci 21(9):965–971. https://doi.org/10.22038/ijbms.2018.25465.6296
130. Valla S, Coucheron DH, Fjaervik E, Kjosbakken J, Weinhouse H, Ross P, Amikam D, Benziman M (1989) Cloning of a gene involved in cellulose biosynthesis in Acetobacter xylinum: complementation of cellulose-negative mutants by the UDPG pyrophosphorylase structural gene. Mol Gen Genet 217(1):26–30. https://doi.org/10.1007/BF00330938
131. Wahid F, Hu X-H, Chu L-Q, Jia S-R, Xie Y-Y, Zhong C (2019) Development of bacterial cellulose/chitosan based semi-interpenetrating hydrogels with improved mechanical and antibacterial properties. Int J Biol Macromol 122:380–387. https://doi.org/10.1016/j.ijbiomac.2018.10.105
132. Walling B, Bharali P, Ramachandran D, Viswanathan K, Hazarika S, Dutta N, Mudoi P, Manivannan J, Manjunath Kamath S, Kumari S, Vishwakarma V, Sorhie V, Gogoi B, Acharjee SA, Alemtoshi (2023) In-situ biofabrication of bacterial nanocellulose (BNC)/graphene oxide (GO) nano-biocomposite and study of its cationic dyes adsorption properties. Int J Biol Macromol 251:126309. https://doi.org/10.1016/j.ijbiomac.2023.126309

133. Wang B, Lv X, Chen S, Li Z, Yao J, Peng X, Feng C, Xu Y, Wang H (2017) Bacterial cellulose/gelatin scaffold loaded with VEGF-silk fibroin nanoparticles for improving angiogenesis in tissue regeneration. Cellulose 24(11):5013–5024. https://doi.org/10.1007/s10570-017-1472-x
134. Wang B, Lv X, Chen S, Li Z, Yao J, Peng X, Feng C, Xu Y, Wang H (2017) Bacterial cellulose/gelatin scaffold loaded with VEGF-silk fibroin nanoparticles for improving angiogenesis in tissue regeneration. Cellulose 24:5013–5024. https://doi.org/10.1007/s10570-017-1472-x
135. Wang B, Lv X, Chen S, Li Z, Yao J, Peng X, Feng C, Xu Y, Wang H (2018) Use of heparinized bacterial cellulose based scaffold for improving angiogenesis in tissue regeneration. Carbohydr Polym 181:948–956. https://doi.org/10.1016/j.carbpol.2017.11.055
136. Wang B, Lv X, Li Z, Zhang M, Yao J, Sheng N, Lu M, Wang H, Chen S (2020) Urethra-inspired biomimetic scaffold: a therapeutic strategy to promote angiogenesis for urethral regeneration in a rabbit model. Acta Biomater 102:247–258. https://doi.org/10.1016/j.actbio.2019.11.026
137. Wang Y, Yuan X, Yu K, Meng H, Zheng Y, Peng J, Lu S, Liu X, Xie Y, Qiao K (2018) Fabrication of nanofibrous microcarriers mimicking extracellular matrix for functional microtissue formation and cartilage regeneration. Biomaterials 171:118–132. https://doi.org/10.1016/j.biomaterials.2018.04.033
138. Wu J, Yin N, Chen S, Weibel DB, Wang H (2019) Simultaneous 3D cell distribution and bioactivity enhancement of bacterial cellulose (BC) scaffold for articular cartilage tissue engineering. Cellulose 26:2513–2528. https://doi.org/10.1007/s10570-018-02240-9
139. Wu Z, Xie S, Kang Y, Shan X, Li Q, Cai Z (2021) Biocompatibility evaluation of a 3D-bioprinted alginate-GelMA-bacteria nanocellulose (BNC) scaffold laden with oriented-growth RSC96 cells. Mater Sci Eng, C 129:112393. https://doi.org/10.1016/j.msec.2021.112393
140. Xun X, Li Y, Zhu X, Zhang Q, Lu Y, Yang Z, Wan Y, Yao F, Deng X, Luo H (2021) Fabrication of robust, shape recoverable, macroporous bacterial cellulose scaffolds for cartilage tissue engineering. Macromol Biosci 21(11):2100167. https://doi.org/10.1002/mabi.202100167
141. Yang L, Zhu X, Chen Y, Wang J (2024) Enhanced bacterial cellulose production in Gluconacetobacter xylinus by overexpression of two genes (bscC and bcsD) and a modified static culture. Int J Biol Macromol 260:129552. https://doi.org/10.1016/j.ijbiomac.2024.129552
142. Yang Y, Zhao J, Zhao Y, Wen L, Yuan X, Fan Y (2008) Formation of porous PLGA scaffolds by a combining method of thermally induced phase separation and porogen leaching. J Appl Polym Sci 109(2):1232–1241. https://doi.org/10.1002/app.28147
143. Yin N, Stilwell MD, Santos TMA, Wang H, Weibel DB (2015) Agarose particle-templated porous bacterial cellulose and its application in cartilage growth in vitro. Acta Biomater 12:129–138. https://doi.org/10.1016/j.actbio.2014.10.019
144. Yu J, Huang TR, Lim ZH, Luo R, Pasula RR, Liao LD, Lim S, Chen CH (2016) Production of hollow bacterial cellulose microspheres using microfluidics to form an injectable porous scaffold for wound healing. Adv Healthc Mater 5(23):2983–2992. https://doi.org/10.1002/adhm.201600898
145. Zeng J, Jia L, Wang D, Chen Z, Liu W, Yang Q, Liu X, Jiang H (2023) Bacterial nanocellulose-reinforced gelatin methacryloyl hydrogel enhances biomechanical property and glycosaminoglycan content of 3D-bioprinted cartilage. Int J Bioprint 9(1):631. https://doi.org/10.18063/ijb.v9i1.631
146. Zhang Q, Lei Z, Peng M, Zhong M, Wan Y, Luo H (2019) Enhancement of mechanical and biological properties of calcium phosphate bone cement by incorporating bacterial cellulose. Mater Technol 34(13):800–806. https://doi.org/10.1080/10667857.2019.1630951
147. Zhang W, Wang X-c, Li X-y, Zhang L-l, Jiang F (2020) A 3D porous microsphere with multistage structure and component based on bacterial cellulose and collagen for bone tissue engineering. Carbohyd Polym 236:116043. https://doi.org/10.1016/j.carbpol.2020.116043
148. Zhang X, Chen H, Zhang H (2007) Layer-by-layer assembly: from conventional to unconventional methods. Chem Commun 14:1395–1405. https://doi.org/10.1002/9783527646746.ch3

149. Zheng L, Li S, Luo J, Wang X (2020) Latest advances on bacterial cellulose-based antibacterial materials as wound dressings. Front Bioeng Biotechnol 8:593768. https://doi.org/10.3389/fbioe.2020.593768
150. Zhong C (2020) Industrial-scale production and applications of bacterial cellulose. Front Bioeng Biotechnol 8:605374. https://doi.org/10.3389/fbioe.2020.605374
151. Zhu X, Chen T, Feng B, Weng J, Duan K, Wang J, Lu X (2018) Biomimetic bacterial cellulose-enhanced double-network hydrogel with excellent mechanical properties applied for the osteochondral defect repair. ACS Biomater Sci Eng 4(10):3534–3544. https://doi.org/10.1021/acsbiomaterials.8b00682

Phasuwit P. Phatchayawat received his B.Sc. in Biotechnology from Kasetsart University, Thailand, in 2018 and his s M. Eng. in Chemical Engineering from Chulalongkorn University, Thailand in 2021. Currently, he is pursuing a Ph.D. in Biomedical Engineering at Chulalongkorn University under the supervision of Prof. Muenduen Phisalaphong. His research focuses on utilizing Thailand's natural resources, such as agricultural biomass and natural biopolymers, to develop high value-added products through advanced biotechnology and biomedical engineering.

Professor Muenduen Phisalaphong has worked in the Department of Chemical Engineering, Faculty of Engineering, Chulalongkorn University since 1989. She received her Ph.D. (Chemical Engineering) from Colorado State University, USA in 1999. She has been working in Biochemical Engineering for more than 30 years. Currently, her research focuses on natural polymer processing and surface modification. Research interest includes technologies/processes for the development of biomass-derived composites for applications in biomedical areas, green catalysis, agricultural and environmental applications, adsorption, packaging and electronics. At present, she has published over 90 research articles.

Cellulose-Based Medical Devices

Sevgi Sözügeçer Çetintaş and Nursel Pekel Bayramgil

Abstract Cellulose is a polysaccharide obtained from generally plant sources. Most preferred cellulose type for medical devices is bacterial cellulose which produced by microorganisms offers higher mechanical strength and water retention capacity. Their provision of moisture to environment resulted from their high water retention capacity which also supports wound healing properties. In relation to this, their usage in areas such as wound dressings and tissue engineering has grown substantially. Additionally, cellulose is an environmentally friendly material due to its biodegradable properties. This feature offers a safe alternative in medical devices. Thanks to their biodegradable properties and biocompatibility, cellulose-based medical devices are cutting-edge area in modern medicine. These areas are such as tissue engineering, wound care, and drug delivery systems. Cellulose-based dressings support the rapid healing of wounds while the risk of infection is also reduced. High water retention capacity catalyzes healing with a moist environment. Cellulose plays a fundamental role in tissue engineering applications by offering structural support: scaffold for cell growth. Recent studies proceed to enhance the performance and safety of cellulose-based medical devices. Studies about bacterial celluloses are the proof of the high potential in medical applications. In addition, variety of studies have demonstrated that new compounds obtained by the modification of cellulose can enhance the performance of medical devices. Hence, cellulose-based medical devices offer both environmentally friendly and effectively suitable option for biomedical applications. Progress in this field is creating opportunities for more sustainable and impactful health solutions in the process of time.

Keywords Bacterial cellulose · Biocompatibility · Hydrogel scaffolds · Tissue engineering · Drug delivery systems · Wound healing · Biodegradability

S. S. Çetintaş · N. P. Bayramgil (✉)
Faculty of Science, Department of Chemistry, Hacettepe University, Beytepe, Ankara, Türkiye
e-mail: nursel@hacettepe.edu.tr

R. Malviya and S. Sundram (eds.), *Sustainable Cellulose Materials for Biomedical Applications*, Sustainable Materials and Technology,
https://doi.org/10.1007/978-981-96-9904-9_15

Abbreviations

ABTS	2,2′-Azino-bis(3-ethylbenzothiazoline-6-sulfonic acid)
BC	Bacterial cellulose
CMC	Carboxymethyl cellulose
CNC	Cellulose nano crystal
CNF	Cellulose nano fibril
EAPap	Electroactive paper
HPC	Hydroxypropyl cellulose
IDT	Interdigitated transducers
IoT	Internet of Things
MEMS	Micro-electro-mechanical systems
ORC	Oxidized regenerative cellulose
PET	Polyethylene terephthalate
PTFE	Polytetrafluoroethylene
SEM	Scanning electron microscopy

1 Introduction

Biocompatible biopolymers have arised as central focus in medical and related fields. A large number of researches have revealed novel systems which can mimic the natural tissues, hereby tailoring them to specific tissue requirements. Their widespread use in biomedical applications is derived from the resemblance of biopolymers to natural tissues. Concurrent advancements in molecular and cellular biology, coupled with polymer chemistry, have driven efforts to enhance the material properties, bioactivity and applicability of biopolymers [11, 82].

Cellulose is one of the most abundant natural polymer. Synthetic and natural are the two classifications of polymers, based on their origin. Cellulose falls under the category of natural polymers. Given its ability to be broken down by biological processes, it is also classified as a biopolymer [61]. When D-glucopyranose monomers linked together by β-1,4-glycosidic bonds and linear chains are formed; cellulose is produced. The structure of cellulose offers numerous hydroxyl groups and potential cross-linking sites [61, 87]. The abundance of intramolecular and intermolecular hydrogen bonds in cellulose molecules leads to a wide variety of structural configurations (Table 1).

Around 30 billion tons of natural biomass containing cellulose is produced yearly [2]. Cellulose, a polysaccharide abundantly found in bacteria and plants, is directly obtained from natural sources including wood, hemp, cotton or can be produced by microorganisms such as bacteria, fungi, or algae [69, 84]. In 1886, Brown's groundbreaking research explored the production of bacterial cellulose. Using glucose as a carbon source, he studied the biosynthesis of cellulose by *Acetobacter xylinum*, an acetic acid bacterium. This bacterium, currently classified as *Gluconacetobacter*

Table 1 Production methods, structural, physical, and mechanical properties of cellulose types [60]

Cellulose type	Production method	Structure	Physical and mechanical properties
Cellulose I	Natural form	Intra and intermolecular hydrogen bonds, parallel-(1,40)-D-glucopyranose chains, triclinic and monoclinic crystal structures	Highest Young's modulus (138 GPa), high stability
Cellulose II	Cellulose I alkali treatment	Hydrogen bonding, Cellulose I subunits are antiparallel	When compared to Cellulose I, lower stability
Cellulose III	Ammonia or amide treatment of Cellulose I and II	Intermolecular hydrogen bonds	Lowest Young's modulus (58 GPa), low stability
PC (Cotton cellulose)	Plant source	Cellulose I structure, hemicellulose residues	Less crystalline, lower water holding capacity
BC (Bacterial cellulose)	Bacterial production	Cellulose I structure, high purity	More crystalline, high water holding capacity, high porosity
CNC (Cellulose nanocrystal)	Obtained from plant cellulose by acid hydrolysis	High crystallinity, short rod-like, no amorphous regions	Highest Young's modulus (100–200 GPa), stiff and rigid
CNF (Cellulose nanofibril)	Cellulose structure subjected to mechanical and chemical processing	Contains both crystalline and amorphous regions, long fibrils	Lower Young's modulus (approximately 100 GPa), more flexible

xylinus, was observed to produce a three-dimensional network of cellulose fibrils when exposed to aerobic conditions. It is now considered the most effective producer of bacterial cellulose [25].

Acid hydrolysis, a type of chemical restructuring, maintains the crystalline nature of cellulose nanocrystals while eliminating amorphous cellulose regions. This restructuring involves chemical and enzymatic treatments that can completely degrade the microfibers within the cellulose structure [52]. Relative to cellulose nanocrystals (CNCs), cellulose nanofibrils (CNFs) possess longer and more flexible chains. This characteristic is due to the lesser degree of fragmentation of cellulose fibrils and the higher retention of amorphous regions during the production of CNFs. CNFs exhibit a more intricate structure comprising either amorphous or crystalline regions, whereas CNCs are primarily crystalline [27]. While CNCs and

CNFs necessitate destructive production methods, bacterial nanocellulose (BC) is biosynthesized by bacteria through processes that involve minimal processing [52]. Bacteria produce glucose chains internally and excrete them through pores in their cell membranes. These extruded glucose chains, upon reaching a diameter of 20–100 nm, self-assemble into microfibrils, thereby forming bacterial nanocellulose [51]. The formation of nanocrystals involves the unidirectional, parallel alignment of cellulose chains within fibrils. This process creates a strong hydroxyl functionality at one end and a reducing end at the other. Glucose units, in contrast, have numerous active hydroxyl (OH) groups. These groups facilitate hydrogen bonding between glucose chains and exhibit reactivity up to ten times greater than typical OH groups [36]. This property allows for the functionalization of nanocellulose to impart specific surface properties [1].

Because of its versatility, cellulose has found applications in various fields including environmental sustainability, catalysis, energy conversion, biosensing, membranes, bioelectronic devices, diagnostic devices, and contact lenses. Thanks to cellulose's unique properties, particularly its similarity to natural tissues, a wide array of innovative medical devices have been developed. These cellulose-based applications span numerous areas, including wound dressings, implants, drug delivery systems, vascular grafts, hemostatic agents, and tissue engineering (Fig. 1) [60, 87]. The subsequent sections will provide in-depth reviews of several key examples of these medical devices.

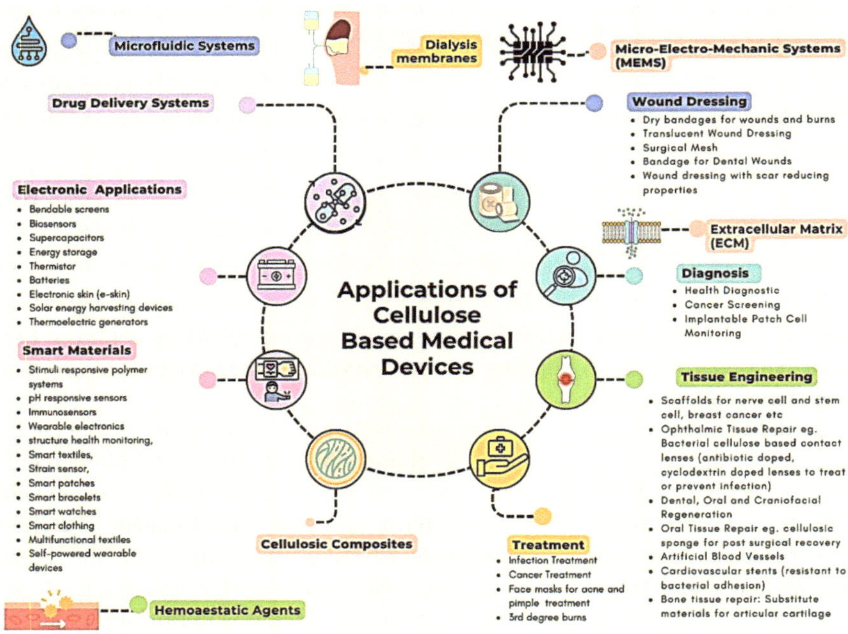

Fig. 1 Applications of cellulose-based medical devices

1.1 Flexible and Wearable Sensors

The biocompatible and biodegradable nature of cellulose makes it a suitable substrate for strain and pressure sensors. Such sensors can be fabricated in various forms, including hydrogels, aerogels, films, paper, and fabrics, and have potential applications in health monitoring, soft robotics, and human–machine interactions.

A biosensor is defined by IUPAC as a device incorporating an integrated receptor transducer, where the receptor component is a biological recognition element [46]. Biosensors rely on a bioreceptor (e.g., enzyme, antibody, living cell, or tissue) and a transducer. The transducer, a key component, detects signals from biological events and converts them into electrical signals for analysis. Cellulose is well-suited for biosensor fabrication due to its unique combination of properties: rapid response, transparency, high sensitivity and accuracy, dimensional stability, low thermal expansion, and affordability. Furthermore, its natural abundance, biodegradability, biocompatibility, and non-toxicity make it an attractive material.

Key challenges in biosensor development include achieving high biocatalytic activity, sensitivity, and selectivity, along with a fast response time, a low detection limit, and environmentally friendly, low-cost production. Bacterial cellulose (BC), with its nanoscale cellulose fibers, offers a promising solution. Its high surface area, porosity, water absorption, biocompatibility, and environmental friendliness enable the creation of innovative sensor materials. These properties can be further tailored and enhanced by incorporating a second phase, such as metal nanoparticles (e.g., Au, Pt, Pd), carbon-based materials (e.g., graphene, carbon nanotubes), or enzymes. Such combinations have been explored to create novel biosensor platforms and improve the electrical and magnetic conductivity of the native bacterial cellulose [81]. Bacterial cellulose (BC)-based biosensors find applications in various fields, including biomedical sensing for humidity, gas, and molecules like glucose and dopamine. Conductive polymer coatings can further improve the performance of cellulose-based actuators used in these biosensors. These cost-effective tools facilitate the detection and monitoring of key biomarkers, such as those associated with Parkinson's disease or α-amylase levels in bodily fluids. The portability, affordability, and disposability of cellulose paper biosensors make them promising alternatives for detecting numerous analytes, including DNA, hydrogen sulfate, and moisture. For instance, researchers have explored magnetite or unsulfated nanocrystalline thin films as oral and dermal glucose biosensors. Their peroxidase-like activity in the presence of 2,2′-azino-bis(3-ethylbenzothiazoline-6-sulfonic acid) (ABTS) and glucose has been demonstrated colorimetrically. These biosensors exhibit high sensitivity, detecting glucose concentrations as low as 5 mM, a level relevant to biological fluids. The catalytic activity of these nanocomposites is significantly enhanced in acidic environments and at ambient temperatures. Consequently, these hybrids can measure glucose levels in slightly acidic bodily fluids like sweat and saliva, even at the low concentrations found in blood. The utilization of paper-based biosensors offers a cost-effective strategy for healthcare applications [53].

For the detection of H_2O_2, dopamine, and glucose, electrochemical biosensors have been created using bacterial cellulose (BC) as a support matrix for immobilized enzymes (laccase, glucose oxidase (GOx)) and heme proteins (horseradish peroxidase) [80]. Nanoscale biomaterials have facilitated the creation of highly sensitive and rapid-response sensors. These advanced biosensors are employed to detect biomarkers associated with diseases like heart disease, cancer, and Alzheimer's. Bacterial cellulose (BC), with its highly porous, interconnected structure, low diffusion barrier, and excellent mechanical properties, provides an ideal platform for biomolecule immobilization and nanomaterial binding.

The biocompatibility, high storage capacity for cells and small molecules, and low interfacial tension at the gel-aqueous solution interface have made cellulose-based materials popular in smart device fabrication. Smart devices, like biochemical sensors, convert chemical information, such as the concentration of a specific sample component, into a comprehensive compositional analysis.

These stimuli-responsive polymer systems enable efficient transduction mechanisms, making them well-suited for sensor applications. Cellulose hydrogel films, for example, can function as highly sensitive, fast-responding pH nanosensors. Beyond biosensors, stimulus-responsive polymers find use in wearable electronics, imaging devices, and smart textiles. The electronic properties and flexibility of cellulose-based materials also enable the production of bendable screens and wearable electronic products [31].

Self-powered, removable skin-mounted electronics are capable of obtaining accurate and reliable biological/physical data, and are called electronic skins. These electronic skins are designed to be attached to any surface and removed without leaving any traces. Flexibility, conductivity, and self-healing properties of cellulose-based materials have been developed for use in electronic devices such as electronic skins [62]. Photonic skins, a novel sensing technology, offer a visual and intuitive way to monitor physical and mechanical stimuli through vibrant colors displayed directly on the target substrate. Despite their potential, their development lags behind that of electronic skins. These skins are constructed from a shape-memory, multi-stage cellulose derivative [83]. One example of a wearable photonic skin comprises a cholesteric hydroxypropyl cellulose photonic sensor bonded to an ultra-conformable amorphous hydroxypropyl cellulose adhesive layer. While the field of soft electronics and nanomaterials has rapidly advanced flexible, wearable, skin-like sensors capable of detecting crucial mechanical and physical signals from various substrates, including human skin, photonic skin development has been slower. Hydroxypropyl cellulose (HPC), a derivative of cellulose, holds significant promise for photonic skin applications due to its biodegradability, renewability, and biocompatibility [92]. Moreover, HPC's ability to spontaneously self-assemble into a chiral nematic liquid crystalline structure, which exhibits vivid colors in aqueous solutions, makes it particularly attractive. This self-assembled liquid crystal mesophase demonstrates stimuli-responsive optical behavior, leading to its use in a range of sensitive photonic devices, including sensors for heat, humidity, and strain.

1.2 Implants and Scaffolds

Bacterial cellulose (BC) is a remarkable biomaterial composed of a dense network of pure cellulose nanofibers produced by bacteria. Its inherent properties—exceptional strength, high water retention, and purity—make it highly biocompatible and well-suited for a range of medical applications.

Tissue engineering, a field that blends life sciences and engineering principles, aims to develop innovative biological solutions for repairing, maintaining, and enhancing damaged tissues and cells [47]. While significant progress has been made, particularly in the design of novel scaffolds, several challenges remain such as ensuring biocompatibility, minimizing cytotoxicity, achieving controlled biodegradability, promoting efficient cell adhesion, proliferation, facilitating nutrient diffusion, scaling up production for widespread use and successfully translating research findings into clinical applications.

Green chemistry and sustainable technologies presents a promising approach to scaffold fabrication which provides advantages such as cost-effectiveness, high biocompatibility, and inherent biodegradability or sustainability, accordingly, reduces toxicity and potential side effects. Cellulose-based scaffolds have showed up as an outstanding choice in tissue engineering due to their desirable mechanical properties, excellent biocompatibility and ability to enhance cell adhesion and proliferation. Cellulose-based scaffolds generally requires specific modifications or functionalization by using variety of polymers, proteins, and solvents in order to improve the catabolic properties and enhance the ability of absorb and bind to other molecules with the aim of optimizing their performance in living systems. Being the most abundant natural occurring biopolymers as cellulose makes it a great sustainable alternative to non-renewable polymer scaffolds. Moreover, the unique properties of cellulose make it an ideal biopolymer for creating intricate three-dimensional (3D) printed structures [30, 38].

Cellulose plays a vital role in tissue engineering by acting as a scaffold that mimics the extracellular matrix, which provides a framework for the growth and development of new tissues [42]. Hydrogel scaffolds have demonstrated significant potential for clinical applications in a wide range of tissue engineering endeavors, including cartilage, bone, muscle, skin, fat, arteries, ligaments, tendons, liver, bladder, and neural tissues. Furthermore, hydrogels can be effectively employed to modify the surface properties of other materials used in tissue growth [6]. The specific biomedical application dictates the desired surface characteristics of a biomaterial. In some cases, the goal is to encourage cell adhesion and the binding of specific proteins and cells, while in others, it may be necessary to discourage these interactions to achieve the desired biological outcomes [45].

Thermosensitive hydrogels offer a unique advantage in cell manipulation by simply lowering the temperature below their lower critical solution temperature that make hydrogels undergo a sol–gel transition and enable easy reshaping of cultured cells. This property ensures the development of injectable hydrogels based on chitosan and poly(*N*-isopropylacrylamide) (PNIPAAm). These injectable hydrogels have shown promise in evaluating chondrogenic differentiation of human mesenchymal stem cells and investigating cartilage formation in vivo through the injection of a cell-hydrogel complex [20].

Cellulose-based biomaterials have been shown to stimulate neuronal regeneration after spinal cord injury. Cellulose nanofibril scaffolds have been utilized in research to promote in vitro neural stem cell differentiation. In vivo, the tunable porosity of cellulose nanofibril scaffolds can facilitate the optimal release of growth factors into injured spinal cord regions. Cellulose-based scaffolds have been employed to deliver and release growth factors for controlled release into the microenvironment surrounding neural stem cells. This is beneficial for the heterogeneous neural differentiation of large stem cell populations and the repair of damaged neural tissue [38].

Cellulose-based materials have emerged as promising candidates for various ophthalmic applications. For instance, composites of bacterial cellulose and polyvinyl alcohol resembles so closely to the natural properties of the cornea. These materials exhibit high light transmittance and flexibility while maintaining a transparent structure with desirable mechanical and thermal characteristics. Moreover, bacterial cellulose-based contact lenses and its lens components can be incorporated with antibiotics, such as ciprofloxacin/γ-cyclodextrin, to prevent infections or treat existing bacterial infections. Research has demonstrated that bacterial biomaterials modified with chitosan and carboxymethyl cellulose to enhance their hydrophilicity can effectively promote the growth of retinal pigment epithelial cells. This finding holds significant promise for the development of novel therapies for various eye diseases, including age-related macular degeneration [58, 85].

Beyond ophthalmic applications, cellulose-based materials have significant potential for oral tissue repair and post-surgical healing. For example, a cellulosic "sponge" is a combination of bacterial cellulose with calcium chloride and sodium alginate which can stimulate the proliferation of gingival fibroblast cells [71]. Bacterial cellulose biomaterials have also demonstrated promising results in root canal surgeries. When used to fill root canals, these biomaterials effectively seal and fill the canal space completely, provides the dual benefit of sterilizing the canal and removing debris [13].

Cellulose-based materials are promising candidates for vascular repair and replacement. Bacterial cellulose, with its inherent capacity to support cellular growth, implements a compelling alternative to traditional synthetic grafts such as polyethylene terephthalate (PET) and polytetrafluoroethylene (PTFE). Particularly, bacterial cellulose-based materials can effectively reduce thrombin levels on their surfaces, mitigate the risk of clot formation which is a significant advantage over conventional grafts [29]. The further advantage in bacterial cellulose-derived composites is enhancing properties. For instance, composites incorporating bacterial cellulose nanocrystals with polyvinyl alcohol demonstrate superior tensile strength, negligible cytotoxicity, and improved suture retention, which makes them attractive candidates for replacing atherosclerotic blood vessels [71].

A critical challenge in implant technology is fostering robust cell-implant interactions. Hydrophilic bacterial cellulose-based biomaterials, modified with polyethylene glycol, have demonstrated enhanced cell adhesion and proliferation, while simultaneously mitigating immune system activation and resisting bacterial colonization [14]. This advancement holds significant promise for the development of novel vascular implants, including antibacterial stent coatings and replacement heart valves.

The mechanical properties of these biomaterials are the most significant for applications in bone tissue engineering and soft tissue repair. Their optimal compressive and tensile strength should effectively support the surrounding tissues [71]. The successful translation of cellulose-based materials into clinically viable tissue engineering scaffolds require very attentive consideration of various factors such as surface chemistry, material thickness, porosity and the mutual attachment of the cellulose nanostructures. Furthermore, cellulose based scaffolds should have such a mechanical characteristic that they could effectively mimic the mechanical properties of natural tissues. These mechanical properties must provide suitable surfaces for cell attachment and proliferation, compatible with cell types such as fibroblasts and osteoblasts. Cellulose derivatives have a wide diversity including hydroxyethyl cellulose, cellulose acetate, cellulose sulfate, methyl cellulose, carboxymethyl cellulose, hydroxypropyl cellulose and ethyl cellulose. They all offer a versatile platform for the advancement of tissue engineering scaffolds [17, 18].

Electrospinning has arisen as a versatile technique for fabricating polymer-based scaffolds including those produced for bone tissue regeneration. According to a recent study, electrospun cellulose nanofibers were used to create scaffolds incorporating nano-hydroxyapatite. While scanning electron microscopy (SEM) analysis revealed a positive correlation between nano-hydroxyapatite incorporated with the average diameter of the nanofibers, results also demonstrated a favorable size distribution closely resemblance to the natural extracellular matrix fibers of bone. Moreover, these electrospun nanofibers exhibited superior mechanical properties, with an elastic modulus of approximately 3.12 GPa and a tensile strength of approximately 70.6 MPa. In vitro biocompatibility assessments demonstrated excellent cell adhesion and proliferation and with the increasing amount of nano-hydroxyapatite the thermal stability of the scaffolds was observed to increased [4]. The nanofibrils structure of bacterial cellulose (BC) exhibits similarities to collagen, which is a primary component of the extracellular matrix in many tissues. Both cellulose and collagen have a mechanical support role within their respective biological contexts. Noticeably, both types exhibit similar dimensions and collagen undergoes extracellular assembly from precursor molecules into polymeric chains [2].

While the potential of BC as a cellular matrix, analogous to collagen, is currently under investigation, it offers several potential advantages. Notably, as a polysaccharide, BC exhibits reduced immunogenicity compared to collagen. A recent study utilizing a dorsal skin fold focal model demonstrated favorable tissue integration and robust extracellular matrix production surrounding a BC implant, further supporting its potential for biomedical applications [28, 68].

1.3 Delivery Systems

Cellulose-based matrices offer a promising platform for the controlled and sustained release of therapeutic agents. A well-designed drug delivery system aims to precisely deliver the requisite amount of medication to the intended target site or organ at the optimal temporal window. The mechanisms governing drug delivery to target cells, tissues, or organs can be modulated by a variety of environmental stimuli, encompassing temperature, light, chemical gradients, pH variations, and the application of electric and magnetic fields [22]. Cellulose, in diverse forms including gels, membranes, spheres, and crystals, has been extensively utilized as an excipient in pharmaceutical tablets to concentrate drug-loaded matrices for oral administration (Table 2). However, nanocellulose represents a relatively novel biomaterial within this domain [37].

Cellulose hydrogels, when employed as substrates in sustained release systems, exhibit a pronounced capacity for water absorption. This phenomenon leads to an increase in the interchain distance between cross-linked polymer chains, thereby facilitating the controlled release of encapsulated drugs into the surrounding tissues and bloodstream. Furthermore, cellulose demonstrates a remarkably slow degradation rate within the human body, rendering it well-suited for localized drug delivery and sustained release profiles. A noteworthy approach leverages the acid instability of Schiff bases formed between aminoglycosides and carboxymethyl cellulose. In this strategy, drug release is orchestrated by the acid produced by bacterial growth, resulting in the cleavage of the Schiff base bond and concomitant drug release. The liberated aminoglycoside antibiotics subsequently interact with bacterial ribosomes, inhibiting protein synthesis and ultimately leading to bacterial cell death. Aminoglycoside hydrogels have demonstrated several advantageous characteristics, including rapid gelation, inherent self-healing capabilities, on-demand drug release, and potent antibacterial activity against both Gram-positive and Gram-negative bacteria. This approach offers a promising and facile strategy for the development of hydrogels for localized, "smart" drug delivery applications. Host–guest interactions provide a versatile platform for modulating the properties of hydrogels to meet the specific demands of diverse medical applications under complex physiological conditions, such as those involving light induction and pH stimulation. The hydrogel undergoes considerable volume expansion and creates ample space for drug diffusion. Furthermore, charges arisen from the repulsive forces contribute to the relaxation of the hydrogel network, which helps drug release easier [48, 86].

Hybrid cellulose-based films exhibit unique attributions such as biodegradability, controlled release of integrated active ingredients and targeted delivery capabilities. These distinctive characteristics position them as promising candidates for transformative advancements in nano-delivery systems. The integration of organic cellulose with inorganic or organic nanomaterials, such as graphene, nanotubes, or nanoparticles, results in the formation of composite materials with significantly enhanced

Table 2 Examples of cellulose-based drug delivery systems: materials and drug delivery applications

Material	Drug delivery applications	Ref. no.
Periodate oxidized cellulose	Immobilization of insulin or p-aminosalicylic acid for sustained drug release	[9]
TEMPO-oxidized cellulose spheres	pH-sensitive sustained release systems for anti-inflammatory drugs	[89]
Amphoteric bacterial cellulose-acrylic acid-chitosan hydrogel	Naproxen delivery carrier, providing more than 24 h of drug release	[43]
Dialdehyde starch cross-linked nanofibril cellulose/gelatin composite gel	High surface area and porosity. Sustains drug release in the intestinal environment for 12 h	[86]
Zinc and copper oxide-doped carboxymethyl cellulose (CMC)-based hydrogel	Zinc and copper oxide-doped carboxymethyl cellulose hydrogel, offering enhanced drug delivery potential through structural modifications involving melamine units	[3]
Hydroxypropyl cellulose and poly(*N*-isopropylacrylamide) hydrogels	Temperature-sensitive polymeric structures used for drug delivery	[55]
Thiolated hydroxypropyl cellulose	Smart polymers for drug delivery systems: redox and temperature-sensitive hydrogels	[77]
Cellulose-based hydrogel derived from Phyllostachys heterocycla	pH/temperature-sensitive system for oral drug delivery	[95]
Poly(L-glutamic acid-co-2-hydroxyethyl methacrylate) and hydroxypropyl cellulose-acrylic acid hydrogel	pH/temperature-sensitive system for controlled release of oral insulin	[8]
IPN and semi-IPN hydrogels of HPC-PAA, PAAm-g-XG, methylcellulose, hydroxyethyl cellulose, and hydroxypropyl cellulose	Controlled release of dextran, ketoprofen, BSA and alaptid like proteins	[44]
Tri-stimulus-responsive hydrogels based on carboxymethyl cellulose and poly(N-isopropylacrylamide)	pH, temperature and redox stimuli for controlled release systems	[40]
Natural cellulose nanocrystals	Holds onto water-soluble drugs like tetracycline and doxorubicin and slowly release them over 24 h	[40]
Cetyltrimethylammonium bromide-modified cellulose nanocrystals	Hydrophobic anticancer drugs, providing 48-h release	[15]

functionality, thereby defining the "hybrid" nature of these systems [57]. The synergistic controlled release, targeted delivery capability, and biodegradability of hybrid films make them suitable for specialized applications of biomedical engineering, environmental sensing, and drug administration.

1.4 Wound Care

Nanocellulose exhibits exceptional biocompatibility and possesses remarkable mechanical strength. Its versatility extends to both dry and wet states, enabling it to mimic the structural and functional characteristics of human bone and the extracellular matrix, respectively. These unique attributes render nanocellulose an ideal candidate for the development of advanced wound care products. Recent advancements in cellulose-based wound dressing technologies have culminated in the development of highly absorbent, antibacterial, and biofilm-disrupting hydrogels. These innovative hydrogels are synthesized through the covalent bonding of ε-poly-L-lysine to carboxyl-modified cellulosic hydrogels, resulting in exceptional bactericidal activity while maintaining excellent compatibility with mammalian cells. This technological innovation provides a highly effective and promising solution for the management of acute and chronic wounds [79].

All injuries, ranging from minor cuts to complex surgical wounds, carry an inherent risk of bacterial infection and subsequent inflammation. The wound healing process can be significantly prolonged by the presence of infection. In immunocompromised individuals, microbial pathogens can initially induce tissue damage, which can subsequently disseminate systemically, leading to organ dysfunction and potentially life-threatening complications [5, 74]. The ideal wound dressing material should fulfill several critical criteria. These include rapid hemostasis, efficient absorption of wound exudate, and the creation of a moist environment conducive to cellular migration and proliferation. Furthermore, a wound dressing must minimize skin irritation in order to prevent secondary damage and provides the controlled delivery of therapeutic agents. A detailed assessment of wound dressing materials requires the evaluation of a wide diversity of performance characteristic such as mechanical properties, rheological behavior, injectable properties, self-healing properties, swelling capacity, porosity, water vapor permeability, biodegradability, adhesion, electrical conductivity, and drug release kinetics [59].

Cellulose-based materials provide physicochemical and biomechanical properties for the development of wound dressings. For instance, hydroxyethyl cellulose-based hydrogels have excellent mechanical flexibility and similarity as mimicking the structural characteristics of the human extracellular matrix. The three-dimensional structure of cellulose hydrogels provide an ideal environment for nutrient exchange, metabolic waste removal, and cellular proliferation, and accordingly efficient wound healing. Noticeably, cellulose hydrogel dressings can be conformed to irregularly shaped wounds, which make it possible to repair of damaged cellular networks without any external manipulation. This adaptability provides them to be particularly well-suited for the management of complex wound geometries [50, 66]. The significant capacity of bacterial nanocellulose to remarkably reduce wound healing times, from an extended period of 240–390 days to a considerably shorter duration of 80 days, has gathered significant attention within the wound care applications field [7]. Moreover, bacterial nanocellulose has minimal toxicity and can effectively stimulate the proliferation of various human cell types including human adipose-derived stem cells. In vivo studies have showed that wound regeneration has been accelerated and capillary formation in tissues are enhanced when treated with bacterial nanocellulose-based biomaterials.

One of the most prominent biomedical applications of bacterial cellulose (BC) lies in its utilization as a topical dressing for the management of severe wounds. Numerous studies have unequivocally demonstrated the superiority of BC over conventional wound dressings in several key aspects, including exudate retention, pain mitigation, accelerated re-epithelialization and overall healing times, reduced rates of wound infection, and diminished scar formation [96]. A significant advantage of BC lies in its inherent conformability, allowing it to seamlessly adapt to the contours of virtually any wound surface. Furthermore, BC serves as an ideal moisturizing agent, effectively maintaining an optimal water balance by dynamically absorbing or releasing fluid in response to the evolving needs of the wound. The unique combination of a highly interconnected network structure, exceptional water content, and favorable mechanical properties positions BC as a promising candidate not only for wound healing but also as a scaffold for the seeding of epithelial cells. In this context, BC can be effectively grafted as a semi-permanent artificial skin, facilitating epithelial healing even in cases of severe tissue damage. Autologous skin grafting currently represents the gold standard for the management of extensive skin defects. However, cell-coated BC-based skin substitutes offer a potentially superior alternative for patients with severe skin damage, minimizing the need for autologous tissue harvesting [26].

BC has also demonstrated significant potential in guided bone regeneration. In neurosurgical procedures, the implantation of barrier membranes that effectively exclude fibroblasts while simultaneously facilitating osteogenesis is crucial for preventing scar formation at the healing site. BC membranes have been successfully employed in this capacity. Another area of active research focuses on the potential of BC for the development of artificial heart valves. Millon and Wan [56], in their comprehensive investigation of BC composites, demonstrated that a biocompatible polyvinyl alcohol-bacterial cellulose composite exhibits mechanical stress–strain properties comparable to those of natural heart valve leaflets. However, this composite demonstrates more rapid relaxation compared to natural valves and exhibits lower residual stress.

1.5 Microfluidic Systems

Recent investigations have highlighted the significant potential of cellulose-based hydrogels for integration within microfluidic systems and advanced imaging platforms. Automated microfluidic systems, leveraging the inherent sensor-actuator properties of hydrogels, offer a unique set of functionalities that remain elusive in conventional systems or those employing alternative actuator materials. The fundamental principles underlying the application of hydrogels within microfluidic devices encompass the design and implementation of microvalves, micropumps, and hydraulic transistors, which constitute the essential building blocks of microfluidic circuitry [66].

Microfluidics has emerged as a transformative technology with immense potential for applications in biomedical analysis and tissue engineering. The successful development of biomicrofluidic systems necessitates the creation of a controlled microenvironment that effectively supports the culture of cellular populations. This necessitates the establishment of conditions that closely mimic the fluid flow dynamics and physiochemical properties of native tissues, thereby imposing unique demands and challenges on the selection and development of suitable chip materials [12].

Cellulose-based hydrogel microchips offer a compelling platform for the realization of sophisticated microfluidic systems. These materials exhibit exceptional capabilities for replicating intricate 2D and 3D microstructures with high fidelity. Furthermore, they possess desirable mechanical properties, exhibit excellent cytocompatibility, facilitating robust cell culture, and demonstrate remarkable dimensional stability even under elevated temperatures [34, 65].

1.6 Dialysis Membranes

Dialyzers filter blood using cellulose-based membranes because these membranes are biocompatible and have filtration capabilities. Extracorporeal immunoadsorption methods can greatly benefit from the creation of immunoadsorbents that are compatible with whole blood. These hemocompatible matrices can be used without the pre-procedure of separating blood cells and plasma, unlike traditional chromatographic media [24, 88]. Traditional hemodialyzers are ideal for this application.

An investigation into the feasibility of creating immunoaffinity supports from regenerated cellulose-based dialyzers was carried out by activating cyanogen bromide and incorporating putative immunochemical ligands, bovine serum albumin, or human immunoglobulin G. Based on the results, it is possible to immobilize 150–300 mg of immunoglobulin per square meter. A reliable carrier material for extracorporeal cleaning systems can be proposed in the form of modified hemodialyzers, which show slightly lower cleaning values and slightly higher ultrafiltration coefficients [33].

1.7 Bioelectronic Devices

Advances in cellulose functionalization have led to the development of new implantable medical devices and healthcare systems. Modern research has led to the creation of flexible supercapacitors with exceptional performance by combining various types of cellulose, including paper, fibers, films, hydrogels, and aerogels. As part of a study aimed at developing innovative ways to use cellulose fibers to make flexible supercapacitors, a metallic paper electrode was created by coating cellulose fibers with metallic nanoparticles such as gold nanoparticles and polyaniline. By coating this metallic paper multiple times, it has been able to create flexible supercapacitors with improved conductivity and structural stability [73].

Most batteries are made from cellulose-based conductive materials because they are very biocompatible, flexible, and have high mechanical strength. Conductive gels made from cellulose increase the electrochemical efficiency of lithium-ion and sodium-ion batteries by creating a stable and effective conductive network. Conductive materials derived from cellulose have excellent electrochemical properties and are versatile, making them excellent possibilities for next-generation battery technologies aimed at improving performance, cycle stability, and energy density. The growing need for green energy storage technologies aligns well with the sustainable sourcing of these materials [83].

1.8 Diagnostic Devices

Bacterial cellulose has emerged as a promising material in the treatment of various cancer types, driven by its exceptional fluid absorption capacity and its suitability for drug loading. In vitro studies employing bacterial cellulose-based scaffolds for cancer cell culture have demonstrated significant enhancements in cell adhesion, proliferation, and the induction of cell differentiation. These findings are particularly noteworthy in the context of breast and ovarian cancer cell lines, where bacterial cellulose scaffolds have been shown to effectively support normal cellular growth [39].

Comprising a network of microfibrils, cellulose-based paper offers a substantial surface area within its intricate microporous structure. This unique architecture enables the efficient storage of analytical reagents and facilitates the capillary-driven transport of specific analytes of interest from liquid samples, enabling rapid on-site detection. The combination of low production costs and ultra-high sensitivity positions these novel devices as a compelling alternative to conventional, sophisticated analytical instruments. These cellulose-based analytical platforms hold immense potential for transformative applications in critical care settings, point-of-care diagnostics, and large-scale early cancer screening programs [70].

1.9 Contact Lenses

Cellulose-based materials have demonstrated significant potential in various ophthalmic applications. Noticeably, bacterial cellulose/polyvinyl alcohol composites have been attracted attention by closely mimicking the structural and functional properties of the native cornea. These composites exhibit exceptional optical clarity, demonstrating high light transmittance and flexibility while maintaining a transparent structure with desirable mechanical and thermal characteristics. Moreover, bacterial cellulose-based contact lenses and their constituent components can be effectively integrated with antimicrobial agents, such as ciprofloxacin/γ-cyclodextrin, and this integrity makes it possible to prevent or treat bacterial infections. A research has shown that bacterial cellulose modificated with chitosan and carboxymethyl cellulose (CMC)-enhanced hydrophilicity and significantly promoted the growth and proliferation of retinal pigment epithelial cells. This result is an evidence for that these materials can be used for the development of novel therapeutic strategies for a range of ocular diseases, including age-related macular degeneration [49, 64].

In order to control the degree of crosslinking effectively within the hydrogel matrix, CMC is used as an alternative to conventional polymers and macromolecules commonly utilized in hydrogel lens fabrication. The inherent hydrophilicity of CMC eliminates the need for incorporation of another hydrophilic structure or apply surface coating in order to add hydrophilic surface. This is how CMC simplifies the manufacturing process and enhances overall performance of cellulosic lens biomaterial [63].

1.10 Micro-Electro-Mechanical Systems (MEMS)

The mechanism of actuation within cellulose electroactive paper (EAPap) is predicated upon the synergistic ion migration and piezoelectric phenomena. A pioneering study demonstrated the feasibility of EAPap actuation by employing microtransfer printing techniques to fabricate intricate microstructures, including rectifier antennas, digital-to-digital converters, dipole rectennas, and meticulously patterned rectangular configurations, onto regenerated cellulose films [75]. This research also comprehenses the microfabrication of a wide diversity of micropatterns and structures. Resistance measurements conducted on interdigitated transducers (IDT) patterned onto cellulose films across varying humidity levels revealed preliminary evidence of linear behavior. This innovative approach has the potential to generate cellulose-based smart devices offering the improvement of flexible microdevices such as acoustic wave sensors, humidity sensors, temperature sensors, and biosensors.

1.11 Extracellular Matrices

Cellulose-based medical devices have gathered significant attention thanks to their exceptional properties such as biocompatibility, biodegradability, and coherance to surface modification. Noticeably, bacterial cellulose (BC) can create a microenvironment that enables to cellular processes, facilitating cell adhesion, proliferation and differentiation. This behavior of BC is points to the inherent structural similarity between BC and the natural extracellular matrix (ECM). These unique properties of it makes BC an ideal biomaterial with diversity for biomedical applications including wound dressings, artificial skin grafts, cardiovascular implants, and tissue engineering. Nevertheless, the combination of exceptional mechanical strength, high moisture retention capacity, and the potential for integration with various bioceramics or polymers enable the utilization of cellulose-based composites in applications such as bone and cartilage tissue engineering. Since excellent integration of material makes it biomedically effective and sustainable, cellulose-based devices allow advanced medical device design and new class for regenerative medicine [41].

1.12 Cellulose-Based Composites

Cellulose-based composites stimulate microenvironment which enhances cellular growth, facilitate with their inherent biocompatibility and desirable mechanical properties. These composites have gathered significant attention within the biomedical field, finding widespread application in cell culture, tissue engineering and wound healing. This widespread possibility of applications can be related to their inherent biocompatibility, mechanical durability and modifiability. These structural characteristics of cellulose provide an ideal substrate for cell adhesion, proliferation, and differentiation, which is why it is a widely preferable material in cell culture applications. With the perspective of tissue engineering, cellulose is strategically combined with a wide range of biomaterials, including collagen, hydroxyapatite, and chitosan, in order to create artificial extracellular matrices that are effectively mimicking the microenvironment of various tissues, including bone, cartilage, and skin. Within the context of wound healing, cellulose-based composites can effectively reduce the infection risk by maintaining a moist at the healing environment and substantially accelerate the wound healing process. Moreover, the nanofibril structure of cellulose, when integrated with natural polymers, transfers inherent biodegradability and bioactivity, and enables composite materials to be designed as actively promote tissue repair. All these attributes position cellulose-based composites as an innovative solution by combining highly efficient and environmentally sustainable materials for the biomedical applications [19, 23, 35, 78].

1.13 Hemoaestatic Agents

Cellulose-based hemostatic agents are the materials used for the control of bleeding due to their capability to absorb fluids and promote clotting. Cellulose-based hemostatic agents offer highly effective and rapid control of bleeding. Efficiency of cellulosic materials as hemoaestatic agents is primarily related to their biocompatible, biodegradable, and hydrophilic properties. Chemically modified cellulose derivatives, especially oxidized regenerative cellulose (ORC), are widely used as hemostatic agents. ORC, when in contact with blood, creates an acidic environment, accelerating the activation of platelets and blood clotting. In addition, cellulose-based materials can provide an antimicrobial effect in the wound area while stopping bleeding and reducing the risk of infection. These agents are generally used in surgical procedures, traumatic injuries, and minimally invasive procedures [94].

Thanks to their tissue-compatible structures, they do not harm the surrounding tissues when placed on the wound and support the tissue healing process. Cellulose-based hemostatic agents can be applied in gauze, pad, or powder form. Thanks to their high absorbency properties, they absorb blood rapidly, promote clot formation, and keep the surgical area clean. At the same time, due to its biodegradable properties, it can be naturally disposed of by breaking down without leaving any residue in the body after surgical interventions [21].

2 New Trends Following Technological Developments

Cellulose emerges as a sustainable and environmentally benign material, characterized by its intrinsic biodegradability, biocompatibility, and abundant natural availability. At the nanoscale, cellulose manifests in distinct structural forms, including nanofibrils and nanocrystals, each exhibiting a unique constellation of properties. These properties encompass flexible surface chemistry, exceptional optical transparency, high elasticity, low thermal expansion coefficients, and inherent anisotropy [16, 90, 91]. The non-toxic nature of cellulose, coupled with its capacity for facile protein conjugation, confers upon it remarkable in vivo stability, rendering it a highly promising material for the development of advanced drug delivery systems. The versatility of cellulose is underscored by its diverse range of applications, spanning from electronics and tissue engineering to the broader realm of biomedicine [10, 17, 18, 32, 54, 67, 72, 76, 93]. The natural origin of cellulose facilitates the development of environmentally sustainable technologies, and its utilization across a wide spectrum of applications, encompassing electronics, energy storage, smart materials, and tissue engineering, has driven the development of cutting-edge technologies, such as electronic skin [93].

3 Expectations for the Future and Problems to Solve

The inherent insolubility of cellulose in aqueous media and its limited susceptibility to natural biodegradation within the human physiological environment constitute significant limitations that restrict its widespread application in biomedical contexts. In the context of lithium-ion battery technology, the utilization of cellulose nanofibril paper as a separator is constrained by restricted ionic transport, a consequence of the inherent porosity limitations imposed by the dense nanofibril network. To circumvent this limitation, the incorporation of cellulose within composite materials is deemed essential. Furthermore, the lack of a well-defined three-dimensional structure presents a significant obstacle to the effective utilization of cellulose in tissue engineering applications. Consequently, the integration of cellulose with other natural or synthetic polymers is frequently necessary. Additionally, the relatively low mechanical properties of pristine cellulose further restrict its applicability in certain biomedical domains. The biodegradation of nanocellulose biomaterials within the physiological environment represents another critical challenge for their successful implementation as scaffolds in tissue engineering applications.

While nanocellulose has been reported to exhibit minimal toxicity, ongoing toxicological investigations are imperative. Animal studies have revealed that inhalation of cellulose nanocrystals can induce pulmonary inflammation, attributed to the self-aggregation of nanocellulose particles and their inherently slow degradation rates. Oxidized cellulose derivatives demonstrate enhanced susceptibility to hydrolysis and, consequently, exhibit increased biodegradability within the human body. Although several studies have explored strategies to augment the biodegradability of non-bacterial cellulose through oxidation protocols, the paucity of in vivo studies necessitates further investigation in this critical area.

Chemical modification emerges as a pivotal strategy for enhancing the biocompatibility of cellulose, thereby expanding its potential for integration within diverse medical devices. Furthermore, chemical modification holds significant promise in advancing the research and development of cellulose-based materials for biomedical applications. The intricate interplay between cell-bacterial cellulose (BC) interactions necessitates further in-depth investigation, encompassing both in vitro and in vivo studies. The development of innovative techniques for the fabrication of BC structures beyond conventional sheet and tubular formats will be crucial for expanding the scope of BC applications. The precise control of porosity within various BC shapes will be paramount for facilitating efficient cell seeding and optimizing the utilization of BC as a versatile tissue engineering scaffold. Moreover, the ability to manipulate BC fiber orientation will unlock new avenues for enhancing the mechanical properties of BC-based materials. The establishment of greater control over cell-BC interactions through the precise manipulation of BC macro and microstructures will be of paramount importance for the successful translation of BC-based technologies into clinical applications. Future research endeavors should prioritize the development of BC-based biosensors for the sensitive detection of biomarkers associated with rare diseases. However, current methodologies for preserving the activity of biorecognition molecules, such as enzymes and proteins, remain limited. To facilitate the practical implementation of these biosensors, the reusability factor must be carefully considered. Future research should also focus on overcoming the current limitations associated with enzyme immobilization onto BC surfaces, thereby expanding the scope of potential applications for BC-based biosensors.

Another significant challenge lies in ensuring the compatibility of cellulose-based materials with a broader spectrum of pharmaceutical compounds. While the successful integration of cellulose-based materials with certain drugs has been demonstrated, their compatibility with other therapeutic agents, particularly protein-based therapeutics, necessitates further comprehensive investigation. The delivery of protein-based therapeutics often requires to keep specific conditions maintained to ensure their stability and biological activity. Additionally, if we take into consideration to the technology which is enduringly evolving, the integration of these materials is being with more sophisticated systems day by day, such as the Internet of Things (IoT), may become increasingly critical. However, security concerns regarding the internet traffic might block the full realization of the potential of IoT technologies within the healthcare domain, particularly within the context of drug delivery systems. If IoT-enabled drug delivery systems are utilized without any implementation of security measure, they carry inherent risks. In order to overcome these security challenges, it is essential to unlock the full potential of IoT technologies in medical drug delivery, ensuring their safe and effective utilization as valuable tools for enhancing patient care.

4 Conclusion

Progress in the cellulose-based medical devices is creating opportunities for more sustainable and impactful health solutions in the process of time. Although the immense potential for targeted drug delivery and a wide diversity of biomedical applications; substantial research and development studies are crucial in order to overcome the significant challenges with the emphasis on patient-specific therapeutic formulations. All these efforts underestimate the necessity of multidisciplinary approaches in particular with pharmacology, materials science, information technology, and artificial intelligence. Collaborative researches across these fields are crucial to define the potential of cellulose-based materials and address the challenges hindering their clinical translation.

References

1. Abushammala H, Mao J (2019) A review of the surface modification of cellulose and nanocellulose using aliphatic and aromatic mono-and di-isocyanates. Molecules 24:2782
2. Adhikari J, Dasgupta S, Barui A, Ghosh M, Saha P (2023) Collagen incorporated functionalized bacterial cellulose composite: a macromolecular approach for successful tissue engineering applications. Cellulose 30:9079–9111
3. Alsaaed FA, El-Lateef HMA, Khalaf MM, Mohamed IM, Al-Omair MA, Gouda M (2022) Drug delivery system based on carboxymethyl cellulose containing metal–organic framework and its evaluation for antibacterial activity. Polymers 14:3815
4. Ao C, Niu Y, Zhang X, He X, Zhang W, Lu C (2017) Fabrication and characterization of electrospun cellulose/nano-hydroxyapatite nanofibers for bone tissue engineering. Int J Biol Macromol 97:568–573
5. Asdaq SMB, Rajan A, Damodaran A, Kamath SR, Nair KS, Zachariah SM et al (2021) Identifying mucormycosis severity in Indian COVID-19 patients: a nano-based diagnosis and the necessity for critical therapeutic intervention. Antibiotics 10:1308
6. Ashammakhi N, GhavamiNejad A, Tutar R, Fricker A, Roy I, Chatzistavrou X et al (2022) Highlights on advancing frontiers in tissue engineering. Tissue Eng B Rev 28:633–664

7. Bacakova L, Pajorova J, Bacakova M, Skogberg A, Kallio P, Kolarova K, Svorcik V (2019) Versatile application of nanocellulose: from industry to skin tissue engineering and wound healing. Nanomaterials 9:164
8. Bai Y, Zhang Z, Zhang A, Chen L, He C, Zhuang X, Chen X (2012) Novel thermo-and pH-responsive hydroxypropyl cellulose-and poly (l-glutamic acid)-based microgels for oral insulin controlled release. Carbohydr Polym 89:1207–1214
9. Bala K, Guha SK, Vasudevan P (1982) P-Amino salicylic acid—oxidized cellulose system: a model for long term drug delivery. Biomaterials 3:97–100
10. Bar-Shai N, Sharabani-Yosef O, Zollmann M, Lesman A, Golberg A (2021) Seaweed cellulose scaffolds derived from green macroalgae for tissue engineering. Sci Rep 11:11843
11. Bharadwaz A, Jayasuriya AC (2020) Recent trends in the application of widely used natural and synthetic polymer nanocomposites in bone tissue regeneration. Mater Sci Eng C 110:110698
12. Bodas D (2025) Fundamentals of microfluidics. In: Microfluidics-aided technologies. Academic Press, New York, NY
13. Cañas-Gutiérrez A, Osorio M, Molina-Ramírez C, Arboleda-Toro D, Castro-Herazo C (2020) Bacterial cellulose: a biomaterial with high potential in dental and oral applications. Cellulose 27:9737–9754
14. Carvalho T, Guedes G, Sousa FL, Freire CS, Santos HA (2019) Latest advances on bacterial cellulose-based materials for wound healing, delivery systems, and tissue engineering. Biotechnol J 14:1900059
15. Casanova F, Pereira CF, Ribeiro AB, Castro PM, Freixo R, Martins E et al (2023) Biological potential and bioaccessibility of encapsulated curcumin into cetyltrimethylammonium bromide modified cellulose nanocrystals. Pharmaceuticals 16:1737
16. Chandel N, Jain K, Jain A, Raj T, Patel AK, Yang YH, Bhatia SK (2023) The versatile world of cellulose-based materials in healthcare: from production to applications. Indus Crops Prod 201:116929
17. Chen C, Xi Y, Weng Y (2022) Recent advances in cellulose-based hydrogels for tissue engineering applications. Polymers 14:3335
18. Chen C, Ding W, Zhang H, Zhang L, Huang Y, Fan M et al (2022) Bacterial cellulose-based biomaterials: from fabrication to application. Carbohydr Polym 278:118995
19. Chiaoprakobkij N, Sanchavanakit N, Subbalekha K, Pavasant P, Phisalaphong M (2011) Characterization and biocompatibility of bacterial cellulose/alginate composite sponges with human keratinocytes and gingival fibroblasts. Carbohydr Polym 85:548–553
20. Cho JH, Kim SH, Park KD, Jung MC, Yang WI, Han SW et al (2004) Chondrogenic differentiation of human mesenchymal stem cells using a thermosensitive poly (N-isopropylacrylamide) and water-soluble chitosan copolymer. Biomaterials 25:5743–5751
21. Cidreira ACM, de Castro KC, Hatami T, Linan LZ, Mei LHI (2021) Cellulose nanocrystals-based materials as hemostatic agents for wound dressings: a review. Biomed Microdev 23:1–23
22. Ciolacu DE, Nicu R, Ciolacu F (2020) Cellulose-based hydrogels as sustained drug-delivery systems. Materials 13:5270
23. da Silva R, Sierakowski MR, Bassani HP, Zawadzki SF, Pirich CL, Ono L, de Freitas RA (2016) Hydrophilicity improvement of mercerized bacterial cellulose films by polyethylene glycol graft. Int J Biol Macromol 86:599–605
24. Daniels CM (2007) Developing and characterization of bioadsorbent for extracorporeal blood purification. Doctoral dissertation, Northwestern University
25. de Oliveira Barud HG, Da Silva RR, da Silva BH, Tercjak A, Gutierrez J, Lustri WR et al (2016) A multipurpose natural and renewable polymer in medical applications: bacterial cellulose. Carbohydr Polym 153:406–420
26. Deng X, Gould M, Ali MA (2022) A review of current advancements for wound healing: biomaterial applications and medical devices. J Biomed Mater Res B Appl Biomat 110:2542–2573
27. Dufresne A (2013) Nanocellulose: a new ageless bionanomaterial. Mater Today 16:220–227
28. Esguerra M (2010) Engineering tissue–bacterial cellulose as a potential biomaterial. Institute of Clincial Sciences Department of Surgery

29. Fink H (2009), Artificial blood vessels—studies on endothelial cell and blood interactions with bacterial cellulose. Institute of Clincial Sciences Department of Surgery
30. Firmanda A, Syamsu K, Sari YW, Cabral J, Pletzer D, Mahadik B et al (2022) 3D printed cellulose based product applications. Mater Chem Front 6:254–279
31. Fu Q, Cui C, Meng L, Hao S, Dai R, Yang J (2021) Emerging cellulose-derived materials: a promising platform for the design of flexible wearable sensors toward health and environment monitoring. Mater Chem Front 5:2051–2091
32. Gil ES, Hudson SM (2004) Stimuli-reponsive polymers and their bioconjugates. Prog Polym Sci 29:1173–1222
33. Godfrey MJ (1998) Immunoaffinity extraction in veterinary residue analysis—a regulatory viewpoint. Analyst 123:2501–2506
34. Hajam MI, Khan MM (2024) Microfluidics: a concise review of the history, principles, design, applications, and future outlook. Biomater Sci 12:218–251
35. Hasan A, Waibhaw G, Saxena V, Pande LM (2018) Nano-biocomposite scaffolds of chitosan, carboxymethyl cellulose and silver nanoparticle modified cellulose nanowhiskers for bone tissue engineering applications. Int J Biol Macromol 111:923–934
36. Hebeish A, Guthrie TJ (2012) The chemistry and technology of cellulosic copolymers, vol 4. Springer Science & Business Media, Cham
37. Heise K, Kontturi E, Allahverdiyeva Y, Tammelin T, Linder MB et al (2021) Nanocellulose: recent fundamental advances and emerging biological and biomimicking applications. Adv Mater 33:2004349
38. Iravani S, Varma RS (2022) Cellulose-based composites as scaffolds for tissue engineering: recent advances. Molecules 27:8830
39. Islam SU, Ul-Islam M, Ahsan H, Ahmed MB, Shehzad A, Fatima A et al (2021) Potential applications of bacterial cellulose and its composites for cancer treatment. Int J Biol Macromol 168:301–309
40. Jackson JK, Letchford K, Wasserman BZ, Ye L, Hamad WY, Burt HM (2011) The use of nanocrystalline cellulose for the binding and controlled release of drugs. Int J Nanomed 6:321–330
41. Jafari A, Al-Ostaz A, Nouranian S (2024) Biomedical potential of cellulose: current trends and future directions. Polyme Adv Technol 35:e6621
42. Janmohammadi M, Nazemi Z, Salehi AOM, Seyfoori A, John JV, Nourbakhsh MS, Akbari M (2023) Cellulose-based composite scaffolds for bone tissue engineering and localized drug delivery. Bioactive Mater 20:137–163
43. Jiang K, Zhou X, He T (2022) The synthesis of bacterial cellulose-chitosan zwitterionic hydrogels with pH responsiveness for drug release mechanism of the naproxen. Int J Biol Macromol 209:814–824
44. Kabir SF, Sikdar PP, Haque B, Bhuiyan MR, Ali A, Islam MN (2018) Cellulose-based hydrogel materials: chemistry, properties and their prospective applications. Prog Biomat 7:153–174
45. Karimi F, O'Connor AJ, Qiao GG, Heath DE (2018) Integrin clustering matters: a review of biomaterials functionalized with multivalent integrin-binding ligands to improve cell adhesion, migration, differentiation, angiogenesis, and biomedical device integration. Adv Healthcare Mater 7:1701324
46. Karunakaran C, Rajkumar R, Bhargava K (2015) Introduction to biosensors. In: Biosensors and bioelectronics. Elsevier, Amsterdam
47. Lanza R, Langer R, Vacanti JP, Atala A (eds) (2020) Principles of tissue engineering. Academic Press, New York, NY
48. Li X, Jiang G, Wang G, Zhou J, Zhang Y, Zhao D (2024) Promising cellulose-based functional gels for advanced biomedical applications: a review. Int J Biol Macromol 260:129600
49. Li Y, Huang C, Yang X, Zhang X (2020) Ofloxacin laden microemulsion contact lens to treat conjunctivitis. J Biomat Sci Polym Ed 31:1566–1579
50. Liang Y, He J, Guo B (2021) Functional hydrogels as wound dressing to enhance wound healing. ACS Nano 15:12687–12722

51. Lin N, Dufresne A (2014) Nanocellulose in biomedicine: current status and future prospect. Eur Polym J 59:302–325
52. Lin SP, LoiraCalvar I, Catchmark JM, Liu JR, Demirci A, Cheng KC (2013) Biosynthesis, production and applications of bacterial cellulose. Cellulose 20:2191–2219
53. Liu X, Huang D, Lai C, Qin L, Zeng G, Xu P et al (2019) Peroxidase-like activity of smart nanomaterials and their advanced application in colorimetric glucose biosensors. Small 15:1900133
54. Marler JJ, Upton J, Langer R, Vacanti JP (1998) Transplantation of cells in matrices for tissue regeneration. Adv Drug Delivery Rev 33:165–182
55. Marsano E, Bianchi E, Viscardi A (2004) Stimuli responsive gels based on interpenetrating network of hydroxy propylcellulose and poly (*N*-isopropylacrylamide). Polymer 45:157–163
56. Millon LE, Wan WK (2006) The polyvinyl alcohol–bacterial cellulose system as a new nanocomposite for biomedical applications. J Biomed Mat Res B Appl Biomat 79:245–253
57. Mishra T, Mandal P, Rout AK, Sahoo D (2022) A state-of-the-art review on potential applications of natural fiber-reinforced polymer composite filled with inorganic nanoparticle. Composites C Open Access 9:100298
58. Mohammadi S, Jabbari F, Babaeipour V (2023) Bacterial cellulose-based composites as vehicles for dermal and transdermal drug delivery: a review. Int J Biol Macromol 242:124955
59. Naghib SM, Amiri S, Mozafari MR (2024) Stimuli-responsive chitosan-based nanocarriers for drug delivery in wound dressing applications: a review. Carbohydr Polym Technol Appl 7:100497
60. Nishino T, Takano K, Nakamae K (1995) Elastic modulus of the crystalline regions of cellulose polymorphs. J Polym Sci B Polym Phys 33:1647–1651
61. Oprea M, Voicu SI (2020) Recent advances in composites based on cellulose derivatives for biomedical applications. Carbohydr Polym 247:116683
62. Pang B, Jiang G, Zhou J, Zhu Y, Cheng W, Zhao D et al (2021) Molecular-scale design of cellulose-based functional materials for flexible electronic devices. Adv Elect Mater 7:2000944
63. Paugh JR, Marsden HJ, Edrington TB, Deland PN, Simmons PA, Vehige JG (2007) A pre-application drop containing carboxymethylcellulose can reduce multipurpose solution-induced corneal staining. Optom Vision Sci 84:65–71
64. Picheth GF, Pirich CL, Sierakowski MR, Woehl MA, Sakakibara CN, de Souza CF et al (2017) Bacterial cellulose in biomedical applications: a review. Int J Biol Macromol 104:97–106
65. Pei Y, Wang X, Huang W, Liu P, Zhang L (2013) Cellulose-based hydrogels with excellent microstructural replication ability and cytocompatibility for microfluidic devices. Cellulose 20:1897–1909
66. Pita-Vilar M, Concheiro A, Alvarez-Lorenzo C, Diaz-Gomez L (2023) Recent advances in 3D printed cellulose-based wound dressings: a review on in vitro and in vivo achievements. Carbohydr Polym 321:121298
67. Radhakrishnan A, Jose GM, Kurup M (2015) PEG-penetrated chitosan–alginate co-polysaccharide-based partially and fully cross-linked hydrogels as ECM mimic for tissue engineering applications. Prog Biomat 4:101–112
68. Rajwade JM, Paknikar KM, Kumbhar JV (2015) Applications of bacterial cellulose and its composites in biomedicine. Appl Microbio Biotech 99:2491–2511
69. Ramezani MG, Golchinfar B (2019) Mechanical properties of cellulose nanocrystal (CNC) bundles: coarse-grained molecular dynamic simulation. J Comp Sci 3:57
70. Ratajczak K, Stobiecka M (2020) High-performance modified cellulose paper-based biosensors for medical diagnostics and early cancer screening: a concise review. Carbohydr Polym 229:115463
71. Raut MP, Asare E, Syed Mohamed SMD, Amadi EN, Roy I (2023) Bacterial cellulose-based blends and composites: versatile biomaterials for tissue engineering applications. Int J Mol Sci 24:986
72. Silvia C, Boffito M, Sartori S (2017) Tissue engineering approaches in the design of healthy and pathological in vitro tissue models. Front Bioeng Biotechnol 5:40

73. Stejskal J, Trchová M, Kovářová J, Prokeš J, Omastová M (2008) Polyaniline-coated cellulose fibers decorated with silver nanoparticles. Chem Pap 62:181–186
74. Stjärne Asplund A (2018) Immunocompromised patients. Infections, diagnostics and nosocomial transmission. Doctoral dissertation, Lund University
75. Suresha KM, Yang SY, Lee MH, Kim JH, Kim J (2008) Actuator, sensor and MEMS devices based on regenerated cellulose. Comp Int 15:679–685
76. Tallawi M, Rosellini E, Barbani N, Cascone MG, Rai R, Saint-Pierre G, Boccaccini AR (2015) Strategies for the chemical and biological functionalization of scaffolds for cardiac tissue engineering: a review. J Royal Soc Interface 12:20150254
77. Tan J, Kang H, Liu R, Wang D, Jin X, Li Q, Huang Y (2011) Dual-stimuli sensitive nanogels fabricated by self-association of thiolated hydroxypropyl cellulose. Polym Chem 2:672–678
78. Tang J, Bao L, Li X, Chen L, Hong FF (2015) Potential of PVA-doped bacterial nano-cellulose tubular composites for artificial blood vessels. J Mat Chem B 3:8537–8547
79. Tavakolian M, Munguia-Lopez JG, Valiei A, Islam MS, Kinsella JM, Tufenkji N, van de Ven TG (2020) Highly absorbent antibacterial and biofilm-disrupting hydrogels from cellulose for wound dressing applications. ACS Appl Mater Int 12:39991–40001
80. Thatikayala D, Ponnamma D, Sadasivuni KK, Cabibihan JJ, Al-Ali AK, Malik RA, Min B (2020) Progress of advanced nanomaterials in the non-enzymatic electrochemical sensing of glucose and H_2O_2. Biosensors 10:151
81. Torres FG, Troncoso OP, Gonzales KN, Sari RM, Gea S (2020) Bacterial cellulose-based biosensors. Med Dev Sens 3:e10102
82. Troy E, Tilbury MA, Power AM, Wall JG (2021) Nature-based biomaterials and their application in biomedicine. Polymers 13:3321
83. Wang DC, Lei SN, Zhong S, Xiao X, Guo QH (2023) Cellulose-based conductive materials for energy and sensing applications. Polymers 15:4159
84. Wang X, Yao C, Wang F, Li Z (2017) Cellulose-based nanomaterials for energy applications. Small 13:1702240
85. Wahid F, Zhong C (2021) Applications of bacterial cellulose. In Nanocellulose: synthesis, structure, properties and applications
86. Wegrzynowska-Drzymalska K, Mlynarczyk DT, Chelminiak-Dudkiewicz D, Kaczmarek H, Goslinski T, Ziegler-Borowska M (2022) Chitosan–gelatin films cross-linked with dialdehyde cellulose nanocrystals as potential materials for wound dressings. Int J Mol Sci 23:9700
87. Wu DT, Munguia-Lopez JG, Cho YW, Ma X, Song V, Zhu Z, Tran SD (2021) Polymeric scaffolds for dental, oral, and craniofacial regenerative medicine. Molecules 26:7043
88. Xiao Y, Xu Y, Wang W, Yin S, Zhao W, Wei R, Zhao C (2024) One-step engineering dual-network reinforced hydrogel microspheres with excellent anti-coagulant and low-density lipoprotein removal. Sep Pur Technol 331:125668
89. Xie F, De Wever P, Fardim P, Van den Mooter G (2021) TEMPO-oxidized cellulose beads as potential pH-responsive carriers for site-specific drug delivery in the gastrointestinal tract. Molecules 26:1030
90. Yan H, Chen X, Feng M, Shi Z, Zhang W, Wang Y et al (2019) Entrapment of bacterial cellulose nanocrystals stabilized Pickering emulsions droplets in alginate beads for hydrophobic drug delivery. Coll Surf B Biointerfaces 177:112–120
91. Yao R, Liu X, Yu H, Hou Z, Chang S, Yang L (2024) Electronic skin based on natural biodegradable polymers for human motion monitoring. Int J Biol Macromol 278:134694
92. Yi H, Lee SH, Ko H, Lee D, Bae WG, Kim TI et al (2019) Ultra-adaptable and wearable photonic skin based on a shape-memory, responsive cellulose derivative. Adv Func Mater 29:1902720
93. Zhang L, Chen L, Wang S, Wang S, Wang D, Yu L et al (2024) Cellulose nanofiber-mediated manifold dynamic synergy enabling adhesive and photo-detachable hydrogel for self-powered E-skin. Nature Comm 15:3859
94. Zhang S, Li J, Chen S, Zhang X, Ma J, He J (2020) Oxidized cellulose-based hemostatic materials. Carbohydr Polym 230:115585

95. Zhou H, Zhu H, Yang X, Zhang Y, Zhang X, Cui K et al (2015) Temperature/pH sensitive cellulose-based hydrogel: synthesis, characterization, loading, and release of model drugs for potential oral drug delivery. BioResources 10:760–771
96. Zubair M, Hussain A, Shahzad S, Arshad M, Ullah A (2024) Emerging trends and challenges in polysaccharide derived materials for wound care applications: a review. Int J Biol Macromol 270:132048

Sevgi Sözügeçer Çetintaş is a dedicated Ph.D. candidate in the Department of Chemistry at Hacettepe University, Türkiye, specializing in polymeric biomedical materials. Her research focuses on the development and characterization of cellulose-based and other polymeric biomedical products, aiming to advance the field of medical devices. In addition to academic work, she is the founder and managing director of a company that provides CE certification, regulatory compliance, and advisory services ensuring that medical devices meet necessary standards, conducting critical evaluations such as device characterization, risk assessment, and the identification of potential health risks associated with medical products. Through a combination of academic research and industry experience, she is committed to bridging the gap between scientific innovation and practical applications in the medical device sector, contributing to the development of safer and more effective healthcare solutions.

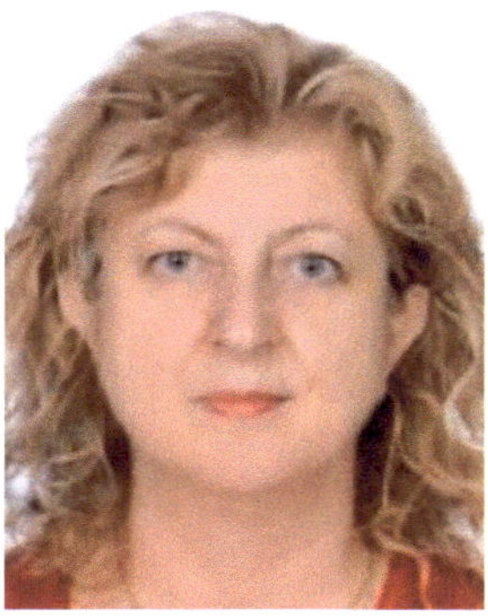

Professor Nursel Pekel Bayramgil is currently working in the Department of Chemistry, Faculty of Science at Hacettepe University, Ankara, Türkiye. She received her M.Sc., and Ph.D. degrees in Physical Chemistry from Hacettepe University in 1994 and 1998, respectively. She worked as a visiting researcher at the Takasaki-Japan Atomic Energy Research Institute in 2002–2003. Dr. Pekel Bayramgil has research experience in cellulose and its derivatives, nanoparticles, nanocomposites, nanofibers, polymer synthesis and characterization, hydrogels, composite films, graphene and its derivatives, drug delivery systems, wound healing systems, personal protective equipment production, and in addition to her own department, she continues her interdisciplinary studies with departments such as medicine, pharmacy, dentistry, and biology. She has numerous projects and scientific articles in her field.

Cellulose-Based Hemostatic Agents: Innovations and Applications in Wound Healing

M. A. Lubna, Akshay R. Thampi, Navami Krishna, and G. K. Rajanikant

Abstract Hemostasis and wound healing are fundamental biological processes crucial for restoring tissue integrity following injury. This chapter presents a comprehensive overview of cellulose-based hemostatic agents, highlighting their innovative applications in modern wound care. Cellulose, a natural polysaccharide, exhibits unique properties such as biodegradability, biocompatibility, and mechanical strength, making it an ideal candidate for enhancing hemostatic efficacy. The chapter discusses the intricate phases of hemostasis, including vasoconstriction, platelet aggregation, and coagulation, alongside the cellular mechanisms involved in inflammation and tissue remodeling. Recent advancements in cellulose technology, particularly the development of nanocellulose composites and hybrid materials, have significantly improved the performance of hemostatic agents. These innovations not only facilitate rapid clot formation but also enhance antimicrobial activity and enable controlled drug delivery. Furthermore, the emergence of smart wound dressings that respond to physiological stimuli exemplifies the potential of cellulose-based materials in personalized medicine. The chapter also addresses the limitations of traditional hemostatic agents, such as biocompatibility issues and biodegradability concerns, emphasizing the advantages of cellulose-derived solutions as sustainable alternatives. By integrating material science with regenerative medicine, cellulose-based hemostatic agents represent a paradigm shift in wound care strategies, offering dynamic solutions for trauma management and chronic wound healing. This work

M. A. Lubna (✉) · A. R. Thampi · N. Krishna · G. K. Rajanikant
Department of Bioscience and Engineering, National Institute of Technology Calicut, Calicut, Kerala, India
e-mail: lubna_p200106bt@nitc.ac.in

A. R. Thampi
e-mail: akshay_m240676bt@nitc.ac.in

N. Krishna
e-mail: navami_p180069bt@nitc.ac.in

G. K. Rajanikant
e-mail: rajanikant@nitc.ac.in

R. Malviya and S. Sundram (eds.), *Sustainable Cellulose Materials for Biomedical Applications*, Sustainable Materials and Technology,
https://doi.org/10.1007/978-981-96-9904-9_16

ultimately underscores the transformative role of cellulose in advancing clinical practices and improving patient outcomes in wound care.

Keywords Cellulose-based hemostatic agents · Wound healing · Hemostasis · Nanocellulose composites · Biocompatibility · Coagulation cascade · Smart wound dressings

Abbreviations

BC	Bacterial cellulose
CMC	Carboxymethyl cellulose
CNF	Cellulose nanofibers
CNC	Cellulose nanocrystals
ECM	Extracellular matrix
EOs	Essential oils
GpIb	Glycoprotein Ib
GpIIb/IIIa	Glycoprotein IIb/IIIa
HPMC	Hydroxypropyl methylcellulose
HEC	Hydroxyethyl cellulose
NPs	Nanoparticles
ORC	Oxidized regenerated cellulose
TEMPO	2,2,6,6-Tetramethylpiperidine-1-oxyl
TXA2	Thromboxane A2
WRV	Water retention value
APS	3-Aminopropyl triethoxysilane

1 Introduction

Hemostasis and wound healing are critical biological processes essential for maintaining homeostasis and restoring tissue integrity following injury. These processes involve intricate mechanisms, beginning with vascular constriction to limit blood loss, followed by platelet aggregation, activation of the coagulation cascade, and subsequent tissue remodeling to achieve complete wound closure. However, in cases of severe trauma, surgical interventions, or chronic wounds, the body's natural healing response may be insufficient, necessitating external support to accelerate and optimize the recovery process.

In recent decades, cellulose, a natural and abundantly available polysaccharide, has emerged as a cornerstone material for developing advanced hemostatic agents and wound care solutions. Its unique characteristics, including excellent mechanical strength, high hydrophilicity, biodegradability, and biocompatibility, make it ideally suited for diverse biomedical applications. Beyond its native form, cellulose can be chemically modified—through processes such as oxidation, esterification, or functionalization—to enhance its interaction with blood components, improve its antimicrobial efficacy, and tailor its properties for specific clinical needs. These modifications enable cellulose-based materials to act not only as physical barriers to bleeding but also as active participants in the wound-healing process by promoting clot formation, absorbing exudates, and fostering tissue regeneration.

Innovative advancements in cellulose-based technologies, such as nanocellulose composites and hybrid materials, have further expanded the scope of these agents. Nanocellulose, with its nanoscale dimensions and high surface area, offers rapid clotting and superior interaction with blood components, making it ideal for hemostatic applications. Hybrid materials that integrate cellulose with bioactive agents or other polymers enhance antimicrobial activity, improve mechanical properties, and enable controlled drug delivery, addressing challenges such as infection prevention and delayed healing. Furthermore, the advent of smart wound dressings—responsive to environmental stimuli such as pH, temperature, or pressure—has redefined the potential of cellulose-based materials in personalized medicine, offering dynamic solutions tailored to the wound's healing state.

As global healthcare systems increasingly emphasize sustainable and eco-friendly medical solutions, cellulose-based hemostatic agents align seamlessly with this vision. Derived from renewable sources, these materials offer an environmentally conscious alternative to synthetic polymers while maintaining exceptional clinical efficacy. By bridging material science and regenerative medicine, cellulose-based innovations represent a paradigm shift in wound care, addressing not only immediate hemostatic needs but also fostering long-term tissue repair and recovery.

This chapter explores the evolution, design, and clinical applications of cellulose-based hemostatic agents, highlighting their transformative role in modern wound care. From traditional applications to the latest technological advancements, this work provides a comprehensive overview of how cellulose is revolutionizing the field, setting the stage for innovative solutions in trauma care, surgery, and chronic wound management.

2 Overview of Hemostasis and Wound Healing

Any wound, internal or external, follows a four-step healing process, the first of which is Hemostasis; the natural ability of the body to stop the bleeding at the site of injury. Hemostasis itself is a multi-step process which occurs in three phases, beginning with vasoconstriction, which is mediated by signaling molecules like endothelin wherein the constriction of blood vessels reduces blood loss. The second phase involves the formation of a platelet plug which occurs after the platelets adhere to the extracellular matrix at the wound site and form a plug on the wall of the injured blood vessel to stop the blood loss. The final step is coagulation, a process which forms a fibrin mesh and stabilizes the platelet plug and secures the wound site [1].

Following hemostasis, the second phase of wound healing is inflammation. This phase begins within 24 h of the injury and lasts around 2 weeks in normal wounds and are mediated by neutrophils, macrophages, and monocytes. The first inflammatory cells that arrive at the wound site are the neutrophils, responsible for phagocytosis [2]. Macrophages play multiple roles in the inflammatory response including phagocytosis, recruitment and activation of more fibroblasts, endothelial cells and more inflammatory cells, promotes angiogenesis, promoting extracellular matrix synthesis to create new tissue.

The third phase of wound healing is Proliferation which begins around 3 days after injury and could last several days or weeks, depending on the severity of the wound. In order to restore the tissue's structure and function, this phase involves replacing the temporary fibrin matrix with a new matrix made of collagen fibers, proteoglycans, and fibronectin [2]. Fibroblasts are the key players in this phase and as they proliferate, new granulation tissue (connective tissue) is formed and is essential for providing structural support as well as promote cell migration and proliferation. Basal epithelial cells (like keratinocytes in skin) migrate across this new tissue to cover the wound area.

The last phase of wound healing is remodeling, and it can take many months or years to complete. Over time, more ordered collagen fibers gradually replace the originally disorderly collagen fibers that were laid down during proliferation. The healed tissue's tensile strength is increased by this restructuring. The granulation tissue formed in the previous phase matures into scars (Fig. 1).

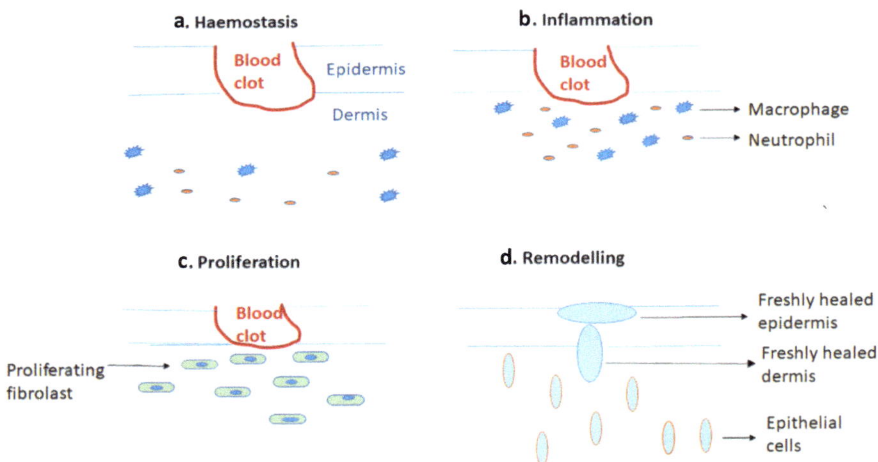

Fig. 1 Phases of cutaneous wound healing. Diagram illustrating the phases of cutaneous wound healing. **a** Hemostasis: Wound formation initiates blood clot development. **b** Inflammation: Recruitment of neutrophils and macrophages to the wound site. **c** Proliferation: Fibroblast proliferation promotes the generation of epithelial cells. **d** Remodeling: Final phase leading to wound closure

3 Mechanisms of Hemostasis

As mentioned earlier, the final step of hemostasis is coagulation; a chain of events triggered in response to vascular injury, known as the coagulation cascade. The platelet plug created during primary hemostasis is stabilized by the gradual activation of clotting factors, which eventually leads to the conversion of fibrinogen to fibrin. The process occurs in three stages and follows intrinsic, extrinsic, and common pathways.

3.1 Coagulation Cascade

Vascular damage during the start of the initiation phase exposes subendothelial components including collagen and von Willebrand factor (vWF). vWF helps platelets adhere to the site of injury by binding to glycoprotein Ib (GpIb) on the platelets [3]. When tissue factor (TF), which is produced by injured endothelial cells, binds with factor VIIa, the extrinsic route is triggered. By activating both factor IX and factor X, this complex connects the intrinsic and extrinsic pathways [4].

Blood components, notably factors XII, XI, IX, and VIII, stimulate the intrinsic pathway during the amplification phase. When negatively charged surfaces, such as exposed collagen, come into touch with factor XII, it is activated, starting a cascade that increases the production of thrombin [4]. By activating factors V and VIII, thrombin (factor IIa) plays a crucial part in this process, further enhancing factor X's activity. The production of enough thrombin to guarantee successful clot formation depends on this amplification [5, 6].

inally, the common pathway, which starts with factor X being activated by both intrinsic and extrinsic pathways, is followed during the propagation phase. Prothrombin is changed into thrombin by the combination of activated factor X (Xa) and factor V, which forms prothrombinase [4, 5]. After thrombin catalyzes the transformation of fibrinogen into fibrin, fibrin strands weave across the platelet plug, stabilizing the clot. Thrombin's capacity to further activate platelets and encourage their aggregation facilitates this process [4].

3.2 Platelet Activation

A critical step in hemostasis that happens along with the coagulation cascade is platelet activation. Platelets stick to exposed collagen after vascular damage with the help of vWF. GpIb binding to vWF mediates this first adhesion [3].

When platelets adhere, they change form and release different granules that contain thromboxane A2 (TXA2), ADP, and other pro-aggregatory chemicals. These chemicals encourage continued activation and bring more platelets to the damage site

[3, 5]. Activated platelets can bind fibrinogen and other adhesive proteins since they have GpIIb/IIIa receptors on their surface. Fibrinogen binding creates a stable platelet plug by joining neighboring platelets [3]. When platelets are activated, calcium ions enter the cell. This is necessary for a number of signaling pathways that promote platelet aggregation and trigger the activation of other clotting factors [3, 5].

3.3 *Factors Affecting Hemostasis*

3.3.1 Blood Flow Dynamics

Blood flow affects how platelets and coagulation factors behave; hence, it is essential for hemostasis.

- **Stress and Shear Rate**: The velocity gradient of blood flow, or shear rate, has a major impact on platelet aggregation and activation. On surfaces exposed to tissue factor (TF), higher shear rates may prevent fibrin formation, but they can improve platelet attachment to the vessel wall [7]. On the other hand, because platelets have more time to aggregate and activate at low shear circumstances, thrombus development may be encouraged.
- **Patterns of Flow**: When laminar flow is disturbed in complicated flow situations, hemostasis takes place. Areas of turbulence produced by abnormal flow patterns, such as those in stenosed arteries, might encourage platelet activation and thrombus formation [7].
- **Vascular Integrity**: Sustaining hemostasis relies on the integrity of the vascular endothelium. Under normal circumstances, endothelial cells release mediators that prevent platelet aggregation, such as prostacyclin and nitric oxide. Increased platelet activity and clot formation result from the loss of these defense mechanisms caused by endothelial degradation [8].

3.3.2 Clotting Disorders

Thrombosis or excessive bleeding can result from several conditions that interfere with normal hemostasis:

- **Congenital Disorders**: Deficits in particular clotting factors (factors VIII and IX, respectively) cause genetic disorders such as hemophilia A and B. Because fibrin production is hindered in people with these conditions, bleeding lasts longer [8, 9].
- **Acquired Diseases**: Coagulation can result from conditions like liver illness that alter the synthesis of clotting components. Pregnancy, hypercoagulable conditions associated with cancer, inflammation, and some drugs, such as anticoagulants, are other acquired variables that affect coagulation [8, 9].

- **Thrombosis**: Hypercoagulable states brought on by acquired illnesses (like antiphospholipid syndrome) or genetic mutations (like the factor V Leiden mutation) can occasionally result in excessive clot formation. Venous thromboembolism is made more likely by certain factors [9].

4 Current Hemostatic Agents in Wound Care

Hemostatic materials help control bleeding after traumatic injuries and surgical procedures. The three primary categories into which they fall are natural, synthetic, and hybrid materials.

4.1 Overview of Existing Hemostatic Agents

4.1.1 Natural Hemostatic Agents

- **Chitosan**: Derived from chitin, chitosan possesses high biocompatibility and hemostatic properties. Hydrogels and sponges that contain chitosan thicken blood and increase platelet aggregation, which aids in clotting. Although they might be difficult to remove, they can be useful for treating a variety of wounds [10].
- **Gelatin**: Hemostatic products based on gelatin, such as gelatin sponges, are distinguished by their capacity to absorb liquids and encourage platelet aggregation. They promote rapid hemostasis by activating the extrinsic and intrinsic coagulation pathways [10, 11].
- **Fibrin Sealants**: These substances are made up of thrombin and fibrinogen, which combine to create a stable clot. They are suitable for a range of surgical applications since they do not require blood to form a clot [11, 12].

4.1.2 Synthetic Hemostatic Agents

Materials created to improve hemostasis through a variety of methods are known as synthetic hemostatic agents.

- **Microfibrillar Collagen**: By acting as a framework for the production of clots, microfibrillar collagen encourages platelet aggregation and improves the natural clotting process. In procedures that call for quick hemostasis, it is frequently utilized.
- **Polymer-Based Agents**: Hydrogels that can absorb blood and encourage clotting can be made using synthetic polymers like polyethylene glycol (PEG). The physical characteristics of these materials allow them to be customized for specific purposes [11].

4.1.3 Hybrid Hemostatic Materials

- **Chitosan/Synthetic Polymer Composites**: These hybrids improve mechanical strength and biocompatibility by combining the natural qualities of chitosan with synthetic polymers. Their efficacy in treating different kinds of wounds can be increased by designing them to react to particular physiological factors [10].
- **Composites Based on Graphene**: In recent years, graphene has been included into hemostatic treatments. These composites have excellent water absorption capabilities and the ability to promote the development of stable fibrin networks which improves hemostasis while lowering the risk of thrombosis [10].

4.2 Limitations of Traditional Hemostatic Agents

Although they are crucial for reducing bleeding in trauma and surgical environments, in practice and during manufacturing, traditional hemostatic agents face two significant challenges, which is ensuring biodegradability and biocompatibility.

4.2.1 Biocompatibility Issues

- **Foreign Body Reactions**: Numerous conventional hemostatic agents are foreign substances that have the potential to trigger immunological reactions. Inflammation, infection, and foreign body reactions may result from this, potentially impeding the healing process of the wound. Hemostatic treatments that contain synthetic materials, for instance, may cause an inflammatory reaction that makes recovery more difficult and lengthens the healing period [13, 14].
- **Risk of Infection**: Hemostatic agents have the potential to spread infections, especially if they are used in non-sterile settings or are not adequately sanitized. This is especially important in emergency scenarios where sterility may be jeopardized. Despite filtration techniques, there remains a chance of viral transmission when materials like fibrin—which is made from pooled human blood—are used [15, 16].
- **Tissue Damage**: When certain hemostatic agents are applied, tissue damage may result. For example, exothermic reactions linked to specific agents, such as zeolite, might result in thermal damage at the wound site, which can impede the healing process and perhaps cause further harm to nearby tissues [15, 16].

4.2.2 Biodegradability Concerns

- **Persistence in Tissues**: Due to their often slow biodegradability, traditional hemostatic medications can remain in tissues for extended periods of time and have adverse effects. Certain synthetic polymers used in hemostatic agents, for instance, may not break down well in the body, which could result in fibrosis or chronic inflammation [13, 16].

- **Incomplete Degradation Products**: Some hemostatic materials may produce incomplete degradation products that are either not biocompatible or may trigger further immune reactions. For example, chemical changes of chitosan derivatives intended to improve hemostatic characteristics have been demonstrated to have harmful effects on normal cells [13, 15].
- **Short Shelf Life and Storage Requirements**: A lot of conventional hemostatic medications have a short shelf life and specific storage requirements (like refrigeration), which makes it difficult to use them in emergency situations like combat medicine or distant locations [13]. When there is an urgent need to stop bleeding, this restriction limits their availability.

5 Cellulose: Structure, Properties, and Biomedical Potential

5.1 Fundamentals of Cellulose

Cellulose ($C_6H_{10}O_5$) is a linear polysaccharide composed of β-D-glucopyranose units interconnected through β-1,4-glycosidic bonds. It represents the most prevalent natural polymer, distinguished by its renewable and biodegradable characteristics. The fundamental repeating unit of cellulose is cellobiose, which comprises two glucose molecules [17]. Each glucose unit in cellulose is oriented 180 degrees relative to its adjacent unit, a configuration that enhances the polymer's tensile strength [18]. The robust hydrogen bonding both within and between cellulose chains results in a rigid and crystalline architecture, positioning cellulose as a critical structural element in plant cell walls. The hydroxyl groups (–OH) present on the glucose units form intra- and intermolecular hydrogen bonds, which are pivotal in maintaining the structure [19]. The biocompatibility, water retention capability, biodegradability, renewability, and modifiability of cellulose render it an ideal biopolymer for applications as a biomaterial [20]. Generally, cellulosic materials are regarded as environmentally friendly and economically advantageous relative to conventional synthetic materials [21].

5.1.1 Molecular Configuration

The molecular structure of cellulose comprises both crystalline and amorphous areas. The crystalline portions are highly ordered and contribute to the strength and insoluble nature of cellulose in water, while the amorphous regions provide flexibility. The degree of polymerization (DP) of cellulose, which ranges from a few hundred to over 10,000 glucose units, varies with the source and influences its mechanical qualities. The crystallinity index, which measures the fraction of crystalline areas, is an essential factor influencing cellulose's physical qualities [22].

5.1.2 Origins and Sources of Cellulose

- **Plant Cellulose**: Sourced from the cell walls of plants, it is particularly abundant in wood, hemp, and cotton. Plant cellulose is embedded within a matrix of lignin and hemicellulose, which separate during the extraction process to obtain pure cellulose. Cotton fibers, predominantly consisting of cellulose, serve as a notable example of high-purity plant cellulose [23].
- **Bacterial Cellulose (BC)**: It is an extracellular polymer produced by various bacterial species, including Gluconacetobacter xylinus (formerly Acetobacter xylinum), Komagataeibacter, Agrobacterium, Achromobacter, and Sarcina. It is distinguished by its high purity and crystallinity. In comparison with plant-derived cellulose, bacterial cellulose demonstrates superior mechanical properties and an enhanced capacity for water retention [24, 25]. In contrast to plant cellulose, which is intertwined with lignin and hemicellulose, bacterial cellulose is synthesized in a pure form, devoid of other polysaccharides and lignin, rendering it highly advantageous for various applications [25].
- **Modified Cellulose**: Chemically altered derivatives of cellulose, including methylcellulose, carboxymethylcellulose, and hydroxypropyl cellulose, exhibit varying solubility, viscosity, and functional properties tailored for specialized biomedical applications. These modifications introduce functional groups that augment the applicability of cellulose in drug delivery, tissue engineering, and a range of other medical applications [26].

5.1.3 Methods of Extraction

The isolation procedures are influenced by the type of plant tissue employed in polymer production. Assessing purity is a crucial aspect of the isolation process, with the required level of purity depending on the intended application of the synthesized polymers. The extraction of cellulose generally entails the removal of lignin, hemicellulose, and other non-cellulosic substances. Commonly employed methods include [27]:

- **Mechanical Pulping**: This method physically separates cellulose fibers through mechanical forces, retaining more lignin and hemicellulose. Consequently, the resulting cellulose is of lower purity, although the process is less chemically intensive [28, 29].
- **Chemical Pulping**: This process employs chemicals, such as sodium hydroxide and sodium sulfide, to degrade lignin and hemicellulose, resulting in cellulose of

higher purity. The Kraft process, a prevalent chemical pulping method, produces robust fibers well-suited for paper manufacturing [28].
- **Enzymatic Treatment**: This approach utilizes cellulases to selectively degrade non-cellulosic components, offering an environmentally sustainable method for producing high-purity cellulose without the use of harsh chemicals [28].

5.1.4 Inherent Biocompatibility

Cellulose's natural origin and lack of toxicity make it biocompatible [30]. It does not trigger substantial immunological responses when implanted in the body, making it appropriate for a variety of biomedical applications. Furthermore, its flexibility to be altered increases its utility in tissue engineering, wound dressings, and drug delivery systems [30]. Cellulose's biocompatibility is boosted by its ability to be treated without compromising structural integrity or functional characteristics.

5.2 Properties Relevant to Hemostasis and Wound Healing

Cellulose derivatives possess favorable properties such as mechanical strength, biocompatibility, and absorbency, making them highly suitable for wound dressing applications [31].

- **Hydrophilicity**: Cellulose is highly hydrophilic due to the presence of hydroxyl groups that can establish hydrogen bonds with water molecules. Cellulose hydrogels are used as wound dressings because of their water-holding property and capacity to keep the wound environment moist [32]. This characteristic enables cellulose to absorb enormous volumes of water and exudate from wounds, resulting in a moist environment beneficial to healing. The water retention capacity of cellulose can be increased by changing its surface characteristics or introducing super-absorbent polymers. Cotton cellulose is used with hydrogels to improve wound-healing characteristics [33]. The hydrophilic property of cellulose allows for the quick uptake of blood and wound exudates, which aids in the development of clots and accelerates the hemostatic process [34].
- **Mechanical Strength**: The strong mechanical strength of cellulose is primarily due to its crystalline structure and abounding hydrogen bonding. The incorporation of crystalline cellulose enhances the mechanical properties and wound-healing efficacy of gelatin-based biocomposite dressings) [35]. This strength ensures that wound dressings maintain structural integrity under mechanical stress, effectively protecting the wound site. The tensile strength of cellulose can be adjusted by regulating the degree of polymerization and crystallinity during production. Additionally, the mechanical properties of cellulose-based wound dressings can be customized through 3D printing, enabling precise control over polymerization and crystallinity [36].

- **Biodegradability**: Cellulose is an ideal material for 3D-printed wound dressings since it is biodegradable, biocompatible, and ability to replicate the extracellular matrix [36].
- Naturally biodegradable, cellulose breaks down into glucose units through enzymatic processes in the body or by environmental microorganisms. Cellulose-based wound dressings are biodegradable and can be employed with antimicrobial agents to address infected chronic wounds [37]. Bacterial cellulose, when combined with cellulase enzymes, is capable of degrading in vitro under conditions similar to wound settings [38]. This characteristic lowers the need to remove cellulose-based wound dressings, as they decompose harmlessly over time, reducing patient discomfort and infection risk. The pace of biodegradation can be affected by the degree of crystallinity and the presence of chemical changes [39, 40]. Biodegradability guarantees that cellulose-based compounds do not remain in the body for longer than necessary, lowering the risk of chronic inflammation or other consequences.
- **Absorbency**: The ability of cellulose to absorb fluids plays a crucial role in managing wound exudates, thereby reducing the risk of infection and fostering an optimal environment for healing [31]. The absorbency of cellulose is commonly quantified by the water retention value (WRV), which serves as a key parameter in the design of wound dressings [41]. Cellulose absorbency, measured by water retention value, is an important factor in the design of 3D-printed cellulose-based wound dressings [33]. Bacterial cellulose has a high water absorption and retention capacity, making it an attractive candidate for wound dressings [42]. Cellulose-based hydrogels are beneficial for wound healing due to their absorbency and ability to maintain adequate moisture levels [43].
- **Tensile Strength**: Tensile strength guarantees that cellulose-based dressings can tolerate mechanical pressures without tearing, resulting in persistent protection. Uniaxial drawing of cellulose films can increase their tensile strength and modulus of elasticity [43]. Fiber orientation, polymerization degree, and cross-linking all have an effect on cellulose's tensile capabilities [44]. The degree of polymerization of cellulose is crucial for producing ultra-strong nanocellulose films [44]. Crystalline cellulose reinforcement increases the tensile strength and other qualities of gelatin-based wound dressings [32].
- **Biocompatibility**: Cellulose's biocompatibility and adaptability make it a viable material for biomedical applications [30]. Cellulose's compatibility with bodily tissues lowers the danger of adverse responses, making it appropriate for direct contact with wounds and tissues. The surface chemistry of cellulose can be altered to improve cell adhesion, proliferation, and differentiation, hence increasing its biocompatibility for tissue engineering purposes [45]. Native cellulose scaffolds are biocompatible and have great potential as surgical biomaterials [46].

Overall, these qualities make cellulose an effective material for hemostatic applications and wound healing, improving patient outcomes and facilitating speedier recovery. The ability to tune these features through chemical and physical alterations enables the creation of sophisticated cellulose-based materials for specialized medicinal applications. Because of its adjustable features that aid in hemostasis and wound healing, cellulose-based biomaterials show great promise as wound-healing materials.

6 Cellulose-Based Hemostatic Agents: Design and Function, Mechanism, Limitations

Cellulose-based hemostatic agents are gaining significant recognition for their efficacy in bleeding control, utilizing the distinctive properties of cellulose to expedite the hemostatic process. These agents are engineered to deliver rapid, efficient, and localized hemorrhage management, particularly in surgical and traumatic wound contexts.

6.1 Design and Function

The design of cellulose-based hemostatic agents emphasizes enhancing their interaction with blood components to accelerate clot formation. Their structure, surface chemistry, and mechanical properties are meticulously engineered to ensure optimal performance. These agents go beyond simply providing a physical barrier to bleeding; they actively participate in the coagulation cascade, promoting platelet aggregation and fibrin network formation, essential for stable clot development. The design and function of cellulose-based hemostatic agents are rooted in their ability to deliver efficient and effective hemostasis through various modifications and types. Because of their variety, efficacy, and biocompatibility, these agents play an important role in current wound treatment and surgical care.

6.2 Development and Modification

For enhancement of the hemostatic characteristics of cellulose, different chemical and physical alterations are used.

6.2.1 Oxidation

Oxidized cellulose is produced by introducing carboxylic acid groups through oxidation, markedly improving its hemostatic properties [47]. This modification enhances the material's interaction with blood components, creating an optimal environment for platelet aggregation and fibrin formation, thereby accelerating the clotting process.

Steps in the Oxidation Process

The selection of oxidizing agents: Sodium periodate ($NaIO_4$) is a common oxidizing agent for cellulose. It selectively converts the hydroxyl groups at the C-2 and C-3 positions of the glucose units in cellulose to aldehydes. Another chemical, 2,2,6,6-Tetramethylpiperidine-1-oxyl (TEMPO), is employed to oxidize the main alcohol groups (C-6) into carboxyl groups [48, 49].

Reaction Conditions: The oxidation reaction is normally carried out in an aqueous media, with the pH and temperature carefully adjusted to produce the required level of oxidation while preserving the cellulose backbone.

Carboxylic acid groups are formed when oxidizing substances transform cellulose's hydroxyl groups into carboxylic acids, aldehydes, or ketones. The addition of carboxylic acid groups increases the material's negative charge, increasing its interaction with positively charged proteins and blood components [49].

Post-treatment: Following oxidation, the cellulose is washed and neutralized to eliminate any remaining oxidizing chemicals. The modified cellulose is subsequently dried and processed into the desired product, such as films, fibers, or powders. Recent study focuses on producing oxidized cellulose nanocrystal composites for enhanced hemostatic materials [47].

Benefits of Oxidized Cellulose

Enhanced Clotting: The carboxyl groups provide sites for calcium ion binding, which is essential in the clotting cascade, hastening the production of blood clots [50].

Improved Absorption: The increased hydrophilicity of carboxylic groups improves the material's ability to absorb wound discharge, resulting in a moist environment suitable for healing. Oxidized cellulose-silk fibroin scaffolds loaded with thrombin have better hemostatic characteristics than empty scaffolds [51].

6.2.2 Functionalization of Cellulose

Functionalization is the process of adding new functional groups or molecules to the cellulose structure to modify its properties to specific purposes. This can be performed using a variety of chemical or physical approaches.

Types of Functionalization

- **Chemical Functionalization**

Grafting: The process of covalently attaching functional groups or molecules to the cellulose backbone. For example, grafting polyethylene glycol (PEG) onto cellulose can increase its biocompatibility while decreasing its immunogenicity [52].

Esterification and etherification: These reactions add ester or ether groups to cellulose, impacting its solubility, hydrophilicity, and interaction with blood proteins [52].

Amine Functionalization: Introducing amine groups can increase the interaction with negatively charged blood components, which improves hemostatic efficiency [52].

- **Physical Functionalization**

Plasma Treatment: When cellulose is exposed to plasma, reactive sites form on its surface, permitting functional molecules to connect later [52]. Surface Coating: Cellulose can be coated with bioactive compounds like growth factors or antibacterial agents to boost its functionality in wound dressings [52].

Benefits of Functionalized Cellulose

Enhanced Biocompatibility: Functional groups can be designed to decrease undesirable immune responses and enhance compatibility with human tissues.

Improved Mechanical Properties: Functionalization can improve the mechanical strength and flexibility of cellulose, making it more suited for a variety of biomedical applications.

Customizable Interactions: By introducing specific compounds, cellulose's interaction with biological systems can be modified, such as boosting its hemostatic or antibacterial activity.

Both oxidation and functionalization of cellulose improve its hemostatic ability, making it a diverse and effective material for wound care and other biomedical applications.

6.3 Types of Cellulose

6.3.1 Native Cellulose

Structure and Properties of Native Cellulose

Native cellulose, a biocompatible and hydrophilic polymer, is extensively utilized in wound care due to its inherent mechanical strength and moisture retention capacity. It has a high level of polymerization, with glucose units creating elongated chains that combine into microfibrils, which contribute to its structural integrity and capacity to sustain a moist wound environment [53]. These microfibrils exhibit a high degree of crystallinity, attributed to extensive intramolecular and intermolecular hydrogen bonding, which enhances the material's mechanical robustness. Native cellulose is a commonly used biomaterial with strong biocompatibility, hydrophilicity, and mechanical strength, making it ideal for wound care applications.

Hydrophilicity

The hydroxyl groups (–OH) on the glucose units endow native cellulose with hydrophilic properties, enabling it to absorb moisture and wound exudate effectively. This moisture retention capability fosters a moist wound environment, which is advantageous for the wound-healing process [42].

Biocompatibility

Inherently biocompatible, native cellulose induces minimal immune response upon application to wounds, making it appropriate for direct contact with human tissue. Its non-toxic and non-irritating properties enhance its usefulness for medical uses [34].

Applications in Wound Dressings

Physical Barrier:

Native cellulose serves as a physical barrier, shielding the wound from external contaminants and preventing further injury. It also reduces bleeding through mechanical pressure and aids in the early phases of wound healing.

Absorption of Exudate:

The hydrophilic nature of native cellulose enables it to absorb and retain wound exudates, managing moisture levels effectively [33]. This property aids in removing excess fluids and reduces the danger of maceration (softening and breaking down of skin due to prolonged exposure to moisture).

Limited Hemostatic Efficiency:

While native cellulose absorbs fluids and concentrates clotting factors, its hemostatic efficiency is lower than modified forms like oxidized cellulose. It lacks functional groups like carboxylic acids, which limits its role in promoting rapid clotting.

Composites with other polymers: Blending with synthetic or natural polymers improves hemostatic properties, tensile strength, and absorbency.

Incorporation of antimicrobial agents: This addition enhances its ability to prevent wound infections.

Advantages of Using Native Cellulose

Cost-Effectiveness:

Due to its natural abundance, native cellulose is inexpensive to source and process, offering a reasonable option for basic wound care applications.

Environmental Sustainability:

As a biodegradable material, native cellulose supports environmentally sustainable practices by reducing ecological impact.

Mechanical Strength:

The fibrous and crystalline structure of native cellulose imparts significant mechanical strength, ensuring durability and resistance to tearing in wound care products.

Limitations

Lower Hemostatic Activity: Native cellulose, unlike modified versions, does not actively participate in blood coagulation, which may restrict its usefulness in reducing severe bleeding.

Potential for Dehydration: While natural cellulose holds moisture, if not correctly controlled, it might cause the wound to dry out, thereby prolonging recovery.

6.3.2 Oxidized Cellulose

Oxidized cellulose is a biodegradable and biocompatible hemostatic substance commonly used in surgical procedures [47].

Hemostatic Mechanism of Oxidized Cellulose

Oxidized cellulose activates the intrinsic coagulation pathway and platelet aggregation, requiring calcium ions and affected by coagulation factors V, VIII, IX, and XII [54]. The carboxylic acid groups on oxidized cellulose provide a surface for platelet and protein adsorption, which is required to initiate clot formation [47]. These groups also increase hydrophilicity, which improves blood absorption and concentrates clotting components near the wound site [47]. The material is created by selectively oxidizing the C6 primary hydroxyl groups of natural cellulose to carboxyl groups, yielding a product with good hemostatic and absorbent characteristics [55]. The material's hemostatic efficacy is attributed to its ability to prolong fibrin gel formation and increase turbidity, depending on its concentration [56]. Furthermore, oxidized cellulose is bactericidal against many pathogens and has antioxidant and wound-healing properties [57]. Thrombin-loaded oxidized cellulose-silk fibroin scaffolds had better hemostatic characteristics than those lacking thrombin. Oxidized cellulose can be processed into a variety of medical products, including sheets, pads, and powders [51].

Biomedical Applications

Surgical Hemostats: Oxidized cellulose is commonly used in surgical settings to cease bleeding. Surgicel, a product consisting of oxidized cellulose, is extensively used in surgeries [47]. These materials are applied directly to bleeding sites to acceleration clotting and successfully reduce hemorrhage [57]. Wound dressing: Oxidized cellulose is used in advanced wound dressings for chronic and bleeding wounds, in addition to surgical applications. Its hemostatic and absorbent qualities aid in maintaining a moist wound environment, managing exudate, and preventing excessive bleeding [58].

Resorbability: Oxidized cellulose is biodegradable and resorbs into the body over time, making it perfect for temporary hemostatic support that does not require removal after healing [59].

Advantages of Oxygenated Cellulose

Enhanced Clotting Efficiency: The inclusion of carboxylic acid groups considerably improves the hemostatic characteristics of cellulose, making it more successful in encouraging blood clot formation than native cellulose.

Biocompatibility: Oxidized cellulose is biocompatible, which reduces the possibility of undesirable immunological responses when employed in the human body [47].

Biodegradability: It dissolves into harmless by-products that the body naturally absorbs, eliminating the need for removal and relieving patient suffering [47].

Versatility: The material can be processed into a variety of forms, including films, sponges, and powders, to meet a wide range of surgical and wound care requirements.

6.3.3 Nanocellulose

Structural and Physical Properties of Nanocellulose

Nanocellulose, generated from plants or microbes, is a versatile material with potential for hemostasis and wound healing. It is created by mechanical, chemical, or enzymatic procedures. Its large surface area and nanoscale dimensions promote fast blood clotting by offering several locations for platelet adhesion and activation [60]. Nanocellulose-based products, including Ca^{2+}-cross-linked hydrogels and oxidized bacterial cellulose composites, have shown improved hemostatic efficacy, quicker clot formation, and lower blood loss in animal models [61]. These materials are also biodegradable, antimicrobial, and biocompatible, which makes them effective for more effective wound management.

High Surface Area: Nanocellulose's increased surface area allows for better interaction with blood components, leading to faster clot formation through protein and platelet adsorption [51].

Nanoscale Dimensions: Cellulose nanofibers (CNFs) and cellulose nanocrystals (CNCs) have unique rheological attributes that help produce gels and films that adhere to wound surfaces. Nanocellulose's high mechanical strength and flexibility make it ideal for long-lasting wound dressings that withstand stress.

Mechanism for Hemostatic Action

Platelet Adhesion and Activation: Nanocellulose's large surface area speeds up the hemostatic process by increasing platelet adhesion and activation [62].

Interaction with Blood Components: It interacts with fibrinogen and other blood components, allowing for faster clot formation than bulk cellulose. Nanocellulose quickly absorbs blood and exudates, concentrating clotting components and speeding up the hemostatic reaction.

Nanocellulose, a flexible cellulose-derived nanomaterial, has received interest for its potential applications in wound care. It can be prepared using a variety of methods, including mechanical, chemical, and biological ones. Nanocellulose-based dressings, such as hydrogels, films, and sponges, are excellent in hemostasis and wound protection. Nanocellulose's biocompatibility and biodegradability allow it to be well tolerated by the body, lowering the chance of adverse reactions.

Hemostatic Dressings: To rapidly prevent bleeding and protect the wound, nanocellulose-based dressings can be utilized in a variety of forms such as hydrogels, films, and sponges [63]. These dressings are especially beneficial in surgical applications, trauma treatment, and military contexts where quick hemostasis is essential.

Controlled Drug Delivery: Nanocellulose can be functionalized to transport hemostatic or antibacterial medicines, providing therapeutic effects beyond simple bleeding control. This dual functionality aids in infection prevention while also encouraging wound healing.

Enhanced Biocompatibility: Because of its natural origin and biocompatibility, nanocellulose is easily accepted by the body, lowering the chance of unpleasant responses. It is also biodegradable, resulting in non-toxic metabolites that the body can easily handle or discard.

The Advantages of Nanocellulose in Hemostasis

Rapid Clot Formation: Nanocellulose-based materials offer great promise for hemostasis and wound-healing applications. These nanoparticles promote fast clot formation, lowering clotting time by up to 88% in vitro biodegradability [64]. The high surface area and contact with blood components promote faster clotting, which is critical for effectively halting bleeding. Ca^{2+}-cross-linked nanocellulose hydrogels have potential as hemostatic wound dressings due to their rapid clot formation, mechanical stability, and biocompatibility. Nanocellulose can be mixed with other materials like chitosan or antibacterial compounds to improve its functionality.

Mechanical Stability: Nanocellulose gives wound dressings structural support, allowing them to remain effective under a variety of situations. These materials are biocompatible, biodegradable, and mechanically stable, making them appropriate for a variety of wound treatment scenarios [65]. Nanocellulose, which is generated from natural cellulose, is biocompatible and less prone to elicit immunological responses.

Biocompatibility: Being derived from natural cellulose, nanocellulose is biocompatible and less likely to cause immune reactions.

Customization: Nanocellulose can be changed or mixed with other materials to enhance its hemostatic capabilities or offer additional functionality such as antibacterial activity.

Limitations and Challenges

Manufacture Complexity: Nanocellulose manufacture is difficult, potentially expensive, and time-consuming [65].

Potential for Over-Swelling: High absorbency has the potential to cause over-swelling, which can compromise dressing adherence [65].

Need for Functionalization: Functionalization may be necessary to increase its hemostatic potential or add other desired qualities.

6.3.4 Modified Cellulose Derivatives—Chemical Structure, Properties, and Applications

Modified cellulose derivatives are chemically changed forms of cellulose that are intended to improve a variety of qualities, including water solubility, absorbency, and interactions with biological components. These alterations are intended to maximize cellulose's natural features for specific biological uses, particularly in hemostasis.

Carboxymethyl Cellulose (CMC)

Carboxymethyl cellulose (CMC) is a widely used cellulose derivative that is highly water soluble and biocompatible [66, 67]. Its structure, featuring carboxyl groups, enables interactions with blood components, enhancing its hemostatic properties. When it comes into contact with blood, it creates gels that act as a physical barrier to bleeding, making it useful for wound dressings and hemostatic treatments [68]. Its variable hydrophilicity and mechanical strength allow for a wide range of uses in the food, pharmaceutical, and biomedical industries. CMC can be converted into nonwoven sheets or hydrogels, which improve clot formation and blood viscosity, hence contributing to effective hemostasis [69]. Chemical Structure: Carboxymethyl groups (–CH_2–COOH) are introduced into cellulose, commonly with chloroacetic acid [69]. Carboxymethyl cellulose (CMC) hydrogels combined with natural antimicrobials such as propolis offer the potential for effective wound care dressings [70].

Properties: Highly water-soluble interacts effectively with blood proteins, generating gels that aid in hemostasis.

Applications: Wound dressings and hemostatic treatments are used to stop bleeding and retain moisture.

Hydroxyethylcellulose (HEC)

Hydroxyethyl cellulose (HEC) is a biocompatible, water-soluble derivative with excellent viscosity and water retention that is ideal for wound dressings [71]. HEC hydrogels can be physically cross-linked or chemically modified to enhance their properties [71]. HEC hydrogels, which are frequently cross-linked or combined with other polymers, have antibacterial properties, absorb exudate, and are biocompatible [71]. Chemical structure contains hydroxyethyl groups (–CH_2CH_2OH), which improves hydrophilicity.

Properties: High viscosity, high moisture retention, ideal for gels and films [31].

Applications: Effective in wound dressings, ensuring an ideal moist environment.

Hydroxypropyl Methylcellulose (HPMC)

Hydroxypropyl methylcellulose (HPMC) is a modified cellulose derivative that forms temperature-sensitive gels for drug delivery and wound treatment.

Chemical structure: Hydroxyl groups were replaced with hydroxypropyl and methyl groups [72].

Properties include temperature-sensitive solubility and gel formation which is useful in drug delivery systems and wound dressings.

Applications: HPMC is used in controlled-release formulations and wound care products that benefit from its gel-forming ability and biocompatibility.

Oxidized Cellulose

Oxidized cellulose undergoes modification to incorporate carbonyl and carboxyl groups, which improves interactions with platelets and clotting factors.

Chemical structure: Oxidation produces carbonyl and carboxyl groups often using agents like sodium periodate or nitric acid.

Characteristics: Improved hemostatic characteristics; bioabsorbable.

Application: Surgical hemostats are used to control bleeding.

Cellulose Acetate (CA)

Cellulose acetate (CA) is well-known for its film-forming capabilities, biocompatibility, and adaptability to hydrogels and nanofibers, making it ideal for wound dressings and surgical masks [73].

Chemical Structure: Acetylation of hydroxyl groups results in acetyl groups ($-COCH_3$).

Properties: Excellent film formation; less hydrophilic than other derivatives.

Applications include protective wound dressings and surgical masks.

Benefits of Modified Cellulose Derivatives in Hemostasis

Improved Solubility: The insertion of carboxyl or hydroxyethyl groups increases cellulose's hydrophilicity, allowing it to dissolve better in water and produce gels that can efficiently interact with blood [31].

Enhanced Absorbency: It manages wound exudates and keeps the surroundings moist, which promotes healing.

Biocompatibility: These derivatives preserve the natural biocompatibility of cellulose, guaranteeing they do not cause unfavorable immune reactions when applied to wounds, making them safe for use in medical applications.

Mechanical qualities: By altering cellulose, its mechanical qualities can be tailored to a variety of applications, ranging from flexible films to sturdy sponges, enabling varied wound care and hemostasis solutions.

Functionalization: The inclusion of functional groups can give cellulose derivatives with additional features, such as antibacterial activity or better interaction with certain blood components, further boosting their hemostatic effectiveness.

Challenges

Production Costs: Chemical modification of cellulose can raise production costs, making certain derivatives more expensive than unmodified cellulose.

Environmental Impact: The chemical procedures used to produce modified cellulose derivatives can have an impact on the environment, necessitating careful waste management and emission control.

Regulatory Approval: As medical goods, these derivatives must go through rigorous testing and approval procedures to verify their safety and efficacy in clinical settings.

7 Different Types of Cellulose-Based Dressings

Cellulose-based dressings have evolved to address a variety of wound care needs. Each variety takes advantage of cellulose's specific qualities, such as biocompatibility, biodegradability, and the ability to retain the environment wet. Cellulose-based dressings are classified into scaffolds, hydrogels, nanofibers, films, foam bandages, and hydrocolloids, as depicted in Fig. 2 and compared in Table 1.

7.1 Scaffold

Scaffolds are three-dimensional structures that model the extracellular matrix (ECM) found in tissues. Cellulose-based scaffolds are typically created by processes such as freeze-drying or 3D printing [31]. Overall, different researches have concluded that 3D scaffolds constructed of cellulose are easy to create, biocompatible and can successfully integrate into healthy tissue [74].

Key Properties:

The porous structure facilitates oxygen exchange and promotes cell proliferation [33]. This can be injected with bioactive chemicals, antibiotics, or growth factors to improve their curative properties [34]. Highly biocompatible and promotes tissue repair.

Applications:

Ideal for chronic wounds, burns, and tissue engineering applications. These active wound dressings defend against germs, keep moisture in, absorb exudates, and help the healing process [34].

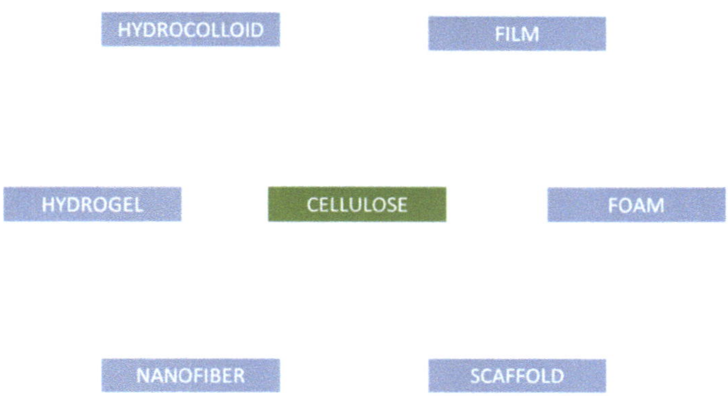

Fig. 2 Different types of cellulose-based dressings

Table 1 Classification of cellulose-based dressings

Type	Type of biomaterial	Application	Benefits	Limitation	References
Scaffold	Cellulose nanofibers Bacterial cellulose Modified cellulose derivative	Surgical wounds, burns, chronic wounds	Facilitates tissue regenerationenhances oxygen supply	Customization may be needed for some applications	[31, 33, 34, 74]
Hydrogel	Hydroxyethyl cellulose Bacterial cellulose, Cellulose nanocrystals	Burns, dry wounds, diabetic ulcers, wounds of the skin and scrapes	High water retention and pain alleviation natural debridement, maintain wet wound environment	Not recommended for highly exuding wounds	[75–78]
Nanofiber	Nanocellulose Electrospun cellulose acetate Carboxymethyl cellulose (CMC) Hydroxypropyl cellulose (HPC)	Burns, chronic wounds, drug delivery systems	Mimics ECM with significant porosity	Higher costs for production	[79, 80]
Film	Bacterial cellulose, Cellulose acetate	Frictional wounds, superficial wounds, skin grafts or contaminated wound chronic wounds, burns, drug delivery systems	Transparent, shielding barrier. Allows for the flow of oxygen and water vapor while preventing the passage of water and bacteria	Limited absorption for bleeding wounds	[81, 82]
Foam	Nanocellulose	Burn, skin graft and skin transplant, pressure ulcers, venous leg ulcers, exuding wounds, chronic wounds	High absorption capacity, effectively absorbs exudates, prevents infiltration, offers thermal insulation, and promotes water and oxygen exchange	Bulkier than films and hydrogels	[83]
Hydrocolloid	Carboxymethyl cellulose	Chronic wounds, minor burns	Gel-forming, seals wound, Provides moist and breathable wound environment. It has good absorption property and is self-adhesive	May fail to stick well to heavily leaking wounds	[31, 84]

Research has established the usefulness of cellulose-based scaffolds in treating chronic wounds, burns, and infected wounds, emphasizing their potential as sophisticated and adaptable structures for wound-healing applications [34].

7.2 *Hydrogel-Based Cellulose Dressings*

Hydrogels are three-dimensional, water-rich networks constructed from modified cellulose (such as carboxymethyl cellulose or hydroxypropyl methylcellulose), which absorb exudates and keep the wound wet.

Key Properties:

High water retention helps to moisturize dry wounds. The use of antibacterial or bioactive substances to promote healing [75]. Non-adhesive and readily removed without harming fresh tissue. According to studies, cellulose hydrogel sheets infused with glycerol improve moisture retention while decreasing wound adhesion, making them excellent for non-adhesive dressings [76].

These hydrogels, which are effective for exuding and granulating wounds, have been shown to speed re-epithelialization and skin regeneration. Recent developments include nanocellulose hydrogels which mimic bacterial cellulose and have soft tissue-like characteristics extending their application in advanced wound treatment [77].

Applications:

Ideal for dry wounds, burns, diabetic ulcers, and surgical wounds [78]. Aquacel® (ConvaTec) is a carboxymethyl cellulose-based hydrogel dressing. 3D-bioprinted cellulose hydrogels improve wound care.

Advantages:

Nonwoven cotton reinforcement is used to regulate moisture, increase air permeability, and absorb exudates [85]. It prevents wound desiccation while also providing pain relief through cooling.

7.3 *Nanofiber-Based Cellulose Dressings*

Nanofiber dressings are made by electrospinning ultra-thin cellulose fibers with a high surface area-to-volume ratio.

Key Properties:

It mimics the extracellular matrix (ECM) to promote cell adhesion and proliferation. High porosity promotes gas exchange and fluid absorption. It can be loaded with medicines, antibacterial agents, or growth hormones for targeted delivery [79].

Applications:

- Ideal for treating chronic wounds and burns, as well as delivering active substances.
- For example, electrospun cellulose acetate nanofibers coated with silver nanoparticles for antibacterial properties. Electrospun nanofibers have many biomedical applications, including wound dressings [80].

Advantages:

- Lightweight and flexible.
- Provides controlled release of medicinal drugs.

7.4 Film-Based Cellulose Dressings

Film dressings are thin, translucent coatings of cellulose or cellulose derivatives (such as bacterial cellulose or hydroxypropyl methylcellulose). These films function as wound-protective barriers.

Key Properties:

- Semi-permeable allows oxygen and moisture exchange while protecting against microorganisms [81].
- Flexible and conforms to the wound's surface.
- They are easily molded, purified, and scaled up, and their nanoporous shape promotes successful wound healing [82].
- Creates a moist wound environment to hasten healing.

Applications:

- Suitable for superficial wounds, including abrasions, cuts, and donor sites.

For example, bacterial cellulose films treated with iodine can help prevent infections.

Advantages:

- Transparent nature allows for wound monitoring without removal.
- Non-adherent and easy to use.

7.5 Foam-Based Cellulose Dressings

Foam dressings are porous cellulose-based structures that absorb large amounts of wound exudate. These dressings are frequently augmented with synthetic polymers to increase durability [83].

Key Properties:

- High absorbency to handle wounds with a lot of exudate.
- Maintains a moist atmosphere while preventing maceration of the surrounding skin.
- Cushioning helps to shield the wound from mechanical trauma.
- The microstructure of foam dressings influences their mechanical qualities, including stiffness, conformability, and durability, which have a substantial impact on clinical performance.

Applications:

- Ideal for pressure ulcers, venous leg ulcers, and postoperative wounds [83]. For example, microfibrillated cellulose foam dressings with antibacterial agents.

Advantages:

- Extended wear period due to high absorption capacity.
- Reduces the frequency of dressing changes, hence reducing patient discomfort.
- The microstructure of foam dressings influences their mechanical qualities, including stiffness, conformability, and durability, which have a substantial impact on clinical performance [83].

7.6 Hydrocolloid-Based Cellulose Dressings

Hydrocolloid dressings are created by combining cellulose derivatives (e.g., carboxymethyl cellulose) with gel-forming substances such as pectin or gelatin. These dressings absorb wound exudates, keep a moist environment, and encourage autolytic debridement [31].

Key Properties:

- Absorbing wound exudates and forming a gel-like substance when in contact with wound fluid.
- Seals the wound, keeping it safe from external pollutants.
- Keeps the wound wet, which promotes autolytic debridement.

Application:

Suitable for treating chronic lesions like pressure sores, diabetic foot ulcers, and small burns [84]. Examples include Comfeel® Plus (Coloplast) and hydrocolloid patches containing silver for antibacterial properties.

Advantages:

- Flexible and sticky, ideal for movable parts of the body.
- Reduces pain and promotes quicker healing.

8 Antimicrobial Properties and Wound-Healing Acceleration

Cellulose-based materials can be developed to have antibacterial properties, which are critical for preventing and treating infections of wounds, especially those associated with chronic conditions [86]. Antimicrobial cellulose dressings safeguard against pathogens, reduce biofilm formation, and enable targeted drug administration, improving wound-healing efficacy while avoiding systemic side effects [37]. These sophisticated dressings can keep the area wet, minimize bacterial invasion, and act as drug carriers [74]. Recent research has yielded good results against drug-resistant bacteria, making cellulose-based dressings a viable option for treating infected chronic wounds [74].

8.1 Role of Antimicrobial Properties in Wound Healing

Infection Prevention: Antimicrobial agents in wound dressings serve primarily to prevent the colonization and multiplication of harmful microorganisms such as bacteria, fungi, and viruses [87]. When loaded with antimicrobial chemicals, cellulose-based dressings form a protective barrier that prevents microbial invasion, resulting in a sterile environment conducive to healing [86].

Reduced Biofilm Formation: Bacterial biofilms, which are resistant to drugs and immunological responses, are a prevalent cause of chronic wound infections. Cellulose-based hemostatic agents with antimicrobial capabilities prevent biofilm formation, ensuring that the wound is free of chronic infections, which can considerably hinder recovery [88]. Composite hydrogels containing bacterial cellulose and chitosan exhibited outstanding anti-biofilm properties, lowering biofilm viability by up to 90% [89].

Localized Antimicrobial Action: Cellulose derivatives can be designed to release antimicrobial agents in a controlled manner just at the wound site [37]. This targeted delivery assures a high local concentration of the antimicrobial agent, which improves efficacy while minimizing systemic adverse effects and lowering the likelihood of antibacterial resistance [90].

8.2 Acceleration of Wound Healing

The presence of antimicrobial compounds in cellulose-based hemostatic agents not only prevents infection, but also assists in accelerating up wound healing through influencing the environment. Maintaining a moist wound environment: Cellulose derivatives naturally promote moisture retention, providing an ideal setting for wound healing. The inclusion of antimicrobial drugs improves this impact by inhibiting microbial development that can cause desiccation or excessive exudate production, both of which can disturb the healing process [87].

Promotion of Cellular Activity: Some antibiotics have bioactive characteristics that stimulate cellular processes necessary for wound healing. For instance, silver nanoparticles and some plant extracts stimulate fibroblast proliferation and keratinocyte migration, both of which are required for tissue regeneration and epithelialization [88, 91].

Balanced Inflammatory Response: Infections can cause prolonged and severe inflammation, delaying wound healing. Antimicrobial cellulose-based dressings assist in controlling the inflammatory response by avoiding infections, allowing the wound to move faster from the inflammatory phase to the proliferative and remodeling phases of healing [92].

8.3 Common Antimicrobial Agents in Cellulose-Based Hemostatic Agents

Several antimicrobial agents have been effectively incorporated into cellulose-based hemostatic agents, each having specific qualities that aid in wound healing.

Silver Nanoparticles: Silver nanoparticles, known for their broad-spectrum antibacterial action, can be integrated into cellulose-based materials to give long-term antimicrobial protection. Their ability to release silver ions over time ensures long-term potency against a variety of diseases. Silver nanoparticle-loaded scaffolds have been demonstrated to increase fibroblast migration and regulate macrophage activation, resulting in better wound healing [93].

Chitosan: Chitosan, a natural polymer with inherent antibacterial qualities, can be combined with cellulose to improve hemostatic and wound-healing properties. Chitosan's ability to stimulate clot formation while inhibiting microbiological growth makes it an ideal candidate for composite wound dressings [94].

Plant-Derived Extracts: Extracts with antibacterial and therapeutic characteristics include aloe vera, propolis, and tea tree oil [95]. Propolis extract coupled with nanocellulose showed antibacterial activity against a variety of pathogens, including Staphylococcus aureus and Pseudomonas aeruginosa [96]. Natural extracts, notably those derived from medicinal plants, have antioxidant and anti-inflammatory characteristics that aid in wound-healing processes [96]. These extracts contain active chemicals, including alkaloids, flavonoids, and phenols, which influence cellular responses throughout the wound-healing process, from hemostasis to remodeling [97]. These natural compounds can be integrated into cellulose matrices to provide additional therapeutic effects, such as reduced inflammation and increased tissue regeneration.

Antibiotics: Loading antibiotics into cellulose-based dressings enables targeted treatment of infected wounds, lowering the risk of systemic complications. This method is especially effective for treating wounds that are prone to serious bacterial infections [34].

8.4 Innovative Applications and Future Directions

The combination of cellulose with antibacterial agents has resulted in the development of sophisticated wound care solutions with increased functionality [37]. Bacterial cellulose, in particular, has great purity, porosity, and mechanical resilience, making it ideal for wound-healing applications [98]. Cellulose derivatives, which are semi-synthetic and water-soluble, provide additional benefits such as thermo-gelling behavior and antimicrobial capabilities [31]. These materials can be functionalized with antimicrobials to combat infections, including those caused by drug-resistant bacteria [74].

Future Advancements Are Anticipated to Concentrate on

Smart Dressings: These are designed to release antimicrobial compounds in response to certain stimuli, such as the presence of bacterial toxins or pH changes, allowing for on-demand infection management.

Biodegradable Dressings: The emphasis on sustainability has led to the development of entirely biodegradable cellulose-based dressings that can give antibacterial protection while remaining environmentally benign.

Personalized Wound Care: Advances in biotechnology and materials science may allow for the creation of customized wound dressings tailored to particular patient needs, including antibacterial medicines unique to the wound's microbiota.

9 Commercially Available Cellulose-Based Hemostatic Agents: Clinical Case Studies and Outcomes

Cellulose-based hemostatic agents are made from several types of cellulose, such as oxidized regenerated cellulose (ORC) and carboxymethyl cellulose (CMC). These agents are intended to enhance hemostasis by impeding blood flow, increasing clot formation, and providing a biocompatible environment for tissue repair. Commercially accessible cellulose-based hemostatic medicines, including Surgicel®, Traumacel®, Celox™, and WoundStat™, have improved bleeding management in various medical fields; see Table 2. These agents not only promote successful hemostasis but also help to improve patient outcomes, making them essential tools in both planned and emergency medical care.

Table 2 Examples of cellulose-based wound dressings available commercially

Brand name	Type of biomaterial	Application
Celmat C®	Nanofiber cellulose	Facial cosmetic treatments, atopic skin condition, chronic wounds, partial cosmetic treatments
Suprasorb® X PHMB	Polyhexamethylene biguanide, nanocellulose,	Arterial ulcers, venous ulcers, diabetic wounds, surgical incisions, skin grafts, and second-degree burns
Nanoderm Ag™	Nanocellulose and silver	Burns, dermabrasions, chronic wounds, and surgical incisions
Surgicel®	Oxidized regenerated cellulose	Maxillofacial surgery and oral
Promogran™	ORC/collagen	Pressure ulcers, diabetic wounds system, surgical wounds, and venous ulcers
Membracel®	Nanocrystalline cellulose	Ulcers, lacerations, and burns
FibDex®	Wood-based cellulose nanofiber	Skin graft, burn wound, and chronic wound
ActCel®	Regenerated cellulose	Capillary wounds, lacerations, punctures, arterial bleeding, and abrasions
BioFill®, BioProcess®, and Nexfill®	Bacterial cellulose	Lesions and burns
Traumacel®	Calcium salt of oxidized cellulose	Recalcitrant ulcers and chronic ulceration
xCell®	Bacterial cellulose	Venous ulcers and other chronic wounds
Oxicel®	Oxidized regenerated cellulose	Surgery, wound
Dermafill™	Nonwoven bacterial cellulose	Pressure ulcer, second-degree burns, abrasions, skin Tears, skin graft, and diabetic wound
Nanoskin®	Bacterial cellulose	Abrasion, facial peeling, and chronic ulcers

9.1 Key Commercial Products—Clinical Case Studies

9.1.1 Surgicel® (Ethicon, Johnson & Johnson)

Composition: Oxidized regenerated cellulose (ORC)

Surgical, a commercial ORC product, is offered in a variety of configurations for diverse uses [47]. Researchers have created improved versions, such as neutralized ORC (NORC), which exhibits higher biocompatibility and hemostatic efficacy [99].

Variants:

Surgicel® Original: The conventional ORC gauze format, suitable for general surgical applications [100].

Surgicel® Fibrillar provides improved conformability, allowing surgeons to mold the material to uneven surfaces [100].

Surgicel® Snow: An ultra-thin format with great absorbency and optimal surface covering. Surgical widely used in cardiovascular, gastrointestinal, and orthopedic operations to control diffuse and capillary bleeding. Advanced Surgicel® products (Fibrillar and SNoW) are more commonly utilized in high-risk bleeding conditions and sophisticated surgeries such as coronary artery bypass grafting and need fewer units per case than Surgicel® Original [100]. The use of enhanced Surgicel® products has been linked to lower hospital expenditures, shorter lengths of stay, and lower ICU costs in brain/cerebral, cardiovascular, and carotid endarterectomy surgeries [101]. Surgicel powder compares to various hemostatic powders in porcine bleeding models. Also effective in neurosurgery and organ transplantation [102]. Surgicel® was highly effective in controlling difficult-to-control venous bleeding in a critical and sensitive surgical situation. In a clinical investigation, a patient undergoing brain surgery for aneurysm repair demonstrated diffuse venous hemorrhage, complicating the process [103]. Surgicel® Original was used on the bleeding location. The outcome is: The product efficiently halted bleeding in minutes, shortening surgery time and reducing tissue handling. There were no reported postoperative problems with the agent. As a result, Surgicel® proved to be highly effective in treating difficult-to-control venous bleeding in a critical and delicate surgical situation.

9.1.2 Traumacel® (Bioster)

Composition: Carboxymethyl cellulose (CMC)

Traumacel® creates a hydrogel upon exposure to blood, adhering to the wound surface and establishing an effective barrier against hemorrhage. Its superior biocompatibility reduces the likelihood of inflammatory responses, guaranteeing safe utilization.

Applications: Traumacel® is specifically intended for surgical operations that involve significant hemorrhage, including gynecological and gastrointestinal surgery. It has also exhibited established effectiveness in emergency trauma management. In thoracic surgery, Traumacel Stapler Seam Protection has demonstrated efficacy in reinforcing stapler sutures and promoting tissue healing, attributable to its bactericidal and hemostatic characteristics [104].

Clinical Evidence: In a case study about liver resections, Traumacel® was employed when traditional hemostatic techniques proved inadequate due to the surgical location or the extent of hemorrhage. In 67 cases of liver resection, the surgical surface was covered with the Traumacel TAF net, which remained in situ. The findings revealed an absence of postoperative hemorrhage in this cohort, and no notable consequences were detected [105].

9.1.3 Celox™ and WoundStat™

Composition: CMC-based hemostatic agents formulated as powders or gauze.

Celox, a hemostatic substance derived from chitosan, has demonstrated considerable effectiveness in managing severe hemorrhage in military and civilian trauma situations. In Afghanistan, Celox effectively attained hemostasis in 18 of 21 soldiers with gunshot wounds upon initial application [106]. A randomized clinical research indicated that Celox-coated gauze decreased the time to hemostasis and blood loss relative to standard pressure dressings in civilian penetrating limb injuries [107]. In a porcine model replicating severe groin injuries, Celox markedly surpassed normal pressure dressings in bleeding management [108]. Clinical case assessments have corroborated Celox's efficacy in swiftly halting severe hemorrhage during emergency trauma scenarios [109]. These trials collectively underscore Celox's efficacy as a hemostatic agent in diverse trauma situations, indicating its potential to save lives in both pre-hospital and therapeutic environments.

9.2 Advantages of Cellulose-Based Commercial Hemostatic Agents

Efficient Hemostasis: Commercial cellulose-based medications swiftly enhance clot formation, proficiently managing hemorrhage in many surgical and traumatic contexts. Their capacity to attain hemostasis within minutes markedly diminishes the likelihood of problems associated with blood loss.

Application Convenience: These agents have been developed for user-friendly application, frequently offered in adaptable formats such as gauze, powders, or gels, facilitating simplicity of use during intricate surgical operations or emergency interventions.

Biocompatibility: Due to their cellulose-based composition, these solutions demonstrate significant biocompatibility, minimizing the likelihood of adverse reactions such as allergies or inflammation, hence facilitating safe application across diverse patient demographics.

Bioresorbable Characteristics: Numerous cellulose-derived hemostatic agents, including those with oxidized regenerated cellulose (ORC), possess resorbable properties, indicating they naturally decompose and are assimilated by the body, hence obviating the necessity for surgical extraction.

Application Versatility: These compounds exhibit adaptability across several medical applications, encompassing surgical procedures in cardiovascular, gastrointestinal, orthopedic, and gynecological domains, in addition to trauma care and military medicine.

Cost-Effectiveness: Their efficacy in hemorrhage management and reduction of procedural complications can result in decreased surgical duration, abbreviated hospital admissions, and diminished overall healthcare expenditures.

Enhanced Patient Outcomes: By effectively controlling hemorrhage and minimizing the likelihood of postoperative complications, cellulose-based hemostatic medicines provide superior clinical results and expedited recovery periods.

10 Advancements in Cellulose-Based Hemostatic Agents

10.1 Emerging Forms: Nanocellulose and Electrospun Cellulose Fibers

Recent developments in cellulose-based materials have led to the emergence of nanocellulose and electrospun cellulose fibers, both of which exhibit enhanced properties tailored for hemostatic applications.

10.1.1 Nanocellulose in Wound-Healing and Biomedical Applications

Nanocellulose, defined by its nanoscale dimensions, offers remarkable mechanical strength, extensive surface area, and biocompatibility, making it an advanced material for biomedical applications, particularly in wound care. Its high aspect ratio and ability to form interconnected networks facilitate rapid blood clot formation, which is critical for effective wound management [110].

Properties and Applications

Nanocellulose has garnered significant attention in recent decades due to its unique combination of biocompatibility, nontoxicity, biodegradability, and superior water absorption capacity. Its incorporation into electrospun nanocomposites enhances the mechanical properties, including Young's modulus and elongation at break, rendering it suitable for diverse biomedical applications [111, 112].

Additionally, its hydrophilicity and extensive hydroxyl (-OH) groups enable strong hydrogen bonding, improving adhesion to other polymers and enhancing water retention in nanocomposites [113].

Biomedical Advances

1. Bacterial cellulose (BC), a prominent form of nanocellulose, has seen extensive commercial use in wound dressings since its introduction by Johnson & Johnson in the 1980s. BC-based products such as Suprasorb X®, Bioprocess®, and XCell® effectively maintain a moist wound environment, absorb exudates, and promote tissue regeneration during the inflammatory phase [114]. These properties highlight BC's role as a reliable material for advanced wound management systems.
2. Nanocellulose aerogels, particularly those derived from dialdehyde nanocellulose fibers combined with collagen, have demonstrated promising results. These aerogels exhibit low density (0.02 g/cm^3), high water absorption (4000%), and excellent biocompatibility. In vitro studies have shown significant cell proliferation, with 96.79% viability of L929 cells over five days. Their high porosity (95%) surpasses the ideal requirements for hemostatic materials, making them viable for tissue engineering and as hemostatic sponges [115, 116].

Classification and Types of Nanocellulose

Nanocellulose is categorized into three main types: cellulose nanofibers (CNFs), cellulose nanocrystals (CNCs), and bacterial cellulose (BC). The primary difference between CNFs and CNCs lies in their dimensions and crystalline structure. CNFs, consisting of both crystalline and amorphous regions, provide flexibility, while CNCs are predominantly crystalline, offering high strength and thermal stability [117]. The production of CNFs involves mechanical treatments such as high-pressure homogenization and microfluidization, often combined with chemical pre-treatments like TEMPO-mediated oxidation to reduce energy consumption and achieve desirable surface chemistry [116].

CNCs, typically isolated through acid hydrolysis, retain crystalline domains resistant to acidic degradation. Sulfuric acid hydrolysis introduces surface sulfate esters, imparting a highly negative charge, enhancing their utility in biomedical applications such as drug delivery and protein immobilization [118]. CNCs derived from diverse sources like plant cell walls, cotton, and bacteria demonstrate high crystallinity (54–88%) and mechanical strength, making them ideal for medical and industrial applications [119].

Bacterial Cellulose (BC)

BC's moldability during biosynthesis allows for customized shapes, enhancing its applicability in wound dressings, tissue engineering scaffolds, and drug delivery systems [110].

Innovations in BC production using static and dynamic fermentation methods, as well as waste-derived carbon sources, have further expanded its potential as a sustainable biomaterial. BC's applications span various biomedical domains, including vascular grafts, dental implants, and scaffolds for corneal and cartilage tissue engineering. Its high tensile strength (200–300 MPa), thermal stability, and resistance to enzymatic degradation make it a reliable base material for next-generation biomaterials [119].

The versatility and exceptional properties of nanocellulose have positioned it as a next-generation material for wound-healing and biomedical applications. Its renewable nature, coupled with its capacity for chemical modification, enables tailored solutions for diverse medical needs. Continued advancements in nanocellulose production and functionalization will likely drive further innovations in healthcare materials.

10.1.2 Electrospun Cellulose Fibers

Electrospun cellulose fibers are produced by creating ultra-fine, porous structures that mimic the extracellular matrix, enhancing cellular adhesion, and proliferation. These fibers also offer controlled drug release, making them suitable for hemostatic applications, especially in complex wounds. The rise of nanotechnology has led to the development of cellulose nanofibers through electrospinning, forming nanocomposite wound dressings that protect wounds and release drugs to prevent postoperative adhesions, promote hemodialysis, and aid tissue repair [120].

Cellulose electrospun nanofibers are attractive due to their fine structure, biocompatibility, biodegradability, hydrophilicity, low density, thermostability, and ease of modification [121]. However, challenges remain in processing cellulose into nanofibers via electrospinning, mainly due to its crystalline structure, rigidity, and strong inter- and intramolecular hydrogen bonding, which hinder its solubility and prevent it from melting [122]. To address these limitations, researchers have used solvents like NMMO, LiCl/DMAc, or ionic liquids (e.g., EmimAc), but these methods are costly and cumbersome, restricting the use of cellulose in large-scale wound-healing applications and increasing the use of cellulose derivatives like cellulose acetate (CA) [123].

The performance of electrospun bioactive dressings depends on the properties of the polymer, drug solubility, drug-polymer interactions, and mat structure. Electrospun mats containing antimicrobial agents outperform films produced by other techniques in water uptake, permeability, drug release rate, and antibacterial properties [124]. Antimicrobial agents such as drugs, nanoparticles, and natural extracts are incorporated into these dressings to reduce infection risks, providing anti-inflammatory, pain-relieving, vasodilation, and antimicrobial effects [125].

espite progress, producing cellulose-based electrospun mats is challenging due to its crystalline structure and rigidity [122]. Solutions include using specific solvents, blending with complementary polymers, or converting cellulose into derivatives. Common solvents for electrospinning cellulose include ionic liquids, alkali/aqueous solvents (NaOH/urea), and polar aprotic solvents with electrolytes (DMAc/LiCl), though these solvents are not volatile and may remain in the final product, limiting large-scale production [123]. Additionally, achieving the correct viscosity for electrospinning requires careful solvent selection and often the use of technologies like melt-electrospinning. Converting cellulose into derivatives like cellulose acetate (CA), cellulose acetate phthalate (CAP), and carboxymethyl cellulose (CMC) simplifies processing, as these derivatives dissolve under various pH conditions, making them ideal for biomedical applications [126].

10.2 Hybrid Cellulose-Based Composites with Polymers and Bioactive Agents for Advanced Wound Care

The integration of cellulose with polymers and bioactive agents has led to the development of hybrid materials with enhanced properties. These composites capitalize on the natural characteristics of cellulose, such as its hydrophilicity and mechanical stability, while incorporating the functionalities of synthetic or natural polymers to improve hemostatic efficacy. Additionally, the incorporation of bioactive agents imparts antimicrobial properties and promotes tissue regeneration.

Hybrid cellulose materials have demonstrated superior swelling capacities, improved adhesion to wound surfaces, and accelerated clotting times, making them suitable for advanced wound care applications. The adaptability of cellulose in forming composites highlights its potential as a versatile platform for designing next-generation hemostatic agents. Nanocellulose, in particular, has shown promise in enhancing drug delivery strategies, as its highly reactive structure enables interactions with other molecules, enzymes, or proteins. This interaction can address the low solubility of many drugs in aqueous environments, thereby improving their bioavailability.

Furthermore, the –OH groups of nanocellulose can be modified through various techniques, such as physical adsorption, surface graft polymerization, and covalent bonding, to improve the performance of incorporated biomolecules. This leads to strong polymer-filler interactions, significantly enhancing the mechanical properties of the resulting materials [127]. However, the in vivo behavior of nanocelluloses remains underexplored. Some studies suggest that toxicity is concentration-dependent, with nanocelluloses not exhibiting toxicity at concentrations below 1 mg/ml, while higher concentrations (2–5 mg/ml) may induce apoptotic toxicity. Notably, anionic nanocelluloses, such as carboxymethylated CNF, have been found to exhibit more cytotoxicity than cationic variants like trimethylammonium-CNF [128].

Among the various types of nanocellulose, bacterial cellulose (BC) is regarded as highly biocompatible and has been successfully applied in wound dressings. However, challenges remain in electrospinning BC due to its structural characteristics [129].

The combination of BC with synthetic and natural polymers has been explored to enhance the morphological, physicochemical, and biological properties of materials. Polymers such as chitosan, polycaprolactone (PCL), polyethylene oxide (PEO), ethylene vinyl alcohol (EVOH), polyvinyl alcohol (PVA), polylactic acid (PLA), polyacrylonitrile (PAN), and silk have been electrospun with BC. Functionalization with 3-aminopropyl triethoxysilane (APS) has further improved cell attachment and antibacterial properties of BC-based electrospun membranes for wound healing. BC membranes grafted with organosilanes and acetylated derivatives have shown improved moisture resistance and hydrophobicity. Naeem et al. [130] developed an in situ BC synthesis process on cellulose acetate (CA)-based electrospun mats through self-assembly to create advanced wound dressings. Despite the promising applications of cellulose nanofibers (CNFs) as reinforcing agents in polymeric composites, there is limited research on incorporating biomolecules into CNF-based fibers. Thus, a deeper exploration of BC and cellulose nanocrystals (CNCs) is warranted.

10.2.1 Drug Loading and Delivery Systems

A variety of hydrophilic and hydrophobic drugs have been incorporated into electrospun polymeric nanofibers. Typically, drugs are added to a polymer solution, ensuring uniform distribution. This method enables the loading of significant drug amounts by adjusting the solution's concentration. However, direct incorporation of drugs into polymer solutions can alter the solution's conductivity, viscosity, and surface tension, affecting the electrospinning process and the morphology of the resulting nanofibers. Furthermore, drugs tend to leach rapidly into aqueous environments, as they are usually concentrated at or near the surface of the fibers. The conventional electrospinning technique allows for partial control over drug release by modulating the pore size and polymer degradation rate, although burst release followed by cytotoxic effects remains a challenge. To overcome these limitations, novel drug delivery systems utilizing multi-fluid, coaxial, and triaxial electrospinning have been developed. These methods facilitate more controlled drug release by isolating a portion of the drug within the fiber core and another near the surface, resulting in a sustained release over time. However, as shown by Yu et al. [131], varying feeding rates during coaxial electrospinning can influence fiber morphology and drug release profiles, demonstrating that optimized parameters are critical for producing high-quality, drug-loaded fibers.

10.2.2 Nanoparticles in Wound Healing

Nanoparticles (NPs) play a critical role in wound healing by promoting antimicrobial activity, angiogenesis, immunomodulation, and cell and drug delivery. NPs can be classified based on their cellular uptake, size, shape, role, and origin (organic or inorganic). Inorganic NPs include carbon-based, metal, metal oxide, semiconducting, and ceramic NPs, while organic NPs are derived from polymers and biomolecules. NPs function as delivery vehicles, protecting and releasing active compounds locally, or by exerting therapeutic effects through their intrinsic properties. The antimicrobial action of NPs arises from their ability to produce reactive oxygen species and interfere with microbial DNA and RNA, inhibiting microbial reproduction [132].

Incorporating NPs into polymeric matrices can enhance the efficacy of wound-healing dressings, with materials such as bioactive glass, gold, copper, cerium, zinc oxide, titanium dioxide, gallium, nitric oxide, and silver NPs (AgNPs) showing significant antimicrobial activity. AgNPs, in particular, have demonstrated efficacy against multidrug-resistant bacteria and have been shown to reduce pro-inflammatory cytokines while enhancing fibroblast migration and keratinocyte proliferation, all of which are crucial for wound healing [133, 134]. However, high concentrations of AgNPs can be toxic to human cells, inhibiting immune cell recruitment and epidermal cell regrowth, potentially hindering wound healing. Therefore, careful dosing is essential to balance antimicrobial effects with biocompatibility.

10.2.3 Natural Extracts in Wound Care

The use of natural biomolecules derived from plant extracts is gaining momentum in wound healing due to concerns over antibiotic resistance and the toxicity of silver-based compounds. Bioactive molecules such as thymol, asiaticoside, curcumin, and gingerol found in plant extracts and possess antimicrobial, antioxidant, and immunomodulatory properties that support tissue regeneration and inflammatory response regulation. Aloe vera, neem (Azadirachta indica), and Panax ginseng have been used traditionally for wound care, with their active compounds showing promise in promoting wound healing [135].

Essential oils (EOs) extracted from aromatic plants have intrinsic antibacterial, antifungal, and insecticidal properties, making them valuable additions to polymeric wound dressings. The hydrophobic nature of EOs enhances their interaction with bacterial cell membranes, increasing membrane permeability and leading to cell death. EO-loaded electrospun nanofibers, particularly those containing oregano, cinnamon, lemongrass, and peppermint oils, have shown potent antimicrobial effects. However, the cytotoxicity of EOs at higher concentrations remains a challenge for their widespread use in wound care [136].

Recent advancements in the controlled release of EOs from electrospun nanofibers have shown promising results in extending the therapeutic efficacy of these natural compounds while minimizing cytotoxicity [137, 138].

10.2.4 Wound Healing and Bioactive Dressings

Wound healing is a complex biological process involving multiple cellular and molecular interactions. Chronic wounds, which fail to heal properly, represent a significant clinical challenge. Traditional wound care treatments often fail to meet the diverse needs of these wounds. As a result, bioactive dressings, incorporating cellulose derivatives and bioactive molecules, have been developed to improve wound-healing outcomes. These dressings are designed to modulate the wound environment, promote tissue regeneration, and reduce the risk of infection.

10.3 Smart Cellulose-Based Wound Dressings

Smart wound dressings have emerged as a significant advancement in utilizing cellulose-based materials for wound care, offering enhanced management of hemostasis and wound healing. These innovative dressings are designed to respond to the dynamic physiological conditions of wounds, including variations in pH, temperature, and moisture content. By incorporating stimuli-responsive components, such as bioactive agents, these dressings facilitate the controlled release of therapeutic substances, creating an ideal environment for wound healing.

Cellulose, with its intrinsic biocompatibility and biodegradability, provides an optimal scaffold for these smart dressings. Noteworthy developments in this area include bacterial cellulose-based dressings, which offer superior management of exudates, and nanocellulose-based materials, capable of releasing bioactive molecules in response to specific stimuli. These innovations contribute to improved patient outcomes, reducing infection risks and accelerating the healing process.

An effective wound dressing must meet a range of requirements tailored to the specific characteristics of the wound. Important considerations include [139–141]:

(a) maintaining a moist wound environment;
(b) facilitating epidermal mobility;
(c) promoting angiogenesis and tissue regeneration;
(d) ensuring oxygen and water vapor permeability;
(e) maintaining appropriate temperature to stimulate blood circulation;
(f) providing protection against microbial pathogens;
(g) being non-adhesive and easily removable;
(h) supporting debridement and promoting leukocyte migration;
(i) being sterile, non-toxic, and hypoallergenic;
(j) offering biocompatibility and effective drug delivery;
(k) demonstrating high absorption capacity for exudates.

10.3.1 pH Responsive Wound Dressings

pH plays a pivotal role in the healing process throughout all stages of wound recovery. The normal pH of healthy skin typically ranges from 4.5 to 6.5, but acute and chronic wounds often exhibit different pH levels. Infection and the production of alkaline by-products can elevate wound pH to values between 7 and 9. Several methods are employed to monitor pH levels during wound healing, including colorimetric and electrochemical techniques; however, these methods can be costly and labor-intensive. The development of sensor-based wound dressings provides a more practical and cost-effective solution. Technologies such as flexible circuit transducers, optical sensor-based dyes, pH-sensitive colorants, electrodes, and carbon dots have been explored as strategies for creating pH-sensitive dressings [140, 142].

10.3.2 Temperature Responsive Wound Dressings

Temperature is another critical factor influencing wound healing, as it affects enzymatic and biochemical processes at the injury site. Normal healing wounds generally maintain a temperature of approximately 37.8 °C, and deviations of as little as 2.2 °C—either an increase or a decrease—can hinder healing, indicating possible ischemia or infection. Temperature monitoring can, therefore, serve as a valuable method for assessing wound status. Various sensor types, including infrared sensors, thermistors, and resistance-based temperature sensors, are employed to measure temperature fluctuations, with materials such as polyurethane nanofibers, graphene, platinum, and gold being incorporated into sensor designs [143, 144].

10.3.3 Pressure Responsive Wound Dressings

Pressure-induced wounds, such as those found in diabetic foot ulcers or pressure ulcers, are commonly treated with therapies aimed at alleviating excessive pressure. Prolonged pressure on the wound site impedes blood circulation, leading to tissue ischemia and hindering the healing process. Flexible pressure sensors integrated into wound dressings allow for real-time monitoring of pressure at the wound site. These sensors can operate based on piezoresistive, capacitive, triboelectric, or piezoelectric principles. The development of advanced flexible pressure sensors continues to progress, improving the ability to monitor wound conditions [145, 146].

10.3.4 Moisture Responsive Wound Dressings

The moisture content of a wound is a crucial factor throughout the healing process. Excess moisture can lead to maceration, while insufficient moisture can slow healing by causing the wound to dry out. Excessive moisture in wounds may result from the natural exudates produced by the injury, as well as external factors like sweating or urinary incontinence. Excessive moisture can lead to frequent dressing changes, which may damage the surrounding skin. In this context, moisture-responsive dressings with integrated sensors offer advantages over traditional dressings, providing real-time feedback on moisture levels. The first commercially available moisture-sensitive wound dressing, "Wound Sense," was introduced in 2016. These advanced dressings are often constructed using composite materials, such as polyvinyl alcohol, carbon nanotubes, and graphene, to improve breathability and moisture regulation [147, 148].

10.3.5 Sustained Drug-Releasing Wound Dressings

Stimuli-responsive dressings enable the targeted, controlled release of therapeutic agents to the wound site, enhancing treatment efficacy. Hydrogels are commonly employed in these dressings, facilitating sustained drug release. A variety of polymers, such as poly lactide-co-glycolide, polyvinyl alcohol, chitosan, and hyaluronic acid, have been studied for their use in drug delivery systems within wound dressings [149, 150].

Moreover, certain wound dressings exhibit self-healing properties, and some are available in injectable or sprayable forms. These multifunctional dressings are designed to address a variety of wound types and healing stages. This review highlights the most recent research on multifunctional wound dressings, emphasizing their potential advantages in promoting wound healing.

The field of cellulose-based wound dressings has witnessed remarkable advancements, particularly with the incorporation of nanocellulose, electrospun fibers, hybrid composites, and smart materials. These innovations harness the unique properties of cellulose to address the complexities of wound healing and hemostasis. As ongoing research further explores these materials, cellulose-based dressings are expected to play an increasingly critical role in modern wound care practices.

11 Regulatory Considerations and Standards

Regional variations exist in the regulatory requirements for cellulose-based hemostatic agents, which are mainly overseen by organizations like the European Medicines Agency (EMA) and the U.S. Food and Drug Administration (FDA). All organizations require clinical and pre-clinical tests as well as post-market surveillance to be conducted to find and deal with potential risks as soon as possible. Depending on its intended purpose, wound dressings may be subject to different regulatory standards (e.g., absorbent dressings vs. antimicrobial dressings). Specific testing procedures could be needed for each kind in order to evaluate performance attributes including fluid absorption rates and antibacterial efficacy.

11.1 FDA Regulations

Classification: Biomaterials are categorized according to their level of risk and intended usage. Cellulose-based products may fall under Class I (low risk), Class II (moderate risk), or Class III (high risk) classifications, which determine the regulatory pathway for approval.

BiocompatibilityTesting: Biocompatibility testing is done in accordance with ISO 10993 standards. Assessments of cytotoxicity, sensitization, irritation, and systemic toxicity are included.

11.2 European Union Regulations (CE Mark)

Compliance with Medical Device Regulation (MDR): The MDR specifies performance and safety standards for medical devices marketed in the European Union and manufacturers are required to perform a conformity assessment under the MDR (EU 2017/745) to show compliance with the law.

Risk Assessment: To find any possible risks of cellulose-based materials, a comprehensive risk assessment needs to be carried out. The testing and documentation needed for CE marking are informed by this assessment.

12 Future Perspectives in Cellulose-Based Hemostatic Research

The field of cellulose-based hemostatic materials holds immense promise for advancing wound care through innovative approaches, particularly in biofabrication, patient-specific customization, and integration with tissue-engineered products. Wound healing, a multifaceted process governed by three distinct phases, can be severely disrupted, leading to non-healing chronic wounds that are challenging to treat. Recent progress has led to the development of advanced therapeutic strategies, including electrospinning, which is highly effective in producing 3D structured dressings that mimic the skin's extracellular matrix. These electrospun dressings exhibit superior characteristics, such as a high surface area-to-volume ratio and porous architecture, facilitating homeostasis, absorption of exudates, gas exchange, cell adhesion, and proliferation while mitigating the risk of infection. Recent advancements have focused on incorporating cellulose and its derivatives into electrospun nanofiber meshes, often modified with specialized biomolecules, offering enhanced functionality for wound management.

To address challenges such as antibiotic resistance and overuse of silver compounds, researchers are exploring novel strategies, including integrating natural plant extracts, alternative drugs, and organic or inorganic nanoparticles with cellulose-based nanofibrous systems. These innovative combinations have shown promising results in accelerating wound healing. Furthermore, nanocellulose has emerged as a valuable material due to its high surface area and abundance of hydroxyl groups, enabling effective binding with proteins and enzymes. This binding capability not only improves solubility but also enhances the mechanical properties of wound dressings. However, challenges persist in processing cellulose, particularly through electrospinning, leading to a shift toward cellulose acetate (CA) as the most widely used derivative for fabricating drug-loaded dressing systems. Despite these advances, in vivo studies are still needed to establish the safety and efficacy of these materials, as concerns regarding their cytotoxicity remain unresolved.

The potential of cellulose-based materials for wound dressings lies in their renewable nature, high water retention capacity, and ability to promote cell attachment, proliferation, and migration, all while demonstrating minimal toxicity. However, the integration of biomolecules into these dressings is not without challenges. Addressing issues such as limited electrospinnability, solvent selection, polymer combinations, and pre-treatments to enhance solubility are crucial for producing effective and reproducible wound dressings. Additionally, surface functionalization with chemical groups to facilitate biomolecule binding offers further opportunities for advancement.

Sustainability is an increasingly critical focus, with efforts directed toward developing eco-friendly production methods, such as employing green solvents, low-energy manufacturing processes, and biodegradable complementary polymers. While these approaches offer promise, large-scale production of sustainable, high-performance wound dressings with minimal environmental impact remains a challenge. Continued research is essential to create "green" wound dressings that not only accelerate healing and prevent infections but also align with the broader goals of environmental stewardship. By integrating advanced biofabrication techniques, sustainable practices, and personalized solutions, cellulose-based hemostatic materials are set to play a transformative role in modern wound care.

13 Conclusion

The advancements in cellulose-based hemostatic agents have revolutionized the field of wound care by leveraging the inherent properties of cellulose and its derivatives. From traditional forms to highly engineered nanomaterials, cellulose has demonstrated remarkable versatility, biocompatibility, and efficacy in controlling bleeding, promoting tissue regeneration, and reducing infection risks. Modifications such as oxidation, functionalization, and hybridization with bioactive agents have further enhanced the hemostatic efficiency, mechanical stability, and therapeutic potential of these materials.

Cellulose-based innovations, including smart wound dressings and nanocellulose composites, hold great promise for addressing current challenges in wound management. These materials provide tailored solutions for a wide range of clinical scenarios, offering benefits such as rapid clot formation, controlled drug delivery, and stimuli-responsive functionalities. Moreover, their biodegradability and environmentally sustainable nature align with the growing emphasis on eco-friendly medical solutions.

As research continues to explore new modifications and applications, the integration of cellulose-based agents into personalized and precision medicine is poised to redefine the standards of care. By bridging the gap between material science and biomedical applications, cellulose-based hemostatic agents represent a transformative step toward improving patient outcomes and advancing the field of regenerative medicine.

References

1. Wallace HA, Basehore BM, Zito PM (2023) Wound healing phases. In: StatPearls. StatPearls Publishing, Treasure Island (FL). Available from: https://www.ncbi.nlm.nih.gov/books/NBK470443/
2. Schultz GS, Chin GA, Moldawer L, Diegelmann RF (2011) Principles of wound healing. In: Fitridge R, Thompson M (eds) Mechanisms of vascular disease: a reference book for vascular specialists. University of Adelaide Press, Adelaide (AU), p 23. Available from: https://www.ncbi.nlm.nih.gov/books/NBK534261/
3. Barmore W, Bajwa T, Burns B (2023) Biochemistry, clotting factors. In: StatPearls. StatPearls Publishing, Treasure Island (FL). Available from: https://www.ncbi.nlm.nih.gov/books/NBK507850/
4. Palta S, Saroa R, Palta A (2014) Overview of the coagulation system. Indian J Anaesth 58(5):515–523. https://doi.org/10.4103/0019-5049.144643
5. Pérez-Gómez F, Bover R (2007) The new coagulation cascade and its possible influence on the delicate balance between thrombosis and hemorrhage. Rev Esp Cardiol 60(12):1217–1219. Available from: https://www.revespcardiol.org/en-the-new-coagulation-cascade-and-its-poss-articulo-13114167
6. Adams RLC, Bird RJ (2009) Coagulation cascade and therapeutics update: relevance to nephrology. Part 1: Overview of coagulation, thrombophilias and history of anticoagulants. Nephrology 14:462–470. https://doi.org/10.1111/j.1440-1797.2009.01128.x
7. Turitto VT, Hall CL (1998) Mechanical factors affecting hemostasis and thrombosis. Thromb Res 92(Suppl 2):S25–S31. https://doi.org/10.1016/s0049-3848(98)00157-1
8. Streiff MB (2023) Overview of hemostasis. MSD Manual. Available from: https://www.msdmanuals.com/professional/hematology-and-oncology/hemostasis/overview-of-hemostasis
9. LaPelusa A, Dave HD (2023) Physiology, hemostasis. In: StatPearls. StatPearls Publishing, Treasure Island (FL). Available from: https://www.ncbi.nlm.nih.gov/sites/books/NBK545263/
10. Guo Y, Wang M, Liu Q, Liu G, Wang S, Li J (2023) Recent advances in the medical applications of hemostatic materials. Theranostics 13(1):161–196. https://doi.org/10.7150/thno.79639IF
11. Gupta R, Mohanty S, Verma D (2023) Current status of hemostatic agents, their mechanism of action, and future directions. J Bioact Compat Polym 38(2):77–105. https://doi.org/10.1177/08839115221147935
12. Kommu SS, McArthur R, Emara AM, Reddy UD, Anderson CJ, Barber NJ, Persad RA, Eden CG (2015) Current status of hemostatic agents and sealants in urologic surgical practice. Rev Urol 17(3):150–159
13. Jamali B, Nouri S, Amidi S (2024) Local and systemic hemostatic agents: a comprehensive review. Cureus 16(10):e72312. https://doi.org/10.7759/cureus.72312
14. Yoon H, Na Y, Choi K, Huh W, Kim J (2019) Wound healing effect of regenerated oxidized cellulose versus fibrin sealant patch: an in vivo study. Arch Craniofac Surg 20:289–296. https://doi.org/10.7181/acfs.2019.00472
15. Yu P, Zhong W (2021) Hemostatic materials in wound care. Burns Trauma 9:tkab019. https://doi.org/10.1093/burnst/tkab019

16. Brown KGM, Solomon MJ (2024) Topical haemostatic agents in surgery. Br J Surg 111(1):znad361. https://doi.org/10.1093/bjs/znad361
17. Gupta PK, Raghunath SS, Prasanna DV, Venkat P, Shree V, Chithananthan C, Choudhary S, Surender K, Geetha K (2019) An update on overview of cellulose, its structure and applications. Cellulose 201(9):84727
18. Gopi S, Balakrishnan P, Chandradhara D, Poovathankandy D, Thomas S (2019) General scenarios of cellulose and its use in the biomedical field. Mater Today Chem 13:59–78
19. Wohlert M, Benselfelt T, Wågberg L, Furó I, Berglund LA, Wohlert J (2022) Cellulose and the role of hydrogen bonds: not in charge of everything. Cellulose 1–23
20. Bhaladhare S, Das D (2022) Cellulose: a fascinating biopolymer for hydrogel synthesis. J Mater Chem B 10:1923–1945
21. Hickey RJ, Pelling AE (2019) Cellulose biomaterials for tissue engineering. Front Bioeng Biotechnol 7:45. https://doi.org/10.3389/fbioe.2019.00045
22. Mattonai M, Pawcenis D, Del Seppia S, Łojewska J, Ribechini E (2018) Effect of ball-milling on crystallinity index, degree of polymerization and thermal stability of cellulose. Bioresour Technol 270:270–277
23. Reis DT, dos Santos Pereira AK, Scheidt GN, Pereira DH (2019) Plant and bacterial cellulose: production, chemical structure, derivatives and applications. Orbital Electron J Chem 321–329
24. Srivastava S, Mathur G (2023) Bacterial cellulose: a multipurpose biomaterial for manmade world. Curr Appl Sci Technol 10:55003
25. Jadczak K, Ochędzan-Siodłak W (2023) Bacterial cellulose: biopolymer with novel medical applications. J Biomater Appl 38(1):51–63
26. Fatema N, Ceballos RM, Fan C (2022) Modifications of cellulose-based biomaterials for biomedical applications. Front Bioeng Biotechnol 10:993711
27. Radotić K, Mićić M (2016) Methods for extraction and purification of lignin and cellulose from plant tissues. In: Sample preparation techniques for soil, plant, and animal samples. pp 365–376
28. Hu M, Lv X, Wang Y, Zhang Y, Dai H (2024) Guideline for the extraction of nanocellulose from lignocellulosic feedstocks. Food Biomacromol 1(1):9–17
29. Magalhães S, Fernandes C, Pedrosa JF, Alves L, Medronho B, Ferreira PJ, Rasteiro MD (2023) Eco-friendly methods for extraction and modification of cellulose: an overview. Polymers 15(14):3138. https://doi.org/10.3390/polym15143138
30. Jafari A, Al-Ostaz A, Nouranian S (2024) Biomedical potential of cellulose: current trends and future directions. Polym Adv Technol 35(12):e6621
31. Tudoroiu EE, Dinu-Pîrvu CE, Albu Kaya MG, Popa L, Anuța V, Prisada RM, Ghica MV (2021) An overview of cellulose derivatives-based dressings for wound-healing management. Pharmaceuticals 14(12):1215
32. Elangwe CN, Morozkina SN, Olekhnovich RO, Krasichkov A, Polyakova VO, Uspenskaya MV (2022) A review on chitosan and cellulose hydrogels for wound dressings. Polymers 14(23):5163
33. Pinho E, Soares G (2018) Functionalization of cotton cellulose for improved wound healing. J Mater Chem B 6(13):1887–1898
34. Abazari MF, Gholizadeh S, Karizi SZ, Birgani NH, Abazari D, Paknia S, Derakhshankhah H, Allahyari Z, Amini SM, Hamidi M, Delattre C (2021) Recent advances in cellulose-based structures as the wound-healing biomaterials: a clinically oriented review. Appl Sci 11(17):7769
35. Bhowmik S, Islam JM, Debnath T, Miah MY, Bhattacharjee S, Khan MA (2017) Reinforcement of gelatin-based nanofilled polymer biocomposite by crystalline cellulose from cotton for advanced wound dressing applications. Polymers 9(6):222
36. Fahma F, Firmanda A, Cabral J, Pletzer D, Fisher J, Mahadik B, Arnata IW, Sartika D, Wulandari A (2023) Three-dimensional printed cellulose for wound dressing applications. 3D Print Addit Manuf 10(5):1015–1035
37. Momin M, Mishra V, Gharat S, Omri A (2021) Recent advancements in cellulose-based biomaterials for management of infected wounds. Expert Opin Drug Deliv 18(11):1741–1760

38. Hu Y, Catchmark JM (2011) Integration of cellulases into bacterial cellulose: toward bioabsorbable cellulose composites. J Biomed Mater Res B Appl Biomater 97(1):114–123
39. Orlando I, Basnett P, Nigmatullin R, Wang W, Knowles JC, Roy I (2020) Chemical modification of bacterial cellulose for the development of an antibacterial wound dressing. Front Bioeng Biotechnol 8:557885
40. Bisaria VS, Ghose TK (1981) Biodegradation of cellulosic materials: substrates, microorganisms, enzymes and products. Enzyme Microb Technol 3(2):90–104
41. Cheng QZ, Wang JX, McNeel JF, Jacobson PM (2010) Water retention value measurements of cellulosic materials using a centrifuge technique
42. He W, Wu J, Xu J, Mosselhy DA, Zheng Y, Yang S (2021) Bacterial cellulose: functional modification and wound healing applications. Adv Wound Care 10(11):623–640
43. Gindl W, Martinschitz KJ, Boesecke P, Keckes J (2006) Changes in the molecular orientation and tensile properties of uniaxially drawn cellulose films. Biomacromolecules 7(11):3146–3150
44. Krässig H, Kitchen W (1961) Factors influencing tensile properties of cellulose fibers. J Polym Sci 51(155):123–172
45. Hickey RJ, Pelling AE (2019) Cellulose biomaterials for tissue engineering. Front Bioeng Biotechnol 7
46. Modulevsky DJ, Cuerrier CM, Pelling AE (2016) Biocompatibility of subcutaneously implanted plant-derived cellulose biomaterials. PLoS ONE 11(6):e0157894
47. Zhang S, Li J, Chen S, Zhang X, Ma J, He J (2020) Oxidized cellulose-based hemostatic materials. Carbohydr Polym 230:115585
48. Coseri S (2017) Cellulose: to depolymerize… or not to? Biotechnol Adv 35(2):251–266
49. Milanović J, Korica M, Nikolić T, Kostić M (2019) OXYCELLULOSE–material for new application areas. Contemp Mater 10(1)
50. Wang Y, Guo Y, Liu Y, Zhao X, Huang Y, Zhang X, Hu X, Mequanint K, Luo G, Xing M (2024) Platelet vesicles synergetic with biosynthetic cellulose aerogels for ultra-fast hemostasis and wound healing. Adv Healthc Mater 2304523
51. Shefa AA, Taz M, Lee SY, Lee BT (2019) Enhancement of hemostatic property of plant-derived oxidized nanocellulose-silk fibroin-based scaffolds by thrombin loading. Carbohydr Polym 208:168–179
52. Bethke K, Palantöken S, Andrei V, Roß M, Raghuwanshi VS, Kettemann F, Greis K, Ingber TT, Stückrath JB, Valiyaveettil S, Rademann K (2018) Functionalized cellulose for water purification, antimicrobial applications, and sensors. Adv Funct Mater 28(23):1800409
53. Poletto M, Pistor V, Zattera AJ (2013) Structural characteristics and thermal properties of native cellulose. Cellulose Fundam Asp 2:45–68
54. Křížová P, Mášová L, Suttnar J, Salaj P, Dyr JE, Homola J, Pecka M (2007) The influence of intrinsic coagulation pathway on blood platelets activation by oxidized cellulose. J Biomed Mater Res A 82(2):274–280
55. Wang L, Xiang BR, Zou QG (2009) Preparation of oxidized cellulose and the related study on its structure and performance. Prog Pharm Sci 8:365–369
56. Rysava J, Masova L, Dyr JE, Suttnar J, Salaj P, Dostalek J, Myska K, Pecka M (2002) Effect of oxidized cellulose on fibrin formation and blood platelets. Cas Lek Cesk 141:50–53
57. Kollár P, Suchý P, Muselik J, Bajerová M, Havelka P, Sopuch T (2008) Hemostatic effects of oxidized cellulose. Ceska Slov Farm 57(1):11–16
58. Rodić-Grabovac B, Sailović P, Lipić N (2020) Medical and pharmaceutical application of oxidized cellulose. J Chem Technol Environ 1(1)
59. Martina B, Kateřina K, Miloslava R, Jan G, Ruta M (2009) Oxycellulose: significant characteristics in relation to its pharmaceutical and medical applications. Adv Polym Technol 28(3):199–208
60. Thomas P, Duolikun T, Rumjit NP, Moosavi S, Lai CW, Johan MRB, Fen LB (2020) Comprehensive review on nanocellulose: recent developments, challenges and future prospects. J Mech Behav Biomed Mater 110:103884

61. Basu A, Hong J, Ferraz N (2017) Hemocompatibility of Ca2+-crosslinked nanocellulose hydrogels: toward efficient management of hemostasis. Macromol Biosci 17(11):1700236
62. Lu X, Li X, Yu J, Ding B (2022) Nanofibrous hemostatic materials: structural design, fabrication methods, and hemostatic mechanisms. Acta Biomater 154:49–62
63. Bacakova L, Pajorova J, Bacakova M, Skogberg A, Kallio P, Kolarova K, Svorcik V (2019) Versatile application of nanocellulose: from industry to skin tissue engineering and wound healing. Nanomaterials 9(2):164
64. Mohamed E, Wang Y, Crispin PJ, Fitzgerald A, Dahlstrom JE, Fowler S, Nisbet DR, Tsuzuki T, Coupland LA (2022) Superior hemostatic and wound-healing properties of gel and sponge forms of nonoxidized cellulose nanofibers: in vitro and in vivo studies. Macromol Biosci 22(10):2200222
65. Du J, Wang J, Xu T, Yao H, Yu L, Huang D (2023) Hemostasis strategies and recent advances in nanomaterials for hemostasis. Molecules 28(13):5264
66. Pettignano A, Charlot A, Fleury E (2019) Carboxyl-functionalized derivatives of carboxymethyl cellulose: towards advanced biomedical applications. Polym Rev 59(3):510–560
67. Rahman MS, Hasan MS, Nitai AS, Nam S, Karmakar AK, Ahsan MS, Shiddiky MJ, Ahmed MB (2021) Recent developments of carboxymethyl cellulose. Polymers 13(8):1345
68. Ohta S, Nishiyama T, Sakoda M, Machioka K, Fuke M, Ichimura S, Inagaki F, Shimizu A, Hasegawa K, Kokudo N, Kaneko M (2015) Development of carboxymethyl cellulose nonwoven sheet as a novel hemostatic agent. J Biosci Bioeng 119(6):718–723
69. Pettignano A, Charlot A, Fleury E (2019) Carboxyl-functionalized derivatives of carboxymethyl cellulose: towards advanced biomedical applications. Polym Rev 59(3):510–560
70. Oliveira RN, McGuinness GB (2018) Blended gels of sodium carboxymethyl cellulose incorporating antimicrobials for absorbance and wound healing applications. Springer, Cham, Switzerland, pp 1–29
71. Luo P, Liu L, Xu W, Fan L, Nie M (2018) Preparation and characterization of aminated hyaluronic acid/oxidized hydroxyethyl cellulose hydrogel. Carbohydr Polym 199:170–177
72. Tundisi LL, Mostaço GB, Carricondo PC, Petri DFS (2021) Hydroxypropyl methylcellulose: physicochemical properties and ocular drug delivery formulations. Eur J Pharm Sci 159:105736
73. Heinze T, Liebert T (2004) Chemical characteristics of cellulose acetate. Macromol Symp 208(1):167–238
74. Momin M, Mishra V, Gharat S, Omri A (2021) Recent advancements in cellulose-based biomaterials for management of infected wounds. Expert Opin Drug Deliv 18(11):1741–1760
75. Orlando I, Roy I (2019) Cellulose-based hydrogels for wound healing. Springer, Heidelberg
76. Kiselioviene S, Baniukaitiene O, Harkavenko V, Babenko NA, Liesiene J (2016) Cellulose hydrogel sheets for wound dressings. Cellul Chem Technol 50:915–923
77. Berglund L, Squinca P, Baş Y, Zattarin E, Aili D, Rakar J, Junker J, Starkenberg A, Diamanti M, Sivlér P, Skog M (2023) Self-assembly of nanocellulose hydrogels mimicking bacterial cellulose for wound dressing applications. Biomacromolecules 24(5):2264–2277
78. Tabatabaei Hosseini BS, Meadows K, Gabriel V, Hu J, Kim K (2024) Biofabrication of cellulose-based hydrogels for advanced wound healing: a special emphasis on 3D bioprinting. Macromol Biosci 24(5):2300376
79. Iacob AT, Drăgan M, Ionescu OM, Profire L, Ficai A, Andronescu E, Confederat LG, Lupaşcu D (2020) An overview of biopolymeric electrospun nanofibers based on polysaccharides for wound healing management. Pharmaceutics 12(10):983
80. Liu M, Duan XP, Li YM, Yang DP, Long YZ (2017) Electrospun nanofibers for wound healing. Mater Sci Eng C 76:1413–1423
81. Thomas S (2008) A review of the physical, biological and clinical properties of a bacterial cellulose wound. J Wound Care 17(8):349–352
82. Horue M, Silva JM, Berti IR, Brandão LR, Barud HDS, Castro GR (2023) Bacterial cellulose-based materials as dressings for wound healing. Pharmaceutics 15(2):424

83. Gefen A, Alves P, Beeckman D, Cullen B, Lázaro-Martínez JL, Lev-Tov H, Santamaria N, Swanson T, Woo K, Söderström B, Svensby A (2024) Fluid handling by foam wound dressings: from engineering theory to advanced laboratory performance evaluations. Int Wound J 21(2):e14674
84. Heyneman A, Beele H, Vanderwee K, Defloor T (2008) A systematic review of the use of hydrocolloids in the treatment of pressure ulcers. J Clin Nurs 17(9):1164–1173
85. Ahmad F, Mushtaq B, Butt FA, Zafar MS, Ahmad S, Afzal A, Nawab Y, Rasheed A, Ulker Z (2021) Synthesis and characterization of nonwoven cotton-reinforced cellulose hydrogel for wound dressings. Polymers 13(23):4098
86. Sarheed O, Ahmed A, Shouqair D, Boateng J (2016) Antimicrobial dressings for improving wound healing. In: Wound healing - new insights into ancient challenges. pp 374–398
87. Simões D, Miguel SP, Ribeiro MP, Coutinho P, Mendonça AG, Correia IJ (2018) Recent advances on antimicrobial wound dressing: a review. Eur J Pharm Biopharm 127:130–141
88. Kim D, Moore J, Buchanan M, Hayes V, Myntti MF, Hakaim A (2018) Clinical assessment of a biofilm-disrupting agent for the management of chronic wounds compared with standard of care: a therapeutic approach. Wounds 30(5):120–130
89. Zmejkoski DZ, Zdravković NM, Trišić DD, Budimir MD, Marković ZM, Kozyrovska NO, Marković BM (2021) Chronic wound dressings–pathogenic bacteria anti-biofilm treatment with bacterial cellulose-chitosan polymer or bacterial cellulose-chitosan dots composite hydrogels. Int J Biol Macromol 191:315–323
90. Zilberman M, Koren E, Guez H, Matsliah L (2019) Controlled release of antimicrobial small molecules. In: Domb AJ, Kunduru KR, Farah S (eds) Antimicrobial materials for biomedical applications. Royal Society of Chemistry, Cambridge (UK), pp 68–112
91. Majhi RK, Mohanty S, Khan MI, Mishra A, Brauner A (2021) Ag@ ZnO nanoparticles induce antimicrobial peptides and promote migration and antibacterial activity of keratinocytes. ACS Infect Dis 7(8):2068–2072
92. Vivcharenko V, Trzaskowska M, Przekora A (2023) Wound dressing modifications for accelerated healing of infected wounds. Int J Mol Sci 24(8):7193
93. You C, Li Q, Wang X, Wu P, Ho JK, Jin R, Zhang L, Shao H, Han C (2017) Silver nanoparticle-loaded collagen/chitosan scaffolds promote wound healing via regulating fibroblast migration and macrophage activation. Sci Rep 7(1):10489
94. Ong SY, Wu J, Moochhala SM, Tan MH, Lu J (2008) Development of a chitosan-based wound dressing with improved hemostatic and antimicrobial properties. Biomaterials 29(32):4323–4332
95. Alven S, Khwaza V, Oyedeji OO, Aderibigbe BA (2021) Polymer-based scaffolds loaded with aloe vera extract for the treatment of wounds. Pharmaceutics 13(7):961
96. Ilomuanya MO, Cardoso-Daodu IM, Ubani-Ukoma UN, Adebona AC (2021) Polymeric biomaterials for wound healing incorporating plant extracts and extracellular matrix components. Recent Adv Wound Healing pp 1–15
97. Raza A, Chohan TA, Zaidi SHH, Hai A, Alzahrani AR, Abida IM, Saleem H (2024) A systematic review on biochemical perspectives on natural products in wound healing: exploring phytochemicals in tissue repair and scar prevention. Chem Biodivers 21(10):e202400615
98. Portela R, Leal CR, Almeida PL, Sobral RG (2019) Bacterial cellulose: a versatile biopolymer for wound dressing applications. Microb Biotechnol 12(4):586–610
99. Wu Y, Wang F, Huang Y (2018) Comparative evaluation of biological performance, biosecurity, and availability of cellulose-based absorbable hemostats. Clin Appl Thromb Hemost 24(4):566–574
100. Stern L, Lee YC, Tao B, Lamerato L, Gangoli G, Kocharian R, Danker III (2017) The real-world utilization pattern of increased utilization of advanced topical adjunctive hemostats in a vertically integrated healthcare system. J Health Econ Outcomes Res 4(2):103
101. Martyn D, Kocharian R, Lim S, Meckley LM, Miyasato G, Prifti K, Rao Y, Riebman JB, Scaife JG, Soneji Y, Corral M (2015) Reduction in hospital costs and resource consumption associated with the use of advanced topical hemostats during inpatient procedures. J Med Econ 18(6):474–481

102. Franceschini G (2021) Use of oxidized regenerated cellulose as a hemostatic agent in neurosurgery: appraisals and recommendations to prevent postoperative complications and facilitate follow-up. Surg Technol Int 38:481–485
103. Ehsaei MR, Sadeghi M, Rahimi S, Gholamveisi B (2021) Reaction to Surgicel in neurosurgery. Med J Tabriz Univ Med Sci 42(6):781–786
104. Habal P, Simek J, Stětina M (2010) Improving treatment safety in emergency thoracic surgery. Rozhl Chir 89(4):261–264
105. Sefr R, Silák J, Ondrák M, Fiala L (2009) Use of local hemostyptic drugs in liver resections. Rozhl Chir 88(6):337–341
106. Pozza M, Millner RW (2011) Celox (chitosan) for haemostasis in massive traumatic bleeding: experience in Afghanistan. Eur J Emerg Med 18(1):31–33
107. Hatamabadi HR, Zarchi FA, Kariman H, Dolatabadi AA, Tabatabaey A, Amini A (2015) Celox-coated gauze for the treatment of civilian penetrating trauma: a randomized clinical trial. Trauma Mon 20(1)
108. Gegel B, Burgert J, Cooley B, MacGregor J, Myers J, Calder S, Luellen R, Loughren M, Johnson D (2010) The effects of BleedArrest, Celox, and TraumaDex on hemorrhage control in a porcine model. J Surg Res 164(1):e125–e129
109. Fourie L, Marshall O (2021) A real-world, case-led evaluation of the haemostatic dressing range, Celox™ and how it can rapidly stop severe haemorrhage to help preserve life in emergency trauma situations. Trauma Cases Rev 7:096. https://doi.org/10.23937/2469-5777/1510096
110. Shi Z, Zhang Y, Phillips GO, Yang G (2014) Utilization of bacterial cellulose in food. Food Hydrocolloids 35:539–545. https://doi.org/10.1016/j.foodhyd.2013.07.012
111. Klemm D, Kramer F, Moritz S, Lindström T, Ankerfors M, Gray D, Dorris A (2011) Nanocelluloses: a new family of nature-based materials. Angew Chem Int Ed 50(24):5438–5466. https://doi.org/10.1002/anie.201001273
112. Moon RJ, Martini A, Nairn J, Simonsen J, Youngblood J (2011) Cellulose nanomaterials review: structure, properties, and nanocomposites. Chem Soc Rev 40(7):3941–3994. https://doi.org/10.1039/C0CS00108B
113. Abitbol T, Rivkin A, Cao Y, Nevo Y, Abraham E, Ben-Shalom T, Lapidot S, Shoseyov O (2016) Nanocellulose, a tiny fiber with huge applications. Curr Opin Biotechnol 39:76–88. https://doi.org/10.1016/j.copbio.2016.01.002
114. Czaja W, Krystynowicz A, Bielecki S, Brown RM Jr (2006) Microbial cellulose—the natural power to heal wounds. Biomaterials 27(2):145–151. https://doi.org/10.1016/j.biomaterials.2005.07.035
115. Li J, Wei X, Xu C, Zhao Y, Li B (2020) Nanocellulose-based aerogels: a promising material for biomedical applications. Carbohydr Polym 231:115744. https://doi.org/10.1016/j.carbpol.2019.115744
116. Lavoine N, Desloges I, Dufresne A, Bras J (2012) Microfibrillated cellulose–its barrier properties and applications in cellulosic materials: a review. Carbohydr Polym 90(2):735–764. https://doi.org/10.1016/j.carbpol.2012.05.026
117. Habibi Y, Lucia LA, Rojas OJ (2010) Cellulose nanocrystals: chemistry, self-assembly, and applications. Chem Rev 110(6):3479–3500. https://doi.org/10.1021/cr900339w
118. Peng BL, Dhar N, Liu HL, Tam KC (2011) Chemistry and applications of nanocrystalline cellulose and its derivatives: a nanotechnology perspective. Can J Chem Eng 89(5):1191–1206. https://doi.org/10.1002/cjce.20554
119. Rojas OJ, Hubbe MA, Lucia LA (2016) Nanocellulose science and technology: from structure to applications. In: Industrial applications of nanocellulose and its nanocomposites. Springer, pp 69–110
120. Alavarse AC, de Oliveira Silva FW, Colque JT, da Silva VM, Prieto T, Venancio EC, Bonvent JJ (2017) Tetracycline hydrochloride-loaded electrospun nanofibers mats based on PVA and chitosan for wound dressing. Mater Sci Eng C 77:271–281. https://doi.org/10.1016/j.msec.2017.03.199

121. Song K, Wu Q, Qi Y, Karki T (2017) Electrospun nanofibers with antimicrobial properties. Woodhead Publishing, Cambridge (UK)
122. Miguel SP, Figueira DR, Simões D, Ribeiro MP, Coutinho P, Ferreira P, Correia IJ (2018) Electrospun polymeric nanofibres as wound dressings: a review. Colloids Surf B Biointerfaces 169:60–71. https://doi.org/10.1016/j.colsurfb.2018.05.011
123. Chen Y, Qiu Y, Chen W, Wei Q (2020) Electrospun thymol-loaded porous cellulose acetate fibers with potential biomedical applications. Mater Sci Eng C 109:110536. https://doi.org/10.1016/j.msec.2019.110536
124. Liakos IL, Holban AM, Carzino R, Lauciello S, Grumezescu AM (2017) Electrospun fiber pads of cellulose acetate and essential oils with antimicrobial activity. Nanomaterials 7:84. https://doi.org/10.3390/nano7040084
125. Abdullah MF, Nuge T, Andriyana A, Ang BC, Muhamad F (2019) Core-shell fibers: design, roles, and controllable release strategies in tissue engineering and drug delivery. Polymers 11:2008. https://doi.org/10.3390/polym11122008
126. Alavarse AC, de Oliveira Silva FW, Colque JT, da Silva VM, Prieto T, Venancio EC, Bonvent JJ (2017) Tetracycline hydrochloride-loaded electrospun nanofibers mats based on PVA and chitosan for wound dressing. Mater Sci Eng C 77:271–281. https://doi.org/10.1016/j.msec.2017.03.199
127. Bacakova L, Pajorova J, Bacakova M, Skogberg A, Kallio P, Kolarova K, Svorcik V (2019) Versatile application of nanocellulose: from industry to skin tissue engineering and wound healing. Nanomaterials 9:164
128. Sheikhi A, Hayashi J, Eichenbaum J, Gutin M, Kuntjoro N, Khorsandi D, Khademhosseini A (2019) Recent advances in nanoengineering cellulose for cargo delivery. J Control Release 294:53–76
129. Ardila N, Medina N, Arkoun M, Heuzey MC, Ajji A, Panchal CJ (2016) Chitosan–bacterial nanocellulose nanofibrous structures for potential wound dressing applications. Cellulose 23:3089–3104
130. Naeem MA, Siddiqui Q, Leroy A, Khan MR, Wei Q (2019) The production and characterization of microbial cellulose—electrospun membrane hybrid nano-fabrics. J Ind Text
131. Yu DG, Yu JH, Chen L, Williams GR, Wang X (2012) Modified coaxial electrospinning for the preparation of high-quality ketoprofen-loaded cellulose acetate nanofibers. Carbohydr Polym 90:1016–1023
132. Sajid M, Płotka-Wasylka J (2020) Nanoparticles: synthesis, characteristics, and applications in analytical and other sciences. Microchem J 154:104623
133. Rubina MS, Said-Galiev EE, Naumkin AV, Shulenina AV, Belyakova OA, Vasil'kov AY (2019) Preparation and characterization of biomedical collagen—chitosan scaffolds with entrapped ibuprofen and silver nanoparticles. Polym Eng Sci 59:2479–2487
134. Seo SB, Dananjaya SH, Nikapitiya C, Park BK, Gooneratne R, Kim TY, Lee J, Kim CH, De Zoysa M (2017) Silver nanoparticles enhance wound healing in zebra fish (Danio rerio). Fish Shellfish Immunol 68:536–545
135. Sridhar R, Lakshminarayanan R, Madhaiyan K, Barathi VA, Lim KH, Ramakrishna S (2015) Electrosprayed nanoparticles and electrospun nanofibers based on natural materials: applications in tissue regeneration, drug delivery, and pharmaceuticals. Chem Soc Rev 44:790–814
136. Liakos I, Rizzello L, Hajiali H, Brunetti V, Carzino R, Pompa PP, Athanassiou A, Mele E (2015) J Mater Chem B 3:1583–1589
137. Suwantong O, Opanasopit P, Ruktanonchai U, Supaphol P (2007) Electrospun cellulose acetate fiber mats containing curcumin and release characteristic of the herbal substance. Polymer 48:7546–7557
138. Phiriyawirut M, Phaechamud T (2012) Gallic acid-loaded cellulose acetate electrospun nanofibers: thermal properties, mechanical properties, and drug release behavior. Open J Polym Chem 21–29

139. Adamu BF, Gao J, Jhatial AK, Kumelachew DM (2021) A review of medicinal plant-based bioactive electrospun nano fibrous wound dressings. Mater Des 209:109942. https://doi.org/10.1016/j.matdes.2021.109942
140. Arafa AA, Nada AA, Ibrahim AY, Sajkiewicz P, Zahran MK, Hakeim OA (2021) Preparation and characterization of smart therapeutic pH-sensitive wound dressing from red cabbage extract and chitosan hydrogel. Int J Biol Macromol 182:1820–1831. https://doi.org/10.1016/j.ijbiomac.2021.05.167
141. Montaser AS, Rehan M, El-Senousy WM, Zaghloul S (2020) Designing strategy for coating cotton gauze fabrics and its application in wound healing. Carbohydr Polym 244:116479. https://doi.org/10.1016/j.carbpol.2020.116479
142. Omidi M, Yadegari A, Tayebi L (2017) Wound dressing application of pH-sensitive carbon dots/chitosan hydrogel. RSC Adv 7:10638–10649. https://doi.org/10.1039/C6RA25340G
143. Tang N, Zheng Y, Jiang X, Zhou C, Jin H, Jin K, Wu W, Haick H (2021) Wearable sensors and systems for wound healing-related pH and temperature detection. Micromachines 12:430. https://doi.org/10.3390/mi12040430
144. Lu SH, Samandari M, Li C, Li H, Song D, Zhang Y, Tamayol A, Wang X (2022) Multimodal sensing and therapeutic systems for wound healing and management: a review. Sens Actuators Rep 4:100075. https://doi.org/10.1016/j.snr.2022.100075
145. Deng WJ, Wang LF, Dong L, Huang QA (2018) LC wireless sensitive pressure sensors with microstructured PDMS dielectric layers for wound monitoring. IEEE Sens J 18:4886–4892. https://doi.org/10.1109/JSEN.2018.2831229
146. Lu L, Zhao N, Liu J, Yang B (2021) Coupling piezoelectric and piezoresistive effects in flexible pressure sensors for human motion detection from zero to high frequency. J Mater Chem C 9:9309–9318. https://doi.org/10.1039/D1TC01894A
147. Messaoud M, Marsiquet C, Revol-Cavalier F, Rat V, Marchand G (2018) Flexible sensors for real-time monitoring of moisture levels in wound dressings. J Wound Care 27:385–391. https://doi.org/10.12968/jowc.2018.27.6.385
148. Tessarolo M, Possanzini L, Gualandi I, Mariani F, Torchia LD, Arcangeli D, Melandri F, Scavetta E, Fraboni B (2021) Wireless textile moisture sensor for wound care. Front Phys 9:722173. https://doi.org/10.3389/fphy.2021.722173
149. Gámez-Herrera E, García-Salinas S, Salido S, Sancho-Albero M, Andreu V, Perez M, Lujan L, Irusta S, Arruebo M, Mendoza G (2020) Drug-eluting wound dressings having sustained release of antimicrobial compounds. Eur J Pharm Biopharm 152:327–339. https://doi.org/10.1016/j.ejpb.2020.05.025
150. Maver U, Gradišnik L, Smrke DM, Kleinschek KS, Maver T (2019) Impact of growth factors on wound healing in polysaccharide blend thin films. Appl Surf Sci 489:485–493. https://doi.org/10.1016/j.apsusc.2019.06.054
152. El Fawal GF, Abu-Serie MM, Hassan MA, Elnouby MS (2018) Hydroxyethyl cellulose hydrogel for wound dressing: fabrication, characterization and in vitro evaluation. Int J Biol Macromol 111:649–659

M. A. Lubna is a dedicated neuroscientist and Senior Research Fellow at the National Institute of Technology, Calicut, where she conducts advanced research under the mentorship of Dr. Rajanikant G. K. Her work focuses on exploring innovative therapeutic approaches in neuroscience, with a particular interest in the neuroprotective potential of bioactive compounds. She has contributed significantly to the field through her scholarly publications, including a book chapter titled "Flavonoid for Neurological Disorders: Prospective and Advances" in Flavonoids for Neurological Disorders: Advances and Future Prospects, published by Apple Academic Press/Taylor & Francis Group. Her research spans neurodegenerative diseases, molecular neuropharmacology, and the translational potential of natural compounds in neurological health. With a strong commitment to scientific discovery and innovation, Lubna actively engages in academic discourse, presenting her findings at national and international conferences. She aims to bridge the gap between fundamental neuroscience and clinical applications, contributing to the development of novel therapeutic strategies for neurological disorders.

Akshay R. Thampi is an emerging innovator at the intersection of biotechnology and entrepreneurship, dedicated to leveraging scientific advancements for real-world impact. Born in 2001 in Kerala, India, he spent his formative years in Qatar, where he developed a global perspective and a deep appreciation for interdisciplinary learning. With a strong academic foundation in Biotechnology from Vellore, Akshay cultivated a passion for scientific research, computational biology, and analytical problem-solving. Currently, he is advancing his expertise with a graduate degree in Bioengineering from the prestigious National Institute of Technology (NIT) Calicut. His research interests span synthetic biology, bioinformatics, and biomedical engineering, with a strong inclination toward translating theoretical knowledge into tangible solutions. Beyond academics, Akshay is driven by an entrepreneurial spirit, aspiring to pioneer biotech innovations that address pressing global challenges in healthcare, sustainability, and biotechnology. His vision is to bridge the gap between research and commercialization, transforming scientific discoveries into accessible and impactful solutions. With a commitment to lifelong learning and a keen eye for technological advancements, Akshay actively engages in scientific writing, innovation-driven collaborations, and knowledge dissemination. As he continues his journey, he aims to make meaningful contributions to the biotech landscape through research, innovation, and enterprise.

Navami Krishna is a Senior Ph.D. Fellow in the Department of Bioscience and Engineering at the National Institute of Technology Calicut (NIT Calicut), India, specializing in neuroscience research with a focus on ischemic stroke treatment. Under the mentorship of Dr. Rajanikant G. K., her research explores cutting-edge therapeutic interventions aimed at mitigating the impact of ischemic strokes. She is a recipient of the prestigious Kerala State Council for Science, Technology, and Environment (KSCSTE) Research Fellowship and the Indian Science, Technology, and Engineering Facilities Map (I-STEM) Catalytic Grant, awarded by the Indian Institute of Science (IISc) in recognition of her significant contributions to the field. Navami has an impressive research portfolio, with two peer-reviewed journal publications in internationally reputed scientific journals, a book chapter published by Taylor & Francis, and five granted Indian patents, reflecting her commitment to translational research and innovation. Academically, she holds a Master's degree in Applied Microbiology from Mahatma Gandhi University, Kerala, India, where she graduated as the university's top-ranked student. Her multidisciplinary expertise in microbiology and neuroscience positions her at the forefront of biomedical research, driving advancements in stroke therapeutics and neuroprotection. Beyond academia, Navami actively engages in scientific outreach, interdisciplinary collaborations, and mentorship programs, contributing to the global scientific community.

Dr. G. K. Rajanikant is a distinguished Professor in the Department of Bioscience and Engineering at the National Institute of Technology Calicut (NITC), India. With over two decades of academic and research excellence, he has made pioneering contributions to neuroscience, particularly in ischemic stroke, neuroprotection, and translational neurobiology. His interdisciplinary expertise spans molecular neurobiology, drug repurposing, bioinformatics, and biosensing technologies, positioning him as a thought leader in biomedical innovation. Currently, Dr. Rajanikant holds key academic and administrative positions at NITC, serving as Chairperson of the Centre for Cultural and Art Relations and the Centre for Public Relations, Information, and Media Exchange. Previously, he played a crucial role in student welfare and institutional development as Dean (Students' Welfare), demonstrating his commitment to fostering academic and research excellence. Dr. Rajanikant's research is supported by national and international grants exceeding ₹2.3 crores (~$275,000 USD), fueling groundbreaking studies in neurotherapeutics, precision medicine, and computational drug discovery. He has authored over 100 high-impact publications, holds multiple patents, and has an impressive h-index of 34, with more than 3,900 citations, underscoring the global impact of his work. His research has contributed to advancing novel therapeutic strategies for neurological disorders, bridging the gap between laboratory discoveries and clinical applications. A recipient of prestigious fellowships, including the Stem Cell Society Singapore Fellowship, Dr. Rajanikant has collaborated with renowned institutions worldwide, such as the University of California, Davis, and Michigan State University. His global research partnerships reflect his commitment to advancing neuroscience through multidisciplinary and cross-border collaborations. He is actively involved in scientific publishing and peer review, serving as an editorial board member of reputed international journals and an expert reviewer for global funding agencies. As a mentor, he has guided numerous young researchers, doctoral candidates, and early-career scientists, nurturing the next generation of innovators in neuroscience and biomedical engineering. Beyond his scientific endeavors, Dr. Rajanikant is a visionary academic leader, mentor, and science communicator. His administrative leadership, research contributions, and commitment to translational neuroscience continue to influence the landscape of biomedical research and healthcare innovation.

Advances in Cellulose-Based Biomaterials for the Sustainable Drug Delivery Systems

Amit Kumar Verma

Abstract Today, in the era of modern medicine, the sustained drug delivery is a major challenge, as conventional delivery approaches like oral, intravenous and transdermal have various shortcomings. Traditional delivery avenues affect the bioavailability, specific target location, immunogenicity, efficacy, cytotoxicity and safety of the drug, which leads to administration of multiple or high amounts of drug doses resulting in low to severe magnitudes of medical complications to the patients. Sustained drug delivery systems using the biopolymers as medium are promising materials as they can overcome the sudden release, enhancing efficacy and safety of drugs. One such polymer is cellulose, which is the most abundant polysaccharide and is procured from different sources like plants, wood, bacteria, algae and chordate animals such as tunicates. Cellulose and its derivatives have charmed the researchers, by several exclusive attributes such as functionality, high mechanical robustness and flexibility, high surface area, optical transparency, low density and biodegradability. The porosity and interconnectivity between cellulose-based polymeric formulations can be controlled for relevant biomedical applications. The commonly existing cellulosic derivatives are cellulose ethers and cellulose esters. Despite several unique features, cellulose has some limitations like low solubility, sensitivity to moisture and susceptibility to microbes, which pose challenges like optimal drug delivery, sustained release, hemocompatibility, immunogenicity, off-target toxicity and biodegradability. This chapter will address the overview of cellulose-based biomaterials, their modifications and characteristics and role in various promising therapeutics in the field of biomedicine.

Keywords Cellulose · Cellulose ethers · Carboxymethyl cellulose · Nanocellulose · Drug delivery · Biodegradability

A. K. Verma (✉)
Department of Biosciences, Jamia Millia Islamia, New Delhi, India
e-mail: averma@jmi.ac.in

R. Malviya and S. Sundram (eds.), *Sustainable Cellulose Materials for Biomedical Applications*, Sustainable Materials and Technology,
https://doi.org/10.1007/978-981-96-9904-9_17

Abbreviations

α-CD	α-Cyclodextrin
β-CD	β-Cyclodextrin
5-FU	5-Fluorouracil
AA	Acrylic acid
ACV	Acyclovir
API	Active pharmaceutical ingredients
BC	Bacteria-based cellulose/bacterial cellulose
BNC	Bacterial nanocellulose
CA	Cellulose acetate
CE	Cellulose ethers
CEC	Carboxyethyl cellulose
CHX	Chlorhexidine
CMC	Carboxymethyl cellulose
CMF	Cellulose microfibers
CNC	Cellulose nanocrystals
CNF	Cellulose nanofibrils
CNW	Cellulose nanowhiskers
CS	Chondroitin sulphate
CTP	Captopril
DEGEE	Diethylene glycol monoethyl ether
EC	Ethyl cellulose
FDA	Food and Drug Administration
GI	Gastrointestinal
HA	Hyaluronan
HEC	Hydroxyethyl cellulose
HPMC	Hydroxypropylmethyl cellulose
HYA	Hyaluronic acid
MBA	N,N′-methylene-bis-acrylamide
MC	Methyl cellulose
MCC	Microcrystalline cellulose
MH	Metformin hydrochloride
NC	Nanocellulose
NCC	Nanocrystalline cellulose
PC	Plant-based cellulose
PLGA	Poly(lactic-*co*-glycolic) acid
PPG	Propylene glycol
PVP	Polyvinylpyrrolidone
TDDS	Transdermal drug delivery systems
WC	Wood cellulose

1 Introduction

Today, in the present time, the human being is rigorously consuming non-renewable and petroleum-based products which give rise to severe environmental pollution, alarming health concerns and abnormal climatic behaviour [85]. So, the modern world is switching from fossil fuel-based or artificial synthetic materials to bio-based renewable and abundant materials from natural sources for a sustainable and eco-friendly environment. Researchers are exploring the novel biomaterials for various advanced biomedical/bioengineering applications with extraordinary features of biodegradability, bioavailability, biocompatibility, mechanical, structural and chemical stability [28]. Among the various biopolymers, cellulose-based biomaterials have attracted incredible consideration of the scientific and industrial community.

Cellulose is the most abundant, cosmopolitan and natural biopolymer on the earth and comprises 40% of all the organic substances and forms 40–50%, which is roughly half of the total biomass of the plants with annual productivity of roughly 10^{10} tonnes [29]. The market demand for such a magnificent biopolymer has increased tremendously over the recent years. Plants, seaweeds, red or brown marine algae, sugarcane bagasse, bacteria and tunicates are the prevalent sources of cellulose [2, 3, 169]. Cellulose is natural, unbranched linear structural homopolymer of anhydrous D-glucose monomers in the structural chair arrangement, interlinked through β-(1 → 4) glycosidic bond with formula of $(C_6H_{10}O_5)n$, where n is equal to 10,000–15,000 pivot on the source involved in the separation of cellulose (Fig. 1) [16, 72, 83]. The abundance of –OH groups in cellulose supports hydrogen bonding leading to fibrillar structure and enabling chemical modifications with high wettability and inertness for water and other chemical reagents. In another way, hydrogen bonding in cellulose gives rise to the development of crystalline and amorphous zones in cellulose fibres. The excessive inter- and intra-molecular hydrogen linkages provide the extraordinary advantage to cellulose such as excellent mechanical (rigidity, high tensile strength, good elastic modulus of 130–150 GPa), physical (crystallinity) and chemical properties (high stability under acidic environment and chirality) and high biodegradability [44, 46, 71, 69, 93, 121].

Plants are an abundant source of cellulose 40–50 wt%, hemicelluloses 20–40 wt% and lignin 20–30 wt%, making them rich natural and renewable reserve on the earth [119]. This natural biopolymer is generally found in the form of microfibrils in the cell wall of plants, wood, algal tissue and epidermal cells of tunicates and the form of nanofibrillar architecture in the bacteria. Cellulose molecules are arranged in macroscopic to nanolevel conformation in the various forms of fibril aggregates, fibrils, nanocrystalline or ordered and amorphous or disordered domains (Fig. 2) [138]. Cellulose is present in the highly complex multi-level conformation formed by aggregates/bundles of fine fibrils. Each fibril has many cellulose chains and possesses repeating units of large ordered crystalline zone and small disordered amorphous zone with cross-sectional dimension of 2–20 nm, depending on the source of synthesis [54]. A single cellulose chain is linked with strong β-(1 → 4) glycosidic

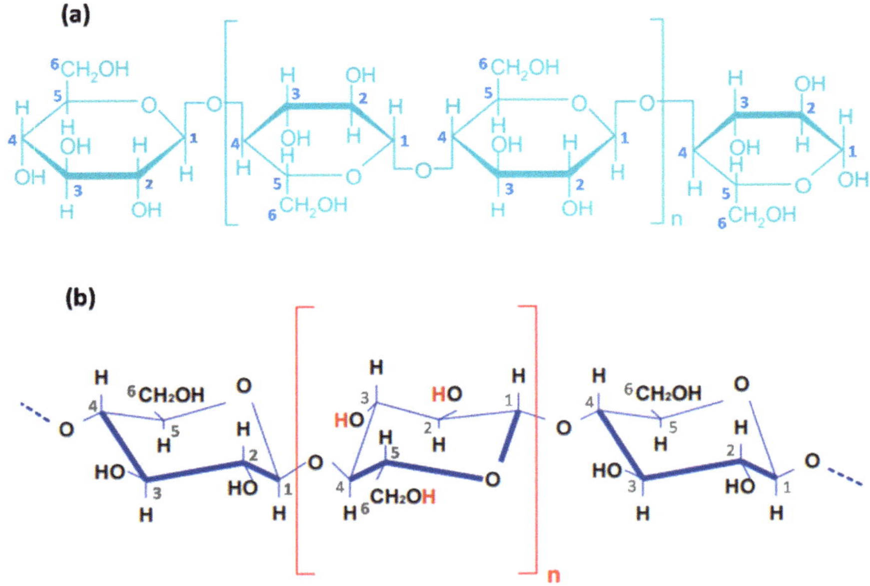

Fig. 1 Cellulose molecular structure: **a** a short segment of cellulose, a linear homopolymer of D-glucose residues with β-(1 → 4) linkage, **b** cellulose polymeric conformation depicted as rigid pyranose rings

linkages and passes through repeated units of crystalline and amorphous domains [79].

Cellulose and its biocompatible polymer derivatives have fascinated the researchers and scientists for a variety of biomedical purposes due to strong mechanical and physical attributes. Due to hierarchical conformation, cellulose inherently developed the functionality, plasticity, physical and mechanical robustness and also has low density, cost-effectiveness and biodegradability [6]. The desired porosity and

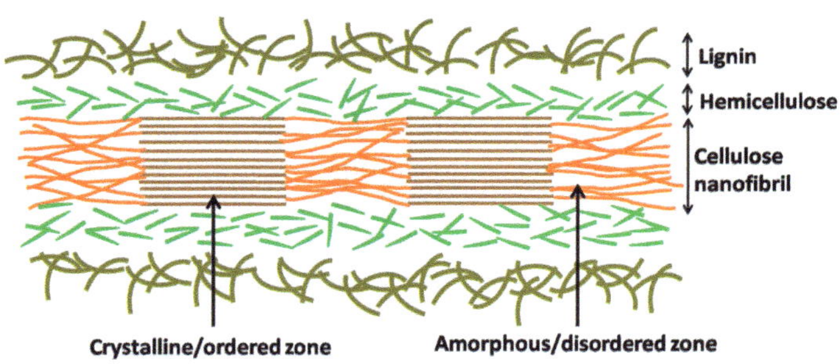

Fig. 2 Cellulose nanofibrils showing crystalline or ordered and amorphous or disordered region embedded in interlinked matrix of hemicelluloses and lignin

interconnected network forming capability makes the cellulose-based materials as a suitable choice for biomedical applications [143]. Apart from the above-mentioned appreciating properties of cellulose, it has some considerable issues like solubility in water and most common solvents, moisture sensitivity and susceptibility to microbial deterioration in the field of biomedical operations [31]. Nonetheless, cellulose can be chemically altered through substitution of intrinsic –OH groups along with other functional chemical groups like oxides, acids, chlorides, nitrates and sulphates to develop the novel and desired features and to reduce the extent of less favourable attributes for biomedical purposes [138].

Since many centuries, cellulose-related derivatives have been commercially used in the textile, paper industry, packaging materials, forest products and other industries, while in the last decades cellulose has attracted the attention for various purposes, one of them is biomedical applications [153]. Cellulose was discovered by French chemist Anselme Payen in 1838, who isolated it from a plant source and determined the chemical formula. After that, cellulose has significantly achieved the prodigious importance over synthetic polymers and has been converted into several water-soluble derivatives. The commonly used typical cellulosic derivatives are carboxymethyl cellulose (CMC), hydroxyethyl cellulose (HEC), methyl cellulose (MC), hydroxypropylmethyl cellulose (HPMC) and cellulose acetate and many more.

Cellulose-based conglomerates with advanced organic and inorganic properties have several biomedical usages [65]. Currently, the cellulosic composites or formulations are used for several types of purposes such as drug transport [56, 71, 81], antibacterial agents [144], wound dressing materials [131], tissue engineering [155], artificial blood vessels [148] and safety from UV light [134]. Cellulose can be formulated into diverse forms like hydrogels, aerogels, membranes and 3D scaffolds with excellent binding properties and potential for medicine [3]. This review will cover the cellulose structure, sources, variety of cellulose-based derivatives based on their chemical nature, as well as their applications in the area of medicine, particularly sustained drug transport systems to develop novel therapeutics.

2 Sources of Cellulose

Generally, cellulose is categorized depending on its origin's source—wood cellulose (WC), plant-based cellulose (PC), bacteria-based cellulose (BC), algae-based cellulose and tunicate-based cellulose (animal origin). Based on affluence, price ratio or inexpensiveness WC and PC are the most frequently used cellulose followed by BC, tunicate-based and algae-based cellulose [79]. The source types and their examples for the separation of cellulose are compiled in Table 1.

Table 1 Major source types and their examples utilized for the extraction of cellulose

Source type	Examples
Softwood	Pine, cedar, spruce, juniper, hemlock, yew, larch
Hardwood	Oak, beech, sakata, *eucalyptus*, aspen, balsa, elm, maple, birch
Annual plants	Cotton, corn, jute, hemp, flax, ramie, oil palm, sisal, *agave*, alfalfa, kenaf, wheat, rice, sunflower, bamboo, canola
Plant residues	Bagasse, cereals straw, soybean straw, cotton seeds, pineapple leaf, potato peels, coconut husk, corn leaf, corn stover, tomato peel, garlic straw, mulberry fibre, mengkuang leaves, oil palm biomass
Algae	*Posidonia oceanica*, *Cladophora glomerata*, *Gelidium elegans*, *Gelidium sesquipedale*, *Laminaria japonica*, *Laminaria hyperborea*, *Ulva lactuca*, *Ulva fasciata*, *Ulva rigida*, *Valonia*, *Rhizoclonium*, *Griffithsia*, *Ceramium*, *Chaetomorpha*, *Chondria*, *Corallina*, *Fucus*, *Saccharina japonica*, *Sargassum fluitans*, *Palmaria palmata*, *Porphyra umbilicalis*, *Gracillaria*, *Gelidiella aceroso*, *Hypnea*, *Halidrys*
Bacteria	*Komagaietibacter xylinus*, *Achromobacter*, *Agrobacterium*, *Salmonella*, *Azotobacter*, *Rhizobium*, *Alkaligenes*, *Pseudomonas*, *Aerobacter*, *Rhodobacter*, *Sarcina*
Animal	*Tunicates*, *Styela clava*, *Halocynthia roretzi*, *Drasche*

2.1 Wood and Plant-Based Cellulose (WC and PC)

Softwood from pine, cedar and spruce and hardwood from oak and beech have been used for the cellulose separation. Apart from soft and hardwood, other plant-based sources like cotton, hemp, jute, flax, corn, bagasse, pineapple leaf, cereal straw, ramie, potato peel waste and palm oil mass are rich in cellulose [138]. Cellulose fibres of wood and plant-based sources are related in structure, chemical composition, along with other polysaccharides, lignin and hemicellulose. WC and PC have complex multi-level architecture and are similar in chemical organization but differ at the microstructural arrangement [100]. In WC and PC, the cellulose fibres are arranged in sheets of nanofibrils and supported via a network of lignin and hemicelluloses outside (Fig. 1). The cellulose content in wood fibres is 40–50%, while in plant fibres is 30–75% (except cotton fibre) and has a different composition of lignin and hemicellulose. The cellulose content in softwood (45–50%) and hardwood (40–50%) is almost similar but lignin (softwood—23–35%; hardwood—18–25%) and hemicellulose (softwood—18–35%; hardwood—24–40%) content are remarkably different. In plant fibres, lignin is 0–20% and hemicellulose is 0–20% [95].

The plant cell wall is mainly composed of primary and secondary walls, and the secondary wall is made up of three layers. The middle layer of secondary cell wall has cellulose microfibers (CMF), and each microfiber is made up of precursor nanofibrils, which are helically coiled and combined by inter- and intra-molecular hydrogen bonds. The cellulose microfibers behave as reinforcement in the cell wall matrix having other polymers of hemicelluloses and lignin, pectin and other organic substances, while lignin acts as the binding substance between hemicellulose and cellulose [156].

2.2 Bacteria-Based Cellulose or Bacterial Cellulose (BC)

The cellulose formed by bacteria is known as bacterial cellulose, microbial cellulose or bacterial nanocellulose (BNC) or biocellulose. In 1886, the British researcher Brown first discovered BC on vinegar fermentation broth's surface like the jelly membrane produced by *Acetobacter xylinum* [13]. BC is synthesized by terminal complex in almost unadulterated form > 90% without contamination of polymers such as lignin and hemicelluloses, which makes the extraction and other processing of BC quite easy followed by less required extensive chemical or other processing in comparison with WC and BC [138]. BC has gained significant concerns due to advantages like high purity, ultra-structured fibrillar morphology, significant crystallinity and high amount of polymerization, mechanical robustness, biocompatibility, biodegradability, water retaining capacity and reliable chemical stability [17]. Other important properties of BC are non-cytotoxicity, non-genotoxicity and high biocompatibility, which attracted the attention for diverse areas of biomedicine [42].

BC is mainly generated as extracellular or exopolysaccharide substance by Gram negative bacteria such as *Komagataeibacter xylinus*, *Agrobacterium*, *Achromobacter*, *Aerobacter*, *Azotobacter*, *Pseudomonas* and *Rhizobium* and only one species of Gram positive bacteria *Sarcina* [43]. *Komagataeibacter xylinus* (formerly *Acetobacter xylinus*) is the most widely utilized species for BC's production at large scale in the variable concentrations of carbon and nitrogen sources in the media [79]. Generally, the carbon source for culture media is agro-industrial waste in the form of rotten fruit, such as pineapple peels, juice and sugar. The harvest of synthesized BC is up to 40% concerning initial carbon source, but the broad-scale production is costly [79]. During synthesis of BC, the molecules are generated in the bacterial cells and then secreted to the outside environment via the small pores of the outer membrane. Normally 10–15 glucan chains are linked together to form subfibrils of roughly 1.5 nm width. Later these subfibrils aggregate side-by-side to form 30–80 nm wide compact cellulose nanofiber, stabilized by inter- and intra-molecular hydrogen bonding. According to various fermentation techniques, BC can be generated in a variety of formulations such as aster/sphere like hydrogels [36].

The prominent features of BC are the large extent of crystallinity (80–90%), large extent of polymerization (7000–16,000 glucose residues) and high mechanical robustness linked with single cellulose nanofibres (like the specific strength of steel or Kevlar) [48]. BC can be shaped into various morphological forms such as fleeces, hollow particles, aster/sphere like hydrogels under agitating/shaking culture conditions. The thin film hydrogels are formed in static culture conditions, which provide versatility for different biomedical materials formulations. Albeit, WC, PC and BC have similar chemical structure but differ by structural architecture [100]. Additionally, WC and PC have lower degradation rate than BC [79]. In some cases, the production of BC consumes a huge water amount of water (more than 99%) and generates durable hydrogels. Due to proper separation and large surface areas, BC nanofibres form an exceptionally porous material. This excellent property of holding a large amount of water displays great elasticity, high wet strength and conformability [42]. The amalgamation of significant characteristics of BC has innovative utilization in the field of tissue engineering and the medical industry as implants, scaffolds, emulsion and hydrogel stabilizers, drug delivery modules, smart artificial skin or wound regeneration therapeutics and enzyme and biomolecules immobilization for augmented activity and greater endurance in vivo [138].

2.3 Algal Cellulose

The cellulose can be processed from the variety of algae sources like red algae (*Gelidium elegans*), brown algae (*Posidonia oceanica*) and green algae (*Cladophora* sp.) [22, 116, 149]. The algal cell walls are highly rich in cellulosic polysaccharides. Red algae have a high concentration of carbohydrates, mainly composed of high energy cellulose and agar as mucilaginous material. Algae are the best alternative source of cellulose as they can extract nutrients from wastewater and can be cultivated on a mass scale, finally becoming eco-friendly. Cellulose from green algae (*Valonia* or *Cladophora*) has significantly higher crystallinity, more than 95%. Cellulose from algae is not as pure as BC, due to the presence of hemicelluloses, proteins and lignin [127]. Cellulose from green algae has tremendous superiority due to high crystallinity (> 70%), low moisture adsorption capability, great porosity and large specific surface area than WC, PC and BC [22, 172]. In nutshell, algal nanocellulose has high potential for biomedicinal purposes due to non-cytotoxicity and easy chemical alterations.

2.4 *Tunicate Cellulose*

The exclusively established animal source of cellulose is the chordate animal-tunicates, around 3000 species, live in the oceans, mostly in shallow waters. In the epidermal cells of tunicates, the plasma membrane with a variety of enzyme complexes is involved in the biosynthesis of cellulose. The pure form of cellulose known as tunicin is received from the outer exoskeleton and complex tissue of tunicate, named tunic [173]. Most of the scientists have centred around the largest class of tunicates—Ascidiacea (sea squirts), comprised of more than 2300 species [170]. In the outer tissue of tunicates, hundreds of cellulose nanofibrils are stuffed together, and the morphology of nanofibrils varies from species to species. The nanofibrils are arranged in layer-by-layer architecture along the surface of epidermis in tissue, and the length of cellulose nanofibrils is 100 nm to several micrometres (generally greater than 2 μm) and width is 10–30 nm, and the aspect ratio is 60–70 [171]. Usually the dry tunic tissue contains nearly 60% cellulose and 27% nitrogenous elements but after the extraction and further processing nearly pure cellulose with high crystallinity (around 95%) and morphologically high aspect ratio fibrils is achieved [170]. The cellulose has a large specific surface area of 150–170 m^2/g and also has reactive surface caused by several –OH groups for further chemical modifications along with reported degree of polymerization of 700–3500 [174, 171].

3 Cellulose Architecture

Cellulose is the high molecular weight, homopolymer of unbranched chain of repeated D-glucose monomers, joined together by β-1,4-glycosidic bonds. Each repeating unit is associated with hydroxyl groups forming inter- or intra-molecular hydrogen bonds and establishing bonds with nearby glucose monomeric units within the same chain and in closeness to other chains [9]. Cellulose has three hydroxyl groups in every glucose monomer, which provides higher polymerization (range—1000–15,000) based on cellulose source and processing. Cellulose molecules are arranged in structured parallel layers of elementary fibrils in plant cell wall, algae cell wall, tunicate epidermal cell and in bacterial cellulose. The cellulose chains in lateral dimension are of 3–5 nm and are strongly bonded through van der Waals and hydrogen bonds. The elementary fibrils are arranged in an ordered crystalline structure and a disordered amorphous region or territory of cellulose microfibrils. Based on source, the microfibrils are of 5–20 nm in diameter and several microns in length [78]. Cellulose is a semi-crystalline polymer and crystallinity relies on the type of source, separation and further processing steps. The extent of crystallinity is 40–60% in case of WC and PC, while 80–100% for the tunicate and bacterial cellulose [106]. The allomorphs of crystalline cellulose are types I, II, III and IV [106]. Cellulose I is arranged as parallel chains and forms sheets and its natural sources are trees, plants, algae, bacteria and tunicates. Cellulose I has two metastable structures—Iα (triclinic) and Iβ (monoclinic), and the ratio of these structures depends on the source. The Iα allomorph is predominantly observed in algal and bacterial cellulose, and Iβ structure with two parallel chains arrangement is mostly observed in plant and tunicate cellulose [161]. The thermodynamically metastable cellulose I can be transformed into both the celluloses II and III. Cellulose II is the most durable structure with antiparallel chain organization along with technical importance for several applications. Through dissolution and recrystallization or mercerization (treatment with NaOH), cellulose II can be produced from cellulose I. Celluloses I and II treatment with amines can produce cellulose III. Cellulose IV can be obtained through treatment of cellulose III with glycerol and high temperature [94].

4 Morphological Variations of Cellulose Molecules

Natural cellulose or extracted from different sources, express typical morphology and shape and forms the basis for categorization (in order of macro-level to nano-dimensions) into cellulose filaments, cellulose crystals, cellulose microfibers/microfibrils and cellulose nanofibrils. Each category of cellulose has unique size, shape, aspect ratio, crystallinity and physicochemical characteristics (mechanical robustness, elastic modulus, tensile strength, solubility and hygroscopic attributes) [138].

4.1 Cellulose Fibres

These fibres are in the macro-level dimensions from a few millimetres to several centimetres. Cellulose fibres are generally occurring in three varieties—strand or long fibres (20–100 cm long), staple or short fibres (60 mm in length) and pulp or very short fibres of 1–10 mm of length [7]. The strand fibres are composed of many cells while staple fibres are an elongated single cell. The very rich source of cellulose, cotton is a staple fibre, 25–45 mm long. Strand and staple fibres from wood or plants are generally obtained via the water retting process while pulp fibres are extracted through various chemical treatment processes from plants like sisal, banana, flax and cotton linters (fibres of few millimetres in length attached to cotton seeds) [7].

4.2 Nanocellulose (NC)

Cellulose polymer with the dimensions at the nanometer scale is commonly known as nanocellulose and it has a diameter of below 100 nm and several micrometres in length [156]. NC can be obtained via the chemical or mechanical processing of microfibers, which would disrupt the inter- or intra-molecular hydrogen bonds. NC has several characteristics like larger specific area, light weight, low density, stiffness higher than Kevlar fibre, strength-to-weight ratio higher than stainless steel and better tensile strength in comparison with cast iron [23, 152]. Sources like plants, crop residues, tunicates, fungi, algae and bacteria provide the nanocellulose. In reports, the agricultural produce such as corncob, banana leaves, cotton, rice husk, sugarcane bagasse, wheat straw, coconut husk and wood are processed for the synthesis of NC [49]. Other reported sources for NC are ginger, coffee grounds, durian rind waste, lemon seeds, pea hull, rubber wood, etc. [156]. The amount of polymerization, polymeric length and features of NC rely on the source and processing methods. The extent of polymerization of wood NC and cotton NC are 10,000 and 15,000, respectively [152]. On the basis of morphological features, nanocellulose can be categorized into three main types—short and rigid cellulose nanocrystals (CNC), long and flexible cellulose nanofibrils (CNF) and highly pure and crystalline BNC, which have the similar chemical content but varies in morphology, size, crystallinity, purity and physicochemical attributes [123]. The morphological and physicochemical properties of variety of cellulose nanoparticles are mentioned in Table 2.

Table 2 Structural, mechanical and physicochemical properties of various types of cellulose nanoparticles [7, 18, 60]

Properties	CNC	CNF	BNC	MCC
Length (nm)	100–2000	> 1000	≥ 1000	< 5000
Diameter (nm)	3–50	2–100	7–10	> 1000
Aspect ratio	10–67	60–100	600–6000	1–2
Degree of polymerization	500–15,000	200–1100	7000–16,000	< 400
Crystallinity (%)	72–80	50–65	80–90	65–83
Functional groups	–OH, –OSO_3H, –COOH	–OH, –COOH, –CHO	–OH	–OH
Young modulus (GPa)	60–200	14–36	15–130	~ 25
Tensile strength (MPa)	7500–7700	~ 1000	200–2000	4.48–6.42
Viscosity	~ 1000	~ 10,000	–	3000–5000

4.2.1 Cellulose Nanocrystals (CNCs)

CNC are rod- or needle-shaped and primarily comprised of crystalline regions with cellulose particle size of 2–20 nm in diameter and length of 100–500 nm [70]. CNCs are also known as nanocrystalline cellulose (NCC), cellulose nanowhiskers or cellulose crystallite and could be generated by acid hydrolysis (HCl, H_2SO_4, HBr, H_3PO_4, $H_2C_2O_4$ (oxalic acid), ionic liquids, metal salt crystals, solid or gaseous acids) of wood or plants (cotton, hemp, flax, wheat straw, mulberry bark, ramie, etc.), tunicate, algae and bacteria. The alkaline treatment and bleaching are implemented for the removal of hemicellulose and lignin, respectively [96]. The acid treatment affects the crystallinity, morphology, dispersibility and physicochemical features of nanocellulose. The H_2SO_4 treatment generates nanocellulose with good dispersibility in hydrophilic solvents but with reduced thermal stability. The processing with popular ionic liquids (1-butyl-3-methylimidazolium hydrogen sulphate (bmimH_2SO_4), 1-allyl-3-methylimidazolium chloride (AmimCl), tetrabutylammonium acetate, 2-hydroxy ammonium hydrogen sulphate) eliminates the disordered amorphous regions, hemicelluloses and lignin and produce high-quality CNCs [41]. The d-block metal salts act as catalysts for the separation of amorphous regions in cellulose [21].

4.2.2 Cellulose Nanofibrils (CNFs)

CNF are elongated fibres of dimension of 1–100 nm in diameter and length of 100 nm to several micrometres. CNFs are obtained from wood, sugar beet, potato, hemp, flax through mechanical pressure, chemical or enzymatic processing [77]. CNFs are also known as microfibrillated cellulose (MFC), nanofibrils, microfibrils or nanofibrillated cellulose (NFC). In morphology, the CNFs have ordered crystalline and disordered amorphous region along with flexibility, hydrophilicity, high aspect ratio and easy chemical modification. CNFs can be generated by mechanical processes like ultrasonication, high-pressure homogenization, steam treatment, grinding, microfluidization or cryocrushing, where pressure causes defibrillation of microfibrils to nanofibrils [70]. The mechanical processing is energy engrossing, so chemical methods use reagents like TEMPO (2,2,6,6-tetramethylpiperidine-1-oxyl) mediated oxidation, carboxymethylation and enzymatic hydrolysis prior to mechanical defibrillation for smooth and easy extraction of nanofibres. TEMPO-treated CNFs have high dispersibility due to high numbers of carboxyl groups, high mechanical stability and crystallinity [110].

4.2.3 Crystalline Cellulose

Suitable mechanical, chemical or enzymatic processing of cellulose microfibrils produces the highly crystalline regions in the form of microcrystalline cellulose

(MCC) with diameter in micrometres or cellulose nanocrystals (CNCs) with diameter at the nanoscale [151]. MCC was first developed by Ranby [128] and is stiff rod-like particles of cellulose chain fractions in nearly perfect crystalline structure. MCC was first introduced into the market under the brand of Avicel® PH and later to the pharmaceutical industry for direct tableting constituents. At present, more than ten companies in the world are involved in the production of MCC. Generally pharmaceutical MCC is obtained from softwood, hardwood and cotton [164]. Commercial MCC has round or rod-like particles with dimensions of 10–200 μm. Partially depolymerised α-cellulose after HCl hydrolysis treatment generates the purified form of MCC. Due to physiological inertness, smooth operation, biodegradability and high abundance, MCC expresses a wide range of compatibility with active pharmaceutical ingredients (API) [111]. Biodegradability, thermal stability, high mechanical robustness, high elasticity, aspect ratio and low density are the advantages of MCC, which makes it suitable biomaterial for filler in biocomposites [120].

5 Cellulose Derivatives

For biomedical purposes, the cellulose has been modified physically, enzymatically or chemically to form the derivatives with enhanced properties of stability, strength, reactivity and biodegradability. The biocompatibility of natural polymers can be increased through surface modification [138]. In the course of surface modification, the outer sheet of polymers for drug transport is altered for the enhanced physiological compatibility, low cytotoxicity, along with reduced side effects [107]. Being the most targeted reactive group, the –OH group is the favoured choice for surface alteration. The chemical methods like etherification, esterification, acetylation, silylation, amination and oxidation, etc. are involved in the surface functionalization of nanocellulose (NC) [5]. The alteration in intra- and inter-chain hydrogen bonding would lower the crystallinity of NC and administration of desired functional features leads the formation of tailored biopolymer for effective drug delivery.

5.1 Cellulose Etherification

Cellulose –OH groups can be partly or completely etherified by various chemicals like epoxides, halogenated carboxylic acids and alkyl halides [66]. In alkaline aqueous conditions, cellulose ethers (CE) have rapid dissolution rate and the compounds such as methyl cellulose (MC), carboxymethyl cellulose (CMC), hydroxyethyl cellulose (HEC), hydroxypropyl methyl cellulose (HPC) and carboxyethyl cellulose (CEC) have been synthesized so far [37]. CMC, MC and HEC have gained high attention for industrial biomedical formulations due to nontoxic properties along with advanced rheological and mechanical features. The solubility of CEs is influenced by acidity or alkalinity of the medium and dissolution rate of CEs in acidic medium is highly reduced as compared to alkaline medium. The high water retention ability and thermo-gelling property of CEs could expedite the wound healing and these valuable features of CEs rely on the chemical structure of the substituent and amount of substitution [35, 137].

5.1.1 Methyl Cellulose (MC)

The very important and simplest CE is the methyl cellulose. MC can be produced in the alkaline environment with methyl chloride or dimethyl sulphate. The various degrees of substitution can be generated through variations in conditions like reaction time or methylating agent [159]. Depending on the extent of replacement the solubility of MC also varies such that if the degree is 1.4–2.0, MC dissolves in water, and if it is 2.4–2.8, MC is usually soluble in water and some organic solvents [158]. MC has thermo-gelling potential and is an emulsifying additive, which is appropriate for drug delivery. MC is commonly used in biomedical purposes of tissue engineering, wound healing and pharmaceutical preparations.

5.1.2 Carboxymethyl Cellulose (CMC)

Commercially available CMC with no known toxicity to humans has exclusive properties of stability, hydrophilicity, high mechanical robustness, chemical stability, biocompatibility, non-cytotoxicity and biodegradability [135]. Chemically CMC can be produced by the reaction of cellulose with monochloroacetic acid, where the carboxymethyl groups replace –OH groups at C_2, C_3 and C_6 positions of glucose monomer. Carboxymethyl groups (–CH_2–COOH) are attached to some of the –OH groups of the cellulose in the CMC chemical structure. The extent of substitution in CMC is mainly in the range of 0.6–1.25 for the commercial grade biomedical products. The molecular weight, extent of replacement and dispersion of carboxymethyl substituents in the polymer are the parameters, on which CMC properties are dependent [147]. CMC is very useful for viscosity enhancing chemical, rheological controlling, binder, stabilizer and film generator in the area of medicine, especially in drug transport and tissue engineering [4, 58, 101]. Alkylation-etherification is the conventional mode of CMC synthesis for long time from various cellulose precursors. The difference in synthesis from various cellulose precursors is in the applied ratio of chemicals to cellulose precursors, along with variation in reaction conditions like temperature, concentration of etherifying agent and alkali, time, and pH.

5.1.3 Hydroxyethyl Cellulose (HEC)

The cellulose precursor and ethylene oxide in alkaline environment would produce the important CE, hydroxyethyl cellulose. HEC solubility in water along with several organic solvents makes it a suitable polymer for various biomedicine purposes. HEC applications in biomedical field are dependent on its non-toxicity, capacity to incorporate large amount of drug and nanoparticles [168]. Mixing of HEC with cellulose provides the enhanced compatibility with less polar matrices, which aid the further applications of HEC in medicine.

5.1.4 Hydroxypropyl Cellulose (HPC)

HPC is generated by the reaction of cellulose precursor and 1,2-propylene oxide in alkaline medium. In the chemical structure of HPC, the free –OH groups are partially or completely substituted by $-CH_2CH(OH)CH_3$ (hydroxypropyl) group. Formation of secondary –OH groups during the synthesis provides further alteration scope for new derivatives for medicine purpose, especially in drug transport and tissue engineering.

5.1.5 Ethyl Cellulose (EC)

The important CE derivative ethyl cellulose is synthesized by the reaction of cellulose precursor and ethyl chloride in alkaline medium at the temperature of 60 °C for several hours. The conversion of some of the –OH groups on the repeating glucose monomers into ethyl ether groups forms the chemical structure of EC [26]. The extent of substitution is 2–2.6 in EC, suitable for various commercial purposes, and it is also the basis for solubility in most of the organic solvents. EC-based composites' physical and performance parameters are dependent on extent of etherification, molecular weight and uniformity at molecular scale. Non-water solubility, biodegradable potential, admirable film-forming ability, water resistance and barrier formation are the significant attributes of EC. Despite having outstanding robustness, the strength of EC reduces with increment in temperature. EC has good scope in medicine, especially in drug delivery modules.

5.2 Cellulose Ester

Cellulose esters can be synthesized by reaction of cellulose with nitric acid, sulphuric acid or acetic anhydride. These esters are thermoplastic biomaterials with commercial potential. In comparison with cellulose, cellulose esters have better solvability in general solvents [32]. These cellulose esters can be produced through less complicated mechanism for applications in medicine. The important cellulose esters are cellulose acetate, cellulose nitrate and cellulose sulphate.

5.2.1 Cellulose Acetate (CA)

In 1865, Schützenberger discovered the thermoplastic and biodegradable polymer cellulose acetate. Conventionally CA is synthesized by acetylation of cellulose precursors with acetic anhydride in the presence of solvent acetic acid and catalyst sulfuric acid. Generally CA is generated by esterification of cellulose precursors like cotton, wood, sugarcane, recycled paper [132]. Due to convenient and easy chemical or mechanical preparation procedures for CA, it is beneficial for biomedical purposes in drug transport, wound healing and tissue engineering.

5.2.2 Cellulose Nitrate

Cellulose nitrate is a semi-synthetic and versatile polymer, also known as celluloid or nitrocellulose, has been in use since 1900 [34]. Cellulose nitrate is produced by the reaction of cellulose with HNO_3, where nitrate groups replace the cellulose –OH groups. Typically 2.2–2.8 nitrate groups per glucose unit in the cellulose nitrate polymer are present. Cellulose nitrate characteristics and applications are dependent on the extent of the nitration. Commercial applications of cellulose nitrate are in explosives, plastics, coating and ink industries. The scope of cellulose nitrate in biomedical purposes is still under inspection.

5.2.3 Cellulose Sulphate

Another cellulose ester derivative, cellulose sulphate can be produced by reaction of cellulose precursor with H_2SO_4, SO_3 and HSO_3Cl. The presence of sulphate groups and higher extent of substitution in polymer contribute properties such as water solubility, antiviral, antibacterial and anti-coagulant. Cellulose sulphate has several advantages like easy synthesis process, cost-effectiveness, large-scale production along with outstanding biocompatibility, film-forming capability and biodegradable potential. These attributes make the cellulose sulphate as potential candidate for biomedical purpose in the field of drug delivery and tissue engineering [115, 142].

6 Advanced Biomaterials Based on Nanocellulose

6.1 Hydrogels

Hydrogels with excellent features have drawn the great consideration, due to advanced progress in the field of nanotechnology in recent years. Hydrogels are playing leading role in the advancement of sustained drug transport modules for variety of biomedical applications [113, 157]. Hydrogels are two or a multicomponent polymeric matrix such as cellulose is formed by 3D crosslinked network of glucose monomers with the capability to absorb water and swell. Hydrogels are formed via physical, chemical and enzymatic methods of crosslinking to augment their physicochemical and biological attributes. The hydrogen bonds, covalent bonds, van der Waals interactions, presence of hydrophilic groups such as –OH, –CONH, $-CONH_2$, $-SO_3H$, capillary effect and osmotic pressure make the hydrogels an excellent entity for biomedical purposes, especially in drug transport [97]. The higher swelling potential, ability to hold substantial amount of fluid soft texture of hydrogels, considered them similar to biological tissues [80]. Hydrogels have several significant advantages for drug transport purposes due to biological compatibility, hydrophilic drugs entrapment and capability for sustained release of drug at specific location [113]. Due to several unique properties, hydrogels have potential to develop various types of biocomposites like blood-contacting hydrogels, contact lenses, wound-healing biosorbents, artificial kidney membranes, artificial skin, vocal cord replacements and artificial tendons [109]. Based on the dimensions of produced particles, hydrogels can be categorized into macro-, micro- or nanogels. The gel particles have dimensions bigger than 100 μm known as macrogels, while the dimensions upto micrometre range, considered as microgels and when the gel particles are smaller than 100 nm known as nanogels.

6.1.1 Nanogels

Nanogels are the type of hydrogels with spherical contour and dimension of 20–200 nm. The hydrogels at nanolevel has enhanced properties of increased solubility of water-soluble drugs, enhanced aggregation of drugs in tumour tissues, reduced cytotoxicity and augmented stability of active ingredients from chemical and enzymatic degradation. Apart from mentioned characteristics, nanogels have other attributes such as enhanced drug capturing ability, chemical stability, excellent mechanical robustness to sustain degradation during transport in biological fluids and responsive nature for rapid and controlled release of drug against the appropriate trigger [87, 103]. With the portfolio of excellent attributes, nanogels have applications in drug delivery systems along with other areas of biomedicine such as chemotherapy, diagnosis of diseases, vaccine delivery, biocatalysis, bioactive scaffolds in tissue regeneration, gene and protein delivery [68, 140, 167]. Various smart nanohydrogels based on NC like injectable hydrogels, shape memory, supramolecular hydrogels, double membrane hydrogels, temperature-sensitive hydrogels have been formulated for different biomedicine purposes [89, 92, 109, 141].

6.2 Nanocomposites

In general, to augment or improve some properties of nanocomposites, they are mixed with nanoparticles or nanofillers to attain desired features. Hydrogels are mixed with nanomaterials to design the nanocomposites with enhanced mechanical and chemical robustness or with antibacterial or magnetic potential [109].

6.2.1 Nanocellulose as Reinforcing Agent

Since few years, NC as a substitute for reinforcing agents in polymeric network has fascinated the scientists. Due to high mechanical robustness, elasticity, surface area, aspect ratio, biodegradability, nanocellulose has upper hand than glass fibres, silica, carbon black and other valuable fillers [133]. Biopolymer-based hydrogels reinforced with NC can be categorized as green nanocomposite materials on the basis of renewability and biodegradability. CNCs have been utilized as reinforcing materials for various types of polymer matrices like poly(vinyl) alcohol, poly(oxyethylene), polyethylene glycol, poly(N-isopropylacrylamide), starch, natural rubber and polyurethane [126]. Due to special morphology of length of fibres in micrometres and width in nanometers, nanofibrillated cellulose (NFC) has advanced attributes for mechanical reinforcement. The high flexibility, aspect ratio and strength are the main parameters to use the CNFs in different applications. BNC has the highest purity among all of the existing nanocellulose materials due to high crystallinity and enhanced biological adhesiveness, and it is excellent reinforcing component for biopolymer composites [52].

6.2.2 Nanocellulose Biomaterials as Matrices

The incorporation of bioactive molecules in biocomposites can improve the biocompatibility, antimicrobial potential and water retaining ability of nanocellulose materials. NC can be used as the substrate for inclusion of inorganic nanoparticles like carbon nanotubes, graphene and graphene oxide for production of hydrogels having antibacterial, antiviral, antifungal potential with magnetic, electrical and high mechanical characteristics. NC acts as support material for metal or metal oxide nanoparticles due to large surface area, decreased count of functional groups and capacity to form aqueous suspension [72]. NC-based materials have applied in the biomedicine field of wound dressings, drug carrier and as scaffold in tissue regeneration. The coating of silver nanoparticles over cellulosic materials have been shown broad antibacterial potential along with low levels of cytotoxicity. The silver nanoparticles were amalgamated with crystalline cellulose through microwave radiation to form the composite with high antibacterial potential for *Escherichia coli* and *Staphylococcus aureus* [84].

7 Application of Cellulosic Derivatives in Drug Delivery Systems

Cellulose, the most abundant natural polymer, is the main component of all the plants and wood with unique features of great mechanical and chemical robustness, thermal stability, biodegradability and low cost which attracted its application in the field of biomedicine and pharmaceutics. The particular disadvantage of cellulose is reduced solubility, which can be superseded by cellulose-based derivatives obtained via etherification, esterification and oxidation [99]. Nanocellulose is equipped with important characteristics of biocompatibility, non-toxicity, antimicrobial potential, wound-healing effects along with enhanced tissue adsorption due to several –OH groups and negative interface charge. The above-mentioned traits of cellulose make the cellulosic hydrogels to be tested in various domains of biomedicine like targeted drug delivery and smart sensors. Cellulose ethers are the most extensively used biomaterials for drug delivery systems in pharmaceutical field and mainly used as excipients, and some of them are methylcellulose, carboxymethylcellulose, ethylcellulose, hydroxyethyl cellulose, hydroxypropylcellulose and hydroxypropyl methyl cellulose [24]. Cellulose-based composites have been applied in drug delivery systems through numerous modes such as oral, dermal, transdermal, dental, ocular, nasal, stomach specific, vaginal, rectal and anti-cancer drug delivery (Fig. 3) [24]. Various cellulose ether-based derivatives have been investigated for the sustained drug delivery modules, and some of them are enlisted in Table 3.

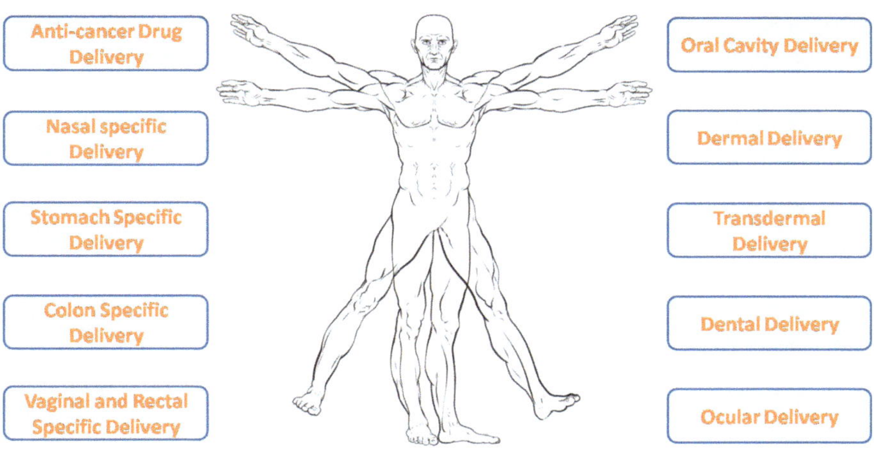

Fig. 3 Various types of drug delivery routes for cellulose-based biopolymers

Table 3 Cellulose-ether-based different derivatives in the form of hydrogels as drug delivery systems

Cellulose derivative	Drug	Additive	Features	Treatment	References
Methylcellulose (MC)	Docetaxel	Non-ionic copolymer surfactant (pluronic F127)	Injectable thermo-responsive gel, increased solubility of lipophilic drug	Cancer therapy	[76]
	Tenofovir	Stearic acid modified	Thermo-sensitive, intravaginal delivery	HIV and vaginal infections	[86]
	Erythropoietin	Hyaluronan (HA)	Epi-cortical administration and minimal invasive delivery	Endogenous stem cell stimulation after brain stroke injury	[162]
	Cyclosporine A	HA, poly(lactic-co-glycolic) acid (PLGA)	Tunable release for hours to weeks epi-cortical delivery	Brain stroke injury	[15]
Carboxymethylcellulose (CMC)	Methotrexate	Graphene oxide	pH-sensitive and controlled release properties	Anti-tumour delivery for colon cancer therapy	[63]
	Chrysin	Guar gum, polyvinyl alcohol, tetraethyl orthosilicate	High stability, swellability	Strong inhibitory effect on MCF-7 cells as anti-cancer	[25]
	Berberine	Hyaluronic acid (HYA)	Anti-adhesion barrier	Anti-inflammatory release delivery	[51]
	Curcumin doxorubicin	Carbon dot, gelatine, folic acid	Photoluminescent bionanogel, spherical morphology, diameter of ~ 70–90 nm	Anti-tumour chemotherapy	[125]
	5-FU	Gum Arabic, Fe(III)	pH triggered swelling of gel	Anti-cancer drug delivery	[40]
	Oral insulin	Poly(acrylic acid)	Biodegradable, pH responsive	Anti-diabetic drug delivery	[38]

(continued)

Table 3 (continued)

Cellulose derivative	Drug	Additive	Features	Treatment	References
	Metformin hydrochloride	Sodium alginate, Fe(III)	Swelling behaviour, encapsulation efficiency, pH sensitive, spherical shape	Anti-diabetic drug delivery	[147]
	Gentamicin	Carboxamide	pH protective, antibacterial cytocompatible	Controlled drug release	[58]
	Diclofenac	AISI 316LVM stainless steel	Corrosion resistant, biocompatible	Controlled drug release, improved osteogenic potential	[101]
	Ibuprofen	Cu-based-metal–organic framework	pH sensitive, low toxicity	In vitro cancer cell delivery	[59]
	Propolis	Polyvinyl alcohol	Antimicrobial, highly absorbent	Drug delivery in wound healing	[112]
	Tetracycline	β-cyclodextrin, cellulose	High compressive strength, swelling ability, superabsorbent, non-cytotoxic	In vitro drug delivery	[62]
	Dexamethasone	Alginate, chitosan	Magnetic hydrogel, high swelling index, pH sensitive	In vitro drug delivery	[61]
	Doxorubicin	Graphene oxide, $FeCl_3 \cdot 6H_2O$	High swelling ratio, pH sensitive	Anticancer drug delivery	[129]
Hydroxyethylcellulose (HEC)	Isoliquiritigenin	HYA	pH sensitive, optimal rheological and adhesive properties, excellent permeability to skin, antimicrobial	Transdermal delivery for skin lesions	[82]
	Eugenol	Poly(β-cyclodextrin)	Super-molecular self-assembled, thermo-stable, bacteriostatic	In vitro drug delivery	[145]
	Metronidazole	Precirol lipid nanoparticles	Enhanced antimicrobial potential	In vitro buccal mucosa drug delivery	[50]

(continued)

Table 3 (continued)

Cellulose derivative	Drug	Additive	Features	Treatment	References
	Doxorubicin Bcl-2 siRNA	Polyethylenimine2k	Low cytotoxicity, higher transfection efficiency, high serum stability	Anticancer drug delivery	[118]
Hydroxypropylcellulose (HPC)	Ibuprofen	Polyvinylpyrrolidone (PVP)	Slower drug release, higher time values	In vitro drug release studies	[33]
	Donepezil hydrochloride	Polyurethane	–	In vitro transdermal delivery	[39]
	Cannabidiol	PEO-b-P(CyCL-co-CL)	Drug solubility enhancer	In vitro anticancer drug delivery	[67]
Hydroxypropyl methylcellulose (HPMC)	Galantamine hydrobromide	Pectin, acrylic acid, methylene bisacrylamide	Good swelling and enhanced release kinetics	In vivo delivery for dementia treatment	[11]
	Benzo-phenone-4	PVP	Bioadhesive, mechanically strong	In vivo drug delivery	[117]
	Ketoprofen	Citric acid, locust bean gum	Increased angle of repose	In vitro drug release	[47]
	Meloxicam	Kollidon	–	In vitro drug release	[57]

7.1 Nanocellulose in Oral Drug Delivery Applications

Several researches have been conducted for the role of NC in oral drug delivery modules and promising findings revealed their superiority. In a study by Bhandari et al. [12], CNF aerogel was used as promising biomaterial for the delivery of model cancer drug bendamustine hydrochloride. The aerogel was characterized for various parameters of shape, mechanical robustness, swelling behaviour at different pH, mucoadhesive detachment force and drug release profile and pharmacokinetics in different pHs. The in vivo findings showed the 3.25 times upsurge in bioavailability of drug. These findings proved that CNFs aerogels are promising gastroretentive drug delivery module with enhanced bioavailability. In a study, CNCs were used as the carrier for various anticancer drugs like etoposide, paclitaxel and docetaxel to evaluate the controlled released behaviour of drugs for few days. The surface morphology of CNC was altered by grafting cationic surfactant, CTAB, leading to concentration guided surge in zeta potential of NC. The findings expressed the sustained release pattern of drugs for few days [14]. In another research by Mohanta et al. [104], thin films of NCC were fabricated for transport of hydrophobic curcumin and hydrophilic anticancer drug doxorubicin hydrochloride in combination with natural cationic polymer chitosan. The $-NH_2$ groups in chitosan showed electrostatic interactions with sulphonate groups of CNC. A layer-by-layer (LbL) method was applied to form multilayered thin films and microcapsules incorporating NCC as base constituent. Drug release experiments were conducted in PBS at physiological pH (7.0) and slightly acidic pH (6.4) due to acidic microenvironment of tumour tissue. The findings showed the controlled release of doxorubicin over nine-hour period with higher release at acidic pH. The curcumin-NCC formulations showed disciplined release of curcumin at pH 7.0. Further, microcapsules containing both curcumin and doxorubicin were synthesized which underscores the adaptability of NCC-based LbL composites for drug delivery purposes. Finally, NCC was the effective transporter system for both hydrophilic and hydrophobic drug delivery in cancer therapies at acidic environment.

In conventional regimen for oral diseases like periodontitis, viral or fungal infections, the drugs have been administered straight into oral cavity. In the oral cavity, due to large amount of salivary fluid, it is pertinent to apply the hydrogels with prolonged adhesion ability for enhanced local drug delivery. CMC is a biopolymer having mucoadhesive features and can be successfully applied for introduction of drugs like miconazole nitrate or lidocaine hydrochloride for transmucosal purposes. Miconazole nitrate provided the buccal-adhesion time of 2.45–3.65 h, while in contrast lidocaine hydrochloride sustained for 6 h at the minimum potent concentration of 4–10 mg/ml in saliva [60].

Cancer is the second dominant reason for global mortality with more than 100 known categories. Few known drugs such as cisplatin, doxorubicin, 5-fluorouracil, methotrexate and curcumin are the major players for the treatment of cancer. The conventional methods like chemotherapy, immunotherapy, radiation therapy and surgery are extensively applied for the remedy of cancer and related diseases [19]. In the research conducted by Poonguzhali et al. [124], the natural biopolymer alginate along with nanocellulose was involved in the development of composite film through solution casting method, loaded with antibiotic ampicillin. Incorporation of nanocellulose into alginate network augmented the mechanical robustness and swelling behaviour of the film and also the controlled drug release profile. The interaction between alginate, nanocellulose and ampicillin was supported through ATR-FTIR analysis. The Alg/NC film exhibited advanced physicochemical features in contrast to films devoid of nanocellulose with extended drug release behaviour. The presence of nanocellulose in in vitro experiments showed the faster release of ampicillin in early stage due to decreased binding interactions between elements and enhanced sponginess of composite film. The controlled release of drug over 500 min focuses the capability of Alg/NC film as efficient drug delivery module in the physiological surroundings. Guo et al. [45] formulated the micro- and nanocellulose/alginate beads in combination of different cellulose sources for the drug release behaviour of metformin hydrochloride (MH). The spheres or beads were generated utilizing various types of cellulose fibres such as cotton linter, MCC and MFC for the improved structural and mechanical robustness along with thermal sturdiness of alginate matrix. The nanolevel MFC with several reactive –OH groups enhanced encapsulation capacity and restricted the fast delivery of MH in imitated gastrointestinal milieu. The cellulose was considered as active alternator for swelling potential, mechanical features and drug delivery pattern of alginate-based delivery arrangements. In imitated gastric fluid (IGF, pH 1.2) and imitated intestinal fluid (IIF, pH 7.4 and pH 6.8), the overall discharge rate of MH was affected by cellulose and its morphological properties. The MFC-Alg beads showed the more constant release of MH in IIF due to hydrogen bonding between MFC and alginate. These results pointing that cellulose-alginate beads could be impressive drug transporter in gastrointestinal surroundings with enhanced durability and customized drug discharge pattern.

BC is a unique and unadulterated crystalline cellulose biomaterial for the applications in biomedicine, especially wound healing, artificial skin and tissue engineering [154]. Along with these applications, the Food and Drug Administration (FDA) has depicted BC as secure and suitable for fabrication of drug transport systems. Badshah et al. [10] developed and examined the single excipient-based BC networks for model drugs famotidine and tizanidine oral delivery. The efficient drug packing and homogeneous dissemination into BC networks were affirmed through SEM and X-ray diffraction. FTIR and thermogravimetric findings expressed the chemical and thermal durability of BC-drug formulations. In vitro dissolution experiments exhibited greater than 80% drug discharge in the first 15 min for all composites, proved the instant discharge parameter. Mohd et al. [105] fabricated and examined the safeness and biological compatibility of BC/acrylamide (AM) pH-susceptible intelligent hydrogels for oral drug transport module for theophylline as model drug. The hydrogels were formulated through microwave irradiation method with the help of N,N′-methylene-bis-acrylamide as crosslinker with soluble BC in NaOH/urea solution. The loaded theophylline compound release profile was analysed at different BC concentrations. The cytotoxicity and acute gastrointestinal toxicity were examined and findings pointed towards the non-cytotoxic behaviour of BC hydrogels.

In an additional study by Supramaniam et al. [146], the scientists developed the ALG/iron oxide modified CNCs hybrids for model drug ibuprofen delivery. The amalgamation of 3% magnetic CNCs (m-CNCs) enhanced the mechanical and swelling index along with excessive drug encapsulation and packaging capacity. The magnetic beads were analysed through techniques like FTIR, XRD, SEM and VSM and showed increased mechanical robustness and swelling magnitude. The ibuprofen release showed the biphasic release profile. The initial rapid release occurred within first 30 min while controlled release phase maintained upto 330 min was affected by m-CNCs power to limit the drug mobility in hydrogel network. Further, inception of Fe nanoparticles escalated the sturdiness in ALG hydrogel and avoided the shattering or fracturing of beads with strengthened target accuracy.

HPC hydrogels are competent for formulation of bioadhesive complexes with combination of poly(acrylic acid) (PAA) and lactose non-adhesive supporting layer. Such types of formulations are commercially accessible as Aftach and are useful in the medication of aphthous ulcers through release of drug triamcinolone acetonide [122]. The in vitro and in vivo drug release profile of drug theophylline was examined through chemically crosslinked hydrogels of cellulose (C) and chondroitin sulphate (CS). For in vitro release in acidic environment, the increase in CS percentage led to reduction in percentage of theophylline release in formulation such as 95% (C), 78% (80/20 C/CS) and 58% (50/50 C/CS). This increment ascertained the reduction of half release time values varies from 14 min for C hydrogels to 60 min for 50/50 C/CS formulation. In vivo release (orally introduction to rats as solution in 0.5% wt CMC) showed the optimal biological compatibility of drug in hydrogel in comparison with alone drug and enhanced release of drug, upto 50 h for 50/50 C/CS matrix [114].

Cellulose ethers are synthesized by etherification reaction of –OH groups with alkyl halide or epoxides. Several types of cellulose ethers till date have been produced, and CMC and HEC are well known for biomedicine purpose. Cellulose ethers are majorly and routinely utilized for film generator in oral drug delivery systems. The additives can be further incorporated to customize the mechanical properties of films. Castro et al. [18] fabricated the hydroxypropyl cellulose as an easily soluble film to treat nausea and vomiting by loading the drug domperidone. The sublingually and buccally administered films retain the promise for magnified onset of action, reduction in doses and refining the overall effectiveness and safety pattern of treatment.

7.2 *Cellulose-Based Formulations in Dermal Applications*

The dermal or skin-related disorders such as irritations, acne, atopic dermatitis, rosacea and various allergies can be treated with the application of cellulosic hydrogels encapsulating the dermal drug. Model dermal drugs such as fluconazole, diclofenac sodium, piroxicam earlier solubilized in emulsion system can be incorporated into cellulose-based hydrogels [160]. Salerno et al. [136] developed drug delivery system having sodium CMC as gel substance, propylene glycol (PPG) as solvent and diethylene glycol monoethyl (DEGEE) ether as penetrating modifier to treat the topical mycosis. The results expressed that microemulsion hydrogels are very effective with antifungal potential in in vitro experiments. DEGEE in hydrogel is successful as standard solution, e.g., < 10 colony forming units (CFU/ml) exhibited the release of complete drug and infiltrating pig skin and four times higher than lipogels and have the potential to retain the drug in skin layers. The sustained discharge profile of an anti-inflammatory drug piroxicam was examined using several MC hydrogels, CMC, HPMC, carbopol 934, carbopol 940 or pluronic F-127. The drug was encapsulated in the microemulsion form in gels along with surfactants Tween 80 and propylene glycol. The in vitro findings pointed that hydrogels 3% MC and 3% HPMC liberated the huge amount of drug 97% and 94%, respectively, after 3 h of experiment. HPMC hydrogel is the most appropriate formulation for topical application of piroxicam in view of rheological features and duration of use of gel [1]. Malik et al. [98] studied the sustained discharge of antiviral drug acyclovir (ACV) through the network of hydrogels composition of CMC, β-cyclodextrin (β-CD), acrylic acid (AA) and N,N′-methylene-bis-acrylamide (MBA). The combination of β-CD and CMC in the solution permits the hydrogels with optimum mechanical robustness and also with upgraded ability to pack and discharge of the drug. The ACV discharge increased with increased concentration of β-CD from 1 to 4% and CMC from 8 to 12%. The hydrogel with effective constitution of 30 g AA/100 g β-CD, 12 g CMC/100 g and 0.6 MBA/100 g, exhibited the optimum maximum release of 96.15% of ACV at physiological pH of 7.4 among all the hydrogels formulations.

7.3 Cellulose-Based Composites in Transdermal Delivery Applications

Transdermal drug delivery systems (TDDSs) represented as an effective and promising delivery alternative instead of oral and hypodermic administration for therapeutic molecules non-invasively into the systemic circulation through skin for enhanced drug release and bioavailability. The important advantage of TDDS is to bypass the liver metabolism and gastrointestinal (GI) tract and providing the medicinal effect at low amount, therefore minimizing or phase out the side effects, especially in GI effect [91]. Due to admirable features such as simple administration, reduced systemic side effects, evasion of liver first-pass effect and ability to bring comfortable sensation for skin in contrast to traditional lotions and patches, the cellulosic hydrogels were researched as efficient TDDS [139]. Skin penetration is reduced with application of high molecular weight or charged active ingredients of drug due to morphology and physicochemical properties. So the transdermal introduction of delivery is achieved by physical methods like iontophoresis, sonophoresis, electroporation and laser radiations [160]. Several studies were conducted based on transdermal delivery of drug using the iontophoresis methods in combination of other physical and chemical methods. In one such case, the clinical performance of iontophoresis method was examined for non-steroidal inflammatory drug celecoxib on various composites with different gelling compounds like sodium alginate, CMC, HPMC and carbopol 934P. HPMC hydrogel showed the higher distribution and enhanced skin holding capacity and maximum release of celecoxib of 41.5% after 5 h. Based on the findings, HPMC is suitable composites for iontophoretic purpose and iontophoretic mode of drug release from hydrogels to rat skin was twice as high as passive flow in in vivo experiment [150].

Kim et al. [75] examined the drug release pattern of captopril (CTP) for antihypertensive therapy, based on pH, temperature and plasticizers. The drug was incorporated into chitosan/carboxymethylcellulose (CS/CMC) biocomposite. In the findings, the lower pH and higher temperature substantially augmented the CTP discharge rate and diffusion rate. The plasticizers like citric acid and glycerol, advanced the physical features of bioformulation and so impacted the faster drug release behaviour. In the release pattern, CTP was expeditiously discharged in first 90 min and totally liberated within 180 min. The citric acid-based formulations showed the fastest release in comparison with without plasticizers and with glycerol. At the temperature of 36.5 °C, the CTP discharge rate was initially escalated and reached to near equilibrium after 120 min. The CTP-CS/CMC biocomposite with citric acid plasticizer showed promising results for TDDS in case of antihypertensive therapy.

In another study, the hydrogel of CMC-poloxamer 407 (P407) was selected for the diffusion pattern of model drug gallic acid through the skin of pig's ear as permeable membrane. The presence of CMC considerably upgraded the texture and morphology of hydrogel matrix, as 4% CMC marginally reduced the pore size from 4.813 to 4.343 μm and lead to noticeable increment in the fraction of pore number around tenfold in contrast to hydrogel without CMC. The noticeable increment of drug permeability through skin was validated via enhanced value of apparent permeability coefficient (Papp) which was backed by the spongy morphology of P407 hydrogel after inclusion of CMC [163]. Chatterjee and Hui [20] reported the formulation of thermo-responsive hydrogel based on CMC and gelatine for the transdermal delivery of drug lidocaine. The hydrogel microspheres had the mean particle size of 5.89–14.60 μm based on the gelatine concentration. All the hydrogels showed the quick discharge of lidocaine throughout the first hour and with stable state profile examined in the next 3 h. The CMC/gelatine (1:1.6) hydrogel having the smaller particle size with higher surface area exhibited the maximum release content of lidocaine (32.3% in 1 h). Several types of nanocellulose-based biocomposites like nanogold-NC and NFC/hydroxypropyl methylcellulose formulations exhibited promising consideration for TDDS. These composites primarily showed the controlled release pattern along with significant success of various nanocellulose materials for TDDS [8].

7.4 Cellulose-Based Composites in Dental Delivery Administration

Dental problems such as caries and periodontal inflammation are the most common infections in humans [108]. Numerous drug transport arrangements have been devised for the treatment of dental diseases such as toothpastes and mouthwashes which are most common. The shortcoming of traditional method is the little oral retention duration due to salivary secretion, water and food ingestion and erosion caused by soft tissue movement [74].

Weyell et al. [165] examined the altered degradation pattern of BC by periodate oxidation to obtain partial biodegradability under buccal physiological milieu along with retention of drug delivery potential. The drug-packaged BC was explored for the applications in dental therapies like dental extraction and mucosal transplantation. A periodate concentration of 0.14 mol/L at temperature of 25 °C and $t = 8$ h proceeded in the material loss of < 10% and significant level of degeneration. Native and oxidized BC was packaged with doxycycline which was examined for prophylaxis against infection. The in vitro MTT assay confirmed the biological compatibility, whereas agar diffusion test showed the antibiotic effectiveness against oral pathogenic bacteria. Drug discharge pattern showed the biphasic behaviour. The study showed the promise for fabrication of customized BC-based biomaterials for dental treatments. In a study by Inoue et al. [53], the researchers prepared the bioabsorbable and bioactive barrier membrane of BC, oxidized with $NaIO_4$ and loaded with model drug chlorhexidine (CHX). For the control release of CHX, the composite of CHX with β-CD was formulated. The membrane oxidation and incorporation of β-CD resulted in tenfold rise in CHX discharge rate in contrast to unaltered cellulose. All the BC-CHX membranes impeded *Staphylococcus aureus*, *Escherichia coli* and *Candida albicans* growth, but DABC (bacterial 2,3-dialdehyde cellulose) + CHX-β-CD exhibited higher zone of inhibition. These findings pointed towards the bioactive and bioabsorbable application of modified BC membranes. Jin et al. [64] fabricated the multifunctional, flexible and biodegradable BC-$Ti_3C_2T_X$ MXene bioaerogel for the precise determination of occlusal force and early diagnosis of periodontal disorders. The resulted bioaerogel showed 3D porous conformation showing pressure stimulated and NH_3 stimulated responses. These findings are promising for diagnosis and oral health and BC-based bioaerogel showed complete biodegradation in the presence of low concentration of H_2O_2 solution providing the eco-friendly and sustainable option.

7.5 Cellulose-Based Composites in Ocular Delivery Applications

The major challenge in ocular disease therapy is to maintain the optimal concentration of drug for required duration to achieve the desired results. To overcome such challenges several viscosity augmenting cellulose-based composites, mucoadhesive polymers and hydrogel systems were developed to delivery desired amount of ocular drug [30]. Various cellulose-based topical ophthalmic formulations such as MC, HEC, HPMC and sodium carboxymethyl cellulose are the most widely used polymer in the present time [166]. An in situ gelling arrangement for ophthalmic drug transport was fabricated based on sodium alginate and MC. At the pH of 4.7, the gel system was in sol state, while increasing the pH upto 7.4, the system experienced the rapid sol–gel transition. The model drug sparfloxacin was incorporated in gel for sustained release. This composite gel system was also examined for corneal permeation on goat eye and showed promising results [73]. Iohara et al. [55] fabricated the thermo-responsive hydrophobically modified composite based on HPMC (HM-HPMC) hydrogel loaded with α-cyclodextrin (α-CD). The formulated gel system exhibited the reversible sol–gel transition at the physiological temperature of 37 °C, while gel system without the drug α-CD expressed temperature dependency. The fabricated gel system showed the quick gelation on the ocular surface and expressive and enhanced assimilation of ocular drug diclofenac sodium.

7.6 Cellulose-Based Composites in Anti-cancer Drug Transport Modules

In the current scenario, the several types of cancer treatments are based on various therapeutic approaches like surgery, chemotherapy and radiotherapy. The regularly applied therapeutic regimen is chemotherapy which involves the administration of drugs with toxicity, off-target specificity and affects the normal body cells [27]. Hydrogels are the appropriate modules for drug transport in cancer therapeutics in sustained and targeted manner. Different categories of hydrogel formulations such as injectable composites, nanoparticles and nanocomposites have been investigated for the regulated drug transport for cancer therapeutics. Nanocelluloses are the very effective formulations for cancer drugs because of great surface area, biodegradability and biocompatibility. In a study by Liu et al. [90], the aerogel of carboxylated cellulose (CC), hydroxyapatite (HAP) and copper nanoparticles (Cu) was developed for the model anticancer drug 5-fluorouracil (5-FU). The CC-HAP-Cu aerogel system enhanced the drug packaging capacity of 181 mg/g for 5-FU and exhibited the controlled discharge over 3 h in highly acidic environment of pH 1.0 and 30 h in neutral condition of pH 7.2. The aerogel showed the maximum compressive stress of 6.70 MPa indicating the mechanical and structural compactness of drug transporter. The low swelling ratio of aerogel further authenticated the controlled release of 5-FU. The MMT biocompatibility assay showed that aerogel has significant compatibility with living cells and relative cell viability exceeded more than 97%. In another study [88], the researchers developed the novel hydrogel drug delivery system by polymerizing polydopamine on cellulose nanofibrils (CNF) followed by incorporation of zeolitic imidazolate frameworks-8 (ZIF-8) nanoparticle. The hydrogel was loaded with hydrophobic curcumin with high loading ratio of 4.5 wt% for the drug. The mechanical robustness of ZIF-8@PCNF hydrogel was 2.75 times greater than CNF hydrogel alone. The hydrogel showed the prolonged and sustained drug release time upto 107 h, higher than other transporter. The release rate was enhanced by acidic environment (up to 90% at pH 2.5) and by near-infrared light (NIR) (84% discharge within 120 min at pH 2.5). The hydrogel showed the promising biological compatibility and distinct anticancer potential against HepG2 cell line.

The stimuli-responsive hydrogels for the cancer chemotherapeutics are examined for entrapment of drug and release triggered via external stimulus. The thermo-responsive hydrogels loaded with anticancer drug can be administered to the target tumour tissue in liquid state and transformed into gel through body temperature [102]. The pH triggered hydrogels have been applied for anticancer drug transport functions on the variation in pH value of tumour milieu (pH 5–6) and normal physiological pH of 7.4. CMC with –COOH groups triggered the discharge of anticancer drug in tumour tissue with pH of 5–6.5. On the other hand dual, temperature/acidic pH-responsive nanogels (DuR-BNGs) were developed by crosslinked polymerization through temperature mediated self-association in aqueous solution. The discharge of encapsulated drug from DuR-BNGs is triggered by higher temperature, and acidic pH can be applied for cancer therapy [130].

8 Conclusion

Cellulose the natural polymer with several advantageous features like abundance, low cost, high biocompatibility and hemocompatibility, cytocompatibility and good biodegradability and chemical, physical or enzymatic derivatization could generate many bioformulations for medical purposes especially in the field of regulated drug delivery, wound dressing, tissue and bone engineering. In particular, carboxymethyl cellulose, hydroxypropyl methylcellulose, methyl cellulose are frequently used in drug delivery systems. Though cellulose-based bioformulations can be easily developed and respond to various types of stimuli with attributes of controlled drug release at target, but some issues have to be resolved for effective and biocompatible drug delivery such as low mechanical and physical robustness of biomaterials, reduced delivery, sudden release of drug, difficult hydrophobic drug release due to solubility in biological fluids, moderate response of biocomposites and the side effects. The various types of delivery routes can affect the potential of biomaterials and their biodegradability, which may lead to various types of side effects.

In recent years, cellulose-based hydrogels have inspired the researchers for drug transport purposes due to possibility to select the suitable polymer type and to fabricate the network for spatial and temporal regulation for various types of pharmaceuticals ingredients release. These hydrogels are loaded with several good features like optimum and sustained discharge of active pharmaceutical ingredient over the long duration, bioadhesiveness to surfaces like skin or tissues, biodegradability to avoid the surgical involvement and expected therapeutic outcomes. The remarkable progress has been achieved for cellulosic hydrogels in the applications of drug transport and other medical purposes. The current, focused and improved developments for hydrogels has shown the promising findings for therapeutic use in several disease treatments.

References

1. Abd-Allah FI, Dawaba HM, Ahmed AM (2010) Preparation, characterization, and stability studies of piroxicam-loaded microemulsions in topical formulations. Drug Discov Ther 4:267–275
2. Abdel-Halim ES (2014) Chemical modification of cellulose extracted from sugarcane bagasse: preparation of hydroxyethyl cellulose. Arab J Chem 7(3):362–371. https://doi.org/10.1016/j.arabjc.2013.05.006
3. Abdelhamid HN, Mathew AP (2022) Cellulose-based nanomaterials advance biomedicine: a review. Int J Mol Sci 23(10):5405. https://doi.org/10.3390/ijms23105405
4. Adeyeye MC, Jain AC, Ghorab MK et al (2002) Viscoelastic evaluation of topical creams containing microcrystalline cellulose/sodium carboxymethyl cellulose as stabilizer. AAPS PharmSciTech 3:16–25. https://doi.org/10.1208/pt030208
5. Aditya T, Allain JP, Jaramillo C et al (2022) Surface modification of bacterial cellulose for biomedical applications. Int J Mol Sci 23(2):610. https://doi.org/10.3390/ijms23020610
6. Ansari F, Sjöstedt A, Larsson PT et al (2015) Hierarchical wood cellulose fiber/epoxy biocomposites—materials design of fiber porosity and nanostructure. Compos Part A Appl Sci 74:60–68. https://doi.org/10.1016/j.compositesa.2015.03.024
7. Ardanuy M, Claramunt J, Toledo Filho RD (2015) Cellulosic fiber reinforced cement-based composites: a review of recent research. Constr Build Mater 79:115–128. https://doi.org/10.1016/j.conbuildmat.2015.01.035
8. Babaei-Ghazvini A, Patel R, Vafakish B et al (2024) Nanocellulose in targeted drug delivery: a review of modifications and synergistic applications. Int J Biol Macromol 278(4):135200. https://doi.org/10.1016/j.ijbiomac.2024.135200
9. Babaei-Ghazvini A, Acharya B (2023) Functional sensory and biomedical applications of cellulose nanocrystals as a sustainable source of material. In: Bressan N (ed) An introduction to veterinary medicine engineering. Springer Nature, Switzerland AG, pp 61–75. https://doi.org/10.1007/978-3-031-22805-6_5
10. Badshah M, Ullah H, Khan SA et al (2017) Preparation, characterization and in-vitro evaluation of bacterial cellulose matrices for oral drug delivery. Cellulose 24(11):5041–5052. https://doi.org/10.1007/s10570-017-1474-8
11. Bashir S, Zafar N, Lebaz N et al (2020) Hydroxypropyl methylcellulose-based hydrogel copolymeric for controlled delivery of galantamine hydrobromide in dementia. Processes 8:1350. https://doi.org/10.3390/pr8111350
12. Bhandari J, Mishra H, Mishra PK et al (2017) Cellulose nanofiber aerogel as a promising biomaterial for customized oral drug delivery. Int J Nanomed 12:2021–2031. https://doi.org/10.2147/IJN.S124318
13. Brown AJ (1886) XLIII.—on an acetic ferment which forms cellulose. J Chem Soc Trans 49:432–439. https://doi.org/10.1039/CT8864900432
14. Burt HM, Jackson JK, Hamad WY (2014) Binding drugs with nanocrystalline cellulose (NCC), 2014. WO/2012/068670
15. Caicco MJ, Cooke MJ, Wang Y et al (2013) A hydrogel composite system for sustained epi-cortical delivery of Cyclosporin A to the brain for treatment of stroke. J Control Release 166:197–202. https://doi.org/10.1016/j.jconrel.2013.01.002

16. Calvino C, Macke N, Kato R et al (2020) Development, processing and applications of bio-sourced cellulose nanocrystal composites. Prog Polym Sci 103:101221. https://doi.org/10.1016/j.progpolymsci.2020.101221
17. Campano C, Balea A, Blanco A et al (2016) Enhancement of the fermentation process and properties of bacterial cellulose: a review. Cellulose 23:57–91. https://doi.org/10.1007/s10570-015-0802-0
18. Castro PM, Fonte P, Oliveira A et al (2017) Optimization of two biopolymer-based oral films for the delivery of bioactive molecules. Mater Sci Eng C 76:171–180. https://doi.org/10.1016/j.msec.2017.02.173
19. Charmsaz S, Collins DM, Perry AS et al (2019) Novel strategies for cancer treatment: highlights from the 55th IACR annual conference. Cancers (Basel) 11(8):1125. https://doi.org/10.3390/cancers11081125
20. Chatterjee S, Hui PC-L (2019) Review of stimuli-responsive polymers in drug delivery and textile application. Molecules 24:2547. https://doi.org/10.3390/molecules24142547
21. Chen YW, Lee HV, Abd Hamid SB (2017) Investigation of optimal conditions for production of highly crystalline nanocellulose with increased yield via novel Cr(III)-catalyzed hydrolysis: response surface methodology. Carbohydr Polym 178:57–68. https://doi.org/10.1016/j.carbpol.2017.09.029
22. Chen YW, Lee HV, Juan JC et al (2016) Production of new cellulose nanomaterial from red algae marine biomass *Gelidium elegans*. Carbohydr Polym 51:1210–1219. https://doi.org/10.1016/j.carbpol.2016.06.083
23. Chu Y, Sun Y, Wu W et al (2020) Dispersion properties of nanocellulose: a review. Carbohydr Polym 250:116892. https://doi.org/10.1016/j.carbpol.2020.116892
24. Ciolacu DE, Nicu R, Ciolacu F (2020) Cellulose-based hydrogels as sustained drug-delivery systems. Materials (Basel) 13(22):5270. https://doi.org/10.3390/ma13225270
25. Dalei G, Das S, Das SP (2021) Low-pressure nitrogen and ammonia plasma treatment on carboxymethyl guar gum/PVA hydrogels: impact on drug delivery, biocompatibility and biodegradability. Int J Polym Mater Polym Biomater 70(2):75–89. https://doi.org/10.1080/00914037.2019.1695204
26. Davidovich-Pinhas M, Barbut S, Marangoni A (2015) The gelation of oil using ethyl cellulose. Carbohydr Polym 117:869–878. https://doi.org/10.1016/j.carbpol.2014.10.035
27. Debele TA, Mekuria SL, Tsai H-C (2016) Polysaccharide based nanogels in the drug delivery system: application as the carrier of pharmaceutical agents. Mater Sci Eng C 68:964–981. https://doi.org/10.1016/j.msec.2016.05.121
28. Deepa B, Abraham E, Cordeiro N et al (2015) Utilization of lignocellulosic biomass for the production of nanocellulose: a comparative study. Cellulose 22:1075–1090. https://doi.org/10.1007/s10570-015-0554-x
29. Du H, Liu W, Zhang M et al (2019) Cellulose nanocrystals and cellulose nanofibrils based hydrogels for biomedical applications. Carbohydr Polym 209:130–144. https://doi.org/10.1016/j.carbpol.2019.01.020
30. Dubashynskaya NV, Poshina D, Raik S et al (2019) Polysaccharides in ocular drug delivery. Pharmaceutics 12:22. https://doi.org/10.3390/pharmaceutics12010022
31. Dumitriu C, Voicu SI, Muhulet A et al (2018) Production and characterization of cellulose acetate–titanium dioxide nanotubes membrane fraxiparinized through polydopamine for clinical applications. Carbohydr Polym 81:215–223. https://doi.org/10.1016/j.carbpol.2017.10.082
32. Edgar KJ, Buchanan CM, Debenham JS et al (2001) Advances in cellulose ester performance and application. Prog Poly Sci 26:1605–1688. https://doi.org/10.1016/S0079-6700(01)00027-2
33. Erkoç T, Sevgili LM, Çavuş S (2022) Hydroxypropyl cellulose/polyvinylpyrrolidone matrix tablets containing ibuprofen: infiltration, erosion and drug release characteristics. ChemistrySelect 30(7):e202202180. https://doi.org/10.1002/slct.202202180
34. Falkowitz A, Piech F (1956) Cellulose lacquers—nitrocellulose lacquers for aircraft. Ind Eng Chem 48:1326. https://doi.org/10.1021/ie50560a606

35. Fidale LC, Heinze T, El Seoud OA (2013) Perichromism: a powerful tool for probing the properties of cellulose and its derivatives. Carbohydr Polym 93:129–134. https://doi.org/10.1016/j.carbpol.2012.06.061
36. Fooladi S, Nematollahi MH, Rabiee N et al (2023) Bacterial cellulose-based materials: a perspective on cardiovascular tissue engineering applications. ACS Biomater Sci Eng 9(6):2949–2969. https://doi.org/10.1021/acsbiomaterials.3c00300
37. Fox SC, Li B, Xu D et al (2011) Regioselective esterification and etherification of cellulose: a review. Biomacromolecules 12:1956–1972. https://doi.org/10.1021/bm200260d
38. Gao X, Cao Y, Song X et al (2013) Biodegradable, pH-responsive carboxymethyl cellulose/poly(acrylic acid) hydrogels for oral insulin delivery. Macromol Biosci 14:565–575. https://doi.org/10.1002/mabi.201300384
39. Gencturk A, Kahraman E, Güngör S et al (2016) Polyurethane/hydroxypropyl cellulose electrospun nanofiber mats as potential transdermal drug delivery system: characterization studies and in vitro assays. Artif Cells Nanomed Biotechnol 45(3):655–664. https://doi.org/10.3109/21691401.2016.1173047
40. Ghawanmeh AA, Tan LL, Ali GAM et al (2024) Optimization of carboxymethyl cellulose-gum Arab-based hydrogel beads for anticancer drugs delivery. J Mol Liq 393:123631. https://doi.org/10.1016/j.molliq.2023.123631
41. Gonçalves AP, Oliveira E, Mattedi S et al (2018) Separation of cellulose nanowhiskers from microcrystalline cellulose with an aqueous protic ionic liquid based on ammonium and hydrogensulphate. Sep Purif Technol 196:200–207. https://doi.org/10.1016/j.seppur.2017.07.054
42. Gorgieva S, Trček J (2021) Bacterial cellulose: production, modification and perspectives in biomedical applications. Nanomaterials 9:1352. https://doi.org/10.3390/nano9101352
43. Gregory DA, Tripathi L, Fricker ATR et al (2021) Bacterial cellulose: a smart biomaterial with diverse applications. Mater Sci Eng R Rep 145:100623. https://doi.org/10.1016/j.mser.2021.100623
44. Grishkewich N, Mohammed N, Tang J et al (2017) Recent advances in the application of cellulose nanocrystals. Curr Opin Colloid Interface Sci 29:32–45. https://doi.org/10.1016/j.cocis.2017.01.005
45. Guo T, Pei Y, Tang K et al (2017) Mechanical and drug release properties of alginate beads reinforced with cellulose. J Appl Polym Sci 134(8). https://doi.org/10.1002/app.44495
46. Gupta PK, Raghunath SS, Prasanna DV et al (2019) An update on overview of cellulose, its structure and applications. In: Pascual AR (ed) Cellulose. IntechOpen. https://doi.org/10.5772/intechopen.84727
47. Hadinugroho W, Martodihardjo S, Fudholi A et al (2023) Hydroxypropyl methylcellulose as hydrogel matrix and citric acid-locust bean gum as negative matrix for controlled release tablet. ACS Omega 8(8):7767–7778. https://doi.org/10.1021/acsomega.2c07432
48. Hallac BB, Ragauskas AJ (2011) Analyzing cellulose degree of polymerization and its relevancy to cellulosic ethanol. Biofuel Bioprod Bioref 5:215–225. https://doi.org/10.1002/bbb.269
49. Hernandez CC, dos Santos Rosa D (2016) Extraction of cellulose nanowhiskers: natural fibers source, methodology and application. In: Mendez-Vilas A (ed) Polymer science: research advances, practical applications and educational aspects. Formatex Research Center S.L., Badajoz, Spain, pp 232–242
50. Ho HN, Le HH, Le TG et al (2022) Formulation and characterization of hydroxyethyl cellulose-based gel containing metronidazole-loaded solid lipid nanoparticles for buccal mucosal drug delivery. Int J Biol Macromol 1(194):1010–1018. https://doi.org/10.1016/j.ijbiomac.2021.11.161
51. Huang Y-C, Huang K-Y, Yang B-Y et al (2016) Fabrication of novel hydrogel with berberine-enriched carboxymethylcellulose and hyaluronic acid as an anti-inflammatory barrier membrane. BioMed Res Int 2016:3640182. https://doi.org/10.1155/2016/3640182
52. Huang Y, Zhu C, Yang J et al (2014) Recent advances in bacterial cellulose. Cellulose 21:1–30. https://doi.org/10.1007/s10570-013-0088-z

53. Inoue BS, Streit S, dos Santos Schneider AL et al (2020) Bioactive bacterial cellulose membrane with prolonged release of chlorhexidine for dental medical application. Int J Biol Macromol 148:1098–1108. https://doi.org/10.1016/j.ijbiomac.2020.01.036
54. Ioelovich M (2008) Cellulose as a nanostructured polymer: a short review. BioResources 3:1403–1418
55. Iohara D, Okubo M, Anraku M et al (2017) Hydrophobically modified polymer/α-cyclodextrin thermoresponsive hydrogels for use in ocular drug delivery. Mol Pharm 14(8):2740–2748. https://doi.org/10.1021/acs.molpharmaceut.7b00291
56. Islam SU, Ul-Islam M, Ahsan H et al (2021) Potential applications of bacterial cellulose and its composites for cancer treatment. Int J Biol Macromol 168:301–309. https://doi.org/10.1016/j.ijbiomac.2020.12.042
57. Jadach B, Misek M, Ferla J (2023) Comparison of hydroxypropyl methylcellulose and alginate gel films with meloxicam as fast orodispersible drug delivery. Gels 9:687. https://doi.org/10.3390/gels9090687
58. Javanbakht S, Nazeri MT, Shaabani A et al (2020) Green one-pot synthesis of multicomponent-crosslinked carboxymethyl cellulose as a safe carrier for the gentamicin oral delivery. Int J Biol Macromol 164:2873–2880. https://doi.org/10.1016/j.ijbiomac.2020.08.168
59. Javanbakht S, Pooresmaeil M, Hashemi H et al (2018) Carboxymethylcellulose capsulated Cu-based metal-organic framework-drug nanohybrid as a pH-sensitive nanocomposite for ibuprofen oral delivery. Int J Biol Macromol 119:588–596. https://doi.org/10.1016/j.ijbiomac.2018.07.181
60. Javanbakht S, Shaabani A (2019) Carboxymethyl cellulose-based oral delivery systems. Int J Biol Macromol 133:21–29. https://doi.org/10.1016/j.ijbiomac.2019.04.079
61. Jeddi MK, Mahkam M (2019) Magnetic nano carboxymethyl cellulose-alginate/chitosan hydrogel beads as biodegradable devices for controlled drug delivery. Int J Biol Macromol 135:829–838. https://doi.org/10.1016/j.ijbiomac.2019.05.210
62. Jeong D, Joo S-W, Hu Y et al (2018) Carboxymethyl cellulose-based superabsorbent hydrogels containing carboxymethyl β-cyclodextrin for enhanced mechanical strength and effective drug delivery. Eur Polym J 105:17–25. https://doi.org/10.1016/j.eurpolymj.2018.05.023
63. Jiao Z, Zhang B, Li C et al (2018) Carboxymethyl cellulose-grafted graphene oxide for efficient antitumor drug delivery. Nanotechnol Rev 7(4):291–301. https://doi.org/10.1515/ntrev-2018-0029
64. Jin X, Li L, Zhao S et al (2021) Assessment of occlusal force and local gas release using degradable bacterial cellulose/$Ti_3C_2T_X$ MXene bioaerogel for oral healthcare. ACS Nano 15(11):18385–18393. https://doi.org/10.1021/acsnano.1c07891
65. Jorfi M, Foster EJ (2015) Recent advances in nanocellulose for biomedical applications. J Appl Polym Sci 132:41719. https://doi.org/10.1002/app.41719
66. Kamel S, Ali N, Jahangir K et al (2008) Pharmaceutical significance of cellulose: a review. Express Polym Lett 2:758–778. https://doi.org/10.3144/expresspolymlett.2008.90
67. Kamenova K, Momekova D, Grancharov G et al (2023) In situ gelling hydroxypropyl cellulose formulation comprising cannabidiol-loaded block copolymer micelles for sustained drug delivery. Int J Mol Sci 24(22):16534. https://doi.org/10.3390/ijms242216534
68. Kandil R, Merkel OM (2019) Recent progress of polymeric nanogels for gene delivery. Curr Opin Colloid Interface Sci 39:11–23. https://doi.org/10.1016/j.cocis.2019.01.005
69. Kargarzadeh H, Ioelovich M, Ahmad I et al (2017) Methods for extraction of nanocellulose from various sources. In: Kargarzadeh H (ed) Handbook of nanocellulose and cellulose nanocomposites. Wiley-VCH Verlag GmbH & Co. KGaA, Weinheim, Germany, pp 1–49
70. Kargarzadeh H, Mariano M, Gopakumar D et al (2018) Advances in cellulose nanomaterials. Cellulose 25:2151–2189. https://doi.org/10.1007/s10570-018-1723-5
71. Karimian A, Parsian H, Majidinia M et al (2019) Nanocrystalline cellulose: preparation, physicochemical properties, and applications in drug delivery systems. Int J Biol Macromol 133:850–859. https://doi.org/10.1016/j.ijbiomac.2019.04.117
72. Kaushik M, Moores A (2016) Review: Nanocelluloses as versatile supports for metal nanoparticles and their applications in catalysis. Green Chem 18:622–637. https://doi.org/10.1039/c5gc02500a

73. Khan N, Aqil M, Imam SS et al (2015) Development and evaluation of a novel in situ gel of sparfloxacin for sustained ocular drug delivery: in vitro and ex vivo characterization. Pharm Dev Technol 20:662–669. https://doi.org/10.3109/10837450.2014.910807
74. Khezri A, Karimi A, Yazdian F et al (2018) Molecular dynamic of curcumin/chitosan interaction using a computational molecular approach: emphasis on biofilm reduction. Int J Biol Macromol 114:972–978. https://doi.org/10.1016/j.ijbiomac.2018.03.100
75. Kim KJ, Hwang MJ, Shim WG et al (2024) Sustained drug release behavior of captopril-incorporated chitosan/carboxymethyl cellulose biomaterials for antihypertensive therapy. Int J Biol Macromol 255:128087. https://doi.org/10.1016/J.IJBIOMAC.2023.128087
76. Kim JK, Won Y-W, Lim KS et al (2011) Low-molecular-weight methylcellulose-based thermo-reversible gel/pluronic micelle combination system for local and sustained docetaxel delivery. Pharm Res 29:525–534. https://doi.org/10.1007/s11095-011-0581-8
77. Klemm D, Cranston ED, Fischer D et al (2018) Nanocellulose as a natural source for groundbreaking applications in materials science: today's state. Mater Today 21:720–748. https://doi.org/10.1016/j.mattod.2018.02.001
78. Klemm D, Heublein B, Fink HP et al (2005) Cellulose: fascinating biopolymer and sustainable raw material. Angew Chem Int Ed 44:3358–3393. https://doi.org/10.1002/anie.200460587
79. Klemm D, Kramer F, Moritz S et al (2011) Nanocelluloses: a new family of nature-based materials. Angew Chem Int Ed 50:5438–5466. https://doi.org/10.1002/anie.201001273
80. Kumar AR, Aeila ASS (2019) Sustained release matrix type drug delivery system: an overview. World J Pharm Pharm Sci 9:470–480
81. Kupnik K, Primožič M, Kokol V et al (2020) Nanocellulose in drug delivery and antimicrobially active materials. Polymers 12:2825. https://doi.org/10.3390/polym12122825
82. Kwon SS, Kong BJ, Park SN (2015) Physicochemical properties of pH-sensitive hydrogels based on hydroxyethyl cellulose–hyaluronic acid and for applications as transdermal delivery systems for skin lesions. Eur J Pharm Biopharm 92:146–154. https://doi.org/10.1016/j.ejpb.2015.02.025
83. Lam E, Hemraz UD (2021) Preparation and surface functionalization of carboxylated cellulose nanocrystals. Nanomaterials (Basel) 11(7):1641. https://doi.org/10.3390/nano11071641
84. Li S-M, Jia N, Ma M-G et al (2011) Cellulose–silver nanocomposites: microwave-assisted synthesis, characterization, their thermal stability, and antimicrobial property. Carbohydr Polym 86:441–447. https://doi.org/10.1016/j.carbpol.2011.04.060
85. Li YY, Wang B, Ma MG et al (2018) Review of recent development on preparation, properties, and applications of cellulose-based functional materials. Int J Polym Sci 8973643:1–18. https://doi.org/10.1155/2018/8973643
86. Li N, Yu M, Deng L et al (2012) Thermosensitive hydrogel of hydrophobically-modified methylcellulose for intravaginal drug delivery. J Mater Sci Mater Med 23:1913–1919. https://doi.org/10.1007/s10856-012-4664-9
87. De Lima CSA, Balogh TS, Varca JPRO et al (2020) An updated review of macro, micro, and nanostructured hydrogels for biomedical and pharmaceutical applications. Pharmaceutics 12:970. https://doi.org/10.3390/pharmaceutics12100970
88. Liu Y, Huo Y, Li M et al (2022) Synthesis of metal–organic-frameworks on polydopamine modified cellulose nanofibril hydrogels: constructing versatile vehicles for hydrophobic drug delivery. Cellulose 29:379–393. https://doi.org/10.1007/S10570-021-04267-X/FIGURES/8
89. Liu S, Jin M, Chen Y et al (2017) High internal phase emulsions stabilised by supramolecular cellulose nanocrystals and their application as cell-adhesive macroporous hydrogel monoliths. J Mater Chem B 5:2671–2678. https://doi.org/10.1039/c7tb00145b
90. Liu L, Wang T, Li M et al (2024) Carboxylated cellulose-based composite aerogel with double filling structure for sustained drug release. Ind Crops Prod 210:118126. https://doi.org/10.1016/J.INDCROP.2024.118126
91. Long W, Ouyang H, Zhou C et al (2021) Simultaneous surface functionalization and drug loading: a novel method for fabrication of cellulose nanocrystals-based pH responsive drug delivery system. Int J Biol Macromol 182:2066–2075. https://doi.org/10.1016/j.ijbiomac.2021.05.193

92. Lu Y, Han J, Ding Q et al (2021) TEMPO-oxidized cellulose nanofibers/polyacrylamide hybrid hydrogel with intrinsic self-recovery and shape memory properties. Cellulose 28:1469–1488. https://doi.org/10.1007/s10570-020-03606-8
93. Lv P, Lu X, Wang L et al (2021) Nanocellulose-based functional materials: from chiral photonics to soft actuator and energy storage. Adv Funct Mater 31:2104991. https://doi.org/10.1002/adfm.202104991
94. Machado B, Costa SM, Costa I et al (2024) The potential of algae as a source of cellulose and its derivatives for biomedical applications. Cellulose 31:3353–3376. https://doi.org/10.1007/s10570-024-05816-w
95. Madsen B, Gamstedt EK (2013) Wood versus plant fibers: similarities and differences in composite applications. Adv Mater Sci Eng 1–14. https://doi.org/10.1155/2013/564346
96. Mahardika M, Abral H, Kasim A et al (2018) Production of nanocellulose from pineapple leaf fibers via high-shear homogenization and ultrasonication. Fibers 6:28. https://doi.org/10.3390/fib6020028
97. Maitra J, Shukla VK (2014) Cross-linking in hydrogels—a review. Am J Polym Sci 4(2):25–31. https://doi.org/10.5923/j.ajps.20140402.01
98. Malik NS, Ahmad M, Minhas MU (2017) Cross-linked β-cyclodextrin and carboxymethyl cellulose hydrogels for controlled drug delivery of acyclovir. PLoS ONE 12(2):e0172727. https://doi.org/10.1371/journal.pone.0172727
99. Marques-Marinho FD, Vianna-Soares CD (2013) Cellulose and its derivatives use in the pharmaceutical compounding practice. In: Van De Ven TGM (ed) Cellulose—medical, pharmaceutical and electronic applications. IntechOpen, London, pp 141–162. https://doi.org/10.5772/56637
100. Martínez-Sanz M, Lopez-Rubio A, Lagaron JM (2011) Optimization of the nanofabrication by acid hydrolysis of bacterial cellulose nanowhiskers. Carbohydr Polym 85:228–236. https://doi.org/10.1016/j.carbpol.2011.02.021
101. Maver U, Xhanari K, Žižek M et al (2020) Carboxymethyl cellulose/diclofenac bioactive coatings on AISI 316LVM for controlled drug delivery, and improved osteogenic potential. Carbohydr Polym 230:115612. https://doi.org/10.1016/j.carbpol.2019.115612
102. Mawad D, Boughton EA, Boughton P et al (2012) Advances in hydrogels applied to degenerative diseases. Curr Pharm Des 18:2558–2575. https://doi.org/10.2174/138161212800492895
103. Miao L, Zhang M, Tu Y et al (2019) Stimuli-responsive cellulose-based hydrogels. In: Mondal MIH (ed) Cellulose-based super-absorbent hydrogels, polymers and polymeric composites: a reference series, 1st edn. Springer, Cham, Switzerland, pp 269–308
104. Mohanta V, Madras G, Patil S (2014) Layer-by-layer assembled thin films and microcapsules of nanocrystalline cellulose for hydrophobic drug delivery. ACS Appl Mater Interfaces 6:20093–20101. https://doi.org/10.1021/am505681e
105. Mohd AMCI, Ahmad N, Pandey M et al (2014) Stimuli-responsive bacterial cellulose-g-poly(acrylic acid-co-acrylamide) hydrogels for oral controlled release drug delivery. Drug Dev Ind Pharm 40(10):1340–1349. https://doi.org/10.3109/03639045.2013.819882
106. Moon RJ, Martini A, Nairn J et al (2011) Cellulose nanomaterials review: structure, properties and nanocomposites. Chem Soc Rev 40:3941–3994. https://doi.org/10.1039/c0cs00108b
107. Mujtaba M, Negi A, King AWT et al (2023) Surface modifications of nanocellulose for drug delivery applications: a critical review. Curr Opin Biomed Eng 28:100475. https://doi.org/10.1016/j.cobme.2023.100475
108. Nguyen S, Hiorth M, Rykke M et al (2011) The potential of liposomes as dental drug delivery systems. Eur J Pharm Biopharm 7(1):75–83. https://doi.org/10.1016/j.ejpb.2010.09.010
109. Nicu R, Ciolacu F, Ciolacu DE (2021) Advanced functional materials based on nanocellulose for pharmaceutical/medical applications. Pharmaceutics 13(8):1125. https://doi.org/10.3390/pharmaceutics13081125
110. de Nooy AEJ, Besemer AC, van Bekkum H (1994) Highly selective TEMPO-mediated oxidation of primary alcohol groups in polysaccharides. Recl Trav Chim Pays-Bas 113:165–166. https://doi.org/10.1002/recl.19941130307

111. Nsor-Atindana J, Chen M, Goff HD et al (2017) Functionality and nutritional aspects of microcrystalline cellulose in food. Carbohydr Polym 172:159–174. https://doi.org/10.1016/j.carbpol.2017.04.021
112. Oliveira RN, Moreira APD, Thiré RMDSM et al (2017) Absorbent polyvinyl alcohol-sodium carboxymethyl cellulose hydrogels for propolis delivery in wound healing applications. Polym Eng Sci 57:1224–1233. https://doi.org/10.1002/pen.24500
113. Onofrei MD, Filimon A (2016) Cellulose-based hydrogels: designing concepts, properties, and perspectives for biomedical and environmental applications. In: Mendez-Vilas A (ed) Polymer science: research advances, practical applications and educational aspects. Formatex Research Center, Badajoz, Spain, pp 108–120
114. Oprea A-M, Nistor MT, Popa MI et al (2012) In vitro and in vivo theophylline release from cellulose/chondroitin sulfate hydrogels. Carbohydr Polym 90:127–133. https://doi.org/10.1016/j.carbpol.2012.05.004
115. Palaninathan V, Raveendran S, Rochani AK et al (2018) Bioactive bacterial cellulose sulfate electrospun nanofibers for tissue engineering applications. J Tissue Eng Regen Med 12:1634–1645. https://doi.org/10.1002/term.2689
116. Pan R, Cheung O, Wang Z et al (2016) Mesoporous *Cladophora* cellulose separators for lithium-ion batteries. J Power Sources 321:185–192. https://doi.org/10.1016/j.jpowsour.2016.04.115
117. Pan P, Svirskis D, Waterhouse GIN et al (2023) Hydroxypropyl methylcellulose bioadhesive hydrogels for topical application and sustained drug release: the effect of polyvinylpyrrolidone on the physicomechanical properties of hydrogel. Pharmaceutics 15:2360. https://doi.org/10.3390/pharmaceutics15092360
118. Park J, Kim S, Kim T (2023) Polyethylenimine-conjugated hydroxyethyl cellulose for doxorubicin/Bcl-2 siRNA co-delivery systems. Pharmaceutics 15(2):708. https://doi.org/10.3390/pharmaceutics15020708
119. Patel R, Dhar P, Babaei-Ghazvini A et al (2023) Transforming lignin into renewable fuels, chemicals, and materials: a review. Bioresour Technol Rep 22:101463. https://doi.org/10.1016/j.biteb.2023.101463
120. Pei A, Zhou Q, Berglund LA (2010) Functionalized cellulose nanocrystals as biobased nucleation agents in poly(l-lactide) (PLLA)—crystallization and mechanical property effects. Comput Sci Technol 70:815–821. https://doi.org/10.1016/j.compscitech.2010.01.018
121. Pelegrini BL, Ré F, de Oliveira MM et al (2019) Cellulose nanocrystals as a sustainable raw material: cytotoxicity and applications on healthcare technology. Macromol Mater Eng 304:1900092. https://doi.org/10.1002/mame.201900092
122. Peppas NA, Bures P, Leobandung W et al (2000) Hydrogels in pharmaceutical formulations. Eur J Pharm Biopharm 50:27–46. https://doi.org/10.1016/S0939-6411(00)00090-4
123. Phanthong P, Reubroycharoen P, Hao X et al (2018) Nanocellulose: extraction and application. Carbon Resour Convers 1:32–43. https://doi.org/10.1016/j.crcon.2018.05.004
124. Poonguzhali R, Khaleel BS, Sugantha KV (2018) Synthesis of alginate/nanocellulose bionanocomposite for in vitro delivery of ampicillin. Polym Bull 75:4165–4173. https://doi.org/10.1007/s00289-017-2253-2
125. Pooresmaeil M, Namazi H (2022) Folic acid-modified photoluminescent dialdehyde carboxymethyl cellulose crosslinked bionanogels for pH-controlled and tumor-targeted co-drug delivery. Int J Biol Macromol 200:247–262. https://doi.org/10.1016/j.ijbiomac.2022.01.002
126. Prusty K, Swain SK (2017) Cellulose-based nanohydrogels for tissue engineering applications. In: Jawaid M (ed) Nanocellulose and nanohydrogel matrices: biotechnological and biomedical applications, 1st edn. Wiley-VCH Verlag GmbH & Co. KGaA, Weinheim, Germany, pp 67–90
127. Rabemanolontsoa H, Saka S (2013) Comparative study on chemical composition of various biomass species. RSC Adv 3:3946–3956. https://doi.org/10.1039/c3ra22958k
128. Ranby B (1949) Aqueous colloidal solutions of cellulose micelles. Acta Chem Scand 3:649–650. https://doi.org/10.3891/acta.chem.scand.03-0649

129. Rasoulzadeh M, Namazi H (2017) Carboxymethyl cellulose/graphene oxide bionanocomposite hydrogel beads as anticancer drug carrier agent. Carbohydr Polym 168:320–326. https://doi.org/10.1016/j.carbpol.2017.03.014
130. Reddy NN, Mohan YM, Varaprasad K et al (2011) Magnetic and electric responsive hydrogel-magnetic nanocomposites for drug-delivery application. J Appl Polym Sci 122:1364–1375. https://doi.org/10.1002/app.34016
131. Resch A, Staud C, Radtke C (2021) Nanocellulose-based wound dressing for conservative wound management in children with second-degree burns. Int Wound J 18:478–486. https://doi.org/10.1111/iwj.13548
132. Rodrigues Filho G, Ribeiro SD, Meireles CdS et al (2011) Release of doxycycline through cellulose acetate symmetric and asymmetric membranes produced from recycled agroindustrial residue: sugarcane bagasse. Ind Crops Prod 33:566–571. https://doi.org/10.1016/j.indcrop.2010.10.037
133. Saba N, Jawaid M (2017) Recent advances in nanocellulose-based polymer nanocomposites. In: Jawaid M (ed) Woodhead publishing series in composites science and engineering, cellulose-reinforced nanofibre composites, 1st edn. Woodhead Publishing, Cambridge, pp 89–112
134. Sadeghifar H, Venditti R, Jur J et al (2017) Cellulose-lignin biodegradable and flexible UV protection film. ACS Sustain Chem Eng 5:625–631. https://doi.org/10.1021/acssuschemeng.6b02003
135. Sahoo PK, Jena DK (2018) Synthesis and study of mechanical and fire retardant properties of (carboxymethyl cellulose-polyacrylonitrile)/montmorillonite biodegradable nanocomposite. J Polym Res 25:260. https://doi.org/10.1007/s10965-018-1659-3
136. Salerno C, Carlucci AM, Bregni C (2010) Study of in vitro drug release and percutaneous absorption of fluconazole from topical dosage forms. AAPS PharmSciTech 11:986–993. https://doi.org/10.1208/s12249-010-9457-1
137. Sanz T, Laguna L, Salvador A (2015) Biscuit dough structural changes during heating: influence of shortening and cellulose ether emulsions. LWT-Food Sci Technol 62:962–969. https://doi.org/10.1016/j.lwt.2015.02.036
138. Seddiqi H, Oliaei E, Honarkar H et al (2021) Cellulose and its derivatives: towards biomedical applications. Cellulose 28:1893–1931. https://doi.org/10.1007/s10570-020-03674-w
139. Simões S, Figueiras A, Veiga F (2012) Modular hydrogels for drug delivery. J Biomater Nanobiotechnol 3:185–199. https://doi.org/10.4236/jbnb.2012.32025
140. Sivaram AJ, Rajitha P, Maya S et al (2015) Nanogels for delivery, imaging and therapy. Wiley Interdiscip Rev Nanomed Nanobiotechnol 7:509–533. https://doi.org/10.1002/wnan.1328
141. Song K, Zhu W, Li X (2020) A novel mechanical robust, self-healing and shape memory hydrogel based on PVA reinforced by cellulose nanocrystal. Mater Lett 260:126884–126887. https://doi.org/10.1016/j.matlet.2019.126884
142. Su T, Wu Q-X, Chen Y et al (2019) Fabrication of the polyphosphates patched cellulose sulfate-chitosan hydrochloride microcapsules and as vehicles for sustained drug release. Int J Pharm 555:291–302. https://doi.org/10.1016/j.ijpharm.2018.11.058
143. Sultan S, Mathew AP (2018) 3D printed scaffolds with gradient porosity based on a cellulose nanocrystal hydrogel. Nanoscale 10:4421–4431. https://doi.org/10.1039/c7nr08966j
144. Sun L, Ding C, Qian X et al (2019) Effective and simple one-step strategy for preparation of "clickable" cellulose modules: support to build antibacterial materials. Cellulose 26:1961–1976. https://doi.org/10.1007/s10570-018-2215-3
145. Sun N, Wang T, Yan X (2017) Self-assembled supermolecular hydrogel based on hydroxyethyl cellulose: formation, in vitro release and bacteriostasis application. Carbohydr Polym 172:49–59. https://doi.org/10.1016/j.carbpol.2017.05.026
146. Supramaniam J, Adnan R, Kaus NHM et al (2018) Magnetic nanocellulose alginate hydrogel beads as potential drug delivery system. Int J Biol Macromol 118:640–648. https://doi.org/10.1016/j.ijbiomac.2018.06.043
147. Swamy BY, Yun Y-S (2015) In vitro release of metformin from iron (III) cross-linked alginate–carboxymethyl cellulose hydrogel beads. Int J Biol Macromol 77:114–119. https://doi.org/10.1016/j.ijbiomac.2015.03.019

148. Tang J, Bao L, Li X et al (2015) Potential of PVA-doped bacterial nano-cellulose tubular composites for artificial blood vessels. J Mater Chem B 3:8537–8547. https://doi.org/10.1039/C5TB01144B
149. Tarchoun AF, Trache D, Klapötke TM (2019) Microcrystalline cellulose from *Posidonia oceanica* brown algae: extraction and characterization. Int J Biol Macromol 138:837–845. https://doi.org/10.1016/j.ijbiomac.2019.07.176
150. Tavakoli N, Minaiyan M, Heshmatipour M et al (2016) Transdermal iontophoretic delivery of celecoxib from gel formulation. Res Pharm Sci 10:419–428
151. Trache D, Hussin MH, Haafiz MM et al (2017) Recent progress in cellulose nanocrystals: sources and production. Nanoscale 9:1763–1786. https://doi.org/10.1039/c6nr09494e
152. Trache D, Tarchoun AF, Derradji M et al (2020) Nanocellulose: from fundamentals to advanced applications. Front Chem 8:392–425. https://doi.org/10.3389/fchem.2020.00392
153. Trache D, Thakur VK, Boukherroub R (2020) Cellulose nanocrystals/graphene hybrids—a promising new class of materials for advanced applications. Nanomaterials 10:1523. https://doi.org/10.3390/nano10081523
154. Ullah H, Badshah M, Mäkilä E et al (2017) Fabrication, characterization and evaluation of bacterial cellulose-based capsule shells for oral drug delivery. Cellulose 24(3):1445–1454. https://doi.org/10.1007/s10570-017-1202-4
155. Ullah S, Chen X (2020) Fabrication, applications and challenges of natural biomaterials in tissue engineering. Appl Mater Today 20:100656. https://doi.org/10.1016/j.apmt.2020.100656
156. Varghese RT, Cherian RM, Chirayil CJ et al (2023) Nanocellulose as an avenue for drug delivery applications: a mini-review. J Compos Sci 7(6):210. https://doi.org/10.3390/jcs7060210
157. Vashist A, Gupta Y, Ahmad S (2012) Interpenetrating biopolymer network based hydrogels for an effective drug delivery system. Carbohydr Polym 87:1433–1439. https://doi.org/10.1016/j.carbpol.2011.09.030
158. Vieira JG, Filho GR, Meireles CdS et al (2012) Synthesis and characterization of methylcellulose from cellulose extracted from mango seeds for use as a mortar additive. Polímeros 22:80–87. https://doi.org/10.1590/s0104-14282012005000011
159. Viera RG, Rodrigues FG, de Assunção RM et al (2007) Synthesis and characterization of methylcellulose from sugar cane bagasse cellulose. Carbohydr Polym 67:182–189. https://doi.org/10.1016/j.carbpol.2006.05.007
160. Vlaia I, Coneac G, Olariu I et al (2016) Cellulose-derivatives-based hydrogels as vehicles for dermal and transdermal drug delivery. In: Majee SB (ed) Emerging concepts in analysis and applications of hydrogels, 1st edn. IntechOpen, London, pp 159–200. https://doi.org/10.5772/63953
161. Wada M, Ike M, Tokuyasu K (2010) Enzymatic hydrolysis of cellulose I is greatly accelerated via its conversion to the cellulose II hydrate form. Polym Degrad Stab 95:543–548. https://doi.org/10.1016/j.polymdegradstab.2009.12.014
162. Wang Y, Cooke MJ, Morshead CM et al (2012) Hydrogel delivery of erythropoietin to the brain for endogenous stem cell stimulation after stroke injury. Biomaterials 33:2681–2692. https://doi.org/10.1016/j.biomaterials.2011.12.031
163. Wang WY, Hui PCL, Wat E et al (2016) Enhanced transdermal permeability via constructing the porous structure of poloxamer-based hydrogel. Polymers 8:406. https://doi.org/10.3390/polym8110406
164. Wardhono EY, Kanani N, Alfirano A (2020) A simple process of isolation microcrystalline cellulose using ultrasonic irradiation. J Disper Sci Technol 41:1217–1226. https://doi.org/10.1080/01932691.2019.1614947
165. Weyell P, Beekmann U, Küpper C et al (2019) Tailor-made material characteristics of bacterial cellulose for drug delivery applications in dentistry. Carbohydr Polym 207:1–10. https://doi.org/10.1016/j.carbpol.2018.11.061
166. Wu Y, Liu Y, Li X et al (2019) Research progress of in-situ gelling ophthalmic drug delivery system. Asian J Pharm Sci 14:1–15. https://doi.org/10.1016/j.ajps.2018.04.008

167. Yan M, Ge J, Liu Z et al (2006) Encapsulation of single enzyme in nanogel with enhanced biocatalytic activity and stability. J Am Chem Soc 128:11008–11009. https://doi.org/10.1021/ja064126t
168. Yang Y, Guo Y, Sun R et al (2016) Self-assembly and β-carotene loading capacity of hydroxyethyl cellulose-graft-linoleic acid nanomicelles. Carbohydr Polym 145:56–63. https://doi.org/10.1016/j.carbpol.2016.03.012
169. Zaki M, Khalil HPSA, Sabaruddin FA et al (2021) Microbial treatment for nanocellulose extraction from marine algae and its applications as sustainable functional material. Bioresour Technol Rep 16:100811. https://doi.org/10.1016/j.biteb.2021.100811
170. Zhao Y, Li J (2014) Excellent chemical and material cellulose from tunicates: diversity in cellulose production yield and chemical and morphological structures from different tunicate species. Cellulose 21:3427–3441. https://doi.org/10.1007/s10570-014-0348-6
171. Zhao Y, Zhang Y, Lindström ME et al (2015) Tunicate cellulose nanocrystals: preparation, neat films and nanocomposite films with glucomannans. Carbohydr Polym 17:286–296. https://doi.org/10.1016/j.carbpol.2014.09.020
172. Zhou S, Nyholm L, Strømme M et al (2019) *Cladophora* cellulose: unique biopolymer nanofibrils for emerging energy, environmental, and life science applications. Acc Chem Res 52:2232–2243. https://doi.org/10.1021/acs.accounts.9b00215
173. Zhu G, Xu H, Dufresne A et al (2018) High-adsorption, self-extinguishing, thermal, and acoustic-resistance aerogels based on organic and inorganic waste valorization from cellulose nanocrystals and red mud. ACS Sustain Chem Eng 6:7168–7180. https://doi.org/10.1021/acssuschemeng.8b01244.s001
174. Šturcová A, Davies GR, Eichhorn SJ (2005) Elastic modulus and stress-transfer properties of tunicate cellulose whiskers. Biomacromolecules 6:1055–1061. https://doi.org/10.1021/bm049291k

Dr. Amit Kumar Verma currently, serving the Department of Biosciences, Jamia Millia Islamia, since 16 years, earned his Master's degree from University of Pune and Ph.D. degree from University of Lucknow and started his career as an Assistant Professor. He has published several impacted research articles and supervised Master's dissertations and Ph.D. theses in the field of protein based biomaterials, molecular biology and cancer biology. Presently, he is actively involved in the teaching at UG, PG and Ph.D. level along with research towards the development of collagen based biomaterials for various biomedical applications. He is an active member of several science societies and departmental and institutional committees and participated in many national and international conferences and seminars.

Innovative Approaches to Corneal Regeneration: From Biomaterials to Clinical Application

Federico Visalli, Caterina Gagliano, Fabiana D'Esposito, Mutali Musa, Daniele Tognetto, and Marco Zeppieri

Abstract Chapter will look at new cutting-edge methods of corneal regeneration, with a focus on the use of bioscaffolds to treat various corneal diseases. Several pathologies can damage the cornea, including infections, hereditary illnesses, degeneration, trauma, and dry eye syndrome. Treatments used, such as corneal transplants and artificial corneas, have severe drawbacks, including donor shortages and immunological rejection. Study investigates the enormous potential of bioscaffolds, which are designed to replicate the native extracellular matrix and assist cell adhesion, proliferation, and differentiation. In addition, bioscaffolds can be employed to transport medications and stem cells. The mechanical characteristics and biocompatibility of many materials are reviewed, including natural polymers such as collagen and

F. Visalli
Department of Ophthalmology, University of Catania, Catania, Italy
e-mail: federico.visalli13@gmail.com

C. Gagliano · F. D'Esposito
Department of Medicine and Surgery, University of Enna "Kore", Enna, EN, Italy
e-mail: caterina.gagliano@unikore.it

F. D'Esposito
e-mail: f.desposito@imperial.ac.uk

C. Gagliano
Eye Center "G.B. Morgagni-DSV", Catania, Italy

F. D'Esposito
Imperial College Ophthalmic Research Group (ICORG) Unit, Imperial College, London, UK

M. Musa
Department of Optometry, University of Benin, Benin City, Edo State, Nigeria
e-mail: mutali.musa@uniben.edu

D. Tognetto · M. Zeppieri
Department of Medicine, Surgery and Health Sciences, University of Trieste, Trieste, Italy
e-mail: tognetto@units.it

M. Zeppieri (✉)
Department of Ophthalmology, University Hospital of Udine, Udine, Italy
e-mail: markzeppieri@hotmail.com

R. Malviya and S. Sundram (eds.), *Sustainable Cellulose Materials for Biomedical Applications*, Sustainable Materials and Technology,
https://doi.org/10.1007/978-981-96-9904-9_18

chitosan, as well as synthetic polymers. Electrospinning, 3D bioprinting, hydrogel formulations, and decellularization are among the fabrication techniques being tested for their ability to produce effective bioscaffolds. The chapter also discusses several functionalization options for improving scaffold performance, such as surface modification and the addition of growth agents and nanoparticles. The integration of stem cells, particularly limbal, mesenchymal, and induced pluripotent stem cells, with bioscaffolds appears to be a potential technique for corneal tissue engineering. Despite advances, other difficulties such as immunogenicity, scaffold design, and the requirement for specific manufacturing methods must also be addressed.

Keywords Bioscaffolds · Corneal endothelial dysfunction · Stem cells · Regenerative medicine · Decellularized scaffolds

1 Introduction to the Cornea

1.1 Corneal Structure

The cornea is the anterior most structure of the human eye. It is the clear and transparent tissue that is responsible for light entering the eye. Within its avascular structure, it is comprised of six layers [1]:

(1) **Epithelium**: The epithelium is its outermost layer. It is a highly mitotic layer capable of remarkable regeneration within short times, allowing it to withstand trauma from external forces. The rigidity of the epithelium is ensured by the interdigitization of different cellular structures.
(2) **Bowman's layer**: It is located below the corneal epithelium. It consists of a fibrous structure. This layer allows on the one hand to stabilize the corneal epithelium and on the other hand guarantees greater protection to the underlying stroma.
(3) **Stroma**: The stroma is located below the Bowman's layer. It represents the thickest corneal layer. It is mainly made up of a collagen scaffold that gives stability and transparency.
(4) **Dua's layer**: The Dua layer is an acellular layer. It represents an important surgical landmark for corneal surgery [2].
(5) **Descemet's membrane**: The descemet membrane is located just below the Dua layer. It is a thin layer that has the task of ensuring stability to the endothelium. This layer is also important for corneal surgery [3]
(6) **Endothelium**: The endothelium is the innermost corneal layer. It plays an important role in the transparency of the cornea. It is a hydrophilic layer, which regulates the flow through pump mechanisms [1].

1.2 Corneal Pathologies

Different pathologies can compromise the correct functioning of the cornea with consequent alteration of corneal transparency:

a. *Infections*: Microorganisms within and outside the eye are responsible for corneal disorders such as ulcers, which may lead to development of corneal scars, preventing light from passing into the eye and causing inflammation [4]. These microbes include bacteria, fungus, viruses and parasites.
b. *Dystrophies*: Dystrophies occur when there is abnormal development and composition of corneal tissue from birth. They can be immediately detectable at birth and progressive, or may lie latent and only manifest after trauma to the cornea at a later date, causing corneal scarring [5].
c. *Degenerations*: As with every tissue in the body, aging can cause the cornea to degenerate, and that can impair its ability to transmit lights.
d. *Trauma*: Any kind of trauma to the cornea will definitely affect his function. Surgical trauma can cause mild scarring at the point of entry into the cornea, while physical trauma can cause loss of transparency and also affects the ability of the cornea to transmit light, thereby leading to blindness.
e. *Dry eye syndrome*: Of increasing importance is dry eye syndrome, which involves one of two mechanisms, or sometimes both. One pathway is caused by an underproduction of tears, while the other pathway is caused by the excessive evaporation of tears. Dry eye can present with inflammation, irritation, and corneal damage.

1.3 Limitations of Traditional Therapy

Presently there are several methods of managing corneal conditions, which include:

a. *Eye drops, anointments*: These are the first line of medication for managing infections and inflammatory processes in the cornea. They might, however, not be sufficient enough for more severe conditions and deep seated infections.
b. *Transplants from donor corneas*: Transplants have been used. To restore lost vision in damaged or diseased corneas, these transplants might be full thickness transplants or partial thickness transplants. The major drawback of transplants is that they are difficult to source and there is the risk of failure after expensive transplant surgery. Another drawback of transplants is the risk of infection before, during, and after surgery.
c. *Artificial corneas*: When tissues are not available, viable options include artificially generated corneas. These are available via 3D printing or by grafting from other tissues. These offer a donor tissue alternative; however, limitations include low integration with tissue host, biocompatibility issues, and long-term functionality [6].

1.4 Stem Cells

Stem cell medicine has evolved rapidly in the past decade in the field of medicine, especially in ocular surgery and cornea wound management. It is not possible to regenerate damage or diseased tissue by leveraging on the ability of these stem cells to differentiate into different kinds of tissues. This novel method can potentially be a last hope for patients suffering from cornea disorders (Mousaei et al. 2022).

2 An Overview of Bioscaffolds

2.1 Overview of Bioscaffolds

Bioscaffolds can be defined as generated materials that are similar to the human extracellular matrix including that of the cornea (Wang 2023). These bioscaffolds help to ensure that the tissue environment is receptive to transplants or stem cell attachments. They can be used to improve proliferation and also catalyze differentiation of stem cells into different layers of corneal tissue [7]. The field of bioscaffold medicine has rapidly developed, by improving the characteristics of these materials to not only support the differentiating and developing cells, but also to act as catalysts and interactive components that help to ensure healing of the surrounding cornea and ocular tissue.

2.1.1 Features of Successful Bioscaffolds

I. *Biocompatibility*: A bioscaffold must be compatible with the tissue it is being ingratiated to be able to achieve its functions and to prevent the tissue for fighting it using the immune system (Mahdavi et al. 2020). This process ensures that the bioscaffold does not pose a threat of infection or inflammation to cornea tissue [8]. Currently, enhancements such as bioactive peptides or laminin proteins can help the bioscaffold to achieve adhesion to the corner tissue, and this folder improves the viability of the process. In addition, sterile scaffolds that do not contain antigens also minimize the risk of an immune response in the grafts.

II. *Cell adhesion and purification*: An effective bioscaffold should enhance cell refrigeration leading to growth of the corner, which in turn would result in a repair of the damage tissues [9]. An example is the catalyzation of scaffold reactions with certain growth factors, which help them to differentiate corneal tissue into epithelial, stromal or endothelial cells, thereby enhancing repairs of the required layer of the cornea.

III. *Optical transparency*: A scaffold must in itself be transparent so as not to compromise normal corneal transparency after being transplanted into the eye [8]. This explains why nanoparticles are preferred for constructing bioscaffolds so as to ensure that they do not interfere with the ability of the cornea to transmit light properly. Electrospun technology, for example, can be used to make stratified collagen tissue, which possibly replicates the objective optical properties of the native cornea and, therefore, ensures transparency.

IV. *Mechanical and structural mimicry*: Scaffolds should closely approximate the anatomical and physiological features of the corners they are grafted onto so as to keep the pressure in the eye normal. This involves ensuring that the scaffold has the required rigidity while also mandating the elasticity needed to act like the native cornea. High-resolution 3D-printed scaffolds have been shown to achieve these characteristics.

V. *Controlled biodegradability*: Bioscaffolds themselves are not expected to be permanent components of the cornea. Therefore, they must be able to degrade over time and be replaced by novel cornea tissue [10]. As this degradation happens, they release pro-regenerative elements that stimulate cornea tissue regeneration while also reducing any unwanted side effects.

2.1.2 Challenges in Bioscaffold Design

The unique composition of the cornea as shown by the structural differences in each of its five layers presents a challenge in designing a compatible bioscaffold. Hence, it is improbable that a single material will be compatible as a bioscaffold for all five layers of the cornea at the same time. Bioscaffold design must, therefore, adopt a layered approach [9]. It is expected that the bioscaffold intended for the epithelium will be able to promote cell adhesion and rapid mitotic differentiation to maintain the cornea's clarity. For the stroma, which is the thickest and most structurally significant layout of the cornea, a bioscaffold must replicate the precise arrangement of the collagen fibrous and also have the durability to withstand trauma from other cornea layers. At the level of the decimate membrane and the cornea endothelium in potential food should be ultra-thin while having rigidity to be able to support and endothelial cells.

2.2 *Materials for Bioscaffolds*

2.2.1 Natural Polymers

Natural polymers are the top-tier materials used for bioscaffolds because of their natural compatibility and degradability. Examples include:

a. **Collagen**: Collagen is the primary protein found in the human cornea, hence a bioscaffold made from similar material would definitely promote easier adhesion and enhance extracellular matrix formation [10]. Collagen-based scaffolds come in a variety of morphologies, including gels, sponges, and membranes. Collagen-based scaffolds have been shown to support the regeneration of different corneal layers. However, collagen has drawbacks that could affect its long-term stability and transparency, such as its low mechanical strength and susceptibility to enzymatic degradation [11].

b. **Gelatin**: Derived from collagen, gelatin is prized for its viscoelastic characteristics and simplicity of manufacturing (J et al. 2023). It is used in various ways and can be functionalized with bioactive peptides to improve cell adherence. It is appropriate for many corneal applications because of its incorporation with bioactive peptides, which enhances its mechanical qualities and regulates degradation rates. Gelatin however has the disadvantage of having and rapidly degenerates and therefore mostly modified before being used for prolonged release scaffolds (Yan 2022).
c. **Chitosan**: It is a derivative from the protein chitin, it has antibacterial properties and it's also very biocompatible with human tissue [12]. Compared to gelatin, it has moderate mechanical 10 size strains but we need to be reinforced before used for certain applications [13]. One problem however is its solubility which makes it difficult to handle. It can be prepared as a gel as a film or as fibers.
d. **Hyaluronic acid**: Hyaluronic acid is a compound which is useful in wetting the corner and promoting cell differentiation and migration from one layer to another. It is present as a hydrogel and can be used to create an environment which enhances tissue support after damage. Hyaluronic acid is hampered by the fact that it has limited tensile strains and it depreciates very fast; hence, it is restricted to applications that do not require significant support.

2.2.2 Synthetic Polymers

Synthetic polymers were created to improve the flexibility and variability in bioscaffold technology [14–16]. Common synthetic bioscaffolds include:

a. **Polylactic acid (PLA)**: This synthetic polymer has buried characteristics and can be used in conjunction with PGA for rapidly degradable scaffolds [17, 18]. PLA is broken down to its metabolize which are acidic in nature and can induce inflammatorily responses thereby rendering it or usable in sensitive tissue.
b. **Polycaprolactone (PCL)**: PCL has slower degradation rates as compared to other polymers. It is therefore suited for use in longer terms. It can also be designed into other smaller structures, but its surface must be altered to make it necessary to integrate into corner tissue [12].
c. **Poly (lactic-co-glycolic acid) (PLGA)**: PLGA combines materials which makeup PLA and PGA thereby incorporating their different properties post-op provides extensive tensile adaptability and is a good option for scaffold design (Yan 2024).

2.2.3 Hybrid Materials

Hybrid materials combine more than one different biographical material thereby harnessing different positive properties into one material. It also combines multiple materials, which can help to exclude the disadvantages of a single material. Hybrid materials can provide custom-made spell scaffold for tissue engineering. This is achieved by combining positive characteristic like bioactivity of one material, which the biocompatibility of another naturally occurring polymer. This may be necessary in cases where complex cornea repairs are required.

One of the major advantages of these hybrid materials is that they can improve the mechanical property in a particular file scaffold. Naturally occurring materials such as collagen and gelatin can provide excellent compatibility, but they often lack the strength required to remove viable over a long period of time the addition of assisted formula such as such as polycaprolactone (PCL) or polyethylene glycol (PEG) to help attenuate these drawbacks. This results in scaffolds which have longer liability and improved flexibility.

One of the primary advantages of hybrid materials is based on the ability to enhance mechanical factors. Natural polymers, such as gelatin and collagen, offer good biocompatibility characteristics, however are limiting in tensile strength, which is required for long-term use. The incorporation of synthetic polymers, such as polyethylene glycol (PEG) or polycaprolactone (PCL), can address these limitations, by providing scaffolds with enhanced flexibility and durability and. Hybrid structures made up of collagen-PCL are capable of maintaining the transparency of the scaffold for a long time. These hybrid structures have mechanical characteristics similar to the native corneal stromal matrix. This also guarantees better stability and resistance.

Some of the characteristics sought by bioscaffolds are represented by a good cell adhesion capacity and the ability to stimulate and support cell growth. In order to improve these characteristics, it is possible to modify some properties of the bioscaffolds. An example is the combination of scaffolds with antimicrobial agents. This combination makes it possible to significantly reduce the risk of infection after transplantation, which is one of the main complications. To improve cell adhesion, you can combine scaffolds with a PEG. This combination provides a hydrophilic surface that improves cell adhesion. In addition, PEG also allows you to improve the mechanical stability of the scaffolds; this translates into greater stability of the scaffold over a long time. The combination of hyaluronic acid with other artificial polymers improves both scaffold hydration and nutrient transfer, both of which are critical for cell generation in the epithelium.

Finally, nanoparticles (such as graphene and silver) can be incorporated into hybrid materials. This improves the stability of the bioscaffold. In addition, these nanomaterials have other advantages such as an important antibacterial activity and increase electrical conductivity. These characteristics can be used as therapies in specific corneal diseases.

Table 1 Summary of advanced technologies in corneal bioscaffold engineering

Technology	Materials	Advantages	Applications
Electrospinning	Collagen, PLA, PCL	High porosity, control over fiber diameter and alignment	Supports cell growth and bioactive compound delivery
3D Bioprinting	Bioinks (cells and biomaterials)	Precise structure, customizable properties	Patient-specific scaffolds with reduced rejection risk
Hydrogel	Hyaluronic acid, PEG	Hydrated environment, controlled release	Tissue regeneration, stem cell delivery
Decellularization	Decellularized corneal tissues	Natural structure, low immunogenicity	Scaffolds for repopulation with patient cells
Nanofabrication	Graphene, silver nanoparticles	Enhanced stability, antibacterial properties	Multifunctional scaffolds

2.3 *Fabrication Techniques*

As can be seen in Table 1, the manufacture of sustainable bioscaffolds depends primarily on the use of advanced technologies. The use of electrospun bioscaffolds or 3D printers allows the structure and characteristics of the scaffold to be adjusted in detail, so as to be able to adapt faithfully to the unique needs of the individual patient [17].

3 Methods for Constricting Corneal Bioscaffolds

3.1 *Electrospinning*

Electrospinning is a process through which a polymer can be spun into fibers of small diameter. Through this technique it is possible to recreate the human extracellular matrix. [19]. This technique is based on the use of a high-voltage electrical charge. The high potential difference that is created allows the polymer to be charged with energy. The electrical discharge generated makes it possible to produce very thin polymeric fibers with variable diameters (Hong et al. 2019). In addition, it is possible to improve the activity of bioscaffolds by combining hydrogels with growth factors or nanoparticles (Yan 2024). Through these modifications, it is possible to create customized scaffolds for the individual patient [20]. Using the electrospinning technique, it is possible to create scaffolds made up of fibers capable of adapting to the tissue environment of the recipient. In fact, several studies suggest that the migraziane process and tissue regeneration are driven by the specific alignment of the fibers within the scaffold [21]. Improved electrospinning techniques have made it possible to create coaxial tissues. This allows you to load the scaffolds with specific medications. In this way, it is possible to create bioscaffolds capable of releasing drugs directly at the site of injury. The bioscaffolds are thus able to actively support the regeneration of damaged tissue [21]. Additionally, multi-jet electrospawn technology now allows multiple layers of scaffolds with different characteristics to be created at once, responding to the different needs of different layers of cornea tissue.

3.1.1 Mechanism and Parameters

To create the substrate nanofibers, an electric field acts on a polymer, which is extruded via a spinneret. These nanofibers are then extracted as deposits on a collector [20]. Several parameters determine the final fiber properties including [21].

a. **Polymer concentration**: The concentration of polymer used as a startup material for the electrospinning process determines the diameter of the resulting fibers and the stability of the polymer jets. Highly viscose polymer would lead to smooth continuous nanofiber extrusion while a low viscosity polymer will lead to beading, which affects the quality after ability of the eventual scaffold
b. **Electric field strengths**: The electric free field controls the conversion of the polymer to nanofibers. The higher the electric field, the more fibers are extracted from the polymer and the increase in speed of the extraction. Excessive electrical energy, however, may damage the fibers and lead to the destruction of the polymers in the same way as inadequate electrical energy may lead to the formation of poor nanofibers and break down of eventual bioscaffold.
c. **Collector distance**: The distance of the collector from the extruded nanofibers can also affect the architecture of the electrospinning results. The distance of the collector is directly related to the quality of the fibers created. If the collector is too close to the electrospun equipment, it may make very thick fibers that might contain some elements of the solvent that have not evaporated. If the collector is placed at the appropriate distances, then the technician can create a well-formed part five of us, which is ideal for transplant. Rotary collecting drums would also result in different nanofibers as compared to a fixed collection drum. The temperature and environmental factors played during the creation of the nanofibers can also affect the composition and the characteristics of the nanofiber created [21].

3.1.2 Application and Advancement

The evolution of electrospun bioscaffolds has seen the impregnation of scaffold tissues with growth catalysts and drugs into them including [20]:

a. **Bioactive protein materials**: Synthesized proteins such as gelatin and collagen have been reported to encourage cell adhesion and multiplication, which in turn results in better corneal tissue regeneration. Stromal cell arrangement can be mimicked with collagen-enhanced bioscaffolds [20]. Its derivative, gelatin, is used as an additive to enhance bioactive activity in the scaffold-tissue zone, thereby catalyzing adhesion.

b. **Nanoparticles**: In electrospun technology, the addition of nanoparticles avails antibacterial properties, thereby allowing them to be used for drug delivery processes. Specifically, silver particles have been known to deter bacteria colonies from forming at the site of grafts [22]. Researchers have shown that particles that are sensitive to magnetic fields can be implanted into these bioscaffolds and allow the graft to be manipulated as healing takes place, thereby enhancing the proper positioning of the transplants [22]. The same can be said about electrospun scaffolds that have been impregnated with growth factors [23].

3.2 3D Bioprinting

3D bioprinting allows one to create cutting-edge scaffolds by forming layers using bioinks, which are made up of sales and biomaterials onto a surface for eventual implantation [24]. This method makes it possible to acutely control the arrangement and location of different cell types within the scaffold, thereby allowing one to closely approximate the scaffold created to that of the structures in the cornea [25] (Zhang 2019). Evolution in the composition of bioinks has also resulted in the enhancement of the kind of corneal scaffolds that can be created. To date, bioinks have demonstrated the capacity to faithfully replicate the structural and biochemical properties of the native extracellular matrix. By incorporating hydrogels such as alginate or collagen, bioinks provide an ideal environment for cell adhesion and proliferation (Wang 2023) [26–28]. The biological activity of scaffolds can be further enhanced by combining them with growth factors and/or nanoparticles.

3.2.1 Procedures and Customization

There are different 3D bioprinting techniques. Each technique has specific advantages and disadvantages. 3D printing techniques have the advantage of being able to be customized for the individual patient.

a. **Layer by layer**: Through this technique, specific bioinks can be printed one layer on top of the other until the final scaffold is formed. This allows you to check all the individual layers precisely. By acting on the individual layers, it will be possible to modify the mechanical characteristics and porosity of the final scaffold. This provides improved biocompatibility, by decreasing chances of tissue rejection and inflammation. Current technological studies have reported bioprinting systems based on multi-materials, which permit the simultaneous deposition of various bioinks and permit the replication of heterogeneous composition of the several layers of the cornea (Zhang 2019). The systems ensure that every scaffold is appropriately adapted to the individual characteristics of the stroma, epithelium, or endothelium, thus providing enhanced similarity to original corneal tissue.

b. **Growth factor incorporation**: Bioscaffolds are usually embedded with growth factors to assist their integration into the whole tissue and minimize immunological rejection (Wang 2023). This controlled production also helps to extend the viability of the scaffold. These factors can enhance healing, thereby allowing the crafted tissue to closely approximate that of the existing tissue, thereby promoting tissue addition [26].

3.2.2 Innovation and Outcomes

Recent researchers suggested using patient-sourced or patient-specific tissue in the scaffold equation to lower the likelihood of immune rejection. Examples include:

a. **Mesenchymal Stem Cells**: 3D-printed scaffolds using corneal stem cells have shown improved corneal regeneration and healing in murine models [28]. The use of these methane stem cells also reduces the events of immune rejection, thereby making them useful in developing specific scaffolds for sensitive patients [26]. Furthermore, the evolution of induced pluripotent stem cell enrichments in bioscaffold medicine has opened new options in patient-specific therapies [29].
b. **Multilayer structures**: Researchers have confirmed that the multiple different layers of the corneas can be recreated using bioscaffolds. This is achieved by electrospinning layers with varying characteristics onto each other, thereby ensuring long-term viability and limiting cytotoxic reactions [27]. Recent advances in this technology have allowed nanocarriers to be impregnated into bioscaffolds, thereby granting anti-inflammatory and antibiotic properties and reducing post-operative negative results [30].

Advanced 3D bioprinting techniques are paving the way for highly customizable and successful corneal scaffolds that are tailored to each patient's specific needs. Creating optically transparent scaffolds with mechanical characteristics that closely resemble those of real corneal tissue is another encouraging outcome of 3D bioprinting. Scaffolds that can retain transparency and have strong tensile strength enough to tolerate intraocular pressure can be produced by adjusting the bioink composition and printing parameters [31]. All these advances are providing possible options for customizable, patient-specific corneal implants.

3.3 Hydrogel Formulation

Hydrogel formulations provide highly hydrated conditions for cell encapsulation and transport [29]. The characteristic of these gels is that they can be generated directly in situ, right in the target region, so as to provide a support matrix for cell proliferation and tissue regeneration, creating an environment favorable to the repair of specific corneal defects and minimizing disruption with surrounding tissues. These gels are capable of faithfully mimicking the extracellular matrix (ECM), thereby offering structural and biochemical support that is essential for cellular viability and function. The high water content of hydrogels, combined with their mechanical properties, makes them ideal formulations for corneal tissue engineering, where hydration, transparency, and biocompatibility are key aspects. Recent technologies have made it possible to expand the use of hydrogels beyond the "simple" function of scaffolding. In fact, the possible incorporation of components sensitive to specific stimuli, such as polymers sensitive to temperature or pH, allows the controlled gelation by these parameters and the release of drugs under specific conditions [29]. Finally, hydrogels can now be functionalized with nanoscale additives, such as nanoparticles or nanofibers, to improve their mechanical strength and bioactivity without compromising their optical clarity.

3.3.1 Composition and Technique

Natural or synthetic polymers can be used to create hydrogels, which typically gel by chemical or physical cross-linking [29]. The selection of polymer type and cross-linking method is critical, as these factors influence the hydrogel's mechanical strength, degradation rate, and compatibility with corneal tissue.

a. **Natural polymers**: Natural polymers such as hyaluronic acid and collagen are characterized by good compatibility with biological tissues. In addition, these polymers are able to support the growth of corneal cells [32]. Hyaluronic acid is a substance characterized by an intrinsic ability to maintain good hydration. Collagen, on the other hand, allows you to generate an environment that is favorable to cell adhesion and growth. Cell adhesion and growth can be further improved by combining natural polymers with bioactive peptides such as RGD (arginine-glycine-aspartic acid) sequences. Studies have shown that this combination allows to faithfully mimic the interactions of the native extracellular matrix (ECM) [33].
b. **Synthetic polymers**: Synthetic polymers have good properties that can be customized, however they need modifications to allow good biocompatibility [33]. For example, to increase the stability of some synthetic polymers such as polyethylene glycol (PEG) and polylactic-co-glycolic acid (PLGA), they are often used in combinations with hydrogel formulations. In addition, synthetic polymers with growth factors or drugs can be integrated into the hydrogel matrix. This integration allows the integrity of the scaffold to be preserved through localized and sustained administration of drugs or growth factors [34].

3.3.2 Clinical Application and Efficacy

In some corneal abnormalities, hydrogels can be injected at the site of the abnormalities to form a scaffold that acts as a support for tissue regeneration. As mentioned above, hydrogels on the one hand act as a scaffold for cell adhesion and on the other hand act as a means of administering bioactive substances. These characteristics make hydrogels effective in the treatment of certain pathological conditions of the cornea [34].

a. **Stem cell encapsulation**: It is possible to incorporate stem cells or growth factors into hydrogels. The incorporation of these factors is an important factor in the regeneration of corneal tissue, as it minimizes the risk of immune rejection. Stem cells play an important role in the regeneration of corneal tissue. The matrix created by hydrogels promotes stem cell function. Mesenchymal stem cells (MSC) cells have immunomodulatory properties capable of reducing inflammation. Less inflammation results in a better tissue repair process. Limbal stem cells (LSCs), on the other hand, have been found to be important in ensuring long-term transparency of the scaffold. These characteristics of stem cells can be used advantageously by incorporating them into scaffolds [35]
b. **Hydrogels**: Hydrogels based on hyaluronic acid and collagen improve the integration of the bioscaffold with the surrounding tissue. Hyaluronic acid is a substance that has hydrophilic properties. Thanks to this characteristic, hyaluronic acid guarantees adequate hydration of corneal cells. Collagen, on the other hand, acts as a support, thus ensuring an ideal biomimetic structure capable of promoting cell adhesion and alignment. The mechanical strength and stability of the bioscaffold can be increased by combining natural hydrogels with agents such as genipin. It follows that these modifications would allow their most suitable use in dealing with more serious corneal defect [36].

Hydrogels therefore have all the potential to represent a good alternative treatment option in patients with more or less severe corneal lesions and other diseases of the ocular surface.

3.4 Decellularization

The technique of decellularization consists of removing cellular components from corneal tissue from the extracellular matrix (ECM), preserving structural and metabolic proteins [37]. Through this technique, it is therefore possible to create biocompatible natural scaffolds ready to be repopulated by the patient's own cells. Decellularization, as it maintains the ECM, allows to preserve important biomechanical and biochemical signals that are crucial for cell adhesion, proliferation, and differentiation [38]. The interest and high potential that this technique offers has made it possible to refine decellurization protocols to try to minimize damage to the ECM during the decellurization process and maximize the amount of cells removed. Standardized and reproducible processes are now available, which allow to improve the clinical potential of decellularized corneal scaffolds.

3.4.1 Techniques and Optimization

There are different techniques of decellurization, chemical, enzymatic, or physical. Each of these techniques has advantages and disadvantages [39]. Therefore, it would be useful to choose the most suitable decellurization technique based on the application.

a. **Chemical methods**: Chemicals such as Triton X-100 and sodium dodecyl sulfate (SDS) are often used in chemical decellurization protocols. These chemicals are capable of effectively lysing cells and removing cellular debris [40]. Although SDS exhibits powerful decellularization capabilities, when used at high concentrations or for excessively long periods, it can compromise the stability and mechanical properties of the scaffold. In addition, prolonged exposure to these chemicals can reduce corneal clarity by altering collagen fibers or degrading glycosaminoglycans. Triton X-100, being a milder chemical than SDS, is often used in conjunction with enzymatic agents [40]. By changing the concentrations of chemical cleaners and the exposure time, it is possible to advantageously balance the ability to remove cells and the destruction of the ECM.
b. **Enzymatic methods**: Using enzymatic decellurization methods, it is possible to preserve ECM components that are essential for cell adhesion and functionalization. The main enzymes used in enzymatic methods are trypsin and nucleases. These enzymes work by selectively degrading certain cellular components. However, these enzymes do not act on the collagen and glycosaminoglycans present in ECM. Compared to chemical methods, enzymatic methods have a lower effect on the final transparency. These methods consequently guarantee a better refractive property to the scaffold. Compared to chemical methods, enzymatic methods also guarantee better rates of repopulation of scaffolds with stem cells or other patient-derived cells (Mendibil et al. 2020). By combining enzymatic techniques with physical techniques, it is also possible to improve cell removal while preserving ECM more. A useful physical technique for this purpose is represented by freeze–thaw cycles. By using physical techniques together with enzymatic methods, it is possible to drastically reduce the cytotoxic effects secondary to the use of chemicals. Ultimately, it should be noted that when necessary it is possible to use treatments that alternate chemical, physical and enzymatic methods (Mendibil et al. 2020). Finally, it is possible to summarize that the decellurization procedures aim to find the best compromise between the efficient removal of cells and the preservation of the ECM.

Scaffolds obtained by decellurization have numerous advantages. Being originating from corneal tissues, they all have characteristics to ensure proper visual function. Due to their high compatibility, the risk of rejection is very low. During the repopulation phase, the high biocompatibility allows to promote cell proliferation. In addition, this process allows the natural structure of the cornea to be preserved. The scaffolds thus created also support the integration of limbal and mesenchymal stem cells. On the other hand, the remaining parts of the natural ECM allow for the transmission of biochemical signals. These natural biochemical signals allow them to guide the proper functioning of cells.

3.5 Nanofabrication

Nanofabrication techniques make it possible to create scaffolds at the nanoscale. The ability to act at the nanoscale allows for improved cell-scaffold interactions and nutrient diffusion. Furthermore, by acting at the nanometer level, it is possible to faithfully imitate the structure of the ECM. Through nanofabrication, it is also possible to create scaffolds containing bioactive nanoparticles. This then allows therapeutic agents to be delivered directly to the target site. The presence of bioactive particles also allows an almost controlled release of the drugs present inside [37].

3.5.1 Methods and Materials

Nanofabrication processes include:

I. **Microfabrication in soft lithography**: This technique acts at the micro and nanoscale level. It therefore allows the characteristics of the native cornea to be replicated. Microfabrication in soft lithography allows you to create anisotropic scaffolds that mimic the orientation of natural corneal cells. It also allows you to precisely control the porosity of the scaffold, which is another important feature.
II. **Solvent casting with particle leaching**: This technique guarantees better control of the prosity. Porosity control is achieved by incorporating "sacrificial" particles inside the scaffold. These particles are then lysed by enzymes, thus allowing the final porosity of the scaffold to be regulated. The porosity of the scaffold is an important feature as it facilitates the diffusion of nutrients on the one hand and allows for better removal of waste products on the other.

As mentioned above, the techniques can be used either individually, but it is also possible to combine them. In this way, it is possible to take advantage of the advantages of the individual techniques [41]. For example, the use of different techniques allows to create nanometric modifications on the surface of electrospun fibers. In this way, porosity can be increased in a controlled manner and consequently improved cell adhesion and proliferation. Similarly, graphene or silver nanoparticles can be incorporated with 3D bioprinted scaffolds. This can improve the clarity and antimicrobial activity of the scaffold [40].

Nanofabrication techniques make it possible to create "smart" scaffolds (Fig. 1). Smart scaffolds are scaffolds that are sensitive to stimuli and adapt to changes in the local tissue environment. Theoretically, these "smart" scaffolds could be able to release specific agents in response to specific factors such as changes in pH or temperature. In this way, it would be possible to act promptly and in a targeted way to change the corneal microenvironment.

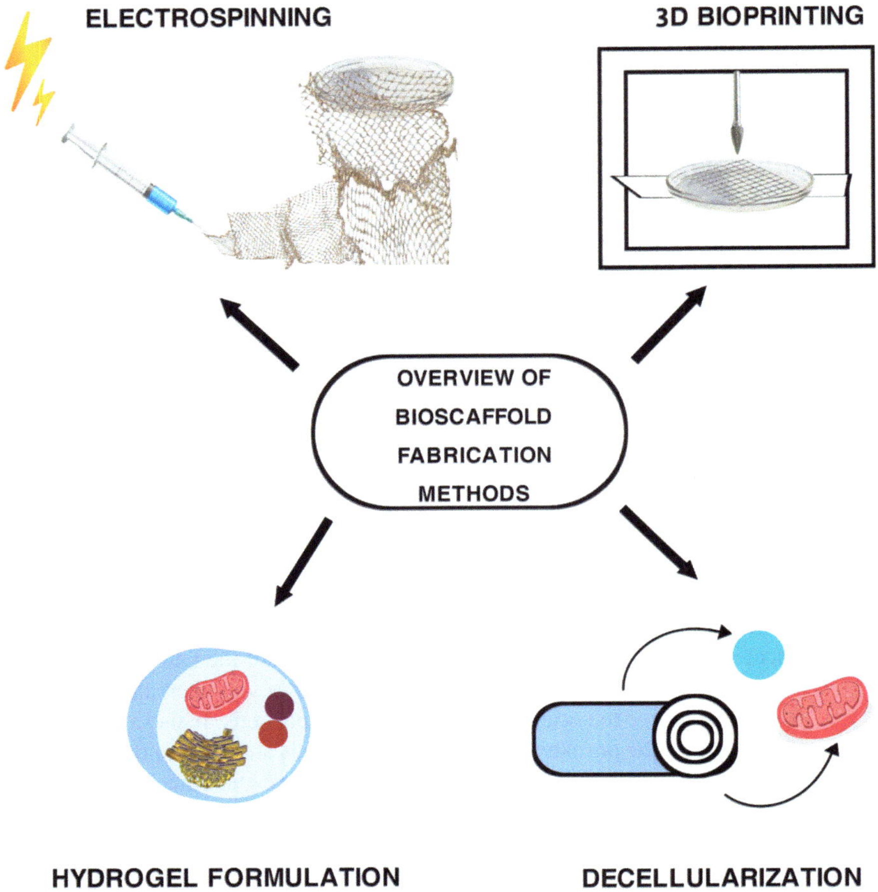

Fig. 1 Overview of bioscaffold fabrication methods

Bioscaffold production techniques. It is possible to observe in each representation the main characteristics of the individual techniques.

3.5.2 Future Directions

The limitations of current bioscaffold manufacturing methods can be overcome by combining different production processes. By combining different techniques, it is possible to create alternative approaches for the treatment of corneal diseases. In addition, the multi-technique approach allows the treatment to be customized to the needs and native corneal structure of the individual patient.

4 Functional Issues

The different cell functionalization techniques have different characteristics and applications (Table 2). Functionalization processes allow to improve scaffold-cell interactions. In this way, it is possible to improve specific biological responses and facilitate tissue integration. The main techniques include [42].

4.1 *Surface Modifications*

It is possible to effect surface modifications of bioscaffolds through chemical, physical, and biological methods [16].

a. **Chemical modification**: Chemical modification is based on the adhesion of chemical compounds to the surface of the scaffold. It is possible to improve cell adhesion by converting fibronectin present in the extracellular matrix into RGD (arginine-glycine-aspartic acid) peptides. The presence of RGD peptides within bioscaffolds allows to significantly improve the adhesion and proliferation of corneal epithelial cells. On the other hand, the migration of epithelial cells can

Table 2 Key functionalization techniques for corneal bioscaffolds

Technique	Method	Description	Application
Surface modifications	Chemical	Bioactive compounds like RGD for cell adhesion	RGD peptides for corneal cell attachment
	Physical	Plasma/UV to increase hydrophilicity	Plasma-treated PCL for better cell growth
	Biological	ECM coatings like collagen to support cell growth	Collagen for limbal stem cell adhesion
Growth factor inclusion	Encapsulation	Slow-release for controlled growth factor delivery	VEGF/bFGF in hydrogels for angiogenesis
	Gradient	Creates signaling gradients for cell guidance	EGF gradient for epithelial migration
Nanoparticles	Drug delivery	Targeted release of anti-inflammatory agents	Reduces inflammation in corneal repair
	Gene delivery	Nanoparticles deliver genes to enhance cell functions	Improves cell proliferation and survival
	Antibacterial	Silver/gold for infection prevention	Silver inhibits bacterial growth

be increased by functionalizing collagen scaffolds with epidermal growth factor (EGF).

b. **Physical treatments**: Physical treatments work by modifying the surface energy and hydrophilicity of the scaffolds. Physical treatments include plasma treatment and UV irradiation.
c. **Biological coatings**: Biological coatings rely on the adhesion of ECM proteins or other bioactive compounds to the scaffold. Limbal stem cell growth can be increased by coating the scaffold with collagen, which are essential for the regeneration of corneal epithelium (Zhang et al. 2022).

4.2 Incorporating Growth Factors

Incorporating growth factors into scaffolds creates an optimal environment for cell differentiation and tissue regeneration. In fact, cell proliferation, differentiation, and migration are regulated by molecular signals provided by growth factors [43]. However, growth factors have the disadvantage of having a short half-life. In fact, these molecules tend to degrade very quickly. To try to increase the half-life of these factors, it is possible to use techniques that allow their slow release, such as the integration of these factors into the scaffold matrix. It is possible to control the release of these factors by integrating them into microspheres or nanoparticles (Wang et al. 2013). The presence of growth factors inside scaffolds allows them to mimic normal tissue signaling pathways. According to a study conducted by Hayashi et al. (2018), using bioscaffolds containing EGF can improve the proliferation of epithelial cells.

4.3 Integration of Nanoparticles

The integration of nanoparticles within bioscaffolds enables drug delivery. In this way, the therapeutic effect of the scaffolds can be improved. In fact, in this way, it is possible to transport and release drugs directly to the site of damage. This allows you to act in a localized manner [44]. As a result, it is possible on the one hand to increase the concentration of drug at the target site, and on the other hand, it is possible to reduce the cytotoxic effects of the drugs themselves. In addition, nanoparticles could also be used to deliver genetic material. This is a very interesting aspect for possible gene therapies [45]. The interest in the development of nanoparticles is mainly due to their small size. The small size allows for precise targeting and controlled release of therapeutic agents. For example, it has been seen that it is possible to reduce post-operative inflammation by incorporating nanoparticles with anti-inflammatory properties inside scaffolds. Similarly, it is possible to prevent the formation of corneal scarring, through nanoparticles that release antifibrotic drugs.

Nanoparticles can also be used as vectors in gene therapy. In fact, it is possible to deliver genetic material embedded in nanoparticles, inside the scaffold. This allows to modify the cellular changes that allow to improve cell survival and proliferation.

Silver and gold nanoparticles are characterized by inherent antibacterial properties. These nanoparticles can consequently help prevent infections. In a study conducted by Sreeharsha et al. (2020), the ability to inhibit bacterial colonizations through the coating of bioscaffolds with silver nanoparticles was highlighted. Again, a controlled release of drugs and not bursting is important. Controlled release at least reduces cytotoxic effects [46].

5 Cells Stem and Bioscaffold

Stem cells are cells capable of differentiating into different types of cell lines. In particular, the use of induced pluripotent stem cells (iPSCs) and mesenchymal stem cells (MSCs) is very promising for corneal regeneration.

I. **Limbal stem cells**: Limbal stem cells (LSCs) are fundamental cells for the corneal epithelium. The possibility of incorporating LSCs into scaffolds and administering them directly to the target site allows for localized promotion of corneal regeneration. In this way, it is possible to counteract the formation of scars [47, 48]. Normally, LSCs are located in the corneal limbus. They have the task of ensuring the continuous turnover of the corneal epithelium. Certain diseases or traumas can compromise LSCs. LSC deficiency causes a state of chronic inflammation of the eye that is unresponsive to the normal drugs used. Oie et al. (2020) showed that in LSCD rabbit models, the use of collagen scaffolds loaded with LSC improves the regeneration of the corneal epithelium and consequently reduces scar formation (M et al. 2019). In another study conducted by Pellegrini et al. (2013), a success rate of approximately 76% in the ability to restore corneal epithelium using fibrin-based scaffolds for LSC transplantation has been reported. These studies illustrate the potential applications and effectiveness of LSC-filled scaffolds. Future research should focus on identifying suitable materials that further improve the survival and function of LSCs, while reducing immunogenicity and inflammation on the other hand.
II. **Mesenchymal stem cells (MSCs)**: MSCs can differentiate into different types of cells, such as corneal, stromal and endothelial cells. This characteristic makes them ideal for the repair of deep corneal tissues. Hydrogel scaffolds are the most widely used for the delivery of MSCs (Zhou et al. 2023). In particular, in a study conducted by Du et al. (2015), it was shown that using MSC-loaded hydrogels can improve corneal regeneration in a rabbit model (Mahmood 2022). In addition, this study also showed that these scaffolds loaded with MSCs also improve integration into the stromal tissue of the recipient. Although clinical studies are needed to evaluate the actual behavior of MSC-loaded bioscaffolds

in vivo; the immunomodulatory capabilities of MSCs and their ability to release trophic substances make them very promising.

III. **Pluripotent stem cells (iPSCs)**: Induced pluripotent stem cells (iPSCs) are cells capable of generating different corneal cells. The scaffolds in this case act as a nest for the differentiation of iPSCs in the different cell lines (epithelial, stromal and endothelial). The ability of these cells to differentiate into different cell types makes them suitable for corneal regeneration. Bioscaffolds can be activated to promote and direct the differentiation of iPSCs. By creating a specific microenvironment within the scaffold, it is possible to promote the maturation of a specific cell line. In a study conducted by Hayashi et al. (2018) in a rabbit model, it was possible to restore endothelial function using decellularized scaffolds loaded with iPSC-derived endothelial cells (Moradi et al. 2019). In this model, the decellularized scaffold allowed to provide adequate MEC for proper cell alignment, while iPSC-derived endothelial cells were able to restore endothelial function.

6 Limitations and Future Studies

Despite the interest shown by bioscaffolds in recent years, their use in clinical practice presents several obstacles. However, new studies should evaluate the actual reduction of immune rejection and long-term outcomes.

6.1 Immunogenicity and Biocompatibility

One of the main challenges to be faced is certainly the immunigenicity of the scaffolds. Immunological responses to scaffolds compromise the stability and transparency of the scaffold. The absence of blood and lymphatic vessels allows keratoplasties to enjoy some immunological privileges. Despite this, however, rejection of the bioscaffold is a possible complication [38]. To reduce the chance of rejection, decellularized bioscaffolds of animal origin or synthetic polymers such as sericin can be used. In addition, to improve compatibility while minimizing the risk of immune reactions, surface modifications can be made to bioscaffolds, such as functionalization of bioactive peptides.

Types of interactions between stem cells and the extracellular matrix (ECM). Figure 2 represents the different types of stem cells and how they interact with the bioscaffold.

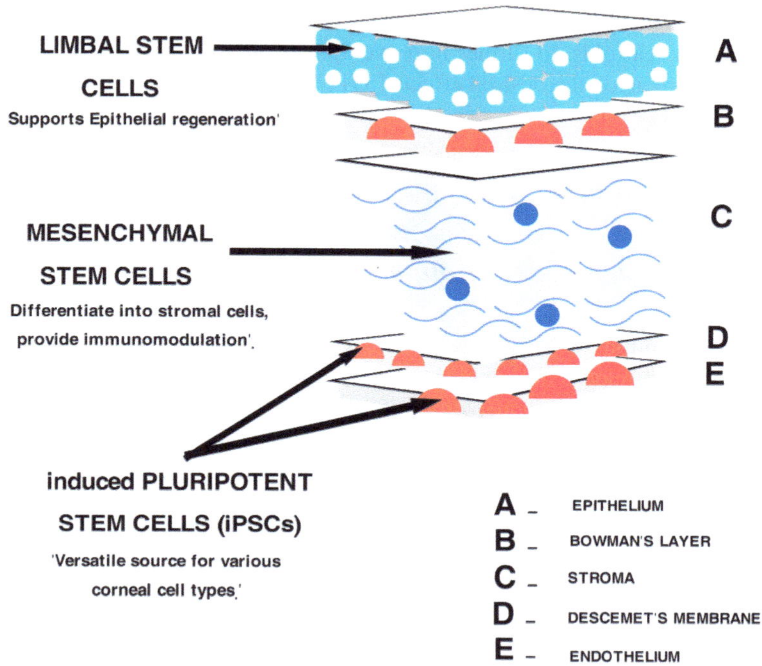

Fig. 2 Stem cells and bioscaffolds for corneal regeneration

6.2 *Scaffold Design and Customization*

To faithfully mimic the corneal tissue of the host, it is essential to design the bioscaffold in a way that is personalized to the recipient patient. 3D bioprinting allows you to fabricate bioscaffolds specific to the needs of the individual recipient patient. In addition, through bioprinting, it is possible to create precise microarchitectures that ensure the best possible alignment. Customizing the scaffold improves the scaffold's fit to the recipient's cornea. This can reduce the risk of scaffolds being rejected [49].

6.3 *Integration of Advanced Technologies*

To further improve the performance of scaffolds, scaffolds can be integrated with gene editing. Through gene editing tools such as CRISPR-Cas9, it is possible to carry out genetic modifications. Through these modifications it is possible to alter some genes fundamental for cell survival, proliferation and immunomodulation within bioscaffolds. CRISPR-Cas9 makes it possible to increase the efficiency of the scaffold in tissue repair, modifying specific genes that on the one hand promote cell adhesion and on the other hand reduce inflammation. Moreover, it would theoretically be possible, through the combination of CRISPR-Cas9 gene editing with iPSCs and decellularized scaffolds, to further improve the performance of scaffolds, directing with greater precision the desired effects [39] (Table 2).

7 Bioscaffolds for Corneal Pathologies

7.1 *Corneal Neovascularization*

A promising option in corneal neovascularization treatment strategies could be bioscaffolds that release anti-angiogenic molecules, e.g., anti-vascular endothelial growth factor. Through bioscaffolds loaded with anti-VEGF, it is possible to administer as many agents directly to the neovessels. This allows both to reduce the risks resulting from prolonged administration and to increase the local concentration of these agents. Some animal studies have shown that scaffolds based on hyaluronic acid and collagen impregnated with anti-VEGF effectively inhibit corneal neovascularization [37].

7.2 *Keratoconus*

Keratoconus is a condition characterized by progressive ectasia and thinning of the cornea. To reduce corneal ectasia, bioscaffolds based on cross-linked collagen can be created. In this way, scaffolds provide mechanical support and are less invasive than laser or surgical treatments. These bioscaffolds can be organized to reflect collagen structure within the corneal stroma, hence granting the thickest corneal layer more biomechanical strength to withstand progressive ecstatic modifications characteristic of keratoconus. Cross-linking increases the scaffold's mechanical strength, making it more resistant to deformation forces. Bioscaffolds also have the advantage of being able to increase corneal thickness [50]. In addition, using scaffolds during cross-linking procedures can increase the penetration and distribution of riboflavin in corneal tissue [51].

7.3 *Corneal Epithelial Disorders*

Limbal stem cell deficiency syndrome (LSCD) still does not have a specific and standardized treatment. Bioscaffolds loaded with stem cells can be used to facilitate the culture and subsequent transplantation of limbal stem cells. Fibrin scaffolds containing limbal stem cells (LSCs) allow for better epithelial restoration and improved ocular surface homeostasis [52].

7.4 Corneal Ulcers

Corneal abrasions and ulcers are quite frequent pathologies in clinical practice. Bioscaffolds have all the features to improve the treatment of these lesions. In fact, through scaffolds, it is possible to stimulate cell regeneration, providing therapeutic substances and providing structural support to the cornea. In addition, the use of scaffolds loaded with growth factors (FGF and EGF) make it possible to reduce the formation of scars. In addition, in the treatment of these lesions, it is possible to convey antifungals, antibiotics, and anti-inflammatories directly to the lesions. This can improve treatment effectiveness and reduce drug resistance (Mahdavi et al. 2020). The higher concentration of drugs directly on the lesion has the advantage of allowing better control of pathogens, reducing the administration of drugs systemically.

7.5 Corneal Endothelial Dysfunction

Some corneal endothelial dysfunctions, such as Fuchs' endothelial dystrophy, lead to an alteration in corneal permeability and consequent corneal transparency. In this sense, bioscaffolds are able to provide a support matrix for the differentiation and proliferation of endothelial cells produced by iPSCs [53]. Preclinical studies have shown that PLGA scaffolds support endothelial cell differentiation and proliferation [54]. Scaffolds represent an ideal substrate for the proliferation and preservation of endothelial cells. Another potential application of bioscaffolds could be to use them to transfer endothelial cells produced from stem cells or cultured human corneal endothelial cells (HCEC) directly onto the patient's damaged cornea [55]. This would represent a unique remedy for the loss of endothelial cells with functional healthy cells.

7.6 Corneal Transplants

When medical therapies fail to resolve corneal pathologies, corneal transplantation is necessary. It is possible to replace the entire cornea or replace only some layers of it. Corneal transplantation is the main disadvantage of the risk of rejection and the limited availability of corneas from donors compared to demand. Bioscaffolds are a very promising alternative that allows you to overcome these limitations. Through scaffolds, it is possible to promote the regeneration and transparency of the transplanted tissue. In addition, thanks to production techniques, such as 3D bioprinting, it is possible to faithfully mimic the biomimetic characteristics of the patient's cornea. In addition, it would be possible to reduce the formation of scars and above all to carry out transplants without corneal sutures. This would allow better patient management during the post-operative period.

8 Ethics and Legal Considerations

Regulatory and ethical hurdles must be addressed for the clinical use of bioscaffolds. Optimal compliance to manufacturing standards must also be upheld to assure safe and successful integration [56]. Rigorous preclinical and clinical testing as well as the approval of national and international health agencies are required. Regulatory frameworks must be adaptable to technical advancements, thus necessitating standards' generation for bioscaffold testing and production (Zocchi 2019).

Some ethical considerations: Transparency during clinical research is key to gaining public trust [57]. Informed consent is critical. It is important that the different scaffold regulatory committees work together to unambiguously examine the potential risks and benefits of scaffolds used in clinical practice. Safety is a priority to ensure the development and use of scaffolds in clinical trials.

High pricing, restricted availability, and difficult procedures are contributory to inequitable access to care. Public policy could be leveraged to promote equitable access and distribution. To confirm the safety of bioscaffolds, large multicentric studies are useful, which must adhere to strict regulatory and ethical guidelines [58]. Longitudinal studies will be key to understanding the true long-term impact of bioscaffolds. Collaboration between different specialists and joint research would accelerate the refinement of scaffolds. In this way, it will be possible to use bioscaffolds as a therapeutic alternative to traditional treatments available today.

9 Conclusions

An alternative to the treatments available for corneal pathologies is the use of biological scaffolds. Traditional therapies have limitations such as immunological rejection, availability of available donors and regeneration capacity. Being able to provide both good structural support and a favorable environment for corneal regeneration, scaffolds functionalized with bioactive peptides represent a revolutionary therapeutic strategy. These therapies have the potential to enable millions of patients to be treated in innovative ways. 3D printing makes it possible to create scaffolds with dimensions, architectures and compositions suitable for the patient's cornea. The personalization of the treatment allows the scaffolds to be adapted to the anatomical and physiological needs of the patient. In this way, it is possible to improve the effectiveness of therapy, minimize the risk of immune rejection and improve functional outcomes. However, in order to be able to move scaffolds from the design laboratory to clinical and surgical practice, it is necessary to overcome the obstacles that still exist today through new inventions and the optimization of existing knowledge. Research should now focus on the design of scaffolds capable of interacting dynamically with the host environment. The use of materials or devices that can respond to stimuli (such as pH and temperature) is a field to be explored with further study and research. The creation of "smart" scaffolds would allow drugs or other agents to be released in response to specific stimuli detected. To give an example, during inflammation, there is an increase in pH levels at the local level; the increase in pH could be detected by pH -sensitive scaffolds that respond by releasing anti-inflammatory agents. This would allow an early and targeted response to be offered. Finally, gene editing combined with scaffolds would make it possible to explore a modality of treatment of corneal diseases that is still to be fully understood and discovered.

References

1. Dawson DG, Geroski DH, Edelhauser HF (2008) Corneal endothelium: structure and function in health and disease. In: Corneal surgery: theory technique and tissue: fourth edition. Elsevier, pp 57–70. https://doi.org/10.1016/B978-0-323-04835-4.50013-6
2. Dua HS, Faraj LA, Said DG, Gray T, Lowe J (2013) Human corneal anatomy redefined: a novel pre-descemet's layer (Dua's Layer). In: Ophthalmology, pp 1778–1785. https://doi.org/10.1016/j.ophtha.2013.01.018
3. Eghrari AO, Riazuddin SA, Gottsch JD (2015) Overview of the cornea: structure, function, and development. In: Progress in molecular biology and translational science. Elsevier B.V., pp 7–23. https://doi.org/10.1016/bs.pmbts.2015.04.001
4. Austin A, Lietman T, Rose-Nussbaumer J (2017) Update on the management of infectious keratitis. Elsevier Inc. https://doi.org/10.1016/j.ophtha.2017.05.012
5. Han SB, Liu YC, Liu C, Mehta JS (2024) Applications of imaging technologies in fuchs endothelial corneal dystrophy: a narrative literature review. MDPI. https://doi.org/10.3390/bioengineering11030271
6. Holland G et al (2021) Artificial cornea: past, current, and future directions. Frontiers Media S.A. https://doi.org/10.3389/fmed.2021.770780
7. Wang P, Sun Y, Shi X, Shen H, Ning H, Liu H (2021) Bioscaffolds embedded with regulatory modules for cell growth and tissue formation: a review. KeAi Communications Co. https://doi.org/10.1016/j.bioactmat.2020.10.014
8. Ahearne M, Fernández-Pérez J, Masterton S, Madden PW, Bhattacharjee P (2020) Designing scaffolds for corneal regeneration. Wiley. https://doi.org/10.1002/adfm.201908996
9. Cai S, Wu C, Yang W, Liang W, Yu H, Liu L (2020) Recent advance in surface modification for regulating cell adhesion and behaviors. Nanotechnol Rev 9(1):971–989. https://doi.org/10.1515/ntrev-2020-0076
10. Watsky MA et al (2002) A collagen-based scaffold for a tissue engineered human cornea: physical and physiological properties. Invest Ophthalmol Vis Sci 43(13):2996
11. Kong B et al (2022) Recombinant human collagen hydrogels with hierarchically ordered microstructures for corneal stroma regeneration. Chem Eng J 428:131012. https://doi.org/10.1016/j.cej.2021.131012
12. Tayebi T et al (2021) Biofabrication of chitosan/chitosan nanoparticles/polycaprolactone transparent membrane for corneal endothelial tissue engineering. Sci Rep 11(1). https://doi.org/10.1038/s41598-021-86340-w
13. Bhattacharyya et al (2023) 3D bioprinting of complex tissue scaffolds with in situ homogeneously mixed alginate-chitosan-kaolin bioink using advanced portable biopen. Carbohydr Polym 317:121046. https://doi.org/10.1016/j.carbpol.2023.121046
14. Jameson JF, Pacheco MO, Nguyen HH, Phelps EA, Stoppel WL (2021) Recent advances in natural materials for corneal tissue engineering. MDPI. https://doi.org/10.3390/bioengineering8110161
15. Nosrati H, Ashrafi-Dehkordi K, Alizadeh Z, Sanami S, Banitalebi-Dehkordi M (2020) Biopolymer-based scaffolds for corneal stromal regeneration: a review. NLM. https://doi.org/10.17219/pim/127653

16. Nosrati H et al (2020) Corneal epithelium tissue engineering: recent advances in regeneration and replacement of corneal surface. Future Medicine Ltd. https://doi.org/10.2217/rme-2019-0055
17. Chi M, Yuan B, Xie Z, Hong J (2023) The innovative biomaterials and technologies for developing corneal endothelium tissue engineering scaffolds: a review and prospect. MDPI. https://doi.org/10.3390/bioengineering10111284
18. Fagerholm P, Lagali NS, Carlsson DJ, Merrett K, Griffith M (2009) Corneal regeneration following implantation of a biomimetic tissue-engineered substitute. Clin Transl Sci 2(2):162–164. https://doi.org/10.1111/j.1752-8062.2008.00083.x
19. Nadaf et al (2022) Recent update on electrospinning and electrospun nanofibers: current trends and their applications. Royal Society of Chemistry. https://doi.org/10.1039/d2ra02864f
20. Patel G, Na K-S, Lee HJ, Koh W-G (2021) Electrospun fibers for corneal regeneration. Curr Ophthalmol Rep 9. https://doi.org/10.1007/s40135-021-00279-9
21. Bhattacharjee P, Ahearne M (2020) Fabrication and biocompatibility of electroconductive silk fibroin/pedot: Pss composites for corneal epithelial regeneration. Polymers (Basel) 12(12):1–7. https://doi.org/10.3390/polym12123028
22. Chen Y, Dong X, Shafiq M, Myles G, Radacsi N, Mo X (2022) Recent advancements on three-dimensional electrospun nanofiber scaffolds for tissue engineering. Adv Fiber Mater 4(5):959–986. https://doi.org/10.1007/s42765-022-00170-7
23. Uhljar LÉ, Ambrus R (2023) Electrospinning of potential medical devices (wound dressings, tissue engineering scaffolds, face masks) and their regulatory approach. MDPI. https://doi.org/10.3390/pharmaceutics15020417
24. Cui H, Nowicki M, Fisher JP, Zhang LG (2017) 3D bioprinting for organ regeneration. Wiley. https://doi.org/10.1002/adhm.201601118
25. Lam EH, Yu F, Zhu S, Wang Z (2023) 3D bioprinting for next-generation personalized medicine. MDPI. https://doi.org/10.3390/ijms24076357
26. Bisht B, Hope A, Mukherjee A, Paul MK (2021) Advances in the fabrication of scaffold and 3D printing of biomimetic bone graft. Ann Biomed Eng 49(4):1128–1150. https://doi.org/10.1007/s10439-021-02752-9
27. Chen Q, Cao P, Advincula R (2018) Mechanically robust, ultraelastic hierarchical foam with tunable properties via 3D printing. Adv Funct Mater 28:1800631. https://doi.org/10.1002/adfm.201800631
28. Jain P, Kathuria H, Dubey N (2022) Advances in 3D bioprinting of tissues/organs for regenerative medicine and in-vitro models. Biomaterials 287:121639. https://doi.org/10.1016/j.biomaterials.2022.121639
29. Lynch CR, Kondiah PP, Choonara YE, Du Toit LC, Ally N, Pillay V (2020) Hydrogel biomaterials for application in ocular drug delivery. Frontiers Media S.A. https://doi.org/10.3389/fbioe.2020.00228
30. Rodríguez-Pombo L, Awad A, Basit AW, Alvarez-Lorenzo C, Goyanes A (2022) Innovations in chewable formulations: the novelty and applications of 3D printing in drug product design. MDPI. https://doi.org/10.3390/pharmaceutics14081732
31. Wu KY, Tabari A, Mazerolle É, Tran SD (2024) Towards precision ophthalmology: the role of 3D printing and bioprinting in oculoplastic surgery, retinal, corneal, and glaucoma treatment. MDPI. https://doi.org/10.3390/biomimetics9030145
32. Ahearne M, Lynch AP (2015) Early observation of extracellular matrix-derived hydrogels for corneal stroma regeneration. Tissue Eng Part C Methods 21(10):1059–1069. https://doi.org/10.1089/ten.tec.2015.0008
33. Ozcelik et al (2014) Biodegradable and biocompatible poly (ethylene glycol)-based hydrogel films for the regeneration of corneal endothelium. Adv Healthc Mater 3(9):1496–1507. https://doi.org/10.1002/adhm.201400045
34. Liang W, Luo Z, Zhou L (2020) Preparation and characterization of the n-HA/PVA/CS porous composite hydrogel. Chin J Chem Eng 28(2):598–602. https://doi.org/10.1016/J.CJCHE.2019.01.029

35. Ghiasi M, Jadidi K, Hashemi M, Zare H, Salimi A, Aghamollaei H (2021) Application of mesenchymal stem cells in corneal regeneration. Elsevier Ltd. https://doi.org/10.1016/j.tice.2021.101600
36. Luo X et al (2023) Research progress of polymer biomaterials as scaffolds for corneal endothelium tissue engineering. MDPI. https://doi.org/10.3390/nano13131976
37. Wilson S, Sidney L, Dunphy S, Rose J, Hopkinson A (2013) Keeping an eye on decellularized corneas: a review of methods, characterization and applications. J Funct Biomater 4(3):114–161. https://doi.org/10.3390/jfb4030114
38. Kasravi M et al (2023) Immunogenicity of decellularized extracellular matrix scaffolds: a bottleneck in tissue engineering and regenerative medicine. Biomater Res 27(1):10. https://doi.org/10.1186/s40824-023-00348-z
39. Isidan A, Liu S, Chen AM, Zhang W, Li P, Smith LJ, Hara H, Cooper DK, Ekser B (2021) Comparison of porcine corneal decellularization methods and importance of preserving corneal limbus through decellularization. PLoS One 16(3):1–12. https://doi.org/10.1371/journal.pone.0243682
40. Neishabouri A, Soltani Khaboushan A, Daghigh F, Kajbafzadeh AM, Majidi Zolbin M (2022) Decellularization in tissue engineering and regenerative medicine: evaluation, modification, and application methods. Frontiers Media S.A. https://doi.org/10.3389/fbioe.2022.805299
41. Wu Z, Kong B, Liu R, Sun W, Mi S (2018) Engineering of corneal tissue through an aligned PVA/collagen composite nanofibrous electrospun scaffold. Nanomaterials 8(2). https://doi.org/10.3390/nano8020124
42. Dave K, Gomes VG (2019) Interactions at scaffold interfaces: effect of surface chemistry, structural attributes and bioaffinity. Elsevier Ltd. https://doi.org/10.1016/j.msec.2019.110078
43. Oliveira ÉR et al (2021) Advances in growth factor delivery for bone tissue engineering. MDPI AG. https://doi.org/10.3390/ijms22020903
44. Madden PW, Lai JNX, George KA, Giovenco T, Harkin DG, Chirila TV (2011) Human corneal endothelial cell growth on a silk fibroin membrane. Biomaterials 32(17):4076–4084. https://doi.org/10.1016/j.biomaterials.2010.12.034
45. Sarkar S, Panikker P, D'Souza S, Shetty R, Mohan RR, Ghosh A (2023) Corneal regeneration using gene therapy approaches. MDPI. https://doi.org/10.3390/cells12091280
46. Hsu CC et al (2015) Stem cell therapy for corneal regeneration medicine and contemporary nanomedicine for corneal disorders. Cogniz Commun Corp. https://doi.org/10.3727/096368914X685744.
47. Iovino C et al (2018) Corneal endothelium features in Fuchs' endothelial corneal dystrophy: a preliminary 3D anterior segment optical coherence tomography study. PLoS One 13(11). https://doi.org/10.1371/journal.pone.0207891
48. Vattulainen M, Ilmarinen T, Viheriälä T, Jokinen V, Skottman H (2021) Corneal epithelial differentiation of human pluripotent stem cells generates ABCB5+ and ΔNp63α+ cells with limbal cell characteristics and high wound healing capacity. Stem Cell Res Ther 12(1). https://doi.org/10.1186/s13287-021-02673-3
49. Roberts C (2005) Biomechanical customization: the next generation of laser refractive surgery. Elsevier Inc. https://doi.org/10.1016/j.jcrs.2004.11.032
50. Pedrotti E et al (2020) New treatments for keratoconus. Int Ophthalmol 40(7):1619–1623. https://doi.org/10.1007/s10792-020-01455-9
51. Dou S, Liu X, Shi W, Gao H (2023) New dawn for keratoconus treatment: potential strategies for corneal stromal regeneration. BioMed Central Ltd. https://doi.org/10.1186/s13287-023-03548-5
52. Nuzzi A, Pozzo Giuffrida F, Luccarelli S, Nucci P (2022) Corneal epithelial regeneration: old and new perspectives. MDPI. https://doi.org/10.3390/ijms232113114
53. Hatou S, Shimmura S (2023) Advances in corneal regenerative medicine with iPS cells. Jpn J Ophthalmol 67(5):541–545. https://doi.org/10.1007/s10384-023-01015-5
54. Jafri M, Colby K (2019) New insights into corneal endothelial regeneration. Curr Ophthalmol Rep 7(1):37–44. https://doi.org/10.1007/s40135-019-00197-x

55. JDelaey J, De Vos L, Koppen C, Dubruel P, Van Vlierberghe S, Van den Bogerd B (2022) Tissue engineered scaffolds for corneal endothelial regeneration: a material's perspective. Royal Society of Chemistry. https://doi.org/10.1039/d1bm02023d
56. Jonlin EC (2022) Ethical considerations in cell therapy. In: Gee AP (ed) Cell therapy: cGMP facilities and manufacturing. Springer, Cham, pp 101–118. https://doi.org/10.1007/978-3-030-75537-9_6
57. King NM, Perrin J (2019) Ethical issues in stem cell research and therapy. [Online]. Available: http://stemcellres.com/content/5/4/85
58. Hunckler MD, Levine AD (2022) Navigating ethical challenges in the development and translation of biomaterials research. Front Bioeng Biotechnol 10. https://doi.org/10.3389/fbioe.2022.949280

Dr. Federico Visalli was born in Catania, Sicily. He obtained his Degree in Medicine and Surgery at the University of Catania with highest honors. He is currently a resident in Ophthalmology, with a profound passion for corneal diseases and corneal surgery. His dedication to the field is driven by an insatiable curiosity and a relentless commitment to innovation in both research and clinical practice. As a passionate researcher, Dr. Visalli actively explores advancements in corneal transplantation, bioengineered scaffolds, and personalized refractive surgery, aiming to enhance surgical outcomes and pioneer new therapeutic approaches. His work is characterized by a strong link between scientific discovery and its real-world application, ensuring that research translates into tangible benefits for patients. He is a proud member of Group of Research in Innovative Discovery in Ophthalmology (GRIDO), where he collaborates with experts to push the frontiers of ophthalmic science. His vision is to contribute to the future of ophthalmology through continuous learning, interdisciplinary collaboration, and a drive for excellence. "There are no limits for those driven by curiosity and passion: every challenge is an opportunity to create something extraordinary."

Caterina Gagliano is a full professor in Ophthalmology at the Kore University Enna (Italy) and is the director of the Ophthalmology Department at the "G.B. Morgagni" Foundation, Catania (Italy). She is the head of the Ocular Immunology and Rare Diseases Unit AOU Policlinico—San Marco Hospital, Catania from 2012 to 2024. She worked as an evaluator for the European Commission "Horizon 2020." She was the president and scientific director of Neurovisual Science Technology srl (NEST), a transfer society that specializes in clinical and preclinical research design. Furthermore, she is the founder of Group of Research in Innovative Discovery in Ophthalmology (GRIDO) with main fields of interest including rare ocular diseases, severe inflammatory eye conditions, and neuro-ophthalmological diseases. Her work is conducted in a multidisciplinary environment, promoting continuous dialogues among colleagues and encouraging the participation of younger coworkers. She works in an environment surrounded by one of the largest gardens in Europe, the Villa Bellini in Catania. This, along with the sunshine of the beautiful Sicily, instills an overwhelming and engaging sense of optimism in her daily life.

Fabiana D'Esposito is an ophthalmologist and a Ph.D. in neuroophthalmology. She qualified in Naples University "Federico II" and trained in London at the Institute of Ophthalmology, Department of Molecular Genetics (UCL). At present, she is a visiting professor of Ophthalmic Genetics at Naples University "Federico II" and honorary senior lecturer at Imperial College. She is a member of ESASO Faculty, board member of the Italian Society of Pediatric Ophthalmology and Strabismus (SIOPS). She is an associate editor of Scientific Reports and European Journal of Ophthalmology. Her current main activity consists in the clinical and molecular diagnosis of genetic eye diseases and in genotype–phenotype correlations.

Mutali Musa is a consultant clinical Optometrist with 15 years of experience, specializing in both academic and primary eye care. He has played a crucial role in establishing and nurturing two of the busiest eye hospitals in southern Nigeria. He has experience in providing end-of-care Low Vision Care and managing a dedicated Glaucoma clinic. His education has taken him from Doctor of Optometry, Masters in Vision Science, and Ongoing Ph.D. at the University of Benin. He also holds a Fellowship in Rehabilitative Optometry and Low Vision Care from the Postgraduate College of Optometrists. He is widely published and cited in top international journals and serves as a guest editor to many. His research interests include: clinical examination and counseling of glaucoma and cataract patients; visual fields, tonometry, and data analysis related to glaucoma disease progression; and measurement of light properties as they interact with the human eye. He is a member of several Research Laboratories and academic societies including the Low Vision Assessment, spectrophotometry, and quality testing of ophthalmic lenses unit at the University of Benin and the ocular biometrics unit at Africa Eye Laser Centre Ltd. He has received multiple scholarships and awards and has presented papers in local and international scientific sessions. He is currently a lecturer at the Department of Optometry, University of Benin, and is also Eyecare Program Consultant to Vision Spring. He is Head of Operations and Senior Consultant Optometrist at Africa Eye Laser Centre Ltd and Centre for Sight Africa Ltd.

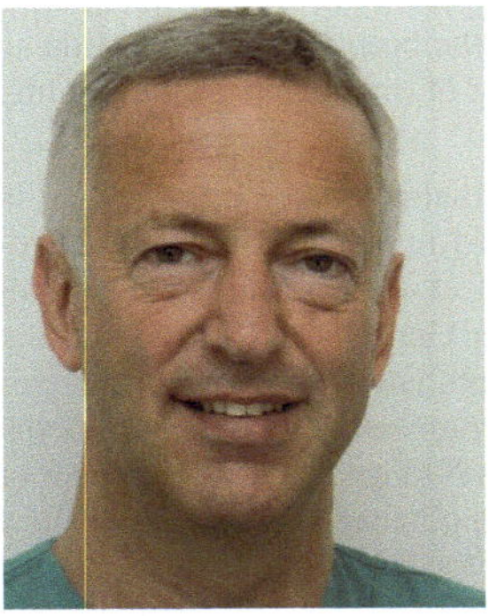

Daniele Tognetto is a full professor of Ophthalmology at the University of Trieste, Italy. He is a chairman of University Eye Clinic of Trieste (Italy) in the Department of Medical Surgical Sciences and Health, in addition to Head of Ophthalmology Course in the Degree Course in Medicine and Surgery at the University of Trieste, and head of Residency Program in Ophthalmology at the University of Trieste. He is an anterior and posterior segment surgeon with more than 30.000 surgical procedures. He has been a speaker in numerous international and national Congresses, including more than 40 live surgeries since 2006. He has membership in several international scientific society and has been the Past President of Italian Society of Vitreoretinal Surgeons. He is the president of Italian Society of Cataract and Refractive Surgeons and President of Triveneta Ophthalmology Society. He is an author of numerous peer-reviewed scientific publications on retina and anterior segment surgery.

Marco Zeppieri was born in Ontario, Canada, and currently lives in Udine, Italy. He graduated with a Degree in Biology from the University of Toronto, Canada, in1992. He obtained his Degree in Medicine and Surgery at the University of Milan-Bicocca, Italy, in 2003. He then obtained his Specialization in Ophthalmology at the University of Udine in 2007. In 2012, he completed his PhD from the University of Udine. He also completed a Post-Doctoral Glaucoma Research Fellowship at Discoveries In Sight, Devers Eye Institute in Portland, Oregon, USA, in 2006. He obtained his National Qualifications as a 2° level University Professor in 2018. Since January 2008 to date, he has worked as an ophthalmologist at the Department of Ophthalmology at Santa Maria della Misericordia Hospital in Udine, Italy. He has won several awards such as the Best Video SOI (2004), AIGS Travel Grant (2007), and ARVO GB Bietti Foundation for Ophthalmology Travel Grant (2007). His scientific interests include new techniques for the morphological and functional diagnosis of ocular hypertension and glaucoma, the use of stem cells for corneal wound lesions, ocular surface and lid diagnostics and treatments and community-based screening procedures. He has participated as invited speaker at numerous national and international ophthalmology conferences. He is a reviewer for numerous international peer-reviewed journals. He is an author of over 200 indexed scientific publications in international scientific journals and book chapters cited on PubMed.

Wound Healing and Cellulose-Based Hemostatic Agent

Asif Karim Neon, Md. Nizam Uddin, Md. Aliahsan Bappy, Md. Mahbubur Rahman, and Md. Moniruzzaman

Abstract Wound healing is a complex, multi-phase process that involves hemostasis, inflammation, proliferation, and remodeling to restore skin integrity and function. Effective wound management, especially in controlling bleeding and promoting tissue regeneration, remains a critical challenge in clinical care. Cellulose-based hemostatic agents have gained attention for their role in wound healing due to their biocompatibility, biodegradability, and ability to form highly absorbent hydrogels. The use of hemostatic agents is crucial in preventing substantial blood loss and reducing the risk of death from excessive bleeding during surgical procedures or in emergency situations. These agents, derived from natural cellulose or modified nanocellulose, exhibit excellent fluid retention, structural stability, and the capacity to promote clotting, making them valuable in bleeding control. Furthermore, oxidized cellulose is a highly effective biodegradable and biocompatible derivative of cellulose, making it one of the most widely used hemostatic agents in surgical procedures. Moreover, the hemostatic properties of cellulose-based materials can be further enhanced through chemical modifications and incorporation of bioactive compounds, leading to improved antibacterial activity and accelerated healing. By creating a moist environment, these agents not only support cell proliferation but also reduce the risk of infection. This chapter reviews the design, mechanisms, and applications of cellulose-based hemostatic agents in wound care, highlighting recent innovations and their impact on enhancing wound healing outcomes.

A. K. Neon · Md. Mahbubur Rahman
Department of Mechanical Engineering, Khulna University of Engineering and Technology, Khulna, Bangladesh

Md. Nizam Uddin (✉)
James C. Morriss Division of Engineering, Texas A&M University-Texarkana, Texarkana, TX, USA
e-mail: muddin@tamut.edu

Md. Aliahsan Bappy
Department of Mechanical Engineering, Lamar University, Beaumont, TX, USA

Md. Moniruzzaman
Department of Industrial and Systems Engineering, Lamar University, Beaumont, TX, USA

R. Malviya and S. Sundram (eds.), *Sustainable Cellulose Materials for Biomedical Applications*, Sustainable Materials and Technology,
https://doi.org/10.1007/978-981-96-9904-9_19

Keywords Hemostatic · Chitosan · Biodegradable · Hydrogel · Wound healing · Oxidized cellulose

Abbreviations

RBCs	Red blood cells
GE	Gelatin
MBG	Mesoporous bioactive glass
MCC	Microcrystalline cellulose
HM-CS	Hydrophobically modified chitosan
NRC	Non-regenerated cellulose
RC	Regenerated cellulose
ONRC	Oxidized non-regenerated cellulose
ORC	Oxidized regenerated cellulose
AgNPs	Silver nanoparticles
OCCSF	Oxidized cellulose carboxylate sodium fiber
OMCC	Oxidized microcrystalline cellulose
TOCNs	TEMPO-oxidized cellulose nanofibers
OCNFs	Oxidized cellulose nanofibers
CNCs	Cellulose nanocrystals
APS	Ammonium persulfate

1 Introduction

Plasma, leukocytes, erythrocytes, and platelets make up blood, which accounts for roughly 7–8% of body weight. The human body uses blood for several vital functions, including delivering nutrients and oxygen to various organs, limiting excessive blood loss, and controlling body temperature [27]. The human body cannot function without blood. One of the main reasons for civilian as well as military deaths is hemorrhage. The hemostatic system in human beings is a complicated system with limited capacity. However, hemostatic materials are necessary to save lives in emergency settings like hospitals and battlefields where the body's natural hemostatic mechanism is unable to properly control bleeding [29]. Multiple organ failure, hypotension, hemorrhagic shock, hypothermia, acidosis, and infections are just a few of the serious consequences that can result from excessive bleeding [28]. For quick hemostasis, especially in emergency conditions, the human body's inherent hemostatic process (Fig. 1) has a limited ability and may require help from hemostatic materials or technologies [74]. The most common techniques for halting bleeding in clinical practice are compression with cotton gauze and wound closure with staples or stitches. Numerous hemostatic materials have recently been developed for the marketplace, including chitosan zeolite, collagen, alginate, gelatin, cyanoacrylate, and cellulose [44]. The ideal hemostatic substance should exhibit quick and lasting hemostatic efficacy, biodegradability, biological compatibility, and solid adherence in moist conditions [17].

Cellulose is used in many kinds of biological applications due to its abundant supply as an organic polymer. It is an organic polymer located in plant cell walls and is a neutral homopolysaccharide composed of D-glucopyranose residues connected by β-(1, 4)-glycosidic linkages. Its low toxicity, biodegradability, and biocompatibility, as well as its superior mechanical, chemical, and physical properties, make it more suitable for biomedical applications [75] (Fig. 2).

It has high-water absorption properties. Where bleeding occurs, it concentrates on platelets and clotting factors. Due to its biocompatibility property, cellulose is non-toxic and doesn't provoke any immune response. Modified form of cellulose supports clot formation, creating an acidic environment. Because of its natural antibacterial qualities, oxidized cellulose lowers the possibility of infection. There are various types of cellulose-based hemostatic agents, including nanocellulose-based hemostatic materials, oxidized regenerated cellulose, chitosan-cellulose composites, and microcrystalline cellulose [66]. The agent assimilates plasma, resulting in the concentration of clotting factors and platelets at the injury location. Oxidized cellulose reduces pH, hence activating coagulation factors. It acts as a block and prevents further blood loss. Cellulose-based hemostatic agents significantly reduce bleeding time and are available in various forms, such as pads, sponges, and powders, for quick and easy applications. They are effective across a wide range of wound types and bleeding conditions, and they also help minimize the risk of infection after application. Its applications include controlling intraoperative bleeding, providing first aid for lacerations and severe wounds, and serving as a portable and effective solution for field injuries. This chapter should include a comprehensive overview of cellulose-based hemostatic agents, specifically focusing on their design, mechanisms, and applications in wound care.

2 Design and Types of Cellulose-Based Hemostatic Agents

Various cellulose derivatives are customized for their hemostatic characteristics by specific chemical and physical alterations.

2.1 Chitosan-Cellulose Composites

Chitosan is a widely common heteropolysaccharide derived from the deacetylation of chitin, a naturally occurring polymer that is second only to cellulose in abundance. The biocompatible, biodegradable, cytocompatible, hemocompatible, and adsorption properties of chitosan are of significant interest to researchers. These properties

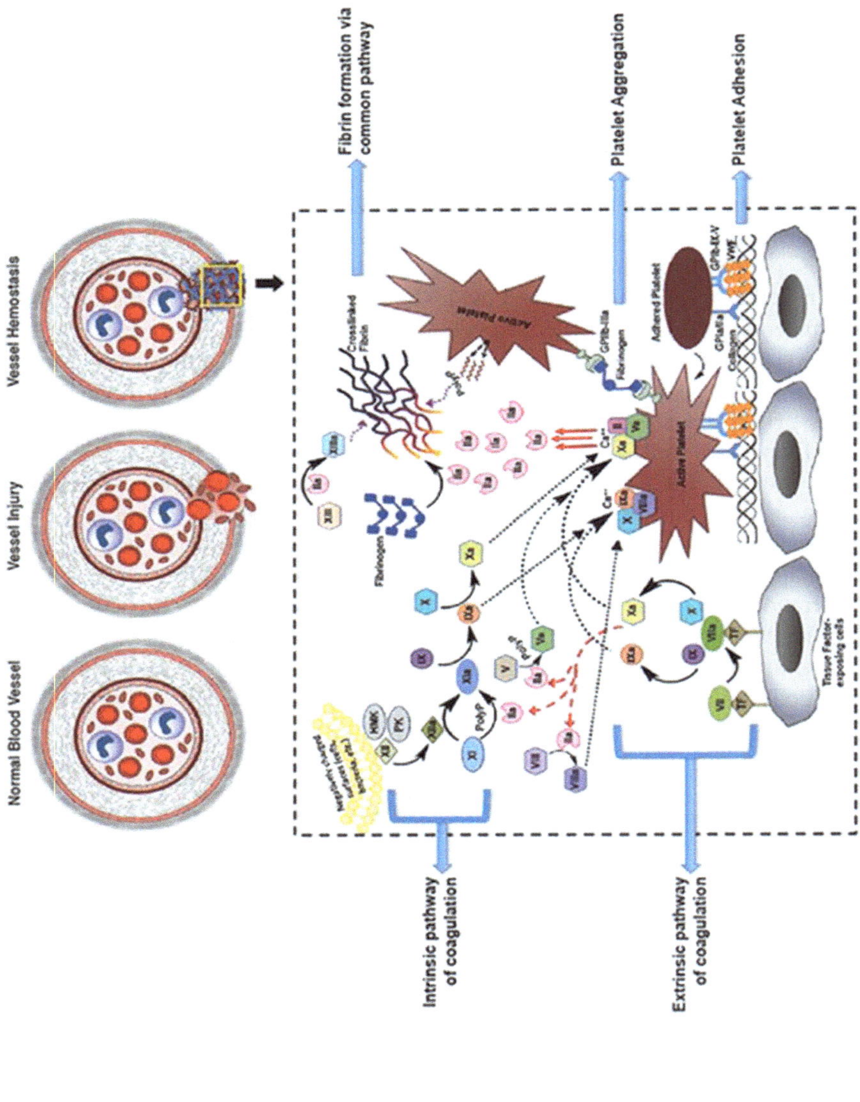

Fig. 1 Schemes of the intrinsic hemostatic mechanisms of the human body [28]

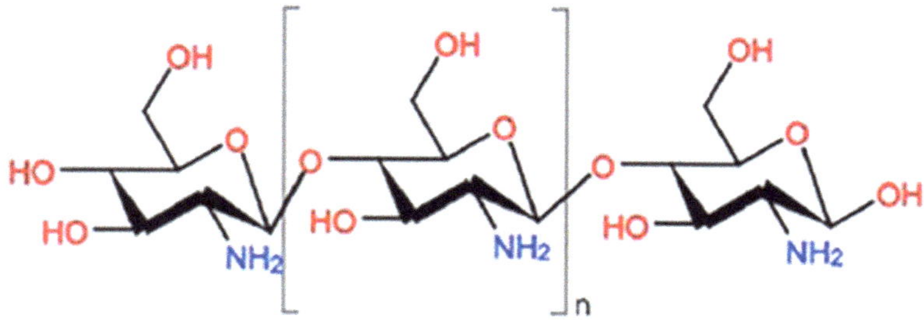

Fig. 2 Chemical structure of cellulose [75]

can be leveraged to develop innovative building blocks with novel functions using Schiff bases [8]. Glucosamine and N-acetylglucosamine units joined by β-(1 → 4) glycosidic linkages form the cationic polymer known as chitosan. Its solubility, mechanical characteristics, and biological activity are all influenced by its degree of deacetylation. Cellulose is a linear polymer made up of D-glucose units connected by β-(1 → 4) bonds. Its crystalline, amorphous, and microfibril sections make up its hierarchical structure [76]. Cellulose is defined by its great tensile strength, chemical stability, and capacity to withstand different chemical changes. It is very compatible with other materials. It makes a suitable matrix for composites. These two polymers are combined using physical or chemical techniques to create chitosan-cellulose composites. Choice depends on the desired applications and properties of the composite. It consists of electrospinning, chemical grafting, physical blending, and hydrogel formation. These composites are biodegradable. The biocompatibility of chitosan-cellulose composites enables their use in biomedical applications such as wound dressings and drug delivery systems. Chitosan's inherent antimicrobial properties are retained in the composites, providing protection against bacterial and fungal infections.

2.2 Hemostatic Mechanism of Chitason

The waterfall theory of blood coagulation, introduced in 1964, established the basis for studying endogenous coagulation pathways. Traditional hemostats promote blood clotting by activating specific components of this theory. However, research has shown that chitosan induces coagulation without triggering the intrinsic pathway, suggesting its hemostatic action operates independently of the classical coagulation cascade. While not yet fully understood, chitosan's hemostatic mechanism involves multiple factors.

2.3 Aggregation of Red Blood Cells

Red blood cells (RBCs), the primary hematocytes in blood, enhance viscosity and facilitate platelet transport to the vascular wall for hemostasis. Their membrane contains negatively charged proteins and glycolipids, preventing aggregation due to electrostatic repulsion [62]. Chitosan, a naturally cationic polysaccharide, carries positively charged $–NH3^{+}$ groups that interact with RBC anions, promoting their aggregation at wound sites to rapidly form clots. The degree of protonation of chitosan's amino groups significantly influences RBC adsorption, with coagulation effectiveness depending on its deacetylation level and the number of protonated amine groups [41]. Additionally, higher molecular weight chitosan strengthens RBC interaction, likely due to increased entanglement and electrostatic forces [35].

2.4 Stimulation of Platelets

Normally, platelets do not adhere to endothelial cells, but their activation and aggregation are crucial for hemostasis during injury. Biopolymers can stimulate this process, which depends on factors like surface chain mobility, chemical composition, hydrogen bonding, charge density, and hydrophobicity [48]. Research suggests that chitosan enhances platelet activation, promoting adhesion and aggregation [31]. Studies indicate that neutralizing chitosan's positive charge reduces platelet adhesion and aggregation, showing that a high positive charge density is essential for its hemostatic effect [18].

2.5 Contact System Activation

When biomaterials encounter blood, they influence coagulation by altering protein functions through adsorption, a process known as the contact activation pathway [78]. This pathway involves coagulation factors XII (FXII), XI (FXI), high-molecular-weight-kininogen, and pre-kallikrein. Contact with functional biomaterials triggers platelet adhesion and activation of the contact system, which together enhance blood coagulation (Fig. 3) [23]. Research indicates that chitosan promotes coagulation in vitro independent of factor VII (FVII) but requires FXI and FXII [37]. Studies also suggest chitosan interacts with plasma proteins and coagulation factors to strengthen blood clots, though some research on its effect on the contact system remains disputed [22].

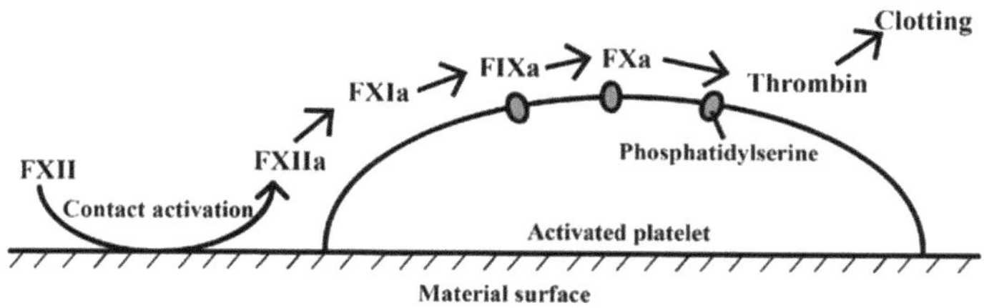

Fig. 3 Simplified scheme of blood clotting reactions on material surface (FXIIa, FXIa, FIXa, and FXa are for activated coagulation factors XII, XI, IX, and X, respectively) [23]

2.6 Development of a Spatial Network Framework

From a molecular perspective, chitosan is a linear glycosaminoglycan, allowing it to form network structures that enhance interactions with blood components and promote clotting. Dowling et al. proposed a gelling hemostatic mechanism, where hydrophobically modified chitosan (HM-CS) self-assembles with blood cells [59]. When HM-CS hydrophobic groups interact with blood cell membranes, they embed into their hydrophobic interiors, creating a three-dimensional gel network. This structure strengthens clot formation and helps stop bleeding (Fig. 4).

2.7 Hemostatic Films Chitosan-Based Composite

Chitosan and gelatin (GE) have become key materials in hemostatic dressings due to their beneficial properties in recent years [16]. Chitosan, a natural alkaline polysaccharide, accelerates blood clotting and promotes wound healing, while GE, a hydrophilic colloid derived from collagen, is widely used in medicine, food, and cosmetics for its unique properties. Li et al. [47] developed ibuprofen-loaded chitosan/GE composite films (Fig. 5) using the solvent casting method, finding that chitosan content influenced tensile strength and elongation. Glutaraldehyde cross-linking enhanced moisture vapor transmission and swelling. Antibacterial tests showed stronger effects against Staphylococcus aureus than *Escherichia coli*. Additionally, cytotoxicity assays confirmed biocompatibility, and hemostatic evaluations demonstrated the films' effectiveness in reducing bleeding during surgery.

For effective hemostasis, selecting appropriate components for powdery hemostats is crucial. Alginate, a natural polysaccharide from brown algae, is widely used in biological carriers and tissue engineering due to its biocompatibility, low toxicity, and affordability. Its strong wound adhesion and high-water absorption make it a valuable hemostatic material [24]. Yunnan Baiyao, a traditional herbal remedy, is an effective surgical sealant but lacks a suitable substrate for direct intraluminal use [14]. Lu et al. [49] developed chitosan/sodium alginate-Yunnan Baiyao composite hemostatic films, which showed superior hemostatic performance in a rat liver bleeding

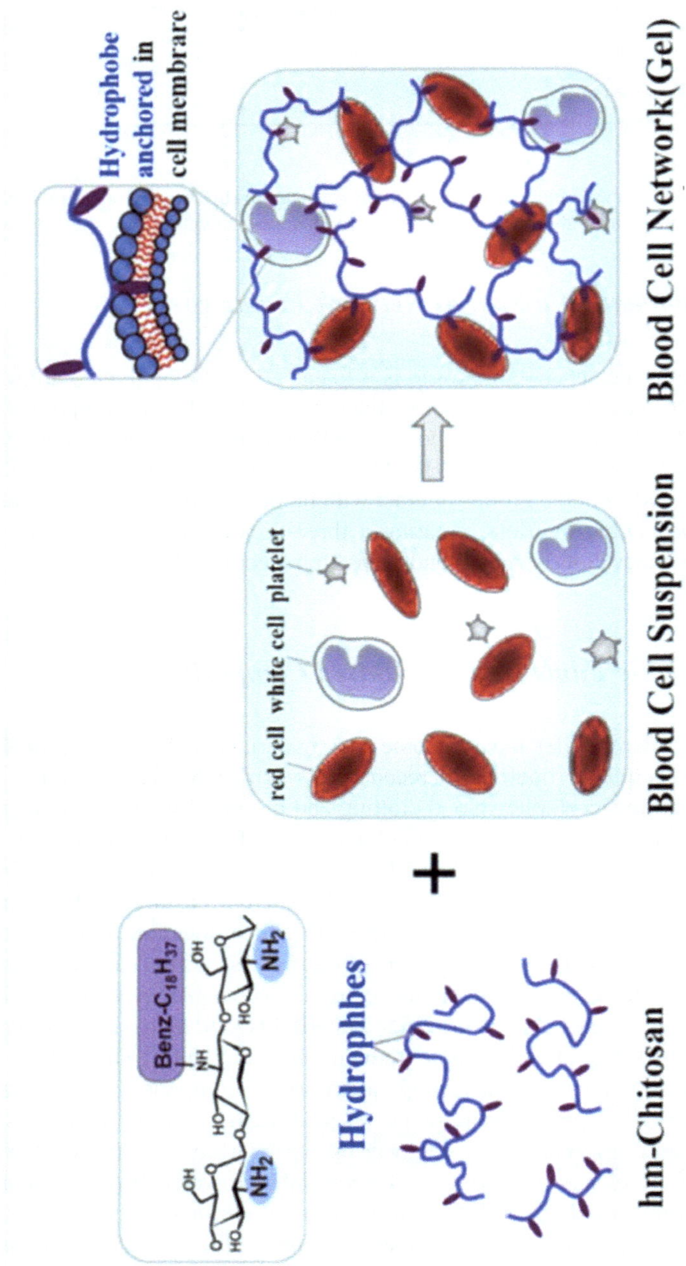

Fig. 4 Mechanism for gelation of blood by hydrophobically modified chitosan [59]

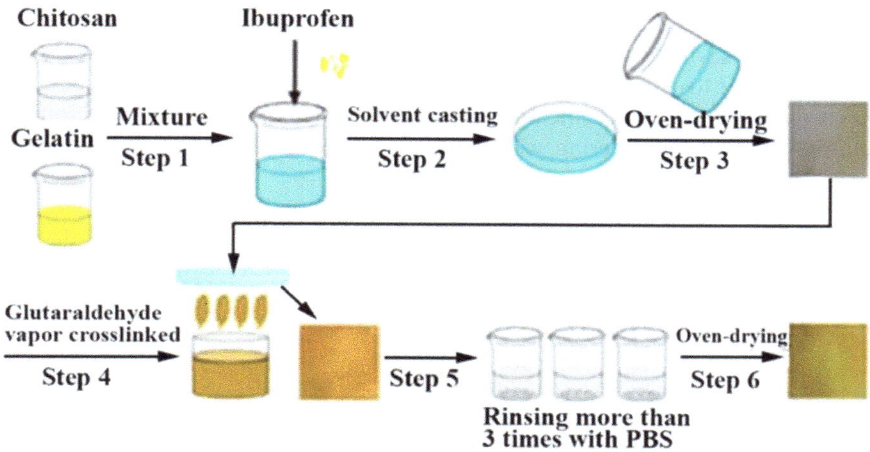

Fig. 5 Preparation scheme for cross-linking of drug-loaded chitosan/GE composite films [47]

model compared to other film and sponge groups [49]. Mesoporous bioactive glass (MBG) is an emerging biomaterial with a uniform nanoscale porous structure, high surface area, and excellent biological activity, making it a promising hemostat [54]. However, some challenges remain in its application. Jia et al. [33] successfully created MBG/chitosan composite porous films through freeze-drying. These highly porous films exhibited strong water absorption, adjustable by modifying MBG-to-chitosan ratios. In rat liver hemorrhage models, increasing MBG content reduced bleeding and clotting time while maintaining good degradability and biocompatibility. Pourshahrestani et al. [57] found that incorporating low concentrations of gallium ions (Ga^{3+}) into MBG improved platelet aggregation, thrombus formation, and overall hemostatic performance. They developed Ga-MBG/chitosan composite scaffolds using lyophilization, which demonstrated superior water uptake, antibacterial properties, and biocompatibility compared to commercial Celox Rapid Gauze. The synergistic effects of chitosan and Ga-MBG made these materials promising candidates for clinical hemostatic applications.

2.8 *Applications for Chitosan-Cellulose-Based Hemostatic Materials*

Chitosan-cellulose composites are widely used as wound dressings due to their hemostatic, antimicrobial, and moisture-retentive properties. They accelerate clot formation, prevent infections, and promote tissue regeneration. These materials are employed in surgical settings to control bleeding during operations. Their biocompatibility and effectiveness make them an attractive alternative to synthetic hemostats. Portable chitosan-cellulose sponges and pads are invaluable in emergency trauma care for controlling bleeding in field conditions. Their ease of use and rapid action are critical for saving lives. Chitosan-cellulose-based hemostatic materials are used in dental surgeries to manage bleeding and protect the surgical site from infections. The biocompatibility and mechanical strength of these composites make them suitable scaffolds for tissue engineering, where controlling bleeding is essential for successful integration.

2.9 Recent Advancements in Chitosan-Cellulose-Based Hemostatic Materials

Nanocellulose and chitosan nanoparticles have been incorporated into composites to enhance their surface area, mechanical strength, and hemostatic efficacy. Nanocomposites exhibit improved platelet adhesion and faster clot formation. Functionalization with bioactive molecules, such as growth factors or antimicrobial agents, has expanded the utility of these materials. Functionalized composites not only control bleeding but also promote wound healing and prevent infections. Advancements in 3D printing technology have enabled the fabrication of custom-shaped hemostatic materials. This approach allows precise control over the material's porosity and architecture, optimizing its performance. Hybrid composites combining chitosan-cellulose with other materials, such as graphene oxide or metallic nanoparticles, have been developed for enhanced hemostatic and antimicrobial properties [49].

3 Oxidized Cellulose

Oxidized cellulose, a significant derivative of cellulose, is produced via the oxidation of natural or regenerated cellulose. Both the physical and chemical characteristics of oxidized cellulose are contingent upon the oxidizing conditions, which include the nature of the agent that oxidizes it. For hemostatic applications, it is most successfully oxidized using nitrogen dioxide gas vapor, resulting in a material with elevated carboxyl content. Typically, oxidized cellulose (6-carboxy cellulose, wherein the alcoholic group on the sixth carbon atom of the anhydroglucose unit is converted to a carboxyl group) having 3–25% carboxyl groups is a suitable and absorbable polymer that can biodegrade within the human body [15].

3.1 Properties of Oxidized Cellulose

Oxidized cellulose possesses many characteristics that make it suitable for hemostatic purposes like biocompatibility, absorbability, antimicrobial activity, and hemostatic efficacy. Oxidized cellulose, a significant derivative of cellulose, is produced via the oxidation of natural or regenerated cellulose. Both the physical and chemical characteristics of oxidized cellulose are contingent upon the oxidizing conditions, which include the nature of the agent that oxidizes it, pH, temperature, and duration of the oxidation process. For hemostatic applications, it is most successfully oxidized using nitrogen dioxide gas vapor, resulting in a material with elevated carboxyl content. Typically, oxidized cellulose (6-carboxy cellulose, wherein the alcoholic group on the sixth carbon atom of the anhydroglucose unit is converted to a carboxyl group) having 3–25% carboxyl groups is a suitable and absorbable polymer that can biodegrade within the human body (Wu et al. 2012).

3.2 Mechanisms of Hemostatic Action

Multiple processes, including a physical barrier, an acidic pH, platelet activation, and thrombin production, are involved in the hemostatic action of oxidized cellulose. Oxidized cellulose acts as a physical barrier when it is applied to a bleeding site. It restricts blood flow and provides scaffold for clot formation. Carboxylic acid groups lower the local pH. It enhances the denaturation of blood proteins and aids clot formation. Platelet adhesion, aggregation, and activation are facilitated by the negatively charged surface of oxidized cellulose. It is essential to start a coagulation cascade. It converts fibrinogen to fibrin. The stabilization of the blood clot results from the interaction of oxidized cellulose with coagulation factors, which speeds up the production of thrombin [63].

3.3 Types of Oxidized Cellulose-Based Hemostatic Materials

Several types of oxidized cellulose are suited to medical requirements. Surgical gauze is made up of soft and malleable sheets that can be used in a wide range of general surgical operations. Fibrous materials are specifically designed for packing irregular wound cavities, while powders and sprays are ideal for use in minimally invasive surgeries or hard-to-reach areas. Composite materials, which combine other hemostatic agents like collagen or chitosan, are formulated to improve overall efficacy [33].

3.4 *Process of Preparing Oxidized Cellulose Through Oxidation*

Cellulose, a widely available biodegradable material, is naturally synthesized by plants and bacteria. It is a neutral homopolysaccharide composed of D-glucopyranose units connected through β-(1,4)-glucosidic bonds [57]. Among its derivatives', oxidized cellulose is particularly valuable and is produced through the oxidation of either natural or regenerated cellulose. For medical applications, particularly in hemostasis, oxidation is most effectively carried out using nitrogen dioxide gas vapor, which results in a material with a high carboxyl content [56]. Typically, oxidized cellulose—also referred to as 6-carboxy cellulose—features an oxidized alcoholic group on the sixth carbon of the anhydroglucose unit, leading to the formation of a carboxyl functional group. This modified cellulose, containing between 3 and 25% carboxyl groups, is recognized as a biocompatible and absorbable polymer that can undergo biodegradation in the human body. Additionally, the rate of degradation and resorption can be adjusted by modifying the oxidation degree and structure of the base cellulose material. Oxidized cellulose with a 6-carboxyl acid content of approximately 16–24% exhibits excellent hemostatic properties, making it a useful agent for bleeding control in various surgical procedures [39]. The oxidation process also increases the acidity of the material due to the high concentration of carboxyl groups. As a result, oxidized cellulose demonstrates antimicrobial activity against numerous Gram-positive and Gram-negative bacteria [77]. Furthermore, the bactericidal efficacy of oxidized cellulose improves with an increased presence of carboxylic acid groups [19]. Generally, different oxidation levels yield oxidized cellulose with modified physicochemical characteristics, influencing its biodegradability, bioabsorbability, and antibacterial properties.

3.5 *Source and Structural Characteristics of Oxidized Cellulose*

The origin and structure of cellulose significantly influence the performance of oxidized cellulose. Oxidation has been studied in both non-regenerated cellulose (NRC) and regenerated cellulose (RC), with NRC fibers having varied diameters and tapered shapes, while RC fibers are more uniform, with lower molecular weight and crystallinity [46]. Despite similar chemical compositions, oxidized non-regenerated cellulose (ONRC) demonstrates superior hemostatic properties compared to oxidized regenerated cellulose (ORC). Additionally, ONRC more effectively inhibits cellular processes involved in wound healing, whereas ORC is preferred for clinical applications due to its oxidation degree and lower cytotoxicity. Bacterial cellulose, while chemically identical to plant cellulose, is purer as it lacks hemicelluloses, lignin, and pectin. Its hydrophilic nature, high permeability, and strong mechanical properties make it suitable for biomedical use. TEMPO-oxidized bacterial cellulose with silver AgNPs (TOBCP/AgNP) has demonstrated strong antibacterial activity against *E. coli* and *S. aureus*, making it a promising wound dressing material. However, while TEMPO-oxidized bacterial cellulose exhibits hemostatic properties, it is not as effective as surgical.

3.6 Safe and Biodegradable Nature of Oxidized Cellulose

Oxidized cellulose is used as hemostatic agents in clinical practice. Research indicates that its biodegradation and absorption at surgical sites occur within a few weeks. Wu et al. [70] examined the breakdown process of oxidized cellulose carboxylate sodium fiber (OCCSF) in vitro, revealing no significant changes in its chemical structure before and after degradation. Over time, OCCSF experiences a decline in polymerization and carboxyl content, suggesting a decarboxylation degradation pathway. Additionally, alterations in fiber color under ultraviolet light indicate an elimination degradation mechanism, supported by studies using 1,2-cyclohexanedione as a model compound. Prior research has confirmed that oxidized cellulose undergoes biosorption through chemical depolymerization and enzymatic hydrolysis [73]. A key benefit of oxidized cellulose is its plant-based origin, eliminating the risk of transmitting viral infections such as HIV. Despite its strong safety record for in vivo use, some adverse effects have been reported. Cases of nerve damage and paralysis have been linked to its use near critical structures like the optic nerve, chiasm, sciatic nerve, and neurocranium. These complications are primarily associated with the material's swelling and acidity, which may lead to nerve compression. However, studies suggest that oxidized regenerated cellulose does not contribute to epidural fibrosis after spinal surgery. To minimize potential risks, it is advisable to remove oxidized cellulose from the surgical site once hemostasis is achieved [6].

3.7 Hemostatic Mechanisms of Oxidized Cellulose

The hemostatic effect of oxidized cellulose operates through both physical and chemical mechanisms. Initially, the physical process begins when oxidized cellulose is applied to a bleeding site. It rapidly absorbs blood fluids, trapping essential blood components such as proteins, platelets, and RBCs [45]. This leads to an increased concentration of coagulation factors, accelerating the clotting process. Following this, oxidized cellulose forms a gel-like "pseudo-clot," acting as a barrier to blood flow and serving as a framework for fibrin clot formation. Chemically, the carboxyl groups present in oxidized cellulose lower the pH, inducing platelet aggregation and facilitating artificial clot formation. Additionally, the acidic environment produced by these groups converts hemoglobin into acid hematin, releasing Fe^{3+} ions that interact with carboxyl groups to further enhance clotting. Despite these known mechanisms, the complete process of hemostasis involving oxidized cellulose-based materials remains not fully understood, and further research is needed to clarify its core functions.

3.8 Various Forms of Oxidized Cellulose-Based Hemostatic Materials

Advancements in material technology have led to the development of oxidized cellulose-based hemostatic materials in different forms, including woven and nonwoven textiles, films, powders, sponges, and hydrogels [10]. Due to the superior tensile strength and flexibility of cellulose fibers, oxidized cellulose is commonly manufactured as gauze, knitted strips, or multilayered filaments. These strips can be tailored to various sizes and shapes, expanding upon contact with blood and forming a clot due to their extensive surface area, which enhances platelet aggregation (Fig. 6) [5]. Woven and nonwoven oxidized cellulose structures are widely used in clinical applications, with studies suggesting that nonwoven ORC achieves hemostasis more rapidly than woven ORC. However, research on the correlation between textile structure and hemostatic efficiency remains limited [20].

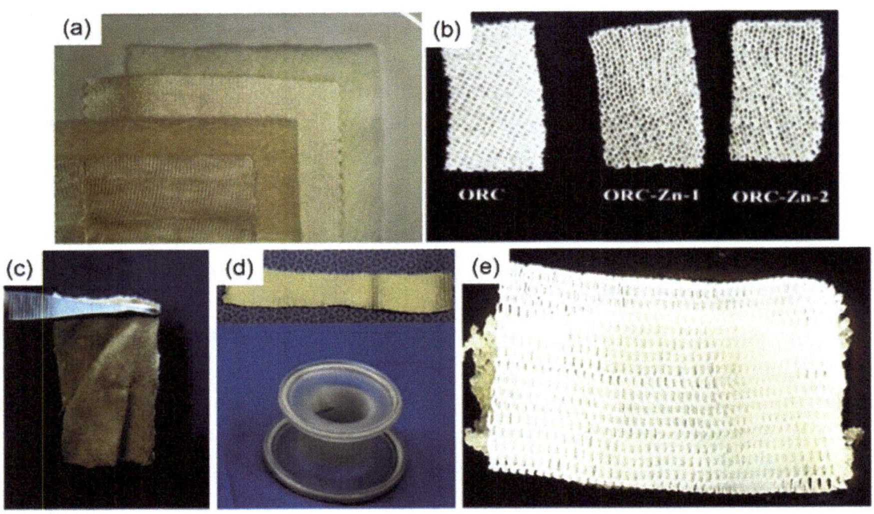

Fig. 6 Images showcasing different structures and applications of oxidized cellulose hemostats: **a** Various woven and nonwoven oxidized cellulose structures. **b** Photographs of oxidized regenerated cellulose (ORC) and zinc-modified oxidized regenerated cellulose gauze (ORC-Zn). **c** Hemostatic experiment using oxidized regenerated cellulose. **d** Oxidized regenerated cellulose sheets cut and rolled around a wound edge protector. **e** Image of OKCEL® H–D, a commercially available oxidized cellulose-based hemostat (Bazghaleh et al. 2016)

Despite their effectiveness, woven and nonwoven oxidized cellulose structures may fail to adhere to surgical instruments, leading to potential rebleeding [46]. To enhance performance, researchers have explored film-based oxidized cellulose materials. For instance, ORC films have been developed using ionic liquids and oxidation systems, demonstrating enhanced hemostatic properties and platelet activation. Furthermore, transparent, hydrophilic films incorporating TEMPO-oxidized cellulose nanofibers (TOCNFs) and calcium phosphate derivatives have been designed for bone surgeries, showing significant blood absorption and clotting potential. However, commercially available oxidized cellulose-based hemostatic films remain scarce [32]

Powdered oxidized cellulose hemostats offer ease of application, making them suitable for controlling bleeding. Researchers have developed potassium and sodium salt forms of ORC, which exhibit bactericidal properties and promote rapid coagulation [38]. Additionally, oxidized cellulose powders can be combined with other hemostatic agents, such as gelatin microparticles, to enhance clotting efficiency. However, powdered hemostats may struggle to adhere effectively to bleeding surfaces, limiting their use for large wounds [10]. To address this, fibrillar ORC powders have been created to form stable blood clots in wet environments [60]. Sponge-based hemostats, known for their porous structure and biocompatibility, offer an effective medium for blood absorption and platelet activation [4]. Recent developments have focused on composite sponges, such as collagen/oxidized microcrystalline cellulose (OMCC) sponges, which enhance intrinsic coagulation pathways. Other innovations include TOCN-alginate sponges, which have demonstrated superior hemostatic efficiency in animal models and biodegrade completely without causing inflammation [68].

Nanotechnology is increasingly being explored for hemostatic and antimicrobial applications in surgery. OCNFs derived from selectively oxidized cellulose are biocompatible, hydrophilic, and capable of forming hydrogels that interact with blood cells to promote clotting [58]. However, their hemostatic efficacy remains relatively low, necessitating the incorporation of additional materials. Researchers have developed OCNF composites with chitosan, thrombin, zinc oxide, and AgNPs, which significantly enhance hemostatic, antibacterial, and mechanical properties. These TOCN-based composite materials hold great promise for future hemostatic applications [11, 12].

3.9 Modification of Oxidized Cellulose-Based Hemostatic Materials

Oxidized cellulose, while widely used as a hemostatic agent, has limitations, such as weak hemostatic properties, especially in severe bleeding or for individuals with blood coagulation disorders like hemophilia. Its low pH (around 3.1) also restricts its use in sensitive tissues such as the nervous and cardiac systems. To overcome these challenges, various modifications have been made to oxidize cellulose-based materials using chemical or physical methods. Neutralizing oxidized cellulose with weak acids or alkali solutions can reduce its cytotoxicity, but this can distort the material's structure and strength [51]. Wu et al. [69, 71] developed a partially water-soluble oxidized regenerated cellulose carboxylate sodium (ORC-Na), which maintains its physical form and tensile strength, showing improved hemostatic properties by forming a gel that seals blood vessels. Surface coating methods, like PEG-coating or chitosan incorporation, have also been explored to enhance hemostatic performance and reduce acidity [69, 71]. Additionally, carbon nanotube functionalization has been used to improve the performance of ORC gauze. Blending or compositing with other materials, such as collagen or gelatin, has shown promise in enhancing the hemostatic properties of oxidized cellulose. For example, collagen/oxidized regenerated cellulose (Col/ORC) composites and gelatin-sodium oxidized cellulose (Gel-NaORC) microparticles have demonstrated excellent hemostatic effects [69, 71]. These modifications not only improve hemostatic efficiency but also enhance biocompatibility, making them suitable for biomedical applications. Furthermore, incorporating thrombin into oxidized cellulose has been shown to improve its hemostatic capabilities, especially for individuals with blood clotting issues [72].

Oxidized cellulose has been developed into various forms such as woven and nonwoven fabrics, films, powders, sponges, and hydrogels [25]. However, oxidized cellulose alone does not have intrinsic hemostatic properties and works by providing a structure that promotes platelet adhesion and aggregation. Its acidic nature can also cause nerve compression and damage. Since single oxidized cellulose does not fully meet the clinical needs, new modifications are being explored. These include adding functional groups or coatings to the surface or combining it with other materials to enhance the coagulation process. These advancements have led to the creation of improved oxidized cellulose-based hemostatic materials, but they are yet to be widely used in clinical practice. The major challenges for these materials' clinical application include the lack of clinical trials to validate their safety and efficacy, low product purity that must meet strict biomedical standards, and the need for specific storage or preparation conditions before use [26]. Addressing these issues will be critical for the future development and clinical adoption of these materials.

3.10 Clinical Applications

In highly vascularized tissues, oxidized cellulose gauze is used to control bleeding. During neurosurgery, it is critical to prevent damage in neural tissues by precise hemostatic control. Oxidized cellulose is frequently used to attain localized hemostasis, avoiding the risk of dislodging. Oxidized cellulose is efficacious in orthopedic interventions. In bone surgeries where bleeding from cancellous bone poses a challenge, oxidized cellulose plays an important role there. For trauma and battlefield injuries, oxidized cellulose powders and dressings allow fast hemostasis in uncontrolled bleeding circumstances. Their lightweight and compact designs enable them to be suitable for incorporation into first-aid kits. Using oxidized cellulose is very common in dental surgery. It manages bleeding from extraction and other oral procedures. Its antibacterial characteristics also assist in reducing postoperative infections [9].

3.11 Advantages, Limitations, and Advances in Oxidized Cellulose Technology

As oxidized cellulose comes from natural sources, there is less chance of allergic responses or toxicity. It is available in different forms and can be applied to various wound types. It is affordable compared to some synthetic agents. It reduces infection risks, which is provided by its antimicrobial properties. It is more sustainable compared to synthetic alternatives as it is biodegradable. Oxidized cellulose also has some limitations too. At massive arterial bleeding, it may not be effective. Its efficacy may be reduced in environments with altered pH. In rare cases, residual oxidized cellulose can cause foreign body reactions. It requires proper storage conditions to maintain its efficacy. Recent progress in material science has improved oxidized cellulose-based hemostatic products. Nanotechnology is incorporated to improve surface area and hemostatic efficiency. For multifunctional applications, gelatin, fibrin, or AgNPs are combined with oxidized cellulose, integrating antibiotics or growth factors for dual hemostatic and therapeutic effects [30].

4 Cellulose Nanocrystal-Based Hemostatic Materials for Wound Dressings

Oxidized cellulose-based hemostatic materials, particularly ORC, are extensively used for controlling internal bleeding (X et al. 2012). ORC is highly effective due to its exceptional blood absorption capacity, which allows it to retain multiple times its weight in blood [21]. Additional benefits of ORC include ease of handling, bactericidal properties, efficient clotting time, and complete degradation within 14–30 days [50]. Additionally, in some cases, ORC left in situ can mimic an abscess in ultrasound imaging, leading to misdiagnosis during postoperative assessments. Recently, nanocellulose-based hemostatic materials have gained attention due to their high surface-to-volume ratio and ability to be integrated into other materials, enhancing their hemostatic performance and wound healing capabilities. Furthermore, nanocelluloses are renewable, widely available, and cost-effective as they can be derived from various biomass sources. Cellulose nanocrystals (CNCs), also referred to as cellulose nanowhiskers, are crystalline nanostructures stabilized by complex hydrogen bonding interactions and dispersed within amorphous cellulose regions. These nanostructures offer excellent biodegradability, biocompatibility, a high elastic modulus, nanoscale dimensions, a high aspect ratio, low density, and an abundance of hydroxyl groups on their surface, making them highly suitable for chemical modifications [36]. Extensive research has been conducted on utilizing cellulose nanocrystal-based materials for wound dressing applications, demonstrating their potential in advancing hemostatic technologies.

4.1 *Sources and Properties of Cellulose Nanocrystals*

CNCs can be derived from numerous sources. The primary sources include lignocellulosic biomass such as wood pulp, agricultural residues, and natural fibers, as well as more refined sources like microcrystalline cellulose (MCC), algal cellulose, and bacterial cellulose [55]. Wood pulp is processed using either the kraft method (sodium hydroxide and sodium sulfide) or sulfite pulping (sulfurous acid salts) to remove lignin from hardwood and softwood. This source yields CNCs with crystallinity ranging from 50 to 83%, with dimensions of 100–300 nm in length and 3–5 nm in diameter [61]. Agricultural waste and natural fibers are gaining attention as eco-friendly and cost-effective alternatives for CNC extraction due to their lower lignin content and loosely bound microfibrils, requiring milder chemical and mechanical treatments. MCC, a purified cellulose form obtained via acid hydrolysis, is commercially available as Avicel®. It is used to produce CNCs through various extraction methods. Algal cellulose, particularly from green algae like Cladophora and Valonia, exhibits a high crystallinity of approximately 95% and possesses unique rosette-like structures that form thin microfibrils. Due to its high structural orientation, it is widely used in biomedical, pharmaceutical, and filtration applications [52]. Bacterial cellulose is synthesized by species such as *Azotobacter*, *Pseudomonas*, *Salmonella*, and *Sarcina ventriculi*. Unlike plant-derived cellulose, BC is free of lignin and hemicellulose, making it a purer form of cellulose. CNCs extracted from BC via acid hydrolysis can achieve high crystallinity depending on reaction conditions.

4.2 *Methods for Preparing CNCs*

Various techniques, including chemical, mechanical, and enzymatic treatments, can be used to extract CNCs by breaking down these disordered regions while preserving the crystalline structure [67]. The amorphous regions are more easily disrupted due to their weaker hydrogen bond interactions. The preparation methods for CNCs are broadly categorized into two main approaches: conventional and green methods, each with distinct processes and advantages.

4.2.1 Conventional CNC Preparation Method

Acid hydrolysis is a widely used technique for extracting CNCs due to its efficiency in breaking C–O bonds in amorphous regions. Protons (H^+) released by the acid interact with glycosidic bonds, leading to chain cleavage and sugar removal in water. Strong acids like sulfuric, hydrochloric, and phosphoric acid can produce CNCs with high crystalline and small sizes. Sulfuric acid improves water dispersion but reduces thermal stability, while hydrochloric acid enhances stability but causes CNC aggregation. Despite its effectiveness, acid hydrolysis is discouraged due to its toxicity, corrosion, high energy use, and environmental risks (Fig. 7) [3, 34].

4.2.2 Green CNC Preparation Methods

The eco-friendly techniques include acid-free approaches such as subcritical and supercritical water hydrolysis, enzymatic hydrolysis, mechanical disintegration, and TEMPO-mediated oxidation (Fig. 8) [55, 65]. Organic acids like citric, maleic, and formic acid offer a sustainable way to extract CNCs with high yield and thermal stability, though they require longer reaction times and high acid concentrations. APS oxidation is another promising method that eliminates the need for lignin and hemicellulose pretreatment, but it has low yields and requires extended reaction times [64]. Subcritical and supercritical water hydrolysis uses high-pressure water as a solvent to break glycosidic bonds, offering a green alternative with lower costs than acid hydrolysis. However, it requires specialized high-pressure reactors, increasing energy demand. Enzymatic hydrolysis, which employs cellulases to break down cellulose into reducing sugars, is a chemical-free and energy-efficient method. However, it has slow reaction rates and requires pre- and post-purification steps. Mechanical treatments like high-pressure homogenization, ball milling, and ultrasonication apply shear forces to break cellulose microfibrils. While they avoid chemical use, they often leave behind amorphous regions, necessitating additional chemical processing for pure CNC isolation.

4.3 Oxidation Treatment of Cellulose

Cellulose is a high-molecular-weight polymer made of 1,4-anhydro-D-glucose units, arranged in microfibrils through strong hydrogen bonding interactions [43]. Due to its rigid structure, native cellulose cannot be broken down into monomers in the human body, requiring chemical modification for biomedical applications. Among these modifications, oxidized cellulose is widely used in wound dressings due to its enhanced coagulation ability, biodegradability, and bactericidal properties.

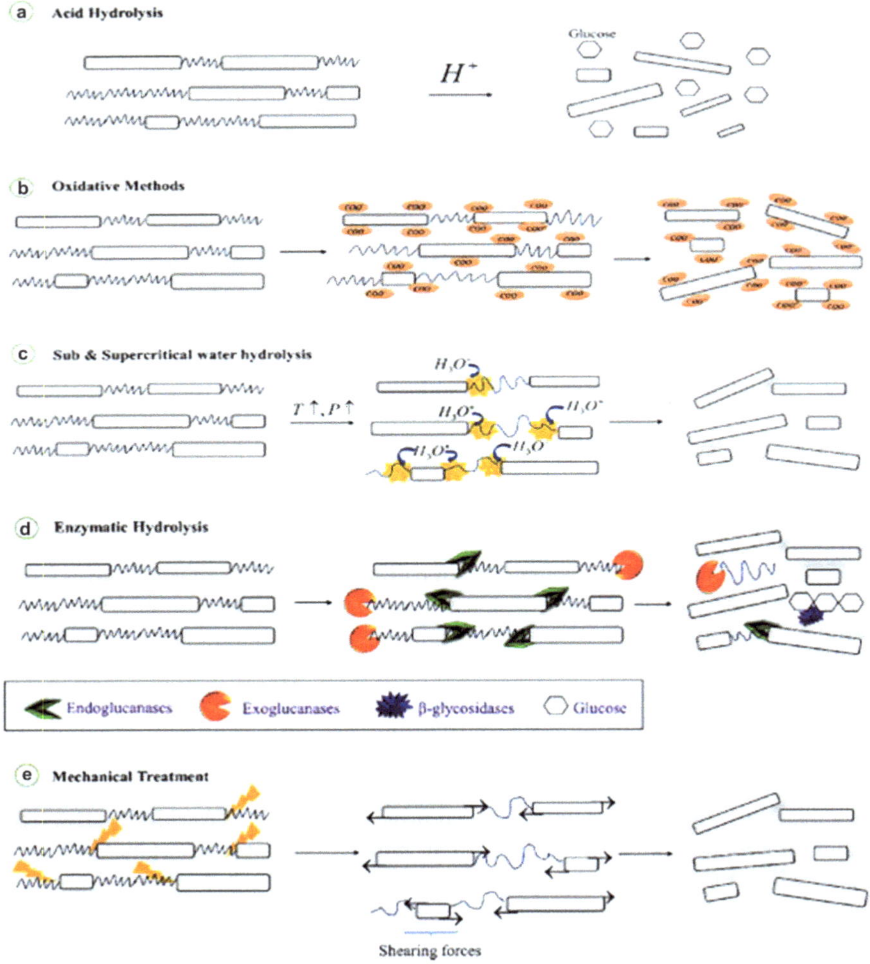

Fig. 7 Schematic representation of different mechanisms used for CNC extraction. **a** Acid hydrolysis-protonated species (H^+) break down amorphous regions, releasing CNCs and degraded sugars. **b** Oxidative methods (e.g., ammonium persulfate (APS) and TEMPO oxidation)—oxidizing agents transform surface hydroxyl groups into carboxyl groups (COO^-), generating repulsive forces that break amorphous chains. **c** Subcritical and supercritical water hydrolysis—High pressure and temperature increase reactive hydronium ion concentration, leading to the breakdown of amorphous regions. **d** Enzymatic hydrolysis—Cellulase enzymes degrade the amorphous phase, releasing crystalline regions. **e** Mechanical treatments—Shearing forces fracture amorphous cellulose chains, isolating CNCs [3, 34]

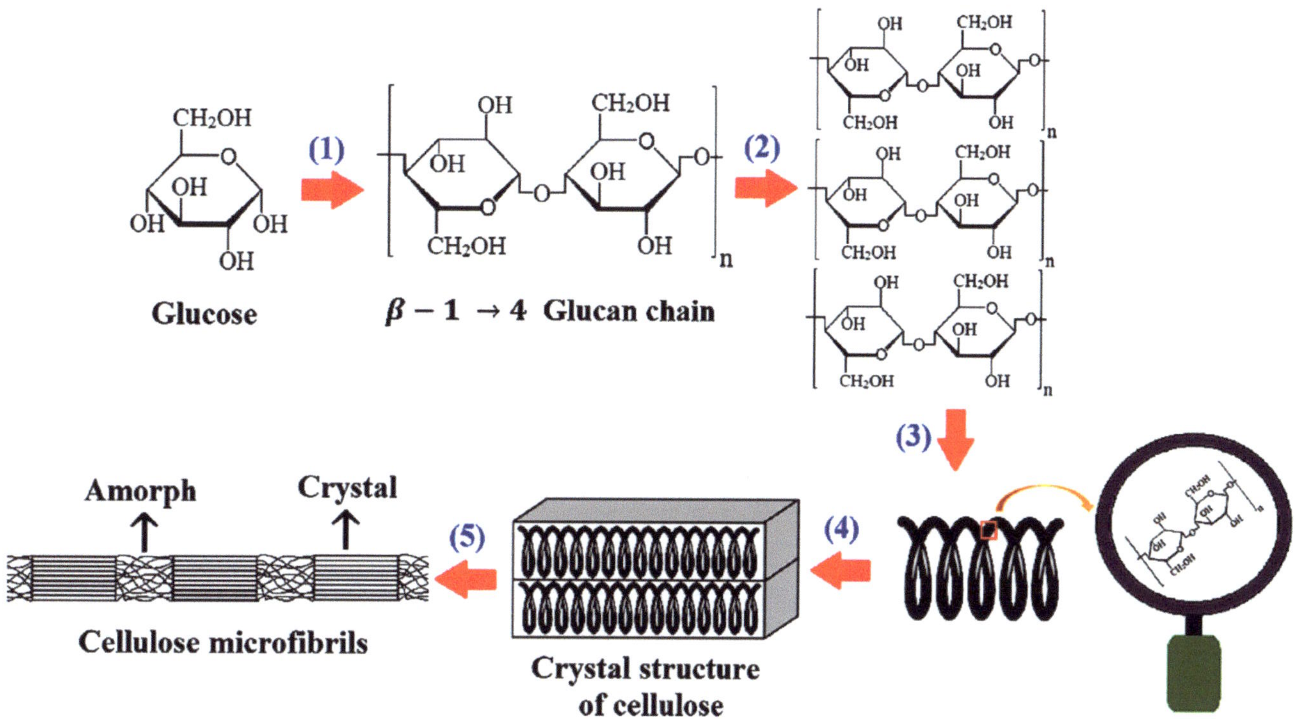

Fig. 8 Illustration of the cellulose synthesis process in various organisms (plants, bacteria, algae, etc.). The steps include: (1) Polymerization of glucose units through β − 1 → 4 linkages. (2) Assembly of glucan chains into cellulose molecules. (3) Formation of thin fibrils through the entanglement of cellulose molecules. (4) Crystallization of cellulose fibrils. (5) Organization of cellulose microfibrils into crystalline and amorphous phases [55]

4.4 TEMPO-Mediated Oxidation

Oxidation introduces carboxyl and aldehyde groups onto the cellulose surface, improving its compatibility, antibacterial activity, and biodegradability. TEMPO-mediated oxidation selectively converts hydroxyl groups at the C6 position into carboxyl groups using NaClO as the oxidant, NaBr, and TEMPO as catalysts under alkaline conditions. This method has been applied to various cellulose sources, including cotton linters, kraft pulp, and bacterial cellulose [77]. The oxidation level depends on factors such as oxidant concentration, reaction time, and cellulose type, influencing biodegradation, hemostatic performance, and antibacterial activity [53]. Methods like conductimetry, NMR, and infrared spectroscopy help quantify carboxyl group content. Studies show that oxidation primarily affects cellulose surfaces without altering its crystalline structure.

4.5 TEMPO-Oxidized Cellulose for Hemostatic Applications

Research comparing oxidized cellulose (OC) and oxidized regenerated cellulose (ORC) indicates that oxidation degree and cellulose type impact hemostatic efficiency and degradation. Higher oxidation levels improve clotting rates, while degradation efficiency varies depending on cellulose structure. Additionally, TEMPO-oxidized cellulose nanocrystals (TOCNCs) have been explored for their improved mechanical and swelling properties in polysaccharide matrices, making them promise for wound dressings [2].

4.6 Nanocomposites with CNCs for Wound Healing

Nanocomposites are materials made by combining a matrix phase with a reinforcing phase in the form of nanoparticles (1–100 nm). These materials enhance properties by integrating small-sized structures. CNCs are widely used as reinforcing agents due to their excellent mechanical strength, ease of chemical modification, high surface area, and renewability [13]. When combined with other biopolymers, CNCs create nanocomposites that improve physical and biological properties, making them ideal for wound healing applications. This section explores common biopolymers used as matrices for CNC-based nanocomposites in wound healing.

5 Conclusions

The primary goal of wound management is to prevent excessive blood loss, minimize infection, and facilitate tissue regeneration. Hemostatic agents play a crucial role in achieving these objectives, particularly in severe wounds where rapid clot formation is essential. Among various biomaterials used in wound healing, cellulose and its derivatives have emerged as highly effective. This study examined different cellulose-based hemostatic agents, including chitosan-cellulose composites, oxidized cellulose-based materials, CNCs, and other bioactive compounds such as chitosan, collagen, gelatin, and alginates. Chitosan-cellulose composites have gained attention due to their synergistic properties. Chitosan, a biopolymer derived from chitin, possesses inherent antimicrobial and hemostatic properties. When combined with cellulose, the resulting composite demonstrates enhanced mechanical strength, fluid absorption, and clotting efficiency. These composites can be used as wound dressings or in surgical applications where rapid blood coagulation is required. This makes chitosan-cellulose composites particularly effective for treating wounds in patients with bleeding disorders.

Oxidized cellulose-based hemostatic materials offer another promising approach to wound healing. The oxidation process enhances cellulose's ability to interact with blood components, increasing clotting speed and efficiency. Oxidized cellulose can be used in the form of gauzes, sponges, or films, making it highly versatile. Additionally, its bactericidal properties help reduce the risk of infection in open wounds. Despite these advantages, oxidized cellulose can sometimes cause irritation or delayed healing in certain patients, necessitating further research into its biocompatibility. CNCs represent a cutting-edge development in hemostatic materials. Due to their nanoscale dimensions, CNCs exhibit a high surface area, which enhances their interaction with blood cells and clotting factors. Additionally, CNCs can be functionalized with bioactive molecules to improve their performance. Their ability to form hydrogels makes them suitable for moist wound healing, which is known to accelerate tissue regeneration. The tunable properties of CNC-based materials make them an exciting avenue for future research and clinical application.

Other bioactive components, such as collagen, gelatin, and alginates, further enhance the wound healing potential of cellulose-based materials. Collagen, a primary structural protein in connective tissue, provides a natural scaffold that supports cell attachment and proliferation. Gelatin, derived from collagen, enhances moisture retention and promotes the migration of fibroblasts, a key cell type in wound repair. Alginates, derived from seaweed, are known for their high absorbency and gel-forming capabilities, making them ideal for treating exudative wounds. In conclusion, cellulose-based hemostatic agents, in combination with other bioactive materials, present a promising approach to wound healing. Their biocompatibility, ability to promote clotting, and potential for customization make them valuable in both clinical and surgical settings. Future research should focus on optimizing these materials for improved efficacy, ensuring they meet the specific needs of different wound types. Additionally, advanced formulations incorporating nanotechnology and bioactive compounds hold great potential for revolutionizing wound care, leading to faster healing and better patient outcomes.

References

1. Abzaeva K, Zelenkov L (2015) Modern topical hemostatic agents and unique representatives of their new generation. Russ Chem Bull 64(6):1233–1239
2. Achneck HE, Sileshi B, Jamiolkowski RM, Albala DM, Shapiro ML, Lawson JH (2010) A comprehensive review of topical hemostatic agents. Ann Surg 251(2):217–228
3. Archana D, Dutta PK, Dutta J (2016) Chitin and chitosan for regenerative medicine. Springer, New Delhi
4. Basagaoglu Demirekin Z, Sezer UA, Ulusoy Karatopuk D, Sezer S (2015) Development of metal ion binded oxidized regenerated cellulose powder as hemostatic agent: a comparative study with in vivo performance. Ind Eng Chem Res 54(18):4906–4914
5. Bazghaleh AA, Dogolsar MA (2019) Preparation of degradable oxidized regenerated cellulose gauze by zinc modification on HNO_3/Cu oxidized viscose fibers. Fibers Polym 20(6):1125–1135
6. Behrens AM, Sikorski MJ, Kofinas P (2014) Hemostatic strategies for traumatic and surgical bleeding. J Biomed Mater Res Part A 102(11):4182–4194
7. Cardoso MJ, Rui RC, Mano JF (2016) Marine origin polysaccharides in drug delivery systems. Mar Drugs 14:34
8. Chai W, Shi Y (2020) Chemical modifications of cellulose and its applications in hemostatic materials. Int J Biol Macromol 164:455–465
9. Chakoli AN, Ahmadi S, He JM, Huang YD (2018) Animated multiwalled carbon nanotubes/oxidized regenerated cellulose nanocomposites for haemostatic applications. Mater Today Proc 5(7):15661–15666
10. Cheng W, He J, Wu Y, Song C, Xie S, Huang Y et al (2013) Preparation and characterization of oxidized regenerated cellulose film for hemostasis and the effect of blood on its surface. Cellulose 20(5):2547–2558
11. Cheng F, Liu C, Wei X, Yan T, Li H, He J et al (2017) Preparation and characterization of 2,2,6,6-tetramethylpiperidine-1-oxyl (TEMPO)-oxidized cellulose nanocrystal/alginate biodegradable composite dressing for hemostasis applications. ACS Sustain Chem Eng 5(5):3819–3828
12. Cheng X, Shao Z, Li C, Yu L, Raja MA, Liu C (2017) Isolation, characterization and evaluation of collagen from jellyfish *Rhopilema esculentum* Kishinouye for use in hemostatic applications. PLoS ONE 12(1):e0169731
13. Conley SP, Littlejohn LF, Henao J, DeVito SS, Zarow GJ (2015) Control of junctional hemorrhage in a consensus swine model with hemostatic gauze products following minimal training. Mil Med 180(11):1189–1195
14. Dai NT, Fu KY, Hsieh PS, Hung YM, Fang YL, Huang NC, Lu TW, Dai LG, Chen SG, Chen TM (2017) A biodegradable hemostatic gelatin/polycaprolactone composite for surgical hemostasis. Ann Plast Surg 78:S124–S128
15. Dai C, Yuan Y, Liu C, Wei J, Hong H, Li X, Pan X (2009) Degradable, antibacterial silver exchanged mesoporous silica spheres for hemorrhage control. Biomaterials 30:5364–5375
16. Dowling MB, Kumar R, Keibler MA, Hess JR, Bochicchio GV, Raghavan SR (2011) A self-assembling hydrophobically modified chitosan capable of reversible hemostatic action. Biomaterials 32:3351–3357
17. Duarte AP, Coelho JF, Bordado JC, Cidade MT, Gil MH (2012) Surgical adhesives: systematic review of the main types and development forecast. Prog Polym Sci 37(8):1031–1050
18. Edwards JV, Howley P, Prevost N, Condon B, Arnold J, Diegelmann R (2009) Positively and negatively charged ionic modifications to cellulose assessed as cotton-based protease-lowering and hemostatic wound agents. Cellulose 16:911–921

19. El-Anwar MW, Nofal AAF, Elmalt A (2017) Surgicel use in control of primary postadenoidectomy bleeding. ENT Ear Nose Throat J 96(9):372–375
20. Elmorsy A, Abdel-Wahab A, Sabry M, Wahby E (2016) Bone wax versus oxidized regenerated cellulose in preventing internal intramedullary bleeding in cardiac surgery. J Egypt Soc Cardio-Thoracic Surg 24(4):302–307
21. Gonzalez ACDO, Andrade ZDA, Costa TF, Medrado ARAP (2016) Wound healing—a literature review. An Bras Dermatol 91:614
22. Hattori H, Amano Y, Nogami Y, Takase B, Ishihara M (2010) Hemostasis for severe hemorrhage with photocrosslinkable chitosan hydrogel and calcium alginate. Ann Biomed Eng 38:3724–3732
23. Hattori H, Ishihara M (2015) Changes in blood aggregation with differences in molecular weight and degree of deacetylation of chitosan. Biomed Mater 10:015014
24. He Q, Gong K, Ao Q, Ma T, Yan Y, Gong Y, Zhang X (2013) Positive charge of chitosan retards blood coagulation on chitosan films. J Biomater Appl 27:1032–1045
25. He J, Wang F, Wu Y, Huang Y, Zhang H (2011) Preparation of the water-soluble chitosan-coated oxidized regenerated cellulose gauze. Cellulose 18(6):1651–1659
26. He JM, Wu YD, Wang FW, Cheng WL, Huang YD, Fu B (2014) Hemostatic, antibacterial and degradable performance of the water-soluble chitosan-coated oxidized regenerated cellulose gauze. Fibers Polym 15(3):504–509
27. Hematology.org (n.d.) Hematology glossary. https://www.hematology.org/education/patients/blood-basics
28. Hickman DSA, Pawlowski CL, Sekhon UDS, Marks J, Sen GA (2018) Biomaterials and advanced technologies for hemostatic management of bleeding. Adv Mater 30(4):1700859. https://doi.org/10.1002/adma.201700859
29. Hong Y, Zhou F, Hua Y, Zhang X, Ni C, Pan D, Zhang Y, Jiang D, Yang L, Lin Q (2019) A strongly adhesive hemostatic hydrogel for the repair of arterial and heart bleeds. Nat Commun 10(1):1–11
30. Hsueh W, Shah-Khan SM, Vallabh H, Cardinal JS, Nasr JY (2017) Endoscopic use of oxidized regenerated cellulose and topical thrombin: a novel approach for control of bleeding pancreatic pseudocyst wall after endoscopic cystogastrostomy. VideoGIE 2(8):193
31. Huang Y, Feng L, Zhang Y, He L, Wang C, Xu J, Wu J, Kirk TB, Guo R, Xue W (2017) Hemostasis mechanism and applications of N-alkylated chitosan sponge. Polym Adv Technol 28:1107–1114
32. Hutchinson RW, George K, Johns D, Craven L, Zhang G, Shnoda P (2013) Hemostatic efficacy and tissue reaction of oxidized regenerated cellulose hemostats. Cellulose 20(1):537–545
33. Jia TB, Chen JY, Feng XX (2011) Fabrication and characterization of chitosan/mesoporous bioactive glasses porous films. J Clin Rehab Tissue Eng Res 15:7877–7880
34. Kargarzadeh H, Mariano M, Gopakumar D, Ahmad I, Thomas S, Dufresne A, Huang J, Lin N (2018) Advances in cellulose nanomaterials. Springer, Netherlands
35. Khan A, Ahmed A (2020) Microcrystalline cellulose: a versatile biopolymer for advanced hemostatic applications. J Biopolym 45(2):123–135
36. Khoshmohabat H, Paydar S, Makarem A, Karami MY, Dastgheib N, Zahraei SAH, Rezaei R, Sadat G, Nezhad M (2019) A review of the application of cellulose hemostatic agent on trauma injuries. Open Access Emerg Med 11:171
37. Klokkevold PR, Fukayama H, Sung EC, Bertolami CN (1999) The effect of chitosan (poly-N-acetyl glucosamine) on lingual hemostasis in heparinized rabbits. J Oral Maxillofac Surg 57:49–52
38. Ko T-M, Lin J-C, Cooper SL (1993) Surface characterization and platelet adhesion studies of plasma-carboxylated polyethylene. J Colloid Interface Sci 156(1):207–217
39. Kordestani SS, Rezaimanesh A, NayebHabib F (2016) The use of oxidized regenerated cellulose (ChitoCell®) in open surgeries. In: 5th international conference on science and technology
40. Kumar S, Dutta PK, Sen P (2010) Preparation and characterization of optical property of crosslinkable film of chitosan with 2-thiophenecarboxaldehyde. Carbohydr Polym 80:564–570

41. Kumar S, Koh J (2014) Physiochemical, optical and biological activity of chitosan-chromone derivatives for biomedical applications. Int J Biol Macromol 65:375–382
42. Kumar S, Nigam N, Ghosh T, Dutta PK, Yadav RS, Pandey AC (2010) Preparation, characterization, and optical properties of a chitosan-anthraldehyde crosslinkable film. J Appl Polym Sci 115:3056–3062
43. Lamichhane S, Anderson JA, Remund T et al (2016) Responses of endothelial cells, smooth muscle cells, and platelets dependent on the surface topography of polytetrafluoroethylene. J Biomed Mater Res Part A 104(9):2291–2304
44. Lan G, Lu B, Wang T, Wang L, Chen J, Yu K et al (2015) Chitosan/gelatin composite sponge is an absorbable surgical hemostatic agent. Colloids Surf B Biointerfaces 136:1026–1034
45. Levy M (2013) Clinical experience with the Surgicel family of absorbable hemostats (oxidized regenerated cellulose) in neurosurgical applications. Wounds 25(6):160–167
46. Lewis PB, Wilson ST, Kentala DR, Barry J, Lewis KM (2016) Computed tomographic characterization of trauma stem—a new oxidized cellulose hemostatic agent. Tomography 2(3):175
47. Li H, Cheng F, Gao S, Wu Z, Dong L, Lin S, Luo Z, Li X (2017) Preparation, characterization, antibacterial properties, and hemostatic evaluation of ibuprofen-loaded chitosan/gelatin composite films. J Appl Polym Sci 134:1–9
48. Li Z, Wang X (2019) Hemostatic efficacy of cellulose-based materials: a comprehensive review. Biomater Sci 7(4):678–692
49. Lu B, Hu S, Zhu X, Zhang LT, Zhang ZL (2011) Experimental study of hemostasis effect of chitosan/sodium-alginate-yunnanbaiyao composite film. Health Res 31:94–96
50. Moeini A, Pedram P, Makvandi P, Malinconico M, d'Ayala GG (2020) Wound healing and antimicrobial effect of active secondary metabolites in chitosan-based wound dressings: a review. Carbohydr Polym 233:115839
51. Nguyen THM, Abueva C, Van Ho H, Lee S-Y, Lee B-T (2018) In vitro and in vivo acute response towards injectable thermosensitive chitosan/TEMPO-oxidized cellulose nanofiber hydrogel. Carbohydr Polym 180:246–255
52. Oksman K, Aitomäki Y, Mathew AP, Siqueira G, Zhou Q, Butylina S, Tanpichai S, Zhou X, Hooshmand S (2016) Review of the recent developments in cellulose nanocomposite processing. Compos Part A Appl Sci Manuf 83:2
53. Park JU, Song EH, Jeong SH, Song J, Kim HE, Kim S (2018) Chitosan-based dressing materials for problematic wound management. In Novel biomaterials for regenerative medicine, pp 527–537
54. Pinkas O, Zilberman M (2017) Novel gelatin-alginate surgical sealants loaded with hemostatic agents. Int J Polym Mater 66:378–387
55. Piozzi GN, Reitano E, Panizzo V, Rubino B, Bona D, Tringali D, Micheletto G (2018) Practical suggestions for prevention of complications arising from oxidized cellulose retention: a case report and review of the literature. Am J Case Rep 19:812–819
56. Pourshahrestani S, Zeimaran E, Kadri NA, Gargiulo N, Jindal HM, Naveen SV, Sekaran SD, Kamarul T, Towler MR (2017) Potency and cytotoxicity of a novel gallium-containing mesoporous bioactive glass/chitosan composite scaffold as hemostatic agents. ACS Appl Mater Inter 9:31381–31392
57. Pourshahrestani S, Zeimaran E, Kadri NA, Gargiulo N, Samuel S, Naveen SV, Kamarul T, Towler MR (2016) Gallium-containing mesoporous bioactive glass with potent hemostatic activity and antibacterial efficacy. J Mater Chem B 4:71–86
58. Qiu Y, Qiu L, Cui J, Wei Q (2016) Bacterial cellulose and bacterial cellulose baccarin membranes for wound healing. Mater Sci Eng C 59:303–309
59. Sagnella S, Kwok J, Marchant RE, Kottke-Marchant K (2001) Shear-induced platelet activation and adhesion on human pulmonary artery endothelial cells seeded onto hydrophilic polymers. J Biomed Mater Res 57:419–431
60. Schonauer C, Tessitore E, Moraci A, Barbagallo G, Albanese V (2005) The use of local agents: bone wax, gelatin, collagen, oxidized cellulose. In: Haemostasis in spine surgery. Springer, pp 89–96

61. Seta FT, An X, Liu L, Zhang H, Yang J, Zhang W et al (2020) Preparation and characterization of high yield cellulose nanocrystals (CNC) derived from ball mill pretreatment and maleic acid hydrolysis. Carbohydr Polym 234:115942
62. Shi B, Shen Z, Zhang H, Bi J, Dai S (2012) Exploring N-imidazolyl-O-carboxymethyl chitosan for high performance gene delivery. Biomacromolecules 13:146–153
63. Shruti S, Salinas AJ, Lusvardi G, Malavasi G, Menabue L, Vallet-Regi M (2013) Mesoporous bioactive scaffolds prepared with cerium-, gallium- and zinc-containing glasses. Acta Biomater 9:4836–4844
64. Singla R, Soni S, Patial V, Kulurkar PM, Kumari A, Mahesh S, Padwad YS, Yadav SK (2017) In vivo diabetic wound healing potential of nanobiocomposites containing bamboo cellulose nanocrystals impregnated with silver nanoparticles. Int J Biol Macromol 105:45
65. Varaprasad K, Jayaramudu T, Kanikireddy V, Toro C, Sadiku ER (2020) Alginate-based composite materials for wound dressing application: a mini review. Carbohydr Polym 236:116025
66. Vasconcelos GF et al (2020) Chitosan-cellulose composite hemostatic agents. Carbohydr Polym 231:115735
67. Vasconcelos NF, Feitosa JPA, da Gama FMP, Morais JPS, Andrade FK, de Souza MDSM, de Freitas Rosa M (2017) Bacterial cellulose nanocrystals produced under different hydrolysis conditions: properties and morphological features. Carbohydr Polym 155:425–431
68. Wang AY, Rafalko J, MacDonald M, Ming X, Kocharian R (2017) Absorbable hemostatic aggregates. ACS Biomater Sci Eng 3(12):3675–3686
69. Wu Y, He J, Cheng W, Gu H, Guo Z, Gao S et al (2012) Oxidized regenerated cellulose-based hemostat with microscopically gradient structure. Carbohydr Polym 88(3):1023–1032
70. Wu Y, He J, Cheng W, Gu H, Guo Z, Gao S et al (2019) Study of the degradation mechanisms of oxidized cellulose-derivative hemostatic agent by using model compound. J Biomater Sci Polym Ed 30(7):580–591
71. Wu YD, He JM, Huang YD, Tang F, Wang FW (2012) Investigation on degradation and stability of oxidized regenerated cellulose. Fiber Polym 13(5):582–586
72. Xu Y, Qiu C, Zhang X, Zhang W (2014) Crosslinking chitosan into H_3PO_4/HNO_3–$NANO_2$ oxidized cellulose fabrics as antibacterial-finished material. Carbohydr Polym 112:186–194
73. Yan T, Cheng F, Wei X, Huang Y, He J (2017) Biodegradable collagen sponge reinforced with chitosan/calcium pyrophosphate nanoflowers for rapid hemostasis. Carbohydr Polym 170:271–280
74. Yang X, Liu W, Li N, Wang M, Liang B, Ullah I et al (2017) Design and development of polysaccharide hemostatic materials and their hemostatic mechanisms. Biomater Sci 5(12):2357–2368
75. Yu P, Zhong W (2021) Hemostatic materials in wound care. Burns Trauma 9:tkab019. https://doi.org/10.1093/burnst/tkab019
76. Zhang Y, Tao L, Li S, Wei Y (2011) Synthesis of multiresponsive and dynamic chitosan-based hydrogels for controlled release of bioactive molecules. Biomacromolecules 12:2894–2901
77. Zhang YP, Wang Y, Chen L, Zheng J, Fan XJ, Xu XL et al (2022) An injectable antibacterial chitosan-based cryogel with high absorbency and rapid shape recovery for noncompressible hemorrhage and wound healing. Biomaterials 285:121546
78. Zhou X, Zhang X, Zhou J, Li L (2017) An investigation of chitosan and its derivatives on red blood cell agglutination. RSC Adv 7:12247–12254

Asif Karim Neon is a dedicated mechanical engineer with a strong academic background and practical experience in the engineering field. He holds a Bachelor of Science degree in Mechanical Engineering from Khulna University of Engineering & Technology (KUET), Bangladesh. Currently he is working as a Junior Executive Engineer at the Rampal Thermal Power Plant, under Bharat Heavy Electricals Limited (BHEL, EPC Contractor), Bangladesh.

Dr. Md. Nizam Uddin is an Assistant Professor of Mechanical Engineering, at Texas A&M University-Texarkana, TX, USA. His research interests are design and advanced manufacturing, nanomaterials, polymer-based nanocomposites, and renewable energy and environmental systems. Throughout his studies, he has published more than 80 peer reviewed articles and conference proceedings. Received 20 honors/awards, and reviewed several manuscripts in international journals and conference proceedings. He has been serving editorial board member; guest editor of several journals.

Md. Aliahsan Bappy currently pursuing an M.Eng. in Mechanical Engineering at Lamar University, Texas, USA, is focused on applying machine learning and nanoelectronics processes to optimize renewable energy technologies. With manufacturing and quality management expertise, he aims to advance solar cell efficiency by developing novel materials and enhancing layer interfaces to reduce energy loss. His research interests are machine learning, product design, development and modification, nanoparticles, energy-efficient engines, solar energy, etc.

Dr. Md. Mahbubur Rahman is currently an Associate Professor in the department of Mechanical Engineering at Khulna University of Engineering & Technology (KUET), Bangladesh. He earned his PhD from the department of Science and Advanced Technology of Saga University, Japan. He has published numerous articles in internationally recognized journals, contributing to the academia community. His research focuses include, but are not limited to, thermal engineering, energy science, materials science and manufacturing processes.

Md Moniruzzaman is currently pursuing a Master of Engineering in Industrial Engineering at Lamar University, TX USA. His expertise includes mechanical product development, thermal systems, and manufacturing process optimization, with hands-on experience in CAD modeling. His contributions also include the development of advanced thermal imaging systems, robotics integration in manufacturing, and Lean Six Sigma-driven process improvements.

Cellulose in 3D Printing for the Development of Biomedical Devices

Deepak Kumar, Rishabha Malviya, Sonali Sundram, and Sathvik Belagodu Sridhar

Abstract Three-dimensional (3D) printing has transformed biomedical applications, offering precise control over complex structures. Among various biopolymers, cellulose has gained significant attention because of its renewability, biocompatibility, and tunable features. Cellulose-based bio-inks exhibit favorable mechanical stability and printability, making them good contenders for wound healing, tissue engineering (TE), and drug delivery. Study aims to explore the improvements in cellulose-based 3D printing, highlighting its biomedical uses. Recent developments in cellulose-derived bio-inks have enhanced their printability and functional performance. Nanocellulose, cellulose derivatives, and composite hydrogels have been utilized to fabricate scaffolds with improved biocompatibility and mechanical integrity. TE applications have established the potential of cellulose-based constructs in regenerating bone, cartilage, and wound tissues. Additionally, 3D-printed cellulose hydrogels have been explored for drug delivery, enabling sustained as well as targeted release of therapeutics. Boundaries like poor mechanical strength, scalability issues, and complex post-processing techniques hinder widespread adoption despite these advancements. Emerging technologies like 4D printing and multi-material bioprinting hold promise for overcoming these challenges. Cellulose-based 3D printing signifies a sustainable and advanced method for application in the biomedical field. Further research on bio-ink formulations, crosslinking strategies, and high-resolution printing techniques is essential to unlock its full potential for clinical and industrial use.

Keywords Cellulose bio-ink · Hydrogel scaffolds · Regenerative medicine · Tissue engineering · Extrusion-based printing · Cellulose derivatives · Scaffold fabrication

D. Kumar · R. Malviya (✉) · S. Sundram
Department of Pharmacy, School of Medical and Allied Sciences, Galgotias University, Greater Noida, U.P., India
e-mail: rishabhamalviya19@gmail.com

S. B. Sridhar
RAK College of Pharmacy, RAK Medical & Health Sciences University, Ras Al Khaimah, UAE

R. Malviya and S. Sundram (eds.), *Sustainable Cellulose Materials for Biomedical Applications*, Sustainable Materials and Technology,
https://doi.org/10.1007/978-981-96-9904-9_20

Abbreviations

2D	Two-dimensional
3D	Three-dimensional
4D	Four-dimensional
AgNPs	Silver nanoparticles
Alg	Alginate
AM	Additive manufacturing
APC	Acetone and acetic acid
APIs	Active pharmaceutical ingredients
BC	Bacterial cellulose
CAD	Computer-aided design
$CaCO_3$	Calcium carbonate
CNC	Cellulose nanocrystals
DDS	Drug delivery system
DLP	Digital light processing
DIW	Direct ink writing
EBP	Extrusion-based printing
FDM	Fused deposition modeling
GelMA	Gelatin methacrylate
HA	Hydroxyapatite
HPMC	Hydroxypropyl methylcellulose
IB	Ibuprofen
LDM	Liquid deposition modeling
MCC	Microcrystalline cellulose
MEX	Material extrusion
NCC	Nanocrystalline cellulose
OPEFB	Oil palm empty fruit bunch
PCL	Polycaprolactone
REPs	Regenerative endodontic procedures
SLA	Stereolithography
SL	Stereolithography
STL	Standard tessellation language
TE	Tissue engineering
TEMPO	2,2,6,6-Tetramethylpiperidine-1-oxyl
TPP	Two-photon polymerization
UV	Ultraviolet

1 Introduction

Three-dimensional (3D) printing, also called additive manufacturing, is a technique that has involved a lot of interest in several domains, including biomedical, aerospace, and structural applications [5]. This method utilizes spatial or layer-by-layer assembly objects with precise measurement [140]. It gets rid of the requirement for lithographic masks, molds, and dies. These innovative manufacturing processes may utilize a diversity of resources, counting metals, alloys, polymers, as well as ceramics, to produce affordable, personalized, and on-demand items with high structural complexity [64, 82, 89].

Over the past several years, there have been appropriate developments in the adaption of 3D bioprinting with drug delivery and cell culture. One advantage of additive manufacturing, including 3D bioprinting, is that technology may be used for producing using live materials and cells [70, 108]. There has been some interest in 3D bioprinting because of the applications of natural items from microorganisms and plants, including lignin, cellulose, alginate, as well as algae. All of these nutrients are readily available from particular plant and animal components, but the extraction procedure involves certain steps [101]. The understanding of chemical and processing technologies of cellulose has advanced greatly [158]. Cellulose, which is derived from plants, bacteria, fungi, and algae. Cellulose has been used as a chemical raw ingredient for more than 150 years worldwide [118]. It has various utilizations, especially as a sustainable substitute for synthetic fibers to be used in 3D Printing [129]. This is because it has several attractive qualities, including being renewable and ecologically friendly, having distinct physicochemical features (such as structure, type, or derivative), being widely available, and being inexpensive [41]. Cellulose along with its derivatives can be utilized to develop hydrogels having desired designs and characteristics, depending on the source, procedure, and modification. Furthermore, cellulose is a great electrical insulator, non-allergic, and heat-resistant. It may be treated to develop pertinent 3D printed structures and microarchitectures. As a result, 3D printed cellulose has an extensive variety of potential uses in various fields [153].

Even though cellulose along with its derivatives have not yet been commercialized for direct public use due to several obstacles, including the need for appropriate solvent development and the fact that cellulose cannot be thermally processed, 3D printed cellulose technology has a lot of potential and has been extensively researched for use in a variety of fields [10, 29]. Numerous properties of cellulose which are used for 3D printing are illustrated in Fig. 1.

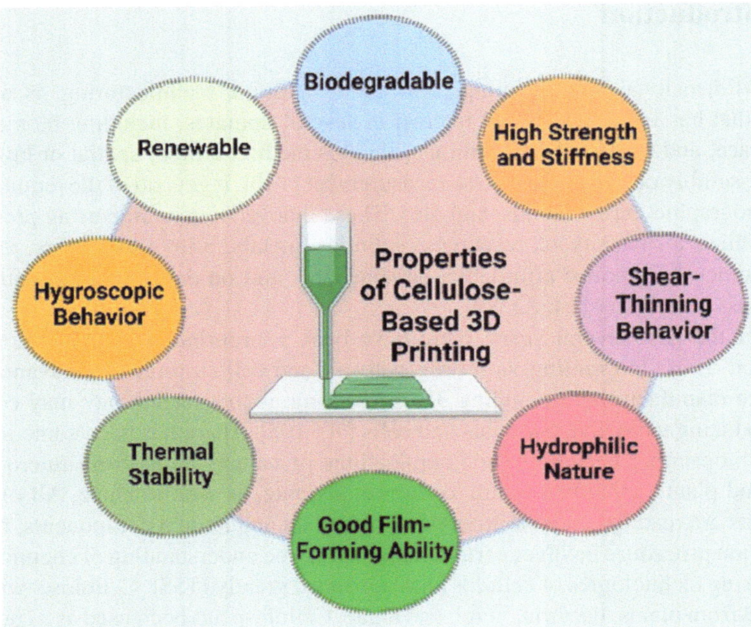

Fig. 1 Various properties of cellulose appropriate for 3D printing

2 Cellulose Derived Bio-ink

Ingredients for 3D bioprinting must be stable in long-term as well as short-term since they will be shared with cells. The long-term durability of materials depends on their biocompatibility. The bio-ink must be biocompatible during the manufacturing procedure till the required phase is attained [92]. Short-term stability is vital to preserve the legitimacy of the item at the first stage by guaranteeing that the appropriate tissue characteristics, stay complete until printing is finished. Before being used in 3D bioprinting to create intricate 3D structures, bio-inks need to have a range of printing and structural characteristics. An ideal bio-ink should have all the necessary physiochemical properties [150]. When it comes to 3D bioprinting, cellulose has become one of the most often used materials for bio-inks. It is a great material for 3D bio-printing because of a number of its intrinsic features [88]. In comparison to 3D bioprinting of cellulose, 3D printing of cellulose has seen substantial growth in the previous five years. A significant difference exists between these two words because 3D bioprinting using cellulose is still very new and has not been well studied [70]. Moreover, in binary ink preparations, they may be printed more, often when combined with another biopolymer, such as gelatin and algae. They are most commonly used as a rheological convertor in these formulations to enhance the printing ability of the bio-ink, extrudability, and shape fidelity [76, 87, 113, 121, 164].

As a result, when 3D bioprinting emerged, scientists worked to create the best bio-ink possible using a variety of materials. These physicochemical and biological characteristics gave cellulose-based bioinks a boost in research.

2.1 Development of Polymeric Cellulose for Bio-ink

Bio-inks based on polymeric along with functionalized cellulose were assessed for their potential use in a diversity of TE applications. Habib et al. assessed a new amalgam hydrogel made of Alg along with CMC in the beginning of 2018 [54]. Additionally, different concentrations were assessed to confirm their cell survival, printability, and form integrity. The shear thinning nature is additionally pronounced in the case of CMC/Alg, indicating a higher likelihood of bio-printing the therapeutically important amalgam hydrogels. The findings show that Alg-CMC sustains cell viability and keeps its shape. In a related study, Morelli et al. used two market grades of cellulose and algae to bioprint umbilical cord matrix stem cells [106]. Commercial grades are chosen because they enhance the rheology and durability of premium industrial paints, water-based paints, and architectural paints and coatings. For 21 days, the in vitro culture was kept alive. The initial findings demonstrated appropriate cell vitality and uniform cell distribution. Passamai et al. used pectin, MCC, and CMC to create a bio-ink. The best square linked channels are shown in pectin/MCC bio-ink (Printability $\approx$ 1). Its potential application as a customized dressing of wound was established by the results [116]. Similarly, different concentrations of aqueous hydrogels depending on CMC were lyophilized and printed. A shear-thinning effect was seen in all combinations of aqueous hydrogels, meaning that as the shear rate augmented, the apparent viscosity reduced. The intestinal drug transport capabilities of the hydrogels were assessed. The findings displayed that the lyophilized hydrogels completely dissolved after 24 h. Non-lyophilized CMC aqueous hydrogel capsules, on the other hand, displayed varying degrees of dissolvability during 24 h [68]. More research is required to ascertain a broader appropriateness in the pharmaceutical industry. Among other methods for using cellulose, Shamsabadi et al. created a method that utilizes EBP to syndicate the xurography of a scaffold made of cellulose [127]. This method was used to generate 3D bioprinted in vitro models by layer-by-layer assembly. Through a suitable 3D distribution of cells, extracellular matrix, and other substances, this innovative technique helps to recreate the intricacies of real tissues while providing the entire structure with mechanical stability as well as direct, informal accessibility to cells.. This approach of the study is fascinating and might be investigated further for usage as an ex vivo model in delivery of drug applications. In a different work, Mohamed et al. created cellulose-based bio-ink functionalized with alkyne and azide using click chemistry [103]. EBP investigating successfully guided the two solutions into Cu (I)/DMSO coagulant solutions and validated the potential use of the clickable cellulose byproducts as precursors to bio-ink. The physical as well as chemical properties of click-gel were identified. The human dermal fibroblast cells (HFB4) showed greater vitality in the biocompatibility examination of the ready clickgels at a concentration of 100 μg/mL [65]. To produce a bio-ink made from TEMPO-oxidized BC, Wei et al. functionalized the cellulose. The bio-ink that is created contains laponite nanoclay, TEMPO-oxidized BC, and algae. As the shear rate varies from 0.01 to 1000 s^{-1}, the generated bio-ink's viscosity decreases, suggesting that non-Newtonian fluids have a shear-thinning impact. The findings of the research demonstrated structural stability as well as release of long-term protein. The 3D bioprinted hydrogel may be a desirable choice for drug release applications because of its release behavior. Ji et al. used two methods to functionalize CMC with norbornene in a different investigation [159]. The first step is to use amide norbornene, followed by carbic functionalized norbornene. According to rheological study, hydrogels made using the aforementioned techniques demonstrated auto-gelation in the absence of UV light. After being exposed to UV light, these bio-inks quickly create cell-rich hydrogels because of the photoinduced thiol-ene crosslinking process. To prove the use of these potential bio-inks in tissue engineering, more research is required.

The mechanical, along with structural stability of polymeric hydrogels is still a challenge; Erkoc et al. combined cellulose with popular gelatin and algae-based bio-ink to address this [37]. According to rheological data, adding cellulose improved the mechanical qualities and made the bioink more printable. According to the results, the presence of cellulose and Alg at the ideal quantities increased the material's durability, homogeneity, and viscoelasticity. It also made 3D printed scaffolds more printable sideways while maintaining form reliability. To determine if these bioprinted scaffolds can be used in TE and medical prosthesis, more research is necessary. Similarly, Gospodinova et al. blended different doses of Alg with hydroxyethylcellulose (HEC) to create HeLa cells that integrated the cervical carcinoma model [50]. Extrusion printing was used to print the constructions. The concentration of Alg is correlated with the viscosity vs. shear speeds. Extrusion printing has an impact on cell survival, as shown by the stress study performed on cells at various stages. The findings showed that HEC helps to enhance molecular permeability, printability, and cell adhesion [85]. In recent years, multicomponent bioinks have received increasing study attention. Gelatin and algae-based bio-inks were examined for their potential in TE [147]. Bioprinted hydrogels have superior cytocompatibility and hemocompatibility, in addition to better shear thinning and viscoelastic characteristics. Geevarghese et al. established a tissue construct using gelatin, algae, cellulose, and collagen peptide. Rheological research shows that the higher the shear rate reduces viscosity, indicating that the created formulation has a shear thinning effect. The printed tissue construct was tested for drug screening purposes. The positive results indicate that multicomponent bioprinting might replace 2D culture techniques for drug development [45]. Several crosslinked hydrogels exhibited viscoelastic activity, as shown by rheological data. The study found that insulin secretion increased by 2–4 times in static cultures. Dynamic culture has shown encouraging outcomes after just 7 days [59]. Using silk fibroin (SF) along with HPCMA, Yan et al. developed a bioprinted scaffold for the purpose of articular cartilage regeneration, as shown in Fig. 2. The rheological investigation of SF-HPCMA showed excellent findings, potentially reducing cell damage from shear stresses and providing mechanical safety

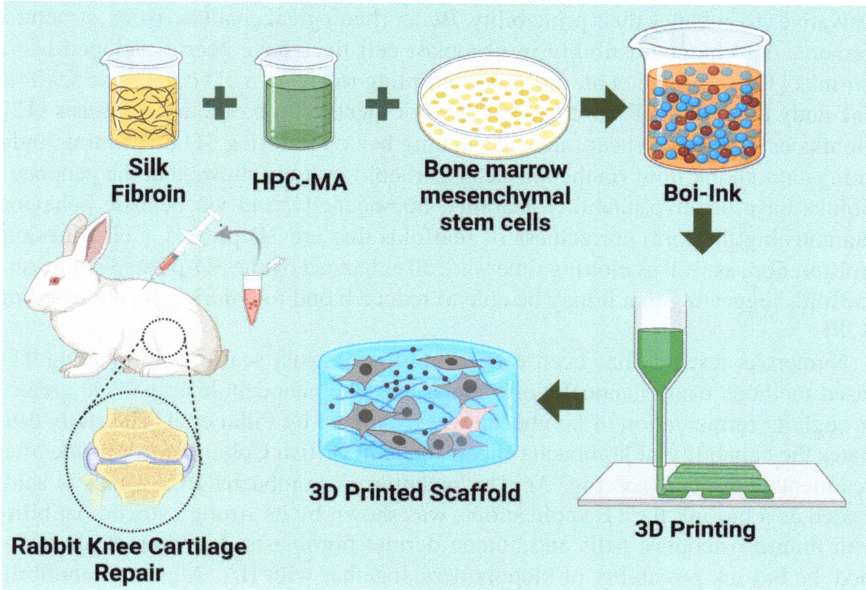

Fig. 2 Diagram illustrating synthesis of silk fibroin with HPC-MA through 3D bioprinting method

throughout 3D bioprinting [165]. Another research used a composite hydrogel made of gelatin methacryloyl, hyaluronic acid methacryloyl, and BC to repair nasal cartilage. Adding BC to hydrogels improved both mechanical properties and printing reliability. Encapsulated ATDC5 cell lines demonstrated higher cell feasibility after 7 days of culture [125].

2.2 *Nanocellulose Reinforced Bio-inks*

Functionalized nanofibers, cellulose nanofibers, and CNCs are examples of cellulosic nanomaterials that have been researched as fillers or functional elements. The bio-inks' physicochemical and biological characteristics were improved by the addition of nanocellulose. For usage in biomedical applications comprising healing of wound, bone, and cartilage TE, and cell-rich 3D printing, nanocellulose-reinforced bio-inks are currently researched [69].

To produce bio-inks that are appropriate for wound dressing, Rees et al. prepared nanocellulose using carboxymethylation and periodate oxidation. The oxidation procedure improved consistency and rheological characteristics, allowing for the production of 3D porous structures [123]. These structures are intriguing prospects for the absorption as well as release of a variety of drugs, proteins, and antibiotics due to their constant porosity. Cellulose nanofibers were added to low-viscosity gelatin derivatives to enhance their printability. Better rheological characteristics, structural elements, and biocompatibility in fibroblast cell lines have been developed using bio-inks [134]. Biranje et al. utilized bioprinting to create a 3D composite scaffold and study the way CNF, casein, and CS work together to promote hemostasis [12]. Bioinks exhibit rising shear rate and thinning behavior during 3D bioprinting, indicating smooth ink flow via the nozzle. The rheology study shows that the generated bioinks have better printability, stability, homogeneity, and viscoelastic behavior, guaranteeing the form correctness of scaffolds that are 3D printed. Cell adhesion, proliferation, as well as clotting time were all enhanced by the 3D-printed composite scaffold, suggesting that it may be able to reduce blood loss during severe bleeding [128].

Numerous research has been carried out to construct scaffolds employing TE-based methods using nanocellulosic materials. The nanocellulosic pellicle created throughout fermentation of kombucha was examined by Pillai et al. The study estimates the capability of kombucha sheets made in British Columbia to extrude after treatment with acid (see Fig. 3). The potential of kombucha BC, which is acid-treated as a bio-ink for TE applications, was shown by its strong cytocompatibility with mouse osteoblast cells and human dermal fibroblasts. Merchan et al. examined the bio-ink possibility of biopolymers, together with HA, Alg, and nanofibrillated cellulose, while also looking at the influence of numerous sterilization techniques. Following autoclaving, β, alongside γ radiation, scaffolds were evaluated for structure, swelling, as well as degradation [80].

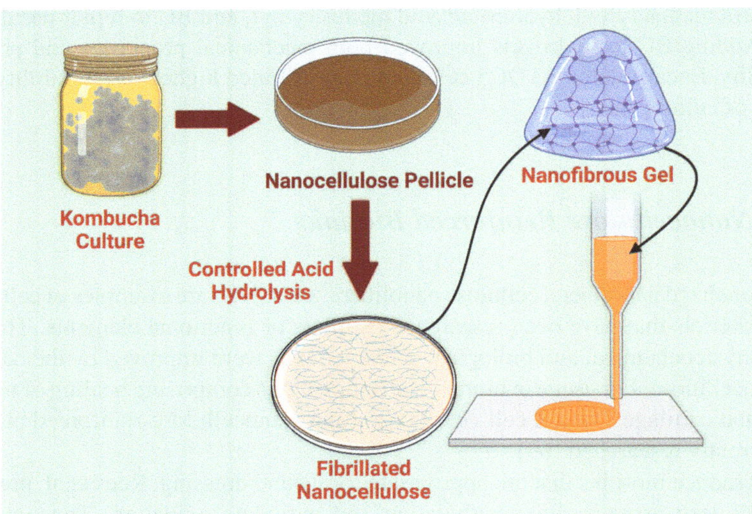

Fig. 3 Diagram representing fabrication of fibrillated nanocellulose through 3D printing

3 3D Bioprinting Technology

In 1974, DE. H. Jones first proposed the idea of 3D printing [36]. The first additive manufacturing (AM) technique was developed in 1981 by Hideo Kodama, used photo-hardening thermoset polymers to produce 3D models [13]. Charles W. Hull established the "stereolithography" technique for 3D printing that includes layer-by-layer assembly of material strata followed by UV light curing [105]. The technique was later used to produce sacrificial resin matrices for the creation of 3D biological scaffolds [26]. 3D bioprinting is now possible in the biomedical sector because of advancements in materials science, nanotechnology, and cell biology. This presents very promising opportunities for future growth in the field of medicine. 3D bioprinting creates constructs that resemble tissue by carefully aligning functional components with tiny units of biochemicals, biomaterials, and living cells [124]. 3D bioprinting offers several benefits over conventional 3D printing, including scalability, cell deposition, accurate cell dispersion, and cost [161]. Nevertheless, there are still barriers preventing 3D bioprinting from being extensively used in a variety of fields, including medicine. Printable biomaterials are very scarce, and printing methods essential to be enhanced for higher scalability and quicker printing rates. Fine printing resolution is ideal for targeting biological activities deprived of affecting mechanical qualities. With the aid of 3D bioprinting, biomimetic complex constructions that precisely distribute different cell types at high densities inside 3D scaffolds may be created. The most popular bioprinting methods for handling bio-inks include extrusion-based, laser-induced forward transfer, acoustic, and inkjet methods [145]. EBP is the most versatile technology for creating tissue-engineered scaffolds, since it may use bio-inks with varying viscosities. To get the optimum printability and performance, bio-inks are still needed that are tailored to a particular application, particularly for bone TE applications [99, 142].

The rapid development of 3D bioprinting in the health field has had a significant influence on both the manufacturing of health supplies and the medical sector today. Intricate organs and tissues have been produced using this method. 3D bioprinting may be divided into a variety of categories based on its methods. The necessary tissue may be produced by using these methods alone or in combination [145].

3.1 Fused Deposition Modeling (FDM)

FDM, often recognized as 3D printing, was the earliest AM method [9]. A thermoplastic polymer is infused into a liquefier, as well as a filament is extruded to generate bio-inks, resulting in a melted thermoplastic base [155]. The properties of FDM products are influenced by several elements. Researchers aim to enhance produced elements by focusing on key components and eventually combining them [32]. Most studies use descriptive methodologies to evaluate the influence of various factors on the accuracy as well as the geometry of printed parts using certain materials along with parameters [2].

To enhance mechanical features, cellulose is joint with other biopolymers, like polylactic acid, to create cellulose-based filaments. To create continuous filaments, the mixture is extruded using a single screw extruder. An accelerating device forces the molten cellulose-polymer mixture through a nozzle. A computerized system controls the nozzle motions, depends on the 3D design of the item being printed [43].

The finished printed item is produced when the nozzle gradually drips the molten cellulose-polymer mixture on the build platform. The building base progressively descends, often by the thickness of the layers. To the previously hardened layer, the nozzle adds a second layer of molten cellulose-polymer mixture. Until the product is finished, the assembly procedure is repeated. The deposited layers cool and change from a liquid to a solid state when they come into contact with the build platform and surroundings. Some FDM printers use cooling fans to speed up the transition and prevent distortion or warping. The thing is fully formed after all of the layers have been applied. Following elimination from the construction platform, post-processing activities including surface polishing and support structure removal are completed. Cellulose is used in FDM filaments to advance the mechanical characteristics of 3D printed goods. To add cellulose fibers to these filaments, several methods have been devised. Fibers and polymers are typically extruded or solvent cast, then melted and compounded to create reinforced filaments. Nanocelluloses have been used as reinforcing materials in the majority of investigations at concentrations between 1 and 40% (15). Stiffness has increased by 190% and tensile strength by 84%. Cellulose, however, did not always enhance mechanical properties. Nevertheless, other benefits were discovered, like enhanced biodegradability and thermomechanical stability [166].

In a work, PLA filaments for 3D printing were reinforced with CNF made by enzymatically hydrolyzing MCC and then homogenizing it at high pressure. The use of polyethylene glycol 600 as a plasticizer led to formulations with up to 5% CNF having acceptable processability. The filaments' failure strain and tensile strength were around 50% and 20% higher, respectively, than those of clean PLA. The addition of 2.5% CNF produced the biggest improvement, whereas 5% CNF produced the least improvement. Recent studies draw attention to difficulties in processing problems, decreased toughness, and nanocellulose agglomeration (heterogeneous dispersion) [43].

Flexibility in material is one advantage of using FDM. Cellulose along with its derivatives, nylon, and additionally low-melting-point metal and ceramic materials may all be utilized with FDM 3D printing. When likened to other printing methods, FDM is more economical. This is because they need less post-processing, FDM printed parts have easily removable support structures. Uneven mechanical characteristics and reduced resilience are the effects of insufficient layer thickness in FDM produced parts. This limitation, then, limits the full use of FDM produced parts [90, 98].

3.2 Stereolithography (SLA)

An improved printing technique based on the polymerization of extremely fine polymers developed in the nineteenth century [63]. Method focuses a light beam onto the liquid photocurable resin's surface using a UV laser and a directed mirror array. The three-dimensional components develop in each layer by doing this along the Z-axis. The method used here is predicated on liquid resin solidifying due to photopolymerization. To achieve targeted polymerization and solidification, a laser is pointed at a certain depth inside a resin vat. Layer by layer, the material solidifies to form a three-dimensional, solid structure. The UV light source is the primary drawback of SL as it causes cell death and encourages skin cancer [13, 35]. Visible light SL is a replacement that is rapidly becoming popular in the bioprinting industry via the optimization as well as stabilization of the structures [156]. Common printing techniques usually produce porous networks through 3D scaffold structures that are poorly built. The structure is being built with the aid of light-sensitive bio-inks, as SL bioprinting depends on a particular light source to control printing. Plane by plane, the bio-inks are then produced [107].

SL offers a number of benefits, including as accuracy, speed, cost-effectiveness, and adaptability. SL is a valuable tool for developing and validating concepts in domains such as medical device design because to its numerous benefits. The capacity to combine medications with photopolymer prior to printing, trapping them in toughened matrices, is a key characteristic of SL printing. The focused laser's beam width is the single factor limiting SL's resolution, which gives it more accuracy than other 3D printing methods. SLA uses a UV laser to toughen a photopolymer typically a resin into the required shape using a CAD file. An extremely accurate 3D printing process is produced by applying a thin layer of liquid resin to the building site and using a laser to solidify it. A number of procedures need to be followed in order to create a prototype utilizing additive manufacturing. As the technique reduces localized heating during printing, it could be more operative for creating dosage forms holding thermo labile materials [61].

3.3 Inkjet Bioprinting

The utilization of 2D desktop inkjet printing machines enables the non-contact technique of inkjet-based bioprinting. They are divided into three types, according to [51]: electrostatic, thermal inkjet bioprinting, and piezoelectric. The bio-ink chamber of piezoelectric inkjet bioprinters is filled with acoustic waves generated by piezoelectric actuators [51]. Pressure is applied among an electrode along with a pressure plate in electrostatic inkjet bioprinters to produce droplets. During thermal inkjet bioprinting, the heat produced in the bio-ink chamber generates pressure [6]. Their popularity has grown in spite of the method of fabrication focus on producing ink droplets because of their reasonable price, compatibility with living organisms, and rapid synthesis of drop bio-inks [126].

A new study suggests using CNC strengthening impacts in conjunction with photoinduced processes to enable DIW of 3D complex structures The effect of CNC on rheological parameters, which show local shear-induced alignment of active reinforcing, permits the DIW of self-standing 3D structures. The printed materials have a longitudinal elastic modulus of 30 MPa and retort to visible light having 30–50% reversible softening and shape memory. Dynamic mechanical responses to external stimuli are now possible because to the Establishing phototunable energy absorption in 3D intricate structures by a blend of geometric as well as photoresponsive effects [110]. 3D printed inks with functionalized CNC have the ability to facilitate quick design for a variety of devices with specific mechanical characteristics, making them ideal for applications requiring dynamic reactions to changes in the environment [95].

3.4 Extrusion-Based Bioprinting (EBP)

The application of EBP, also known as direct writing, in TE and biofabrication has grown significantly in recent years [66]. The extruded bio-inks are widely dispersed as a strand using two fundamental techniques: mechanical force and pneumatical force. In actuality, a stage regulator controls a dispenser system while turning on a robotic stage. The bio-inks are put on to a construction substrate using a piston-driven deposition apparatus, and a screw-driven device may regulate the bio-ink overflow as shown in Fig. 4 [109]. After deposition, the bioink changes from a liquid state to a solid state, temperature changes, exposure to UV radiation, or chemical crosslinking can all help in this process. The entire strength and integrity of the manufactured edifice depend heavily on this solidification process. The construct may undergo further procedures after printing, such as sterilizing, eliminating support materials, or growing in a bioreactor to encourage tissue growth and cell proliferation [130]. EBP is extensively utilized in TE and regenerative medicine because to its ability to accurately and modify complex cell-laden structures. This is an vital subject in the deposition of very viscous biomaterials. As an STL file format, a CAD file is snuck into manufacturing and printed using the designated structure. But because to advancements in machinery, numerous printheads may now deposit various bio-inks, which have the benefit of controlling the created components' porosity, form, and cell distribution [109]. EBP has access to a range of printable materials as well as permits the deposition of high cell densities [172].

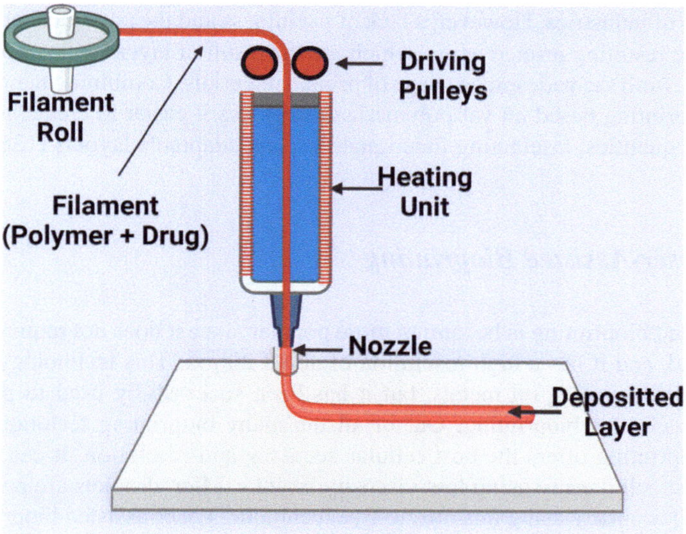

Fig. 4 Diagram illustrating the fused deposition modeling in 3D bioprinting

3.5 Vat Polymerization

Vat polymerization involves a variety of processes, such as two-photon polymerization and digital light processing. DLP accurately crosslinks the photosensitive resin by shining a range of UV light over its surface [25]. Two-photon polymerization (TPP) creates incredibly complicated and convoluted three-dimensional structures by initiating polymerization at a specified target in the resin using focused ultrashort laser pulses [136].

The vat polymerization method is based on solution of polymer with a laser-solidification mechanism [169]. The two primary vat polymerization techniques that have been presented thus far are SL and DLP. Although there are many advantageous biomaterials, there are also significant technological obstacles that need to be overcome. For instance, the UV light used in SL aims of polymerize layers in a certain 2D pattern [112, 160]. Despite some commercial or scholarly interest, this approach has a number of disadvantages, such as the cytotoxic nature of photoinitiators and the shortage of photosensitive polymers [86]. In one work, MAcr-CMC formulations were used to print hydrogels with complex geometries utilizing DLP and TPP technologies. To produce a hydrogel scaffold with the right size to support the cell proliferation necessary for soft-tissue regeneration, both procedures were tested. TPP was suggested as a substitute vat polymerization method for printing MAcr-CMC aqueous formulations into small hydrogels with intricate shapes and high resolutions. The first and most popular 3D printing method, vat polymerization, offers several benefits, such as quick printing speeds and high-precision production facilities. This technique is often used to create complexly structured structures that are very useful in a range of industries. However, a lack of usefulness and the poor mechanical qualities of the resulting printed items, which are the result of layer-by-layer production processes, limit the widespread usage of printed materials. Combining nanomaterials with 3D printing based on vat polymerization makes it easier to create goods with improved qualities, fascinating functionalities, and adaptable layouts [136].

3.6 Laser-Assisted Bioprinting

Laser-based bioprinting is becoming more popular since it does not require a nozzle or scaffold, and it has a high resolution of about 20 μm. This technology was first developed for writing on metals, but it has been successfully used to print DNA or organ cells for bioprinting. Out of all the many bioprinting techniques, laser-based bioprinting offers the best cellular accuracy and resolution. It can also print the highest cell density using low-viscosity bioinks. Microdroplets are printed with single-cell accuracy using this nozzle-free technique. Laser-assisted bioprinting has several problems, one of which being the heat death of cells due to nanosecond laser irritation [24].

4 Post-processing of Printed Object

Post-processing is a crucial step in additive manufacturing using nonthermoplastic feedstocks. For the benefit of the current discussion, post-processing refers to any procedures or treatments used to produce the final structure after material extrusion or printing. Long print durations provide a barrier for all-cellulose feedstocks, as this process is often undertaken after layer deposition to maintain structural integrity and shape fidelity [115]. Printing processes, particularly material extrusion, may be used with a variety of cellulose feedstocks as well as applications. However, the post-processing approach chosen depends on the feedstock qualities and desired application. This might involve air drying, coagulation, or a mix of these methods [3]. Overview of various types of post processing techniques after 3D printing is shown in Fig. 5.

4.1 Solvent Evaporation

Air-drying at ambient temperature or solvent evaporation is the simplest technique and the one that is most similar to conventional thermal MEX printing. This technique has mostly been established to work with feedstocks that contain cellulose derivatives that dissolve in frequently used NCC suspended in water to aid in drying and evaporation. Solution-/suspension-based printing differs from a melt-solidification

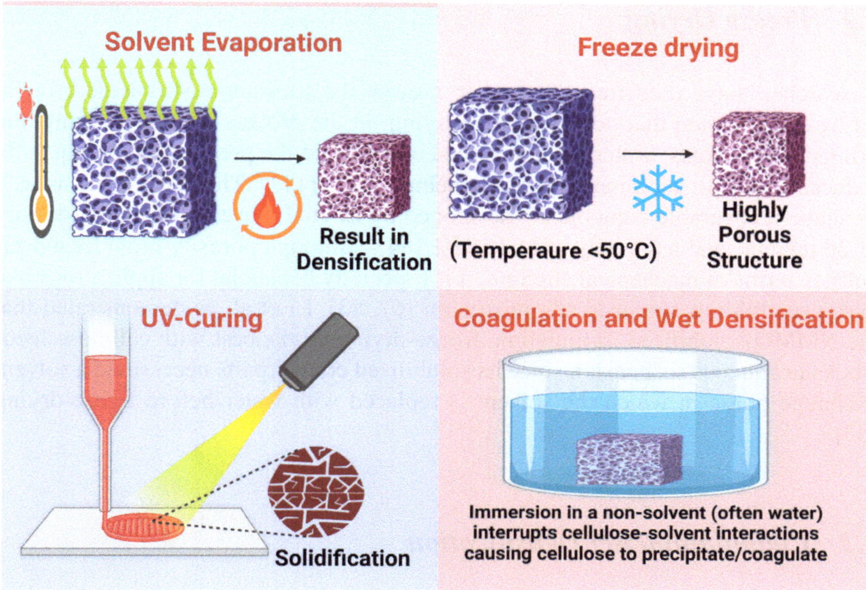

Fig. 5 Diagram illustrating post processing after 3D printing

process, however, because of the effect that solvent incorporation has on the density and form integrity of the finished print. An important factor to consider when using these formulas is post-print shrinkage brought on by densification as the solvent evaporates. Tenhunen et al. compared the use of acetone as well as acetic acid as solvents for CA and APC and discovered that it was better for CA-based feedstocks to utilize acetic acid for slower evaporation since it improved printability and showed reduced shrinkage [141]. Nonetheless, shrinkage may be measured in controlled drying circumstances (temperature, relative humidity), and then taken into consideration in the initial digital design and printing settings to produce a component with the required dimensions [77]. Pattinson et al. were able to create structurally robust, high-density CA prints by allowing each layer to fully dry thereafter to prevent shrinking and promote inter-layer adhesion [117], this limited print efficiency by requiring up to 60 s among layers for higher CA concentrations (35 wt%). As anticipated, as the amount of evaporating liquid decreases, higher cellulose concentrations lead to improved form fidelity (less shrinkage) [77, 117]. Consequently, non-derivatized cellulose feedstocks, which frequently have a cellulose concentration of less than 10%, are more likely to shrink after drying than highly concentrated derivatized cellulose feedstocks [111].

4.2 Freeze Drying

Researchers have used freeze-drying to prevent the substantial shrinkage brought on by densification that occurs with airdrying. In the process of freeze-drying, the printed components' water content is ice-covered, and the pressure is subsequently reduced to permit sublimation, leaving behind pores [111]. The deformable as well as shape-recoverable components developed by Li et al. have a volumetric density of 26 mg/cm^3 and a porosity of up to 98% [84]. Although porosity provides a difficulty in terms of mechanical strength, it is especially beneficial for applications like tissue scaffolding and drug administration [67, 83]. Li et al. as demonstrated that the NMMO-solubilized formulation, freeze-drying works best with cellulose feedstocks in aqueous suspension since its solubilized counterparts necessitate a solvent exchange phase in which the solvent is replaced with water before freeze-drying [83].

4.3 Coagulation/Wet Densification

This processing technique works by substituting a material that interacts with cellulose less strongly for the solvent through which cellulose is dissolved. This increases the interactions between molecules of cellulose, which promotes densification. A suitable solvent is one whose cohesive energy density, which is the increase in volumetric internal energy that occurs when intermolecular interactions are removed, is comparable to that of the particles in solution. Specifically, a more homogeneous solution is produced because particle–particle interactions and solvent–particle interactions have similar energetics. Liquid-rich phases and particle-rich phases would arise from a mismatch in CED [56]. Through the formation of a hydrodynamic shell around the solvent ions, a nonsolvent (water) disrupts the interaction between cellulose and the ionic liquid in this solute displacement process. This enables the hydrogen bonds in cellulose to rebuild, leading to coagulation [168]. As proven by Gunasekera et al., this may be achieved by immersing printed objects in water, such as those deposited using MJT. Even so, they used submersion periods of up to two hours to guarantee complete solvent clearance [53]. Markstedt et al. discovered that using an atomizing nozzle to spray water in between MEX layer depositions in an attempt to avoid long submersion durations was unproductive since it led to poor inter-layer adhesion [97]. To enable water to constantly seep into the printed section from the bottom, they also tried printing within an agar mold. This technique was only effective on items that were less than a certain size, beyond that the diffusion process was too slow to allow subsequent layers to gel properly. The primary disadvantage of this solidification method for larger prints is the amount of time required to accomplish total solvent replacement. A little differently, Hausmann et al. used organic solvents including ethanol, acetone, and acetonitrile to progressively replace the water in their aqueous feedstocks, which is extremely polar and interacts with hydrophilic cellulose [56]. Inter-cellulose hydrogen bonding was less disturbed by the less polar and cellulose-interactive replacement solvents, which produced a denser end product. Coagulation/cellulose regeneration techniques often result in significant shrinkage and loss of form integrity, similar to solvent evaporation or air-drying. However, these illustrations show crucial basic ideas and useful components that will propel the development of cellulose-based 3D printing technology [39].

4.4 UV-Curing

UV-curing is the process of promoting polymerization using UV light, which causes the print structure to crosslink and solidify. It is quicker and, when utilized as the printing method, allows for high-resolution printing [149]. The cellulose component of the formulations in the examples given here is derivatized cellulose, typically acrylated, to aid in photopolymerization. Since standard acrylic and epoxide resins are derivative from fossil fuels and may not be suitable for use in medical applications owing to their composition, the completed prints' sustainability and biocompatibility may be restricted by the introduction of extra photoactive/crosslinking additives [27]. Cafiso et al. demonstrated how UV-curing can be used very well for DLP printing as well as post-curing, attaining a 27 μm print resolution (height ~ 1–2 cm) with 12–15 s cures in among layers and a 3 min post-cure [22]. Additionally, Chan et al. observed that the as-printed microstructure is preserved by continuous UV irradiation during printing, together with the corresponding rapid crosslinking and gelation, allowing for additional structural control of the finished print [23]. Although there are several adjustable factors, including UV light energy and intensity, the polymerization process may also be influenced by the curing environment. For instance, to prevent oxygen from inhibiting the photo-induced polymerization process, Sieira et al. carried out UV-curing in a nitrogen environment. UV-curing enables incredibly accurate, effective printing, but it also necessitates careful feedstock preparation. It usually a derivatization of the cellulose component [135].

5 3D-Printed Cellulose Applications

3D printed cellulose has several applications beyond hydrogel the production process. 3D printed hydrogel derivatives have several applications, including plastics, smart-paper, textiles, biomedical, electronics, living ink, and sensors. Wang et al. utilized FDM 3D printing to create a polylactic/polyethylene glycol/cellulose nanofiber (CNF) biocomposite. Potential uses for composites exist in the clothing and textile sectors [154]. Biodegradable cellulose promotes recyclable goods and energy-efficient printing in ecologically friendly products. Adding varying amounts of CNF can result in a 3D printed object with unique qualities and functionality. The

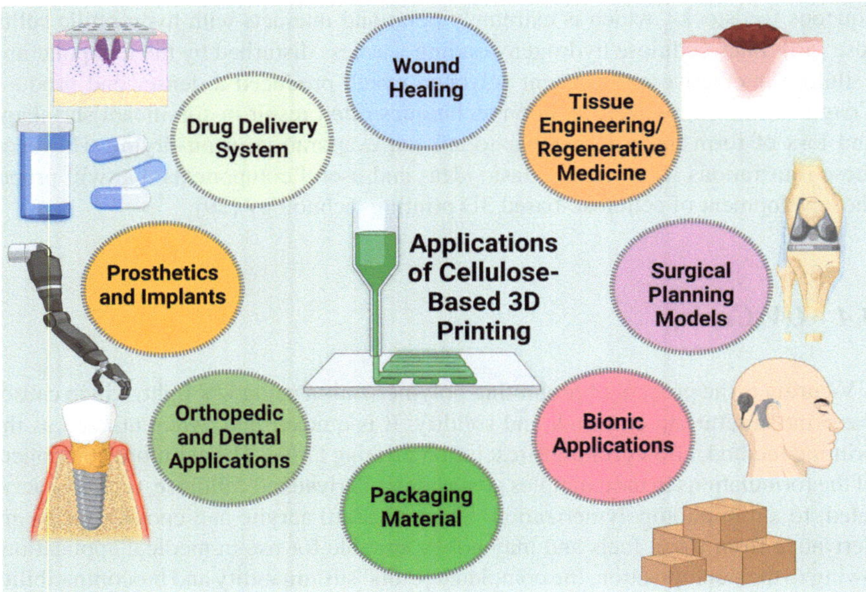

Fig. 6 Various biomedical applications of cellulose-based 3D printing

scaffold may also have uses in architecture, interior design, and packaging. Figure 6 summarizes many applications for 3D printed cellulose.

5.1 Biomedical Applications

Nanocellulose fibers are the natural building blocks of cellulose, a polymer. Because of its mechanical properties, low toxicity, and biodegradability, cellulose along with its derivatives have possibilities of use in biomedicine. Among the many medical applications of nanocellulose are biomedical TE, implants, healing of wounds, drug delivery, and diagnostics [52]. Recent research shows that cellulose along with its derivatives have potential medical use. The researchers looked at hydrogels based on cellulose as well as its derivatives for use in biomedical uses like bioimaging, wound dressings, scaffolds for TE, and sensors. They looked at long-term prospects, physicochemical characteristics, hydrogel production technologies, and preparation techniques. Cellulose-based 3D printing technology was utilized until recently to develop a quick and affordable method for medicinal purposes [34, 42, 49]. This study will provide instances of 3D-printed cellulose for biological applications. Various patents related to cellulose-based 3D printing in biomedical applications are shown in Table 1.

Table 1 Overview of patient related to the inventions in cellulose-based 3D printing

Patent number	Title	Inventor	Publication date	Note	References
US20180305569	Preparation and applications of biocompatible conductive inks based on cellulose nanofibrils for 3d printing of conductive biomedical devices and for use as models for study of neurodegenerative disorders and connection between brain/neurons and communication or other electronic devices	Paul Gatenholm, Erdem Karabulut	25.10.2018	The invention uses nanocellulose fibrils or crystals, along with carbon nanotubes, as a biocompatible and conductive ink for 3D printing electrically conductive patterns. These devices are biocompatible, have excellent mechanical properties, and can be used as bioassays for neurodegenerative diseases, study brain function, and link the human brain to electronic devices	[143]
CN106344527	3D printing type ginkgo biloba leave extract dispersible tablet and preparation method thereof	Liu Qiang, Cao Siwei, An Baichao, Xiang Shijian, Hong Junhui, Ruan Shifa, Zhu Hongxia, Weng Lidong, Chen Huoji	25.01.2017	The invention outlines a 3D printing method for a dispersible tablet of ginkgo biloba leaf extract. The process involves mixing the extract, microcrystalline cellulose (25–40%), mannitol, povidone, and gel in a powder box, filling an ink cartridge with adhesive, modeling the tablet, and controlling the 3D printing rapid forming machine	[15]
CN118436589	Preparation method of nanocellulose aerogel drug carrier capable of being used for 3D printing	Liu Zhongming, Kong Fangong Wang Shoujuan, Li Yang	06.08.2024	The invention outlines a 3D printing method for preparing nano cellulose aerogel drug carriers, utilizing agricultural and forestry straws. The method provides regular shape, high adsorption capacity, and controlled release behavior, making it suitable for drug sustained release research	[21]

(continued)

Table 1 (continued)

Patent number	Title	Inventor	Publication date	Note	References
WO2023227831	Ink formulation, method of producing the same and uses of the ink formulation	Wang Xiaoju, Brusentsev Yury, Xu Wenyang, Xu Chunlin, Wågberg Lars, Backman Oskar, Willför Stefan	30.11.2023	The innovation focuses on methacrylated cellulose nanofibril-based ink compositions for 3D printing. These formulations are made up of a hydrogel that contains nanofibrils and a cross-linking agent; biological material may also be included. They are used in drug screening and cell culture	[152]
CN113925837	Method for preparing ibuprofen by applying 3D printing technology	Shang Jianping, Fan Huajun, Qin Xiaoping, Zhao Bin, Hao Shixiong, Tian Haiyang, Fan Beibei, Xue Zhiqi	14.01.2022	The invention utilizes biological 3D printing technology for biomedicine, specifically for preparing ibuprofen tablets. The optimal ratio of medicine, including 10% of IB, 20% of HPMC, and 70% of PCL, is optimized, ensuring no chemical changes among medicines, auxiliary materials, and wires	[19]
CN106511285	3D printing breviscapine orally disintegrating tablet and preparation method thereof	Liu Qiang, Cao Siwei, An Baichao, Xiang Shijian, Hong Junhui, Ruan Shifa, Zhu Hongxia, Weng Lidong, Chen Huoji	22.03.2017	The invention outlines a 3D printing method for disintegrating breviscapine orally. The process involves mixing breviscapine, dextrin, cellulose, povidone, silica powder, and corrigent, and then printing the tablet	[16]

(continued)

Table 1 (continued)

Patent number	Title	Inventor	Publication date	Note	References
KR1020220021048	Bioink for 3d bioprinting including microfibrillated cellulose and guar gum, and 3d printing method	Young Hoon Jung, Jun Tae Kim, and Gi Eon Park	22.02.2022	The invention describes a 3D bioink using microfibrillated cellulose and guar gum, offering excellent printability, rapid crosslink formation, and mechanical properties. This bioink is suitable for biomaterials, drug delivery systems, and tissue engineering supporters	[71]
WO2019232114	Injectable thermoresponsive hydrogels as a combinatory modality for controlled drug delivery, biomaterial implant and 3d printing bioink	Benhabbour Soumya Rahima and Maturavongsadit Panita	05.12.2019	This invention presents pharmaceutical compositions containing active agents, cellulose nanocrystals, thermogelling polymers, and gelling agents for 3D printing, treating bone disorders, Paget's disease, cancer, infectious diseases, and regenerative medicine, including osteoporosis	[151]
US20210205459	Injectable thermoresponsive hydrogels as a combinatory modality for controlled drug delivery, biomaterial implant and 3d printing bioink	Soumya Rahima Benhabbour, Panita Maturavongsadit	08.07.2021	Pharmaceutical compositions containing active agents, cellulose nanocrystals, thermogelling polymers, and gelling agents are designed for 3D printing, treating bone disorders, Paget's disease, cancer, infectious diseases, and regenerative medicine	[144]
CN117883400	Reszopiclone pulse timed release tablet and preparation method thereof	Li Zhengqi, Xu Heng, Geng Congcong, Xu Chao, Fu Jian, Wang Jiawen	16.04.2024	The invention outlines a pulse timed release tablet. The controlled release layer is made of modified cellulose and a water-soluble polymer. The tablet's preparation method involves wet granulation, 3D printing, and coating the controlled release layer	[20]

(continued)

Table 1 (continued)

Patent number	Title	Inventor	Publication date	Note	References
CN106511416	3D printed radix notoginseng total saponins orally disintegrating tablet and preparation method thereof	Liu Qiang, Cao Siwei, An Baichao, Xiang Shijian, Hong Junhui, Ruan Shifa, Zhu Hongxia, Weng Lidong, Chen Huoji	22.03.2017	The invention outlines a 3D printed radix notoginseng total saponins orally disintegrating tablet preparation method. The process involves mixing radix notoginseng total saponins, cellulose, mannitol, povidone, silica gel, and flavoring agent. The tablet is then printed using a 3D printing and fast-forming machine	[17]
CN112043676	Three-dimensional printed loratadine oral disintegrating tablet as well as raw material composition and preparation method thereof	Yi Shaoling, Chen Lingli, Yu Zejun, and Wu Zepeng	08.12.2020	The invention relates to a three-dimensional printed loratadine oral disintegrating tablet. Solid component containing loratadine, ethyl cellulose, sodium carboxymethyl starch, mannitol, and flavoring agent	[18]

5.1.1 Tissue Engineering Application

Developing biomaterials with characteristics similar to natural tissue poses a substantial challenge in creating a 3D microenvironment for complex tissue regeneration. Tissue engineering has made extensive use of cellulose, both as nanomaterials and as its derivatives [58]. Biomedical interest is drawn to nanocellulose due to its great surface area, biodegradability, and surface chemistry. The potential of nanocellulosic materials for 3D bioprinting is enormous because of their shear-thinning behavior, support for the development of live cells, and printability the ability to create and preserve repeatable 3D objects from bioink/biomaterial ink using the bioprinting process. The inherent characteristics, dimensions, and makeup of native cellulose which may be altered chemically or by building hybrid structures using other natural or synthetic polymers are the one gives it this potential [40]. Nanocellulose can be added to printable pastes or hydrogels to help in deposition and the development of a stable structure. 3D printed hydrogels made of cellulose have already been reported by various researchers [7, 55, 91, 138].

One of the key challenges in 3D printing is producing stable hydrogel inks that are either cell-free or cell-charged, have the best physicochemical properties, and are biocompatible. The properties of 3D-printed hydrogels may be enhanced by carboxylated cellulose nanocrystals for tissue engineering applications. In combination with xanthan gum, Kumar et al. evaluated the sodium alginate hydrogel matrix's rheological behavior. In human skin fibroblast cells, this hydrogel showed high vitality, suitable rheological properties, dynamic mechanical capabilities, and post-printing fidelity [79]. Additionally, Palaganas et al. utilized 3D printing stereolithography to create hydrogels with a complex architecture and enhanced material characteristics appropriate for tissue engineering using nanocrystalline cellulose [114].

The rheological characteristics of material are correlated with printing behaviors including stability, accuracy, and resolution. The cellulose nanofiber's diameters alter the rheological characteristics via influencing the fiber attachment. Pseudoplastic fluid behavior occurs when the tension caused by moving the nozzle in the matrix instantly lowers the viscosity of the fluid. The concentration along with physical features of cellulose nanofibers impact printing accuracy by affecting values such as angled line sharpness, cross-sectional ratio, surface roughness, and multi-line writing. These values are closely related to the rheological features of matrix [133].

5.1.2 3D-Printed Wound Dressing

Nanocellulose has exceptional mechanical qualities that enable the development of transparent films. In order to promote healing of wound and create elastic gels with bioresponsive properties, nanocellulose can also provide moisture to wounds. A 3D nanocellulose–alginate hydrogel was produced by Leppiniemi et al. for use as a wound treatment. Requirements for moisturization, absorption, barrier, and support must be met by 3D nanocellulose printed structures [81]. In other words, they should be 3D printed, non-irritating, sterile, as well as applied painlessly. Additionally, they should maintain their shape to prevent maceration, be mechanically stable under moderate compression and shear stresses, and be able to remove exudate and keep the wound moist. Additionally, the build needs to be on-demand and enable the patient to carry out everyday duties [41].

For use as wound dressings, Xu et al. investigated the strength of a 3D-printed hydrogel that contained 1% nanocellulose. The twofold cross-linking method in situ physical Ca^{2+} crosslinking as well as chemical cross-linking by 1,4-butanediol diglycidyl ether was used to mold the wound dressing. Consequently, the proliferation and stiffness of fibroblast cells were enhanced by 3D printed wound dressings. These qualities support wound healing and are favorable for cellular functions, tissue repair, and regeneration [162]. Kanjou et al. employed BNC for 3D printing in situations involving chronic wounds. Hyaluronic acid and chondroitin sulfate were used in place of the fermentation medium to create the bacterial cellulose, which was then cross-linked using calcium chloride as well as sodium alginate. Developing 3D printed scaffolds appropriate for cell viability was the goal of this procedure [72].

For applications involving 3D printed wound dressings, Rees et al. employed TEMPO nanocellulose, a mixture of carboxymethylation as well as periodic oxidation. A homogenous mixture including short nanofibrils (<20 nm broad and <200 nm long) was created by the mixture of carboxymethylation along with periodic oxidation. As a biomaterial for ink, the nanofibrils' tiny size can lower the viscosity of the nanocellulose, giving it favorable rheological characteristics. Furthermore, nanocellulose creates an open-porous, three-dimensional structure that may release antimicrobials. Wound dressings made of nanocellulose can shield wounds from pathogenic germs (micrometer-sized) because of its nanometer-sized pore structure and extra antibacterial qualities [123].

3D-printed bandages offer a custom fit by being printed directly onto the skin, enhancing wound protection and healing. Gerstenhaber et al. developed a soy protein fiber bandage using electrospinning, designed to function like a second skin. Additionally, nano-cellulosic materials can be used for 3D-printed wound dressings, providing mechanical strength, permeability, and absorption. However, cellulose lacks antimicrobial properties, requiring active ingredients for microbial protection and antioxidant effects. Its nanostructure aids in microbial filtration, and its properties influence drug release in wound dressings [60]. Overview of 3D-printed cellulose-based materials for wound healing applications is mentioned in Table 2.

5.1.3 Controlled Drug Delivery Systems

APIs are supplied via drug delivery systems (DDS) to provide a therapeutic impact in either humans or animals. This mechanism regulates the drug release rate, timing, and target to guarantee optimum drug distribution and absorption and to maximize efficacy and safety [33].

Table 2 Various 3D-printed cellulose-based materials for wound healing applications

Primary materials	3D-printed type	Description	Key features	Applications	References
Carboxymethyl cellulose, nano-based electrospun material	Electrospinning	Provides a controlled drug-release model for wound treatment	Drug-delivery potential for healing	Drug-eluting wound dressings	[44, 100]
TEMPO-oxidized cellulose nanofibrils, alginate, calcium chloride	Extrusion-based 3D printing	Aerogel-based material demonstrates excellent water absorption, ideal for wound dressings	High water retention for moist wound healing	Moisture-retentive wound dressings	[38]
Mesoporous silica (SBA-15) and cellulose membrane	Extrusion-Based 3D Printing	Composite (P-CM-SBA 30%) demonstrates antibacterial activity, tensile strength	Meets key wound dressing criteria	Advanced wound care materials	[41, 131]
Cellulose, mesoporous silica, and $CaCO_3$	Extrusion-based 3D printing	Composite membranes offer antibacterial properties against *S. aureus* and *E. coli*, with favorable water vapor permeability and swelling	Promising for wound healing applications	Antibacterial wound dressings	[132, 146]
Sodium alginate, gelatin, and cellulose nanocrystal hydrogel	Extrusion-based 3D printing	Bio-based scaffolding for tissue regeneration that has regulated nanocrystal alignment, gradient structure, and pore size	Supports tissue engineering and cell growth	Scaffold development for tissue regeneration	[137]
Nanofibril cellulose	Extrusion-based 3D printing	Alginate and Ca^{2+} influence dimensional stability, enhancing potential wound care applications	Structural improvement for wound treatment	Structural support in wound dressings	[96]
Silver nanoparticles (AgNPs), alginate, and nanocrystalline cellulose	Extrusion-based 3D printing	Exhibits antimicrobial effects against *S. aureus* and *P. aeruginosa*; AgNPs show cytotoxic effects on HepG2 cell cultures	Antibacterial with potential cytotoxicity concerns	Antimicrobial wound dressings	[11, 119]
Cellulose nanofibers or nanocrystals	Hydrogel (direct ink writing)	Supports the use of bioprinting technology to promote the growth of human mesenchymal stem cells	Cell-friendly hydrogel for tissue engineering	TE and regenerative medicine	[88, 139]

(continued)

Table 2 (continued)

Primary materials	3D-printed type	Description	Key features	Applications	References
TEMPO-oxidized cellulose nanofibrils and GelMA	Extrusion-based 3D printing	GelMA enhances fibroblast proliferation, making it a viable option for wound healing	Promotes cell growth and regeneration	Cell-proliferative wound dressings	[14, 163]
Cellulose nanofibrils, cellulose nanocrystals, and alginate	Extrusion-based 3D printing	Alginate contributes to mechanical properties but does not impact the ink's viscoelastic behavior	Structural enhancement for 3D-printed biomaterials	Mechanical reinforcement in bioprinted structures	[57]

Zhang et al. used hot-melt extrusion to develop zero-order release APIs, which is a revolutionary way to make tablets with 3D printed components. The model medication used was acetaminophen. Soluplus® and BenecelTM hydroxypropyl methylcellulose (HPMC) E5 were used in the 3D-printed tablet formulation. The results showed that the rate of drug release may be effectively and efficiently controlled by the 3D structure design. The outer shell structure was thick and stiff, acting as a barrier during in vitro drug release trials. Hot-melt extrusion combined with FDM-based 3D printing provides a novel, cost-effective, and effective technique for complicated production, customized dosage forms, and a more regulated release profile [170]. Infanger et al. made use of a powder bed for 3D printing. To make 3D objects, a liquid binder was sprayed over the powder using inkjet printing. In order to bind APIs (70% caffeine), hydroxypropyl cellulose was used in a dropon-powder deposition approach. This ink was made with a blend of ethanol and water. The findings demonstrated that the brittleness of the tablets was influenced by the disintegration time as well as particle size of the binder, while the viscosity of the binder affected the dissolving characteristics. Slower dissolution and disintegration were the results of a high viscosity binder, and vice versa. HPC can be utilized as a solid binding for 3D drop-on-powder printing, according to this study [62]. Overview of 3D printing cellulose-based material in drug delivery is demonstrated in Table 3.

5.2 Other Application

5.2.1 Packaging Applications

Product packaging is typically used by manufacturers or distributors to preserve the product's quality. Paper, plastic, metal, glass, and other materials can be used for packaging, depending on their properties. Technology for 3D printing was created to create 3D packaging. It is now feasible to print product packaging in the required size thanks to this technology [31].

The plastic industry is becoming more ecologically conscious, combining biodegradable materials and composites with synthetic polymers to reduce non-biodegradable plastic waste. Kudahettige-Nilsson et al. created a 3D printing ink of glycerol or ethylene glycol, cellulose, and citric acid, which showed promise for plastic applications due to its excellent tensile qualities and water absorption characteristics [78]. Similarly, Cisneros-López et al. studied 3D-printed bioplastics made by combining polylactic acid, epoxy, and microcrystalline cellulose. Using fused deposition modeling, they discovered that epoxy and Joncryl efficiently regulated the MFI of biocomposites made from recycled PLA. The inclusion of epoxy increased impact strength, whereas microcrystalline cellulose improved the mechanical characteristics of 3D-printed samples by 88%, 127%, and 11%, respectively. In contrast, injection-molded samples outperformed 3D-printed samples in terms of mechanical performance [28]. Further research by Zander et al. explored the potential of recycled plastics and renewable materials in 3D printing. They produced green composite

Table 3 3D-printed cellulose-based materials in drug delivery applications

Cellulose material	Drug	3D printing technique	Key findings	References
$CaCO_3$-nanocellulose composite from oil palm empty fruit bunch (OPEFB)	5-Fluorouracil	Liquid deposition modeling (LDM)	Developed a medium for controlled release of 5-fluorouracil, a chemotherapeutic agent used in treating various cancers	[104]
Hydroxypropyl methylcellulose (HPMC)	Acetaminophen	Fused deposition modeling	Created 3D-printed tablets with zero-order release kinetics, achieving controlled acetaminophen release by designing specific tablet geometries	[171]
Cellulose acetate	Diltiazem	Fused deposition modeling	Fabricated osmotically controlled dosage forms with a cellulose acetate shell, enabling delayed and controlled release of diltiazem	[46]
Hydroxypropyl cellulose	Caffeine	Powder bed 3D printing with Inkjet Binder	Produced 3D objects by spraying a liquid binder over a powder bed, demonstrating that binder viscosity influences dissolution characteristics and that hydroxypropyl cellulose can serve as a solid binder in 3D printing	[62]
Cellulose acetate	Propranolol HCl	Extrusion-based 3D printing	Established a coated shell system to regulate propranolol HCl release from immediate-release tablets, resulting in a profile of prolonged drug release	[4]
Hydroxypropyl methylcellulose (HPMC)	Multiple drug	Fused deposition modeling	Evaluated drug release behavior from 3D-printed cellulose implants, demonstrating controlled release over an extended period	[47]
Ethylcellulose	Acetaminophen	Inkjet printing	Displayed hallmarks of zero-order drug release	[8]
Hydroxypropyl cellulose	Cinnarizine	Fused deposition modeling, hot-melt extrusion	Drug release from floating tablets followed pseudo-zero-order kinetics	[148]

(continued)

Table 3 (continued)

Cellulose material	Drug	3D printing technique	Key findings	References
HPMC 2208, Methocelt, microcrystalline cellulose, and cellulose acetate	Captopril, nifedipine, and glipizide	Hot-melt extrusion	Zero-order drug release was shown by captopril. On the other hand, depending on the active/excipient ratio, glipizide and nifedipine showed either Korsmeyer-Peppas release kinetics or first-order release kinetics	[30]
Hydroxypropyl cellulose	Theophylline	Hot-melt extrusion	The kinetics of release were zero order	[48]
Hydroxypropyl methylcellulose, MCC	Guaifenesin	Extrusion-based 3D printing	Drug release by Fickian diffusion via a layer of hydrated hydroxypropyl methylcellulose gel	[74]
Cellulose acetate, and hydroxypropyl methylcellulose	Aspirin, hydrochlorothiazide, ramipril, pravastatin sodium, and atenolol	Extrusion-based 3D printing	The profile of immediate and sustained release was influenced by the ratio of active to excipient	[73]

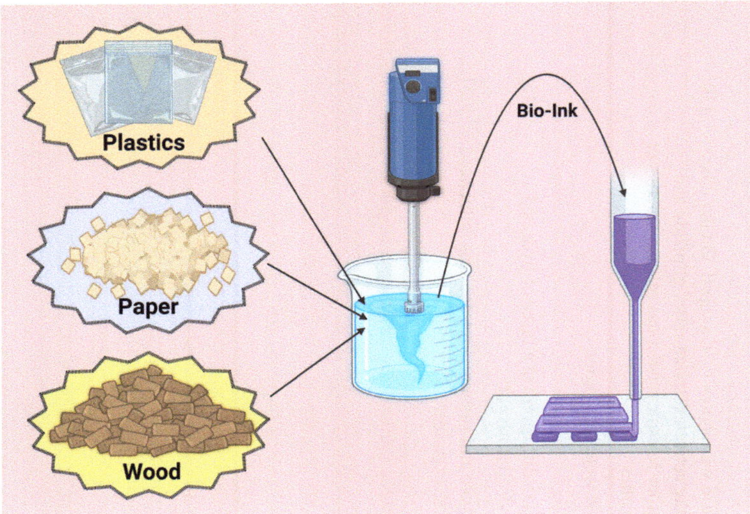

Fig. 7 Diagram illustrating utilization of cellulose related waste for 3D printing application

materials by incorporating recycled polypropylene with waste cellulose from paper, cardboard, and wood powder as shown in Fig. 7. These composites, created through extrusion-based polymer additives and shear crushing processes, exhibited increased filler content compared to pure polymer composites. Thermal gravimetric analysis confirmed their stability, while dynamic mechanical analysis revealed that cellulose addition improved storage modulus by 20–30%. However, increasing cellulose content to 10% by weight did not significantly enhance tensile strength [167].

5.2.2 Bionic Applications

Bionic architectures can facilitate the multifaceted integration of useful abiotic nanomaterials with live microorganisms [102]. Wangpraseurt et al. looked at the potential uses of cellulose-based bionics for the development of living cells. To support the alive microalgae, they created 3D printed corals. These live hybrid bionic corals can grow algal cell densities of up to 109 cells mL^{-1}. Different algae strains may be directly encapsulated using this 3D bioprinting technique, which also facilitates their development in a specially designed optical micro-habitat. Large-scale bionic corals might be developed using this research to create spatially effective photobioreactors for algae growth. This method has the potential to be used in research on algal-coral symbiosis, coral reef conservation, and algae biotechnology [157].

Optimally adjustable photosynthetic materials that mimic the structure of coral tissue and skeletons may be 3D printed with micron-level precision using the bioprinting process. Through the use of photopolymerizable gelatin–methacrylate/nanocellulose and a combination of nanocrystalline cellulose and polyethylene glycol diacrylate, an artificial skeleton is 3D printed utilizing the bioink solution for the symbiosis of microalgae (*Symbiodinium* sp.) to form coral tissue. Stability and the capacity to modify the tissue's scattering characteristics are provided by this substance. Bionic corals can extend photon residence duration in algal cultures by analyzing the development of green microalgae from the *Chlorellaceae* family, which produce fatty acids for biofuel. The microalgae might develop to an algal cell density of $> 8 \times 10^8$ cells mL^{-1} on day 12 as concrete aggregates. In the process of building tissues, algae produce oxygen through photosynthesis [120].

6 Cellulose-Based 4D Printing

4D printing has arisen as a ground-breaking technique in biomedical applications, allowing for the fabrication of smart materials that may transform in reply to external stimuli like moisture, heat, pH, and electric or magnetic fields. Unlike conventional 3D printing, 4D printing has a time-dependent component, allowing structures to modify their shape or function following production. This technique has great promise in the biomedical arena, where shape-memory polymers (SMPs), hydrogels, and nanocellulose-based composites play critical roles [122].

Most important uses of 4D-printing in biomedicine are tissue engineering, in which smart scaffolds may dynamically adapt to biological surroundings, boosting cell growth and regeneration. Cellulose nanocrystals (CNCs) are frequently used in 4D-printed biomedical materials because of their biocompatibility, biodegradability, and outstanding mechanical qualities [93]. These CNC-based materials are utilized for healing of wounds, bone scaffolding, and drug delivery systems. For example, CNCs paired with hydrogels allow for regulated drug release, improving therapeutic efficacy. Furthermore, 4D-printed biomaterials can be utilized to develop self-healing medical devices, like wearable sensors and implants that respond to physiological changes. Smart bandages, for example, may constrict or expand in reaction to humidity, resulting in ideal wound healing conditions. Similarly, 4D-printed bioelectronics, which include soft robotics and flexible sensors, provide real-time health monitoring and diagnostics [94]. Overview of various 4D printing along with various stimuli is shown in Fig. 8.

The use of stimuli-responsive CNCs in 4D-printing has opened up new paths for personalized medicine, allowing for the development of patient-specific implants and prostheses that adapt to the motions of the body. As research advances, the combination of 4D printing, nanotechnology, and bioengineering will further transform biomedical applications, resulting in more efficient, adaptable, and sustainable healthcare solutions [75].

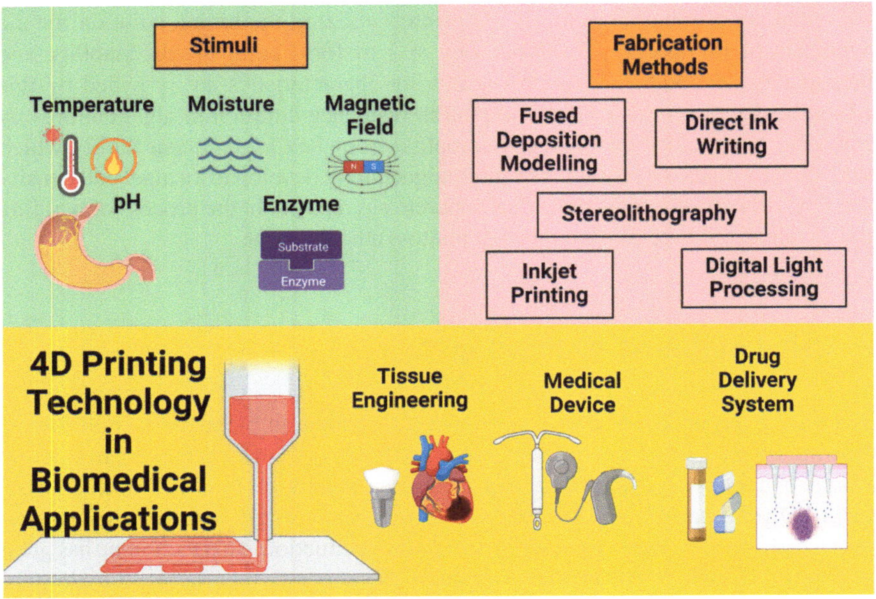

Fig. 8 Diagram illustrating the overview of 4D printing in biomedical application

7 Challenges and Limitations

Cellulose-based 3D printing faces numerous challenges and limits that prevent widespread commercialization. Solvent development is a major challenge for printing on cellulose, which cannot be thermally treated. Printed cellulose structures have low mechanical characteristics, making them unsuitable for load-bearing applications in biomedicine and industry. The layer-by-layer deposition approach causes anisotropy in 3D-printed materials, which impacts their strength and durability. Another issue is biocompatibility and stability. Although cellulose is considered friendly, maintaining structural integrity and stability in biological applications remains an issue. Printability and resolution of cellulose-based bio-inks are limited, making it difficult to achieve high-resolution objects with complicated geometries. Additionally, post-processing activities including solvent evaporation, freeze-drying, and curing increase manufacturing time and complexity. The high manufacturing costs and lack of commercially accessible cellulose-based filaments limit the technology's scalability for general industrial application. Overcoming these problems will need ongoing study into material composition, processing processes, and structural reinforcing solutions.

8 Conclusion

The integration of cellulose into 3D printing technology has demonstrated significant potential for biomedical applications, particularly in tissue engineering, wound healing, and drug delivery. Cellulose-based bio-inks exhibit favorable biocompatibility, mechanical stability, and printability, making them viable candidates for the development of advanced biomaterials. The unique properties of nanocellulose, including its rheological behavior and reinforcement capabilities, further enhance the performance of printed scaffolds and hydrogels. Various studies have explored cellulose derivatives for controlled drug delivery, with promising outcomes in sustained drug release and targeted therapy.

Despite its advantages, several challenges remain, including optimizing print fidelity, enhancing mechanical strength, and developing suitable post-processing techniques. Addressing these limitations requires advancements in crosslinking strategies, bio-ink formulations, and multi-material printing approaches. Moreover, incorporating 4D printing with stimuli-responsive cellulose-based materials presents new opportunities for dynamic biomedical applications. Future research should focus on refining cellulose-based inks for high-resolution bioprinting, exploring scalable manufacturing methods, and investigating regulatory pathways for clinical translation. The continued innovation in cellulose-based 3D printing holds great promise for sustainable and personalized healthcare solutions.

References

1. 3D Printed Bandages Conform to Skin and Help Heal Wounds—3DPrint.com | The Voice of 3D Printing/Additive Manufacturing. https://3dprint.com/209620/3d-printed-bandages/. Accessed 26 Feb 2025
2. Acierno D, Patti A (2023) Fused deposition modelling (FDM) of thermoplastic-based filaments: process and rheological properties—an overview. Materials 16
3. Albelo I, Raineri R, Salmon S (2024) Materials and methods for all-cellulose 3D printing in sustainable additive manufacturing. Sustain Chem 5:98–115. https://doi.org/10.3390/SUSCHEM5020008
4. Algahtani MS, Mohammed AA, Ahmad J, Saleh E (2020) Development of a 3D printed coating shell to control the drug release of encapsulated immediate-release tablets. Polymers (Basel) 12. https://doi.org/10.3390/polym12061395
5. Arif ZU, Khalid MY, Noroozi R et al (2022) Recent advances in 3D-printed polylactide and polycaprolactone-based biomaterials for tissue engineering applications. Int J Biol Macromol 218
6. Assad H, Assad A, Kumar A (2023) Recent developments in 3D bio-printing and its biomedical applications. Pharmaceutics 15
7. Athukoralalage SS, Balu R, Dutta NK, Choudhury NR (2019) 3D bioprinted nanocellulose-based hydrogels for tissue engineering applications: a brief review. Polymers (Basel) 11
8. Awad A, Fina F, Trenfield SJ, et al (2019) 3D printed pellets (Miniprintlets): a novel, multi-drug, controlled release platform technology. Pharmaceutics 11. https://doi.org/10.3390/pharmaceutics11040148

9. Baca D, Ahmad R (2020) The impact on the mechanical properties of multi-material polymers fabricated with a single mixing nozzle and multi-nozzle systems via fused deposition modeling. Int J Adv Manuf Technol 106. https://doi.org/10.1007/s00170-020-04937-3
10. Baranov A, Sommerhoff F, Curnow OJ, Staiger MP (2023) Prospects for manufacturing of complex 3D-shaped all-cellulose composites. Compos Part A Appl Sci Manuf 173. https://doi.org/10.1016/j.compositesa.2023.107627
11. Bergonzi C, Remaggi G, Graiff C et al (2020) Three-dimensional (3D) printed silver nanoparticles/alginate/nanocrystalline cellulose hydrogels: study of the antimicrobial and cytotoxicity efficacy. Nanomaterials 10. https://doi.org/10.3390/nano10050844
12. Biranje SS, Sun J, Cheng L et al (2022) Development of cellulose nanofibril/casein-based 3D composite hemostasis scaffold for potential wound-healing application. ACS Appl Mater Interfaces 14. https://doi.org/10.1021/acsami.1c21039
13. Biswas A, Kumar Singh A, Das D (2024) Stereolithography-based polymer additive manufacturing process for microfluidics devices. Adv Addit Manuf 237–268. https://doi.org/10.1002/9781394238316.CH15
14. Brusentsev Y, Yang P, King AWT et al (2023) Photocross-linkable and shape-memory biomaterial hydrogel based on methacrylated cellulose nanofibres. Biomacromolecules 24. https://doi.org/10.1021/acs.biomac.3c00476
15. CN106344527A—3D printing type ginkgo biloba leave extract dispersible tablet and preparation method thereof. Google Patents. https://patents.google.com/patent/CN106344527A/en?oq=CN106344527. Accessed 21 Feb 2025
16. CN106511285A—3D printing breviscapine orally disintegrating tablet and preparation method thereof. Google Patents. https://patents.google.com/patent/CN106511285A/en?oq=CN106511285. Accessed 21 Feb 2025
17. CN106511416A—3D printed radix notoginseng total saponins orally disintegrating tablet and preparation method thereof. Google Patents. https://patents.google.com/patent/CN106511416A/en?oq=CN106511416. Accessed 21 Feb 2025
18. CN112043676A—3D printing loratadine orally disintegrating tablet and raw material composition and preparation method thereof. Google Patents. https://patents.google.com/patent/CN112043676A/en?oq=CN112043676. Accessed 21 Feb 2025
19. CN113925837A—A kind of 3D printing technology is applied to the preparation method of ibuprofen. Google Patents. https://patents.google.com/patent/CN113925837A/en?oq=CN113925837. Accessed 21 Feb 2025
20. CN117883400A—Eszopiclone pulse time release tablet and preparation method thereof. Google Patents. https://patents.google.com/patent/CN117883400A/en?oq=CN117883400. Accessed 21 Feb 2025
21. CN118436589A—A method for preparing a 3D-printable nanocellulose aerogel drug carrier. Google Patents. https://patents.google.com/patent/CN118436589A/en?oq=CN118436589. Accessed 21 Feb 2025
22. Cafiso D, Septevani AA, Noè C et al (2022) 3D printing of fully cellulose-based hydrogels by digital light processing. Sustain Mater Technol 32. https://doi.org/10.1016/j.susmat.2022.e00444
23. Chan CLC, Lei IM, van de Kerkhof GT et al (2022) 3D printing of liquid crystalline hydroxypropyl cellulose—toward tunable and sustainable volumetric photonic structures. Adv Funct Mater 32. https://doi.org/10.1002/adfm.202108566
24. Chang J, Sun X (2023) Laser-induced forward transfer based laser bioprinting in biomedical applications. Front Bioeng Biotechnol 11
25. Chekkaramkodi D, Jacob L, Muhammed Shebeeb C et al (2024) Review of vat photopolymerization 3D printing of photonic devices. Addit Manuf 86:104189. https://doi.org/10.1016/J.ADDMA.2024.104189
26. Chinga-Carrasco G (2018) Potential and limitations of nanocelluloses as components in biocomposite inks for three-dimensional bioprinting and biomedical devices. Biomacromolecules 19

27. Chiulan I, Heggset EB, Voicu ŞI, Chinga-Carrasco G (2021) Photopolymerization of bio-based polymers in a biomedical engineering perspective. Biomacromolecules 22
28. Cisneros-López EO, Pal AK, Rodriguez AU et al (2020) Recycled poly (lactic acid)-based 3D printed sustainable biocomposites: a comparative study with injection molding. Mater Today Sustain 7–8. https://doi.org/10.1016/j.mtsust.2019.100027
29. Dai L, Cheng T, Duan C et al (2019) 3D printing using plant-derived cellulose and its derivatives: a review. Carbohydr Polym 203
30. Deng GY, Xiang LY, Wei DH et al (2007) Tablets with material gradients fabricated by three-dimensional printing. J Pharm Sci 96. https://doi.org/10.1002/jps.20864
31. Dermol Š, Borin B, Gregor-Svetec D et al (2024) The development of a bacterial nanocellulose/cationic starch hydrogel for the production of sustainable 3D-printed packaging foils. Polymers (Basel) 16:1527. https://doi.org/10.3390/POLYM16111527/S1
32. Dey A, Yodo N (2019) A systematic survey of FDM process parameter optimization and their influence on part characteristics. J Manuf Mater Process 3
33. Ding L, Agrawal P, Singh SK et al (2024) Polymer-based drug delivery systems for cancer therapeutics. Polymers 16:843. https://doi.org/10.3390/POLYM16060843
34. Du H, Liu W, Zhang M et al (2019) Cellulose nanocrystals and cellulose nanofibrils based hydrogels for biomedical applications. Carbohydr Polym 209
35. Dupont L, Xu J, Harasek M, Gföhler M (2025) From soft lithography to 3D printing: current status and future of microfluidic device fabrication. Polymers 17:455. https://doi.org/10.3390/POLYM17040455
36. Dwijaksara NLB, Andromeda S, Gusti AWR et al (2024) Innovations in bioink materials and 3D bioprinting for precision tissue engineering. Metta: Jurnal Ilmu Multidisiplin 4:101–118. https://doi.org/10.37329/METTA.V4I3.3520
37. Erkoc P, Uvak I, Nazeer MA et al (2020) 3D printing of cytocompatible gelatin-cellulose-alginate blend hydrogels. Macromol Biosci 20. https://doi.org/10.1002/mabi.202000106
38. Espinosa E, Filgueira D, Rodríguez A, Chinga-Carrasco G (2019) Nanocellulose-based inks—effect of alginate content on the water absorption of 3D printed constructs. Bioengineering 6. https://doi.org/10.3390/bioengineering6030065
39. Etale A, Onyianta AJ, Turner SR, Eichhorn SJ (2023) Cellulose: a review of water interactions, applications in composites, and water treatment. Chem Rev 123
40. Finny AS, Popoola O, Andreescu S (2021) 3D-printable nanocellulose-based functional materials: Fundamentals and applications. Nanomaterials 11
41. Firmanda A, Syamsu K, Sari YW et al (2022) 3D printed cellulose based product applications. Mater Chem Front 6
42. Fu LH, Qi C, Ma MG, Wan P (2019) Multifunctional cellulose-based hydrogels for biomedical applications. J Mater Chem B 7
43. Gauss C, Pickering KL, Muthe LP (2021) The use of cellulose in bio-derived formulations for 3D/4D printing: a review. Compos Part C: Open Access 4
44. Gaydhane MK, Sharma CS, Majumdar S (2023) Electrospun nanofibres in drug delivery: advances in controlled release strategies. RSC Adv 13
45. Geevarghese R, Somasekharan LT, Bhatt A et al (2022) Development and evaluation of a multicomponent bioink consisting of alginate, gelatin, diethylaminoethyl cellulose and collagen peptide for 3D bioprinting of tissue construct for drug screening application. Int J Biol Macromol 207. https://doi.org/10.1016/j.ijbiomac.2022.02.191
46. Gioumouxouzis CI, Tzimtzimis E, Katsamenis OL et al (2020) Fabrication of an osmotic 3D printed solid dosage form for controlled release of active pharmaceutical ingredients. Eur J Pharm Sci 143. https://doi.org/10.1016/j.ejps.2019.105176
47. Giri BR, Poudel S, Kim DW (2021) Cellulose and its derivatives for application in 3D printing of pharmaceuticals. J Pharm Investig 51
48. Giri BR, Song ES, Kwon J et al (2020) Fabrication of intragastric floating, controlled release 3D printed theophylline tablets using hot-melt extrusion and fused deposition modeling. Pharmaceutics 12. https://doi.org/10.3390/pharmaceutics12010077

49. Gopi S, Balakrishnan P, Chandradhara D et al (2019) General scenarios of cellulose and its use in the biomedical field. Mater Today Chem 13
50. Gospodinova A, Nankov V, Tomov S et al (2021) Extrusion bioprinting of hydroxyethylcellulose-based bioink for cervical tumor model. Carbohydr Polym 260. https://doi.org/10.1016/j.carbpol.2021.117793
51. Gudapati H, Dey M, Ozbolat I (2016) A comprehensive review on droplet-based bioprinting: past, present and future. Biomaterials 102
52. Gumrah Dumanli A (2016) Nanocellulose and its composites for biomedical applications. Curr Med Chem 24. https://doi.org/10.2174/0929867323666161014124008
53. Gunasekera DHAT, Kuek S, Hasanaj D et al (2016) Three dimensional ink-jet printing of biomaterials using ionic liquids and co-solvents. Faraday Discuss 190. https://doi.org/10.1039/c5fd00219b
54. Habib A, Sathish V, Mallik S, Khoda B (2018) 3D printability of alginate-carboxymethyl cellulose hydrogel. Materials 11. https://doi.org/10.3390/ma11030454
55. Halib N, Ahmad I, Grassi M, Grassi G (2019) The remarkable three-dimensional network structure of bacterial cellulose for tissue engineering applications. Int J Pharm 566. https://doi.org/10.1016/j.ijpharm.2019.06.017
56. Hausmann MK, Siqueira G, Libanori R et al (2020) Complex-shaped cellulose composites made by wet densification of 3D printed scaffolds. Adv Funct Mater 30. https://doi.org/10.1002/adfm.201904127
57. Heggset EB, Strand BL, Sundby KW et al (2019) Viscoelastic properties of nanocellulose based inks for 3D printing and mechanical properties of CNF/alginate biocomposite gels. Cellulose 26. https://doi.org/10.1007/s10570-018-2142-3
58. Hickey RJ, Pelling AE (2019) Cellulose biomaterials for tissue engineering. Front Bioeng Biotechnol 7
59. Hospodiuk-Karwowski M, Chi K, Pritchard J, Catchmark JM (2022) Vascularized pancreas-on-a-chip device produced using a printable simulated extracellular matrix. Biomed Mater (Bristol) 17. https://doi.org/10.1088/1748-605X/ac8c74
60. https://temple-news.com/professor-develops-individual-3d-printed-bandages/. Accessed June 2025
61. Husna A, Ashrafi S, Tomal AA et al (2024) Recent advancements in stereolithography (SLA) and their optimization of process parameters for sustainable manufacturing. Hybrid Adv 7:100307. https://doi.org/10.1016/J.HYBADV.2024.100307
62. Infanger S, Haemmerli A, Iliev S et al (2019) Powder bed 3D-printing of highly loaded drug delivery devices with hydroxypropyl cellulose as solid binder. Int J Pharm 555. https://doi.org/10.1016/j.ijpharm.2018.11.048
63. Jang J, Park HJ, Kim SW et al (2017) 3D printed complex tissue construct using stem cell-laden decellularized extracellular matrix bioinks for cardiac repair. Biomaterials 112. https://doi.org/10.1016/j.biomaterials.2016.10.026
64. Javaid M, Haleem A (2018) Additive manufacturing applications in medical cases: a literature based review. Alexandria J Med 54. https://doi.org/10.1016/j.ajme.2017.09.003
65. Ji S, Abaci A, Morrison T et al (2020) Novel bioinks from UV-responsive norbornene-functionalized carboxymethyl cellulose macromers. Bioprinting 18. https://doi.org/10.1016/j.bprint.2020.e00083
66. Ji S, Guvendiren M (2017) Recent advances in bioink design for 3D bioprinting of tissues and organs. Front Bioeng Biotechnol 5
67. Jia C, Bian H, Gao T et al (2017) Thermally stable cellulose nanocrystals toward high-performance 2D and 3D nanostructures. ACS Appl Mater Interfaces 9. https://doi.org/10.1021/acsami.7b08760
68. Jiang X, Huang Y, Cheng Y et al (2021) Effects of lyophilization on the release profiles of 3d printed delivery systems fabricated with carboxymethyl cellulose hydrogel. Polymers (Basel) 13. https://doi.org/10.3390/polym13050749
69. Joseph B, Sagarika VK, Sabu C et al (2020) Cellulose nanocomposites: fabrication and biomedical applications. J Bioresources Bioproducts 5. https://doi.org/10.1016/j.jobab.2020.10.001

70. Jusoh WNLW, Sajab MS, Abdul PM, Kaco H (2022) Recent advances in 3D bioprinting: a review of cellulose-based biomaterials ink. Polymers (Basel) 14
71. KR1020220021048 bioink for 3D bioprinting including microfibrillated cellulose and guar gum, and 3D printing method. https://patentscope.wipo.int/search/en/detail.jsf?docId=KR353451468&_cid=P22-M7EL5Y-13336-1. Accessed 21 Feb 2025
72. Kanjou M, Abdulhakim H, de Olyveira GM, Basmaji P (2019) 3-D print cellulose nanoskin: future diabetic wound healing. J Biomater Nanobiotechnol 10. https://doi.org/10.4236/jbnb.2019.104011
73. Khaled SA, Burley JC, Alexander MR et al (2015) 3D printing of five-in-one dose combination polypill with defined immediate and sustained release profiles. J Control Release 21. https://doi.org/10.1016/j.jconrel.2015.09.028
74. Khaled SA, Burley JC, Alexander MR, Roberts CJ (2014) Desktop 3D printing of controlled release pharmaceutical bilayer tablets. Int J Pharm 461. https://doi.org/10.1016/j.ijpharm.2013.11.021
75. Khalid MY, Arif ZU, Noroozi R et al (2023) 3D/4D printing of cellulose nanocrystals-based biomaterials: additives for sustainable applications. Int J Biol Macromol 251
76. Kim SH, Kim K, Kim BS et al (2020) Fabrication of polyphenol-incorporated anti-inflammatory hydrogel via high-affinity enzymatic crosslinking for wet tissue adhesion. Biomaterials 242. https://doi.org/10.1016/j.biomaterials.2020.119905
77. Klar V, Pere J, Turpeinen T et al (2019) Shape fidelity and structure of 3D printed high consistency nanocellulose. Sci Rep 9. https://doi.org/10.1038/s41598-019-40469-x
78. Kudahettige-Nilsson RL, Ullsten H, Henriksson G (2019) Plastic composites made from glycerol, citric acid, and forest components. Bioresources 13. https://doi.org/10.15376/biores.13.3.6600-6612
79. Kumar A, I Matari IA, Han SS (2020) 3D printable carboxylated cellulose nanocrystal-reinforced hydrogel inks for tissue engineering. Biofabrication 12. https://doi.org/10.1088/1758-5090/ab736e
80. Lafuente-Merchan M, Ruiz-Alonso S, Espona-Noguera A et al (2021) Development, characterization and sterilisation of nanocellulose-alginate- (hyaluronic acid)-bioinks and 3D bioprinted scaffolds for tissue engineering. Mater Sci Eng C 126. https://doi.org/10.1016/j.msec.2021.112160
81. Leppiniemi J, Lahtinen P, Paajanen A et al (2017) 3D-printable bioactivated nanocellulose-alginate hydrogels. ACS Appl Mater Interfaces 9. https://doi.org/10.1021/acsami.7b02756
82. Li J, Wu C, Chu PK, Gelinsky M (2020) 3D printing of hydrogels: rational design strategies and emerging biomedical applications. Mater Sci Eng R: Rep 140
83. Li L, Zhu Y, Yang J (2018) 3D bioprinting of cellulose with controlled porous structures from NMMO. Mater Lett 210. https://doi.org/10.1016/j.matlet.2017.09.015
84. Li VCF, Mulyadi A, Dunn CK et al (2018) Direct ink write 3d printed cellulose nanofiber aerogel structures with highly deformable, shape recoverable, and functionalizable properties. ACS Sustain Chem Eng 6. https://doi.org/10.1021/acssuschemeng.7b03439
85. Li X, Deng Q, Wang S et al (2021) Hydroxyethyl cellulose as a rheological additive for tuning the extrusion printability and scaffold properties. 3D Print Addit Manuf 8. https://doi.org/10.1089/3dp.2020.0167
86. Li Y, Zhang X, Zhang X et al (2023) Recent progress of the vat photopolymerization technique in tissue engineering: a brief review of mechanisms, methods, materials, and applications. Polymers (Basel) 15
87. Li Z, Wang D, Bai H et al (2020) Photo-crosslinking strategy constructs adhesive, super-absorbent, and tough PVA-based hydrogel through controlling the balance of cohesion and adhesion. Macromol Mater Eng 305. https://doi.org/10.1002/mame.201900623
88. Lin L, Jiang S, Yang J et al (2023) Application of 3D-bioprinted nanocellulose and cellulose derivative-based bio-inks in bone and cartilage tissue engineering. Int J Bioprint 9. https://doi.org/10.18063/ijb.v9i1.637
89. Liu G, Zhang X, Chen X et al (2021) Additive manufacturing of structural materials. Mater Sci Eng R: Rep 145

90. Lu Y, Song J, Yao X et al (2021) 3D printing polymer-based bolus used for radiotherapy. Int J Bioprint 7. https://doi.org/10.18063/ijb.v7i4.414
91. Luo H, Cha R, Li J et al (2019) Advances in tissue engineering of nanocellulose-based scaffolds: a review. Carbohydr Polym 224
92. Lv H, Deng G, Lai J et al (2023) Advances in 3D bioprinting of biomimetic and engineered meniscal grafts. Macromol Biosci 23
93. Magalhães MI, Almeida APC (2023) Nature-inspired cellulose-based active materials: from 2D to 4D. Appl Biosci 2. https://doi.org/10.3390/applbiosci2010009
94. Mahmud MAP, Tat T, Xiao X et al (2021) Advances in 4D-printed physiological monitoring sensors. Exploration 1. https://doi.org/10.1002/EXP.20210033
95. Maity K, Mondal A, Saha MC (2022) Cellulose nanocrystal-based all-3D-printed pyro-piezoelectric nanogenerator for hybrid energy harvesting and self-powered cardiorespiratory monitoring toward the human-machine interface. ACS Appl Mater Interfaces. https://doi.org/10.1021/acsami.2c21680
96. Mallakpour S, Azadi E, Hussain CM (2021) State-of-the-art of 3D printing technology of alginate-based hydrogels—an emerging technique for industrial applications. Adv Colloid Interface Sci 293
97. Markstedt K, Sundberg J, Gatenholm P (2014) 3D bioprinting of cellulose structures from an ionic liquid. 3D Print Addit Manuf 1. https://doi.org/10.1089/3dp.2014.0004
98. Mathew A, Kishore SR, Tomy AT et al (2023) Vapour polishing of fused deposition modelling (FDM) parts: a critical review of different techniques, and subsequent surface finish and mechanical properties of the post-processed 3D-printed parts. Prog Addit Manuf 8
99. Maturavongsadit P, Narayanan LK, Chansoria P et al (2021) Cell-laden nanocellulose/chitosan-based bioinks for 3D bioprinting and enhanced osteogenic cell differentiation. ACS Appl Bio Mater 4. https://doi.org/10.1021/acsabm.0c01108
100. Maver T, Smrke DM, Kurečič M et al (2018) Combining 3D printing and electrospinning for preparation of pain-relieving wound-dressing materials. J Solgel Sci Technol 88. https://doi.org/10.1007/s10971-018-4630-1
101. Mehanny S, Abu-El Magd EE, Ibrahim M et al (2021) Extraction and characterization of nanocellulose from three types of palm residues. J Mater Res Technol 10. https://doi.org/10.1016/j.jmrt.2020.12.027
102. Mei X, Liu C, Li Z (2024) Research progress on functional, structural and material design of plant-inspired green bionic buildings. Energy Build 316:114357. https://doi.org/10.1016/J.ENBUILD.2024.114357
103. Mohamed AL, Soliman AAF, Ali EAB et al (2020) Hydrogel bioink based on clickable cellulose derivatives: synthesis, characterization and in vitro assessment. Int J Biol Macromol 163. https://doi.org/10.1016/j.ijbiomac.2020.07.068
104. Mohan D, Khairullah NF, How YP et al (2020) 3D printed laminated $CaCO_3$-nanocellulose films as controlled-release 5-fluorouracil. Polymers (Basel) 12. https://doi.org/10.3390/POLYM12040986
105. Moharana S, Acharya B, Behera A, Maharana T (2024) 3D printing in advanced materials from used PPEs—the state-of-the-art technology. Mater Horizons: Nat Nanomat Part F 3376:107–134. https://doi.org/10.1007/978-981-97-4692-7_6
106. Morelli G, Pescara T, Greco A et al (2023) Utilization of a commercial 3D printer for the construction of a bio-hybrid device based on bioink and adult human mesenchymal cells. Energies (Basel) 16. https://doi.org/10.3390/en16010374
107. Morris VB, Nimbalkar S, Younesi M et al (2017) Mechanical properties, cytocompatibility and manufacturability of chitosan:PEGDA hybrid-gel scaffolds by stereolithography. Ann Biomed Eng 45. https://doi.org/10.1007/s10439-016-1643-1
108. Mukherjee P, Rani A, Saravanan P (2019) Polymeric materials for 3D bioprinting. In: 3D printing technology in nanomedicine
109. Munaz A, Vadivelu RK, St. John J et al (2016) Three-dimensional printing of biological matters. J Sci: Adv Mater Dev 1

110. Müller LAE, Zingg A, Arcifa A et al (2022) Functionalized cellulose nanocrystals as active reinforcements for light-actuated 3d-printed structures. ACS Nano 16. https://doi.org/10.1021/acsnano.2c05628
111. Nowak D, Jakubczyk E (2020) The freeze-drying of foods—the characteristic of the process course and the effect of its parameters on the physical properties of food materials. Foods 9
112. Obed OZ, Uthman TO, Surgun S (2025) A review on lignin-derived resins for sustainable vat polymerization in additive manufacturing. Green Energy Environ Technol 4. https://doi.org/10.5772/GEET.20240068
113. Ojansivu M, Rashad A, Ahlinder A et al (2019) Wood-based nanocellulose and bioactive glass modified gelatin-alginate bioinks for 3D bioprinting of bone cells. Biofabrication 11. https://doi.org/10.1088/1758-5090/ab0692
114. Palaganas NB, Mangadlao JD, De Leon ACC, et al (2017) 3D printing of photocurable cellulose nanocrystal composite for fabrication of complex architectures via stereolithography. ACS Appl Mater Interfaces 9. https://doi.org/10.1021/acsami.7b09223
115. Parhi R (2021) A review of three-dimensional printing for pharmaceutical applications: quality control, risk assessment and future perspectives. J Drug Deliv Sci Technol 64
116. Passamai VE, Katz S, Rodenak-Kladniew B et al (2023) Pectin-based inks development for 3D bioprinting of scaffolds. Journal of Polymer Research 30. https://doi.org/10.1007/s10965-022-03402-x
117. Pattinson SW, Hart AJ (2017) Additive Manufacturing of Cellulosic Materials with Robust Mechanics and Antimicrobial Functionality. Adv Mater Technol 2. https://doi.org/10.1002/admt.201600084
118. Piras CC, Fernández-Prieto S, De Borggraeve WM (2017) Nanocellulosic materials as bioinks for 3D bioprinting. Biomater Sci 5
119. Pita-Vilar M, Concheiro A, Alvarez-Lorenzo C, Diaz-Gomez L (2023) Recent advances in 3D printed cellulose-based wound dressings: a review on in vitro and in vivo achievements. Carbohydr Polym 321
120. Pu X, Wu Y, Liu J, Wu B (2024) 3D bioprinting of microbial-based living materials for advanced energy and environmental applications. Chem Bio Eng 1:568–592. https://doi.org/10.1021/CBE.4C00024
121. Radeke C, Pons R, Mihajlovic M et al (2023) Transparent and cell-guiding cellulose nanofiber 3D printing bioinks. ACS Appl Mater Interfaces 15. https://doi.org/10.1021/ACSAMI.2C16126
122. Ramezani M, Mohd Ripin Z (2023) 4D printing in biomedical engineering: advancements, challenges, and future directions. J Funct Biomater 14:347. https://doi.org/10.3390/JFB14070347
123. Rees A, Powell LC, Chinga-Carrasco G et al (2015) 3D bioprinting of carboxymethylated-periodate oxidized nanocellulose constructs for wound dressing applications. Biomed Res Int 2015. https://doi.org/10.1155/2015/925757
124. Ribezzi D, Pinos R, Bonetti L et al (2023) Design of a novel bioink suitable for the 3D printing of lymphoid cells. Front Biomater Sci 2. https://doi.org/10.3389/fbiom.2023.1081065
125. Sang S, Ma Z, Cao Y et al (2023) BC enhanced photocurable hydrogel based on 3D bioprinting for nasal cartilage repair. Int J Polym Mater Polym Biomater 72. https://doi.org/10.1080/00914037.2022.2052727
126. Sears NA, Seshadri DR, Dhavalikar PS, Cosgriff-Hernandez E (2016) A review of three-dimensional printing in tissue engineering. Tissue Eng Part B Rev 22
127. Shahin-Shamsabadi A, Selvaganapathy PR (2019) ExCeL: combining extrusion printing on cellulose scaffolds with lamination to create in vitro biological models. Biofabrication 11. https://doi.org/10.1088/1758-5090/ab0798
128. Shahriari-Khalaji M, Sattar M, Cao R, Zhu M (2023) Angiogenesis, hemocompatibility and bactericidal effect of bioactive natural polymer-based bilayer adhesive skin substitute for infected burned wound healing. Bioact Mater 29. https://doi.org/10.1016/j.bioactmat.2023.07.008

129. Sharma C, Raza MA, Purohit SD et al (2025) Cellulose-based 3D printing bio-inks for biomedical applications: a review. Int J Biol Macromol 305:141174. https://doi.org/10.1016/J.IJBIOMAC.2025.141174
130. Shashikumar U, Saraswat A, Deshmukh K et al (2024) Innovative technologies for the fabrication of 3D/4D smart hydrogels and its biomedical applications—a comprehensive review. Adv Colloid Interface Sci 328:103163. https://doi.org/10.1016/J.CIS.2024.103163
131. Shen Z, Cai N, Xue Y et al (2020) Porous SBA-15/cellulose membrane with prolonged antimicrobial drug release characteristics for potential wound dressing application. Cellulose 27. https://doi.org/10.1007/s10570-020-02967-4
132. Shen Z, Cai N, Xue Y et al (2019) Engineering sustainable antimicrobial release in silica-cellulose membrane with $CaCO_3$-aided processing for wound dressing application. Polymers (Basel) 11. https://doi.org/10.3390/polym11050808
133. Shin S, Hyun J (2021) Rheological properties of cellulose nanofiber hydrogel for high-fidelity 3D printing. Carbohydr Polym 263. https://doi.org/10.1016/j.carbpol.2021.117976
134. Shin S, Park S, Park M et al (2017) Cellulose nanofibers for the enhancement of printability of low viscosity gelatin derivatives. Bioresources 12. https://doi.org/10.15376/biores.12.2.2941-2954
135. Siqueira G, Kokkinis D, Libanori R et al (2017) Cellulose nanocrystal inks for 3D printing of textured cellular architectures. Adv Funct Mater 27. https://doi.org/10.1002/adfm.201604619
136. Soullard L, Schlepp A, Buret R et al (2024) Towards the 3D printing of innovative hydrogel scaffolds through vat polymerization techniques using methacrylated carboxymethylcellulose aqueous formulations. Prog Addit Manuf 1–13. https://doi.org/10.1007/S40964-024-00744-4/FIGURES/5
137. Sultan S, Mathew AP (2018) 3D printed scaffolds with gradient porosity based on a cellulose nanocrystal hydrogel. Nanoscale 10. https://doi.org/10.1039/c7nr08966j
138. Sultan S, Siqueira G, Zimmermann T, Mathew AP (2017) 3D printing of nano-cellulosic biomaterials for medical applications. Curr Opin Biomed Eng 2
139. Tabatabaei Hosseini BS, Meadows K, Gabriel V et al (2024) Biofabrication of cellulose-based hydrogels for advanced wound healing: a special emphasis on 3D bioprinting. Macromol Biosci 24
140. Tareq MS, Rahman T, Hossain M, Dorrington P (2021) Additive manufacturing and the COVID-19 challenges: an in-depth study. J Manuf Syst 60. https://doi.org/10.1016/j.jmsy.2020.12.021
141. Tenhunen TM, Moslemian O, Kammiovirta K et al (2018) Surface tailoring and design-driven prototyping of fabrics with 3D-printing: an all-cellulose approach. Mater Des 140. https://doi.org/10.1016/j.matdes.2017.12.012
142. Tripathi S, Mandal SS, Bauri S, Maiti P (2023) 3D bioprinting and its innovative approach for biomedical applications. MedComm (Beijing) 4
143. US20180305569A1—Preparation and applications of biocompatible conductive inks based on cellulose nanofibrils for 3d printing of conductive biomedical devices and for use as models for study of neurodegenerative disorders and connection between brain/neurons and communication or other electronic devices. Google Patents. https://patents.google.com/patent/US20180305569A1/en?oq=US20180305569. Accessed 21 Feb 2025
144. US20210205459A1—Injectable thermoresponsive hydrogels as a combinatory modality for controlled drug delivery, biomaterial implant and 3d printing bioink. Google Patents. https://patents.google.com/patent/US20210205459A1/en?oq=US20210205459. Accessed 21 Feb 2025
145. Vanaei S, Parizi MS, Vanaei S et al (2021) An overview on materials and techniques in 3D bioprinting toward biomedical application. Eng Regener 2
146. Vaseghi A, Sadeghizadeh M, Herb M et al (2024) 3D printing of biocompatible and antibacterial silica-silk-chitosan-based hybrid aerogel scaffolds loaded with propolis. ACS Appl Bio Mater. https://doi.org/10.1021/ACSABM.4C00697/SUPPL_FILE/MT4C00697_SI_001.PDF

147. Vel R, Bhatt A, Priyanka A et al (2022) DEAE—cellulose-based composite hydrogel for 3D printing application: physicochemical, mechanical, and biological optimization. Mater Today Commun 33. https://doi.org/10.1016/j.mtcomm.2022.104335
148. Vo AQ, Zhang J, Nyavanandi D et al (2020) Hot melt extrusion paired fused deposition modeling 3D printing to develop hydroxypropyl cellulose based floating tablets of cinnarizine. Carbohydr Polym 246. https://doi.org/10.1016/j.carbpol.2020.116519
149. Voet VSD, Guit J, Loos K (2021) Sustainable photopolymers in 3D printing: a review on biobased, biodegradable, and recyclable alternatives. Macromol Rapid Commun 42
150. Vrana NE, Gupta S, Mitra K et al (2022) From 3D printing to 3D bioprinting: the material properties of polymeric material and its derived bioink for achieving tissue specific architectures. Cell Tissue Bank 23
151. WO2019232114A1—Injectable thermoresponsive hydrogels as a combinatory modality for controlled drug delivery, biomaterial implant and 3d printing bioink. Google Patents. https://patents.google.com/patent/WO2019232114A1/en?oq=WO2019232114. Accessed 21 Feb 2025
152. WO2023227831A1—Ink formulation, method of producing the same and uses of the ink formulation. Google Patents. https://patents.google.com/patent/WO2023227831A1/en?oq=+WO2023227831. Accessed 21 Feb 2025
153. Wajahat M, Kim JH, Ahn J et al (2020) 3D printing of Fe_3O_4 functionalized graphene-polymer (FGP) composite microarchitectures. Carbon N Y 167. https://doi.org/10.1016/j.carbon.2020.05.045
154. Wang Q, Ji C, Sun L et al (2020) Cellulose nanofibrils filled poly (lactic acid) biocomposite filament for FDM 3D printing. Molecules 25. https://doi.org/10.3390/molecules25102319
155. Wang S, Ma Y, Deng Z et al (2020) Effects of fused deposition modeling process parameters on tensile, dynamic mechanical properties of 3D printed polylactic acid materials. Polym Test 86. https://doi.org/10.1016/j.polymertesting.2020.106483
156. Wang Z, Abdulla R, Parker B et al (2015) A simple and high-resolution stereolithography-based 3D bioprinting system using visible light crosslinkable bioinks. Biofabrication 7. https://doi.org/10.1088/1758-5090/7/4/045009
157. Wangpraseurt D, You S, Azam F et al (2020) Bionic 3D printed corals. Nat Commun 11. https://doi.org/10.1038/s41467-020-15486-4
158. Waryono T (2019) RINGKASAN Puding merah (Gruptophyllum pictum L Griff). Jurnal Farmasi Galenika (Galenika J Pharmacy) (e-Journal) 1
159. Wei J, Wang B, Li Z et al (2020) A 3D-printable TEMPO-oxidized bacterial cellulose/alginate hydrogel with enhanced stability via nanoclay incorporation. Carbohydr Polym 238. https://doi.org/10.1016/j.carbpol.2020.116207
160. Whyte DJ, Doeven EH, Sutti A et al (2024) Volumetric additive manufacturing: a new frontier in layer-less 3D printing. Addit Manuf 84:104094. https://doi.org/10.1016/J.ADDMA.2024.104094
161. Xie Z, Gao M, Lobo AO, Webster TJ (2020) 3D bioprinting in tissue engineering for medical applications: the classic and the hybrid. Polymers (Basel) 12
162. Xu C, Zhang Molino B, Wang X et al (2018) 3D printing of nanocellulose hydrogel scaffolds with tunable mechanical strength towards wound healing application. J Mater Chem B 6. https://doi.org/10.1039/c8tb01757c
163. Xu W, Molino BZ, Cheng F et al (2019) On low-concentration inks formulated by nanocellulose assisted with gelatin methacrylate (GelMA) for 3D printing toward wound healing application. ACS Appl Mater Interfaces 11. https://doi.org/10.1021/acsami.8b21268
164. Xu W, Wang X, Sandler N et al (2018) Three-dimensional printing of wood-derived biopolymers: a review focused on biomedical applications. ACS Sustain Chem Eng 6
165. Yan K, Zhang X, Liu Y et al (2023) 3D-bioprinted silk fibroin-hydroxypropyl cellulose methacrylate porous scaffold with optimized performance for repairing articular cartilage defects. Mater Des 225. https://doi.org/10.1016/j.matdes.2022.111531
166. Zaaba NF, Jaafar M, Ismail H (2021) Tensile and morphological properties of nanocrystalline cellulose and nanofibrillated cellulose reinforced PLA bionanocomposites: a review. Polym Eng Sci 61

167. Zander NE, Park JH, Boelter ZR, Gillan MA (2019) Recycled cellulose polypropylene composite feedstocks for material extrusion additive manufacturing. ACS Omega 4. https://doi.org/10.1021/acsomega.9b01564
168. Zavrel M, Bross D, Funke M et al (2009) High-throughput screening for ionic liquids dissolving (ligno-)cellulose. Bioresour Technol 100. https://doi.org/10.1016/j.biortech.2008.11.052
169. Zhang AP, Qu X, Soman P et al (2012) Rapid fabrication of complex 3D extracellular microenvironments by dynamic optical projection stereolithography. Adv Mater 24. https://doi.org/10.1002/adma.201202024
170. Zhang J, Yang W, Vo AQ et al (2017) Hydroxypropyl methylcellulose-based controlled release dosage by melt extrusion and 3D printing: structure and drug release correlation. Carbohydr Polym 177. https://doi.org/10.1016/j.carbpol.2017.08.058
171. Zhang P, Xu P, Chung S et al (2022) Fabrication of bilayer tablets using hot melt extrusion-based dual-nozzle fused deposition modeling 3D printing. Int J Pharm 624. https://doi.org/10.1016/j.ijpharm.2022.121972
172. Zieliński PS, Gudeti PKR, Rikmanspoel T, Włodarczyk-Biegun MK (2023) 3D printing of bio-instructive materials: toward directing the cell. Bioact Mater 19. https://doi.org/10.1016/j.bioactmat.2022.04.008

Deepak Kumar has completed B. Pharmacy from Galgotias University, Greater Noida. Currently he is pursuing his M. Pharmacy from Galgotias University, Greater Noida. His area of interest includes nano formulations, formulation optimization, cancer nanomedicine, targeted drug delivery system, pharmaceuticals, and green nanotechnology. He has authored 3 book chapters. He has published many good Scopus index papers with reputed publishers like Elsevier and Bentham Science. He has attended more than 10 international conferences.

Rishabha Malviya completed B. Pharmacy from Uttar Pradesh Technical University and M. Pharmacy (Pharmaceutics) from Gautam Buddha Technical University, Lucknow, Uttar Pradesh. His PhD (Pharmacy) work was in the area of Novel formulation development techniques. He has 15 years of research experience and is presently working as Associate Professor in the Department of Pharmacy, School of Medical and Allied Sciences, Galgotias University for past 8 years. He has edited/editing 85 books with Springer Nature, CRC Press, Taylor and Francis, Apple Academic Press, Wiley, etc. His area of interest includes formulation optimization, nanoformulation, targeted drug delivery, artificial intelligence in healthcare, localized drug delivery and characterization of natural polymers as pharmaceutical excipients. He has authored more than 150 research/review papers for national/international journals of repute. He is an editorial board member/reviewer of more than 40 journals. More than 22 patent grants from different countries are in his credit, while 40 patents are published/under evaluation.

Sonali Sundram completed B. Pharm & M. Pharm (pharmacology) from AKTU, Lucknow. She has worked as a research scientist on the project of ICMR at King George's Medical University, Lucknow, after that she joined BBDNIIT, and currently, she is working at Galgotias University, Greater Noida. Her PhD (Pharmacy) work was in the area of Neurodegeneration and Nano formulation. Her area of interest is Neurodegeneration, clinical research, and artificial Intelligence. She has edited 14 books (Wiley, Springer Nature, CRC Press/Taylor and Francis, River Publisher, and IOP Publishing). She has attended as well, and organized more than 15 national and international seminars/conferences/workshops. She has more than 8 patents, national and international, in her credit.

Index

R. Malviya and S. Sundram (eds.), *Sustainable Cellulose Materials for Biomedical Applications*, Sustainable Materials and Technology,
https://doi.org/10.1007/978-981-96-9904-9

T

U

V

W

Z

MIX
Papier aus verantwortungsvollen Quellen
Paper from responsible sources
FSC® C105338

If you have any concerns about our products,
you can contact us on
ProductSafety@springernature.com

In case Publisher is established outside the EU,
the EU authorized representative is:
Springer Nature Customer Service Center GmbH
Europaplatz 3, 69115 Heidelberg, Germany

Printed by Libri Plureos GmbH
in Hamburg, Germany